# ECOSYSTEMS OF THE WORLD 17B

# MANAGED GRASSLANDS
## ANALYTICAL STUDIES

Edited by

R.W. Snaydon

*Department of Agricultural Botany*
*University of Reading*
*Whiteknights*
*Reading RG6 2AS (United Kingdom)*

# ELSEVIER
Amsterdam — Oxford — New York — Tokyo 1987

ELSEVIER SCIENCE PUBLISHERS B.V.
Sara Burgerhartstraat 25
1055 KV Amsterdam, The Netherlands

*Distributors for the United States and Canada:*

ELSEVIER SCIENCE PUBLISHING COMPANY INC.
52 Vanderbilt Avenue
New York, N.Y. 10017, U.S.A.

**Library of Congress Cataloging-in-Publication Data**
Main entry under title:

Managed grasslands.

(Ecosystems of the world ; 17A)
Includes bibliographies and indexes.
1. Pastures. 2. Pasture ecology. I. Snaydon, R. W.
II. Series.
SB199.M29 1986      633.2'02      86-541
ISBN 0-444-42565-9 (U.S.)

ISBN 0-444-42565-9 (Vol. 17B)
ISBN 0-444-41702-8 (Series)

Copyright © 1987 by Elsevier Science Publishers B.V., Amsterdam

All rights reserved. No part of this publication may be reproduced, stored in a retrieval system or transmitted in any form or by any means, electronic, mechanical, photocopying, recording or otherwise, without the prior written permission of the publisher, Elsevier Science Publishers B.V./Science & Technology Division, P.O. Box 330, 1000 AH Amsterdam, The Netherlands.

Special regulations for readers in the U.S.A. – This publication has been registered with the Copyright Clearance Center Inc. (CCC), Salem, Massachusetts. Information can be obtained from the CCC about conditions under which photocopies of parts of this publication may be made in the U.S.A. All other copyright questions, including photocopying outside of the U.S.A., should be referred to the publisher.

Printed in The Netherlands

# PREFACE

Most volumes in this series deal with unmanaged ecosystems and, in general, with ecosystems only slightly influenced by man. This volume deals uncompromisingly with ecosystems that are intensively managed by man. Ecological justifications for studying agricultural ecosystems are considered in Chapter 1 but, in addition, such studies are justified by the large area of land which is used agriculturally, especially in the more heavily populated areas of the world. Leaving aside the large areas of barren desert, tundra and mountains, most land is managed by man to some extent. On a worldwide scale, about 75% of the habitable land is used for agriculture. The use of such a large area is hardly surprising, in view of the world's growing population, and the fundamental importance of food to man. Within the area used for agriculture, almost twice as much land is used for grassland as for crops.

The large areas of land used for agriculture are rarely studied by ecologists, in spite of agriculture's claim to be the oldest and most important form of applied ecology. The provision of food has always been, and will continue to be, man's most fundamental requirement. To satisfy that requirement, man has had increasingly to exploit and manipulate plant and animal communities and, more recently, his environment. At first he gathered food and hunted, like many other animals in natural ecosystems. However, as the population density increased, man was forced to impose greater control and management over ecosystems to obtain the necessary food supply.

Although the requirement for food is paramount, food is not the only resource required by man, nor is agriculture the only user of land. Production of other resources, such as water, timber, and minerals, requires the management and exploitation of land. In addition, man's recreational and aesthetic demands also require the use and management of land. These various requirements are frequently in conflict, one with another. In particular, agriculture has increasingly come into conflict with biological and landscape conservation.

The major role of the ecologist, it seems to me, is two-fold: (1) to provide the necessary information on ecosystems, and to provide an intellectual framework into which that information can be placed; and (2) to take a broad, objective and balanced view of the conflict between the various uses of land and other resources, recognizing the relative importance of supplies of food, timber, water, minerals and recreational facilities to man. To fulfil these roles the ecologist must have a clear understanding of the ecological processes occurring within ecosystems, whether natural or managed. He must also have a clear understanding of the objectives and functions of each of these enterprises. Seen from an agricultural viewpoint, there is need for a clear understanding of the ecological processes occurring in agricultural systems, if those systems are to be managed rationally and effectively.

It is surprising that a systematic ecological approach has not been used more extensively in agriculture, in view of the fact that agriculture is, to a considerable extent, applied ecology. Perhaps the main reason for this is that ecology is a very young science, while agriculture is an old technology. As a result, there has been little opportunity for an ecological perspective to be assimilated into agriculture. Indeed, ecology has probably had little to offer agriculture, and the flow of ideas and information has tended to be in the opposite direction (Chapter 1). However, ecological concepts have been applied for a considerable time, especially to agricultural grasslands (e.g. Levy, 1951; Davies, 1952); more recently, Spedding (1971, 1975) and Spedding et al. (1981) in particular have applied these concepts more widely.

There is obviously considerable scope for developing a more ecological approach to agriculture. If this volume induces more ecologists and agriculturalists to view agriculture in an ecological perspective, it will have largely succeeded in its objective.

Like several previous volumes, this volume is presented in two parts; Part A deals descriptively with types of managed grasslands world-wide, whereas this part gives a more detailed analysis of ecosystem processes, mainly using temperate grasslands as examples.

## REFERENCES

Davies, W., 1952. *The Grass Crop*. Span, London, 1st ed.

Levy, E.B., 1951. *Grasslands of New Zealand*. Government Printer, Wellington, 1st ed.

Spedding, C.R.W., 1971. *Grassland Ecology*. Oxford University Press, Oxford, 221 pp.

Spedding, C.R.W., 1975. *The Biology of Agricultural Systems*. Academic Press, London, 261 pp.

Spedding, C.R.W., Walsingham, J.M. and Hoxey, A.M., 1981. *Biological Efficiency in Agriculture*. Academic Press, London, 383 pp.

# LIST OF CONTRIBUTORS TO VOLUME 17

J. ANCHORENA
D. Norte 547
1035 Buenos Aires (Argentina)

G.W. ARNOLD
CSIRO Division of Wildlife and Rangelands
Private Bag, P.O.
Midland, W.A. 6056 (Australia)

J. ARES
Centro Nacional Patagonico
28 de Julio
9120 Puerto Madryn (Argentina)

J.A. BAARS
Ruakura Soil and Plant Research Station
Private Bag
Hamilton (New Zealand)

N.J. BARROW
CSIRO Division of Animal Production
Private Bag, P.O.
Wembley, W.A. 6014 (Australia)

A.M. BEESKOW
Centro Nacional Patagonico
28 de Julio
9120 Puerto Madryn (Argentina)

M. BERTILLER
Centro Nacional Patagonico
28 de Julio
9120 Puerto Madryn (Argentina)

E.F. BIDDISCOMBE
CSIRO Division of Groundwater Research
Private Bag, P.O.
Wembley, W.A. 6014 (Australia)

H.A. BIRRELL
Pastoral Research Institute
P.O. Box 180
Hamilton, Vic. 3300 (Australia)

A.I. BREYMEYER
Institute of Geography and Spatial Organization
Polish Academy of Sciences
Krakowskie Przedmiescie 30
00927 Warsaw (Poland)

J.L. BROCK
DSIR Grasslands Division
Private Bag
Palmerston North (New Zealand)

L. BULLA
Instituto de Zoologia Tropical
Universidad Central de Venezuela
Apartado 59058
Caracas 104 (Venezuela)

K.R. CHRISTIAN
CSIRO Division of Plant Industry
P.O. Box 1600
Canberra City, A.C.T. 2601 (Australia)

G. DEFOSEE
Centro Nacional Patagonico
28 de Julio
9120 Puerto Madryn (Argentina)

C.J. DOYLE
The Animal and Grassland Research Institute
Hurley, Maidenhead SL6 5LR (United Kingdom)

# LIST OF CONTRIBUTORS

F.X. DUNIN
CSIRO Division of Plant Industry
P.O. Box 1600
Canberra City, A.C.T. 2601 (Australia)

A. DZIEWULSKA
Institute of Geography and Spatial Organization
Polish Academy of Sciences
Krakowskie Przedmiescie 30
00927 Warsaw (Poland)

P.J. EDWARDS
Dept. of Agriculture and Fisheries
University of Natal
Private Bag X9059
Pietermaritzburg 3200 (Republic of South Africa)

F. FILLAT
Instituto Pirenaico de Ecologia
Apartado 64
Jaca (Spain)

M.J.S. FLOATE
Invermay Agricultural Research Centre
Private Bag
Mosgiel (New Zealand)

N. FRENCH
Box 557
Star Rt. 89038
Las Vegas, Nev. 89124 (U.S.A.)

A.G. GILLINGHAM
Whatawhata Hill Country Research Station
Private Bag
Hamilton (New Zealand)

J.O. GREEN
26 Spinfield Lane
Marlow, Bucks SL7 2LB (United Kingdom)

W. HARRIS
Botany Division, D.S.I.R.
Private Bag
Lincoln (New Zealand)

J.H. HOGLUND
DSIR Grasslands Division
Private Bag
Lincoln (New Zealand)

C.W. HOLMES
Department of Animal Science
Massey University
Palmerston North (New Zealand)

W. HOLMES
Wye College
Nr. Ashford, Kent TN25 5AH (United Kingdom)

M. IRISARRI
Centro Nacional Paragonico
28 de Julio
9120 Puerto Madryn (Argentina)

I. ITO
Grassland Research Laboratory
Tohoku University
Kawatabi, Narugo, Tamatsukuri
Miyagi 989-67 (Japan)

D.I.H. JONES
Welsh Plant Breeding Station
Plas Gogerddan
Nr. Aberystwyth SY23 3EB (United Kingdom)

A. KAJAK
Institute of Ecology, Polish Academy of Sciences
05150 Dzienkanów Leśny, p-ta Łomianki (Poland)

C.J. KORTE
Ministry of Agriculture and Fisheries
Private Bag
Gisborne (New Zealand)

P. LOISEAU
INRA Station d'Agronomie
12 Avenue de l'Agriculture
63100 Clermont-Ferrand (France)

C. MARSHALL
School of Plant Biology
University College of North Wales
Bangor, Gwynedd LL57 2UW (United Kingdom)

# LIST OF CONTRIBUTORS

C. MERINO
Centro Nacional Patagonico
28 de Julio
9120 Puerto Madryn (Argentina)

D.J. MINSON
CSIRO Division of Tropical Crops and Pastures
306 Carmody Road
St. Lucia, Qld. 4067 (Australia)

F.X. de MONTARD
INRA Station d'Agronomie
12 Avenue de l'Agriculture
63100 Clermont-Ferrand (France)

P. MONTSERRAT
Instituto Pirenaico de Ecologia
Apartado 64
Jaca (Spain)

G. MORALES
Instituto de Zoologia Tropical
Universidad Central de Venezuela
Apartado 59058
Caracas 104 (Venezuela)

J. MORRISON
The Animal and Grassland Research Institute
North Wyke
Okehampton EX20 2SB (United Kingdom)

H. OKRUSZKO
Institute of Land Reclamation and Grassland Farming
IMUZ-Falenty
05–550 Raszyn (Poland)

J. PACHECO
Instituto de Zoologia Tropical
Universidad Central de Venezuela
Apartado 59058
Caracas 104 (Venezuela)

J.E. RADCLIFFE
Agricultural Research Division
P.O. Box 24
Lincoln, Canterbury (New Zealand)

P.V. RATTRAY
Whatawhata Hill Country Research Station
Private Bag
Hamilton (New Zealand)

G. RICOU
INRA, Laboratoire de Recherche sur les Ecosystèmes Prairiaux
16 rue Dufay
71600 Rouen (France)

M. ROSTAGNO
Centro Nacional Patagonico
28 de Julio
9120 Puerto Madryn (Argentina)

J.P. SHILDRICK
The Sports Turf Research Institute
Bingley BD16 1AU (United Kingdom)

R.W. SNAYDON
Department of Agricultural Botany
University of Reading
Whiteknights
Reading RG6 2AS (United Kingdom)

C.R.W. SPEDDING
Department of Agriculture and Horticulture
University of Reading
Early Gate
Reading RG6 2AT (United Kingdom)

K.W. STEELE
Invermay Agricultural Research Centre
Private Bag
Mosgiel (New Zealand)

N.M. TAINTON
P.O. Box 375
Department of Grassland Science
University of Natal
Pietermaritzburg (Republic of South Africa)

T.A. THOMAS
Welsh Plant Breeding Station
Plas Gogerddan
Nr. Aberystwyth SY23 3EB (United Kingdom)

A.A. TITLYANOVA
Institute of Soil Science and Agrochemistry
U.S.S.R. Academy of Sciences
18 Sovietskaya
Novosibirsk 630099 (U.S.S.R)

J.M. WALSINGHAM
Department of Agriculture and Horticulture
University of Reading
Earley Gate
Reading RG6 2AT (United Kingdom)

D.H. WHITE
Animal Research Institute
Princes Highway
Werribee, Vic. 3030 (Australia)

R. ZLOTIN
Institute of Geography
U.S.S.R. Academy of Sciences
Staromonetny 29
109017 Moscow (U.S.S.R.)

# CONTENTS OF VOLUME 17A

*Chapter 1.* DISTRIBUTION AND PRODUCTIVITY OF GRASSLANDS IN EUROPE
by A. Dziewulska

*Chapter 2.* THE DISTRIBUTION AND MANAGEMENT OF GRASSLANDS IN THE BRITISH ISLES
by J.O. Green

*Chapter 3.* THE SYSTEMS OF GRASSLAND MANAGEMENT IN SPAIN
by P. Montserrat and F. Fillat

*Chapter 4.* GRASSLANDS IN UPLAND AREAS: THE MASSIF CENTRAL, FRANCE
by P. Loiseau, F.X. de Montard and G. Ricou

*Chapter 5.* MANAGED GRASSLANDS IN SOUTH AFRICA
by P.J. Edwards and N.M. Tainton

*Chapter 6.* MANAGED GRASSLANDS IN JAPAN
by I. Ito

*Chapter 7.* STRUCTURAL AND DYNAMIC CHARACTERISTICS OF OVERGRAZED LANDS OF NORTHERN PATAGONIA, ARGENTINA
by J. Ares, A.M. Beeskow, M. Bertiller, M. Rostagno, M. Irisarri, J. Anchorena, G. Defosee and C. Merino

*Chapter 8.* SEASONALLY FLOODED NEOTROPICAL SAVANNA CLOSED BY DIKES
by L. Bulla, J. Pacheco and G. Morales

*Chapter 9.* GRASSLANDS ON DRAINED PEATS IN POLAND
by A. Kajak and H. Okruszko

*Chapter 10.* THE USE OF TURFGRASSES IN TEMPERATE HUMID CLIMATES
by J. Shildrick

*Chapter 11.* CHANGES IN STRUCTURE AND FUNCTION OF TEMPERATE ZONE GRASSLANDS UNDER THE INFLUENCE OF MAN
by A. Titlyanova, N. French and R. Zlotin

*Chapter 12.* MANAGED GRASSLANDS AND ECOLOGICAL EXPERIENCE
by A. Breymeyer

*Chapter 13.* THE FUTURE ROLE OF MANAGED GRASSLANDS
by C.R.W. Spedding

SYSTEMATIC LIST OF GENERA
AUTHOR INDEX
SYSTEMATIC INDEX
GENERAL INDEX

# CONTENTS OF VOLUME 17B

| | |
|---|---|
| PREFACE . . . . . . . . . . . . . . . V | |
| References . . . . . . . . . . . . . . VI | |
| LIST OF CONTRIBUTORS . . . . . . . . VII | |
| **Section I. Primary productivity** . . . . . . . . 1 | |

*Chapter 1.* GENERAL INTRODUCTION
by R.W. Snaydon . . . . . . . . 3

Managed and unmanaged ecosystems . . . . . . 3
Ecological aspects . . . . . . . . . . . . 3
Fundamental and applied studies . . . . . . . 4
Status of managed grasslands . . . . . . . . 4
Scope of the volume . . . . . . . . . . . 4
References . . . . . . . . . . . . . . . 5

*Chapter 2.* THE PRODUCTIVITY OF TEMPERATE GRASSLANDS
by J.E. Radcliffe and J.A. Baars . . . . 7

Introduction . . . . . . . . . . . . . . . 7
Pasture production . . . . . . . . . . . . 7
Factors which affect pasture growth . . . . . . 10
Potential production from pasture . . . . . . . 13
Hill pastures . . . . . . . . . . . . . . . 14
Conclusions . . . . . . . . . . . . . . . 15
Acknowledgements . . . . . . . . . . . . 15
References . . . . . . . . . . . . . . . 15

*Chapter 3.* THE PRODUCTIVITY OF MEDITERRANEAN AND SEMI-ARID GRASSLANDS
by E.F. Biddiscombe . . . . . . . . 19

Introduction . . . . . . . . . . . . . . . 19
Regional pasture productivity . . . . . . . . 19
Limitations to primary productivity . . . . . . 23
Improvements of pasture productivity . . . . . 24
Conclusions and future research . . . . . . . 24
Acknowledgements . . . . . . . . . . . . 25
References . . . . . . . . . . . . . . . 25

*Chapter 4.* PHYSIOLOGICAL ASPECTS OF PASTURE GROWTH
by C. Marshall . . . . . . . . . . 29

Introduction . . . . . . . . . . . . . . . 29
The assimilation of carbon . . . . . . . . . 30
The utilization of carbon . . . . . . . . . . 32
Environmental and developmental influences on growth . . . . . . . . . . . . . . . 34
Conclusions . . . . . . . . . . . . . . . 41
References . . . . . . . . . . . . . . . 42

*Chapter 5.* MODELLING PASTURE GROWTH
by K.R. Christian . . . . . . . . 47

The pasture system . . . . . . . . . . . . 47
The choice of models . . . . . . . . . . . 47
Interception of radiation . . . . . . . . . . 49
Photosynthesis . . . . . . . . . . . . . . 50
Respiration . . . . . . . . . . . . . . . 50
Water relations . . . . . . . . . . . . . . 51
Allocation of assimilate . . . . . . . . . . . 52
Shoot growth and phenology . . . . . . . . 52
Root growth . . . . . . . . . . . . . . . 53
Nitrogen . . . . . . . . . . . . . . . . 53
Plant senescence and decomposition . . . . . . 55
Plant competition . . . . . . . . . . . . . 55
Grazing . . . . . . . . . . . . . . . . . 56
Conclusions . . . . . . . . . . . . . . . 56
References . . . . . . . . . . . . . . . 57

*Chapter 6.* EFFECTS OF NITROGEN FERTILIZER
by J. Morrison . . . . . . . . . . 61

Introduction . . . . . . . . . . . . . . . 61
Yield response . . . . . . . . . . . . . . 62
Factors affecting response . . . . . . . . . . 63
Effects on herbage composition . . . . . . . . 67
Recovery of fertilizer nitrogen in herbage . . . . 67
Postscript . . . . . . . . . . . . . . . . 68
References . . . . . . . . . . . . . . . 68

*Chapter 7.* EFFECTS OF GRAZING AND CUTTING
by C.J. Korte and W. Harris . . . . 71

Introduction . . . . . . . . . . . . . . . . . 71
Effects of defoliation . . . . . . . . . . . . 71
Differences between cut and grazed swards . . . . . 75
Conclusion . . . . . . . . . . . . . . . . . 76
References . . . . . . . . . . . . . . . . . 77

*Chapter 8.* THE BOTANICAL COMPOSITION OF PASTURES
by R.W. Snaydon . . . . . . . . . 81

Introduction . . . . . . . . . . . . . . . . . 81
Factors affecting composition . . . . . . . . . . 81
Control of composition . . . . . . . . . . . . 81
Sward age and composition . . . . . . . . . . 82
Sward age and yield . . . . . . . . . . . . . 83
Herbage yield of species . . . . . . . . . . . . 83
Herbage yield of mixtures . . . . . . . . . . . 84
Seasonal pattern of production . . . . . . . . . 85
Herbage quality . . . . . . . . . . . . . . . 85
Conclusions . . . . . . . . . . . . . . . . . 86
References . . . . . . . . . . . . . . . . . 86

**Section II. Secondary productivity** . . . . . . . . 89

*Chapter 9.* BEEF PRODUCTION FROM MANAGED GRASSLANDS
by W. Holmes . . . . . . . . . . . 91

Introduction . . . . . . . . . . . . . . . . . 91
Production of herbage . . . . . . . . . . . . 91
Utilization of herbage . . . . . . . . . . . . 92
Responses to management . . . . . . . . . . . 96
The biological efficiency of beef production systems . 97
Models of beef systems . . . . . . . . . . . . 98
References . . . . . . . . . . . . . . . . . 99

*Chapter 10.* MILK PRODUCTION FROM MANAGED GRASSLANDS
by C.W. Holmes . . . . . . . . . 101

Introduction . . . . . . . . . . . . . . . . . 101
Productivity of grassland dairy farms . . . . . . . 101
Factors affecting productivity . . . . . . . . . . 103
Conclusions . . . . . . . . . . . . . . . . . 109
References . . . . . . . . . . . . . . . . . 110

*Chapter 11.* SHEEP PRODUCTION FROM MANAGED GRASSLANDS
by P.V. Rattray . . . . . . . . . 113

Introduction . . . . . . . . . . . . . . . . . 113
Annual production . . . . . . . . . . . . . . 113
Matching feed requirements and pasture growth . . 114
Breed and reproductive rate . . . . . . . . . . 116
Pasture conditions . . . . . . . . . . . . . . 117
Grazing management . . . . . . . . . . . . . 118
Critical periods of the reproductive cycle . . . . . 118
Sheep as converters . . . . . . . . . . . . . . 120
Conclusions . . . . . . . . . . . . . . . . . 120
References . . . . . . . . . . . . . . . . . 120

*Chapter 12.* OTHER DOMESTICATED ANIMALS
by J.M. Walsingham . . . . . . . 123

Introduction . . . . . . . . . . . . . . . . . 123
Ruminants . . . . . . . . . . . . . . . . . 123
Non-ruminants . . . . . . . . . . . . . . . . 126
Conclusion . . . . . . . . . . . . . . . . . 127
References . . . . . . . . . . . . . . . . . 127

*Chapter 13.* GRAZING BEHAVIOUR
by G.W. Arnold . . . . . . . . . 129

Introduction . . . . . . . . . . . . . . . . . 129
The circadian patterns of behaviour . . . . . . . 129
Time spent feeding . . . . . . . . . . . . . . 129
Grazing time and management . . . . . . . . . 130
The choice of diet . . . . . . . . . . . . . . 132
Animal factors influencing preferences . . . . . . 133
References . . . . . . . . . . . . . . . . . 134

*Chapter 14.* PLANT FACTORS AFFECTING INTAKE
by D.J. Minson . . . . . . . . . 137

Introduction . . . . . . . . . . . . . . . . . 137
Forage intake by stall-fed ruminants . . . . . . . 137
Grazing factors controlling intake . . . . . . . . 141
Conclusion . . . . . . . . . . . . . . . . . 142
References . . . . . . . . . . . . . . . . . 142

*Chapter 15.* MINERALS IN PASTURES AND SUPPLEMENTS
by D.I.H. Jones and T.A. Thomas . . . 145

Introduction . . . . . . . . . . . . . . . . . 145
Mineral deficiency in pasture . . . . . . . . . . 145
Pasture minerals and nutritional requirement . . . . 147
The diagnosis of mineral deficiencies . . . . . . . 148
Factors influencing the mineral content
  of pastures . . . . . . . . . . . . . . . . 149
Methods of mineral supplementation . . . . . . . 151
Conclusion . . . . . . . . . . . . . . . . . 152
References . . . . . . . . . . . . . . . . . 152

*Chapter 16.* REPRODUCTION, LIFESPAN AND EFFICIENCY OF PRODUCTION
by J.M. Walsingham . . . . . . . 155

Introduction . . . . . . . . . . . . . . . . . 155
Reproduction . . . . . . . . . . . . . . . . 155
Life-span . . . . . . . . . . . . . . . . . . 157
Conclusion . . . . . . . . . . . . . . . . . 158
References . . . . . . . . . . . . . . . . . 159

**Section III. Nutrient cycling** . . . . . . . . . . 161

*Chapter 17.* NITROGEN CYCLING IN MANAGED GRASSLANDS
by M.J.S. Floate . . . . . . . . 163

Introduction . . . . . . . . . . . . . . . . . 163
Ecosystem components . . . . . . . . . . . . 164
Nitrogen transformations and flows between
  components . . . . . . . . . . . . . . . . 164

| | |
|---|---|
| Inputs of nitrogen | 165 |
| Losses of nitrogen | 166 |
| Control of nitrogen cycling | 167 |
| The dimensions of ecosystems | 167 |
| Equilibrium, steady state and balance in ecosystems | 168 |
| Case studies | 169 |
| Control and manipulation of grassland ecosystems | 170 |
| Modelling the nitrogen cycle | 171 |
| References | 171 |

*Chapter 18.* PHOSPHORUS CYCLING IN MANAGED GRASSLANDS by A.G. Gillingham . . . 173

| | |
|---|---|
| Introduction | 173 |
| Soil phosphorus | 173 |
| Plant uptake of phosphorus | 175 |
| Animal intake of phosphorus | 175 |
| Returns of phosphorus | 176 |
| Above-ground losses of phosphorus | 177 |
| Modelling of the phosphorus cycle | 177 |
| References | 178 |

*Chapter 19.* RETURN OF NUTRIENTS BY ANIMALS by N.J. Barrow . . . 181

| | |
|---|---|
| Introduction | 181 |
| Distribution of nutrients between faeces and urine | 181 |
| Spatial distribution of the return | 183 |
| Availability of plant nutrients in faeces | 183 |
| Availability of plant nutrients in urine | 184 |
| References | 185 |

*Chapter 20.* NITROGEN FIXATION IN MANAGED GRASSLANDS by J.H. Hoglund and J.L. Brock . . . 187

| | |
|---|---|
| Introduction | 187 |
| Nitrogen fixation in grassland ecosystems | 187 |
| Nitrogen fixation, legume growth and soil mineral nitrogen | 188 |
| Nitrogen fixation and other edaphic factors | 190 |
| Nitrogen fixation and competition | 191 |
| Effects of defoliation | 193 |
| Conclusion | 193 |
| References | 194 |

*Chapter 21.* NITROGEN LOSSES FROM MANAGED GRASSLAND by K.W. Steele . . . 197

| | |
|---|---|
| Introduction | 197 |
| Gaseous losses | 197 |
| Wind erosion | 200 |
| Fire | 200 |
| Solution losses | 200 |
| Animal retention, products and transfer | 200 |
| Management of grasslands to reduce losses of nitrogen | 201 |
| Concluding comments | 202 |
| References | 202 |

*Chapter 22.* RUN-OFF AND DRAINAGE FROM GRASSLAND CATCHMENTS by F.X. Dunin . . . 205

| | |
|---|---|
| Introduction | 205 |
| A perspective on grassland hydrology | 205 |
| Management influences on grassland hydrology | 207 |
| Hydrologic models of grassland management | 210 |
| Concluding remarks | 211 |
| References | 211 |

**Section IV. Systems management** . . . 215

*Chapter 23.* ECONOMIC CONSIDERATIONS IN THE PRODUCTION AND UTILIZATION OF HERBAGE by C.J. Doyle . . . 217

| | |
|---|---|
| Introduction | 217 |
| Valuing increased grass production | 217 |
| Optimal level of inputs | 219 |
| Optimizing the use of herbage as a feed | 220 |
| Valuing economic benefits arising at different times | 222 |
| Risk and uncertainty | 223 |
| Conclusions | 225 |
| Acknowledgements | 225 |
| References | 225 |

*Chapter 24.* STOCKING RATE by D.H. White . . . 227

| | |
|---|---|
| Introduction | 227 |
| Stocking rate and pasture | 227 |
| Stocking rate and soil properties | 228 |
| Stocking rate and animal nutrition | 228 |
| Quantifying animal output in relation to stocking rate | 229 |
| Economic responses to increasing stocking rate | 231 |
| Interaction of stocking rate with other management practices | 233 |
| Conclusions | 234 |
| References | 235 |

*Chapter 25.* FERTILIZER INPUTS AND BOTANICAL COMPOSITION by R.W. Snaydon . . . 239

| | |
|---|---|
| Introduction | 239 |
| Interactions with stocking rate | 240 |
| Fertilizer use | 240 |
| Botanical composition | 243 |
| Conclusions | 245 |
| References | 245 |

*Chapter 26.* HERBAGE CONSERVATION AND SUPPLEMENTS by H.A. Birrell . . . 247

| | |
|---|---|
| Introduction | 247 |
| Forage conservation | 247 |

Regional use of conservation . . . . . . . . . . 248
Effects of conservation . . . . . . . . . . . 249
Forage crops . . . . . . . . . . . . . 251
Conclusions . . . . . . . . . . . . . . 251
References . . . . . . . . . . . . . . 252

GLOSSARY . . . . . . . . . . . . . . 255

SYSTEMATIC LIST OF GENERA . . . . . . . 257

AUTHOR INDEX . . . . . . . . . . . . 259

SYSTEMATIC INDEX . . . . . . . . . . . 271

GENERAL INDEX . . . . . . . . . . . . 275

Section I

# PRIMARY PRODUCTIVITY

The chapters in this section cover various aspects of the primary productivity of managed temperate grasslands. Chapters 2 and 3 deal with the overall pattern of primary productivity in temperate and mediterranean environments respectively. Attention is focussed especially on geographical variation in productivity, and on the environmental factors determining that variation. The seasonal pattern of production is also considered, because of its subsequent importance in secondary productivity (Chapters 9–12). Chapters 4 and 5 deal with the physiological processes underlying pasture growth, whereas Chapters 6–8 deal with some of the more important management variables which influence grassland productivity.

The abundance of information, set against the constraints of space, has severely limited the number of topics that could be covered in this section. In particular, various important limiting factors, such as water supply and the availability of various macro-nutrient and micro-nutrient elements, other than nitrogen, have not been included in spite of their importance as management variables, and the large amount of information available. Pests and pathogens are also not considered though, in this case, information is more scanty and their importance is less clear.

The chapters in this section clearly indicate the breadth and intensity of the studies that have been made on temperate managed grasslands. In general, the intensity of study has been considerably greater than that of comparable "natural" ecosystems. The approach has also been more experimental, compared with the more descriptive approach to "natural" ecosystems (Chapter 1). It is also worth noting the contribution that studies of managed grasslands have made to studies of "natural" plant communities. In particular, important techniques used in measuring productivity, and such techniques as growth analysis, the point quadrat technique and computer modelling (Chapter 5), have been used first for managed grasslands. Not only have these techniques been used in later studies of "natural" communities, but also the principles first defined for managed grasslands have usually been applied elsewhere, and some have contributed substantially to our understanding of "natural" communities. In other cases, such as the physiological analysis of growth (Chapter 4), the developments in studies of managed grasslands have, as yet, not been widely applied to "natural" communities.

Chapter 1

# GENERAL INTRODUCTION

R.W. SNAYDON

### MANAGED AND UNMANAGED ECOSYSTEMS

Traditionally, ecologists have studied plants and animals in habitats, such as sand dunes, salt marshes, mountain tops and natural forest, that are largely unaffected by man. They have tended to shun habitats that are affected by man, and have rarely studied those managed by man. However, there are several good ecological reasons for choosing to study habitats that are managed by man. Firstly, such managed ecosystems are frequently more simple ecologically than "natural" ecosystems, and are therefore easier to study and to understand. Secondly, managed ecosystems have been treated in a diversity of ways (e.g. by the use of fertilizers and grazing management) giving, in effect, pre-existing experimental treatments; by comparing different fields or farms, it is possible therefore to deduce, more accurately than can be done for "natural" ecosystems, the effects of individual factors on components of the ecosystem. Thirdly, more information has been obtained on managed ecosystems, largely because of their economic importance. As a result, there is a plethora of information on managed grasslands, especially in temperate environments; a volume such as this cannot fully do justice to that information. Finally, research on managed ecosystems is usually more clearly focussed than that on unmanaged ecosystems because of the clearer objectives in applied studies.

There is also a large difference in the nature of information collected from managed and "natural" ecosystems. On the one hand, much of the information concerning managed ecosystems has been obtained from experimental studies. Some experimental studies, such as the Park Grass Experiment at Rothamsted, which was designed to study the effects of various mineral nutrients on grassland, were started over a century ago. By contrast, few experimental studies have been made of "natural" ecosystems, because of fear of disturbance; consequently, most studies of "natural" ecosystems have been descriptive, resulting in much slower progress. Although the primary reason for studying managed ecosystems is to increase the output from those systems, and to increase the economic efficiency of production, most of the information obtained is of ecological value. In some cases, for instance in plant population biology (Harper, 1977; Snaydon, 1980a), studies of managed ecosystems have laid the foundations for major developments in ecological theory.

### ECOLOGICAL ASPECTS

Many ecologists seem to assume that managed ecosystems are not proper subjects for ecological study. It is not clear whether this is because they believe that ecological principles do not apply to managed ecosystems, or whether it is simply a continuation of a prejudice against cultivated species, already apparent a century ago; Darwin (1868) commented that "Botanists have generally neglected cultivated varieties as beneath their notice". I have argued elsewhere (Snaydon, 1980b) that ecological principles are just as applicable to managed ecosystems as to "natural" ecosystems. This is true of each level of ecological organization, whether it be the dynamics of plant populations, succession within plant communities, or energy flow in ecosystems. Plants and animals respond to environmental variables, and to each other, irrespective of whether the ecological variables are manipulated by man or not. Man can often use ecological principles to his advantage in manipulating ecosystems, but he can rarely bypass or prevent them.

Effective, rational management of ecosystems, whether for agriculture, forestry, recreation or conservation, depends on an understanding of the ecological processes that occur within them. Much can be achieved by an empirical and pragmatic approach to management, based upon past experience or trial and error, but this commonly leads to rules of thumb and so to dogma (Snaydon, 1980b). Such empirical management is usually site-specific, and often season-specific, and can only be modified by further trial and error. However, empirical and rule of thumb management is still widely practiced, not only in agriculture but also in forestry, recreation and conservation. However, the trend is towards a more rational approach, based on theoretical principles, though theory currently lags behind practice and, as a result, pure science has more often been a spin-off from technology and applied science than vice versa (Snaydon, 1984).

## FUNDAMENTAL AND APPLIED STUDIES

We need to recognize that fundamental and applied studies have different objectives; fundamental studies are undertaken solely to increase knowledge and understanding, but applied studies are undertaken to improve our ability to manipulate systems, and so to increase output or economic efficiency. Such improvements in management depend on an ability to predict, usually quantitatively, the outcome of any change in management. This predictive ability, it might be argued, should be improved by increasing knowledge and understanding — that is, by fundamental study. In the long term this may be true, but fundamental studies tend to be at lower levels of biological organization, and made on rare or exceptional species and environments, and it is rarely possible to relate or extrapolate these to agricultural conditions, largely because of our inability to extrapolate from one level to another (Snaydon, 1984) and because of the highly specific ecological responses of organisms. In the shorter term, therefore, information and understanding for predictive purposes must be obtained at the relevant level of organization, using the relevant organisms and environment.

## STATUS OF MANAGED GRASSLANDS

Intensively managed grasslands are mostly manmade, as opposed to less intensively managed rangelands (e.g. steppe and savannah), many of which are modified natural grassland. Most of the managed grasslands were created by the removal of natural forest; this has occurred in temperate, mediterranean and tropical regions (Moore, 1964; Snaydon, 1981). Without grazing, cutting or fire, these man-made grasslands would revert to forest — that is, they are plagioclimaxes (Davies, 1959; Levy, 1970). As a result, man-made grasslands are ecologically unstable; the botanical composition changes rapidly (Chapter 8), particularly in response to grazing or cutting management (e.g. Jones, 1933; Brougham, 1960) and to fertilizer use (e.g. Brenchley and Warington, 1958; Elberse et al., 1983).

Large areas of managed grassland have not been ploughed for decades, and sometimes centuries; these swards have reached dynamic equilibrium with the environmental conditions and management imposed. Such swards are usually referred to as *permanent pasture*. Other pastures, termed *leys*, have been ploughed and resown at intervals, usually as part of an arable rotation. These recently sown swards are highly unstable and change quickly with age (Chapter 8). Large numbers of species are usually found in permanent pastures, but only a few of these species are normally sown in leys; there has been a recent tendency to sow fewer species, and sometimes only one grass species.

## SCOPE OF THE VOLUME

The scope of this volume has been purposely restricted, by dealing predominantly with temperate managed grasslands. However, there is still a plethora of information; I have therefore focussed attention on two important ecological aspects of ecosystems, energy flow and nutrient cycling, and on the factors which influence those processes.

Chapters 2 to 16 essentially deal with energy flow in managed grassland ecosystems, the processes involved in energy flow, and the factors which determine its magnitude, giving special attention to those management variables that can be effectively and economically varied. Chapters 1 to 8 deal with primary productivity, whereas Chapters 9 to 16 deal with secondary productivity.

Chapters 17 to 22 deal with nutrient cycling and some of the more important processes involved. Nitrogen and phosphorus are considered, because of their agricultural importance, though considerable information is also available for other nutrients (e.g. potassium and sulphur).

The final group of chapters (Chapters 23 to 26) deals with the management of systems as a whole, considering management variables, such as stocking rate and fertilizer use, which influence many aspects of the system, such as energy flow, including both primary and secondary productivity, and nutrient cycling.

Some chapters (e.g. 9 to 12 and 23 to 26) largely present the agricultural context of systems, providing an essential background to the ecological approach of most other chapters.

I am very conscious of many gaps in the cover, and especially of the relatively scant attention given to soil water and rainfall, in the context of both productivity and nutrient cycling. The space given to modelling aspects of the grassland ecosystem also does not do justice to the considerable progress that has been made in the past decade. Indeed, a volume four times as big would be needed to do justice to the wealth of information available on ecosystem analysis of managed temperate grasslands.

## REFERENCES

Brenchley, W.E. and Warington, K., 1958. *The Park Grass Tests at Rothamsted, 1856–1949*. Rothamsted Experimental Station, Harpenden.

Brougham, R.W., 1960. The effect of frequent hard grazing at different times of year on the productivity and species yield of a grass-clover pasture. *N.Z. J. Agric. Res.*, 3: 125–136.

Darwin, C., 1868. *Variation of Animals and Plants Under Domestication*. Murray, London.

Davies, W., 1959. *The Grass Crop*. Span, London, 2nd ed.

Elberse, W.T., Van den Bergh, J.P. and Dirven, J.G.P., 1983. Effect of use and mineral supply on the botanical composition and yield of old grassland on heavy-clay soil. *Neth. J. Agric. Sci.*, 31: 63–88.

Harper, J.L., 1977. *Population Biology of Plants*. Academic Press, London, 892 pp.

Jones, M.G., 1933. Grassland management and its influence on the sward. *Empire J. Exper. Agric.*, 1: 43–57.

Levy, E.B., 1970. *Grasslands of New Zealand*. Government Printer, Wellington, 3rd ed., 324 pp.

Moore, C.W.E., 1964. Distribution of grasslands. In: C. Barnard (Editor), *Grasses and Grasslands*. MacMillan, London, pp. 182–205.

Snaydon, R.W., 1980a. Plant demography in agricultural systems. In: O.T. Solbrig (Editor), *Demography and Evolution in Plant Populations*. Blackwell, Oxford, pp. 131–160.

Snaydon, R.W., 1980b. Ecological aspects of management — a perspective. In: I.H. Rorison and R. Hunt (Editors), *Amenity Grassland: An Ecological Perspective*. Wiley, Chichester, pp. 219–231.

Snaydon, R.W., 1981. The ecology of grazed pastures. In: F.H.W. Morley (Editor), *Grazing Animals*. Elsevier, Amsterdam, pp. 13–31.

Snaydon, R.W., 1984. Plant demography in an agricultural context. In: R. Dirzo and J. Sarukhan (Editors), *Perspectives on Plant Population Ecology*. Sinauer, Sunderland, pp. 389–407.

Chapter 2

# THE PRODUCTIVITY OF TEMPERATE GRASSLANDS

J.E. RADCLIFFE and J.A. BAARS

## INTRODUCTION

About 30% of the world's grasslands occur in the temperate region, and temperate grasslands maintain approximately 35% of the world's ruminant livestock (Reid and Jung, 1982), mainly sheep, beef cattle and dairy cattle.

In this chapter, we present information on the variation and seasonal patterns of pasture production, drawing mainly on data collected in New Zealand and Britain. Most of the data presented here refer to lowland pastures, but data for hill pastures are also considered.

New Zealand and Britain have been selected for study, because standardized experimental techniques have been used, which permit valid comparisons between sites within each country, and because of the contrasting conditions in the two countries. In New Zealand, grass-clover swards are grazed by livestock throughout the year, and little fertilizer nitrogen is applied; nitrogen fixation by clover provides the nitrogen supply, but growth of the most productive pastures can be restricted by nitrogen deficiency at some times (Ball et al., 1979). In Britain, intensive pastoral farming is largely based on grass swards which receive nitrogenous fertilizer, currently averaging 120 kg N ha$^{-1}$ yr$^{-1}$ (Chapter 25). Livestock are usually removed from pastures in winter, when they receive conserved fodder and concentrates. In both countries, perennial ryegrass (*Lolium perenne*) is the most widely sown grass (Langer, 1963; Hopkins, 1979), though most of the existing grasslands contain a wide range of species (Chapter 8).

No direct comparisons can be made of grassland production in New Zealand and Britain, since different defoliation and fertilizer regimes were used in the two countries. It is known that annual yields of pastures are substantially affected by the length of the interval between *defoliations*, and by the height and severity of defoliations (Chapter 6; Davidson, 1968), though these effects vary with season (Brougham, 1970). The effects of plant nutrients, water, plant species, insect pests and pathogens on pasture production are not considered in detail here; some are the subjects of subsequent chapters (Chapters 6–8).

## PASTURE PRODUCTION

### New Zealand

#### Measurement

The pattern of lowland pasture production in New Zealand, considered in this chapter, is based on a series of standardized experiments conducted on grazed pastures. The experiments provide basic information on pasture production and seasonal patterns of growth at various sites. They also provide a reference base for comparing species, for interpreting results of other agronomic and livestock experiments, and for defining periods of likely feed shortage, or abundance, in specified livestock management systems (Chapters 9–11).

Grazed pastures, dominated by perennial ryegrass (*L. perenne*) and white clover (*Trifolium repens*), which were at least two years old, were studied at 20 lowland sites (Fig. 2.1). These were generally below 200 m elevation, except those in Central Otago, which were on elevated valley floors at 300 to 400 m, and those on a volcanic plateau at about 400 m elevation in central North Island. Several of the South Island sites were irrigated.

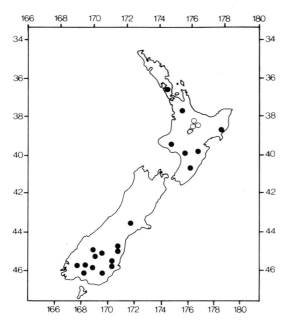

Fig. 2.1. Location of study sites in New Zealand (● = lowland sites; ○ = cooler elevated plateau).

Growth was measured at each site for at least five years, and often for much longer. Pastures received fertilizer as required (mostly phosphorus), but no fertilizer nitrogen was supplied, nitrogen being supplied solely through the clover.

Seasonal patterns of growth were assessed by a "trim" cutting technique (Radcliffe, 1974), in which previously grazed pasture was trimmed to a height of 2.5 cm, caged, and then harvested to the same height, usually after fourteen days, to simulate intensive sheep grazing. Cages were then moved to a new area and pasture trimmed in readiness for the next harvest. The cutting interval was extended in some winters and in droughts when no growth occurred. In determining seasonal growth patterns, yields were apportioned to standard dates (Radcliffe, 1974).

### Annual production

Annual yields averaged 10.4 t DM ha$^{-1}$ (Table 2.1), excluding very droughty sites which produced less than 6 t DM ha$^{-1}$ (e.g. Radcliffe and Cossens, 1974; Rickard and Radcliffe, 1976).

Annual production among North Island sites was rather greater, and much more variable, than among South Island sites, which generally experienced less moisture stress. Within each group, the variation in annual yield from year to year at a site was generally greater than the variation in annual means among sites. Some sites in different climatic regions (Robertson, 1959) had similar mean annual yields and growth patterns (Radcliffe, 1975a, b).

### Seasonal pattern

Proportionately more growth occurred in winter, and less in summer, in the North Island than in the South Island (Table 2.1).

TABLE 2.1

Annual production (t DM ha$^{-1}$) and seasonal distribution of pasture growth in New Zealand[1]

|  | Mean | | Site extremes | |
|---|---|---|---|---|
|  | North Island | South Island | North Island | South Island |
| *Annual yield* | 10.9 | 9.9 | 10.0–12.8 | 8.3–11.5 |
| Variation[2] | 3.4 | 2.3 |  |  |
| *Percentage of annual yield* |  |  |  |  |
| Winter (June–Aug.) | 14 | 4 | 12–19 | 0–7 |
| Spring (Sept.–Nov.) | 39 | 38 | 35–44 | 33–49 |
| Summer (Dec.–Febr.) | 26 | 39 | 24–30 | 31–47 |
| Autumn (March–May) | 21 | 19 | 18–25 | 15–22 |

[1] Grazed perennial ryegrass–white clover pastures cut at 14-day intervals. North Island data from 7 sites, on average 11 years per site, South Island data from 13 sites, on average 10 years per site.
[2] 90% confidence half range within site.

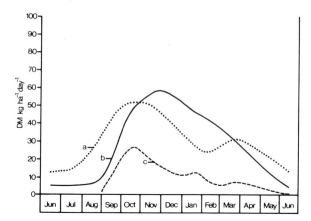

Fig. 2.2. Average patterns of seasonal pasture growth rates in New Zealand: $a$ = North Island (seven sites which have warmer winters and drier summers); $b$ = South Island (thirteen sites, which have colder winters and autumns with less summer moisture stress than the North Island sites); $c$ = a very dry site in Central Otago, South Island, with severe moisture stress over most of the growing season.

Sites in the North Island had a pronounced spring peak of pasture growth (Fig. 2.2, line $a$) followed by lower growth rates over summer, then an autumn "flush" of growth in response to late season rains and to a reduced evaporative demand. The South Island sites, which included four irrigated sites, had a shorter growing season (Fig. 2.2, line $b$) with a peak in November–December (late spring–early summer). Growth was probably curtailed at the beginning and end of the growing season by cool temperatures. The droughty site (Fig. 2.2, line $c$) had lower growth rates, with a marked spring peak.

## Britain

### Measurement

The response of ryegrass swards to fertilizer nitrogen was measured at 21 sites (Fig. 2.3) over four years (Morrison et al., 1980). Perennial ryegrass monocultures were established, and measurements began one year after sowing. The swards were cut six times, to a height of 5 cm, during the growing season to simulate grazing. Swards received regular nitrogen applications (up to 750 kg N ha$^{-1}$) and other mineral fertilizers as necessary.

### Annual production

Annual yields, from swards receiving 450 kg N ha$^{-1}$ yr$^{-1}$ (in six dressings each of 75 kg N),

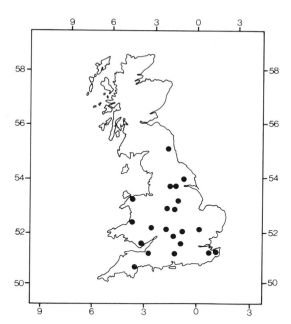

Fig. 2.3. Location of study sites in Britain (● = lowland sites).

averaged 11.4 t DM ha$^{-1}$, with a range from 6.1 to 14.3 t DM ha$^{-1}$ among sites (Morrison, 1980). As in New Zealand, the year-to-year variation within British sites was at least equal to the variation between site means (Morrison, 1980).

### Seasonal pattern

Nearly half the annual production came from the first two cuts in April and May (Table 2.2). Growth rates at this time were undoubtedly enhanced by plentiful soil moisture, the addition of fertilizer nitrogen to vegetative tillers, the longer cutting interval and the longer stubble (compared with the New Zealand series); these are all factors conducive to grass growth. Mid-season growth

TABLE 2.2

Percentage of annual yield produced in four seasons in Britain[1]

| Season | Month | Mean (%) | Site extremes |
|---|---|---|---|
| Winter | November–February | 0 | |
| Spring | March–May (2 cuts) | 48 | 42–65 |
| Summer | June–August (3 cuts) | 41 | 31–47 |
| Autumn | September–October (2 cuts) | 11 | 3–16 |

[1]Average of 21 sites, adapted from Morrison et al. (1980).

contributed on average 41% of the annual yield, while late-season growth contributed 11%, but this varied widely between sites.

## FACTORS WHICH AFFECT PASTURE GROWTH

### Defoliation

Both height and frequency of defoliation affect subsequent growth (Chapter 7). Experiments in New Zealand have shown that a 28-day interval between defoliations gave 22% more annual production than a 14-day interval (Baars, 1982), although the seasonal pattern was only slightly affected (Fig. 2.4). Pastures defoliated to 7.5 cm stubble in summer regrew much faster than those defoliated more severely to 2.5 cm stubble (Brougham, 1970).

Models have been used to simulate the pasture growth rates of irrigated pastures, assuming rotation lengths and stubble heights typical for dairy cows or for breeding sheep in Canterbury New Zealand (Baars, 1980). The longer rotation intervals and laxer defoliation of the dairying management gave much higher pasture growth rates in summer, but lower growth rates in autumn (Fig. 2.5). The more frequent defoliation intervals and shorter stubble heights, normally used for sheep management, encouraged more early-season and late-season growth than the rigid 14-day interval between defoliations used in the standard cutting trial (Fig. 2.5).

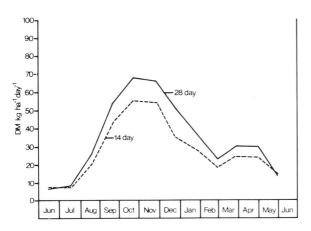

Fig. 2.4. Effect of cutting interval on seasonal pattern of pasture growth rates at Takapau, New Zealand.

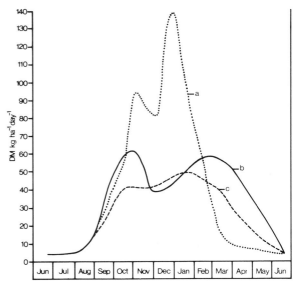

Fig. 2.5. Effect of management on seasonal pattern of pasture growth rates under irrigation in Canterbury, New Zealand: $a$ = simulated results of dairy cow management with a 15-day rotation in mid-spring increasing to 35 days in late autumn (residual dry matter ranging from 1800 kg DM ha$^{-1}$ in spring to 2200 kg DM ha$^{-1}$ in summer); $b$ = simulated results of sheep management with set stocking in early spring, rotation length increasing from 20 days in mid-spring to 40 days in autumn and 80 days in winter (residual dry matter 1200 kg DM ha$^{-1}$ in spring, 700 kg DM ha$^{-1}$ in summer, 1200 kg DM ha$^{-1}$ in autumn and 300 kg DM ha$^{-1}$ in winter); $c$ = actual yield of plots cut at 14-day intervals to 2.5 cm. Mean over thirteen years (from Rickard and Radcliffe, 1976).

### Climate

#### Radiation and daylength

Radiation receipt, rather than temperature, governs pasture growth rate when pasture is in the linear phase of growth — that is, from the third to tenth week following defoliation to 2.5 cm (Brougham, 1959b). Baars and Waller (1979) found a strong linear relationship between the growth rate of irrigated pasture and the logarithm of radiation, under infrequent lax defoliation with beef cattle. Alberda (1962) also concluded that, provided the supply of water and mineral nutrients was adequate, the rate of pasture growth was principally determined by radiation, after the canopy closed.

In pastures defoliated every fourteen days, as in the present New Zealand series, growth remained mostly in the exponential phase, so that maximal growth rates were not usually attained before defoliation. Under these conditions, radiation is not such an important determinant of pasture growth rate (Fitzgerald, 1978).

According to Behaeghe (1974) the effects of daylength may differ between vernalized and non-vernalized swards. However, his experimental conditions were extreme, and further study is required to see if the results apply under field conditions. In experiments sited from 51° to 69°N, Deinum et al. (1981) found that the greatest daily growth rates (135 to 250 kg DM ha$^{-1}$) were associated with the longest daylengths. However, in the absence of moisture stress, with plentiful nutrients and a cutting interval of fourteen days, there was no indication of a daylength effect in mid-latitudes (36° to 46°) in New Zealand.

### Temperature

Experiments with young, spaced plants of temperate grass species, grown in controlled environments, have shown an optimum temperature for growth of about 20°C (Mitchell, 1956), but extrapolating from climate laboratory data to the field is fraught with difficulties. Peacock (1975a, b) found that, in the field, the temperature at the level of the stem apex was the most important factor determining leaf extension rates, but such temperature measurements are rarely available. We have found a close correlation ($R^2 = 0.75$) between spring (August–October) pasture growth rate and soil temperature at 10 cm depth (Fig. 2.6), and a similar correlation ($R^2 = 0.70$) between autumn growth rate and soil temperature (Fig. 2.6).

### Rainfall

In both New Zealand and Britain, soil water supplies are usually replenished over winter, so early season growth is little affected by moisture shortages. As a result, spring growth rates, at many lowland sites in New Zealand, show little variation from year to year (Radcliffe, 1979); mid-season growth (December–April) usually shows the greatest variation in response to fluctuations in effective rainfall (Fig. 2.7). Large areas of grassland in both countries have insufficient soil water in summer for maximum grass growth (Coulter, 1975; Wilkins et al., 1981). In addition, summer droughts can affect subsequent response to autumn rains, though this depends on species and cultivar (Norris and Thomas, 1982).

In the British study (Morrison et al., 1980), where the annual rainfall of sites ranged between 450 and 1150 mm, more than 60% of the variation in annual yield between sites was attributed to differences in the available soil water, caused by differences in rainfall over the growing season and by the water-holding capacity of the soil (Morrison et al., 1980).

In the New Zealand study, spring and summer rainfall (September–February) accounted for at least 60% of the variation in annual yield. When sites were grouped into broad climatic regions, herbage production increased by about 13.5 kg DM ha$^{-1}$ for every millimetre increase in rainfall, for sites receiving up to 550 mm of rain in spring and summer, but only by about 4 kg DM ha$^{-1}$ for every millimetre at wetter sites (Fig. 2.8). The smaller response to spring and summer rainfall among the six sites in group 2 (Fig. 2.8), compared to group 3, can be attributed to their higher

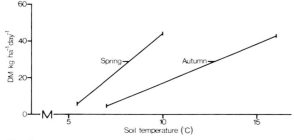

Fig. 2.6. Relationships between pasture growth rate and soil temperature (10 cm depth) in spring (mean of four sites) and in autumn (mean of two sites) in North Island of New Zealand.

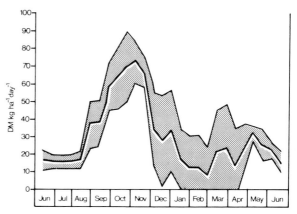

Fig. 2.7. Seasonal pattern of pasture growth rates at Masterton, New Zealand, over five years, mean ± standard error (from Radcliffe, 1975b).

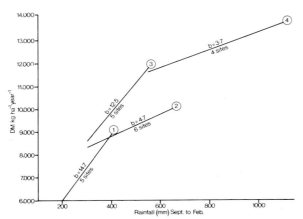

Fig. 2.8. Relationships between annual yield of pasture, and rainfall from September to February in New Zealand for four groups of sites. Group 1: South Island sites, with cold winters and hot summers. Group 2: eastern North Island sites, with moderately cold winters and very warm summers. Group 3: windy coastal sites with cold winters and warm summers. Group 4: northern North Island sites, with mild winters and warm humid summers.

radiation receipts and hence greater evaporation rates, which diminish effective rainfall. The effect of summer rainfall on summer yield, at the various sites in New Zealand was consistently about 8 kg DM ha$^{-1}$ for every millimetre increase in rainfall, when mean summer rainfall was between 130 and 400 mm. The water-holding capacity of soils also affected pasture growth. For example, a volcanic soil of low water-holding capacity in Central North Island, New Zealand, produced 5.7 t DM ha$^{-1}$, whereas a more moisture-retentive volcanic soil nearby produced 9.0 t DM ha$^{-1}$ yr$^{-1}$ (Baars et al., 1975).

Water supply affects seasonal growth patterns as well as annual yields; an example of dryland and irrigated pastures on similar soils in the South Island of New Zealand is given in Fig. 2.9A, and simulated data (Baars, 1980) for North Island sites are given in Fig. 2.9B. In summer-dry areas of South Island, New Zealand, irrigation doubled or even trebled annual production (Radcliffe and Cossens, 1974; Rickard and Radcliffe, 1976; Scott and Maunsell, 1981). At moderately dry sites in both New Zealand and Britain, irrigation over the summer and autumn also increased yield (Fig. 2.9B and C).

## Plant species

Temperate grass species and cultivars often have different flowering times, and thus have rather

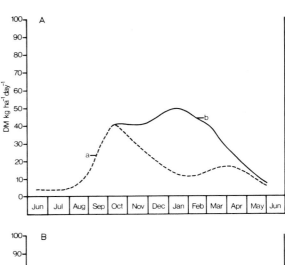

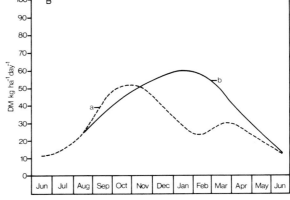

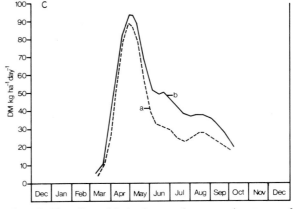

Fig. 2.9. The effect of irrigation on the seasonal patterns of pasture growth. A. Canterbury Plains site, New Zealand, mean of 13 years: $a$ = without added water; $b$ = irrigated with approx. 600 mm yr$^{-1}$. (Redrawn from Rickard and Radcliffe, 1976.) B. New Zealand, North Island, mean of five sites with average of thirteen years: $a$ = without added water; and $b$ = simulated for same sites and years with soil water non-limiting. C. Perennial ryegrass (pasture) at Hurley, Britain, mean of thirteen years: $a$ = without added water; and $b$ = irrigated. (Redrawn from Corrall, 1978).

different patterns of growth (Fig. 2.10). Subtropical grasses have a very different pattern of growth, being summer-active. Subtropical grasses, such as paspalum (*Paspalum dilatatum*) and kikuyu (*Pennisetum clandestinum*), form a major component of pasture in the warmer temperate environment of Northland, New Zealand (Baars, 1976; Lambert et al., 1977; Piggot et al., 1978; Percival et al., 1979). In mixtures with temperate species, these grasses give more pasture growth in summer and autumn than do pastures based on ryegrass and white clover. As a result, ryegrass–paspalum pastures average 17.2 t DM ha$^{-1}$ yr$^{-1}$ (Baars, 1976), whereas estimated yields of perennial ryegrass–white clover pastures under similar management average only 11.0 t DM ha$^{-1}$ yr$^{-1}$.

### Effect of flowering

A short-term decline in grass growth occurs in summer, during the late stages of flowering, when the reproductive tillers mature and few vegetative tillers remain. The reduction in growth occurs even if defoliation prevents flowering and if water and nitrogen supplies are adequate. It is accompanied by a halt in leaf expansion and hence dry weight harvested (Anslow, 1965; Leafe et al., 1974). Such developments affect seasonal growth patterns.

## POTENTIAL PRODUCTION FROM PASTURE

### New Zealand

From studies under controlled environments, Mitchell (1963) has estimated that the potential annual production from grassland in New Zealand ranges from 28 t DM ha$^{-1}$ in Northland, to 18 t DM ha$^{-1}$ in Southland. In central New Zealand, Brougham (1959b) estimated potential yields of 25 t DM ha$^{-1}$ yr$^{-1}$, when pasture was continuously maintained in the linear growth phase. Yields of 18 to 19 t DM ha$^{-1}$ yr$^{-1}$ are attained under dairy cow management in the warmer North Island, and 15 to 16 t DM ha$^{-1}$ yr$^{-1}$ in the cooler South Island. Under sheep management, yields of 14 t DM ha$^{-1}$ yr$^{-1}$ in the North Island and 12 t DM ha$^{-1}$ yr$^{-1}$ in the South are attained. Clearly, the best achieved annual yields are still much less than the calculated potential.

Pasture growth rates are greatest when the critical leaf area necessary for complete light interception is attained. Data obtained by Brougham (1957, 1958) indicate that the critical leaf area index for ryegrass–white clover pastures is attained with a herbage mass of 2.2 t DM ha$^{-1}$ in early summer and about 1.0 t DM ha$^{-1}$ in winter. Under these conditions, maximal daily growth rates can be 220 kg DM ha$^{-1}$ (Brougham, 1959a; Harris et al., 1973; Geenty, 1983).

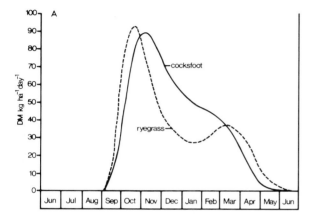
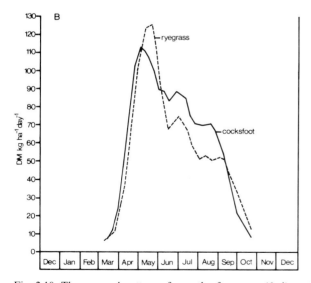

Fig. 2.10. The seasonal pattern of growth of ryegrass (*Lolium perenne*) and cocksfoot (*Dactylis glomerata*). A. "Grasslands Apanui" cocksfoot and "Grasslands Ruanui" perennial ryegrass grown in the MacKenzie Plains, South Island, New Zealand, with irrigation and fertilizer (redrawn from Scott and Maunsell, 1981). B. S 37 cocksfoot and S 23 perennial ryegrass grown at Hurley, Britain (redrawn from Corrall, 1978).

### Europe

Potential yields from densely tillering grasses, harvested at four-week to five-week intervals and supplied with ample water and mineral fertilizers (up to 600 kg N ha$^{-1}$ yr$^{-1}$), have been estimated at 20 to 25 t DM ha$^{-1}$ yr$^{-1}$ (Cooper, 1970; Voightlander and Voss, 1980). These yields agree with estimates of 20 t DM ha$^{-1}$ obtained from calculations based on measures of photosynthesis and respiration (Leafe, 1978). However, yields from experiments receiving up to 600 kg N ha$^{-1}$ yr$^{-1}$ have averaged only 12.5 t DM ha$^{-1}$ under infrequent cutting, and 11.9 t DM ha$^{-1}$ under more frequent cutting (Morrison, 1980). Using these data, Morrison et al. (1980) have developed relationships between herbage yield and available soil water, summer rainfall and fertilizer nitrogen. Using these relationships, Wilkins et al. (1981) have estimated that current grassland yields in lowland Britain could be at least doubled with more use of fertilizer nitrogen.

In Holland, ryegrass pastures reach a critical leaf area index with a herbage mass of 1.5 t DM ha$^{-1}$ (Alberda and Sibma, 1968). Under these conditions, and with optimal water and mineral supplies, daily growth rates reached 200 kg DM ha$^{-1}$ with little variation between years or between soil types (Alberda and Sibma, 1968; Sibma, 1968). This maximum growth rate is similar to that attained in England (Leafe et al., 1974) and in New Zealand (see above).

## HILL PASTURES

Herbage production on hill lands varies widely, due to variations in both soil and climatic conditions. In both Britain (Munro and Davies, 1974) and New Zealand (N.Z. DSIR, 1980), acid soils are widespread in upland areas, giving soils of low mineral fertility, which severely curtail pasture growth. However, only climatic factors will be considered here.

Slopes of different aspects receive differing amounts of solar radiation and wind, which affect precipitation and moisture losses, heat balances, and soil-forming processes, and so affect plant growth (Lambert and Roberts, 1978). These differences also cause stock preferentially to graze or rest on certain areas, so depleting or enriching, respectively, the pool of soil nutrients.

In Britain, rough and hill grazings constitute more than one-third of agricultural land (Lazenby, 1981). The upland climate is characterized by sustained high wind speeds, large variations in the amount and duration of rain and snow, and a marked altitudinal variation in temperature (Gloyne, 1968). Compared with New Zealand hill country, whose climate is not yet well documented, radiation levels are considerably lower, and the grass growing season much shorter (Munro, 1973).

Altitude may restrict pasture production, mainly because of cooler spring temperatures. There is an average reduction of 0.6°C for every 100 m increase in altitude (Smith, 1976), and exposure to wind also increases with altitude. In northwestern Lancashire, on fertilized soils, Morris and Thomas (1972) reported a reduction in the annual yield of grasses from 7.0 t DM ha$^{-1}$ at 50 m to 5.5 t DM ha$^{-1}$ at 300 m. In eastern Scotland, flowering of perennial ryegrasses was delayed by three to ten days for every 100 m increase in altitude up to 600 m, though annual yields were little affected (Hunter and Grant, 1971); better soil water conditions in mid-season at higher altitudes compensated for cooler temperatures and slower growth in spring. Elsewhere in Scotland, yields were also little affected by altitude (250 m to 450 m) or by slope, and the effect of aspect was not consistent (Burnham et al., 1970). In Wales, severe winters can reduce production in the subsequent year through cold damage. These factors reduced annual yields from 10.1 t DM ha$^{-1}$ at 30 m to 4.4 t DM ha$^{-1}$ at 300 m in Wales in 1969 (Munro and Davies, 1973).

New Zealand hill pasture lands, which cover 4.5 million ha, are extremely diverse in lithology, structure and geology (N.Z., DSIR, 1980). Their existing and potential production has been little investigated, as compared with lowland pastures. The effect of altitude is evident in central North Island, where cooler spring temperatures at about 400 m above sea level can appreciably delay spring growth (Fig. 2.11). In Otago, South Island, low temperatures at 1100 m curtail the growing season to 170 days, with nearly half the annual yield of 2.5 t DM ha$^{-1}$ being produced in the spring months of October and November (Cossens and Brash, 1981). Growth rates during winter tend to be slightly higher on the warmer north-facing slopes, whereas growth rates in summer tend to be slightly higher on south-facing slopes, which retain more soil moisture. However, sometimes the overall

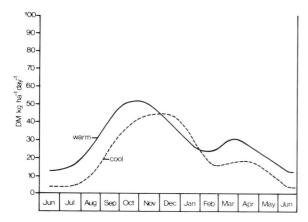

Fig. 2.11. Seasonal patterns of pasture growth at warmer lowland sites (<100 m) in the North Island of New Zealand (mean of seven sites), and cooler inland sites of the central North Island (mean of three sites at 380 m elevation).

pattern of growth is similar on nearby north-facing and south-facing hillsides (Gillingham, 1973; Radcliffe, 1982). At least five studies in New Zealand (Radcliffe et al., 1968; Gillingham, 1973; Suckling, 1975; Grant and Lambert, 1979; Radcliffe, 1982) have found that annual pasture production within single hill paddocks can differ up to three-fold, depending on topography, slope and deposition of stock excreta.

## CONCLUSIONS

Pasture yields under lowland conditions in New Zealand and Britain averaged about 10 to 11 t DM $ha^{-1}$ $yr^{-1}$ under standardized, frequent defoliations. Year-to-year variation within sites was usually greater than differences between site means.

Available soil water during the growing season, and soil temperature at the beginning and near the end of the growing season, were the overriding factors determining both annual yields and the seasonal pattern of pasture growth.

Frequency and severity of defoliation were also important factors determining annual yield; these effects varied with season. With infrequent lax defoliation, radiation receipt and daylength determined maximum growth rates and hence yield.

Less information is available on hill pastures, though, altitude, aspect and slope all affect microclimate and so influence pasture growth.

## ACKNOWLEDGEMENTS

We thank our many colleagues in the New Zealand Ministry of Agriculture and Fisheries, who have obtained and summarized pasture data.

## REFERENCES

Alberda, T., 1962. Actual and potential production of agricultural crops. *Neth. J. Agric. Sci.*, 10: 325–333.

Alberda, T. and Sibma, L., 1968. Dry matter production and light interception of crop surfaces. 3. Actual herbage production in different years as compared with potential values. *J. Br. Grassl. Soc.*, 23: 206–216.

Anslow, R.C., 1965. Grass growth in midsummer. *J. Br. Grassl. Soc.*, 20: 19–26.

Baars, J.A., 1976. Seasonal distribution of pasture production in New Zealand. 8. Dargaville. *N.Z. J. Exper. Agric.*, 4: 151–156.

Baars, J.A., 1980. Development of a simulation model for pasture growth. *Annu. Conf. Agron. Soc. N.Z.*, 10: 103–106.

Baars, J.A., 1982. Variation in grassland production in the North Island with particular reference to Taranaki. *Proc. N.Z. Grass. Assoc.*, 31: 81–107.

Baars, J.A. and Waller, J.E., 1979. Effects of temperature on pasture production. *Annu. Conf. Agron. Soc. N.Z.*, 9: 101–104.

Baars, J.A., Radcliffe, J.E. and Brunswick, L., 1975. Seasonal distribution of pasture production in New Zealand. 7. Wairakei, pasture and lucerne production. *N.Z. J. Exper. Agric.*, 3: 253–258.

Ball, P.R., Brougham, R.W., Brock, J.L., Crush, J.R., Hoglund, J.H. and Carran, R.A., 1979. Nitrogen fixation in pasture. 1. Introduction and general methods. *N.Z. J. Exper. Agric.*, 7: 1–5.

Behaeghe, T.J., 1974. Experiments on the seasonal variations of grass growth. In: *Proc. 12th Int. Grassl. Congr., Moscow*, Vol. 1, Part 1, pp. 76–85.

Brougham, R.W., 1957. Some factors that influence the rate of growth of pasture. *Proc. N.Z. Grassl. Assoc.*, 19: 109–116.

Brougham, R.W., 1958. Interception of light by the foliage of pure and mixed stands of pasture plants. *Aust. J. Agric. Res.*, 9: 39–52.

Brougham, R.W., 1959a. The effects of frequency and intensity of grazing on the productivity of a pasture of short rotation ryegrass and red and white clover. *N.Z. J. Agric. Res.*, 2: 1232–1248.

Brougham, R.W., 1959b. The effects of season and weather on the growth rate of a ryegrass and clover pasture. *N.Z. J. Agric. Res.*, 2: 283–296.

Brougham, R.W., 1970. Frequency and intensity of grazing and their effects on pasture production. *Proc. N.Z. Grassl. Assoc.*, 32: 137–144.

Burnham, C.P., Court, M.N., Jones, R.J.A. and Tinsley, J., 1970. Effect of soil parent material, elevation, aspect and fertiliser treatment on upland grass yields. *J. Br. Grassl. Soc.*, 25: 272–277.

Cooper, J.P., 1970. Potential production and energy conversion in temperate and tropical grasses. *Herbage Abstr.*, 40: 1–15.

Corrall, A.J., 1978. The effect of genotype and water supply on the seasonal pattern of grass production. In: *Constraints to Grass Growth and Grassland, Pap. Europ. Grassl. Fed., 7th Gen. Meet., Belgium*, pp. 2.23–2.31.

Cossens, G.G. and Brash, D.W., 1981. Seasonal distribution of pasture production in New Zealand. 15. The Higher Otago Plateau: Rock and Pillar Range. *N.Z. J. Exper. Agric.*, 9: 73–78.

Coulter, J.D., 1975. The climate. In: G. Kuschel (Editor), *Biogeography and Ecology in New Zealand*. W. Junk, The Hague, pp. 87–138.

Davidson, J.L., 1968. Growth of grazed plants. *Proc. Aust. Grassl. Conf.*, 1a: 1–10.

Deinum, B., De Beyer, J., Nordfeldt, P.H., Kornher, A., Østgård, O. and Van Bogaert, G., 1981. Quality of herbage at different latitudes. *Neth. J. Agric. Sci.*, 29: 141–150.

Fitzgerald, P.D., 1978. The relationship between irrigated pasture production and radiation. *Nordic Hydrol.*, 9: 263–266.

Geenty, K.G., 1983. *Influence on Nutrition and Body Composition on Milk Production in the Grazing Ewe*. Thesis, Lincoln College, Lincoln, 205 pp.

Gillingham, A.G., 1973. Influence of physical factors on pasture growth on hill country. *Proc. N.Z. Grassl. Assoc.*, 35: 77–85.

Gloyne, R.W., 1968. Some climatic factors affecting hill-land productivity. *Occas. Symp. Br. Grassl. Soc.*, 4: 9–15.

Grant, D.A. and Lambert, M.C., 1979. Nitrogen fixation in pasture. 5. Unimproved North Island hill country, 'Ballantrae.' *N.Z. J. Exper. Agric.*, 7: 19–22.

Harris, A.J., Brown, K.R., Turner, J.D., Johnston, J.M., Ryan, D.L. and Hickey, M.J., 1973. Some factors affecting pasture growth in Southland. *N.Z. J. Exper. Agric.*, 1: 139–163.

Hopkins, A., 1979. The botanical composition of grasslands in England and Wales: an appraisal of the role of species and varieties. *J. R. Agric. Soc. Eng.*, 140: 140–150.

Hunter, R.F. and Grant, S.A., 1971. The effect of altitude on grass growth in East Scotland. *J. Appl. Ecol.*, 8: 1–19.

Lambert, M.C. and Roberts, E., 1978. Aspect differences in an improved hill country pasture. *N.Z. J. Agric. Res.*, 21: 255–260.

Lambert, J.P., Rumball, P.J. and Christie, A.J.R., 1977. Comparison of ryegrass and Kikuyu grass pastures under mowing. *N.Z. J. Exper. Agric.*, 5: 71–77.

Langer, R.H.M., 1963. Grass species and strains. In: R.H.M. Langer (Editor), *Pastures and Pasture Plants*. A.H. and A.W. Reed, Wellington, pp. 65–83.

Lazenby, A., 1981. British grasslands; past, present and future. *Grass Forage Sci.*, 36: 243–266.

Leafe, E.L., 1978. Physiological, environmental and management factors of importance to maximum yield of the grass crop. In: J.K.R. Gasser and B. Wilkinson (Editors), *Maximising Yields of Crops*. HMSO, London, pp. 37–49.

Leafe, E.L., Stiles, W. and Dickinson, S.E., 1974. Physiological processes influencing the pattern of productivity of the intensively managed grass sward. In: *Proc. 12th Int. Grassl. Congr., Moscow, Part 1*, pp. 442–457.

Martin, T.W., 1960. The role of white clover in grassland. *Herbage Abstr.*, 30: 159–164.

Mitchell, K.J., 1956. Growth of pasture species under controlled environment. 1. Growth at various levels of constant temperature. *N.Z. J. Sci. Technol.*, 38A: 203–216.

Mitchell, K.J., 1963. Production potential of New Zealand pasture land. *Proc. N.Z. Inst. Agric.*, 9: 80–96.

Morris, R.M. and Thomas, J.G., 1972. The seasonal pattern of dry-matter production of grasses in the North Pennines. *J. Br. Grassl. Soc.*, 27: 163–172.

Morrison, J., 1980. The influence of climate and soil on the yield of grass and its response to fertilizer nitrogen. In: *The Role of Nitrogen in Intensive Grassland Production, Pap. Int. Symp. Europ. Grassl. Fed.* PUDOC, Wageningen, pp. 51–57.

Morrison, J., Jackson, M.V. and Sparrow, P.E., 1980. *The Response of Perennial Ryegrass to Fertilizer Nitrogen in Relation to Climate and Soil*. Grassland Research Institute Technical Report, 27. Grassland Research Institute, Hurley, 27 pp.

Munro, J.M.M., 1973. Potential pasture production in the uplands of Wales. 1. Climatic variation. *J. Br. Grassl. Soc.*, 28: 59–67.

Munro, J.M.M. and Davies, D.A., 1973. Potential production in the uplands of Wales. 2. Climatic limitations on production. *J. Br. Grassl. Soc.*, 28: 161–169.

Munro, J.M.M. and Davies, D.A., 1974. Potential pasture production in the uplands of Wales. 5. The nitrogen contribution of white clover. *J. Br. Grassl. Soc.*, 29: 213–223.

New Zealand D.S.I.R., 1980. *New Zealand Department of Scientific and Industrial Research, Discussion Paper 3. Land Alone Endures*. Government Printer, Wellington, 286 pp.

Norris, I.B. and Thomas, H., 1982. Recovery of ryegrass species from drought. *J. Agric. Sci., Cambridge*, 98: 623–628.

Peacock, J.M., 1975a. Temperature and leaf growth in *Lolium perenne*. 1. The thermal microclimate: its measurement and relation to crop growth. *J. Appl. Ecol.*, 12: 99–114.

Peacock, J.M., 1975b. Temperature and leaf growth in *Lolium perenne*. 2. The site of temperature perception. *J. Appl. Ecol.*, 12: 115–123.

Percival, N.S., Lambert, J.P., Christie, A.R.J. and McClintock, M.M., 1979. Evaluation of *Paspalum* selections. 2. Productivity under grazing. *N.Z. J. Exper. Agric.*, 7: 65–73.

Piggot, G.H., Baars, J.A., Cumberland, G.L.B. and Honore, E.N., 1978. Seasonal distribution of pasture production in New Zealand. 13. South Kaipara, Northland. *N.Z. J. Exper. Agric.*, 6: 43–46.

Radcliffe, J.E., 1974. Seasonal distribution of pasture production in New Zealand. 1. Methods of measurement. *N.Z. J. Exper. Agric.*, 2: 337–340.

Radcliffe, J.E., 1975a. Seasonal distribution of pasture production in New Zealand. 4. Westport and Motueka. *N.Z. J. Exper. Agric.*, 3: 239–246.

Radcliffe, J.E., 1975b. Seasonal distribution of pasture production in New Zealand. 7. Masterton (Wairarapa) and Maraekakaho (Hawke's Bay). *N.Z. J. Exper. Agric.*, 3: 259–265.

Radcliffe, J.E., 1979. *Climatic and Aspect Influences on Pasture Production in New Zealand*. Thesis, Lincoln College, Lincoln, 365 pp.

Radcliffe, J.E., 1982. Effects of aspect and topography on pasture production in hill country. *N.Z. J. Agric. Res.*, 25: 485–496.

Radcliffe, J.E. and Cossens, G.G., 1974. Seasonal distribution of pasture production in New Zealand. 3. Central Otago. *N.Z. J. Exper. Agric.*, 2: 349–358.

Radcliffe, J.E., Dale, W.R. and Viggers, E., 1968. Pasture production measurements on hill country. *N.Z. J. Agric. Res.*, 11: 658–700.

Reid, R.L. and Jung, G.A., 1982. Problems of animal production from temperate pastures. In: J.R. Hacker (Editor) *Nutritional Limits to Animal Production from Pastures. Pap. Int. Symp. CSIRO Division of Tropical Crops and Pastures, St. Lucia, Qld., 1981*, pp. 21–43.

Rickard, D.S. and Radcliffe, J.E., 1976. Seasonal distribution of pasture production in New Zealand. 12. Winchmore, Canterbury Plains dryland and irrigated pastures. *N.Z. J. Exper. Agric.*, 4: 329–335.

Robertson, N.G., 1959. The climate of New Zealand. In: A.H. McLintock (Editor), *Descriptive Atlas of New Zealand*. Government Printer, Wellington, pp. 19–23.

Scott, D. and Maunsell, L.A., 1981. Pasture irrigation in the Mackenzie Basin. 1. Species comparison. *N.Z. J. Exper. Agric.*, 9: 279–290.

Sibma, L., 1968. Growth of closed green crop surfaces in the Netherlands. *Neth. J. Agric. Sci.*, 16: 211–216.

Smith, L.P., 1976. *The Agricultural Climate of England and Wales*. Ministry of Agriculture, Fisheries and Food, Technical Bulletin 35, London, HMSO, 147 pp.

Suckling, F.E.T., 1975. Pasture management trials on unploughable hill country at Te Awa. 3. Results for 1959–69. *N.Z. J. Exper. Agric.*, 3: 351–436.

Voightlander, G. and Voss, N., 1980. Factors maximizing pasture growth. In: *Proc. Int. Meet. Animal Production from Temperate Grassland, Dublin*, pp. 390–413.

Wilkins, R.J., Morrison, J. and Chapman, P.E., 1981. Potential production from grasses and legumes. In: J.L. Jollans (Editor), *Grassland in the British Economy*. CAS Paper 10. Centre for Agricultural Strategy, Reading, pp. 390–413.

Chapter 3

# THE PRODUCTIVITY OF MEDITERRANEAN AND SEMI-ARID GRASSLANDS

E.F. BIDDISCOMBE

## INTRODUCTION

In this chapter the plant productivity of Mediterranean and semi-arid grasslands of four world regions is considered as background for detailed discussion of the processes involved and of the important management variables, which are considered in subsequent chapters. I shall concentrate on recent research papers and reviews, as far as possible, and attempt to assess the current value of some lines of study. Finally, some potentially useful topics for future research on the role of plant productivity in animal production will be suggested.

*Total annual dry matter production* (ADM) and pasture growth rate are useful starting points for estimates of primary productivity. Measurement of ADM requires full-season exclosure from grazing (Chapter 1). A disadvantage is that exclosure per se may increase growth within small cages during periods of slow winter growth (Heady, 1957), or may induce botanical composition different from that of grazed pasture outside (Rossiter and Pack, 1956). Few data on ADM under grazing have been published and, indeed, grassland ecologists have for long questioned their value (Rossiter, 1966). More measurements have been made of the periodic amounts of *dry matter on offer* to animals (DMO) and of the botanical composition of the material. These measures are important in that they frequently determine intake; however, DMO is not a good index of pasture productivity since it depends on stocking rate and season, and often includes green and dead material. DMO is used by many workers for modelling and analyzing relationships between pasture and animal production, sometimes in association with intake data obtained from oesophageal, rumen or faecal samples. In this chapter both ADM and DMO data for the regions are presented if possible. DMO at seasonal peak, under a conservative stocking rate, will be quoted extensively to compare the potential productivity of regions, mainly because of the availability of data.

## REGIONAL PASTURE PRODUCTIVITY

### Mediterranean Basin

The productivity of annual and annual + perennial pastures of Israel is known from recent work on both ungrazed and grazed grasslands. Gutman (1978) recorded a range of 2.5 to 3.8 t ha$^{-1}$ in the primary production (ADM) of herbaceous vegetation of northern Israel sampled through eight years, and with an average rainfall of about 600 mm. DMO is normally between 2 and 4 t ha$^{-1}$ (Table 3.1), but may be less than 2 t ha$^{-1}$ at Kare Deshe with moderate stocking of cattle, and less than 1 t ha$^{-1}$ under drought in the northern Negev. Cattle-grazed pastures at Neve Yaar (32°44'N, 38°10'E) produce up to 8 t ha$^{-1}$ given abundant, well-distributed rainfall, weed control and adequate fertilizer. In the early part of the growing season of all zones (October to December) DMO values of 0.1 to 0.2 t ha$^{-1}$ are common (Noy-Meir, 1975; E. Arnold, 1980); this is unsufficient for the maintenance of lambing ewes and cattle (Seligman and Gutman, 1978). The depression of liveweight of adult sheep normally continues until February in the presence of 0.25 to 0.28 t green herbage ha$^{-1}$ (Tadmor et al., 1974). After the pasture peaks in late April there is a rapid loss

TABLE 3.1

Yearly peak DMO of Mediterranean regional grassland

| Zone | Average rainfall (mm) | Stocking rate (ha$^{-1}$) | DMO (t ha$^{-1}$) | Reference |
|---|---|---|---|---|
| *Israel* | | | | |
| Northern Negev | 250 | 0.6–1.0 ha sheep | 2–4 | Tadmor et al (1974), Noy-Meir (1975) |
| Neve Yaar | 550 | *ca.* 1.8 cattle | 2.4 | Naveh (1982) |
| Kare Deshe | 600 | 0.8 cattle (January to June) | 1.1–1.6 | Seligman and Gutman (1978) |
| *Southern France* | | | | |
| Corsica | 500–750 | 0.4–7.9 sheep (over 3 years) | 0.3, 1.3 3.3 over 3 years | Long et al. (1978) |

of 20 to 30% of the dry matter, followed by a steady decrease through summer.

Unfertilized "garrigue" communities in southern France, cleared of inedible shrubs by burning and/or mulching, have an ADM of 0.3 to 0.6 t dry matter ha$^{-1}$, of which herbaceous species contribute over 50% (Long et al., 1978); the annual rainfall here is 500 to 750 mm. Given all the recommended treatments, including moderate fertilizer application, ADM increases at least fourfold and provides 90% herbaceous material after seven years. In Corsican maquis, pasture improved by the foregoing techniques produces DMO values (Table 3.1) which enable a multiple increase in sheep stocking rate over several years.

Le Houérou and Hoste (1977) reviewed data from eight countries of the Mediterranean Basin and concluded that ADM with 300 mm annual rainfall was about 1.2 t ha$^{-1}$, of which only about 0.4 t was consumed by various animals. The estimates for 700 mm rain were 1.9 t ADM and 0.8 t eaten. Naveh (1982) agreed with these rainfall/productivity relationships for unimproved Tabor oak grassland of the East Mediterranean.

In Aegean Turkey (690 mm rainfall) ungrazed annual and perennial pastures have a peak yield of 3.1 to 4 t ha$^{-1}$ of green shrub-free dry matter one year after sowing (Pringle and Cornelius, 1968). Primary productivity of a grass + forb stand in the lowlands of northern Greece was 2.6 t ha$^{-1}$ from 480 mm rain (Papanastasis, 1981), though peak biomass fluctuates by 50% between years, depending on rainfall. Ungrazed swards of subterranean clover (*Trifolium subterraneum*) at Elvas, Portugal, with 600 mm rainfall produce about 8 t ha$^{-1}$ (Crespo, 1970). Data on ADM in other parts of the Basin are scarce, and inadequate to estimate stock-carrying capacity.

E. Arnold (1980) recorded that the dry matter content of prominent grasses of the semi-arid northern Negev increased rapidly from 45 to 55%, and that of forbs increased to 40%, when they were nearing the reproductive stage in late winter. At the same stage, N content of mixed pasture was 2 to 3%.

**California**

**Productivity.** Pitt and Heady (1978) reported a 19-year average of 2.2 t ha$^{-1}$ at maturity (1 June) in exclosures of annual grassland in the central coastal ranges of northern California (39°N, 123°W) with an annual rainfall of 890 mm, though the yearly variation was between 1.1 and 5.6 t ha$^{-1}$. Mean ADM of winter annual pasture in the Sierra Foothills, where annual rainfall averages 750 mm and varies between 250 and 1100 mm, is recorded as 4.2 t ha$^{-1}$ (Raguse et al., 1980). Green forage is rarely present until well into autumn, increasing slowly without grazing to 0.8 t ha$^{-1}$ by 1 March (range 0.2 to 2.2 t between years). In the same area, Evans et al. (1975) showed that between 0.35 and 0.65 t ha$^{-1}$ growth of herbage was produced in autumn, following a clearing defoliation; this was 10 to 31% of the maximum spring biomass.

At the San Joaquin Field Station (rainfall 490 mm), Wagnon et al. (1958) measured peak DMO means between 1.7 and 3.1 t ha$^{-1}$ on unfertilized plots grazed at an average of 73 steer days per hectare.

### Botanical composition

Plant cover of pastures in the central coastal ranges of northern California in March average 35% annual grasses and 12% legumes (*Medicago, Trifolium, Vicia* spp.) (Pitt and Heady, 1978), but composition varies greatly. For example, the cover of *Bromus mollis* varies from 5 to 31% between years; the percentage of annual legumes tending to vary inversely.

## South Africa

**Productivity.** Sown winter pastures of grass + legume are used commonly in three coastal subregions of the Western Cape Province, often in rotation with cereals (Joubert, 1980). Wassermann and Wicht (1972) found that annual legumes (*Trifolium, Medicago* spp.), mown periodically, yielded between 1.7 and 3.2 t ha$^{-1}$ in drier years (320–370 mm) and 4.9 to 8.6 t ha$^{-1}$ in a wet year (510 mm rainfall).

Carrying capacity of low-rainfall (350 mm per year or less) veld of the Province is only about 0.2 small stock unit ha$^{-1}$, based on ADM of 2.8 t ha$^{-1}$ per year or less (Walter and Volk, 1955). The Succulent Form of Karroid Broken Veld (Acocks, 1975) has an above-ground biomass of 7 to 8 t ha$^{-1}$ (Rutherford, 1981). Using the estimate by Rodin et al. (1974) that the annual increment is 20 to 55% of the total biomass, this indicates an annual increment of 1.5 to 4 t ha$^{-1}$.

**Botanical composition.** The components of Karroid Veld are usually assessed by ground cover, rather than biomass, and Joubert (1980) concluded that palatable species comprise about 40% of the total number of species and 30% of total cover. Grasses contribute about 30% of the biomass of the Succulent Karroid Veld (Rutherford, 1981).

Composite samples of edible plants of Broken Veld in winter contain 27% fibre, 8.7% crude protein, and only 0.09% phosphorus which is deficient for grazing animals (data quoted by Rutherford, 1981).

## Southern Australia

Grassland of the region is mainly used for sheep production in areas where rainfall is about 300 to 700 mm and mainly for cattle production where rainfall is about 700 to 1200 mm.

### Southeastern and southern zones

**Productivity.** ADM of exclosed annual pasture in northern Victoria varies from 2.1 t ha$^{-1}$ in a drought year (325 mm rainfall) to 8.8 t ha$^{-1}$ in above-average rainfall (880 mm) (White et al., 1980), with about 4 to 5 t ha$^{-1}$ yr$^{-1}$ in an average rainfall (580 mm) (McGowan et al., 1983). This production is only achieved if phosphate requirements are met. Perennial ryegrass + subterranean clover pasture in 850 mm rainfall at Hamilton, southwestern Victoria, has an estimated ADM of 16 to 21 t ha$^{-1}$ (Cayley et al., 1980), though this includes 59 and 34% respectively of dead herbage.

Silsbury et al. (1979) obtained ADM values of 10.0 t ha$^{-1}$ for swards of *Medicago truncatula* sown at high densities in South Australia. The best approximations of ADM for the area are those of M.V. Smith (1970), Carter and Day (1970) and Brown (1976), based on the peak DMO of swards grazed at low stocking rate, but ungrazed for the final four to six weeks of the growing season. Smith (1970) studied lucerne + annual species swards on deep sand and obtained values of 1.2 t ha$^{-1}$ for a drought year (200 mm rain) and 3.0 t ha$^{-1}$ for a year of above-average rainfall (470 mm). Corresponding values obtained by Carter and Day (1970) for a perennial ryegrass + subterranean clover pasture on Kangaroo Island were 2.5 t ha$^{-1}$ (380 mm rain) and 4.3 t ha$^{-1}$ (690 mm). Brown (1976) reported a range of 0.5 to 3.2 t ha$^{-1}$ from annual pasture receiving similar rainfall extremes but higher stocking rates than those of the other workers. In dry years, only 20% of the peak production of these grazing systems is achieved by mid winter.

**Botanical composition.** A common feature of swards grazed at low stocking rates is the dominance of grasses, which may constitute 90% of ground cover in spring, especially in dry years. Subterranean clover usually recovers to contribute 30 to 60% in subsequent wet years.

## Southwestern zone

**Productivity.** Greenwood et al. (1967) found that a clover + annual grass pasture gave an ADM of 5.9 t ha$^{-1}$, when given N fertilizer and 550 mm rainfall. About 16% of the yearly production was achieved by late winter.

Table 3.2 indicates the peak DMO of subterranean clover-based annual pastures in this zone. The results were obtained from swards adequately fertilized with P, continuously grazed and measured over a range of annual rainfall; stocking rates were about average for the district. In the range 460 to 620 mm rainfall, and 5 to 11.7 sheep ha$^{-1}$, peak DMO varied between 3.1 and 6.7 t ha$^{-1}$, depending mainly on rainfall.

At a mean stocking rate of 12.3 sheep ha$^{-1}$, Mann et al. (1966) found peak DMO of between 3.4 and 5.7 t ha$^{-1}$ with clover + grass pastures in an area where rainfall fluctuated around 750 mm. A *Dactylis glomerata* + annual species pasture during an eight-year experiment, with mean rainfall of 830 mm, had a maximum DMO of 1.8 to 3.6 t ha$^{-1}$ under rotational grazing by 2.3 steers ha$^{-1}$ (Nicholas, 1979; D.A. Nicholas, pers. commun., 1982). Corresponding values for mixed-annuals pasture were between 1.6 and 3.1 t ha$^{-1}$, with less DMO than *Dactylis* pastures in spring. Allowing for probable consumption during the growing season, the apparent primary productivity in this region is within the range quoted by Cayley et al. (1980) for a similar area of rainfall in southwestern Victoria.

DMO of annual species in the autumn–early winter period in the region is <0.5 t ha$^{-1}$ in most years (Table 3.2); this is below the level required to maintain sheep liveweight (R.C.G. Smith et al., 1972). This contrasts with southwestern Victoria, where the rains at late March usually maintain adequate herbage supply (0.7–0.8 t DMO ha$^{-1}$) through autumn and winter for 15 to 20 non-breeding sheep ha$^{-1}$ (Birrell et al., 1980).

McKeown and Smith (1970) and Rossiter (1966) calculated that only about 50% of the ADM from sown pastures in the region is eaten. This indicates that the primary productivity is about 8.5 kg ha$^{-1}$ of consumable dry matter mm$^{-1}$ rainfall; this is about double the average

TABLE 3.2

Peak DMO of annual-type pastures of grazing experiments in southwestern Australia

| Average annual rainfall (mm) | Main sward component(s)[1] | Stocking rate (sheep ha$^{-1}$ yr$^{-1}$) | DMO in successive years of experiment | | Reference |
|---|---|---|---|---|---|
| | | | autumn/ early winter (t ha$^{-1}$) | spring peak (t ha$^{-1}$) | |
| 400 | *Erodium*, grass, clover | 8.2 | 0.46, 0.54 | 2.1, 1.9 | White et al. (1966) |
| 460 | clover | 8.2 | 0.50 (mean of 3 years) | 6.8 (mean of 3 years) | McKeown and Smith (1970), Experiment A |
| 475 | clover (sprayed to control fungal attack) | 8 | 0.31, 1.5, n.d. | 3.1, 3.2, 4.5 | Anderson et al. (1982) |
| 500 | clover, annual grasses | 11.7 | 0.29, 0.5, 1.0 | n.d., 6.7, n.d. | Rogers et al. (1982), Experiment 2 |
| 550 | clover, *Bromus mollis* (no N fertilizer) | 10.4 (mean of 8.6, 12.3) | 0.15, 0.28 | 5.4, 6.0 | Greenwood et al. (1967) |
| 620 | clover alone | 10 | 0.39, 0.47, 0.32 | 3.3, 3.5, 4.8 | Biddiscombe et al. (1976), Experiment A |
| 620 | clover alone | 10 | 0.48, 1.2 | 5.0, 5.7 | Biddiscombe et al. (1976), Experiment B |
| 620 | clover alone | 8.75 | 3.1 | 5.4 | Biddiscombe et al. (1980) |
| 620 | 3 annual clover species | 5 | n.d. | 4.3, 6.4 | Rossiter et al. (1972) |

[1]Clover = *Trifolium subterraneum*. n.d. = not determined.

amount for the Mediterranean Basin (Le Houérou and Hoste, 1977). However, intensively improved grassland in oak woodland areas yields 6.7 kg mm$^{-1}$ (Naveh, 1982), which is comparable with the McKeown and Smith figure, if allowance is made for the high stocking rate (1.8 steers ha$^{-1}$) imposed for most of the year in Naveh's study.

**Botanical composition.** Rossiter (1966, pp. 9, 15, 16) listed floristic dominants of pastures in the region; these include clovers on acidic soils, medics on alkaline soils, clover with low applied P, medic with high applied P. The average clover content of sown pasture is sharply reduced by applications of N fertilizer, and the content of sown grasses increases correspondingly (Greenwood et al., 1967). The pasture is later invaded by volunteer grasses and herbs, which constitute an average of 50% in the fourth year, though rather less when N fertilizer has been applied.

The nutritive value of subterranean clover was assessed by Galbraith et al. (1980). The N content of herbage remains steady at 4.8% until early winter, then falls rapidly to 3% by early spring and maintains at 2% thereafter. Digestibility is also high (78–80%) until early winter, then falls steadily to 60% at peak biomass.

## LIMITATIONS TO PRIMARY PRODUCTIVITY

The length of the growing season — that is, the period when adequate soil water is available — generally determines the primary productivity of mediterranean plant communities (Mooney and Vieira da Silva, 1981). Either total annual rainfall (Le Houérou and Hoste, 1977), annual and particular monthly rainfalls (Duncan and Woodmansee, 1975), or progressive seasonal totals combined with winter minimum temperatures (Pitt and Heady, 1978; Naveh, 1982) may be correlated with herbage yield. Le Houérou and Hoste recognized a relationship over the range of 200 to 1000 mm annual rainfall, and an asymptotic relationship at higher rainfalls. Duncan and Woodmansee (1975), using multiple linear regression, found that the ADM of Californian rangelands was most highly correlated with annual rainfall combined with April, November and January rain. Pitt and Heady (1978) showed that yield in June was significantly correlated with rainfall in the previous September to November period and by the number of days $<0°C$ in October. Naveh (1982) found that rainfall for six months beginning in October, together with minimum temperatures in December–January, account for high proportions of the variance in ADM; legume yields were more dependent on annual rainfall, whereas grass yields were more dependent on early-season rains. Greater productivity ($>0.5$ t DMO ha$^{-1}$, Table 3.2) in autumn–early winter is invariably associated with above-average, persistent autumn rainfall according to the respective authors. Gutman (1978) also explained differences in autumn production between years on the basis of autumn rainfall.

Naveh (1982) emphasized the resilient nature of mediterranean grassland in persisting through wide fluctuations in yearly weather patterns, whereas Arnold and Bennett (1975) noted the stable level of peak DMO under low stocking rate, over a nine-year period with many years of poor rainfall.

The inhibiting effect of low winter temperatures and low radiation receipt on growth is evident for a number of prominent mediterranean species: annual ryegrass (Donald, 1951), subterranean clover (Silsbury and Fukai, 1977; Galbraith et al., 1980) and *Medicago truncatula* (Silsbury et al., 1979), but also for steppe vegetation in Israel (Gutman, 1978), and annual pastures in southern Victoria (Birrell et al., 1980).

Volunteer, early-maturing annual species, such as *Hordeum murinum*, *Vulpia myuros* and *Erodium botrys*, which may dominate annual pastures, tend to reduce the productivity of pastures, because they mature too early to take full advantage of spring rainfall for leaf production (D.F. Smith, 1968; Biddiscombe et al., 1976). These species also subsequently pose problems because of their low-quality herbage.

Productivity is often limited by deficiency of major nutrients, such as N, P, K and S (Rossiter, 1966). On the podzolic and sandy soils of Southern Australia, there are responses to lime and sulphur (Barrow, 1965, 1966), nitrogen (Greenwood et al., 1967) and to applied phosphate (Carter and Day, 1970; Ozanne and Howes, 1971). Nitrogen and sulphur applications also affect the productivity of California rangeland (Martin and Berry, 1970). Both legumes and grasses of pastures of northern

Israel are responsive to N and P dressings (Naveh, 1982). Soil N is depleted in semi-arid agro-ecosystems of Israel (Noy-Meir and Harpaz, 1977). The yield of herbaceous plants of the garrigue of southern France is increased by applied N, P and K (Long et al., 1978). Acid soils of the cereal belt of the Western Cape, South Africa, require liming for satisfactory growth of annual pasture legumes (Wassermann and Wicht, 1972).

Nitrogen appears to be needed more widely than other nutrients. The soil fertility that develops during two years under subterranean clover + annual grass + herb pasture ($\geqslant 50\%$ clover content) is adequate for the subsequent growth of cereal crops (Watson, 1963). The clover component of the ley is also desirable as a nutritive summer feed for stock (Wilson and Hindley, 1968).

Grazing management may limit plant productivity. Total and net pasture productivity may decrease with increasing stocking rate (Cayley et al., 1980). In some cases, however, low stocking rates allow a large carry-over of dry herbage which depresses pasture productivity in the following year (Hooper and Heady, 1970). Defoliation per se appears to have only a temporary effect on the growth rate of an emerging pasture (Greenwood et al., 1967; Greenwood and Arnold, 1968). Resting the young pasture in autumn and winter considerably increases DMO in winter (Brown, 1976) and throughout the year (R.C.G. Smith et al., 1973), compared with continuous grazing. Arnold (1968) pointed out that no single management system, such as continuous, rotational or strip grazing, or any combination of these, gives a consistent response in terms of animal productivity; the various systems show only small differences in long-term productivity.

## IMPROVEMENTS OF PASTURE PRODUCTIVITY

The semi-natural grasslands of most of the Mediterranean regions of the world contain relatively unpalatable, and often unproductive, species in abundance. Replacement of species within the grass genera *Bromus*, *Hordeum*, *Poa*, *Aristida*, *Festuca* and *Vulpia* and the herb genera *Erodium*, *Echium* and *Hypochoeris* is usually recommended in most regions. The introduction of subterranean clover to Australia (Rossiter, 1966) and to Portugal (Crespo, 1970) has generally increased pasture yield. Sowing of wheat grasses and annual *Medicago* in central California (Cornelius and Burma, 1970) has succeeded similarly. Joubert (1980) advocated the introduction of *Atriplex* and *Chrysanthemoides* species to Karroid Veld types for increased plant productivity. A perennial grass in place of annual grass may extend the growing season, but the evidence to date from southern Australia indicates that little extra animal production results from the presence of the perennial grass in the medium- to low-rainfall zones ($< 750$ mm per year) (Reed, 1970; Rogers et al., 1982). Longer-term advantage from the use of perennial grass seems to be restricted to high-rainfall areas and longer-growing seasons (Nicholas, 1979; Arkell et al., 1981).

An ability to predict plant productivity is useful in manipulating stocking rate and management, and in planning large-scale movements of stock to favoured zones. Regression analyses, based on early-season weather factors, can be used to predict plant productivity or botanical composition (Duncan and Woodmansee, 1975; Pitt and Heady, 1978; Naveh, 1982). Similarly, simulation models, based on precipitation, temperature, evapotranspiration and photosynthetic season, can improve prediction of regional and zonal patterns of net primary productivity (Lieth, 1974). Simulation of plant productivity, under Australian conditions, has pointed to the importance of seed pools (Galbraith et al., 1980), initial plant density (R.C.G. Smith and Williams, 1973) and plant mortality rates (Bowman et al., 1982). Modelling of plant growth is treated in more detail in Chapter 5.

## CONCLUSIONS AND FUTURE RESEARCH

Peak DMO is within the range of 3500 to 5500 kg ha$^{-1}$ in about half of the years in southwestern Australia; primary productivity of sheep-grazed pastures is about 28% greater than this (Biddiscombe et al., 1980). Thus the primary productivity of improved, clover-based annual pastures in these conditions lies within the bottom half of the range for temperate grasslands given by Whittaker (1975). The critical period for animal production is the autumn–winter period; at this time, sheep may have to be removed and given supplementary feed (Arnold and Bennett, 1975).

Large year-to-year variations in herbage production and botanical composition restricts the full use of managed grasslands in mediterranean regions. Continued research is needed to improve both the stability and level of animal productivity. In particular, the early growth of pasture in autumn and the quality of dry herbage in summer deserve special emphasis. This may be achieved, in part, by introducing grass species with earlier autumn establishment (R.C.G. Smith et al., 1972). Breeding programmes may also help to stabilize plant productivity by developing resistance to drought, disease and insects. Further exploitation should be made of the high nutritive value of those legumes which have a low level of oestrogenic activity.

There appears also to be a need for further tuning of models of plant productivity of mediterranean grassland (Galbraith et al., 1980), including a widening of the data bases for extrapolation to other regions (Morley, 1982). Existing models, such as that of M.V. Smith (1974), need to be validated. Further information is also needed on nitrogen cycling, the attributes which determine the success of species (Rossiter, 1978), the physiology of response to defoliation (Arnold, 1977) and the effects of weather variables (Pitt and Heady, 1978) on productivity.

## ACKNOWLEDGEMENTS

I am grateful to my colleague Dr. G.W. Arnold for help with the literature search.

## REFERENCES

Acocks, J.P.H., 1975. *Veld Types of South Africa. Mem. Bot. Surv. S. Afr.*, No. 40: 128 pp.

Anderson, W.K., Parkin, R.J. and Dovey, M.D., 1982. Relations between stocking rate, environment and scorch disease on grazed subterranean clover pasture in Western Australia. *Aust. J. Exper. Agric. Anim. Husb.*, 22: 182–189.

Arkell, P.T., Glencross, R.N., Nicholas, D.A., Paterson, J.G., Anderson, G.W., Biddiscombe, E.F. and Rogers, A.L., 1981. *Perennial Pasture Grasses in South-Western Australia. 3. Productivity Under Grazing. Tech. Bull. No. 57,* Western Australian Department of Agriculture, Perth, W.A., 11 pp.

Arnold, E., 1980. *Diet Selection by Oesophageal-Fistulated Sheep Grazing in the Semi-Arid Northern Negev.* Pamphlet No. 225, Division of Scientific Publications, The Volcani Center, Bet Dagan, 65 pp.

Arnold, G.W., 1968. Pasture management. In: *Proc. 4th Australian Grasslands Conference, Perth, W.A., 1968*, pp. 189–211.

Arnold, G.W., 1977. Effect of herbivores on arid and semi-arid rangelands. Defoliation and growth of forage plants. In: *Proc. of the 2nd United States/Australia Rangeland Panel.* Australian Rangeland Society, Perth, W.A., pp. 57–72.

Arnold, G.W. and Bennett, D., 1975. The problem of finding an optimum solution. In: G.E. Dalton (Editor), *Study of Agricultural Systems.* Applied Science Publishers, London, pp. 129–173.

Barrow, N.J., 1965. Further investigations of the use of lime on established pastures. *Aust. J. Exper. Agric. Anim. Husb.*, 5: 441–449.

Barrow, N.J., 1966. The residual value of the phosphorus and sulphur components of superphosphate on some Western Australian soils. *Aust. J. Exper. Agric. Anim. Husb.*, 6: 9–16.

Biddiscombe, E.F., Boundy, C.A.P. and Southey, I.N., 1976. *Comparative Grazing Value of Annual Pasture Species.* Tech. Memo. 76/1, CSIRO Division of Land Resources Management, Perth, W.A., 13 pp.

Biddiscombe, E.F., Arnold, G.W., Galbraith, K.A. and Briegel, D.J., 1980. Dynamics of plant and animal production of a subterranean clover pasture grazed by sheep. Part 1 — Field measurements for model calibration. *Agric. Systems*, 6: 3–22.

Birrell, H.A., Reed, K.F.M. and Bird, P.R., 1980. Seasonal limitations to the nutrition of sheep and beef cattle in the high rainfall areas of South-Eastern Australia. *Proc. Aust. Soc. Anim. Prod.*, 13: 32–36.

Bowman, P.J., White, D.H., Cayley, J.W.D. and Bird, P.R., 1982. Predicting rates of pasture growth, senescence and decomposition. *Proc. Aust. Soc. Anim. Prod.*, 14: 36–37.

Brown, T.H., 1976. Effect of deferred autumn grazing and stocking rate of sheep on pasture production in a Mediterranean-type climate. *Aust. J. Exper. Agric. Anim. Husb.*, 16: 181–188.

Carter, E.D. and Day, H.R., 1970. Interrelationships of stocking rate and superphosphate rate on pasture as determinants of animal production. 1. Continuously grazed old pasture land. *Aust. J. Agric. Res.*, 21: 473–491.

Cayley, J.W.D., Bird, P.R. and Chin, J.F., 1980. Effect of stocking rate of steers on net and true growth rates of perennial pasture. *Proc. Aust. Soc. Anim. Prod.*, 13: 468.

Cornelius, D.R. and Burma, G.D., 1970. Seeding and seedbed ridging to improve dry grazing land in central California. In: *Proc. XI Int. Grassland Congr.*, pp. 107–111.

Crespo, D.G., 1970. Some agronomic aspects of selecting subterranean clover (*Trifolium subterraneum* L.) from Portuguese ecotypes. In: *Proc. XI Int. Grassland Congr.*, pp. 207–210.

Donald, C.M., 1951. Competition among pasture plants. 1. Intra-specific competition among annual pasture plants. *Aust. J. Agric. Res.*, 2: 355–376.

Duncan, D.A. and Woodmansee, R.G., 1975. Forecasting forage yield from precipitation in California's annual rangeland. *J. Range Manage.*, 28: 327–329.

Evans, R.A., Kay, B.L. and Young, J.A., 1975. Microenvironment of a dynamic annual community in relation to range improvement. *Hilgardia*, 43(3): 102 pp.

Galbraith, K.A., Arnold, G.W. and Carbon, B.A., 1980. Dynamics of plant and animal production of a subterranean clover pasture grazed by sheep. Part 2 — Structure and validation of the pasture growth model. *Agric. Systems*, 6: 23–43.

Greenwood, E.A.N. and Arnold, G.W., 1968. The quantity and frequency of removal of herbage from an emerging annual grass sward by sheep in a set-stocked system of grazing. *J. Br. Grassland Soc.*, 23: 144–148.

Greenwood, E.A.N., Davies, H. Lloyd and Watson, E.R., 1967. Growth of an annual pasture on virgin land in southwestern Australia including effects of stocking rate and nitrogen fertiliser. *Aust. J. Agric. Res.*, 18: 447–459.

Gutman, M., 1978. Primary production of transitional Mediterranean steppe. In: *Proc. I Int. Rangeland Congr.*, pp. 225–228.

Heady, H.F., 1957. Effect of cages on yield and composition in the California annual type. *J. Range Manage.*, 10: 175–177.

Hooper, J.F. and Heady, H.F., 1970. An economic analysis of optimum rates of grazing in the California annual-type grassland. *J. Range Manage.*, 23: 307–311.

Joubert, J.G.V., 1980. Veld and pasture in animal production systems in the Western Cape and Southwest Africa/Namibia. *S. Afr. J. Anim. Sci.*, 10: 299–303.

Le Houérou, H.N. and Hoste, C.H., 1977. Rangeland production and annual rainfall relations in the Mediterranean Basin and in the African Sahelo-Sudanian Zone. *J. Range Manage.*, 30: 181–189.

Lieth, H., 1974. Comparative productivity in ecosystems — the primary productivity. In: *Proc. I Int. Congr. Ecol.*, p. 36.

Long, G.A., Etienne, M., Poissonet, P.S. and Thiault, M.M., 1978. Inventory and evaluation of range resources in "maquis" and "garrigues" (French Mediterranean Area): Productivity levels. In: *Proc. I Int. Rangeland Congr.*, pp. 505–509.

Mann, P.P., Gorddard, B.J., Glencross, R.N. and Fitzpatrick, E.N., 1966. Stocking rate and rate of superphosphate in a higher rainfall area. *J. Dep. Agric. W. Aust.*, 7: 482–488.

Martin, W.E. and Berry, L.J., 1970. Use of nitrogenous fertilisers on California rangeland. In: *Proc. XI Int. Grassland Congr.*, pp. 817–822.

McGowan, A.A., Cameron, I.H. and White, D.H., 1983. Effect of quantity and frequency of application of superphosphate on the seasonal yield of annual pasture. *Aust. J. Exper. Agric. Anim. Husb.*, 23: 64–72.

McKeown, N.R. and Smith, R.C.G., 1970. Seasonal pasture production, liveweight change and wool growth of sheep in a Mediterranean environment. In: *Proc. XI Int. Grassland Congr.*, pp. 873–876.

Mooney, H.A. and Vieira da Silva, J., 1981. Photosynthesis and allocation. In: N.S. Margaris and H.A. Mooney (Editors), *Components of Productivity of Mediterranean-Climate Regions — Basic and Applied Aspects*. W. Junk, The Hague, p. 39.

Morley, F.H.W., 1982. The role of modelling in sheep production. *Proc. Aust. Soc. Anim. Prod.*, 14: 45–46.

Naveh, Z., 1982. The dependence of the productivity of a semi-arid Mediterranean hill pasture ecosystem on climatic fluctuations. *Agric. Environ.*, 7: 47–61.

Nicholas, D.A., 1979. Perennial pasture measures up. *J. Dep. Agric. W. Aust.*, 20: 57–59.

Noy-Meir, I., 1975. *Primary and Secondary Production in Nomadic and Sedentary Grazing Systems in the Semi-Arid Regions: Analysis and Modelling*. Final Research Report to Ford Research Foundation. Hebrew University, Jerusalem, 332 pp.

Noy-Meir, I. and Harpaz, Y., 1977. Agro-ecosystems in Israel. *Agro-Ecosystems*, 4: 143–167.

Ozanne, P.G. and Howes, K.M.W., 1971. The effects of grazing on the phosphorus requirements of an annual pasture. *Aust. J. Agric. Res.*, 22: 81–92.

Papanastasis, V.P., 1981. Species structure and productivity in grasslands of Northern Greece. In: N.S. Margaris and H.A. Mooney (Editors), *Components of Productivity of Mediterranean-Climate Regions — Basic and Applied Aspects*. W. Junk, The Hague, pp. 205–217.

Pitt, M.D. and Heady, H.F., 1978. Responses of annual vegetation to temperature and rainfall patterns in Northern California. *Ecology*, 59(2): 336–350.

Pringle, W.L. and Cornelius, D.R., 1968. Grazing potential in Aegean Turkey. *J. Range Manage.*, 21: 151–154.

Raguse, C.A., Hull, J.L. and Delmas, R.E., 1980. Perennial irrigated pastures. III. Beef calf production from irrigated pasture and winter annual range. *J. Range Manage.*, 72: 493–499.

Reed, K.F.M., 1970. Variation in liveweight gain with grass species in grass-clover pastures. In: *Proc. XI Int. Grassland Congr.*, pp. 877–880.

Rodin, L.E., Bazilevich, N.I. and Rozov, N.N., 1974. Primary productivity of the main world ecosystems. In: *Proc. I Int. Congr. Ecol.*, pp. 176–181.

Rogers, A.L., Biddiscombe, E.F., Barron, R.J.W. and Briegel, D.J., 1982. Comparative productivity of perennial and annual pastures under continuous grazing by sheep. *Aust. J. Exper. Agric. Anim. Husb.*, 22: 364–372.

Rossiter, R.C., 1966. Ecology of the Mediterranean annual-type pasture. *Adv. Agron.*, 18: 1–56.

Rossiter, R.C., 1978. The ecology of subterranean clover-based pastures. In: J.R. Wilson (Editor), *Plant Relations in Pastures*. CSIRO, Canberra, A.C.T., pp. 325–339.

Rossiter, R.C. and Pack, R.J., 1956. The effect of protection from grazing on the botanical composition of an annual-type subterranean clover pasture. *J. Aust. Inst. Agric. Sci.*, 22: 71–73.

Rossiter, R.C., Taylor, G.B. and Anderson, G.W., 1972. The performance of subterranean, rose, and cupped clovers under set-stocking. *Aust. J. Exper. Agric. Anim. Husb.*, 12: 608–613.

Rutherford, M.C., 1981. Biomass structure and utilisation of the natural vegetation in the winter rainfall region of South Africa. In: N.S. Margaris and H.A. Mooney (Editors), *Components of Productivity of Mediterranean-Climate Regions — Basic and Applied Aspects*. W. Junk, The Hague, pp. 135–149.

Seligman, N.G. and Gutman, M., 1978. Cattle and vegetation responses to management of Mediterranean rangeland in Israel. In: *Proc. I Int. Rangeland Congr.*, pp. 616–618.

Silsbury, J.H. and Fukai, S., 1977. Effects of sowing time and sowing density on the growth of subterranean clover at Adelaide. *Aust. J. Agric. Res.*, 28: 427–440.

Silsbury, J.H., Adem, L., Baghurst, P. and Carter, E.D., 1979. A quantitative examination of the growth of swards of *Medicago truncatula* cv. Jemalong. *Aust. J. Agric. Res.*, 30: 53–63.

Smith, D.F., 1968. The growth of barley grass (*Hordeum leporinum*) in annual pasture. 1. Germination and establishment in comparison with other annual species. *Aust. J. Exper. Agric. Anim. Husb.*, 8: 478–483.

Smith, M.V., 1970. Effects of stocking rate and grazing management on the persistence and production of dryland lucerne on deep sands. In: *Proc. XI Int. Grassland Congr.*, pp. 624–628.

Smith, M.V., 1974. Stochastic influences on the profitability of resowing rundown pastures with perennial species. In: *Proc. XII Int. Grassland Congr.*, pp. 388–394.

Smith, R.C.G. and Williams, W.A., 1973. Model development for a deferred-grazing system. *J. Range Manage.*, 26: 454–460.

Smith, R.C.G., Biddiscombe, E.F. and Stern, W.R., 1972. Evaluation of five Mediterranean annual pasture species during early growth. *Aust. J. Agric. Res.*, 23: 703–716.

Smith, R.C.G., Biddiscombe, E.F. and Stern, W.R., 1973. Effect of spelling newly sown pastures. *Aust. J. Exper. Agric. Anim. Husb.*, 13: 549–555.

Tadmor, N.H., Eyal, E. and Benjamin, R.W., 1974. Plant and sheep production on semi-arid annual grassland in Israel. *J. Range Manage.*, 27: 427–432.

Wagnon, K.A., Bentley, J.R. and Green, L.R., 1958. Steer gains on annual-plant range pastures fertilised with sulphur. *J. Range Manage.*, 11: 177–187.

Walter, H. and Volk, O.H., 1955. *Grundlagen der Weiderwirtschaft in Süd-West Africa*. Ulmer Verlag, Stuttgart, 218 pp.

Wassermann, V.D. and Wicht, J.E., 1972. A preliminary evaluation of rose clover (*Trifolium hirtum* All.) and cupped clover (*Trifolium cherleri* L.) as pasture legumes in the cereal areas of the winter rainfall region. *Proc. Grassland Soc. S. Afr.*, 7: 93–97.

Watson, E.R., 1963. The influence of subterranean clover pastures on soil fertility. 1. Short-term effects. *Aust. J. Agric. Res.*, 14: 796–807.

White, D.H., McConchie, B.J., Curnow, B.C. and Ternouth, A.H., 1980. A comparison of levels of production and profit from grazing Merino ewes and wethers at various stocking rates in Northern Victoria. *Aust. J. Exper. Agric. Anim. Husb.*, 20: 296–307.

White, L.D., Lightfoot, R.J. and Glencross, R.N., 1966. The Avondale stocking rate experiment. *J. Dep. Agric. W. Aust.*, 7: 442–455.

Whittaker, R.H., 1975. *Communities and Ecosystems*. MacMillan, New York, N.Y., 2nd ed., 385 pp.

Wilson, A.D. and Hindley, N.L., 1968. The value of the seed, pods and dry tops of subterranean clover (*Trifolium subterraneum*) in the summer nutrition of sheep. *Aust. J. Exper. Agric. Anim. Husb.*, 8: 168–171.

Chapter 4

# PHYSIOLOGICAL ASPECTS OF PASTURE GROWTH

C. MARSHALL

## INTRODUCTION

In an intensively managed grass sward, with adequate supplies of mineral nutrients and water, the production of dry matter is determined by the efficiency with which leaves intercept light and utilize it in the assimilation of carbon. In addition, the total biomass production is dependent on the length of the growing season, whereas the economic (or harvested) yield depends on the pattern of allocation of carbon from the leaves to the various sinks of the plant.

In temperate grasses, herbage production is characterized by a marked seasonal pattern in growth rate (Chapter 2). Since this is related more to the transition from vegetative to reproductive growth than to the pattern of incoming radiation (Fig. 4.1A) or temperature (Corrall and Fenlon, 1978), there would appear to be considerable potential for increasing herbage production, especially in the climatically favourable conditions, immediately after flowering.

The physiological features of pasture growth, where leaf material is harvested at intervals, contrast greatly with those of annual crops, where a reproductive sink is harvested at the end of the lifecycle. In grassland, leaves are both sources and sinks; their removal inevitably results in a period in which sward growth is limited by assimilate supply. As new leaf area is produced, after cutting or grazing, photosynthesis increases and so the growth rate of the sward increases until the accumulated dry matter is sufficient to be harvested again (Fig. 4.1B). As a result, carbon assimilation and tissue production by the sward, and each of its individual component plants, follow a cyclical pattern over the growing season between succes-

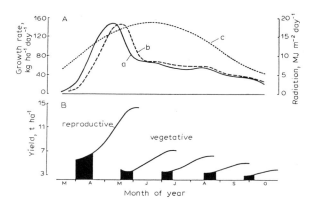

Fig. 4.1. A. Seasonal pattern of herbage production of S24 (*a*) and S23 (*b*) perennial ryegrass grown at the Grassland Research Institute, Hurley, Berkshire, England, together with the seasonal pattern of radiation receipt (*c*). B. Above-ground dry weight of infrequently cut swards of S24 during two-month periods following cutting at the Grassland Research Institute, Hurley; shaded areas indicate periods of incomplete light interception (after Robson, 1981).

sive cuts, though the rate of regrowth is variable, due to seasonal factors (Leafe, 1978). By contrast, in a continuously grazed sward, herbage dry matter shows only slight changes with time throughout the growing season (M.B. Jones et al., 1982).

Most studies of the physiology of growth have been concerned with the assimilation of carbon and its utilization in leaf growth and tiller production. Particular attention has been paid to the effect of environmental and seasonal factors. Such studies have established certain underlying physiological principles of growth, particularly in the infrequently cut sward; this knowledge may be applied to increase productivity, by modifying management techniques, or may be used to define selection criteria in the breeding of improved cultivars. In physiological terms, the productivity of intensively managed, as opposed to unmanaged, grassland is based on manipulating the key physiological activities of the component species so as to optimize performance throughout the growing season.

## THE ASSIMILATION OF CARBON

### Light interception and canopy photosynthesis

Leaf and tiller production establishes the leaf surface which intercepts light. Observations on a wide range of crops indicate that crop growth rate, and the total production of biomass, is a linear function of the amount of radiation intercepted by the crop (Monteith, 1977, 1981). The growth rate of a sward recovering from defoliation, or recently established, is therefore determined primarily by the rate at which leaf area is produced, and by its arrangement in the canopy, but also by the photosynthetic capacity of the leaves and by their persistence. The amount of light intercepted at any one time establishes the upper limit of photosynthesis of the canopy but, if the photosynthetic efficiency of leaves becomes limited by disease, drought or other environmental factors, then photosynthesis will fall short of this potential. Many aspects of the relationships between light and carbon assimilation of herbage grasses have been discussed in recent reviews (Rhodes, 1973; D. Wilson, 1973; Rhodes and Stern, 1978; D. Wilson et al., 1980a; Robson, 1980) and will not be considered comprehensively in this chapter.

The now classical relationship between growth, light distribution and the ratio of canopy photosynthesis to respiration in the grass sward was first described by Donald (1961). The proportion of the incident radiation that is intercepted increases with time after defoliation. Initially, all leaves make a positive contribution to the dry-weight increase of the sward; however, as *leaf area index* (LAI) increases and shading within the canopy increases, the photosynthetic contribution of the lower leaves greatly declines so that the rates of photosynthesis and respiration in the canopy become more similar with time. Donald (1961) considered that there was an optimum LAI value at which the pasture growth rate was maximal; further increases in LAI would reduce pasture growth rate, because the respiration rate would increase, due to the increased accumulation of non-photosynthetic tissue with time. The results of some experiments support this view, but others show that the rates of canopy photosynthesis, canopy respiration and crop growth all tend to reach a plateau with increasing LAI — that is, there is no optimal LAI (Yoshida, 1972; Robson, 1973a; Wilson, 1973). For example, in miniswards of *Lolium perenne*, Robson (1973a) found that the rate of photosynthesis per unit area of sward reached a peak after six weeks, at the time when virtually all the light was intercepted; but thereafter the photosynthetic rate remained more or less constant while LAI, dark respiration and herbage yield continued to increase steadily for a further three weeks before reaching their maximum values. Ceiling yield was achieved at this point because the senescence of old and poorly illuminated leaves matched the production of new leaf tissue. These results suggest that the pattern of growth following defoliation is determined, on the one hand, by the changing pattern of light interception, caused by the increase in LAI with time, as predicted by Donald (1961) and, on the other hand, by leaf mortality, rather than by canopy respiration as proposed by Donald (1961). The latter point is emphasized in a recent model of sward growth, in which the rate of tissue death steadily increases from the time that the rate of dry matter production becomes maximal, whereas the respiration rate remains more or less constant (Fig. 4.2) (Parsons et al., 1983). Leaf senescence in swards is considered in more detail later in this chapter.

The spatial arrangement of leaf area within the canopy has a profound effect on the amount of light intercepted, and therefore on the photosynthetic potential of the sward. Horizontally displayed leaves absorb more light at low LAI than erect leaves. As a result, communities with horizontal leaves will be more productive at low irradiance, or when frequently cut, whereas those with erect leaves will be more productive at high irradiance, or when infrequently cut (Rhodes, 1973, 1975). Selection in *Lolium perenne*, for the combination of canopy characters that are associated with a high growth rate under infrequent defoliation has produced lines yielding up to 30% more than the original populations (Rhodes and Mee,

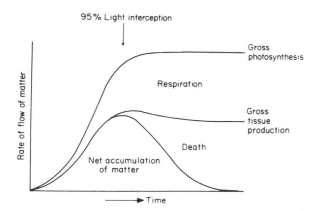

Fig. 4.2. Predicted sward growth following cutting, based on a model, showing the balance between the rate of gross photosynthesis, the rate of loss of matter in respiration and death, and the rate of net accumulation of matter (Parsons et al., 1983).

1980). Similarly, differences in pasture growth rate between species and cultivars are more closely related to differences in canopy structure than to leaf photosynthetic rates (Sheehy and Cooper, 1973). However, marked differences in canopy structure in grasses are not always correlated with differences in growth rate (Sheehy and Peacock, 1977).

**Individual leaf photosynthesis**

As the growth rate of a sward is related more to the total amount of photosynthesis than to the rate of light-saturated net photosynthesis of individual leaves, selection for increased rates of leaf photosynthesis is unlikely to result in faster pasture growth. Indeed, there appears to be an inverse correlation between rate of leaf photosynthesis and mean leaf size in grasses and cereals (Evans, 1976). However, if the rate of leaf photosynthesis departs from its optimum, then the growth rate will decline. For example, nitrogen deficiency or water stress greatly reduces the photosynthetic rate of leaves, and correspondingly reduces canopy photosynthesis and growth rate (Robson and Parsons, 1978; M.B. Jones et al., 1980a). The effects of environmental factors on the photosynthetic activity of leaves have been well documented (D. Wilson, 1973; Cooper, 1975; Milthorpe and Moorby, 1979) and will not be considered further.

The photosynthetic rate of a leaf changes greatly with age both directly, by a decline in metabolic potential with time, and indirectly, by its change of location within the canopy (Jewiss and Woledge, 1967; Leafe, 1972; Woledge and Leafe, 1976). Leafe (1972) studied the regrowth of a sward of *L. perenne* after cutting and found that, when most of the light was intercepted, 80 to 90% of the interception was by the two youngest fully expanded leaves of tillers — that is, the leaves with the highest photosynthetic potential. Thus, although the photosynthetic rate of leaves declines steadily soon after appearance, the overall photosynthesis of the canopy is not greatly affected, because new leaves soon take over their function. However, another factor is also important in this respect. Leaves of grasses that emerge in partial shade have a far lower photosynthetic potential than leaves that emerge in full sunlight (Woledge, 1973; Woledge and Leafe, 1976; Parsons and Robson, 1981a). As a result, as the LAI increases, the photosynthetic potential of successive leaves decreases (Fig. 4.3) and the photosynthetic potential of the sward declines during regrowth. Robson (1973a) concluded that 30% less carbon was fixed by a minisward of *L. perenne* than would have been the case if the photosynthetic rate of young leaves at the end of the growth period was equal to that of leaves produced earlier. Successive leaves produced by elongating flowering tillers do not show this decline in photosynthetic potential, perhaps because they do not emerge into a shaded environment (Woledge and Leafe, 1976; Woledge, 1979), though Parsons and Robson (1981a) noted that even leaves produced before stem elongation maintained their photosynthetic capacity.

In view of the factors already considered, it

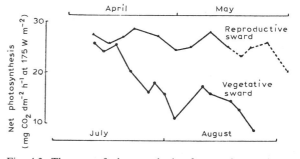

Fig. 4.3. The rate of photosynthesis of successive newly expanded leaves of S24 *Lolium perenne*: in a vegetative sward of increasing LAI (July and August) and a reproductive sward (April and May); ----- represents the flag leaf only. (From Woledge and Leafe, 1976.)

would seem that the decline in photosynthetic potential could be prevented if the sward was kept less dense, for example by frequent grazing or cutting; however, this would be offset by reduced light interception at the lower LAI. Attempts are now being made to overcome this conflict by selecting genotypes that are relatively insensitive to the effects of shading. For example, D. Wilson (1981) has shown that there is considerable variation within *L. perenne* in the degree to which photosynthetic potential is affected by low irradiance.

## THE UTILIZATION OF CARBON

### Assimilate partitioning and growth

Growth represents the conversion of assimilated carbon into leaf, stem and root tissue; the overall pattern of growth depends on the relative partition of carbon from each expanded source-leaf to the various meristematic sinks. In seedling plants of *Lolium multiflorum*, $^{14}C$-assimilate moves freely within the plant from each expanded main shoot leaf to the developing leaves of the main shoot and tillers and to every seminal and nodal root (Marshall, 1967). Emerging tillers are therefore dependent on the main shoot for carbohydrate (Fig. 4.4). When the first primary tillers (tillers arising from buds in the axils of main shoot leaves) appear and produce an expanded leaf, they also export assimilate to all meristematic regions of the plant (Marshall and Sagar, 1968; Clifford et al., 1973). There is thus a mutual exchange of newly assimilated carbohydrate between the main shoot and daughter tillers, and between sister tillers; the main shoot and tillers can therefore be regarded as interdependent. As each tiller increases in size and in assimilatory activity, the degree of assimilate transfer between it and adjacent shoots gradually declines, so it becomes more or less independent in terms of its carbon economy. Similar patterns of assimilate flow between the main shoot and developing tillers have been observed in various grasses (Williams, 1964; Nyahoza et al., 1973, 1974; Rogan and Smith, 1974). In miniswards the first primary tiller (T1) of a plant of *L. perenne*, arising from the first leaf of the main shoot, essentially becomes independent when it reaches a dry weight of about 25 mg (that is, at about the two-leaf stage), even though it still imports some assimilate from the main shoot until it is double that weight (Colvill and Marshall, 1981). Primary tillers appearing later, when the sward is denser, import assimilate from the main shoot for a much longer period, relative to their weight, than do the first primary tillers (Fig. 4.4B). The first secondary tillers (tillers arising from buds in the axils of leaves on primary tillers) are initially supported by assimilate from

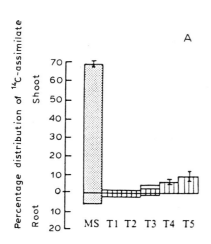
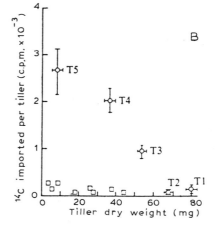

Fig. 4.4. A. The percentage distribution of $^{14}C$-labelled assimilate from the main shoot of plants of *L. perenne* after eight weeks of growth in a sward (*MS* = main shoot; *T1–T5* indicates individual primary tillers). B. The relationship between the import of $^{14}C$-assimilate from the main shoot and the tiller dry weight for results presented in A (○ = primary tillers; □ = secondary tillers) (after Colvill and Marshall, 1981).

the main shoot, as well as from their parental primary tiller; however, later-appearing secondary tillers are supplied with assimilate only by the primary tiller (Colvill and Marshall, 1981). Each developing family of tillers, the offshoots of a single primary tiller, becomes a distinct physiological component of the plant, in which the primary tiller functions in a similar manner to the main shoot in supplying assimilate to its developing daughter (secondary) tillers. Thus, at any one time, the vegetative grass plant consists of a main shoot and independent tillers, with a localized pattern of assimilate export, and a range of developing tillers that import assimilate from specific sources.

If all the tillers except the main shoot are severely defoliated, the pattern of $^{14}$C-assimilate distribution becomes greatly modified (Marshall and Sagar, 1965, 1968; Forde, 1966; Gifford and Marshall, 1973; Rogan and Smith, 1974). Under these conditions, all the defoliated tillers are supplied with carbon from the intact main shoot; the more severe the defoliation the greater is the degree and duration of support. The translocated $^{14}$C-assimilate moves to the meristematic and elongating zones of the cut leaves of each tiller, contributing to the re-establishment of leaf area. As leaf growth recovers, the degree of support by the main shoot declines and assimilate is again distributed only to the younger primary tillers. In *Poa pratensis*, which is rhizomatous, assimilate is transported along rhizomes from the main shoot to defoliated tillers (Nyahoza et al., 1973, 1974). In addition, when only the T1 family of tillers are defoliated, assimilate may be supplied to these tillers from a newly established primary tiller (T6) (Nyahoza et al., 1973). Similar assimilate transport occurs if a previously independent tiller is shaded (Forde, 1966; Rogan and Smith, 1974). A heavily shaded tiller can thus be maintained for up to five weeks (Ong and Marshall, 1979). These various experimental studies show that grass plants function as a family of physiologically integrated tillers, in which no one tiller becomes dominant, but young or stressed tillers are supported at the expense of established tillers or root systems; the increased demand for assimilate may actually stimulate the rate of photosynthesis of the main shoot and established tillers (Gifford and Marshall, 1973). However, when the whole plant is shaded (Ong, 1978; Ong and Marshall, 1979), or when flowering places an overwhelming demand on resources (Ong et al., 1978a), severely stressed tillers are not supported and die.

Productivity of the sward depends upon the partitioning of resources between shoots and roots. This partitioning depends on species (Ryle, 1970), on reproductive development (Ryle and Powell, 1972; Parsons and Robson, 1981b), and on irradiance and nitrogen supply (Powell and Ryle, 1978). During the main phase of reproductive growth, only about 5% of the current assimilate is allocated to the root system (Parsons and Robson, 1981b; Colvill and Marshall, 1984), compared with about 25% during seedling growth, and about 10% during vegetative growth. This reflects the high rate of above-ground biomass production during flowering. Thus, there is little scope for increasing annual yield by diverting a greater proportion of the assimilate to shoot growth (Robson, 1980).

**Respiratory loss**

Respiratory activity is essential for the maintenance of mature tissues and for growth itself. Respiration accounts for about 50% of the carbon assimilated in photosynthesis by a sward (Robson, 1973a). Approximately half of this is associated with maintenance, though D. Wilson (1975a) reported considerable variation between genotypes of *L. perenne* in the rate of mature leaf respiration. Selection for low respiration rates has led to an increase in growth of the order of 25% (D. Wilson, 1982; D. Wilson and Jones, 1982; Robson, 1982). The selected lines have a greater tiller density, and leaf area is developed more rapidly after cutting. Selection for low rates of maintenance respiration seems to offer a realistic opportunity to increase pasture productivity, though long-term effects of stress on genotypes with low respiration have not yet been investigated.

**Carbohydrate reserves**

A significant proportion of current assimilate is not immediately utilized in growth, but is laid down in carbohydrate reserves and used later. Fructans — that is, polymers of fructose — are accumulated in temperate grasses, especially during the reproductive stage and during autumn and winter (Waite and Boyd, 1953; Smith, 1973; Pollock and Jones, 1979). During flowering, fructans are temporarily stored in stem internodes, and subsequently mobilized to developing seeds and to new tillers formed in the post-flowering period (Colvill and Marshall, 1984). These tillers then themselves accumulate fructans during autumn and winter, when growth is restricted by a decline in mean air temperature. Carbohydrate reserves are mobilized during spring, and this may partially account for the early flush of leaf growth in spring, at the time when the potential for leaf extension rapidly increases (Pollock and Jones, 1979; Parsons and Robson, 1980). Fructans may also be of particular significance in unpredictable environments, such as montane regions, acting as a readily available source of sugars of low molecular weight that play a part in drought-tolerance or cold-tolerance mechanisms (Levitt, 1980; Atkinson and Farrar, 1983). A greater understanding of the control of mobilization may provide an opportunity to manipulate, by selection or perhaps by the use of a plant growth regulator, the utilization of carbohydrate reserves and thereby influence the earliness and degree of growth in the spring.

## ENVIRONMENTAL AND DEVELOPMENTAL INFLUENCES ON GROWTH

As well as influencing the supply of carbon for growth, environmental factors exert a direct effect on growth by regulating the processes of cell division and cell elongation. In particular, temperature significantly affects the rate of developmental processes in meristematic tissues. Thus, although there is a close correlation between light interception and dry-matter production of the sward, the rate of production and development of individual components, such as leaves, is more closely related to temperature than to the amount of assimilate available for growth (Monteith, 1977; Kemp and Blacklow, 1980). On the other hand, internal factors, such as inter-sink competition for resources, or the inhibitory influence of certain endogenous growth substances, restrict growth and development despite favourable environmental conditions for growth. Both external and internal influences on growth are considered further here in relation to the production, growth and maintenance of leaves and tillers.

## Growth and development of leaves

### Leaf production

In temperate grasses, the rate of leaf primordium initiation is usually greater than the rate of leaf development, so that primordia tend to accumulate at the shoot apex with time. As each primordium starts to elongate, its meristematic activity becomes localized at the base so that distinct lamina and sheath intercalary meristems are formed. Cell division ceases in the lamina meristem when the ligule is differentiated, but division continues in the leaf sheath meristem until the ligule appears above the previous sheath (Langer, 1979). Cell enlargement is confined to the unemerged part of the leaf; thus both cell division and cell elongation have ceased in newly emerged leaf tissue (Begg and Wright, 1962). Cell enlargement makes a greater contribution to the overall increase in length of the lamina than the production of new cells, so measurements of leaf elongation predominantly reflect cell extension. In most grasses only a limited number of leaves grow at any one time; as one leaf ceases to extend the next leaf appears, and another leaf starts to elongate at the base of the tiller.

There have been relatively few studies of the effects of environmental factors on the rate of leaf primordium production in herbage grasses, but temperature, irradiance and photoperiod are known to influence the rate of leaf appearance (Anslow, 1966; Ryle, 1966; Silsbury, 1970). Temperature has the greatest effect and, over the range 5 to 25°C, leaves appear more frequently with increasing temperature (Mitchell, 1953; Cooper, 1964; Robson, 1972), so there is a very marked seasonal pattern in the rate of leaf appearance. In the field, a new leaf appears on tillers of *L. perenne* approximately every 15 days from the end of April to the beginning of September, and approximately every 52 days from the end of October to mid-February (Vine, 1983) (Fig. 4.5). There is a clear correlation between the rate of leaf appearance per tiller and both soil temperature (Thomas and Norris, 1977; Vine, 1983) and air temperature (Davies and Simons, 1979); the effect is greater for reproductive tillers in the spring than for vegetative tillers in summer and autumn. In addition, Vine (1983) observed that vegetative tillers on reproductive plants had a reduced rate of leaf appearance in May and June, compared with reproductive tillers;

# PHYSIOLOGICAL ASPECTS OF PASTURE GROWTH

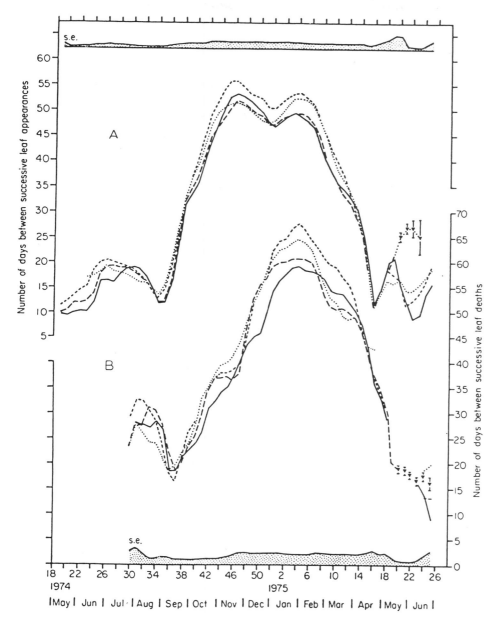

Fig. 4.5. The rate of leaf appearance (A) and leaf death (B) in cut swards of S24 perennial ryegrass receiving 95 kg N ha$^{-1}$ (······), 185 kg N ha$^{-1}$ (-----), 275 kg N ha$^{-1}$ (---) or 365 kg N ha$^{-1}$ (——). In May–June 1975, data are also presented for leaf appearance on vegetative tillers in swards left to grow uncut in 1975 (▼···▼). (After Vine, 1983.)

this possibly reflects the outcome of competition for assimilate or nutrients between vegetative and reproductive tillers.

## Leaf growth and temperature

Day-to-day variation in leaf extension rates are closely correlated with mean daily air temperatures (Williams and Biddiscombe, 1965; Peacock, 1975a; Keatinge et al., 1979) (Fig. 4.6), and the rate of leaf extension is most closely correlated with the temperature of the basal meristem of the elongating leaf (Peacock, 1975b). As the basal meristem of vegetative tillers is situated close to the surface of the soil, leaf growth may be determined more by

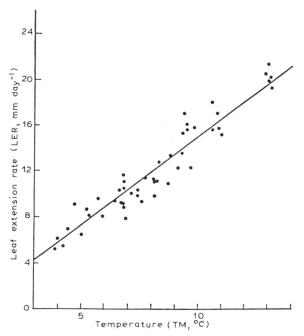

Fig. 4.6. The relationship between leaf extension rate and daily mean air temperature in a sward of *L. perenne* cv. Perma grown in Northern Ireland (after Keatinge et al., 1979).

soil than air temperatures (Peacock, 1975b; Thomas and Norris, 1977). The overall response to temperature is reduced by water stress (see below) but, in the absence of drought, the daily air-temperature mean can be used to predict productivity, since leaf extension is a major determinant of sward productivity (Thomas, 1975). In the field, the growth response to temperature is greater in the spring than in the autumn (Peacock, 1975c; Thomas, 1977; Thomas and Norris, 1977), and the development of the spring response coincides with the onset of floral induction (Parsons and Robson, 1980). Emerging leaves of reproductive tillers are more responsive to temperature than leaves of tillers from vegetative plants; this difference is reflected in the seasonal pattern of sward productivity. Firstly, greater response to temperature in spring allows active early growth when radiation is high but at a time when temperatures are relatively low; the utilization of accumulated fructans may be involved in this response (Pollock and Jones, 1979). Secondly, the smaller temperature response in the vegetative phase may be an important factor restricting productivity in the post-flowering period, when radiation and temperature are both favourable for rapid growth.

There is considerable intraspecific variation in response to temperature. The leaf extension of Mediterranean populations of *Lolium perenne* and *Dactylis glomerata* is more responsive to temperature, at low temperatures, than that of populations from Scandinavia or Britain; however, Mediterranean populations have poor survival at sub-zero temperatures and tend to become summer-dormant (Cooper, 1964; Robson and Jewiss, 1968; Eagles and Østgård, 1971; Østgård and Eagles, 1971). Crosses between Mediterranean and Scandinavian populations give rise to hybrids that can be selected for leaf extension at low temperature, cold-tolerance and summer-production (Eagles and Othman, 1978; D. Wilson et al., 1980a); selection within indigenous species may also give rise to lines with a similar combination of characters (Ollerenshaw et al., 1976). Selections which grow at low temperatures usually display higher rates of both photosynthesis and dark respiration than those of other populations (Ollerenshaw et al., 1976).

Greater leaf growth at low temperature may reflect changes in the supply of endogenous growth substances from root to shoot (Atkin et al., 1973), and the supply of gibberellins may be especially significant as they have a marked effect on cell extension in grass leaves, through their effect on cell-wall plasticity (Goodwin, 1978; R.L. Jones, 1980). Supplying gibberellic acid ($GA_3$) can offset the effects of low soil temperature on leaf extension in *D. glomerata* (Menhenett and Wareing, 1976), and can lead to increased pasture growth rates over winter (Arnold et al., 1967). Differences between genotypes in response to low temperatures may thus be related to differences in the production of gibberellins or perhaps to differences in the sensitivity of elongating cells to gibberellins. The different responses of leaf extension to temperature in spring and autumn may also be related to differences in the hormonal balance of tillers in vegetative and reproductive plants (R.L. Jones, 1973).

**Leaf growth and water stress**

Leaf growth is very sensitive to water deficits (Wardlaw, 1969; Ludlow and Ng, 1976). As a result, sward growth and productivity are significantly reduced by periods of water stress (Turner and Begg, 1978; M.B. Jones et al., 1980a). Cell expansion is more sensitive to water stress than other processes (Hsaio, 1973), so that the rate of leaf extension is reduced even by mild water stress. This sensitivity arises because the uptake of water into growing cells provides the mechanical force for cell enlargement. Leaf water potential, which depends on both water supply and water loss, displays a characteristic diurnal fluctuation in the field, in both well-irrigated and stressed plants (M.B. Jones et al., 1980b; Hanson and Hitz, 1982). Correspondingly, if cell turgor declines with the reduction in leaf water potential, the rate of leaf extension must also decline but, if turgor can be partially buffered, some growth will be sustained. The maintenance of turgor by osmotic adjustment (Turner and Begg, 1978; Turner and Jones, 1980) is a particular feature of the metabolic adaptation to water deficits in the field, where stress slowly increases with time. Inorganic ions ($K^+$ and $Cl^-$), organic ions, soluble carbohydrates, amino acids (especially proline), and quaternary ammonium compounds gradually accumulate in growing and fully expanded tissues, and thereby stabilize turgor against an overall background of declining leaf water potential (J.R. Wilson et al., 1980b; Munns and Weir, 1981; Hanson and Hitz, 1982). Maintenance of turgor by osmotic adjustment therefore requires metabolic expenditure to synthesize and transport materials; this is likely to occur at the expense of growth. However, this response does permit some photosynthetic activity to be maintained, at least in the short term, since it allows stomata to remain more open than if there was no osmotic adjustment (Turner and Begg, 1978; Ludlow, 1980). In addition, some leaf growth and, more significantly, some root growth can be maintained. As well as osmotic adaptation, it is clear that abscisic acid (ABA) also plays a major role in the physiological response to water stress (Aspinall, 1980; W.J. Davies et al., 1980). ABA rapidly accumulates in water-stressed leaves, perhaps in response to the decline in cell turgor (Pierce and Raschke, 1980), and induces stomatal closure. The characteristic reduction in leaf size in water-stressed plants can be reproduced in well-watered plants by repeated foliar applications of ABA (Quarrie and Jones, 1977).

A period of water stress usually reduces leaf area expansion of a sward (Turner and Begg, 1978; M.B. Jones et al., 1980a and b), so that less light is intercepted and growth is reduced (Table 4.1). On rewatering, the rate of leaf extension may significantly exceed that of unstressed plants (Lawlor, 1972; Ludlow and Ng, 1977; Norris and Thomas, 1982a). This may simply reflect the utilization of carbohydrate or other substrates that accumulate at the leaf base during stress, but is more likely to be the result of the simultaneous growth of a large number of cells that have accumulated over the period of water stress, as cell division appears to be less sensitive to water stress than cell elongation (Hsaio, 1973; Clough and Milthorpe, 1975).

There is considerable variation, both between and within grass species, in response to water stress (D. Wilson et al., 1980a; Norris and Thomas, 1982a, b). In *Lolium perenne*, D. Wilson (1975b, c) has shown that selection for fewer, shorter stomata and less leaf ridging gave plants that maintained leaf extension rates and active photosynthesis for a longer period during water stress; these selections also showed increased growth in the field during dry conditions (D. Wilson et al., 1980a). Further improvement might be achieved by selection for

TABLE 4.1

Above-ground dry-matter yield, LAI and tiller density of irrigated (*I*) and stressed (*S*) field swards of perennial ryegrass, following a cut on 9 July 1976 (from M.B. Jones et al., 1980a)

| Date of sampling: | 25 July | | 1 August | | 8 August | | S.E. of means |
|---|---|---|---|---|---|---|---|
| | I | S | I | S | I | S | |
| Dry matter ($g\ m^{-2}$) | 424.2 | 437.6 | 479.0 | 441.7 | 641.6 | 509.5 | ±31.8 |
| LAI | 2.7 | 2.0 | 3.0 | 1.6 | 4.8 | 2.4 | ±0.4 |
| Tiller density ($m^{-2}$) | 13242 | 12498 | 11532 | 8047 | 12409 | 9523 | ±1241 |

more efficient osmoregulation (Blum et al., 1983), or more rapid release of ABA (W.J. Davies et al., 1980) at the onset of water stress. Such selections, by their capacity to buffer cell water relations during stress, might also be expected to have a rapid growth recovery when the water supply is restored, similar to that occurring in populations of *L. perenne* and *L. multiflorum* from dry regions (Norris and Thomas, 1982b).

## Leaf growth and mineral nutrition

The growing leaf is a major sink for nitrogen, phosphorus and potassium; these nutrients are supplied directly by uptake from soil, or indirectly by retranslocation from senescing leaves. Nutrient supply has little effect on the rate of leaf appearance (Langer, 1959; Ryle, 1964; McIntyre, 1965) (see also Fig. 4.5), though low nutrient supply may reduce the number of leaves per flowering shoot. There have been few direct observations of the effect of mineral nutrition on the rate of leaf extension, though Robson and Deacon (1978) showed that nitrogen deficiency greatly reduced the rate of elongation of main shoot leaves of *L. perenne*. The major effect of nutrient supply is on the size of individual leaves; this is particularly evident for nitrogen (Langer, 1959, 1966; Ryle, 1964). Leaf length and width are both increased by an increased supply of nitrogen. Both cell number and cell size are increased, though it is not clear whether both the rate and duration of cell division are affected. Nitrogen thus exerts a major effect on the leaf area per tiller and also stimulates tillering (p. 39); as a result it plays a significant role in determining the LAI of the sward. High inputs of nitrogen fertilizer are therefore essential to maximize grassland productivity (see Chapter 6).

## Leaf life-span

There have been few observations of the life-span of leaves of herbage species, or of the factors regulating their senescence and death. Robson (1973b) observed that, in a minisward of *L. perenne* grown from seed in a controlled environment, six leaves were produced by the main shoot before the oldest leaf died (Fig. 4.7). Thereafter, for a short period, the production of each new leaf was accompanied by the death of the oldest leaf on the shoot. As tillering commenced and the LAI of the sward increased, the rate of leaf death rapidly

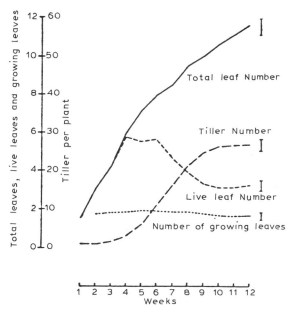

Fig. 4.7. The production and survival of leaves on the main shoot, and the tiller production per plant by *L. perenne* growing in a minisward (from Robson, 1973b).

increased until ceiling yield was reached, when leaf death matched the rate of leaf production. At this point, each shoot consisted of about three fully expanded leaves, which is typical of tillers of *L. perenne* in a field sward, and a new leaf was produced every nine to ten days, each with a life-span of about four weeks.

In field swards of *L. perenne* that are cut periodically, leaves have longer life-spans. Leaves appearing during the summer survive for just under two months (Bean, 1964; M.B. Jones et al., 1982; Vine, 1983), and those appearing during the autumn and early winter survive for approximately four months (Vine, 1983). However, there are marked differences between species in mean leaf longevity; for example, leaves of *Festuca ovina* live longer than those of *Dactylis glomerata* (Williamson, 1976). It appears that species with long-lived leaves may be characterized by low relative growth rates (Grime and Hunt, 1975). Although under favourable conditions the oldest leaf on a tiller dies as a new leaf appears, so that the number of live leaves per tiller remains more or less constant (Bean, 1964; Hunt and Brougham, 1966; M.B. Jones et al., 1982), leaf appearance and leaf death are not precisely matched throughout the year. Vine (1983) observed that between August and December death exceeds appearance, and between January and May appearance exceeds death (Fig. 4.5).

As a result of leaf death, a large proportion of the above-ground biomass of grass swards is lost as dead material — almost 30% in an uncut seedling minisward (Robson, 1973b), and about 50% of the recovery growth in a seven-week period after defoliation in the field (Hunt, 1965). Such large losses are obviously of importance in determining sward productivity, and have been considered by many investigators (Hunt, 1965; Hodgson et al., 1981; Parsons et al., 1983). Some have suggested that selection for greater leaf life-span, ideally without reducing the rate of leaf appearance, might increase ceiling yield (Robson, 1980, 1981) but, since photosynthetic capacity declines with age (Jewiss and Woledge, 1967; Woledge, 1972) and with shading, this approach seems unlikely to be successful. The rapid senescence of leaves as soon as they become shaded might be the best strategy, because of the fact that mineral nutrients and nitrogenous compounds are retranslocated from senescing leaves and utilized in growing tissues (Brady, 1973; Thomas and Stoddart, 1980); their sequestration in leaves with only marginal photosynthetic potential might be counterproductive.

## Growth and development of tillers

### Environmental regulation of tillering

The tiller is the leaf-producing unit of the sward, so the factors influencing the production and survival of tillers play a critical role in establishing and maintaining the growth potential of the sward. Tiller primordia are initiated in sequence at the shoot apex in the axil of each successive leaf primordium, soon after it appears, and develop into tiller buds. In general, the production of tiller buds is not greatly influenced by the environment, but the development and emergence of tillers is highly dependent on external conditions. In an optimal environment, the development of tiller buds occurs relatively freely in vegetative plants, but tillering is suppressed in adverse conditions to a far greater degree than is leaf production (Jewiss, 1966). Tiller buds therefore accumulate under such conditions, but may develop later if the environment improves. This suggests that tiller buds are in competition with other parts of the plant, especially expanding leaves, for substrates such as carbohydrate and nitrogen.

Low irradiance and low nutrient supply both have a marked effect on tiller production (Mitchell, 1953; Troughton, 1968; Ong et al., 1978b). From studies in controlled environments, and shading experiments in the field, the main effect of reducing irradiance on growth and development in grasses is a reduction in tiller production (Mitchell, 1953; Langer, 1963; Bean, 1964; Spiertz and Ellen, 1972). Tillering is also reduced by reducing the supply of nitrogen (Langer, 1959; McIntyre, 1965; Fletcher and Dale, 1974) and, to a lesser extent, by reducing the supply of potassium and phosphorus (Langer, 1959). The pattern and duration of tillering is also greatly modified by the continuity of nutrient supply (Aspinall, 1961). In young barley plants, where a delay in supplying nitrogen retards the outgrowth of the first tiller bud (Fletcher and Dale, 1974), the application of cytokinins through the roots stimulated tiller bud growth in the absence of nitrogen (Sharif and Dale, 1980). Similarly the application of cytokinins directly to tiller buds stimulates their out-growth (Jewiss, 1972; Langer et al., 1973; Clifford and Langer, 1975; Jinks and Marshall, 1982). These results suggest that the influence of nitrogen on tiller bud development may, at least in part, be due to a hormonal effect.

The density of the plant population exerts a major effect on the production of tillers (Kirby and Faris, 1972; Kays and Harper, 1974); this seems related to a reduction in both irradiance and nutrient supply. In favourable conditions, in both spaced plants and plants establishing at high density, the production of tillers tends to be initially exponential, but the rate declines as self-shading and mutual shading increases. This suggests that the supply of assimilate is no longer sufficient to initiate or maintain the extension of all tiller buds. In *Lolium perenne* grown in miniswards at 500 to 5000 plants $m^{-2}$, Colvill and Marshall (1981) observed that there was very little difference in leaf production per main shoot after twelve weeks, but that tiller number at high density was only one-third of that at low density. This difference was characterized by the production of fewer secondary tillers and by the greatly reduced weight of the main shoot and tillers at the higher density.

Cutting or grazing also greatly affects tillering. M.B. Jones et al. (1982) reported that tiller density increased five-fold in a sward of *L. perenne* managed by continuous grazing, as compared with an infrequently cut sward. This difference developed over one growing season, and was characterized by the production of small tillers in the continuously grazed sward. It would be interesting to know if such differences in tiller density reflect increased tillering *per se*, or an increased survival of young tillers that would normally die in a flowering sward as a result of shading (see p. 39).

Other environmental factors, such as temperature, day-length and water stress, also influence tillering but this seems primarily to be by their influence on the rate of leaf production of established shoots. For example, the rate of tiller production has a fairly low temperature optimum in temperate grasses; this seems to reflect directly the effect of temperature on the rate of leaf appearance and rate of tiller emergence (Langer, 1963; Evans et al., 1964).

### Internal regulation of tillering

Tillering is greatly affected by reproductive development. During the early phase of stem elongation and inflorescence development, the growth of tiller buds is suppressed but bud growth may be resumed after inflorescence emergence (Aspinall, 1961; Jewiss, 1972). Competition for assimilates and nutrients may be a factor in this, but there is considerable evidence that the growth of tiller buds may be more directly controlled by growth substances (Jewiss, 1972; Langer et al., 1973; Clifford, 1977; Jinks and Marshall, 1982). For example, application of the auxin transport inhibitor TIBA (tri-iodobenzoic acid) stimulates tillering during this period (Jewiss, 1972; Langer et al., 1973). Developing inflorescences or elongating stem internodes may be the source of auxin, since their removal results in rapid tillering (Laidlaw and Berrie, 1974; Harrison and Kaufman, 1980). Similarly, excision of the developing inflorescence promotes the movement of $^{14}$C-assimilate to tiller buds, but application of indole acetic acid to decapitated plants reduces it (Clifford, 1977). However, in view of the previously described effects of cytokinins on tiller bud outgrowth (p. 39), it seems unlikely that auxin alone regulates tiller bud development. Langer et al. (1973) described a changing pattern of response to applied growth substances at different stages of development in wheat, but overall it was clear that bud inhibition was a function of auxin, and also of assimilate supply, and that these factors could be overcome by supplying cytokinin. Harrison and Kaufman (1980) concluded that the ratio of cytokinin to auxin was of particular importance in regulating tiller bud development in oats, with outgrowth promoted by a high ratio of cytokinin to auxin. There is considerable evidence that the source of cytokinin for bud growth is the root system (Van Staden and Davey, 1979).

### Tiller dynamics and life histories

The seasonal pattern of tillering, and the longevity of individual tillers, has been studied in detail in several temperate grasses. Langer (1956) found that tillers appeared and died in all months of the year, and concluded that the plant was a dynamic population of short-lived tillers. The perenniality of the individual plant, and therefore the sward, depends on the capacity to replace dying tillers, and this is complicated by the very marked seasonal flushes in tiller production and death, especially those associated with flowering. If this replacement fails to occur, the plant will die. Colvill and Marshall (1984) found that the death of plants in a sward of *Lolium perenne* was characterized by a progressive decline in the number of live tillers; the tillers died in an apparently random order, not related to age or position. Since the sward was not grazed, death probably resulted from competition with adjacent plants. Similar observations on infrequently cut swards of *Phleum pratense* and *L. perenne* have shown that the number of live tillers per plant and per unit area increased rapidly (that is, rate of appearance > rate of death) during late winter and spring and reached a peak value in April (Fig. 4.8); this can be attributed to the improved light and temperature conditions compared with those prevailing over winter (Langer et al., 1964; Garwood, 1969; Hebblethwaite, 1977; Colvill and Marshall, 1984). However, the tiller number declined rapidly (tiller death > tiller appearance) from April to June, when flowering occurred; as a result, the live tiller population fell by as much as 50%. Tiller numbers increased soon after flowering; the tillers produced at this time are particularly important, since most survive the winter and form the majority of the flowering tiller population in the following year

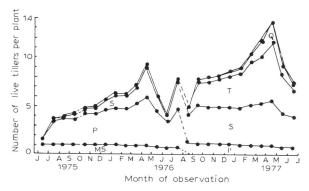

Fig. 4.8. The mean number of live tillers per plant of *L. perenne* over a two-year period after seedling establishment in a field plot ($MS$ = main shoot; $P$ = primary tillers; $S$ = secondary tillers; $T$ = tertiary tillers; $Q$ = quaternary tillers; ----, period of the annual cut) (after Colvill and Marshall, 1984).

(Jewiss, 1966; Lambert and Jewiss, 1970; Colvill and Marshall, 1984). In contrast, only a small proportion of the tillers appearing between October and May survive and flower. Very few tillers therefore live for more than a year in *L. perenne* and *P. pratense* though, in frequently cut swards of *Festuca pratensis*, some tillers survive for more than three years (Jewiss, 1966).

The dynamic nature of tillering in *L. perenne* is emphasized by the degree and rapidity of tiller turnover just before and during stem elongation in the spring. The majority of the flush of new tillers, mainly second-order tillers in a second-year sward, are ephemeral and form the bulk of the tillers that die during stem elongation (Hill and Watkin, 1975; Colvill and Marshall, 1984) (Fig. 4.8). Thus most dead tillers are young, vegetative and small (Ong et al., 1978a) and, even though up to half the tillers may die at this time, there is little effect on dry-matter accumulation (Hebblethwaite, 1977).

Irrigation has little effect on the death rate of tillers (Garwood, 1969; Hebblethwaite, 1977), but the addition of nitrogen fertilizer in spring increases the proportion of tillers that die, and causes earlier death (Hebblethwaite, 1977; Ong et al., 1978a). It seems likely that tiller mortality is caused by increased shading, especially by the elongation of flowering tillers (Langer et al., 1964; Spiertz and Ellen, 1972). This view is supported by the very poor assimilatory capacity of small, basal vegetative tillers, prior to senescence and death (Ong et al., 1978a). These tillers are particularly at risk, because they are not supplied with assimilate from the main shoot or from other elongating flowering tillers (Ong et al., 1978a; Colvill and Marshall, 1984), as is the case for shaded tillers in vegetative plants (Ong and Marshall, 1979).

Reproductive tillers die after flowering. In *L. perenne* reproductive tillers represent about 20 to 30% of the peak live tiller number in spring (Hebblethwaite, 1977; Colvill and Marshall, 1984). The combined loss of vegetative and reproductive tillers leaves few vegetative tillers immediately after flowering, so that the growth rate of the sward is low (Jewiss, 1972). Recovery primarily depends on the production of new tillers by the surviving vegetative tillers, and by the out-growth of suppressed tiller buds. Any treatment that stimulates tiller production at this stage is likely to improve pasture growth after flowering. Plant growth regulators might be used to break tiller bud dormancy, but selection for reduced dormancy might also result in plants with a more rapid post-flowering recovery. The supply of carbohydrate may also be an important factor regulating the production of new tillers, especially after cutting, but there is evidence that the bases of flowering tillers accumulate carbohydrate reserves which are readily mobilized to newly developing tillers after the period of reproductive suppression (Smith, 1973; Colvill and Marshall, 1984).

## CONCLUSIONS

The grass sward is a dynamic population of short-lived tillers of different ages. Each tiller is itself a dynamic population of leaves of limited life-span. Growth and productivity therefore depend on the continuous production of new leaves and tillers to replace those that die or are harvested. The life-span of individual leaves or tillers is closely related to their time of appearance but, under conditions of stress such as shading, leaf death is accelerated and may lead to the death of whole tillers and even whole plants. Nevertheless, the tillers of an individual plant are physiologically integrated and tend to share, rather than compete for, resources in times of transient stress.

Temperature plays a critical role in the rate of production and growth of leaves and tillers, and so determines the canopy structure and sets the potential for light interception by the sward. Selection of genotypes with faster leaf growth at low temperatures could increase productivity, by lengthening the growing season; such genotypes would also have to be frost-tolerant.

The production of leaves in infrequently harvested swards is seriously restricted by tiller bud dormancy, especially after cutting during the flowering period. The selection of genotypes with less bud dormancy might improve productivity, by accelerating regrowth after cutting.

For an established leaf canopy, growth depends upon the area and arrangement of leaves for light interception, and upon their photosynthetic potential and respiratory activity. Selection has been carried out for each of these attributes, with beneficial effects on sward productivity. The challenge for the plant breeder is how to incorporate all these attributes in a single genotype.

The key to increasing productivity with existing cultivars is the establishment of a harvesting strategy in which the degree and frequency of cutting or grazing reduces to a minimum the time when the sward has less than complete light interception. A far more physiologically based approach to pasture management is needed if this is to be widely achieved.

**REFERENCES**

Anslow, R.C., 1966. The rate of appearance of leaves on tillers of the Gramineae. *Herbage Abstr.*, 36: 149–155.

Arnold, G.W., Bennett, D. and Williams, C.N., 1967. The promotion of winter growth in pastures through growth substances and photoperiod. *Aust. J. Agric. Res.*, 18: 245–257.

Aspinall, D., 1961. The control of tillering in the barley plant. I. The pattern of tillering and its relation to nutrient supply. *Aust. J. Biol. Sci.*, 14: 493–505.

Aspinall, D., 1980. Role of abscisic acid and other hormones in adaptation to water stress. In: N.C. Turner and P.J. Kramer (Editors), *Adaptation of Plants to Water and High Temperature Stress*. Wiley, New York, N.Y., pp. 155–172.

Atkin, R.K., Barton, G.E. and Robinson, D.K., 1973. Effect of root-growing temperature on growth substances in xylem exudate of *Zea mays*. *J. Exper. Bot.*, 24: 475–487.

Atkinson, C.J. and Farrar, J.F., 1983. Allocation of photosynthetically-fixed carbon in *Festuca ovina* L. and *Nardus stricta* L. *New Phytol.*, 95: 519–531.

Bean, E.W., 1964. The influence of light intensity upon the growth of an S.37 Cocksfoot (*Dactylis glomerata*) sward. *Ann. Bot.*, 28: 427–443.

Begg, J.E. and Wright, M.J., 1962. Growth and development of leaves from intercalary meristems in *Phalaris arundinacea* L. *Nature*, 194: 1079–1098.

Blum, A., Mayer, J. and Gozlan, G., 1983. Associations between plant production and some physiological components of drought resistance in wheat. *Plant, Cell Environ.*, 6: 219–225.

Brady, C.J., 1973. Changes accompanying growth and senescence and effect of physiological stress. In: G.W. Butler and R.W. Bailey (Editors), *Chemistry and Biochemistry of Herbage, 2*. Academic Press, London, pp. 317–351.

Clifford, P.E., 1977. Tiller bud suppression in reproductive plants of *Lolium multiflorum* Lam. cv. Westerwoldicum. *Ann. Bot.*, 41: 605–615.

Clifford, P.E. and Langer, R.H.M., 1975. Pattern and control of distribution of $^{14}C$-assimilates in reproductive plants of *Lolium multiflorum* Lam. cv. Westerwoldicum. *Ann. Bot.*, 39: 403–411.

Clifford, P.E., Marshall, C. and Sagar, G.R., 1973. The reciprocal transfer of radiocarbon between a developing tiller and its parent shoot in vegetative plants of *Lolium multiflorum* Lam. *Ann. Bot.*, 37: 777–785.

Clough, B.F. and Milthorpe, F.L., 1975. Effects of water deficit on leaf development in tobacco. *Aust. J. Plant Physiol.*, 2: 291–300.

Colvill, K.E. and Marshall, C., 1981. The patterns of growth, assimilation of $^{14}CO_2$ and distribution of $^{14}C$-assimilate within vegetative plants of *Lolium perenne* at low and high density. *Ann. Appl. Biol.*, 99: 179–190.

Colvill, K.E. and Marshall, C., 1984. Tiller dynamics and assimilate partitioning in *Lolium perenne* L., with particular reference to flowering. *Ann. Appl. Biol.*, 104: 543–557.

Cooper, J.P., 1964. Climatic variation in forage grasses. I. Leaf development in climatic races of *Lolium* and *Dactylis*. *J. Appl. Ecol.*, 1: 45–61.

Cooper, J.P., 1975. Control of photosynthetic production in different environments. In: J.P. Cooper (Editor), *Photosynthesis and Productivity in Different Environments*. Cambridge University Press, Cambridge, pp. 593–612.

Corrall, A.J. and Fenlon, J.S., 1978. A comparative method for describing the seasonal distribution of production from grasses. *J. Agric. Sci., Cambridge*, 91: 61–67.

Davies, A. and Simons, R.G., 1979. Effect of autumn cutting regime on developmental morphology and spring growth of perennial ryegrass. *J. Agric. Sci., Cambridge*, 92: 457–469.

Davies, W.J., Mansfield, T.A. and Wellburn, A.R., 1980. A role for abscisic acid in drought endurance and drought avoidance. In: F. Skoog (Editor), *Plant Growth Substances 1979*. Springer-Verlag, Berlin, pp. 242–253.

Donald, C.M., 1961. Competition for light in crops and pastures. In: F.L. Milthorpe (Editor), *Mechanisms in Biological Competition*. Symposium of the Society for Experimental Biology 15. Cambridge University Press, Cambridge, pp. 283–313.

Eagles, C.F. and Østgård, O., 1971. Variation in growth and development in natural populations of *Dactylis glomerata* from Norway and Portugal. I. Growth analysis. *J. Appl. Ecol.*, 8: 367–381.

Eagles, C.F. and Othman, O.B., 1978. Physiological studies of a hybrid between populations of *Dactylis glomerata* from contrasting climatic regions. I. Inter-population differences. *Ann. Appl. Biol.*, 89: 71–79.

Evans, L.T., 1976. Physiological adaptation to performance as crop plants. *Phil. Trans. R. Soc., Lond., Ser. B*, 275: 71–83.

Evans, L.T., Wardlaw, I.F. and Williams, C.N., 1964. Environmental control of growth. In: C. Barnard (Editor), *Grasses and Grasslands*. Macmillan, London, pp. 102–125.

Fletcher, G.M. and Dale, J.E., 1974. Growth of tiller buds in barley: effects of shade treatment and mineral nutrition. *Ann. Bot.*, 38: 63–76.

Forde, B.J., 1966. Translocation in grasses. II. Perennial ryegrass and couch grass. *N.Z. J. Bot.*, 4: 496–514.

Garwood, E.A., 1969. Seasonal tiller populations of grass and grass/clover swards with and without irrigation. *J. Br. Grassland Soc.*, 24: 333–344.

Gifford, R.M. and Marshall, C., 1973. Photosynthesis and assimilate distribution in *Lolium multiflorum* Lam. following differential tiller defoliation. *Aust. J. Biol. Sci.*, 26: 517–526.

Goodwin, P.B., 1978. Phytohormones and growth and development of organs of the vegetative plant. In: D.S. Letham, P.B. Goodwin and T.J.V. Higgins (Editors), *Phytohormones and Related Compounds II*. Elsevier/North Holland Biomedical Press, Amsterdam, pp. 31–173.

Grime, J.P. and Hunt, R., 1975. Relative growth rate: its range and adaptive significance in a local flora. *J. Ecol.*, 63: 393–422.

Hanson, A.D. and Hitz, W.D., 1982. Metabolic responses of mesophytes to plant water deficits. *Annu. Rev. Plant Physiol.*, 33: 163–203.

Harrison, M.A. and Kaufman, P.B., 1980. Hormonal regulation of lateral bud (tiller) release in oats (*Avena sativa* L.). *Plant Physiol.*, 66: 1123–1127.

Hebblethwaite, P.D., 1977. Irrigation and nitrogen studies in S.23 ryegrass grown for seed. I. Growth, development, seed yield components and seed yield. *J. Agric. Sci., Cambridge*, 88: 605–614.

Hill, M.J. and Watkin, B.R., 1975. Seed production studies on perennial ryegrass, timothy and prairie grass. I. Effect of tiller age on tiller survival, ear emergence and seedhead components. *J. Br. Grassland Soc.*, 30: 63–71.

Hodgson, J., Bircham, J.S., Grant, S.A. and King, J., 1981. The influence of cutting and grazing management on herbage growth and utilization. In: C.E. Wright (Editor), *Plant Physiology and Herbage Production*. British Grassland Society, Maidenhead, pp. 51–62.

Hsaio, T.C., 1973. Plant responses to water stress. *Annu. Rev. Plant Physiol.*, 24: 519–570.

Hunt, L.A., 1965. Some implications of death and decay in pasture production. *J. Br. Grassland Soc.*, 20: 27–31.

Hunt, L.A. and Brougham, R.W., 1966. Some aspects of growth in an undefoliated stand of Italian ryegrass. *J. Appl. Ecol.*, 3: 21–28.

Jewiss, O.R., 1966. Morphological and physiological aspects of growth of grasses during the vegetative phase. In: F.L. Milthorpe and J.D. Ivins (Editors), *The Growth of Cereals and Grasses*. Butterworths, London, pp. 39–54.

Jewiss, O.R., 1972. Tillering in grasses — its significance and control. *J. Br. Grassland Soc.*, 27: 65–82.

Jewiss, O.R. and Woledge, J., 1967. The effect of age on the rate of apparent photosynthesis in leaves of tall fescue (*Festuca arundinacea* Schreb.). *Ann. Bot.*, 31: 661–671.

Jinks, R.L. and Marshall, C., 1982. Hormonal regulation of tiller bud development and internode elongation in *Agrostis stolonifera* L. In: J.S. McLaren (Editor), *Chemical Manipulation of Crop Growth and Development*. Butterworths, London, pp. 525–542.

Jones, M.B., Leafe, E.L. and Stiles, W., 1980a. Water stress in field-grown perennial ryegrass. I. Its effect on growth, canopy photosynthesis and transpiration. *Ann. Appl. Biol.*, 96: 87–101.

Jones, M.B., Leafe, E.L. and Stiles, W., 1980b. Water stress in field-grown perennial ryegrass. II. Its effect on leaf water status, stomatal resistance and leaf morphology. *Ann. Appl. Biol.*, 96: 103–110.

Jones, M.B., Collett, B. and Brown, S., 1982. Sward growth under cutting and continuous stocking managements: sward canopy structure, tiller density and leaf turnover. *Grass Forage Sci.*, 37: 67–73.

Jones, R.L., 1973. Gibberellins: their physiological role. *Annu. Rev. Plant Physiol.*, 24: 571–598.

Jones, R.L., 1980. The physiology of gibberellin-induced elongation. In: F. Skoog (Editor), *Plant Growth Substances 1979*. Springer-Verlag, Berlin, pp. 188–195.

Kays, S. and Harper, J.L., 1974. The regulation of plant and tiller density in a grass sward. *J. Ecol.*, 62: 79–105.

Keatinge, J.D.H., Stewart, R.H. and Garrett, M.K., 1979. The influence of temperature and soil water potential on the leaf extension rate of perennial ryegrass in Northern Ireland. *J. Agric. Sci., Cambridge*, 92: 175–183.

Kemp, D.R. and Blacklow, W.M., 1980. Diurnal extension rates of wheat leaves in relation to temperatures and carbohydrate concentrations of the extension zone. *J. Exper. Bot.*, 31: 821–828.

Kirby, E.J.M. and Faris, D.G., 1972. The effect of plant density on tiller growth and morphology in barley. *J. Agric. Sci., Cambridge*, 78: 281–288.

Laidlaw, A.S. and Berrie, A.M.M., 1974. The influence of expanding leaves and the reproductive stem apex on apical dominance in *Lolium multiflorum*. *Ann. Appl. Biol.*, 78: 75–82.

Lambert, D.A. and Jewiss, O.R., 1970. The position in the plant and the date of origin of tillers which produce inflorescences. *J. Br. Grassland Soc.*, 25: 107–112.

Langer, R.H.M., 1956. Growth and nutrition of Timothy (*Phleum pratense*). I. Life history of individual tillers. *Ann. Appl. Biol.*, 44: 166–187.

Langer, R.H.M., 1959. Growth and nutrition of Timothy (*Phleum pratense*). IV. The effect of varying nutrient supply on growth during the first year. *Ann. Appl. Biol.*, 47: 211–221.

Langer, R.H.M., 1963. Tillering in herbage grasses. *Herbage Abstr.*, 33: 141–148.

Langer, R.H.M., 1966. Mineral nutrition of grasses and cereals. In: F.L. Milthorpe and J.D. Ivins (Editors), *The Growth of Cereals and Grasses*. Butterworths, London, pp. 213–226.

Langer, R.H.M., 1979. *How Grasses Grow*. Edward Arnold, London, 2nd ed., 66 pp.

Langer, R.H.M., Prasad, P.C. and Laude, H.M., 1973. Effects

of kinetin on tiller bud elongation in wheat (*Triticum aestivum* L.). *Ann. Bot.*, 37: 565–571.
Langer, R.H.M., Ryle, S.M. and Jewiss, O.R., 1964. The changing plant and tiller populations of timothy and meadow fescue swards. I. Plant survival and the pattern of tillering. *J. Appl. Ecol.*, 1: 197–208.
Lawlor, D.W., 1972. Growth and water use of *Lolium perenne*. II. Plant growth. *J. Appl. Ecol.*, 9: 99–105.
Leafe, E.L., 1972. Micro-environment, carbon dioxide exchange and growth in grass swards. In: A.R. Rees, K.E. Cockshull, D.W. Hand and R.G. Hurd (Editors), *Crop Processes in Controlled Environments*. Academic Press, London, pp. 157–174.
Leafe, E.L., 1978. Physiological, environmental and management factors of importance to maximise yield of the grass crop. In: *Maximizing Yields of Crops*. HMSO, London, pp. 37–49.
Levitt, J., 1980. *Responses of Plants to Environmental Stresses Volume 1: Chilling, Freezing, and High Temperature Stresses*. Academic Press, London, 2nd ed., 497 pp.
Ludlow, M.M., 1980. Adaptive significance of stomatal responses to water stress. In: N.C. Turner and P.J. Kramer (Editors), *Adaptation of Plants to Water and High Temperature Stress*. Wiley-Interscience, New York, N.Y., pp. 123–138.
Ludlow, M.M. and Ng, T.T., 1976. Effect of water deficit on carbon dioxide exchange and leaf elongation rate of the $C_4$ tropical grass *Panicum maximum* var. *trichoglume*. *Aust. J. Plant Physiol.*, 3: 401–413.
Ludlow, M.M. and Ng, T.T., 1977. Leaf elongation rate in *Panicum maximum* var. *trichoglume* following removal of water stress. *Aust. J. Plant Physiol.*, 4: 263–272.
Marshall, C., 1967. The use of radioisotopes to investigate organization in plants, with special reference to the grass plant. In: *Isotopes in Plant Nutrition and Physiology*. I.A.E.A., Vienna, pp. 203–216.
Marshall, C. and Sagar, G.R., 1965. The influence of defoliation on the distribution of assimilates in *Lolium multiflorum* Lam. *Ann. Bot.*, 29: 365–370.
Marshall, C. and Sagar, G.R., 1968. The distribution of assimilates in *Lolium multiflorum* Lam. following differential defoliation. *Ann. Bot.*, 32: 715–719.
McIntyre, G.I., 1965. Some effects of nitrogen supply on the growth and development of *Agropyron repens* L. Beauv. *Weed Res.*, 5: 1–12.
Menhenett, R. and Wareing, P.F., 1976. Effects of soil temperature on the growth and hormone content of *Dactylis glomerata* L. (cocksfoot) in controlled environments. *J. Exper. Bot.*, 27: 1259–1267.
Milthorpe, F.L. and Moorby, J., 1979. *An Introduction to Crop Physiology*. Cambridge University Press, Cambridge, 2nd ed., 244 pp.
Mitchell, K.J., 1953. Influence of light and temperature on the growth of ryegrass (*Lolium* spp). I. Pattern of vegetative development. *Physiol. Plant.*, 6: 21–46.
Monteith, J.L., 1977. Climate and the efficiency of crop production in Britain. *Philos. Trans. R. Soc., Lond., Ser. B*, 281: 277–294.
Monteith, J.L., 1981. Does light limit crop production? In: C.B. Johnson (Editor), *Physiological Processes Limiting Plant Productivity*. Butterworths, London, pp. 23–38.

Munns, R. and Weir, R., 1981. Contribution of sugars to osmotic adjustment in elongating and expanded zones of wheat leaves during moderate water deficits at two light levels. *Aust. J. Plant Physiol.*, 8: 93–105.
Norris, I.B. and Thomas, H., 1982a. Recovery of ryegrass species from drought. *J. Agric. Sci., Cambridge*, 98: 623–628.
Norris, I.B. and Thomas, H., 1982b. The effect of droughting on varieties and ecotypes of *Lolium*, *Dactylis* and *Festuca*. *J. Appl. Ecol.*, 19: 881–889.
Nyahoza, F., Marshall, C. and Sagar, G.R., 1973. The interrelationship between tillers and rhizomes of *Poa pratensis* L. — an autoradiographic study. *Weed Res.*, 13: 304–309.
Nyahoza, F., Marshall, C. and Sagar, G.R., 1974. Assimilate distribution in *Poa pratensis* L. — a quantitative study. *Weed Res.*, 14: 251–256.
Ollerenshaw, J.H., Stewart, W.S., Gallimore, J. and Baker, R.H., 1976. Low temperature growth in grasses from Northern Latitudes. *J. Agric. Sci., Cambridge*, 87: 237–239.
Ong, C.K., 1978. The physiology of tiller death in grasses. I. The influence of tiller age, size and position. *J. Br. Grassland Soc.*, 33: 197–203.
Ong, C.K. and Marshall, C., 1979. The growth and survival of severely-shaded tillers in *Lolium perenne* L. *Ann. Bot.*, 43: 147–155.
Ong, C.K., Marshall, C. and Sagar, G.R., 1978a. The physiology of tiller death in grasses. 2. Causes of tiller death in a grass sward. *J. Br. Grassland Soc.*, 33: 205–211.
Ong, C.K., Marshall, C. and Sagar, G.R., 1978b. The effects of nutrient supply on flowering and seed production in *Poa annua* L. *J. Br. Grassland Soc.*, 33: 117–121.
Østgård, O. and Eagles, C.F., 1971. Variation in growth and development of natural populations of *Dactylis glomerata* from Norway and Portugal. II. Leaf development and tillering. *J. Appl. Ecol.*, 8: 383–392.
Parsons, A.J. and Robson, M.J., 1980. Seasonal changes in the physiology of S24 perennial ryegrass (*Lolium perenne* L.). 1. Response of leaf extension to temperature during the transition from vegetative to reproductive growth. *Ann. Bot.*, 46: 435–444.
Parsons, A.J. and Robson, M.J., 1981a. Seasonal changes in the physiology of S24 perennial ryegrass (*Lolium perenne* L.). 2. Potential leaf and canopy photosynthesis during the transition from vegetative to reproductive growth. *Ann. Bot.*, 47: 249–258.
Parsons, A.J. and Robson, M.J., 1981b. Seasonal changes in the physiology of S24 perennial ryegrass (*Lolium perenne* L.). 3. Partition of assimilates between root and shoot during the transition from vegetative to reproductive growth. *Ann. Bot.*, 48: 733–744.
Parsons, A.J., Leafe, E.L., Collett, B., Penning, P.D. and Lewis, J., 1983. The physiology of grass production under grazing. II. Photosynthesis, crop growth and animal intake of continuously-grazed swards. *J. Appl. Ecol.*, 20: 127–139.
Peacock, J.M., 1975a. Temperature and leaf growth in *Lolium perenne*. I. The thermal microclimate: its measurement and relation to crop growth. *J. Appl. Ecol.*, 12, 99–114.
Peacock, J.M., 1975b. Temperature and leaf growth in *Lolium*

perenne. II. The site of temperature perception. *J. Appl. Ecol.*, 12: 115–123.

Peacock, J.M., 1975c. Temperature and leaf growth in *Lolium perenne*. III. Factors affecting seasonal differences. *J. Appl. Ecol.*, 12: 685–697.

Pierce, M. and Raschke, M., 1980. Correlation between loss of turgor and accumulation of abscisic acid in detached leaves. *Planta*, 148: 174–182.

Pollock, C.J. and Jones, T., 1979. Seasonal patterns of fructan metabolism in forage grasses. *New Phytol.*, 83: 9–15.

Powell, C.E. and Ryle, G.J.A., 1978. Effect of nitrogen deficiency on photosynthesis and the partitioning of $^{14}$C-labelled leaf assimilate in unshaded and partially shaded plants of *Lolium temulentum*. *Ann. Appl. Biol.*, 90: 241–248.

Quarrie, S.A. and Jones, H.G., 1977. Effects of abscisic acid and water stress on development and morphology of wheat. *J. Exper. Bot.*, 28: 192–203.

Rhodes, I., 1973. Relationship between canopy structure and productivity in herbage grasses and its implications for plant breeding. *Herbage Abstr.*, 43: 129–133.

Rhodes, I., 1975. The relationships between productivity and some components of canopy structure in ryegrass (*Lolium* spp.). IV. Canopy characters and their relationship with sward yield in some intra-population selections. *J. Agric. Sci., Cambridge*, 84: 345–351.

Rhodes, I. and Mee, S.S., 1980. Changes in dry matter yield associated with selection for canopy characters in ryegrass. *Grass Forage Sci.*, 35: 35–39.

Rhodes, I. and Stern, W.R., 1978. Competition for light. In: J.R. Wilson (Editor), *Plant Relations in Pastures*. C.S.I.R.O., Melbourne, Vic., pp. 175–189.

Robson, M.J., 1972. The effect of temperature on the growth of S.170 tall fescue (*Festuca arundinacea*). I. Constant temperature. *J. Appl. Ecol.*, 9: 643–653.

Robson, M.J., 1973a. The growth and development of simulated swards of perennial ryegrass. II. Carbon assimilation and respiration in a seedling sward. *Ann. Bot.*, 37: 501–518.

Robson, M.J., 1973b. The growth and development of simulated swards of perennial ryegrass. I. Leaf growth and dry weight change as related to the ceiling yield of a seedling sward. *Ann. Bot.*, 37: 487–500.

Robson, M.J., 1980. A physiologist's approach to raising the potential yield of the grass crop through breeding. In: R.G. Hurd, P.V. Biscoe and C. Dennis (Editors), *Opportunities for Increasing Crop Yields*. Pitmans, London, pp. 33–49.

Robson, M.J., 1981. Potential production — what is it and can we increase it? In: C.E. Wright (Editor), *Plant Physiology and Herbage Production*. British Grassland Society, Maidenhead, pp. 5–18.

Robson, M.J., 1982. The growth and carbon economy of selection lines of *Lolium perenne* cv. S23 with differing rates of dark respiration. I. Grown as simulated swards during a regrowth period. *Ann. Bot.*, 49: 321–329.

Robson, M.J. and Deacon, M.J., 1978. Nitrogen deficiency in small closed communities of S24 ryegrass. II. Changes in weight and chemical composition of single leaves during their growth and death. *Ann. Bot.*, 42: 1199–1213.

Robson, M.J. and Jewiss, O.R., 1968. A comparison of British and North African ecotypes of tall fescue (*Festuca arundinacea*). II. Growth during winter and survival at low temperatures. *J. Appl. Ecol.*, 5: 179–190.

Robson, M.J. and Parsons, A.J., 1978. Nitrogen deficiency in small closed communities of S24 ryegrass. I. Photosynthesis, respiration, dry matter production and partition. *Ann. Bot.*, 42: 1185–1197.

Rogan, P.G. and Smith, D.L., 1974. Patterns of translocation of $^{14}$C-labelled assimilates during vegetative growth of *Agropyron repens* (L.) Beauv. *Z. Pflanzenphysiol.*, 73: 405–414.

Ryle, G.J.A., 1964. A comparison of leaf and tiller growth in seven perennial grasses as influenced by nitrogen and temperature. *J. Br. Grassland Soc.*, 19: 281–290.

Ryle, G.J.A., 1966. Effect of photoperiod in growth cabinets on the growth of leaves and tillers in three perennial grasses. *Ann. Appl. Biol.*, 57: 269–279.

Ryle, G.J.A., 1970. Partition of assimilates in an annual and a perennial grass. *J. Appl. Ecol.*, 7: 217–227.

Ryle, G.J.A. and Powell, C.E., 1972. The export and distribution of $^{14}$C-labelled assimilates from each leaf on the shoot of *Lolium temulentum* during reproductive and vegetative growth. *Ann. Bot.*, 36: 363–375.

Sharif, R. and Dale, J.E., 1980. Growth-regulating substances and the growth of tiller buds in barley; effects of cytokinins. *J. Exper. Bot.*, 31: 921–930.

Sheehy, J.E. and Cooper, J.P., 1973. Light interception, photosynthetic activity, and crop growth rate in canopies of six temperate forage grasses. *J. Appl. Ecol.*, 10: 239–250.

Sheehy, J.E. and Peacock, J.M., 1977. Microclimate, canopy structure and photosynthesis in canopies of three contrasting temperate forage grasses. I. Canopy structure and growth. *Ann. Bot.*, 41: 567–578.

Silsbury, J.H., 1970. Leaf growth in pasture grasses. *Trop. Grasslands*, 4: 17–36.

Smith, D., 1973. The nonstructural carbohydrates. In: G.W. Butler and R.W. Bailey (Editors), *Chemistry and Biochemistry of Herbage 1*. Academic Press, London, pp. 105–155.

Spiertz, J.H.J. and Ellen, J., 1972. The effect of light intensity on some morphological and physiological aspects of the crop perennial ryegrass (*Lolium perenne* L. var. "Cropper") and its effects on seed production. *Neth. J. Agric. Sci.*, 20: 232–246.

Thomas, H., 1975. The growth responses to weather of simulated vegetative swards of a single genotype of *Lolium perenne*. *J. Agric. Sci., Cambridge*, 84: 333–343.

Thomas, H., 1977. The influence of autumn cutting regime on the response to temperature of leaf growth in perennial ryegrass. *J. Br. Grassland Soc.*, 32: 227–230.

Thomas, H. and Norris, I.B., 1977. The growth responses of *Lolium perenne* to the weather during winter and spring at various altitudes in mid-Wales. *J. Appl. Ecol.*, 14: 949–964.

Thomas, H. and Stoddart, J.L., 1980. Leaf senescence. *Annu. Rev. Plant Physiol.*, 31: 83–111.

Troughton, A., 1968. Influence of genotype and mineral nutrition on the distribution of growth within plants of *Lolium perenne* L. grown in soil. *Ann. Bot.*, 32: 411–423.

Turner, N.C. and Begg, J.E., 1978. Response of pasture plants to water deficits. In: J.R. Wilson (Editor), *Plant Relations in Pastures*. C.S.I.R.O., Melbourne, Vic., pp. 50–66.

Turner, N.C. and Jones, M.M., 1980. Turgor maintenance by osmotic adjustment: A review and evaluation. In: N.C. Turner and P.J. Kramer (Editors), *Adaptation of Plants to Water and High Temperature Stress.* Wiley, New York, N.Y., pp. 87–103.

Van Staden, J. and Davey, J.E., 1979. The synthesis, transport and metabolism of endogenous cytokinins. *Plant, Cell Environ.*, 2: 93–106.

Vine, D.A., 1983. Sward structure changes within a perennial ryegrass sward: leaf appearance and death. *Grass Forage Sci.*, 38: 231–242.

Waite, R. and Boyd, J., 1953. The water soluble carbohydrates of grasses. I. Changes occurring during the normal life cycle. *J. Sci. Food Agric.*, 4: 197–204.

Wardlaw, I.F., 1969. The effect of water stress on translocation in relation to photosynthesis and growth. II. Effect during leaf development in *Lolium temulentum* L. *Aust. J. Biol. Sci.*, 22: 1–16.

Williams, C.N. and Biddiscombe, E.F., 1965. Extension growth of grass tillers in the field. *Aust. J. Agric. Res.*, 16: 14–22.

Williams, R.D., 1964. Assimilation and translocation in perennial grasses. *Ann. Bot.*, 28: 419–425.

Williamson, P., 1976. Above-ground primary production of chalk grassland allowing for leaf death. *J. Ecol.*, 64: 1059–1075.

Wilson, D., 1973. Physiology of light utilization by swards. In: G.W. Butler and R.W. Bailey (Editors), *Chemistry and Biochemistry of Herbage 2.* Academic Press, London, pp. 57–101.

Wilson, D., 1975a. Variation in leaf respiration in relation to growth and photosynthesis of *Lolium*. *Ann. Appl. Biol.*, 80: 323–328.

Wilson, D., 1975b. Leaf growth, stomatal diffusion resistances and photosynthesis during droughting of *Lolium perenne* populations selected for contrasting stomatal length and frequency. *Ann. Appl. Biol.*, 79: 67–82.

Wilson, D., 1975c. Stomatal diffusion resistances and leaf growth during droughting of *Lolium perenne* plants selected for contrasting epidermal ridging. *Ann. Appl. Biol.*, 79: 83–94.

Wilson, D., 1981. The role of physiology in breeding herbage cultivars adapted to their environment. In: C.E. Wright (Editor), *Plant Physiology and Herbage Production.* British Grassland Society, Maidenhead, pp. 95–108.

Wilson, D., 1982. Response to selection for dark respiration rate of mature leaves in *Lolium perenne.* L. and its effects on growth of young plants. *Ann. Bot.*, 49: 303–312.

Wilson, D. and Jones, J.G., 1982. Effect of selection for dark respiration rate of mature leaves on crop yields of *Lolium perenne* c.v. S23. *Ann. Bot.*, 49: 313–320.

Wilson, D., Eagles, C.F. and Rhodes, I., 1980. The herbage crop and its environment — exploiting physiological and morphological variations to improve yields. In: R.G. Hurd, P.V. Biscoe and C. Dennis (Editors), *Opportunities for Increasing Crop Yields.* Pitmans, London, pp. 21–32.

Wilson, J.R., Ludlow, M.M., Fisher, M.J. and Schulze, E.-D., 1980. Adaptation to water stress of the leaf water relations of four tropical forage species. *Aust. J. Plant Physiol.*, 7: 207–220.

Woledge, J., 1972. The effect of shading on the photosynthetic rate and longevity of grass leaves. *Ann. Bot.*, 36: 551–561.

Woledge, J., 1973. The photosynthesis of ryegrass leaves grown in a simulated sward. *Ann. Appl. Biol.*, 73: 229–237.

Woledge, J., 1979. Effect of flowering on the photosynthetic capacity of ryegrass leaves grown with and without natural shading. *Ann. Bot.*, 44: 197–207.

Woledge, J. and Leafe, E.L., 1976. Single leaf and canopy photosynthesis in a ryegrass sward. *Ann. Bot.*, 40: 773–783.

Yoshida, S., 1972. Physiological aspects of grain yield. *Annu. Rev. Plant Physiol.*, 23: 437–464.

Chapter 5

# MODELLING PASTURE GROWTH

K.R. CHRISTIAN

## THE PASTURE SYSTEM

Managed pastures show similarities with unmanaged grasslands on the one hand and with crops on the other. Compared with native or unimproved grasslands, managed grasslands receive higher inputs of mineral nutrients, and are subject to closer control of grazing frequency and intensity. As a result, herbage production is usually greater in managed grasslands but fewer plant species occur. Compared with crops, pastures are usually more permanent, even when they consist largely of annual species. Their agricultural value lies in vegetative growth, whereas the value of most crops lies in their reproductive output. In addition, the indeterminate nature of pasture growth, and the presence of more than one species, results in a heterogeneous structure in space, and complex patterns of growth and composition in time. These features, together with selective grazing and non-uniform treading (Chapter 7) and excretion by animals (Chapter 19), make the task of describing and modelling pasture growth a much more formidable task than that for crop production.

In spite of the differences considered above, there is no hard-and-fast distinction between managed grasslands and natural grasslands or crops. At the one extreme, many agricultural grasslands are so little managed as to be, to all intents and purposes, natural grassland while, at the other extreme, many crops are used wholly or in part as forage. Despite the different purposes to which the vegetation is put, the principles governing its growth differ in degree rather than in kind. It is therefore reasonable, in examining pasture models, to consider relevant studies on other herbaceous plant communities. Indeed, in view of the comparative paucity of systems studies of managed pastures, it is essential to consider models that have been developed for crops, since these have sometimes been adapted, with little modification, for grasslands (Dayan et al., 1981; Wight and Hanks, 1981).

## THE CHOICE OF MODELS

The term "model" has been used in such a variety of contexts, over the past two decades, that it has become virtually meaningless, unless used with some qualification. In the present context, a model may be regarded as a mathematical representation of physical or biological processes. Plant growth models are as varied as the purposes for which they have been constructed, and cover an enormous range of plant activities in various degrees of detail. Only a few examples can be mentioned here, and I shall concentrate on the agronomic or whole-crop point of view, which is concerned with the response of the system to management and environmental factors, and focuses on the effects, rather than on the mechanisms, of the biophysical processes involved. Plant-physiological models have been extensively reviewed by Singh et al. (1980).

The classification of plant growth models by Baier (1979) is based on the usual dichotomies (*static/dynamic*, *deterministic/stochastic*, *discrete/continuous*), and indicates the variety of approaches that have been adopted. However, an examination of the literature suggests that most models currently being developed belong to one of three main types: empirical, analytical and simulation. These terms are somewhat relative, since

model can be constructed entirely from first principles, and parameter values for the relevant processes and their interactions must be obtained either from the literature, or by experiment, or inferred from values for other species. It must be appreciated that all models, at their lowest level of organization, are based on descriptive and empirical functions. It is the way in which these functions are used and combined that determines whether the model is predominantly empirical or mechanistic.

Empirical models are expressly designed to fit data, using regression-type equations, and the relationships between environmental variables and plant growth are associative rather than causal. Models of this type will generate no more than the data fed into them, and are consequently of little general interest, even though they may have greater predictive ability than mechanistic or theoretically based models. A notable example is the regression model of Baier (1973), which predicts crop yield as the product of quadratic terms in climatic variables (see also chapter 2), each term having as coefficient a fourth-order polynomial in a time scale defined by phenological stage; it would not seem feasible to go any further than this in fitting a static model to a dynamic situation.

Although mechanistic models rely on experimental data for their construction, they offer the opportunity of synthesizing the known information to predict the behaviour of the system under any given combination of environmental conditions, thereby providing a much broader base than would be physically possible by experimentation alone. In bringing several processes together into the same framework, models allow, and often compel, the testing of evidence from a variety of sources for consistency. They therefore have the salutary effect of drawing attention to deficiencies in knowledge of plant functions and their interactions.

Analytical models are generally composed of a set of differential equations describing the transformation of energy and material components. They are particularly useful for dealing with a segment of the system or a single process that can be isolated, with some degree of realism, from the rest of the system for the period of study, especially where a result may be obtained in integrated form. In this way, a great deal of information may be stored concisely. However, when assumptions regarding the constancy of parameters are made, usually to obtain tractable equations, the model is cast into a static mould, and the generality of the solution is correspondingly reduced. Though conceptually of interest, analytical models are of limited value for studying plant behaviour under the continually changing conditions that are obtained in the field.

Compartmental simulation models, using finite-difference equations to update state variables, have proved by far the most popular in studies of plant growth. Perhaps the main reasons for this are the flexibility with which they can be constructed and modified, their suitability for dealing with a system that is subject to continually changing external influences, and the facility with which they enable plant responses to be adjusted over time, according to the current status of the system. In addition, they generate a time sequence of results which not only can be readily compared with actual experimental performances, but which can also make it relatively easy to understand how the system is functioning, and in what way any anomalous behaviour is occurring. In essence, a simulation model appeals to most biologists because it can give a much clearer picture of what is happening than all but the simplest of analytical models.

The plant system has been depicted as a hierarchical organization of crop, single plant, organ, cell and molecule — corresponding very broadly to the scientific disciplines of agronomy, plant physiology, and molecular biology — with the effects at each level being transmitted to the next highest level. Since simulation in greater detail requires smaller time-steps, it is not practicable to incorporate more than one or two levels within a single model. However, detailed examination of plant processes frequently introduces further complexity, rather than enabling the net effect to be summarized in a form suitable for inclusion in a model of lower resolution.

Comprehensive models of the transfer of energy and matter in the plant–environment system have been constructed (e.g. Paltridge, 1970; Shawcroft et al., 1974; Christian and Milthorpe, 1981), but these cannot be viewed as practicable methods for calculating growth and productivity over long periods. This is partly because of the computing time required, but mainly because of the detailed input data which they demand. Their main application lies in testing physical and biological principles, for which purpose they must be supported by large numbers of precise, intensive measurements. Observations of this nature are seldom practicable in the field, and there is a popular tendency towards developing models that require as input only standard meteorological readings, the water-holding capacity of the soil at various depths, and the values of a few readily-measured plant parameters. It is also preferable, in many respects, to disregard the individual plant and treat the sward as a population of tillers, leaves and roots, either singly or collectively.

In general, plant growth models are built up from a consideration of the component processes, with the most basic form being a function of the type:

$$\frac{1}{W} \cdot \frac{dW}{dt} = R X_1 X_2 X_3 \dots \qquad (1)$$

where $W$ is plant mass per unit area, and where $R$, the relative growth rate under optimal conditions, is multiplied by the coefficients $X_1, X_2, X_3 \dots$, each ranging between 0 and 1, to give a rate which is adjusted for the effects of such environmental variables as temperature, water stress and mineral supply. The widely-quoted growth index of Fitzpatrick and Nix (1970) is of this form, and the concept has been used in a variety of pasture growth models (Smith and Stephens, 1976; Cale, 1979; Loomis et al., 1979; Selirio and Brown, 1979; Fick, 1980). Interactions between environmental factors have received little study, and it is problematical whether such multiplicative functions are more appropriate than the alternative of considering $R$ to be modified only by the single most limiting factor.

The typical growth curve is sigmoidal, with exponential growth up to a critical leaf area index, followed by a period of linear growth until a maximum plant mass, depending on solar light receipt, is reached (e.g. Fick, 1980). The equation for growth constrained by environmental variables may be expressed in integral form (Wallach, 1975), but at the cost of flexibility.

The major component processes of growth will now be considered from the point of view of their incorporation into current models.

## INTERCEPTION OF RADIATION

In a comprehensive review, Lemeur and Blad (1974) commented that formulae dealing with light interception were so complex that they obstructed communication between the modeller and the crop ecologist; the situation is still much the same today.

The nature of grassland canopies is such that a statistical approach has been universally adopted. Data are required on leaf area and inclination at various heights within the canopy; leaf orientation has generally been regarded as unimportant. Canopy structure changes during growth (C.T. de Wit, 1965), and further complexity is added in mixed swards by the presence of species with different growth habit. Ross et al. (1972) suggested that the vertical pattern of leaf areas of grasses and legumes could be assumed to be represented by a normal distribution about their respective mean heights or, alternatively, that all grass leaf area occurred above that of legume; however, not all swards may conform to either of these patterns.

Almost all models of light penetration through canopies are based on the premise that radiation intensity, $I$, declines exponentially with depth:

$$I = I_0 e^{-KL} \qquad (2)$$

where $L$ is the leaf area index above the given height, $I_0$ is the radiation intensity at the top of the canopy, and $K$ a coefficient which ranges from 0.3 to 0.8, according to canopy structure. The complications caused by transmission through leaves, and reflection from them, are probably insignificant, compared with the errors involved in establishing a mean value for $K$. The equation has the advantage that it can be integrated over the whole canopy and combined with the photosynthesis function, as discussed in the next section.

Detailed models divide the canopy into vertical layers, usually each of unit leaf area (Shawcroft et al., 1974; Goudriaan, 1977; Christian and Milthorpe, 1981), thus allowing for changes in leaf characteristics with height. Since such models also often incorporate a heat-balance component, the behaviour of near infra-red and long-wave radiation are conveniently handled at the same time, though calculations need to be made at least hourly to accommodate the changing geometry. Since the photosynthetic response to light is not linear, particularly in $C_3$ species, it is important to distinguish the leaf area receiving both direct and diffuse light from that receiving diffuse light only. The distribution of diffuse light may be approximated by mean values (Anderson and Denmead, 1969), or the amounts arriving from different parts of the sky may be summed (C.T. de Wit, 1965), whereas light conditions under various types of cloud cover have been examined by O'Rourke and Terjung (1981).

Kimes et al. (1980) proposed the use of Monte Carlo methods for simulating light interception in multi-component canopy structures.

## PHOTOSYNTHESIS

The increase in knowledge concerning the metabolic pathways associated with photosynthesis and photorespiration over the past decade has been accompanied by a greater sophistication in physiological and biological models (Singh et al., 1980; Tenhunin and Westrin, 1979). However, there seems little point in using complex equations for simulating crop photosynthesis, since it is generally accepted that gross photosynthetic rate $P$ is adequately described by the equation:

$$P = \left(\frac{1}{tC} + \frac{1}{aI}\right)^{-1} \qquad (3)$$

where $a$ is described as photochemical efficiency, and is the inverse of the initial slope of the response curve, $t$ is the carbon dioxide conductance, and $C$ the ambient concentration of carbon dioxide (Acock et al., 1976). In $C_4$ pastures, all leaves may receive light intensities within the linear part of the response curve, if incoming radiation is not too high, and the expression may be further simplified (Connor et al., 1974).

Monteith (1965) developed a simple binomial model, in which the leaves that received effective radiation were classed as either "sunlit" or "once shaded". When the total areas of these fractions were combined separately with the simple equation for $P$ given above, the resulting expression could be integrated over the whole daylight period to give total daily photosynthesis on clear days. The method has found wide acceptance in various forms in pasture growth models (Curry, 1971; Ross et al., 1972; Acock et al., 1976; Sheehy et al., 1979; Ludlow et al., 1982). On cloudy days, photosynthesis may be approximated by comparing observed with clear-day solar radiation and partitioning the daylight hours into clear and overcast portions, assuming these to be normally distributed (C.T. de Wit et al., 1978).

As an alternative to the Monteith model, Goudriaan and Van Laar (1978) have developed a set of equations, based on the pioneer work of C.T. de Wit (1965), to calculate daily totals of gross carbon dioxide assimilation, assuming that the response to light intensity is exponential and that the effect of leaf angle can be neglected. Despite the simplicity of these methods, it is claimed that their accuracy compares well with that of more complex formulae. Such models may well prove more useful than attempts to account for the effects of penumbra and the anisotropism of diffuse radiation.

There is an increasing awareness that $P'_{max}$ is not a fixed quantity for a given species, but varies according to current leaf temperature and physiological leaf age (Angus and Wilson, 1976; Detling et al., 1979). Sheehy et al. (1980) derived $P'_{max}$ as a function of irradiance and temperature at the day of leaf emergence, and leaf age. However, these factors may interact — for instance, the rate of ageing may depend on the microclimate of the leaf.

## RESPIRATION

Mitochondrial or dark respiration has been separated by Penning de Vries (1975) into a growth component, resulting from biosynthesis and largely independent of temperature, and a maintenance component, associated with protein turnover and energy-requiring functions. Sheehy et al. (1980) described respiration as consisting of a short intense efflux of carbon dioxide, associated with biosynthesis and a slower efflux proportional to the remaining carbohydrate, associated with the maintenance of metabolic activity.

Typical values for respiration, used in models, are 0.03, 0.015 and 0.01 kg kg$^{-1}$ dry matter per day for leaf, stem, and root or storage organs respectively at 25°C, with a $Q_{10}$ of 2 (Van Keulen et al., 1982). Wide variations can be expected in these estimates, and Sheehy et al. (1980) found that the temperature relationship was critical in governing the rate of death in leaves. Protein turnover is almost certainly greater in expanding than in senescing leaves, and maintenance respiration may be more closely associated with the maximum rate of photosynthesis than with leaf weight (Angus and Wilson, 1976). Parton et al. (1978) suggested that root respiration was reduced by suberization and soil water deficit, though the experimental evidence is meagre.

A knowledge of plant chemical composition at different stages of growth is necessary to make a precise assessment of the carbon requirements for growth (C.T. de Wit et al., 1978) but, since the main changes during maturation involve the replacement of protein with structural carbohydrate and lignin, the energy content does not alter substantially. As a result, the conversion efficiency (0.7 kg kg$^{-1}$) should remain fairly constant throughout the year (Van Keulen et al., 1982).

## WATER RELATIONS

### Water in the soil

Because of the importance of soil water, many simulation models of soil water content have been developed, with simple water budgets (e.g. Keig and McAlpine, 1974) being the most popular. Finite-difference methods for simulating vertical water flow in soils, based on the Darcy–Richards equation (Richards, 1931), require such small time-steps that they are often unsuitable for plant growth studies, though approximate methods may prove adequate (Richter, 1980). Alternatively, infiltration can be simulated by the "parametric" method, in which each succeeding soil layer is "filled" to field capacity (e.g. Stroosnijder, 1982). Redistribution between layers, notably during surface evaporation, is not so readily handled. Rainfall interception by vegetation can be related to the plant cover, while run-off during storms can be estimated from long-term records (Stroosnijder, 1982). Run-off and drainage are considered in more detail in Chapter 20; Molz (1981) provides an excellent review of models of water extraction by roots. He concludes that further refinement is unwarranted until values for the basic parameters have been established with greater confidence. In particular, the question of the relative magnitudes of resistances to flow in soils and in roots remains unresolved.

Water extraction from various soil depths should be proportional to effective root densities and soil water potentials (e.g. Selirio and Brown, 1979). However, it is difficult to measure the distribution of roots in the profile, and even more difficult to determine which are effective in water uptake. Rather than attempt to work from first principles, it is common practice to insert empirical relationships, relating the ratio of actual to potential evaporation to available soil water (Molz and Remson, 1970; Saxton et al., 1974).

### Evapotranspiration and plant growth

Water deficit is often the greatest limitation to pasture growth (see Chapters 2 and 3), with yield proportional to the ratio of actual to potential evapotranspiration (A.J. de Wit, pers. comm., 1980); this has formed the basis for a number of crop models.

Although there is increasing understanding of water flow in the plant in relation to growth (e.g. Fiscus et al., 1983), the processes are unfortunately too complex to be incorporated in agronomic models. Stomatal and aerodynamic resistances can be predicted from calculations of wind speed, energy balance and water potential gradients within the canopy (Waggoner and Reifsnyder, 1968; Shawcroft et al., 1974; C.T. de Wit et al., 1978; Christian and Milthorpe, 1981), but iterative solutions are needed to achieve stable and balanced energy budgets. Sinclair et al. (1976) simplified the system by condensing the canopy into a single leaf layer and obtained good agreement with the multi-layer model, but experienced difficulty in estimating effective canopy and aerodynamic resistances.

Of the various methods used for estimating potential evapotranspiration, $E_0$, the Penman formula (Penman, 1963), has generally proved to be the most reliable. Adjustments may be made for leaf surface wetness (Jensen et al., 1971), for stage of growth or season (Saxton et al., 1974) and for the height or weight of vegetation (Russo and Knapp, 1976). In many circumstances, advection undoubtedly causes appreciable bias. The ratio of actual evapotranspiration, $E_a$, to potential evapotranspiration, $E_0$, depends on soil water availability (Denmead and Shaw, 1962). The relationships generally used in models are evidently based largely on experience of the soil characteristics, and usually show $E_a$ equal to $E_0$ above a critical percentage of soil water availability, with a linear decline to zero at wilting point (McCown, 1973; R.C.G. Smith and Johns, 1975; Selirio and Brown, 1979).

The contribution of evaporation from the soil to potential evapotranspiration depends not only on plant cover, sometimes expressed as plant height (De Jong and Cameron, 1979), but also on the plant litter layer. Various formulae have been used to account for soil evaporation, but the method of Ritchie et al. (1976) seems to have been the most widely adopted.

## ALLOCATION OF ASSIMILATE

This remains perhaps the most unsatisfactory aspect of plant growth models, particularly those that are governed by production at the source of the supply, rather than by the requirements of the various sinks. In the absence of general principles, most models distribute assimilates in somewhat arbitrary fashion, in order to follow the observed changes in the weight of plant fractions. The procedures adopted often appear to have little in common. For example, Detling et al. (1979) reserved a constant proportion of the photosynthesis of mature leaves for their maintenance, and allocated the remainder to the growth of young expanding leaves, apart from a minimum allowance for the roots. By contrast, Sheehy et al. (1980) allocated more to the roots and less to the leaves when the total amount of photosynthate increased, distributing the remainder to the stems. The net effects might not be very different, under normal conditions for growth, but the approaches suggest little insight into the processes involved.

Davidson (1969) has suggested that the ratio of shoot mass to root mass is proportional to the ratio of their respective specific activities. However, since shoot-specific activity is defined as rate of carbon uptake per unit mass, and root-specific activity as rate of uptake of the mineral nutrient in limiting supply (usually nitrogen), the expression is no more than a condition for equilibrium growth. Furthermore, "specific activity", as used here, is highly dependent on other environmental factors, and does not define the functional capacity of the plant. Nevertheless, the idea has been accepted by some (e.g. Charles-Edwards, 1982) as a basic biological principle, and made the starting point for mathematical manipulations relating to shoot/root partitioning. The usefulness of such exercises is doubtful, since they require the introduction of hypothetical parameters whose magnitudes and stability are unknown.

The fact that growth is increased both by light and carbon dioxide enrichment has often been taken as evidence that growth is limited by photosynthesis; however, Sheehy and Peacock (1975) and others have shown that growth is poorly related to photosynthetic rate. Photosynthetic capacity may not be a major limiting factor, since evidence suggests that the amount of material partitioned to leaf area may be mainly responsible for changes in relative growth rate (Potter and Jones, 1977). Although negative feedback mechanisms have not been conclusively demonstrated, examples of responses in photosynthetic activity to increased plant demand are well known, and it would appear that the photosynthetic rate observed under normal conditions is not a maximum value, but the rate at which assimilate can be currently metabolized by the plant. The emphasis on source-directed models has tended to obscure the obvious fact that the problem of allocation can only be resolved by a better understanding of the changes in sink sizes and strengths during ontogeny.

## SHOOT GROWTH AND PHENOLOGY

Although the course of pasture growth is profoundly modified by phenology, few models of pasture growth, as opposed to crop growth, have considered growth stage. Parton et al. (1978), modelling shortgrass prairie, defined six phenological stages, starting from first-shoot appearance and proceeding through to senescence. Rate of development was dependent upon cumulative heat summation, modified by the assumption that water stress retarded development during vegetative growth, but hastened it during the reproductive stages.

The regeneration of shoots, following drought or winter, must occur through the allocation of assimilates originating from below-ground reserves. Detling et al. (1979) assumed that translocation of labile carbohydrate from the crowns and upper roots during spring growth was dependent upon the degree of shading by dead leaves, and was inversely related to the quantity of assimilate in the new leaves. Regrowth following defoliation (Chapter 7) also seems to be partly determined by reallocation of assimilates, with growth rates often more rapid than leaf area indices would suggest. In the model of Sheehy et al. (1979), assimilates are transferred from the stem to new leaves over a four-day period.

The date of initiation of reproductive growth is influenced by responses to temperature, daylength and vernalization, and varies widely according to species and genotype. Modelling the main effects or interactions between these various factors is frustrated by lack of knowledge of the mechanisms involved. It has often been suggested that each species and genotype has evolved a strategy which optimizes reproductive effort; and the model of King and Roughgarden (1983) suggested that this may have been achieved by the timing of the switch from vegetative to reproductive growth, but it is likely that, where seasonal variation is high, a number of different strategies could be equally successful.

The transition from the vegetative to the reproductive state is accompanied by changes in allocation priorities (Chapter 4). Sheehy et al. (1979) found, in ryegrass swards, that assimilate partitioning to stems increased at the expense of root growth, and this may be generally the case. They also observed that the leaves formed during reproduction occurred near the top of the canopy, thus maintaining the photosynthetic capacity of the sward.

## ROOT GROWTH

Although roots may comprise over half the total weight of the plant, the difficulties of measurement have meant that the pattern of root growth is much more poorly understood than that of shoots. The problem is exacerbated in established swards, where the death of roots and their replacement by others leads to a confused agglomeration of living and dead roots.

Under these circumstances, simulation of the growth of individual roots may not be particularly informative. Page and Gerwitz (1974) treated the movement of ryegrass roots into soil as analogous to diffusion from a plane surface into a semi-infinite liquid medium, a concept that was extended by Hillel and Talpaz (1976), who modelled root growth in terms of initiation, proliferation and extension into deeper layers.

Root growth depends both on soil conditions and on the allocation of carbohydrate from the shoot, though these factors are obviously interdependent. In the model of Gilmanov (1977), shoot growth took place at the expense of root growth when water supply was plentiful but, as the soil dried out, carbohydrate supply to the roots, and consequently root growth, increased until mechanical impairment of root penetration and water conductivity by the soil became restrictive. This could occur if shoot growth was inhibited by water stress sooner than photosynthesis or root growth. Similarly, increases in root/shoot ratios, when mineral nutrients are limiting (Sheehy et al. 1979), can be attributed to a restriction in shoot growth, rather than to any enhancement of root growth. The optimum temperatures for root growth differ among plant species, and differences between air and soil temperatures can influence root/shoot relationships (Davidson, 1969).

Little is known regarding the seasonal development and phenology of the root system. Parton et al. (1978) assumed that juvenile root growth in grassland was initiated by transfer of carbohydrate from suberized roots, and that juvenile roots aged at a constant rate to mature non-suberized roots, which became suberized at a rate governed by soil temperature. Death of roots can be expected to be greatest when conditions for root growth are most unfavourable (Gilmanov, 1977), and the model of Parton et al. (1978) showed an increased loss of roots at extreme temperatures and at lower soil water potentials. What proportion of the roots remain alive at the end of the growing season, or following severe drought or defoliation, is usually a matter for conjecture, though it will certainly vary greatly according to environment and plant species.

## NITROGEN

Numerous models have been constructed on widely different aspects of nitrogen in grassland

and crop ecosystems, including the fate of fertilizer nitrogen in soils (Saxton et al., 1977), leaching (Addiscott, 1981), microbial processes (McGill et al., 1981), grass–legume associations (Ross et al., 1972), and distribution of nitrogen in the plant (Seligman et al., 1975). Simulation of nitrogen cycling (Chapter 17) is heavily reliant on other state variables of the system, and sub-models of carbohydrate metabolism and soil moisture are prerequisites.

Absorption of nitrogen by plant roots is assumed to occur largely in the form of nitrate, and ion intake has been related to the concentration at the root surface. Reuss and Innis (1977) used a double Michaelis–Menten curve to account for the fact that uptake is independent of concentration over a large part of the range. Movement of ions to the root takes place both by mass flow and by diffusion (Nye and Tinker, 1977), and with decreasing water potential uptake falls steadily to zero (Greenwood et al., 1974; Reuss and Innis, 1977). Seligman et al. (1975) calculated that mass flow alone was inadequate to account for the rapid growth of a closed grass canopy, and concluded that the difference must be made up by diffusion, although Greenwood et al. (1974) assumed a concentration at the root surface equal to that in the bulk soil. Uptake from each soil level has been assumed to be proportional to the product of water uptake and nitrate concentration. However, the total uptake is also evidently under the control of plant requirements, and Reuss and Innis (1977) found it necessary to introduce a term for reverse flow to prevent excess accumulation in the roots, whereas Seligman et al. (1975) assumed that nitrogen uptake by the plant was proportional to the difference between maximum and current nitrogen contents of the various plant organs, according to sink size.

The general metabolism of nitrogen within the plant itself has not been extensively studied by modellers. Pate et al. (1979) have used an empirical model to describe in detail the transport and distribution of carbon and symbiotically fixed nitrogen in a legume, while Greenwood and Barnes (1978) have examined a mechanism for the changes in organic nitrogen content of plants with increasing size and age. The effects of nitrogen deficiency have usually been expressed directly in terms of growth; for example, Morris (1982) considered growth responses of leaves to nitrogen concentration to be hyperbolic, like the photosynthetic response to radiation, while Sheehy et al. (1979) assumed that partitioning of assimilate between leaf, stem and root changed from proportions of 10:40:50 under low nitrogen supply to 25:65:10 under high nitrogen treatment and irrigation. Model building has evidently not yet reached the stage where such specific effects as reduction in photosynthetic leaf area or increase in specific leaf area can be readily incorporated.

Few models have been developed for the important grass–legume association. The model of Ross et al. (1972) treats the effect of available nitrogen on plant growth and the transfer of legume nitrogen to the available fraction, but more detailed studies are needed.

The vertical movement of nitrate through the soil is governed by water flow, and has been calculated by diffusion–convection equations (Wagenet, 1981), although Saxton et al. (1977) found that a simple finite-difference equation, based on concentration and water flow, was adequate, except when the soil was well above field capacity. Nitrate ions have been assumed either not to be absorbed by the soil (Saxton et al., 1977) or to be partitioned in dynamic equilibrium between a mobile and a retained phase (Addiscott, 1981).

Grasslands are subject to large inputs (Chapter 6) and losses (Chapter 22) of nitrogen (Jones and Woodmansee, 1979), with grazing animals playing an important part (Chapter 19). Deposition of faeces and urine produces large local variability in soil nitrogen content; perhaps for that reason, it has received little attention from modellers. The total quantity of nitrogen excreted, and its distribution between faeces and urine, would have to be determined either by direct collection or by the application of an animal nutrition model, which estimated intakes and the *digestibility* or the nitrogen content of the material selected. Data regarding the subsequent fate of excreted nitrogen are highly variable; in particular, the amount of urinary urea–nitrogen lost as ammonia (Chapter 22) is largely dependent on environmental conditions (Jones and Woodmansee, 1979).

The breakdown of organic nitrogen to ammonium is intimately linked with the growth requirements of decomposing organisms. Mineralization rates increase with temperature, soil water potential, and nitrogen content; at high C:N ratios, nitrogen immobilization predominates (Reuss and Innis, 1977), and hence both processes occur concurrently in different regions, according to the nature of the substrate. The conversion of ammonium to nitrate ion takes place rapidly in well-drained soils, and has been expressed in terms of the rate-limiting step, the oxidation to nitrite, as a Michaelis–Menten equation (Reuss and Innes, 1977; McGill et al., 1981). In saturated soils, the rate is reduced to zero, while anaerobic conditions may cause denitrification, which requires the modelling of oxygen diffusion through water-filled pores from the atmosphere for its description (Luxmoore et al., 1970; Leffelaar, 1979).

The turnover rates of other mineral nutrients show obvious similarities to that of nitrogen, both conceptually and in compartmental structure (Chapter 18) though the dynamics are very different. Phosphorus cycling in semi-arid grasslands has been modelled in detail by Cole et al. (1977), and conceptually in a grazing system by Katznelson (1977), whereas soil transformations of phosphorus and potassium are described in the model of O.L. Smith (1979).

## PLANT SENESCENCE AND DECOMPOSITION

The loss of function and the death of plant material usually receives little attention in models of managed grassland, though it is often treated in considerable detail in models of natural grasslands. Even in heavily stocked swards, a large proportion of the primary production dies, as opposed to being eaten (Hutchinson, 1971). This may be regarded as a waste of animal feed, but it serves an important role in recycling mineral nutrients.

Detling et al. (1979) assumed, in their model, that leaf death was a function of soil water potential and minimum temperature as well as age, but they ensured a constant leaf area index by relating senescence to the production of new leaves. Sheehy et al. (1980) assumed that ryegrass leaves died when 50% of their weight was lost through respiration, so that there were always about three live leaves per tiller, even under varying temperatures. The rate of conversion of standing dead material to fallen dead matter or litter depends upon plant species and weather conditions.

Dead litter disappears through decay and by being broken down to smaller particles, under the combined influence of weather and soil fauna. O.L. Smith (1979) developed a very extensive model of microbial processes involved in the decomposition of organic matter, whereas Cale and Waide (1980) described a simpler model, in which decomposers were aggregated into functional groups. Hunt (1977) assumed that decomposition was independent of microbial numbers, and considered the heterogeneous plant material to consist of a labile and a resistant fraction, depending on their C:N ratios, with highly contrasting decomposition rates. In all of these models, decomposition rates depended on temperatures, soil water potentials, and nitrogen.

## PLANT COMPETITION

The density and botanical composition of pastures continually change (Chapter 8). In most pasture models, plants are regarded as uniform in habit, vigour and phenological development, a simple condition that is rarely apparent in practice.

Competition within monocultures may be intense, depending on density; empirical relationships between density and yield have received considerable attention (Willey and Heath, 1969). The reciprocal relationship between stand yield per unit area and density is an interesting analogy of the Langmuir isotherm relating adsorption of gas molecules to the available surface area:

$$Y = \frac{Y_{max}}{1 + k\rho} \qquad (4)$$

where $Y$ is the yield, $Y_{max}$ the maximum yield, $\rho$ the density of sowing and $k$ a constant, so that yield increases monotonically with density. Aikman and Watkinson (1980) have modelled the effects of density, using a logistic growth model, and have simulated self-thinning effects, whereas Ford and Diggle (1981) have used Monte Carlo simulation to model effects of spatial distribution on individual plant size.

The basic models of plant species competition have been summarized by Antonovics (1978), who pointed out some of the implications of selection for competitive ability. Torssell and Nicholls (1978) have modelled the effects of inter-specific competition at different developmental stages on total productivity, whereas Torssell et al. (1976) assessed the relative dominance of two tropical legumes in mixtures. Berendse (1979) examined the situation where one population could draw on a source of some limiting factor not available to the other, for instance by greater rooting depth, or fixation of nitrogen.

Few models have dealt with both density and species composition. In the model of Scott et al. (1978), the growth of one species was related to the weight and competitive ability of the competing species. However, since competitive ability can vary with phenological stage or environment, a more fundamental approach is clearly desirable. Byrne et al. (1976) used a process-governed model, relating the growth of each species to relative rates of transpiration. Wright (1981) extended the relationship between stand yield and density, derived for single-species stands, to the case of binary mixtures. Similarly, Weiner (1982), using an analogy with the reciprocal-yield law, developed a model relating the reproductive output of a individual plant to the numbers and competitive ability of neighbours lying within concentric radii.

The growth of weeds in pastures has been given scant attention by modellers. One of the few exceptions is the model by Medd and Smith (1978) of the likely spread of *Carduus nutans* (nodding thistle) in Australia.

## GRAZING

Although the general effects of animals on pastures are well known (Watkin and Clements, 1978; Chapter 7), quantitative estimates are extremely difficult to obtain, and it is all too convenient to simplify the problem by considering only cut or ungrazed pasture. Not only growth rates but the growth habit and the competitive ability of pasture species can be altered by contrasting grazing regimes (Alexander and Thompson, 1982). Defoliation commonly results in cessation of root growth, followed by a period of rapid tillering, but the net effect of defoliation on pasture production depends on the relative growth rate of the sward, as well as on the intensity and frequency of grazing (Hilbert et al., 1981). Much of the dead material in laxly-grazed swards comes from the death of defoliated residual herbage (Korte and Sheath, 1979), and may represent a considerable wastage. Following intense defoliation, on the other hand, young leaves expand in conditions of high light intensity, unshaded by older vegetation, though much of the light may also be intercepted by leaf sheaths, which have a low photosynthetic efficiency (Parsons et al., 1983). Preserving the optimal leaf area for light interception may in fact be less important than controlling the reproductive development of plants and maintaining a leafy sward (Korte et al., 1982). Effects such as these cannot be readily modelled, but nevertheless merit more detailed investigation.

A further consideration regarding pastures is that growth must be specified in terms that are meaningful to the animal, and a mere figure for dry matter is inadequate. Keeping an account of leaf ages, as in the model of Sheehy et al. (1980), is a useful step in this direction, although the difficulties of predicting even the mean digestibility of successive harvests are indicated by the regression model of Edelsten and Corrall (1979).

## CONCLUSIONS

The most important requirement when modelling grassland systems is a sense of proportion. Faced with a welter of factors and their interactions, it is highly tempting to become immersed in those detailed aspects with which one is most familiar, while dismissing the rest of the system in a few lines of perfunctory code. When dissecting individual processes, it is important to keep in mind that they do not operate in isolation or without reference to one another. In paying great attention to detail, models may fail to treat the plant as the highly integrated organism that we otherwise acknowledge it to be. Even if nitrogen is believed to be the main limiting factor, it is still a mistake to produce a model that contains little else besides nitrogen metabolism, because such a model can never indicate that plant growth is directed by anything but nitrogen.

The nature of a model is, to a large extent, a reflection of the data available, but much also depends on the way this information is used. Models must provide a reduced description encapsulating the essence of the processes involved. Cale and Waide (1980) claimed, on the basis of their results, that models could be constructed on theoretical principles, with the aid of a relatively small base, which would not only simulate the natural process but also provide insight into its operation. On the other hand, biologists may be permitted to question the utility of such a model as that of Johnson et al. (1983), which requires the creation of more than a dozen parameters and considerable algebraic calisthenics to simulate a data set that could be fitted equally well by two straight lines.

It should be a matter of concern that each person constructing a new model feels obliged to start afresh, despite the number of models already published. There is a need for standard packages, well-documented and efficiently coded. Even those who are interested only in a specific area, such as soil microbiology, need routines to calculate temperature and moisture profiles, and much time can be wasted if they have to construct sub-models to do this.

Since the growth of pasture is largely at the mercy of weather conditions, forecasts for seasonal growth cannot generally be regarded as a realistic objective, but further development should enable predictions to be made of the suitability of particular climatic regions and local environments for a given species or genotype. In agricultural research programmes, the main purpose of models is in complementing and interpreting experimental studies. In particular, they may be expected to become increasingly important in investigating the effects on productivity of different grazing patterns, irrigation, fertilizer application and other management practices, the effect of specific environmental variables on growth, the performance of various combinations of species, and the development of more productive cultivars.

The goals achieved by pasture models to date have been limited, though they have proved a useful accessory to research. Much has been learned on both sides. Modellers have come to appreciate that more than band-aid algorithms are needed to cope with problems that have concerned plant physiologists for decades, whereas biologists have become increasingly aware of the need to think quantitatively and to view their results within the context of the whole plant–environment system.

## REFERENCES

Acock, B., Hand, D.W., Thornley, J.H.M. and Warren Wilson, J., 1976. Photosynthesis in stands of green peppers. An application of empirical and mechanistic models to controlled-environment data. *Ann. Bot.*, 40: 1293–1307.

Addiscott, T.M., 1981. Leaching of nitrate in structured soils. In: M.J. Frissel and J.A. van Veen (Editors), *Simulation of Nitrogen Behaviour of Soil–Plant Systems*. PUDOC, Wageningen, pp. 245–253.

Aikman, D.P. and Watkinson, A.R., 1980. A model for growth and self-thinning in even-aged monocultures of plants. *Ann. Bot.*, 45: 419–427.

Alexander, K.I. and Thompson, K., 1982. The effect of clipping frequency on the competitive interaction between two perennial grass species. *Oecologia (Berl.)*, 53: 251–254.

Anderson, M.C. and Denmead, O.T., 1969. Short wave radiation on inclined surfaces in model plant communities. *Agron. J.*, 61: 867–872.

Angus, J.F. and Wilson, J.H., 1976. Photosynthesis of barley and wheat leaves in relation to canopy models. *Photosynthetica*, 10: 367–377.

Antonovics, J., 1978. The population genetics of mixtures. In: J.R. Wilson (Editor), *Plant Relations in Pastures*. CSIRO, Melbourne, Vic., pp. 233–252.

Baier, W., 1973. Crop weather analysis model: review and model development. *J. Appl. Meteorol.* 12: 937–947.

Baier, W., 1979. Note on the terminology of crop-weather models. *Agric. Meteorol.*, 20: 137–145.

Berendse, F., 1979. Competition between plant populations with different rooting depths. I. Theoretical considerations. *Oecologia (Berl.)*, 43: 19–26.

Byrne, G.F., Torssell, B.W.R. and Sastry, P.S.N., 1976. Plant growth curves in mixtures and climatological response. *Agric. Meteorol.*, 16: 37–44.

Cale, W.G., 1979. Modelling grassland primary productivity using piecewise stationary, piecewise linear mathematics. *Ecol. Modelling*, 7: 107–123.

Cale, W.G. and Waide, J.B., 1980. A simulation model of decomposition in a shortgrass prairie. *Ecol. Modelling*, 8: 1–14.

Charles-Edwards, D.A., 1982. *Physiological Determinants of Plant Growth*. Academic Press, Sydney, N.S.W., 161 pp.

Christian, K.R. and Milthorpe, F.L., 1981. A systematic approach to the simulation of short-term processes in the plant–environment complex. *Plant, Cell Environ.*, 4: 275–284.

Cole, C.V., Innis, G.S. and Stewart, J.W.B., 1977. Simulation of phosphorus cycling in semiarid grasslands. *Ecology*, 58: 1–15.

Connor, D.J., Brown, L.F. and Trlica, M.J., 1974. Plant cover, light interception, and photosynthesis of shortgrass prairie: a functional model. *Photosynthetica*, 8: 18–27.

Curry, R.B., 1971. Dynamic simulation of plant growth — part I. Development of a model. *Trans. ASAE*, 14: 946–949.

Davidson, R.L., 1969. Effect of root/leaf temperature differentials on root/shoot ratios in some pasture grasses and clover. *Ann. Bot.*, 33: 561–569.

Dayan, E., Van Keulen, H. and Dovrat, A., 1981. Experimental evaluation of a crop growth simulation model. A case study with Rhodes grass. *Agro-Ecosystems*, 7: 113–126.

De Jong, R. and Cameron, D.R., 1979. Computer simulation model for predicting soil water content profiles. *Soil Sci.*, 128: 41–48.

Denmead, O.T. and Shaw, R.H., 1962. Availability of soil water to plants as affected by soil moisture content and meteorological conditions. *Agron. J.*, 54: 385–389.

Detling, J.K., Parton, W.J. and Hunt, H.W., 1979. A simulation model of *Bouteloua gracilis* biomass dynamics on the North American shortgrass prairie. *Oecologia (Berl.)*, 38: 167–191.

De Wit, C.T., 1958. *Transpiration and Crop Yields.* Versl. Landbouwk. Onderz. (Agric. Res. Rep.) 64.6. PUDOC, Wageningen, 88 pp.

De Wit, C.T., 1965. *Photosynthesis of Leaf Canopies.* Versl. Landbouwk. Onderz. (Agric. Res. Rep.) 663. PUDOC, Wageningen, 57 pp.

De Wit, C.T., et al., 1978. *Simulation of Assimilation, Respiration and Transpiration of Crops.* Wiley, New York, N.Y.

Edelsten, P.R. and Corrall, A.J., 1979. Regression models to predict herbage production and digestibility in a non-regular sequence of cuts. *J. Agric. Sci. (Cambridge)*, 92: 575–585.

Fick, G.W., 1980. A pasture production model for use in a whole farm simulator. *Agric. Systems*, 5: 137–161.

Fiscus, E.L., Klute, A. and Kaufmann, M.R., 1983. An interpretation of some whole plant water transport phenomena. *Plant Physiol.*, 71: 810–817.

Fitzpatrick, E.A. and Nix, H.A., 1970. The climatic factor in grassland ecology. In: R.M. Moore (Editor), *Australian Grasslands.* ANU Press, Canberra, A.C.T., pp. 3–26.

Ford, E.D. and Diggle, P.J., 1981. Competition for light in a plant monoculture modelled as a spatial stochastic process. *Ann. Bot.*, 48: 481–500.

Gilmanov, T.G., 1977. Plant submodel in the holistic model of a grassland ecosystem (with special attention to the belowground part). *Ecol. Modelling*, 3: 149–163.

Goudriaan, J., 1977. *Crop Micrometeorology: a Simulation Study.* PUDOC, Wageningen, 249 pp.

Goudriaan, J. and Van Laar, H.H., 1978. Calculation of daily totals of the gross $CO_2$ assimilation of leaf canopies. *Neth. J. Agric. Sci.*, 26: 373–382.

Greenwood, D.J. and Barnes, A., 1978. A theoretical model for the decline in the protein content of plants during growth. *J. Agric. Sci. (Cambridge)*, 91: 461–466.

Greenwood, D.J., Wood, J.T. and Cleaver, T.J., 1974. A dynamic model for the effects of soil and weather conditions on nitrogen response. *J. Agric. Sci. (Cambridge)*, 82: 455–467.

Hilbert, D.W., Swift, D.M., Detling, J.K. and Dyer, M.I., 1981. Relative growth rates and the grazing optimization hypothesis. *Oecologia (Berl.)*, 51: 14–18.

Hillel, D. and Talpaz, H., 1976. Simulation of root growth and its effect on the pattern of soil water uptake by a non-uniform root system. *Soil Sci.*, 121: 307–312.

Hunt, H.W., 1977. A simulation model for decomposition in grasslands. *Ecology*, 58: 469–484.

Hutchinson, K.J., 1971. Productivity and energy flow in grazing/fodder conservation systems. *Herbage Abstr.*, 41: 1–10.

Jensen, M.E., Wright, J.L. and Pratt, B.J., 1971. Estimating soil moisture depletion from climate, crop and soil data. *Trans. ASAE*, 14: 954–959.

Johnson, I.R., Ameziane, T.E. and Thornley, J.H.M., 1983. A model of grass growth. *Ann. Bot.*, 51: 599–609.

Jones, M.B. and Woodmansee, R.G., 1979. Biogeochemical cycling in annual grassland ecosystems. *Bot. Rev.*, 45: 111–144.

Katznelson, J., 1977. Phosphorus in the soil–plant–animal ecosystem. *Oecologia (Berl.)*, 26: 325–334.

Keig, G. and McAlpine, J.R., 1974. WATBAL. A computer system for the estimation and analysis of soil moisture regimes from climatic data. *CSIRO Tech. Mem.*, 74/4: 44 pp.

Kimes, D.S., Ranson, K.J. and Smith, J.A., 1980. A Monte Carlo calculation of the effects of canopy geometry on PhAR absorption. *Photosynthetica*, 14: 55–64.

King, D. and Roughgarden, J., 1983. Energy allocation patterns of the California grassland annuals *Plantago erecta* and *Clarkia rubicunda*. *Ecology*, 64: 16–24.

Korte, C.J. and Sheath, G.W., 1979. Herbage dry matter production: the balance between growth and death. *Proc. N.Z. Grassland Soc.*, 40: 152–161.

Korte, C.J., Watkin, B.R. and Harris, W., 1982. Use of residual leaf area index and light interception as criteria for spring-grazing management of a ryegrass-dominant pasture. *N.Z. J. Agric. Res.*, 25: 309–319.

Leffelaar, P.A., 1979. Simulation of partial anaerobiosis in a model soil in respect to denitrification. *Soil Sci.*, 128: 110–120.

Lemeur, R. and Blad, B.L., 1974. A critical review of light models for estimating the shortwave radiation regime of plant canopies. *Agric. Meteorol.*, 14: 255–286.

Loomis, R.S., Rabbinge, R. and Ng, E., 1979. Explanatory models in crop physiology. *Annu. Rev. Plant Physiol.*, 30: 339–367.

Ludlow, M.M., Stobbs, T.H., Davis, R. and Charles-Edwards, D.A., 1982. Effect of sward structure in two tropical grasses with contrasting canopies on light distribution, net photosynthesis and the size of bite harvested by grazing cattle. *Aust. J. Agric. Res.*, 33: 187–201.

Luxmoore, R.J., Stolzy, L.H. and Letey, J., 1970. Oxygen diffusion in the soil–plant system. I. A model. *Agron. J.*, 62: 317–322.

McCown, R.L., 1973. An evaluation of the influence of available soil water storage capacity on growing season length and yield of tropical pastures using simple water balance models. *Agric. Meteorol.*, 11: 53–63.

McGill, W.B., Hunt, H.W., Woodmansee, R.G., Reuss, J.O. and Paustian, K.H., 1981. Formulation, process controls, parameters and performance of PHOENIX: a model of carbon and nitrogen dynamics in grassland soils. In: M.J. Frissel and J.A. van Veen (Editors). *Simulation of Nitrogen Behaviour of Soil–Plant Systems.* PUDOC, Wageningen, pp. 171–188.

Medd, R.W. and Smith, R.C.G., 1978. Prediction of the potential distribution of *Carduus nutans* (nodding thistle) in Australia. *J. Appl. Ecol.*, 15: 603–612.

Molz, F.J., 1981. Models of water transport in the soil–plant system: a review. *Water Resour. Res.*, 17: 1245–1260.

Molz, F.J. and Remson, I., 1970. Extraction-term models of

soil–moisture use by transpiring plants. *Water Resour. Res.*, 6: 1346–1356.

Monteith, J.L., 1965. Light distribution and photosynthesis in field crops. *Ann. Bot.*, 29: 27–38.

Morris, J.T., 1982. A model of growth responses by *Spartina alterniflora* to nitrogen limitation. *J. Ecol.*, 70: 25–42.

Nye, P.H. and Tinker, P.B., 1977. *Solute Movement in the Soil–Root System*. Blackwell, Oxford, 342 pp.

O'Rourke, P.A. and Terjung, W.H., 1981. Modelling of influence of cloud amounts and types on leaf net photosynthetic rates inside a mature canopy. *Photosynthetica*, 15: 317–329.

Page, E.R. and Gerwitz, A., 1974. Mathematical equations, based on diffusion equations, to describe root systems of isolated plants, row crops, and swards. *Plant Soil*, 41: 243–254.

Paltridge, G.W., 1970. A model of a growing pasture. *Agric. Meteorol.*, 7: 93–130.

Parsons, A.J., Leafe, E.L., Collett, B. and Stiles, W., 1983. The physiology of grass production under grazing. I. Characteristics of leaf and canopy photosynthesis of continuously-grazed swards. *J. Appl. Ecol.*, 20: 117–126.

Parton, W.J., Singh, J.S. and Coleman, D.C., 1978. A model of production and turnover of roots in shortgrass prairie. *J. Appl. Ecol.*, 15: 515–542.

Pate, J.S., Layzell, D.B. and McNeil, D.L., 1979. Modelling the transport and utilization of carbon and nitrogen in a nodulated legume. *Plant Physiol.*, 63: 730–737.

Penman, H.L., 1963. *Vegetation and Hydrology*. Commonwealth Agricultural Bureau, Farnham Royal, 124 pp.

Penning de Vries, F.W.T., 1975. The cost of maintenance processes in plant cells. *Ann. Bot.*, 39: 77–92.

Potter, J.R. and Jones, J.W., 1977. Leaf area partitioning as an important factor in growth. *Plant Physiol.*, 59: 10–14.

Reuss, J.O. and Innis, G.S., 1977. A grassland nitrogen flow simulation model. *Ecology*, 58: 379–388.

Richards, L.A., 1931. Capillary conduction of liquids through porous mediums. *Physics*, 1: 318–333.

Richter, J., 1980. A simple numerical solution for the vertical flow equation of water through unsaturated soils. *Soil Sci.*, 129: 138–144.

Ritchie, J.T., Rhoades, E.D. and Richardson, C.W., 1976. Calculating evaporation from native grassland watersheds. *Trans. ASAE*, 19: 1098–1103.

Ross, P.J., Henzell, E.F. and Ross, D.R., 1972. Effects of nitrogen and light in grass–legume pastures: a systems analysis approach. *J. Appl. Ecol.*, 9: 535–556.

Russo, J.M. and Knapp, W.W., 1976. A numerical simulation of plant growth. *Int. J. Biometeorol.*, 20: 276–285.

Saxton, K.E., Johnson, H.P. and Shaw, R.H., 1974. Modelling evaporation and soil moisture. *Trans. ASAE*, 17: 673–677.

Saxton, K.E., Schuman, G.E. and Burwell, R.E., 1977. Modelling nitrate movement and dissipation in fertilized soils. *Soil Sci. Soc. Am. J.*, 41: 265–271.

Scott, H.D., Griffis, C.L., Brewer, D.W. and Oliver, L.R., 1978. Simulation of plant competition. *Trans. ASAE*, 21: 813–821.

Seligman, N.G., Van Keulen, H. and Goudriaan, J., 1975. An elementary model of nitrogen uptake and redistribution by annual plant species. *Oecologia (Berl.)*, 21: 243–261.

Selirio, I.S. and Brown, D.M., 1979. Soil moisture-based simulation of forage yield. *Agric. Meteorol.*, 20: 99–114.

Shawcroft, R.W., Lemon, E.R., Allen, L.H., Stewart, D.W. and Jensen, S.E., 1974. The soil–plant–atmosphere model and some of its predictions. *Agric. Meteorol.*, 14: 287–307.

Sheehy, J.E. and Peacock, J.M., 1975. Canopy photosynthesis and crop growth rate of eight temperate forage grasses. *J. Exp. Bot.*, 26: 679–691.

Sheehy, J.E., Cobby, J.M. and Ryle, G.J.A., 1979. The growth of perennial ryegrass: a model. *Ann. Bot.*, 43: 335–354.

Sheehy, J.E., Cobby, J.M. and Ryle, G.J.A., 1980. The use of a model to investigate the influence of some environmental factors on the growth of perennial ryegrass. *Ann. Bot.*, 46: 343–365.

Sinclair, T.R., Murphy, C.E. and Knoerr, K.R., 1976. Development and evaluation of simplified models for simulating canopy photosynthesis and transpiration. *J. Appl. Ecol.*, 13: 813–829.

Singh, J.S., Trlica, M.J., Risser, P.G., Redmann, R.E. and Marshall, J.K., 1980. Autotrophic subsystems. In: A.I. Breymeyer and G.M. van Dyne (Editors), *Grasslands, Systems Analysis and Man*. Cambridge University Press, Cambridge, pp. 59–200.

Smith, O.L., 1979. An analytical model of the decomposition of soil organic matter. *Soil Biol. Biochem.*, 11: 585–606.

Smith, R.C.G. and Johns, G.G., 1975. Seasonal trends and variability of soil moisture under temperate pasture on the Northern Tablelands of New South Wales. *Aust. J. Exp. Agric. Anim. Husb.*, 15: 250–255.

Smith, R.C.G. and Stephens, M.J., 1976. Importance of soil moisture and temperature on the growth of improved pasture on the Northern Tablelands of New South Wales. *Aust. J. Agric. Res.*, 27: 63–70.

Stroosnijder, L., 1982. Simulation of the soil water balance. In: F.W.T. Penning de Vries and H.H. van Laar (Editors), *Simulation of Plant Growth and Crop Production*. PUDOC, Wageningen, pp. 175–193.

Tenhunin, J. and Westrin, S., 1979. Development of a photosynthesis model with an emphasis on ecological applications. IV· WHOLEPHOT — Whole leaf photosynthesis in response to four independent variables. *Oecologia (Berl.)*, 41: 145–162.

Torssell, B.W.R., Ive, J.R. and Cunningham, R.B., 1976. Competition and population dynamics in legume-grass swards with *Stylosanthes hamata* (L.) Taub. (sens. lat) and *Stylosanthes humilis* (H.B.K.). *Aust. J. Agric. Res.*, 27: 71–83.

Torssell, B.W.R. and Nicholls, A.O., 1978. Population dynamics in species mixtures. In: J.R. Wilson (Editor), *Plant Relations in Pastures*. CSIRO, Melbourne, Vic., pp. 217–232.

Van Keulen, H., Penning de Vries, F.W.T. and Drees, E.M., 1982. A summary model for crop growth. In: F.W.T. Penning de Vries and H.H. van Laar (Editors), *Simulation of Plant Growth and Crop Production*. PUDOC Wageningen, pp. 87–97.

Wagenet, R.J., 1981. Simulation of soil-water and nitrogen movement. In: M.J. Frissel and J.A. van Veen (Editors). *Simulation of Nitrogen Behaviour of Soil–Plant Systems*. PUDOC, Wageningen, pp. 67–77.

Waggoner, P.E. and Reifsnyder, W.E., 1968. Simulation of the temperature, humidity and evaporation profiles in a leaf canopy. *J. Appl. Meteorol.*, 7: 400–409.

Wallach, D., 1975. The effect of environmental factors on the growth of a natural pasture. *Agric. Meteorol.*, 15: 231–244.

Watkin, B.R. and Clements, R.J., 1978. The effects of grazing animals on pastures. In: J.R. Wilson (Editor), *Plant Relations in Pastures*. CSIRO, Melbourne, Vic., pp. 273–289.

Weiner, J., 1982. A neighborhood model of annual-plant interference. *Ecology*, 63: 1237–1241.

Weiner, J. and Conte, P.T., 1981. Dispersal and neighborhood effects in an annual plant competition model. *Ecol. Modelling*, 13: 131–147.

Wight, J.R. and Hanks, R.J., 1981. A water-balance, climate model for range herbage production. *J. Range Manage.*, 34: 307–311.

Willey, R.W. and Heath, S.B., 1969. The quantitative relationships between plant population and crop yield. *Adv. Agron.*, 21: 281–321.

Wright, A.J., 1981. The analysis of yield–density relationships in binary mixtures using inverse polynomials. *J. Agric. Sci. (Cambridge)*, 96: 561–567.

Chapter 6

# EFFECTS OF NITROGEN FERTILIZER

J. MORRISON

## INTRODUCTION

Vigorous grass growth is dependent on an adequate supply of available nitrogen in the soil throughout the growing season. Since grass is harvested in a succession of vegetative crops, the amount of nitrogen needed to support maximum yield is greater than that for most other crops. The amount of nitrogen supplied by the decomposition of soil organic matter (Chapter 17), biological fixation by legumes (Chapter 21), and excretal return (Chapter 19) is rarely adequate to achieve the potential yield in most temperate grasslands (Chapter 2). Fertilizer application is a convenient way of supplying nitrogen to grassland, since the amount and timing can be closely controlled. If the farmer is to use fertilizer nitrogen effectively, he needs to know when and in what quantities to apply fertilizer, and to know the likely response to those applications in his particular conditions of management and environment.

Although the response of grassland to nitrogen fertilizer has been widely recognized, it is only in the Northern Hemisphere, principally northern Europe and the humid parts of the United States of America, that fertilizer nitrogen is extensively used on grassland. This is a consequence of the need for a sound industrial base to provide the input, combined with a high domestic demand and relatively high prices for products, particularly dairy products. The intensive use of fertilizer nitrogen on grassland has developed only in the last 25 years. Current use in some European countries is shown in Table 6.1. In some other countries, such as New Zealand and parts of Australia, intensive grassland farming is based on nitrogen fixation by legumes, rather than fertilizer nitrogen. In most of the tropical world and developing countries, grassland production is less intensive and little nitrogen fertilizer is used on grassland.

There have been numerous studies of the effects of nitrogen fertilizer on grassland production. Whitehead (1970), in an excellent review, summarized the results for temperate grassland. There is no such comprehensive review for tropical or semi-arid grassland, but Bogdan (1977) cites many references concerning the response of tropical grasses to nitrogen fertilizer.

This chapter is selective, and largely based on studies in Europe, especially the United Kingdom and the Netherlands. It examines the effect of nitrogen fertilizer on the yield and composition of grassland, and how this is modified by environment, soil nitrogen, frequency of defoliation, species, season and the presence of a legume. In considering the response of grassland to fertilizer nitrogen, it is assumed that other nutrients are not limiting. In most farm conditions, deficiencies of other major nutrient elements (potassium, phosphorous and sulphur) must be corrected to obtain the maximum response to nitrogen. Normally these elements are applied in proportion to the amount of nitrogen applied.

The effect of form of nitrogen fertilizer is not

TABLE 6.1

Average fertilizer nitrogen use (kg N ha$^{-1}$ yr$^{-1}$) on grassland in selected countries in Europe, 1980 (after Lee, 1983)

| | |
|---|---|
| The Netherlands | 265 |
| United Kingdom | 120 |
| Federal Republic of Germany | 100 |
| Belgium | 120 |
| France | 30 |

considered here. The most commonly used forms are soluble granular fertilizers, such as ammonium nitrate, calcium ammonium nitrate and urea, or as aqueous or anhydrous ammonia applied by soil injection. Although slight differences occur, the response to the various forms is broadly proportional to the amount of nitrogen applied (van Burg et al., 1982).

## YIELD RESPONSE

Most studies of response to nitrogen fertilizer have been carried out under cutting regimes; few have measured the response under grazing. In most studies, particularly those before 1970 and in tropical areas, relatively small amounts of fertilizer ($<200$ kg N ha$^{-1}$ yr$^{-1}$) have been applied, and it is therefore not possible to construct a full response curve.

The shape of the response curve of grassland to fertilizer nitrogen is characteristic (Fig. 6.1), and similar to that of most other crops. Response is linear at low nitrogen applications, reaching a maximum yield and subsequently declining at high rates of application. Reid (1970), who studied the response over the full range of rates, fitted a polynomial regression to define the response.

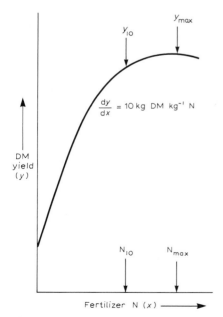

Fig. 6.1. General form of the response of grassland to nitrogen fertilizer (Morrison et al., 1980).

The most comprehensive series of nitrogen fertilizer trials on grassland was undertaken in the United Kingdom. Two experiments were carried out at each of 21 contrasting sites, over a period of four years. There was marked variation between the various sites and between years (Morrison et al., 1980). Several statistical models were fitted, but an inverse quadratic model fitted the result best in most cases (Sparrow, 1979).

Attempts have been made to define an economic optimum rate of application but, since the value of grass depends on its quality and use, an optimum can only be defined for a specific system of management (Chapter 24). Reid (1970) estimated that a response of at least 5.7 kg DM per kg N would be needed to be economic for grass fed to dairy cows in the United Kingdom; Prins et al. (1980) proposed a figure of 7.5 kg DM per kg N for the Netherlands. Morrison et al. (1980) chose an arbitrary value of 10 kg DM per kg N on the basis that this was the point where the response began to decline markedly and where yield was about 90% of the maximum, yet required only about 60% of the nitrogen needed to achieve maximum yield. The nitrogen supply to achieve that point was termed $N_{10}$.

In the study by Morrison et al. (1980), there was a wide variation in the yields between sites at a given nitrogen application, and in the response to nitrogen fertilizer (Table 6.2), though the average yields and responses were similar to those described in previous studies (Whitehead, 1970; Van Steenbergen, 1977).

TABLE 6.2

Range of variation in dry-matter yield and response to nitrogen of cut swards of *Lolium perenne* at 21 contrasting sites, United Kingdom (after Morrison, 1980b)

|  | Range | Mean |
| --- | --- | --- |
| Maximum yield (t ha$^{-1}$) | 6.5–15.0 | 11.9 |
| $N_{max}$[1] (kg ha$^{-1}$) | 540–678 | 624 |
| $N_{10}$[2] (kg ha$^{-1}$) | 250–530 | 386 |
| Optimum yield[3] (t ha$^{-1}$) | 5.2–14.3 | 10.8 |
| Response between $N_0$ and $N_{10}$ (kg DM kg$^{-1}$ N) | 15–26 | 21 |

[1] $N_{max}$ = N application rate giving maximum yield.
[2] $N_{10}$ = N application rate at the point where the response is 10 kg DM per kg N.
[3] Optimum yield = yield at $N_{10}$.

## FACTORS AFFECTING RESPONSE

### Climate

The yield response of pastures to nitrogen is very dependent on growing conditions. Periods of active growth are also periods of greatest response to fertilizer nitrogen, and seasonal responses to nitrogen reflect this. Few papers dealing with the response of grassland to fertilizer nitrogen fully describe climatic conditions and it is therefore difficult to assess quantitative relationships between response to fertilizer nitrogen and environmental factors.

In the United Kingdom, the annual yield of grass and the response to fertilizer nitrogen are predominantly dependent on the amount of water available over the growing season, which is determined by summer rainfall and the available water capacity of the soil (Morrison et al., 1980). Soil water deficits restrict growth from mid-May onwards (Garwood and Stiles, 1974). The amount and seasonal distribution of rainfall, therefore, greatly affect nitrogen uptake and yield response. Summer rainfall and soil characteristics have been used to predict fertilizer nitrogen requirements (Thomas and Young, 1982; Morrison and Russell, 1980). In a classic study, Garwood and Williams (1967) showed that grass growth was severely limited by soil water deficits too small to restrict transpiration, mainly because dry soil conditions prevented the uptake of nitrogen from the upper 10 cm of soil; if nitrogen was injected deeper into the soil, nitrogen uptake occurred and growth continued.

Growth and response to fertilizer nitrogen are also limited by temperature and radiation, especially in spring and autumn. Consequently, the annual yield and response to fertilizer nitrogen in upland conditions in the United Kingdom are less than those in lowland conditions (Davies and Munro, 1974).

In mediterranean, semi-arid and sub-tropical regions, the periods of active growth are also determined by the periods of effective rainfall. Henzell and Stirk (1963) and Vicente-Chandler et al. (1959) have demonstrated the effects of rainfall on the yield and nitrogen response of tropical grasslands (see below).

### Soil nitrogen

Although the amount of nitrogen contained in soil organic matter is large (2–20 t ha$^{-1}$) in developed grassland soils, the rate of turnover is slow (Lazenby, 1983). The rate of mineralization is dependent on the C:N ratio, pH, aeration, temperature and the water content of the soil. The amount of nitrogen released annually from the soil organic matter, as measured by the amount of nitrogen harvested in herbage when no fertilizer nitrogen is applied, is in the range of 20 to 275 kg ha$^{-1}$ in the United Kingdom, with a median value of 80 to 100 kg ha$^{-1}$ (Richards, 1977). Similar ranges have been recorded in the Netherlands (Van der Meer, 1983) and in tropical grasslands (Henzell, 1963; Vicente-Chandler et al., 1959). The amount of nitrogen mineralized depends on soil type, age of grassland, and whether the grassland is grazed or not.

Grassland yield is related to total nitrogen supply, which includes nitrogen from decomposing soil organic matter, from fertilizer and from excreta (Richards, 1977). Morrison et al. (1980) considered that the effects of fertilizer and soil nitrogen are additive, so that fertilizer nitrogen requirements are inversely related to soil nitrogen (Table 6.3). The likely contribution of soil nitrogen needs to be considered in determining the optimum rate of fertilizer nitrogen, but standard techniques of measurement are inadequate; Whitehead et al. (1981) and Whitehead (1981) have proposed an extraction technique, with an adjustment for water availability, for estimating the mineralization of soil nitrogen.

### Seasonal effects

Grass growth in any given climatic zone has characteristic seasonal patterns. In temperate conditions, growth in spring is initially slow; it accelerates to a peak in early summer, declines sharply to a midsummer trough, increases in July and August and then declines slowly after August (Anslow and Green, 1967). The midsummer trough is most evident when water is limiting. The yield response to fertilizer nitrogen might be expected to follow a similar pattern. Cowling and Lockyer (1970) found this, to some extent, but Van Burg (1970), Prins and Van Burg (1979) and Morrison (1980a) found much smaller seasonal differences in response to

TABLE 6.3

The influence of soil nitrogen ($N_s$) on the optimum fertilizer nitrogen rate ($N_{10}$) for *Lolium perenne* swards at pairs of sites with similar growing conditions and yield but contrasting soil nitrogen status (after Morrison et al., 1980)

| Location | Yield at $N_{10}$ (t ha$^{-1}$) | $N_s$ (kg ha$^{-1}$) | $N_{10}$ (kg ha$^{-1}$) |
|---|---|---|---|
| Aylesbury, Buckinghamshire | 12.2 | 102 | 338 |
| Pluckley, Kent | 12.6 | 16 | 475 |
| Bangor, Gwynedd | 13.2 | 136 | 337 |
| Wenvoe, South Glamorganshire | 14.3 | 28 | 530 |

nitrogen; response at any one time was also affected by previous management and fertilization.

Fertilizer nitrogen was originally used in early spring to obtain "early bite", but responses were small and unreliable. If nitrogen fertilizer is applied too early there is little yield response and a risk of leaching loss; if nitrogen is applied too late, production is lost. In northern Europe, the first application has usually been sometime in March, when active grass growth begins. Farmers in the Netherlands have recently been advised to make the initial application when the accumulated day degrees reach 200°C (T200) (Van Burg et al., 1980), but results from the United Kingdom have been inconsistent.

Fertilizer nitrogen is usually applied at intervals during the growing season. Residual effects from previous applications are usually small, and only measurable following high rates of application (Prins et al., 1980). The recommended pattern of application in Europe is to decrease application from May onwards, on the basis that this would give the greatest annual response. However, this pattern exaggerates the seasonality of growth and, though it is suitable if pasture is to be cut for conservation, it is not ideal for grazing. Provided that the annual application is not greater than the sward's ability to respond, the annual yield is not greatly affected by the pattern of distribution, though the seasonal pattern of production can be altered (Wolton et al., 1971; Morrison et al., 1980; Morrison, 1980b).

When grass is to be cut for conservation (Chapter 26), the seasonal pattern of production is not critical, and a pattern of diminishing doses for each successive cut is satisfactory. In intensive systems, whatever the strategy of conservation used, there is likely to be a surplus of herbage production in May and a deficit in June (Thomas and Young 1982), hence some modification of mid-season production would be useful. For grazing, increasing the amount of fertilizer applied in mid-season, and reducing early application, can increase mid-season production (Table 6.4) to produce a pattern of production better matched to the grazing animals' requirements (Chapters 9–11).

**Species composition**

Species differ, to some extent, in their response to fertilizer nitrogen. In the United Kingdom most nitrogen trials have been carried out on sown swards of *Lolium* spp., whereas in the Netherlands and eastern Europe most trials have been carried out on permanent grassland. There have been few direct comparisons of permanent grassland and sown swards, and there is little information on the response of individual "unsown" or "indigenous" grasses.

Cowling and Lockyer (1965) compared the responses of *Dactylis glomerata, Festuca pratensis, Lolium perenne, Phleum pratense*, which are commonly sown grasses and *Agrostis tenuis*, which is rarely sown. Responses of the four cultivated grasses were similar over the linear phase, although *Dactylis glomerata* tended to be the most responsive; cultivated grasses were slightly more responsive than *Agrostis tenuis*. Haggar (1976) compared the response of *Lolium perenne* and four grasses that are rarely sown, *Agrostis stolonifera, Festuca rubra* and *Holcus lanatus*, under infrequent cutting. *Agrostis stolonifera* responded much less than the other three species, whereas *Lolium perenne* responded more than the other grasses early in the

# EFFECTS OF NITROGEN FERTILIZER

TABLE 6.4

Effect of various patterns of application of 300 kg N ha$^{-1}$ yr$^{-1}$ on the seasonal distribution of yield of *Lolium perenne*, cut six times a year

| Applied N (kg ha$^{-1}$) | | | | | | April–May | June–July | Aug.–Oct. | Total |
|---|---|---|---|---|---|---|---|---|---|
| Cuts 1 | 2 | 3 | 4 | 5 | 6 | | | | |
| 75 | 75 | 38 | 37 | 38 | 37 | 5.2 | 2.5 | 2.0 | 9.7 |
| 50 | 50 | 50 | 50 | 50 | 50 | 4.4 | 2.8 | 2.4 | 9.6 |
| 38 | 37 | 75 | 75 | 38 | 37 | 3.9 | 3.4 | 2.4 | 9.7 |

Each value is the mean of four years and of 21 sites in the United Kingdom (after Morrison et al., 1980).

season, though less in mid-season. Frame (1984) also showed that *L. perenne* responded more than eight indigenous species.

The response of permanent grassland, which is often dominated by grasses considered to be inferior, has been similar to that of cultivated grasses (Whitehead, 1970; Reid, 1970; Mudd, 1971; Van Steenbergen, 1977; Van Burg et al., 1980; Morrison et al., 1980). In infertile upland conditions, however, swards of *Festuca ovina—Agrostis tenuis* and *Molinia caerulea* responded only one quarter as well as sown swards (Davies and Munro, 1974).

Applications of fertilizer nitrogen generally increase the content of *Lolium perenne* in swards (De Vries and Kruyne, 1960; Garstang, 1981). However, on sandy soils in the Netherlands, high rates of fertilizer nitrogen, combined with infrequent cutting for silage, cause increases in *Elymus repens, Poa annua*, and *Rumex* spp. (De Gooijer, 1973).

The potential dry-matter production of tropical grassland is substantially greater than that of temperate grasses, because of the longer growing season and more suitable environmental conditions. As a result, the nitrogen requirements are greater, and so the response to nitrogen may be greater. Tropical and temperate grasses have similar responses over the linear phase, when cut at the same frequency (Fig. 6.2), but the tropical species, under suitable conditions, can continue to respond up to nitrogen applications of 1000 kg ha$^{-1}$ yr$^{-1}$ or more. Frequent and close defoliation greatly reduces the yield of tropical grasses, and also reduces their response to nitrogen fertilizer (see below).

## Frequency of defoliation

The dry-matter yield of grassland decreases with increasing frequency of defoliation (Chapter 7).

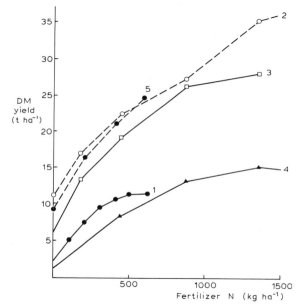

Fig. 6.2. Response of some tropical grasses to nitrogen fertilizer, compared with that of a temperate grass (*Lolium perenne*). *1* = *Lolium perenne* (Morrison et al., 1980); *2* = *Pennisetum purpureum*, Costa Rica (Vicente-Chandler et al., 1959); *3* = *Digitaria decumbens*, Costa Rica (Vicente-Chandler et al., 1959); *4* = *Cynodon dactylon*, U.S.A. (Fisher and Caldwell, 1959); *5* = *Pennisetum purpureum*, Kenya (Thairu et al., 1968).

It is, therefore, not surprising that the absolute response to nitrogen decreases with increasing frequency of cutting (Table 6.5), though the proportionate response is actually increased. The optimum rate of nitrogen application ($N_{10}$) is increased with frequent cutting (Table 6.6). Similar effects have been reported by Vicente-Chandler et al. (1959) and Fisher and Caldwell (1959) for tropical grasses (Table 6.7).

Table 6.5

Effect of cutting frequency on the response (t DM ha$^{-1}$) of *Lolium perenne* to fertilizer nitrogen (after Cowling and Lockyer, 1963)

| Cutting frequency | Fertilizer N (kg ha$^{-1}$) | | | |
|---|---|---|---|---|
| | 0 | 140 | 280 | 560 |
| Twice weekly | 1.3 | 3.4 | 4.7 | 7.7 |
| Six cuts/year | 1.4 | 3.5 | 6.1 | 9.3 |
| Hay plus aftermath | 2.3 | 6.3 | 9.3 | 11.7 |

TABLE 6.6

The effects of cutting frequency on the yield and response to fertilizer nitrogen of *Lolium perenne*

| | Cuts per year | |
|---|---|---|
| | 4 | 6 |
| Optimum yield (t ha$^{-1}$) | 11.9 | 10.8 |
| $N_{10}$ (kg ha$^{-1}$) | 281 | 386 |
| Response $N_0$–$N_{10}$ (kg DM kg N$^{-1}$) | 27 | 21 |

Each value is the mean of four years and 17 sites in the United Kingdom (after Morrison, 1980a).

TABLE 6.7

Effect of nitrogen fertilizer and cutting frequency on the yield (t ha$^{-1}$) of *Pennisetum purpureum* in Puerto Rico (after Vicente-Chandler et al., 1959)

| Fertilizer (N kg ha$^{-1}$) | Cutting frequency (days) | | |
|---|---|---|---|
| | 40 | 60 | 90 |
| 0 | 10.7 | 17.0 | 34.1 |
| 178 | 17.0 | 27.6 | 47.9 |
| 448 | 22.7 | 41.1 | 63.3 |
| 897 | 27.4 | 49.9 | 85.9 |
| 1345 | 35.4 | 52.2 | 77.9 |
| Response between 0 and 897 kg DM kg$^{-1}$ N | 18.6 | 36.7 | 57.6 |

## Grazing

Grazing has two effects which influence the response of grassland to fertilizer nitrogen: first, nitrogen is returned in excreta and, secondly, treading can damage pastures on some soils and at high stocking rates.

Richards (1977) postulated that, at low applica-

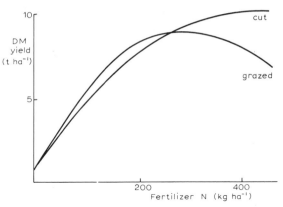

Fig. 6.3. The response of cut and grazed swards to nitrogen fertilizer (after Richards, 1977).

tions, response to nitrogen fertilizer is greater under grazing than cutting (Fig. 6.3), because of nitrogen returned in excreta, but that the maximum yield may be less, and reached at a lower rate of fertilizer nitrogen, because of increased treading damage. This has been confirmed by Baker and Large (1983). Jackson and Williams (1979) confirmed that yield was reduced under grazing, and that response to nitrogen was also less under grazing than cutting.

## Presence of legumes

Grasslands which contain an appreciable proportion of legumes respond less to fertilizer nitrogen than pure grass swards (Whitehead, 1970). Only *Trifolium repens*, the most common legume in temperate grassland, is considered here, but other legumes have a similar effect. Studies by Cowling (1961), Chestnutt and Lowe (1970), Reid (1970) and others have shown responses similar to those in Fig. 6.4. The response of grass–clover swards, over the linear phase, is less than that of pure grass swards. On average, grass–clover swards receiving no nitrogen fertilizer yield about the same as grass receiving 150 to 200 kg N ha$^{-1}$, though this can range from about 50 to 300 kg N ha$^{-1}$ (Morrison, 1981).

Applications of nitrogen fertilizer to grass–clover swards reduce the legume content, especially under sheep grazing (Brockman and Wolton, 1963; Curll, 1981). Since quite small applications of fertilizer nitrogen (100–200 kg N ha$^{-1}$ yr$^{-1}$) can reduce the legume content substantially, it is diffi-

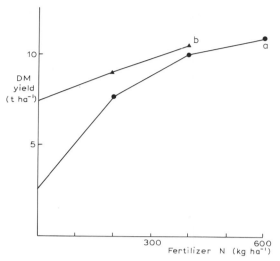

Fig. 6.4. Response to nitrogen fertilizer of a pure *Lolium perenne* sward (*a*), and a mixture of *Lolium perenne* and *Trifolium repens* (*b*) (after Morrison, 1981).

cult to use fertilizer nitrogen effectively on swards containing clover. However, early spring applications of fertilizer nitrogen can increase early growth without substantially reducing subsequent production (Cowling, 1966; Morrison et al., 1983; Laissus, 1983).

## EFFECTS ON HERBAGE COMPOSITION

A number of aspects of the chemical composition of herbage can affect animal intake and production (Chapter 14), several of these are in turn affected by applications of nitrogen.

### Digestibility

There is no evidence that the application of fertilizer nitrogen significantly affects digestibility (Spedding and Diekmahns, 1972). However, Prins and Van Burg (1979) have pointed out that fertilizer nitrogen allows more frequent harvesting and hence indirectly increases herbage digestibility.

### Soluble carbohydrate content

Soluble carbohydrate content is particularly important in silage fermentation; a minimum content of 12 to 15% is needed for a stable fermentation without additives (M.A.F.F., 1976). Soluble carbohydrate content is reduced by fertilizer nitrogen (e.g. Nowakowski, 1962), most of the reduction being in the fructosans.

### Nitrogen content

The total nitrogen content of herbage is consistently increased by the application of fertilizer nitrogen (Vicente-Chandler et al., 1959; Henzell, 1963; Whitehead, 1970). Typical effects are shown in Table 6.8.

Fertilizer nitrogen increases the organic nitrogen content of herbage, including the proportion in soluble forms, such as peptides, amides and amino acids, as opposed to proteins (Nowakowski, 1962).

Most inorganic nitrogen in herbage occurs as nitrate. Nitrogen fertilizer increases the nitrate content of herbage, which reaches a peak one or two weeks following the application of fertilizer nitrogen (Wilman, 1965). When nitrogen fertilizer is applied above optimal rates, the concentration of nitrate in the plant may exceed 0.5%, which may be toxic to animals (Van Burg, 1966; Deinum and Sibma, 1980).

## RECOVERY OF FERTILIZER NITROGEN IN HERBAGE

Nitrogen cycling and losses are dealt with elsewhere (Chapters 17 and 22). It is appropriate here, however, to consider the recovery of applied nitrogen in herbage. The apparent recovery of applied nitrogen is the proportion of the nitrogen applied that is harvested in herbage, allowing for the amount of nitrogen contributed by the soil. This recovery can range from 40 to 90%, with an average of about 60% (Vicente-Chandler et al.,

TABLE 6.8

The effect of nitrogen fertilizer application on the total nitrogen content (% of dry matter) of *Lolium multiflorum* (Wilman, 1965)

| N application (kg N ha$^{-1}$) | Harvested after weeks | | | | | |
|---|---|---|---|---|---|---|
| | 1 | 2 | 3 | 4 | 5 | 6 |
| 28 | 4.7 | 4.0 | 2.7 | 1.9 | 1.5 | 1.3 |
| 140 | 5.7 | 6.7 | 5.6 | 3.8 | 3.0 | 2.4 |

1959; Whitehead, 1970; Williams and Jackson, 1976; Dowdell et al., 1980). The recovery seems to be greater under grazing than cutting (Richards, 1977), probably because of recycling.

The losses of the remaining nitrogen, in soil organic matter and through denitrification, volatilization and leaching, are considered elsewhere (Chapter 22). Leaching losses to groundwater (Chapter 20) are of major concern, but current evidence suggests that this occurs mainly when fertilizer nitrogen is applied above optimum rates (Garwood and Tyson, 1973; Dowdell et al., 1980).

## POSTSCRIPT

It is of interest to speculate on likely future trends in the use of fertilizer nitrogen on grassland. In areas of intensive grassland production, in temperate countries, the use of fertilizer nitrogen on grassland is already a firmly established practice; the amount used in these areas is likely to increase steadily, since the average application is considerably less than the estimated economic optimum (Chapter 25). However, economic constraints, and public concern about water pollution, put an onus on the farming industry to make the most efficient use of fertilizer nitrogen applied to grassland. The emphasis of research will probably be on the response of grazed grassland to fertilizer nitrogen and the fate of the nitrogen applied. Nevertheless, at the same time, there is increased research interest in alternative sources of nitrogen, and especially in the potential contribution from legumes.

In other regions, some increase in the use of fertilizer nitrogen on grassland can be expected, but it is likely that the increase will be restricted to intensive and irrigated systems, possibly mainly for milk production, and for strategic use to extend the period of forage availability, for instance in spring and autumn, or to grassland intended for hay or silage. Legumes already have a major role (Chapters 8 and 25) in improving the nitrogen status, yield and herbage quality in many of these areas, such as New Zealand, but the full potential of legumes in many areas, particularly the tropics, has yet to be achieved.

## REFERENCES

Anslow, R.C. and Green, J.O., 1966. The seasonal growth of pasture grasses *J. Agric. Sci. Cambridge*, 68: 109–122.

Baker, R.D. and Large, R.V., 1983. *Annual Report Grassland Research Institute, Hurley 1982*, pp. 113–115.

Bogdan, A.V., 1977. *Tropical Pasture and Fodder Plants*. Longman, London.

Brockman, J.S. and Wolton, K.M., 1963. The use of nitrogen on grass/white clover swards. *J. Br. Grassland Soc.*, 18: 7–13.

Chestnutt, D.M.B. and Lowe, J., 1970. Review of the agronomy of white clover/grass swards, white clover research. *Br. Grassland Soc., Occas. Symp.*, No. 6: 191–213.

Cowling, D.W., 1961. The effect of white clover and nitrogenous fertilizer on the production of a sward. L. Total annual production. *J. Br. Grassland Soc.*, 16: 281–290.

Cowling, D.W., 1966. The effect of time of cutting and early application of nitrogenous fertilizer on the yield of ryegrass/white clover swards. *J. Agric. Sci. Cambridge*, 66: 413–431.

Cowling, D.W. and Lockyer, D.R., 1963. *Grassland Research Institute Experiments in Progress*, 15: 17–18.

Cowling, D.W. and Lockyer, D.R., 1965. A comparison of the reaction of different grass species to fertilizer nitrogen and to growth in association with white clover. 1. Yield of dry matter. *J. Br. Grassland Soc.*, 20: 197–204.

Cowling, D.W. and Lockyer, D.R., 1970. The response of perennial ryegrass to nitrogen in various periods in the growing season. *J. Agric. Sci., Cambridge*, 75: 539–546.

Curll, M.L., 1981. *The Effects of Grazing by Set-Stocked Sheep on a Perennial Ryegrass/White Clover Pasture*. PhD Thesis, Reading University.

Davies, D.A. and Munro, J.M.M., 1974. Potential pasture production in the uplands of Wales. 4. Nitrogen response from sown and natural pastures. *J. Br. Grassland Soc.*, 29: 149–158.

De Gooijer, H.H., 1973. Wat gebeurt er met ons grasland? *Stikstof*, 73: 552–556.

Deinum, B. and Sibma, L., 1980. Nitrate content in relation to fertilizer and management. In: *Proc. Int. Symp. European Grassland Fed. on the Role of Nitrogen in Intensive Grassland Production, Wageningen, 1980*. PUDOC, Wageningen, pp. 95–102.

De Vries, D.M. and Kruyne, A.A., 1960. The influence of nitrogen fertilization on the botanical composition of grassland. *Stikstof*, 4: 26–36.

Dowdell, R.J., Morrison, J. and Hood, A.E.M., 1980. The rate of fertilizer nitrogen applied to grassland: uptake by plants, immobilization into soil organic matter and losses by leaching and denitrification. In: *Proc. Int. Symp. European Grassland Fed. on The Role of Nitrogen in Intensive Grassland Production, Wageningen, 1980*. PUDOC, Wageningen, pp. 129–136.

Fisher, F.L. and Caldwell, A.G., 1959. The effects of continued use of heavy rates of fertilizers on forage production and quality of coastal bermuda grass. *Agron. J.*, 51: 99–102.

Frame, J., 1984. Natural grasses come in from the cold. *W. Scotland Coll. Agric. Agron. Publ.*, No. 780: 8 pp.

Garstang, J.R., 1981. The long term effects of the application of different amounts of nitrogenous fertilizer on dry matter yield, sward composition and chemical composition of a permanent grassland sward. *Exper. Husb.*, 37: 117–132.

Garwood, E.A. and Stiles, W., 1974. *Soil Water and Irrigation*, Silver Jubilee Report, Grassland Research Institute, Hurley, pp. 89–95.

Garwood, E.A. and Tyson, K.C., 1973. Losses of nitrogen and other nutrients to drainage from soil under grass. *J. Agric. Sci., Cambridge*, 80: 303–312.

Garwood, E.A. and Williams, T.E., 1967. Growth, water use and nutrient uptake from the subsoil by grass swards. *J. Agric. Sci. Cambridge*, 69: 125–130.

Haggar, R.J., 1976. The seasonal productivity, quality and response to nitrogen of four indigenous grasses compared with *Lolium perenne*. *J. Br. Grassland Soc.*, 31: 197–207.

Henzell, E.F., 1963. Nitrogen fertilizer responses of pasture grasses in south eastern Queensland. *Aust. J. Exper. Agric. Anim. Husb.*, 3: 290–299.

Henzell, E.F. and Stirk, G.B., 1963. Effects of nitrogen deficiency and soil moisture stress on growth of pasture grasses at Samford in south eastern Queensland. *Aust. J. Exper. Agric. Anim. Husb.*, 300–306.

Jackson, M.V. and Williams, T.E., 1979. Response of grass swards to fertilizer N under cutting or grazing. *J. Agric. Sci., Cambridge*, 92: 549–562.

Laissus, R., 1983. How to use nitrogen fertilizers on a grass/white clover sward. In: A.J. Corrall (Editor), *Efficient Grassland Farming. Proc. 9th Gen. Meet., European Grassland Fed., Reading 1982. Occas. Symp. Br. Grassland Soc.*, No. 14: 223–226.

Lazenby, A., 1983. Nitrogen relationships in grassland ecosystems. In: *Proc. XIV Int. Grassland Congr. Lexington, 1981*, pp. 56–62.

Lee, J., 1983. The spatial pattern of grassland production in Europe. In: A.J. Corrall (Editor), *Efficient Grassland Farming. Proc. 9th Gen. Meet. European Grassland Fed., Reading, 1982. Occas. Symp. Br. Grassland Soc.*, No. 14: 11–20.

M.A.F.F. (Ministry of Agriculture, Fisheries and Food), 1976. *Liscombe Experimental Husbandry Farm, Annual Review, 1976*. Ministry of Agriculture, Food and Fisheries, London, pp. 18–35.

Morrison, J., 1980a. The growth of *Lolium perenne* and response to fertilizer N in relation to season and management. In: *Proc. XIII Int. Grassland Congr., Leipzig, 1977*, pp. 943–946.

Morrison, J., 1980b. The influence of climate and soil on the yield of grass and its response to fertilizer nitrogen. In: *Proc. Int. Symp. European Grassland Fed. on the Role of Nitrogen in Intensive Grassland Production, Wageningen, 1980*. PUDOC, Wageningen, pp. 51–57.

Morrison, J., 1981. The potential of legumes for forage production. In: *Winter Meeting, on Legumes and Fertilizers in Grassland Systems*. British Grassland Society, pp. 1.1–1.8.

Morrison, J. and Russell, R.D., 1980. Fertilizer recommendations for grassland. *Chem. Ind.*, 17: 686–688.

Morrison, J., Jackson, M.V. and Sparrow, P.E., 1980. *The Response of Perennial Ryegrass to Fertilizer Nitrogen in Relation to Climate and Soil*. Grassland Research Institute, Hurley, Tech. Rep., No. 27: 71 pp.

Morrison, J., Denehy, H. and Chapman, P.F., 1983. Possibilities for the strategic use of fertilizer N on white clover/grass swards. In: A.J. Corrall (Editor), *Efficient Grassland Farming. Proc. 9th Gen. Meet., European Grassland Fed., Reading, 1982. Occas. Symp. Br. Grassland Soc.*, No. 14: 227–231.

Mudd, C.H., 1971. Yields of natural and artificial grassland under five levels of fertilizer treatment. In: *Proc. European Grassland Fed. 4th Gen. Meet., Lausanne, 1971*, pp. 69–72.

Nowakowski, T.Z., 1962. Effects of nitrogen fertilizers on total nitrogen, soluble nitrogen and soluble carbohydrate contents of grass. *J. Agric. Sci., Cambridge*, 59: 387–392.

Prins, W.H. and Van Burg, P.F.J., 1979. The seasonal response of grassland to different levels of nitrogen pre-treatment. 1. Experiments, 1972 and 1973. *Neth. Nitrogen Tech. Bull.*, No. 11: 11–33.

Prins, W.H., Van Burg, P.F.J. and Wieling, H., 1980. The seasonal response of grassland to nitrogen at different intensities of nitrogen fertilization with special reference to response measurements. In: *Proc. Int. Symp. European Grassland Fed. on the Role of Nitrogen in Intensive Grassland Production, Wageningen, 1980*. PUDOC, Wageningen, pp. 35–49.

Reid, D., 1970. The effects of a wide range of nitrogen application rates on the yields of a perennial ryegrass sward with and without white clover. *J. Agric. Sci., Cambridge*, 74: 227–240.

Richards, I.R., 1977. Influence of soil and sward characteristics on the response to nitrogen. In: *Proc. Int. Meet. on Animal Production from Temperate Grassland. Ir. Grassland Anim. Prod. Assoc., Dublin, 1977*, pp. 45–49.

Sparrow, P.E., 1979. The comparison of five mathematical models which can represent the relationship between the dry matter yield of grass herbage and fertilizer nitrogen. *J. Agric. Sci., Cambridge*, 93: 513–520.

Spedding, C.R.W. and Diekmahns, E.G. (Editors), 1972. Grasses and legumes in British Agriculture. *Common. Bur. Pastures Field Crops, Bull.*, 49: 483 pp.

Thairu, D.M., Morrison, J. and Keya, N.C.O., 1968. *Kenya Ministry of Agric. Annu. Rep. Part II, 1968*.

Thomas, C. and Young, J.W.O., 1982. *Milk from grass*. I.C.I. Agricultural Division, Billingham, and Grassland Research Institute, Hurley: 104 pp.

Van Burg, P.F.J., 1966. Nitrate as an indicator of the nitrogen nutrition status. In: *Proc. 10th Int. Grassland Congr., Helsinki*, pp. 257–262.

Van Burg, P.F.J., 1970. The seasonal response of grassland herbage to nitrogen. *Neth. Nitrogen Tech. Bull.*, No. 8: 6–25.

Van Burg, P.F.J., Hart, M.L. and Thomas, H., 1980. Nitrogen and grassland; past and present situation in the Netherlands. In: *Proc. Int. Symp. European Grassland Fed. on the Role of Fertilizer Nitrogen in Intensive Grassland Production, Wageningen, 1980*. PUDOC, Wageningen, pp. 15–33.

Van Burg, P.F.J., Dilž, K. and Prins, W.H., 1982. Agricultural value of various nitrogen fertilizers in the Netherlands and elsewhere in Europe. *Neth. Nitrogen Tech. Bull.*, No. 13: 27–32.

Van der Meer, H.G., 1983. Effective use of nitrogen on grassland farms. In: A.J. Corrall (Editor), *Efficient grass-*

land Farming. Proc. 9th Gen. Meet. European Grassland Fed., Reading, 1982. Occas. Symp. Br. Grassland Soc., No. 14: 61–68.

Van Steenbergen, T., 1977. The influence of soil and year on the effect of nitrogen fertilizer on the yield of grassland. Stikstof, 20: 29–35.

Vicente-Chandler, R., Silva, S. and Figarella, J., 1959. The effect of nitrogen fertilization and frequency of cutting on the yield and composition of three tropical grasses. Agron. J., 51: 202–206.

Whitehead, D.C., 1966. Nutrient minerals in grassland herbage. Rev. Ser. 1/1966. Common. Bur. Pastures Field Crops.

Whitehead, D.C., 1970. The role of nitrogen in grassland productivity. Common. Bur. Pastures Field Crops, Bull., 48: 202 pp.

Whitehead, D.C., 1981. An improved chemical extraction method for predicting the supply of available soil nitrogen. J. Sci. Field Agric., 32: 359–365.

Whitehead, D.C., Barnes, R.J. and Morrison, J., 1981. An investigation of analytical procedures for predicting soil nitrogen supply to grass in the field. J. Sci. Field Agric., 32: 211–218.

Williams, T.E. and Jackson, M.V., 1976. The recovery of fertilizer N in herbage and soil. Min. Agric., Food and Fish., Tech. Bull., No. 32. H.M.S.O., London, pp. 145–152.

Wilman, D., 1965. The effect of nitrogenous fertilizer on the rate of growth of Italian ryegrass. J. Br. Grassland Soc., 20: 248–254.

Wolton, K.M., Brockman, J.S. and Shaw, P.G., 1971. The effect of time and rate of N application on the productivity of grass swards in two environments. J. Br. Grassland Soc., 26: 123–131.

Chapter 7

# EFFECTS OF GRAZING AND CUTTING

C.J. KORTE and W. HARRIS

## INTRODUCTION

Grasslands are utilized both by grazing and cutting. Cut herbage is usually conserved and used to supplement animals during periods when there is insufficient available pasture (Chapter 26), though it is sometimes fed fresh. The main effect of grazing animals on the productivity and botanical composition of pastures is through defoliation, though they also have an effect through treading, nutrient returns and seed dispersal (Donald, 1941; Sears, 1956; Watkin and Clements, 1978; Curll, 1982). This chapter will mainly deal with the effects of defoliation, but the other effects of grazing animals on pastures will also be considered, in so far as they cause differences in response to cutting and grazing. Nutrient returns are considered further in Chapter 19.

## EFFECTS OF DEFOLIATION

Defoliation is usually defined in terms of three parameters (Harris, 1978): (1) intensity, i.e. the proportion of herbage removed, usually measured by residual mass, height, or leaf area index; (2) frequency, i.e. the time interval between successive defoliations; (3) timing, i.e. the date or developmental stage of the sward at the time of defoliation.

The effects of defoliation can be measured in terms of total herbage yield, herbage composition, seasonal distribution and changes in plant species composition.

### Herbage yield

The effects of defoliation have normally been measured by: (1) the amount of herbage dry matter (DM) per unit area removed by cutting (herbage harvested); (2) the amount removed by grazing (herbage consumed); and (3) the amount accumulated between successive grazings (herbage accumulation). These three measures may differ from each other, and none directly measures pasture growth rate (Hodgson et al., 1981), which has seldom been measured in cut or grazed pasture.

### Frequency and intensity of defoliation

The response of pasture to defoliation has been the subject of considerable research, and of several reviews (Donald, 1941; Jameson, 1963; Alcock, 1964; Harris, 1978). Herbage DM production is usually reduced by more frequent and more intensive defoliation, whether by cutting or grazing. The adverse effects of more intense defoliation are often reduced by less frequent defoliation. For example, intensive grazing (38 ewes $ha^{-1}$) reduced pasture production compared with lax grazing (25 ewes $ha^{-1}$) with 7-day rotational grazing (Table 7.1). However, at a given intensity (38 ewes $ha^{-1}$), a 28-day grazing rotation gave more herbage production than a 7-day rotation. Recent detailed studies of the effects of continuous grazing (Fig. 7.1) show that higher stocking rates reduced photosynthesis and shoot growth of swards, probably as a result of reduced leaf area index. However, increasing the stocking rate increased herbage consumption per unit area, except at the highest stocking rates, where greater efficiency of harvesting could not compensate for the reduction in shoot growth.

In some cases, more frequent or intensive defoliations have increased DM production (Alcock, 1964). In particular, closer cutting has given higher yields in a number of experiments, for example with ryegrass (Reid, 1959, 1962; Appadurai and

TABLE 7.1

Herbage accumulation (t DM ha$^{-1}$ yr$^{-1}$) from perennial ryegrass–white clover pastures rotationally grazed by sheep at 7- or 28-day intervals and stocked at 25 or 38 ewes ha$^{-1}$ (from Campbell, 1969)

| Ewes ha$^{-1}$: | 25 | 38 | 38 |
|---|---|---|---|
| Regrowth (days): | 7 | 7 | 28 |
| First year | 11.2 | 9.6 | 14.5 |
| Second year | 13.4 | 10.0 | 14.9 |
| Mean | 12.3 | 9.8 | 14.7 |

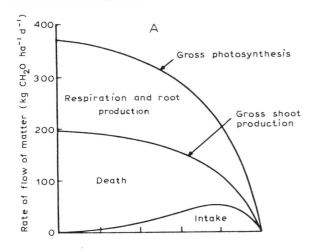

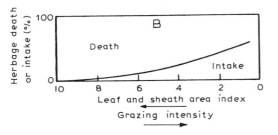

Fig. 7.1.A. The effect of grazing intensity, and leaf area index, on gross photosynthesis, gross herbage (shoot) production, herbage death and herbage intake by sheep in continuously grazed perennial ryegrass pastures. B. The relationship in A redrawn to show the proportions of herbage production that are eaten or are lost by death and decay. Reproduced from Parsons et al. (1983).

Holmes, 1964; Binnie and Harrington, 1972; Binnie et al., 1974, 1980). Tayler and Rudman (1966) concluded that greater utilization of herbage from the base of the sward, rather than increased growth with closer cutting, was a major reason for higher yields with more intensive defoliation. Differences in crop utilization can also influence yield responses to cutting frequency (Anslow, 1967).

In other situations, various combinations of defoliation frequency and intensity have had relatively little effect on harvested yield (e.g. Hodgson et al., 1981). In these cases, although photosynthesis and herbage growth were reduced by more frequent and intense defoliation, death and decay of unharvested herbage was also reduced. However, very intensive or very frequent defoliation did reduce yield.

The effect of more frequent and intense defoliation has been attributed to: (1) reduced light interception by photosynthetic tissue; (2) depletion of metabolic reserves; (3) reduced uptake of nutrients and water; and (4) damage to meristems or depletion of seed reserves. The relative importance of these factors depends on environmental conditions and pasture species (Harris, 1978). For example, under rangeland conditions in North America, the level of reserve carbohydrates has been a useful indicator of need for rest from grazing (Dahl and Hyder, 1977). By contrast, the effects of light interception have been stressed in New Zealand, where deficiencies of nutrients and water are less important.

**Light interception.** Under favourable environmental conditions, the growth rate of pasture increases with increasing leaf area and associated light interception (Brown and Blaser, 1968), the maximum growth rate being reached when 95 to 100% of light is intercepted. More frequent and intense defoliation reduces leaf area and so reduces light interception and growth.

**Reserves.** After defoliation, metabolites for the production of new shoot and root structures come either from current photosynthesis or metabolic reserves. Reserves are likely to be of greater importance when the residual leaf area is small (Booysen and Nelson, 1975), or where the residual leaf area has a low photosynthetic efficiency (Langer and Keoghan, 1970). Frequent intense defoliation, by reducing the opportunity for replenishment of reserves, can result in slower pasture regrowth. The importance of reserves varies considerably with species and environment, reserves being most important where climatic extremes of drought or low temperature seriously reduce growth for prolonged periods (Harris, 1978).

**Uptake of nutrients and water.** Frequent and intense defoliation could reduce uptake of nutrients and water from soil by three possible mechanisms (Harris, 1978): (1) reduction in root growth following defoliation, so limiting exploitation of nutrients and water; (2) reduction in transpiration, so restricting absorption of nutrients; and (3) reduction in assimilate levels, from reserves and photosynthesis, limiting active ion uptake.

**Meristems and seeds.** Apical meristems, from which shoot growth originates, may be removed by defoliation. Species which are resistant to frequent and intense defoliation tend to branch or tiller prolifically, have a high ratio of vegetative to reproductive shoots, and usually grow in a manner such that their apical meristems are inaccessible to grazing animals (Booysen et al., 1963; Dahl and Hyder, 1977). Intense defoliation may also deplete seed reserves in the soil, by reducing the number of flowering heads that mature and set seed (Archer and Rochester, 1982; Sheath and Boom, 1985). This reduction in seed reserves in the soil can be particularly important for sward recovery after drought.

### Timing of defoliation

The most important aspect of timing of defoliation is its relation to stage of reproductive development (Stapledon, 1924; Bird, 1943; Austenson, 1963; Lawrence and Ashford, 1966, 1969; Sheard and Winch, 1966; Bonin and Tomlin, 1968a, b; Gillet, 1970; Winch et al., 1970; Corrall, 1974; Mislevy et al., 1974, 1977; Corrall et al., 1979; Gervais and St-Pierre, 1979; Hunt et al., 1979; Matches, 1979; Binnie et al., 1980; Omaliko, 1980; Mason and Lachance, 1983). Greater annual yields of herbage are usually obtained when reproductive development is allowed to occur uninterrupted. For example, Corrall (1974) obtained yields of 17 t DM ha$^{-1}$ from perennial ryegrass swards when the first cut was taken after anthesis, but only 14 t DM ha$^{-1}$ when the first cut was taken at the start of stem elongation; proportionately larger differences have been obtained by others (e.g. Bird, 1943; Bonin and Tomlin, 1968a).

The effect of defoliation depends on season. For example, Brougham (1960) found that frequent close grazings of a mixed-species sward in autumn and winter increased herbage yield, but frequent close grazing in spring and summer reduced annual yield. The results reflected species tolerances to different grazing intensities at various stages in their annual growth cycles, as well as the effects of defoliation on herbage growth under different seasonal conditions.

### Seasonal distribution of yield

Grazing animals require a continuous supply of food, so changes in seasonal pattern of pasture production can have important consequences on animal production where all-year-round grazing is practised (Chapters 9–11). The importance of seasonal pattern of herbage production depends upon whether forage can be conserved (Chapter 26), the type of animal production enterprise (such as milk, wool or lamb production) (Chapters 9–11), and the degree to which body reserves of grazing animals can act as a buffer against variations in food supply.

Generally, manipulation of defoliation to modify the seasonal pattern of herbage production results in some loss in total annual DM yield. More frequent defoliation (Humphreys and Robinson, 1966; Harris et al., 1980) and more intensive defoliation (Korte et al., 1982) reduces seasonal variation in herbage production. This occurs largely because more frequent or intense defoliation interrupts reproductive development earlier, reducing the early summer peak of production, but it can increase post-flowering growth (Winch et al., 1970; Mason and Lachance, 1983) and autumn production (Korte, 1982).

### Pasture composition

#### Herbage components

Animals generally prefer the leaf fraction of pastures, and avoid dead herbage in preference to green herbage (Minson, 1983; Chapter 14). It is therefore important to distinguish the effect of defoliation on these components. More frequent defoliation generally increases the proportion of leaf material relative to "stem" material, and the proportion of green relative to dead herbage (Stapledon, 1924; Burton et al., 1963; Humphreys and Robinson, 1966; Wilman et al., 1976a, b; Omaliko, 1980; Wilman and Asiegbu, 1982). The interval between defoliations that maximizes leaf yield is

TABLE 7.2

Effect of cutting interval, over a 30-week period, on the yield (kg DM ha$^{-1}$) of herbage components, and the percentage and yield of digestible organic matter (DOM) from perennial ryegrass swards (from Wilman et al., 1976b)

| Cutting interval (weeks): | 3 | 4 | 5 | 6 | 8 | 10 |
|---|---|---|---|---|---|---|
| *Herbage components* | | | | | | |
| Green leaf blade | 6500 | 7400 | 6270 | 6320 | 4710 | 2010 |
| Dead leaf | 100 | 160 | 460 | 490 | 960 | 1400 |
| "Stem" | 1130 | 1890 | 2880 | 3690 | 5210 | 4070 |
| Emerged inflorescence | – | – | 10 | 60 | 180 | 210 |
| Total ryegrass | 7730 | 9450 | 9620 | 10560 | 11060 | 7690 |
| Unsown species | 2500 | 2560 | 3050 | 3080 | 4030 | 7860 |
| Total DM yield | 10230 | 12010 | 12670 | 13640 | 15090 | 15550 |
| *Digestible organic matter* | | | | | | |
| Percentage | 68.7 | 68.4 | 67.4 | 67.3 | 65.6 | 63.9 |
| DOM yield | 6910 | 8350 | 8730 | 9010 | 9770 | 9740 |

"Stem" included true stem, leaf sheath, unemerged leaf blade and unemerged inflorescence.

usually considerably less than that which maximizes total DM yield (Table 7.2), though the interval is dependent on factors such as time of year, nitrogen availability, and plant species or cultivar. Frequent defoliation also increases the proportion of dead leaf to green leaf material (Table 7.2). Longer periods of uninterrupted growth permit greater culm development, and much of the increase in annual yield caused by infrequent defoliation is caused by increasing amounts of "stem" (Table 7.2; Austenson, 1963).

As dead leaves generally accumulate in the base of swards (Jackson, 1976), more intensive (closer) defoliation might be expected to increase the dead matter content of harvested herbage. However, when herbage has been measured to ground level, closer defoliation has usually been found to reduce dead herbage production, probably because less herbage is left unharvested and available to die (Hodgson et al., 1981; Korte et al., 1982; Parsons et al., 1983).

### Digestibility

In general, less frequent cutting reduces the digestibility of harvested herbage but, because it increases herbage yield (see above), it usually increases the annual yield of digestible matter (Table 7.2; Wilman et al., 1976a, b; Bartholomew and Chestnutt, 1977; Chestnutt et al., 1977; Reid, 1978; Griffin and Watson, 1982; Monson and Burton, 1982). The quality of harvested herbage appears to be influenced more by cutting frequency than by cutting height (Binnie and Harrington, 1972; Binnie et al., 1974; Briseno and Wilman, 1981).

Infrequent cutting affects pasture quality most during reproductive development, since digestibility declines most rapidly after ear emergence in grasses (Minson et al., 1960), but timing of cutting has little effect on annual digestible DM yield (Bonin and Tomlin, 1968a, b; Winch et al., 1970; Corrall, 1974; Mislevy et al., 1974, 1977; Corrall et al., 1979; Mason and Lachance, 1983).

### Botanical composition

Harris (1978) has reviewed the effect of defoliation on the botanical composition of pastures. Species differ in their response to defoliation, but the effects depend on environmental conditions and competition from other species.

Prostrate plants, with rhizome, stolon or basal rosettes, are usually more resistant to frequent intensive defoliation than species with a twining, scrambling or erect habit (Harris, 1978; Humphreys, 1981; Table 7.3). Differences in response to defoliation occur because of differences in removal of photosynthetic area and meristems, bud regeneration, flowering, seed production, soil seed reserves and seedling regeneration.

Defoliation at a particular time of year may change species dominance, because plants may be at different stages in their annual physiological or

TABLE 7.3

Effects of six years of grazing on the botanical composition of (% of turf cores with species present) ryegrass–white clover swards (from Harris and Brougham, 1968)

| Species | Moderate grazing (%) | Intensive grazing (%) |
|---|---|---|
| *Lolium* spp. | 68 | 95 |
| *Trifolium repens* (stoloniferous) | 62 | 69 |
| *Agrostis tenuis* (rhizomatous) | – | 35 |
| *Poa annua* | – | 45 |
| *P. trivialis* (stoloniferous) | 1 | 23 |
| *Oxalis corniculata* (prostrate stolons) | – | 41 |
| *Hydrocotyle moschata* (prostrate stolons) | – | 47 |
| *Sagina procumbens* (prostrate stolons) | – | 20 |
| *Taraxacum officinale* (rosette) | – | 11 |

The moderately grazed sward was rotationally grazed down to a height of 5 to 7 cm each time the sward reached a height of 15 to 20 cm. The intensively grazed sward was grazed continuously to a height of 3 to 5 cm.

morphological cycle. For example, Jones (1933) demonstrated that either white clover or ryegrass could be induced to dominate pasture by hard or light spring grazings respectively. Other examples of timing of defoliation changing species dominance are given by Brougham (1960), Lawrence (1973) and Baars et al. (1979).

Changes in botanical composition can influence total and seasonal distribution of herbage production or forage quality. At the extreme, over-grazing can lead to invasion of pasture by unpalatable or poisonous weed species. In terms of yield, the most beneficial changes in botanical composition occur in pasture communities where defoliation encourages species of different growth periodicities to dominate the mixture during the part of the year when they make maximum growth (Harris, 1978).

## DIFFERENCES BETWEEN CUT AND GRAZED SWARDS

Grazing experiments usually require larger areas and are more expensive to carry out than cutting experiments. As a result, comparisons of large numbers of species, cultivars, fertilizers or grazing management treatments are usually carried out in cutting experiments. It is therefore important to know if the responses observed under cutting can be safely extrapolated to grazed swards. In general, different effects of treatments on herbage production and botanical composition have been found under cutting than under grazing (Sears, 1951; Elliot and Lynch, 1958; Taylor et al., 1960; Bryant and Blaser, 1961, 1968; Wolton, 1963; Cuykendall and Marten, 1968; Matches, 1968; Frame, 1965, 1976; Calder et al., 1970; Frame and Hunt, 1971; Richards et al., 1976; Wilman, 1977; Jackson and Williams, 1979; Briseno and Wilman, 1981; Harris, 1983). However, similar results have sometimes been found (e.g. Aldrich and Elliott, 1974; Camlin and Stewart, 1975). Differences in response can be attributed to the return of nutrients in animal excreta, the effects of treading, and selective defoliation. Each of these animal influences on pasture are therefore separately reviewed.

### Nutrient return

Dung and urine are potentially significant sources of N, P, K, S, Mg and Ca for pastures (Ch. 19; Watkin and Clements, 1978). Although responses to dung and urine have been variable, many studies have shown that the return of excreta to pastures can markedly stimulate herbage production (Table 7.4) and affect botanical composition. Most responses can be attributed to N and, less often, to K and P (Wolton, 1963; Cuykendall and Marten, 1968; Calder et al., 1970; Frame and Hunt, 1971; Frame, 1976).

In cut swards, nutrients are continually removed from the sward by removal of herbage. Attempts have been made to prevent differences in nutrient availability between cut and grazed swards by: (1)

TABLE 7.4

The effect of defoliation ($D$), treading ($T$) and excreta return ($E$) at two stocking rates on herbage accumulation (kg DM ha$^{-1}$) from a perennial ryegrass–white clover pasture (from Curll and Wilkins, 1983)

| | Sheep ha$^{-1}$ | |
|---|---|---|
| | 25 | 50 |
| $D$ | 8990 | 5420 |
| $D+T$ | 8640 | 4870 |
| $D+T+E$ | 10880 | 7440 |
| Mean | 9500 | 5910 |

fertilizing cut swards, at a rate either sufficient to prevent deficiency (Cuykendall and Marten, 1968) or calculated to return nutrients removed in herbage (McNeur, 1953); (2) returning clippings after cutting (Lynch, 1947; Wolton, 1963); and (3) applying animal excreta to cut swards (Sears, 1951; Wolton, 1963). Each of these approaches has inherent difficulties. Plant nutrients may be more or less readily available from artificial fertilizers or clippings than from excreta. Clippings can smother swards and it may be difficult to accurately simulate the patchy pattern of dung and urine return. Elliot and Lynch (1958) showed that when nutrients were returned in the form of clippings or excreta, mown swards still differed from grazed swards both in yield and botanical composition. They concluded that selective grazing was the main reason for these differences.

### Defoliation pattern

Grazing is usually much less uniform than cutting. In mixed swards, legumes (Sears, 1951; Taylor et al., 1960; Briseno and Wilman, 1981) and more palatable grasses and weeds (Elliot and Lynch, 1958) are often more intensively defoliated in grazed than mown swards.

Mowing can, however, cause selective defoliation, because dead herbage, leaf, stem and floral tissues are not uniformly distributed vertically within swards (Sato, 1973; Jackson, 1976). As a result, the harvested yield of these components may not be in proportion to their relative abundance in pasture. For example, little dead herbage is harvested by cutting, because dead herbage is concentrated below cutting height. Similarly, the foliage of different plant species is seldom uniformly distributed within the canopy (Clark et al., 1974), so mowing can selectively defoliate different species.

Differences in defoliation pattern between grazed and cut swards have been considered to contribute to differences in yield. Within grazed swards, some areas are more closely grazed than others, and overgrazing may occur to such an extent that pasture regrowth is reduced, compared to a cut sward (Bryant and Blaser, 1968). In other instances, plants or plant parts that are partially grazed, or that escape grazing entirely, may increase the regrowth of the grazed sward (Matches, 1968).

In set-stocked systems, the seasonal pattern of defoliation may be dictated by the grazing animals' requirement for food. It has been suggested by Harris (1983) that this is one reason why higher yields have been obtained from cutting than grazing. Defoliation by cutting can be more readily controlled to match herbage production.

### Treading

Treading by sheep and cattle can reduce pasture growth directly, through damage to plant growing points and photosynthetic tissue, or indirectly, through soil compaction and puddling (Watkin and Clements, 1978). The reduction in yield increases with stocking rate, the reduction depending mainly on pasture species (Edmond, 1964) and the water content of the soil (Edmond, 1962).

Richards et al. (1976) suggested that only at high levels of herbage production, and associated high stocking rates, would the damaging effects of treading outweigh the beneficial effects of excretal returns, so reducing the yields of grazed compared with cut swards. However, Curll and Wilkins (1983) have recently shown that, even at extremely high stocking rates, the benefits of excreta return can more than compensate for yield reductions caused by treading (Table 7.4). It is therefore concluded that treading is probably the least significant animal effect causing differences between cut and grazed swards, except perhaps on wet poorly drained soils (Scott, 1963).

Cutting machinery can also damage pasture plants. For example, Frame (1983) reported yield losses up to 30% in red clover swards caused by machinery wheels, the damage being worst in wet conditions, and on sloping fields. Red clover is very sensitive to treading damage (Edmond, 1964), but damage may occur in other pasture species.

## CONCLUSION

More frequent and intense defoliation, whether by cutting or grazing, generally reduces pasture growth, but may not always reduce the amount of forage harvested. Reductions in growth with more frequent and intense defoliation may be wholly or partially compensated for by improved utilization, and therefore by reduced losses of unharvested herbage.

Forage quality and the proportion of green leaf are generally increased by more frequent defoliation. More frequent and intense defoliation often improves the seasonal pattern of herbage production, usually at the expense of total annual yield.

The effects of defoliation on botanical composition are briefly reviewed.

Although the general conclusions obtained from cutting and grazing experiments are similar, important differences have been observed. It is concluded that the results of cutting experiments, even where they incorporate methods to simulate other aspects at grazing, such as nutrient returns, may not be applicable to grazing conditions. It is therefore important to involve the grazing animal in pasture experimentation as soon as possible to adequately account for the factors which are linked with defoliation by ruminant livestock.

## REFERENCES

Alcock, M.B., 1964. The physiological significance of defoliation on the subsequent regrowth of grass-clover mixtures and cereals. In: B.J. Crisp (Editor), *Br. Ecol. Soc., Symp.*, No. 4: 25–41.

Aldrich, D.T.A. and Elliot, C.S., 1974. A comparison of the effects of grazing and of cutting on the relative herbage yields of six varieties of perennial ryegrass (*Lolium perenne*). In: *Proc. XII Int. Grassland Congr., Moscow*, 5: 11–17.

Anslow, R.C., 1967. Frequency of cutting and sward production. *J. Agric. Sci., Cambridge*, 68: 377–384.

Appadurai, R.R. and Holmes, W., 1964. The influence of stage of growth, closeness of defoliation, and moisture on the growth and productivity of a ryegrass-white clover sward. I. Effect on herbage yields. *J. Agric. Sci. Cambridge*, 62: 327–332.

Archer, K.A. and Rochester, I.J., 1982. Numbers and germination characteristics of white clover seed recovered from soils on the Northern Tablelands of New South Wales after drought. *J. Aust. Inst. Agric. Sci.*, 48: 99–101.

Austenson, H.M., 1963. Influence of time of harvest on yield of dry matter and predicted digestibility of four forage grasses. *Agron. J.*, 55: 149–153.

Baars, J.A., Percival, N.S., Goold, G.J. and Weeda, W.C., 1979. The use of grazing pressure to manipulate the balance of paspalum/ryegrass-based pasture. *Proc. N.Z. Grassland Assoc.*, 41: 89–95.

Bartholomew, P.W. and Chestnutt, D.M.B., 1977. The effect of a wide range of fertilizer nitrogen application rates and defoliation intervals on the dry matter production, seasonal response to nitrogen, persistence and aspects of chemical composition of perennial ryegrass (*Lolium perenne* cv. S.24). *J. Agric. Sci., Cambridge*, 88: 711–721.

Binnie, R.C. and Harrington, F.J., 1972. The effect of cutting height and cutting frequency on the productivity of Italian ryegrass sward. *J. Br. Grassland Soc.*, 27: 177–182.

Binnie, R.C., Harrington, F.J. and Murdoch, J.C., 1974. The effect of cutting height and nitrogen level on yield, *in vitro* digestibility and chemical composition of Italian ryegrass swards. *J. Br. Grassland Soc.*, 29: 57–62.

Binnie, R.C., Chestnutt, D.M.B. and Murdoch, J.C., 1980. The effect of time of initial defoliation on the productivity of perennial ryegrass swards. *Grass Forage Sci.*, 35: 267–273.

Bird, J.N., 1943. Stage of cutting studies; I. Grasses. *J. Am. Soc. Agron.*, 33: 845–861.

Bonin, S.G. and Tomlin, D.C., 1968a. Effects of nitrogen on herbage yields of timothy harvested at various developmental stages. *Can. J. Plant Sci.*, 48: 501–509.

Bonin, S.G. and Tomlin, D.C., 1968b. Effects of nitrogen on herbage yields of reed canarygrass harvested at various developmental stages. *Can. J. Plant Sci.*, 48: 511–517.

Booysen, P. de V. and Nelson, C.J., 1975. Leaf area and carbohydrate reserves in regrowth of tall fescue. *Crop Sci.*, 15: 262–266.

Booysen, P. de V., Tainton, N.M. and Scott, J.D., 1963. Shoot-apex development in grasses and its importance in grassland management. *Herbage Abstr.*, 33: 209–213.

Briseno de la Hoz, V.M. and Wilman, D., 1981. Effects of cattle grazing, sheep grazing, cutting and sward height on a grass-white clover sward. *J. Agric. Sci., Cambridge*, 97: 699–709.

Brougham, R.W., 1960. The effects of frequent hard grazing at different times of year on the productivity and species yields of grass-clover pastures. *N.Z. J. Agric. Res.*, 3: 125–136.

Brown, R.H. and Blaser, R.E., 1968. Leaf area index in pasture growth. *Herbage Abstr.*, 38: 1–9.

Bryant, H.T. and Blaser, R.E., 1961. Yields and stands of Orchardgrass compared under clipping and grazing intensities. *Agron. J.*, 53: 9–11.

Bryant, H.T. and Blaser, R.E., 1968. Effects of clipping compared to grazing of Ladino clover–orchardgrass and alfalfa–orchardgrass mixtures. *Agron. J.*, 60: 165–166.

Burton, G.W., Jackson, J.E. and Hart, R.H., 1963. Effects of cutting frequency on yield, *in vitro* digestibility, and protein, fibre, and carotene content of coastal bermudagrass. *Agron. J.*, 55: 500–502.

Calder, F.W., Nicholson, W.G. and Carson, R.B., 1970. Effect of actual versus simulated grazing on pasture productivity and chemical composition of forage. *Can. J. Anim. Sci.*, 50: 475–482.

Camlin, M.S. and Stewart, R.H., 1975. Reaction of Italian ryegrass cultivars under grazing as compared with cutting. *J. Br. Grassland Soc.*, 30: 121–129.

Campbell, A.G., 1969. Grazing interval, stocking rate, and pasture production. *N.Z. J. Agric. Res.*, 12: 67–74.

Chestnutt, D.M.B., Murdoch, J.C., Harrington, F.J. and Binnie, R.C., 1977. The effect of cutting frequency and applied nitrogen on pasture production and digestibility of perennial ryegrass. *J. Br. Grassland Soc.*, 32: 177–183.

Clark, J., Kat, C. and Santhirasegaram, K., 1974. The dry-matter production, botanical composition, *in vitro* digestibility and protein percentage of pasture layers. *J. Br. Grassland Soc.*, 29: 179–184.

Corrall, A.J., 1974. The effect of interruption of flower development on yield and quality of perennial ryegrass. *Vaxtodling*, 29: 39–43.

Corrall, A.J., Lavender, R.H. and Terry, C.P., 1979. Grass species and varieties. Seasonal patterns of production and relationships between yield, quality and date of first harvest. *Grassland Res. Inst., Hurley, Tech. Rep.*, No. 26: 23 pp.

Curll, M.L., 1982. Grass and clover content of pastures grazed by sheep. *Herbage Abstr.*, 52: 403–411.

Curll, M.L. and Wilkins, R.J., 1983. The comparative effects of defoliation, treading and excreta on a *Lolium perenne–Trifolium repens* pasture grazed by sheep. *J. Agric. Sci., Cambridge*, 100: 451–460.

Cuykendall, C.H. and Marten, G.C., 1968. Defoliation by sheep-grazing versus mower-clipping for evaluation of pasture. *Agron. J.*, 60: 404–408.

Dahl, B.E. and Hyder, D.N., 1977. Developmental morphology and management implications. In: R.E. Sosebee (Editor), *Rangeland Plant Physiology*. Society for Range Management, Denver, Colo., pp. 257–290.

Donald, C.M., 1941. *Pastures and Pasture Research*. University of Sydney, Sydney, N.S.W., 108 pp.

Edmond, D.E., 1962. Effects of treading pasture in summer under different soil moisture levels. *N.Z. J. Agric. Res.*, 5: 389–395.

Edmond, D.E., 1964. Some effects of sheep treading on the growth of 10 pasture species. *N.Z. J. Agric. Res.*, 7: 1–16.

Elliot, I.L. and Lynch, P.B., 1958. Techniques for measuring pasture production in fertilizer trials. *N.Z. J. Agric. Res.*, 1: 498–521.

Frame, J., 1965. The effect of cutting and grazing technique on the productivity of grass/clover swards. In: *Proc. X Int. Grassland Congr., São Paulo*, pp. 1155–1516.

Frame, J., 1976. A comparison of herbage production under cutting and grazing (including comments on deleterious factors such as treading). In: J. Hodgson and D.K. Jackson (Editors), *Br. Grassland Soc., Occas. Symp.*, No. 8: 39–49.

Frame, J., 1983. The effect of wheel tracking on red clover (*Trifolium pratense*) swards. In: A.J. Corrall (Editor), *Br. Grassland Soc., Occas. Symp.*, No. 14: 313–315.

Frame, J. and Hunt, I.V., 1971. The effects of cutting and grazing systems on herbage production from grass swards. *J. Br. Grassland Soc.*, 26: 163–171.

Gervais, P. and St-Pierre, J.C., 1979. Influence du stade de croissance à la première récolte sur le rendement, la composition chimique et les réserves nutrives de la fléole des prés. *Can. J. Plant Sci.*, 59: 177–183.

Gillet, M., 1970. Physiology of some temperate forage grasses and cutting date in spring. In: *Proc. XI Int. Grassland Congr., Australia*, pp. 545–548.

Griffin, J.L. and Watson, V.H., 1982. Production and quality of four bermudagrasses as influenced by rainfall patterns. *Agron. J.*, 74: 1044–1047.

Harris, W., 1978. Defoliation as a determinant of the growth, persistence and composition of pasture. In: J.R. Wilson (Editor), *Plant Relations in Pastures*. CSIRO, Melbourne, Vic., pp. 67–85.

Harris, W., 1983. Simulation by cutting of stocking rate and rotational and continuous management. 1. Herbage removal, standing herbage mass, and net herbage accumulation rate. *N.Z. J. Agric. Res.*, 26: 15–27.

Harris, W. and Brougham, R.W. 1968. Some factors affecting change in botanical composition in a ryegrass–white clover pasture under continuous grazing. *N.Z. J. Agric. Res.*, 11: 15–38.

Harris, W., Pineiro, J. and Henderson, J.D., 1980. Performance of mixtures of ryegrass cultivars and prairie grass with red clover cultivars under two grazing frequencies. *N.Z. J. Agric. Res.*, 23: 339–348.

Hodgson, J., Bircham, J.S., Grant, S.A. and King, J., 1981. The influence of cutting and grazing management on herbage growth and utilization. In: C.E. Wright (Editor), *Br. Grassland Soc., Occas. Symp.*, No. 13: 51–62.

Humphreys, L.R., 1981. *Environmental Adaptation of Tropical Pasture Plants*. Macmillan, London, 261 pp.

Humphreys, L.R. and Robinson, A.R., 1966. Subtropical grass growth. 1. Relationship between carbohydrate accumulation and leaf area in growth. *Qld. J. Agric. Anim. Sci.*, 23: 211–259.

Hunt, I.V., Frame, J. and Harkess, R.D., 1979. The effect of date of harvest of primary growth and levels of nitrogen and potassium fertilizers on the dry matter production of timothy (*Phleum pratense*). *Grass Forage Sci.*, 34: 131–137.

Jackson, D.K., 1976. The influence of patterns of defoliation on sward morphology. In: J. Hodgson and D.K. Jackson (Editors), *Br. Grassland Soc., Occas. Symp.*, No. 8: 51–60.

Jackson, M.V. and Williams, T.E., 1979. Response of grass swards to fertiliser N under cutting or grazing. *J. Agric. Sci., Cambridge*, 92: 549–562.

Jameson, D.A., 1963. Response of individual plants to harvesting. *Bot. Rev.*, 29: 532–594.

Jones, M.J., 1933. Grassland management and its influence on the sward. II. The management of a clovery sward and its effects. *Empire J. Exper. Agric.*, 1: 122–128.

Korte, C.J., 1982. Grazing management of perennial ryegrass/white clover pasture in late spring. *Proc. N.Z. Grassland Assoc.*, 43: 80–84.

Korte, C.J., Watkin, B.R. and Harris, W., 1982. Use of residual leaf area index and light interception as criteria for spring-grazing management of ryegrass dominant pasture. *N.Z. J. Agric. Res.*, 25: 309–319.

Langer, R.H.M. and Keoghan, J.M., 1970. Growth of lucerne following defoliation. *Proc. N.Z. Grassland Assoc.*, 32: 98–107.

Lawrence, T., 1973. Productivity of intermediate wheatgrass as influenced by date of initial cutting, height of cutting, and N fertilizer. *Can. J. Plant Sci.*, 53: 295–301.

Lawrence, T. and Ashford, R., 1966. The productivity of intermediate wheatgrass as affected by initial harvest dates and recovery periods. *Can. J. Plant Sci.*, 46: 9–15.

Lawrence, T. and Ashford, R., 1969. Effect of stage and height of cutting on the dry matter yield and persistence of intermediate wheatgrass, bromegrass, and reed canarygrass. *Can. J. Plant Sci.*, 49: 321–332.

Lynch, P.B., 1947. Methods of measuring the production from grassland. *N.Z. J. Sci. Technol.*, 28A: 385–405.

Mason, W. and Lachance, L., 1983. Effects of initial harvest date on dry matter digestibility and protein in timothy, tall fescue, reed canarygrass and Kentucky bluegrass. *Can. J. Plant Sci.*, 63: 675–685.

Matches, A.G., 1968. Performance of four pasture mixtures defoliated by mowing or grazing with sheep or cattle. *Agron. J.*, 60: 281–285.

Matches, A.G., 1979. Management. In: R.C. Buckner and L.P. Bush (Editors), *Tall Fescue*. American Society of Agronomy, Madison, Wisc., pp. 171–199.

McNeur, A.J., 1953. Pasture measurement techniques as applied to strain testing. *Proc. N.Z. Grassland Assoc.*, 15: 157–165.

Minson, D.J., 1983. Forage quality: assessing the plant animal complex. In: *Proc. XIV Int. Grassland Congr., Lexington, U.S.A.*, pp. 23–29.

Minson, D.J., Raymond, W.F. and Harris, C.E., 1960. Studies in the digestibility of herbage. VIII. The digestibility of S37 cocksfoot, S23 ryegrass and S24 ryegrass. *J. Br. Grassland Soc.*, 15: 174–180.

Mislevy, P., Washko, J.B. and Harrington, J.D., 1974. Effects of different initial cutting treatments on the production of Climax timothy and reed canarygrass. *Agron. J.*, 66: 110–112.

Mislevy, P., Washko, J.B. and Harrington, J.D., 1977. Influence of plant stage at initial harvest and height of regrowth at cutting on forage yield and quality of timothy and orchardgrass. *Agron. J.*, 69: 353–356.

Monson, W.G. and Burton, G.W., 1982. Harvest frequency and fertilizer effects on yield, quality and persistence of eight bermudagrasses. *Agron. J.*, 74: 371–374.

Omaliko, C.P.E., 1980. Influence of initial cutting date and cutting frequency on yield and quality of star, elephant and Guinea grasses. *Grass Forage Sci.*, 35: 139–145.

Parsons, A.J., Leafe, E.L., Collett, B., Penning, P.D. and Lewis, J., 1983. The physiology of grass production under grazing. II. Photosynthesis, crop growth and animal intake of continuously-grazed swards. *J. Appl. Ecol.*, 20: 127–139.

Reid, D., 1959. Studies on the cutting management of grass-clover swards. I. The effect of varying the closeness of cutting on the yields from an established grass-clover sward. *J. Agric. Sci., Cambridge*, 53: 299–312.

Reid, D., 1962. Studies on the cutting management of grass-clover swards. III. The effects of prolonged close and lax cutting on herbage yields and quality. *J. Agric. Sci., Cambridge*, 59: 359–368.

Reid, D., 1978. The effects of frequency of defoliation on the yield response of a perennial ryegrass sward to a wide range of nitrogen application rates. *J. Agric. Sci., Cambridge*, 90: 447–457.

Richards, I.R., Wolton, K.M. and Ivins, J.D., 1976. A note on the effect of sheep grazing on the yield of grass swards. *J. Agric. Sci., Cambridge*, 87: 337–340.

Sato, K., 1973. Comparison among four temperate grasses in production structure at each heading stage. *J. Jap. Soc. Grassland Sci.*, 19: 208–214.

Scott, R.S., 1963. The effect of mole drainage and winter pugging on grassland production. *Proc. N.Z. Grassland Assoc.*, 25: 119–127.

Sears, P.D., 1951. The technique of pasture measurement. *N.Z. J. Sci. Technol.*, 33A: 1–29.

Sears, P.D., 1956. The effect of the grazing animal on pasture. In: *Proc. VII Int. Grassland Congr., N.Z.*, pp. 92–103.

Sheard, R.W. and Winch, J.E., 1966. The use of light interception, grass morphology and time as criteria for harvesting of timothy, smooth brome and cocksfoot. *J. Br. Grassland Soc.*, 21: 231–237.

Sheath, G.W. and Boom, R.C., 1985. Effects of November–April grazing pressure on hill country pastures. 2, Pasture species composition. *N.Z. J. Exper. Agric.*, 13: 329–340.

Stapledon, R.G., 1924. Seasonal productivity of grasses. *Bull. Welsh Plant Breeding Stn.*, H3: 5–84.

Tayler, J.C. and Rudman, J.E., 1966. The distribution of herbage at different heights in "grazed" and "dung patch" areas of sward under two methods of grazing management. *J. Agric. Sci., Cambridge*, 66: 29–39.

Taylor, T.H., Washko, J.B. and Blaser, R.E., 1960. Dry matter yield and botanical composition of an orchardgrass–ladino white clover mixture under clipping and grazing conditions. *Agron. J.*, 53: 217–220.

Watkin, B.R. and Clements, R.J., 1978. The effects of grazing animals on pastures. In: J.R. Wilson (Editor), *Plant Relations in Pastures*. CSIRO, Melbourne, Vic., pp. 273–289.

Wilman, D., 1977. The effect of grazing compared with cutting, at different frequencies, on a lucerne-cocksfoot ley. *J. Agric. Sci., Cambridge*, 88: 483–492.

Wilman, D. and Asiegbu, J.E., 1982. The effects of clover variety, cutting interval and nitrogen application on herbage yields, proportions and heights in perennial ryegrass–white clover swards. *Grass Forage Sci.*, 37: 1–13.

Wilman, D., Drougsiotis, D., Koocheki, A., Lwoga, A.B. and Shim, J.S., 1976a. The effect of interval between harvests and nitrogen application on the proportion and yield of crop fractions in four ryegrass varieties in the first harvest year. *J. Agric. Sci., Cambridge*, 86: 189–203.

Wilman, D., Koocheki, A. and Lwoga, A.B., 1976b. The effect of interval between harvests and nitrogen application on the proportion and yield of crop fractions and on the digestibility and digestible yield of two perennial ryegrass varieties in the second harvest year. *J. Agric. Sci., Cambridge*, 87: 59–74.

Winch, J.E., Sheard, R.W. and Mowat, D.N., 1970. Determining cutting schedules for maximum yield and quality of bromegrass, timothy, lucerne and lucerne/grass mixtures. *J. Br. Grassland Soc.*, 25: 44–52.

Wolton, K.M., 1963. An investigation into the simulation of nutrient returns by the grazing animal in grassland experimentation. *J. Br. Grassland Soc.*, 18: 213–219.

Chapter 8

# THE BOTANICAL COMPOSITION OF PASTURES

R.W. SNAYDON

## INTRODUCTION

The botanical composition of pastures varies greatly. On a worldwide scale, extreme differences exist between the major climatic zones, mainly as a result of differences in temperature regime and water availability (Snaydon, 1981a). This chapter deals with variation on a smaller scale, using temperate grasslands as an example. The main aims are to consider the factors responsible for differences in botanical composition, and then to consider the possible effects of these differences on the productivity and quality of pastures. The effects of botanical composition on animal performance are considered in Chapter 25.

## FACTORS AFFECTING COMPOSITION

One of the most comprehensive surveys investigating the factors affecting the botanical composition of temperate pastures was carried out in the Netherlands by De Vries (reviewed by Kruijne et al., 1967), though King (1962), Rogers and King (1972), Grime and Lloyd (1973) and Grant and Brock (1974) have carried out more restricted surveys. Some of the studies (e.g. Kruijne et al., 1967; Grime and Lloyd, 1973) have considered only the distribution of individual species, but others have attempted to classify stands, either objectively by multivariate analysis (e.g. King, 1962; Lloyd, 1972), or subjectively (e.g. O'Sullivan, 1983).

Differences in botanical composition, in the various surveys, were largely associated with differences in soil conditions (pH, nutrient status and water content) and differences in grazing/cutting management (frequency and intensity). Some species, such as *Dactylis glomerata*, *Lolium perenne*, *Phleum pratense*, *Poa annua*, *P. trivialis* and *T. repens* occur most abundantly on soils with a high nutrient status and a pH near neutrality; these pastures are usually the most productive. Other species, such as *Agrostis stolonifera*, *Cynosurus cristatus* and *Holcus lanatus*, occur most abundantly on soils of intermediate fertility, whereas *A. tenuis* and *Festuca rubra* occur most abundantly on rather more infertile soil. *Agrostis canina*, *Briza media*, *Festuca ovina* and *Nardus stricta* occurred almost entirely on very infertile acid soils, where productivity is usually extremely low. Some species, such as *Arrhenatherum elatius*, *Leucanthemum vulgare* and *Dactylis glomerata* occur predominantly in infrequently cut swards, whereas other species such as *Bellis perennis*, *Lolium perenne* and *Trifolium repens* occur predominantly in heavily grazed swards (e.g. Kruijne et al., 1967; Elberse et al., 1983).

## CONTROL OF COMPOSITION

The botanical composition of pastures can be controlled directly, by sowing, but is also indirectly affected by various management practices, such as fertilizer use. Direct control, by sowing the desired species and cultivars, is usually only effective in the short term. As other species invade the sward, the proportion of the sown species in temperate pastures declines, on average, from about 75% after one year to about 40% after twenty years (Morrison, 1979; Green and Baker, 1981).

Most sown grasslands are part of a rotation system, where grassland is periodically ploughed and sown with arable crops, usually to prevent the build-up of weeds, pests and pathogens, or to exploit the soil fertility that develops under pasture. The practice of sowing grassland has tended to decline in Europe, as herbicides, pesticides, fungicides and fertilizers have been progressively used to overcome these constraints, and continuous arable production has been practised. Some of the sown grassland, however, replaces older grassland which has been heavily invaded by unsown species. However, this practice has also tended to decline because of increasing costs and doubts about its efficacy (see below).

Indirect control of the botanical composition of pastures occurs largely as a by-product of management practices, such as fertilizer application or grazing/cutting regime, which are imposed mainly to manipulate the yield, quality or utilization of pasture. Fertilizer applications greatly affect the botanical composition of lowland pastures (e.g. Brenchley and Warington, 1958; Williams, 1978; Elberse et al., 1983) and upland pastures (e.g. Milton, 1940; Luscombe et al., 1981); these effects have been reviewed by Rabotnov (1977). Similarly, grazing management causes rapid changes in the botanical composition of swards (e.g. M.G. Jones, 1933; Brougham, 1960).

## SWARD AGE AND COMPOSITION

Only one year after sowing, about 25% of the sward is composed of invading species, especially *Poa annua*, *P. trivialis* and dicotyledonous annuals, such as *Stellaria media* (Morrison and Idle, 1972). These species are probably successful invaders in the early stages because of the abundance of their seed in the soil (Champness and Morris, 1948; Champness, 1949; Howe and Chancellor, 1983), and their requirement for high fertility conditions (see above). Between two and twenty years after sowing, swards are progressively invaded by such species as *Agrostis* spp., *Festuca rubra* and *Holcus lanatus* (Fig. 8.1A). These species generally have lower requirements for soil fertility (see above). After about thirty years, most swards have reached dynamic equilibrium, the botanical composition being dependent upon soil nutrient status, soil water content and grazing/cutting management (see above).

Although this general pattern of changing composition with age occurs on farms, different results have sometimes been obtained under experimental conditions. For example, Garwood and Tyson (1979) found only slight changes in botanical composition over a twenty-year period, though Smith (1979) found large changes within ten years. These differences, between farm and experimental conditions and between experiments, are probably due to different fertilizer inputs and cutting/grazing management. For example, the use of fertilizer (N, P and K) on farms declines as the sward ages, so that on average the use of fertilizer is twice as great on swards less than seven years old as on swards

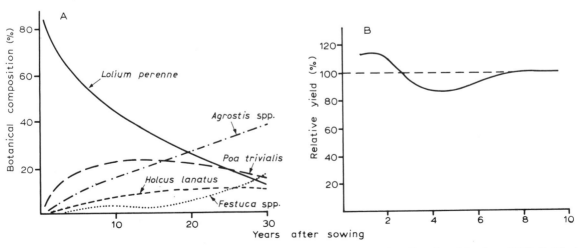

Fig. 8.1. A. The botanical composition of sown pastures at various times after sowing (data from Morrison, 1979). B. The herbage yields of sown pastures at various times after sowing.

more than seven years old (Church and Leech, 1983). By contrast, experimental plots usually receive constant fertilizer inputs, regardless of age, and usually receive higher inputs than the farm average. Another difference is that experimental plots are normally cut whereas most farm swards are grazed.

## SWARD AGE AND YIELD

The usually accepted pattern of changes in yield with sward age is shown in Fig. 8.1B, though results have been somewhat conflicting (Smith, 1979). The large yield in the first harvest year can be attributable to the release of nutrients when the old sward is ploughed in (e.g. Watson, 1963; Davies, 1964; Clement, 1971). The subsequent decline in yield has been attributed to the fixation of nutrients in the soil organic matter; for example, Garwood and Tyson (1979) found that about 75 kg N ha$^{-1}$ yr$^{-1}$ was fixed during the first eight years after sowing. After about the fifth year, the amount of nutrients fixed is increasingly balanced by the amount released by mineralization (Garwood and Tyson, 1979), so the yield increases. If this interpretation is correct, the pattern of change in yield with sward age should be changed, or eliminated, by applying sufficient fertilizer; Smith (1979) provided some evidence to support this.

There seems to be little indication that the magnitude of changes in yield with age reflects the magnitude of changes in botanical composition. For example, Garwood and Tyson (1979) found large changes in yield but little change in botanical composition, whereas Smith (1979) found little change in yield but large changes in botanical composition. However, even if such correlations existed, it would be dangerous to infer a causal relationship between yield and botanical composition — that is, that botanical composition determines yield. The comparative yield of species, and of communities, must be compared directly, under uniform conditions.

## HERBAGE YIELD OF SPECIES

Most of the grass and legume species normally sown in temperate pastures (e.g. *Dactylis glomerata, Lolium perenne, Medicago sativa* and *Trifolium repens*) are reputed to be high-yielding, whereas invading species (e.g. *Agrostis* spp., *Holcus lanatus, Lotus corniculatus, Poa* spp.) are reputedly low-yielding. These reputations seem to be based largely on field observation; species which normally occur in low-fertility sites, where productivity is inherently low, have reputations for low yield, whereas species from high-fertility sites, with high productivity, have a reputation for high yields. However, there are other reasons to expect that species from low-fertility conditions might indeed be lower-yielding. Bradshaw et al. (1964) found that species from infertile soil were slower growing in sand culture. They concluded that slow growth rate was one aspect of adaptation to low-fertility conditions. Grime and Hunt (1975) have confirmed this in more extensive studies of the seedling growth of a wide range of species. However, as Grime and Hunt (1975) pointed out, these differences in relative growth rate under controlled conditions might not be reflected in higher yields in closed stands in the field, so sward comparisons are required.

The yields of various grass species have been compared in the field, both in pure stands and in simple mixtures with legumes. In addition, complex naturally occurring mixtures (permanent pastures) have been compared with sown swards (leys), usually simple mixtures. The results of these various studies have been reviewed by Snaydon (1978, 1979) and by Dibb and Haggar (1979).

The most extensive comparisons of pure stands have been made for commonly sown species, and have been reviewed by Snaydon (1979). Although the larger-seeded species (e.g. *Lolium* spp.) establish more rapidly, and therefore yield more in the first harvest year, thereafter there are no consistent differences between species (Snaydon, 1979). Any differences between species are small compared with the effects of fertilizer use, cutting frequency or sites, and are inconsistent, depending on fertilizer use, grazing/cutting regime, site and year — that is, there are large species × environment interactions compared with species differences.

Where sown species have been compared with indigenous species in pure stands (e.g. Stapleton and Milton, 1932; Morris and Thomas, 1972; Haggar, 1976; Frame, 1983), commonly sown grasses, such as *Dactylis glomerata, Lolium perenne* and *Phleum pratense* have been among the most productive species, but some indigenous species, such as *Alopecurus pratensis*, *Anthoxanthum odoratum*, *Festuca rubra* and *Holcus lanatus*, have been at least as productive, and sometimes more productive. However, some indigenous species, such as *Poa trivialis* and *Trisetum flavescens*, have usually been less productive, mainly because of poor establishment or poor persistence, while species indigenous to highly infertile sites tend to be less productive. Once again, the large-seeded species, such as *Lolium* spp., have established more rapidly, and yielded more in the first harvest year, than small-seeded species, such as *Agrostis tenuis*. As a result, the period over which yield is measured can greatly affect apparent performance. In addition, climatic conditions (e.g. Morris and Thomas, 1972) and fertilizer use (e.g. Cowling and Lockyer, 1965; Frame, 1983) can greatly affect relative performance. The frequency and height of cutting or grazing might also be expected to affect the relative performance of species, though few comparisons have been made. One reason for expecting differences in response to cutting/grazing management is that grass species differ in resource allocation between vegetative propagules (stolons and rhizomes) and leaves (Snaydon, 1984); tall upright species (e.g. *L. perenne*, *D. glomerata* and *H. lanatus*) have a greater proportion of shoot material above the cutting height than do short spreading species (e.g. *Agrostis stolonifera*). As a result, tall upright species may appear to be more productive under infrequent lax cutting but are usually less productive, and less persistent, under frequent close cutting or intense grazing. This is one of the reasons why the proportion of sown species in swards declines with time (see above). However, there are also differences between sown species in response to cutting/grazing management. In particular, *Lolium perenne* is more tolerant of frequent close defoliation than *Dactylis glomerata* (e.g. Brougham, 1960; Jones, 1983).

**HERBAGE YIELD OF MIXTURES**

Most agricultural grasslands are composed of complex mixtures of species, so any differences in productivity between species, measured in pure stands, have little relevance in practice unless these differences are also expressed in mixtures. Most of the available evidence indicates that differences in productivity between grass species are normally reduced, and often disappear entirely when those grasses are grown with a legume, such as *Trifolium repens* (Fig. 8.2). This would be expected if the most productive grasses were also the most competitive against clover, so reducing nitrogen fixation and hence productivity. When grass species are grown in mixtures, the yield of the mixture is usually slightly more than the mean of the component species when grown in pure stands (Trenbath, 1974). There is a tendency for mixtures to be similar in yield, regardless of their composition.

When other species invade pure stands of a sown species there is usually little effect on the yield of the stand (e.g. Smith, 1979). However, yield may sometimes be slightly reduced (e.g. Camlin and Stewart, 1976), apparently because the death of sown plants leaves bare ground, which is not immediately invaded by other species. On the other hand, in a few cases, invasion by other species may actually increase yield, even if the invading species is itself lower-yielding in pure stands (e.g. Wells and Haggar, 1974).

Surprisingly few valid comparisons of yield have

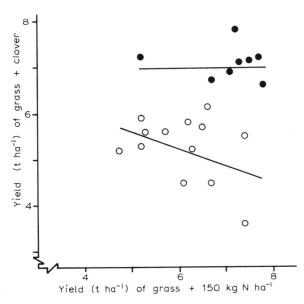

Fig. 8.2. Comparisons of the yield of various grass species and cultivars grown in pure stands receiving 150 kg N ha$^{-1}$ and grown in mixtures with *Trifolium repens*. Data from W. Murphy (pers. comm.) (●), and from Holmes and MacLusky (1955) (○).

been made between complex mixtures of indigenous species (permanent pasture) and pure stands or single mixtures of selected species (leys). Comparison is made difficult because of the fact that, under normal farm conditions, leys receive about twice as much fertilizer as permanent pasture (see above) and, in many experimental studies, fertilizer use has not been standardized for the two sward types. Where similar fertilizer inputs have been used (e.g. Kleter and Bakhuis, 1972; Grant et al., 1981), no significant difference in productivity has been detected. Leys yield less in the year of sowing, more in the first harvest year, when soil nutrients are released (see above), and about the same thereafter (e.g. Mudd and Meadowcroft, 1964).

## SEASONAL PATTERN OF PRODUCTION

While the total annual production of a pasture sets an upper limit to the feed available for grazing animals, the amount that is actually utilized is determined more by stocking rate (Chapter 24), by herbage quality (Ch. 14), and by the correspondence between the seasonal pattern of herbage production and the seasonal pattern of herbage requirement (Chapters 9–11). The match between herbage production and requirement can be manipulated to a considerable extent. On the one hand, the pattern of production can be varied, for example by varying the timing of nitrogen application (Chapter 6), but a more common method of manipulating the supply of herbage is by conserving herbage (e.g. as silage and hay) in periods of excess to be used in periods of deficiency (Chapter 26). On the other hand, the pattern of herbage requirement can also be varied by varying the type of livestock and the time of reproduction. For example, dairy cows and, to a lesser extent lambing ewes, have a peak of requirement soon after parturition, whereas non-reproductive animals (e.g. steers and wethers) have a steadily increasing feed requirement as their weight increases. The timing of the peak of herbage requirement for reproductive animals (dairy cows and ewes) can be varied by changing the time of parturition.

The seasonal pattern of herbage production varies to some extent between grass species. Most of the commonly sown species (e.g. *Dactylis glomerata*, *Lolium perenne* and *Phleum pratense*) have a very marked peak of production at flowering time in early summer (Anslow and Green, 1967). *Anthoxanthum odoratum*, *Cynosurus cristatus*, *Holcus lanatus* and *Poa trivialis* have similar marked peaks in early summer, but *Agrostis* spp., *Festuca ovina* and *F. rubra* have a more even pattern of production throughout the growing season (Morris and Thomas, 1972; Haggar, 1976). The relative value of these patterns of production will depend on the system of animal production (see above). In theory, a more even distribution of production throughout the growing season can be obtained by growing species in mixtures, but few field comparisons have been made to confirm this.

## HERBAGE QUALITY

The intake of herbage by grazing animals, and hence animal performance, is greatly affected by herbage quality (Chapter 14). The most commonly used measure of herbage quality is digestibility, though other chemical and physical attributes are also important (Chapter 14).

There are large differences in digestibility between the various parts of the plant (e.g. leaf lamina, leaf sheath and stem): even for a given plant part there is considerable variation with age, season and environmental conditions (Hacker and Minson, 1981). These differences make valid comparisons between species difficult, though some consistent differences in digestibility between species have been found. In particular, panicoid grasses generally have lower digestibility than pooid grasses (Minson and McLeod, 1970). There is also considerable variation among the temperate pooid grasses, when swards are grown under identical conditions and cut similarly. The digestibility of *Dactylis glomerata*, *Festuca ovina* and *F. rubra* is consistently about 5% less than that of *Lolium perenne* (Thomas and Morris, 1973; Haggar, 1976; Wilson and Collins, 1980), whereas the digestibility of *Holcus lanatus* is consistently about 5% greater than *L. perenne*. The results for *Agrostis* spp. and *Poa trivialis* have been variable (Thomas and Morris, 1973; Haggar, 1976; Wilson and Collins, 1980), but *Agrostis* spp. tend to be slightly less digestible and *P. trivialis* tends to be more digestible than *L. perenne*. There is therefore variation between both sown species, such as *L. perenne* and *D. glomerata*, and between indigenous species such as *H. lanatus* and *F. rubra*, but no consistent difference between the two groups of species. In general, however, species indigenous to extremely infertile soils tend to be less digestible than either sown species or species indigenous to reasonably fertile soils (Newbould, 1979).

## CONCLUSIONS

Managed grasslands are man-made plagioclimaxes; as a result, the botanical composition of these grasslands is extremely dynamic, changing rapidly in response to such factors as grazing/cutting management and fertilizer use.

Both the yield and botanical composition of temperate pastures are greatly affected by soil conditions (nutrient status, pH and water content) and by grazing/cutting management. The resulting correlation between yield and composition has often been interpreted as causal, but most of the available experimental evidence indicates that, in general, most grass species differ only slightly in productivity, when grown under identical conditions. However, there is considerable species × environment interaction; for example, species from fertile conditions usually yield more than those from low fertility conditions under high fertilizer inputs and vice versa under low fertilizer inputs. Species also differ in response to grazing/cutting management. Management (fertilizer use and grazing/cutting regime) has a much greater effect on productivity than does changing the pasture composition. As a result, resowing a pasture can only have a lasting effect on yield if it is accompanied by improved management.

Most of the sown temperate grasses, and many of the indigenous grasses, have similar patterns of seasonal production, with a large peak of production in early summer. Some indigenous species have a more uniform seasonal distribution of production. The relative value of these patterns depends on the use to which the pasture is put.

The difference in herbage quality between species, when they are treated identically, is not as great as is usually assumed. However, species from extremely infertile soils tend to have lower herbage yield and quality.

## REFERENCES

Anslow, R.C. and Green, J.O., 1967. The seasonal growth of pasture grasses. *J. Agric. Sci., Cambridge*, 68: 109–122.

Bradshaw, A.D., Chadwick, M.J., Jowett, D. and Snaydon, R.W., 1964. Experimental investigations in the mineral nutrition of several grass species. IV. Nitrogen. *J. Ecol.*, 52: 665–676.

Brenchley, W.E. and Warington, K., 1958. *The Park Grass Plots at Rothamsted 1856–1949*. Rothamsted Experimental Station, Harpenden, 144 pp.

Brougham, R.W., 1960. The effect of frequent hard grazing at different times of the year on the productivity and species yield of a grass-clover pasture. *N.Z. J. Agric. Res.*, 3: 125–136.

Camlin, M.S. and Stewart, R.H., 1976. The assessment of persistence and its application to the evaluation of early perennial ryegrass cultivars. *J. Br. Grassland Soc.*, 31: 1–6.

Champness, S.S., 1949. Notes on the buried seed populations beneath different types of leys in their seeding year. *J. Ecol.*, 37: 51–56.

Champness, S.S. and Morris, K., 1948. The population of buried viable seed in relation to contrasting pasture and soil types. *J. Ecol.*, 36: 149–173.

Church, B.M. and Leech, P.K., 1983. *Survey of Fertilizer Practice*. Rothamsted Experimental Station, Harpenden.

Clement, C.R., 1971. Effect of cut and grazed swards on the supply of nitrogen to subsequent arable crops. *Min. Agric., Food Fish., Tech. Bull.*, 20: 166–171.

Cowling, D.W. and Lockyer, D.R., 1965. A comparison of the reaction of different grass species to fertilizer nitrogen and to growth with white clover. I. Yield of dry matter. *J. Br. Grassland Soc.*, 20: 197–204.

Davies, W.E., 1964. The yield and composition of lucerne, grass and clover under different systems of management. VI. Residual fertility. *J. Br. Grassland Soc.*, 19: 358–361.

Dibb, C. and Haggar, R.J., 1979. Evidence of sward changes on yield. In: A.H. Charles and R.J. Haggar (Editors), *Changes in Sward Composition and Productivity*. British Grassland Society, Hurley, pp. 11–20.

Elberse, W.Th., Van den Bergh, J.P. and Dirven, J.G.P., 1983. Effects of use and mineral supply on the botanical composition and yield of old grassland on heavy-clay soil. *Neth. J. Agric. Sci.*, 31: 63–88.

Frame, J., 1983. *The Response of Indigenous Species to Fertilizer Nitrogen*. Agron. Publ. 744. W. Scotland Agric. Coll., Ayr, 10 pp.

Garwood, E.A. and Tyson, K.C., 1979. Productivity and botanical composition of a grazed ryegrass–white clover sward over 24 years as affected by soil conditions and weather. In: A.H. Charles and R.J. Haggar (Editors), *Changes in Sward Composition and Productivity*. British Grassland Society, Hurley, pp. 41–46.

Grant, D.A. and Brock, J.L., 1974. A survey of pasture composition in relation to soils and topography on a hill country farm in the Southern Ruahine Range, New Zealand. *N.Z. J. Exper. Agric.*, 2: 243–250.

Grant, D.A., Luscombe, P.C. and Thomas, V.J., 1981. Responses of ryegrass, browntop, and an unimproved resident pasture in hill country to nitrogen, phosphorus and potassium fertilisers. 1. Pasture production. *N.Z. J. Exper. Agric.*, 9: 227–236.

Green, J.O. and Baker, R.D., 1981. Classification, distribution and productivity of UK grasslands. In: J.L. Jollans (Editor), *Grassland in the British Economy*. University of Reading, Reading, pp. 237–247.

Grime, J.P. and Lloyd, P.S., 1973. *An Ecological Atlas of Grassland Plants*. Arnold, London, 192 pp.

Grime, J.P. and Hunt, R., 1975. Relative growth-rate: its range and adaptive significance in a local flora. *J. Ecol.*, 63: 393–422.

Hacker, J.B. and Minson, D.J., 1981. The digestibility of plant parts. *Herbage Extr.*, 51: 459–482.

Haggar, R.J., 1976. The seasonal productivity, quality and response to nitrogen of four indigenous species compared with *Lolium perenne*. *J. Br. Grassland Soc.*, 31: 197–207.

Holmes, W. and McLusky, D.S., 1955. The intensive production of herbage for crop drying. VI. *J. Agric. Sci. Cambridge*, 46: 269–286.

Howe, C.D. and Chancellor, R.J., 1983. Factors affecting the viable weed content of soils beneath lowland pastures. *J. Appl. Ecol.*, 20: 915–922.

Jones, E.L., 1983. The production and persistence of different grass species cut at different heights. *Grass Forage Sci.*, 38: 79–87.

Jones, M.G., 1933. Grassland management and its influence on the sward. *Empire J. Exper. Agric.*, 1: 42–57.

King, J., 1962. The *Festuca–Agrostis* grassland complex in South-East Scotland. *J. Ecol.*, 50: 321–355.

Kleter, H.J. and Bakhuis, J.A., 1972. The effect of white clover on the production of young and older grassland compared to that of nitrogen fertilizer. *J. Br. Grassland Soc.*, 27: 229–239.

Kruijne, A.A., De Vries, D.M. and Mooi, H., 1967. Bijdragen tot de oecologie van de Nederlandse graslandplanten. *Versl. Landbouwkundige Onderzoekingen*, 696: 65 pp.

Lloyd, P.S., 1972. The grassland vegetation of the Sheffield region. II. Classification of grassland types. *J. Ecol.*, 60: 759–776.

Luscombe, P.C., Grant, D.A. and Thomas, V.J., 1981. Responses of ryegrass, browntop and unimproved resident pasture in hill country, to nitrogen, phosphorus and potassium fertilisers. II. Species composition of resident pastures. *N.Z. J. Exper. Agric.*, 9: 237–241.

Milton, W.E.J., 1940. The effect of manuring, grazing and cutting on the yield and botanical composition of natural hill pasture. I. Yield and botanical composition. *J. Ecol.*, 28: 326–356.

Minson, D.J. and McLeod, M.N., 1970. The digestibility of temperate and tropical grasses. In: *Proc. XI Int. Grassland Congr.*, 719–722.

Morris, R.M. and Thomas, J.G., 1972. The seasonal pattern of dry-matter production of grasses in the North Pennines. *J. Br. Grassland Soc.*, 27: 163–172.

Morrison, J., 1979. Botanical change in agricultural grassland in Britain. In: A.H. Charles and R.J. Haggar (Editors), *Changes in Sward Composition and Productivity*. British Grassland Society, Hurley, pp. 5–10.

Morrison, J. and Idle, A.A., 1972. *A Pilot Survey of Grassland in S.E. England*. *Tech. Bull.*, 10. Grassland Research Institute, Hurley, 76 pp.

Mudd, C.H. and Meadowcroft, S.C., 1964. Comparison between the improvement of pastures by the use of fertilisers and by reseeding. *Exper. Husbandry*, 10: 66–84.

Newbould, P., 1979. Soils and vegetation of the hills and their limitations. In: *Science and Hill Farming*. Hill Farming Research Organisation, Edinburgh, pp. 9–21.

O'Sullivan, A.M., 1983. The lowland grasslands of Ireland. *J. Life Sci., R. Dublin Sci.*, 4: 131–142.

Rabotnov, T.A., 1977. The influence of fertilizers on the plant communities of mesophytic grasslands. In: W. Kraus (Editor), *Applications of Vegetation Science to Grassland Husbandry*. Junk, The Hague, pp. 461–497.

Rogers, J.A. and King, J., 1972. The distribution and abundance of grassland species in hill pasture in relation to soil aeration and base status. *J. Ecol.*, 60: 1–18.

Smith, A., 1979. Changes in botanical composition and yield in a long-term experiment. In: A.H. Charles and R.J. Haggar (Editors), *Changes in Sward Composition and Productivity*. British Grassland Society, Hurley, pp. 69–75.

Snaydon, R.W., 1978. Indigenous species in perspective. In: *Proc. 1979 Crop Protection Conference (Weeds)*, pp. 905–913.

Snaydon, R.W., 1979. Selecting the most suitable species and cultivars. In: A.H. Charles and R.J. Haggar (Editors), *Changes in Sward Composition and Productivity*. British Grassland Society, Hurley, pp. 179–189.

Snaydon, R.W., 1981a. The ecology of grazed pastures. In: F.H.W. Morley (Editor), *Grazing Animals*. Elsevier, Amsterdam, pp. 13–31.

Snaydon, R.W., 1981b. How important is the botanical composition of pastures? *J. Agric. Soc. Univ. Coll., Wales*, 62: 126–139.

Snaydon, R.W., 1984. Plant demography in an agricultural context. In: R. Dirzo and J. Sarukhan (Editors), *Perspectives in Plant Population Ecology*. Sinauer, Sunderland, pp. 389–407.

Stapledon, R.G. and Milton, W.E.J., 1932. Yield, palatability and other studies on strains of various grass species. *Welsh Plant Breed. Stn. Bull.*, H10.

Thomas, J.G. and Morris, R.M., 1973. Seasonal patterns of digestible organic matter and protein production from grasses in the North Pennines. *J. Br. Grassland Soc.*, 28: 31–40.

Trenbath, B.R., 1974. Biomass productivity of mixtures. *Adv. Agron.*, 26: 177–210.

Watson, E.R., 1963. The influence of subterranean clover pastures on soil fertility. I. Short-term effects. *Aust. J. Agric. Res.*, 14: 796–807.

Wells, G.J. and Haggar, R.J., 1974. Herbage yields of ryegrass swards invaded by *Poa* species. *J. Br. Grassland Soc.*, 29: 109–111.

Williams, E.D., 1978. *Botanical Composition of the Park Grass Experiment at Rothamsted, 1856–1976*. Rothamsted Experimental Station, Rothamsted, 61 pp.

Wilson, R.K. and Collins, D.P., 1980. Chemical composition of silage made from different grass genera. *Irish J. Agric. Res.*, 19: 75–84.

Section II

# SECONDARY PRODUCTIVITY

The structure of this section essentially parallels that of section I. The first chapters (9–12) deal with overall patterns of secondary productivity in managed grasslands, considering the more important animal species. The aim in these chapters is to set the agricultural context in which animal production occurs. Attention is focussed on the use of herbage by the various animal species and the conversion of herbage into products (meat, milk and wool) used by man. Consideration is given to the efficiency with which herbage is used and converted, and the way in which the seasonal pattern of food requirement relates to the seasonal pattern of herbage production. The second group of chapters (13–16) deal with some of the more important factors and processes that affect secondary productivity. The ways in which pasture attributes (e.g. yield and composition) affect the intake and utilization of herbage, and hence affect animal performance, are given special attention. The converse effects of animals on pastures receive less attention. Apart from stocking rate, which is considered in Chapter 24, probably the most important attributes of animals which affect the efficiency of conversion of herbage to animal products are reproduction rate and lifespan; these are considered in Chapter 16.

Although the information presented in Chapters 9–12 does not appear to be directly relevant to "natural" ecosystems, most of the ecological principles in these chapters, and even more those in Chapters 13 to 16, are applicable to "natural" ecosystems. As in the case of primary productivity, there has been much more detailed study of managed grassland ecosystems than of "natural" ecosystems. As a result, the techniques (e.g. in behaviour studies, feed requirements, or herbage quality analysis) have been more fully developed, and the principles more firmly established. Studies of managed grassland ecosystems may therefore indicate ways in which studies of animal performance in "natural" ecosystems may be intensified and extended.

Chapter 9

# BEEF PRODUCTION FROM MANAGED GRASSLANDS

W. HOLMES

## INTRODUCTION

Previous chapters (Chapters 2, 3 and 5) in this volume have shown that the primary productivity of managed grassland is heavily dependent on soil conditions and climate, and can also be greatly affected by management. Variations in these factors result in variations in annual production of grassland herbage between 2.5 and 25 t of dry matter (DM) per hectare in temperate conditions and to over 50 t DM ha$^{-1}$ in the wet tropics. This is equivalent to approximately 45 to 450 GJ of gross energy per hectare and 100 to 3750 kg of crude protein (N × 6.25) per ha in temperate conditions.

Grass and forage crops are not normally suitable for human food and, although methods of extraction of high-quality nutrients have been developed (Wilkins, 1977), they have not been widely adopted. Herbivores digest forage through the activity of the micro-organisms in their digestive tracts (Hungate, 1966). With some herbivores, such as Equidae, fermentation occurs in the hind-gut or caecum, while ruminants such as the Bovidae depend on fermentation in the fore-gut. Ruminants retain the forage for a prolonged period (c. 50 h) in the rumen; they include cattle, sheep and goats, which are the most important species used by man to convert forage products into meat, milk and fibre. A few non-ruminants, such as horses and rabbits, are used to a lesser extent to exploit forage resources (Ch. 12).

Beef cattle are found typically in pastoral areas of moderate soil fertility, and at a greater distance from markets than are animals kept for milk production (Ch. 10). They require less uniform nutrition and less frequent attention than milking cows, and the product can be transported more easily. Beef production occurs either as a single enterprise, often large-scale, or as a complementary enterprise in mixed farming systems. In mixed farming systems, pastures grazed by sheep (Ch. 11) or beef cattle may form part of a cropping rotation.

Winrock International (1978) have reviewed the role of ruminants on a world scale, and shown the contribution of beef cattle to human diets. Meat from beef cattle is an important component of the diet in North and South America, Oceania and Western Europe, where the combination of climate and economic factors has led to a dominance of milk and beef production from grass (Allen et al., 1982). This chapter reviews beef production and considers its biological efficiency. Chapter 10 of Volume 21 deals with beef production under more intensive conditions, in feed lots.

## PRODUCTION OF HERBAGE

### Seasonal variability in the supply and quality of herbage

Variations in the growth rate of herbage are discussed in detail in Chapter 2, and also by Anslow (1967) and Morrison et al. (1980) for temperate regions and Vicente-Chandler et al. (1964) for the wet tropics. Seasonal variations in pasture growth rate, from 0 to 120 kg DM ha$^{-1}$ day$^{-1}$, pose special problems in the efficient utilization of pastures; variations from year to year are also important.

Species, and cultivars within the species, differ in agronomic characteristics, but there is little evidence that the total annual yield or the seasonal distribution of yield of species adapted to a particular environment varies widely (Chapter 8). Differences in tolerance of the vagaries of weather and management, and differences in persistency, are probably the agronomic characters of greatest importance, although possible future breeding for greater herbage production should not be ignored.

Although sown pastures form an important element in some pastoral systems, semi-natural permanent pastures occupy an even greater area in many countries. Recent studies in the United Kingdom (Forbes et al., 1980) have shown that they respond in a similar manner to management as do sown pastures (Chapter 8).

The dominant feature influencing the feeding value of any pasture species is the stage of development. In general, leafy vegetative material is of higher digestibility, and is more readily harvested, than is stemmy material produced in the reproductive phase of development (Osbourn, 1982). Management is generally organized so that at least a proportion of the pasture is maintained in leafy condition.

## UTILIZATION OF HERBAGE

### The conservation of surplus forage

Whereas the rate of growth of grass and forage crops varies widely with the season, the nutrient requirements of the grazing animals are less variable. Several special management methods are used to overcome this; for example, the sale of **finished stock** is timed to coincide with declining pasture production, whereas additional stock are introduced to consume excess herbage at peak growing periods. However, the most important method is to conserve forage when production exceeds requirements, and to use this when production is less than requirement (Chapter 26). Hay-making and silage-making are both widely practised in temperate regions. The need for conservation is less in the wet tropical areas, where seasonal variations are slight. Fig. 9.1 compares typical herbage growth rates of temperate pasture with requirements of beef cattle.

### The harvesting of grass by the grazing animal

The feed requirements of beef cattle depend on the size and physiological activity of the animal — that is, whether it is lactating, pregnant or growing.

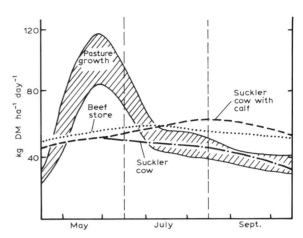

Fig. 9.1. The seasonal variation in herbage requirements of beef cattle in relation to pasture growth (from Holmes, 1982).

Animals will endeavour to satisfy this requirement by grazing sufficient herbage from the pasture. Typically beef cattle consume herbage dry matter to the extent of 1.5 to 2.0% of their body weight daily. A beef animal of 400 kg body weight thus consumes about 8 kg organic matter daily. Daily intake depends on bite size, bite frequency and grazing time. These are all affected by the available herbage mass per unit area and per animal (Fig. 9.2). Cattle can, to a limited extent, compen-

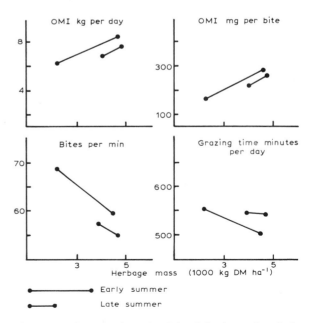

Fig. 9.2. Herbage intake and grazing behaviour of cattle in relation to herbage mass on offer (Zoby and Holmes, 1983).

sate for declining herbage mass by prolonging grazing time, as can sheep (Bircham, 1981; Hodgson, 1982). The provision of supplementary feed may replace pasture and so reduce grazing time. If herbage intake is to be maximized, the animal must be able to take large bites. This occurs when tiller density is high (c. 15 000 tillers $m^{-2}$ for *Lolium perenne*), when the pasture is tall (6–12 cm), when the available herbage is in the range 2.5 to 3.5 t DM $ha^{-1}$, and when the available herbage per animal exceeds 0.5 t. These conditions are difficult to achieve with sparse or stemmy pastures, so short leafy pasture is essential for maximum intake.

Hodgson et al. (1981) and Bircham and Hodgson (1983) have considered the optimal herbage mass, height and leaf area index for maximal herbage production. They found that, for sheep on continuously stocked pasture, net herbage accumulation was maximal at a herbage mass of about 1200 to 1500 kg $ha^{-1}$ then declined slowly to 2000 kg $ha^{-1}$. Recent estimates by the author on cattle pastures, which are normally utilized at higher herbage masses of 2000 to 3000 kg $ha^{-1}$ with continuous stocking, have shown that net herbage accumulation is greater at 2000 than at 3000 kg DM $ha^{-1}$.

## The utilization of grass and other feeds by beef animals

### The diversity of systems

There are two major sources of calves for beef production: (1) beef-breeding herds, where the calf suckles the cow for six to nine months and is then weaned and fed until it reaches slaughter weight; and (2) surplus bull calves from dairy herds, which are usually raised on other farms, often intensively (Vol. 21, Chapter 10). In addition, some beef is produced from "culled" beef and dairy cows beyond their peak of productivity. In the United Kingdom, these categories account for 29, 43 and 24% respectively of the total home-produced beef [MLC (Meat and Livestock Commission), 1983].

Cattle can calve at any time of the year, though cows in well-managed herds calve in groups, either in the spring or in the autumn. Calves born in spring are normally weaned during summer and autumn and fed on conserved grass and other feeds during the following winter (Fig. 9.3). If well fed they may reach finished weight (400–500 kg) at the end of that winter, but are more commonly finished on pasture in their second year. In the latter case, they are on a low plane of nutrition in winter and show compensatory growth in spring and summer. In continental Europe, cattle are often kept several years and killed at 600 to 750 kg weight (Micol and Beranger, 1984).

The calves of dairy cows may follow many routes to beef (Fig. 9.3). Rapid beef-producing systems involve feeding animals on high-quality concentrates (largely cereals and agro-industrial by-products); slower beef-producing systems involve grazing, or feeding on silage and hay; 18-month and 24-month beef systems normally require about 1 t of concentrated feed per animal, in addition to grazing and cut forage.

Cull cows are usually fed on grass or high-quality silage, often with limited supplements of concentrate feeds.

Early-maturing smaller breeds, such as the Angus and Hereford, require only about 90% of the time required by late-maturing and larger breeds, such as Charolais and Simmental, to reach finished condition. Pure dairy breeds tend to be late-maturing when used for beef production, but crosses of beef with dairy breeds may be early- or late-maturing depending on the beef breed used in the cross. In general, early-maturing types of beef animal reach the finished condition more rapidly and can derive a greater proportion of their total nutrient from grazing and forage.

### Suckler systems

In these systems the sole product is the calf, so the number of calves per cow per year and growth rate of the calf are the major components of efficiency. The age at first calving, and the longevity of the cow, will affect the number of replacements required and so affect efficiency (Holmes, 1977). The implications of some of these variables are shown in Fig. 9.4 (see also Chapter 16).

In more extensive conditions, where feed requirement must be more closely related to the natural growth cycle of the grass, calving is usually in early spring. Since the combined feed requirement of cow and calf in the autumn will exceed feed supply, the weaned calves are normally sold for finishing elsewhere.

In more favoured areas, where either fresh or conserved winter forage is available, calving may

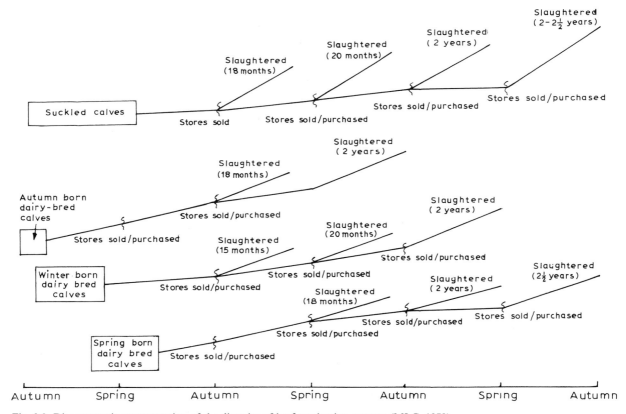

Fig. 9.3. Diagrammatic representation of the diversity of beef production systems (MLC, 1978).

occur in autumn; limited quantities of concentrated feeds may be used during winter, and the calves are weaned in summer or early autumn.

In Britain, autumn-born lowland calves are weaned at about 320 kg while spring-born hill calves are weaned at about 240 kg. Autumn calving gives a greater annual output per cow, albeit at a higher cost.

Efficiency can be greatly increased if cows produce twins or suckle an additional calf. Twinning is rare and, although it has been induced experimentally, this is not a common practice. If an additional calf is produced or fostered, the calf weight produced per hectare can be raised from 450 kg to 800 kg (MLC, 1981) (see also Fig. 9.4).

### Weaned calves

The diet of calves after weaning depends on the season of birth. Spring-born calves, weaned in the autumn, are usually maintained on a diet of conserved forage in the first winter, and then grow rapidly on grass in the second summer to finish at about twenty months of age. The autumn-born calf grows well on grass during the first summer, then normally receives a high-quality diet of forage plus concentrate in the second winter to finish at fifteen months of age.

Although breeds differ in rate of weight gain, the efficiency of feed use is very similar. For example, the following quantities of feed were consumed per kilogram weight gain during winter: dairy breed crossed with Aberdeen Angus, 10.5; with Charolais 11.0; with Hereford 10.2; with Limousin 10.6; with Sussex 10.6 (MLC, 1981). During summer, Angus, Charolais, Hereford and Simmental cross calves all consumed 10.3 kg feed per kilogram gain, while Sussex crosses consumed 9.8 kg. All these differences are probably within the range of experimental error.

### Dairy beef systems

A feature of dairy beef systems is the diversity of possible growth patterns (Fig. 9.3). Calves can be

# BEEF PRODUCTION

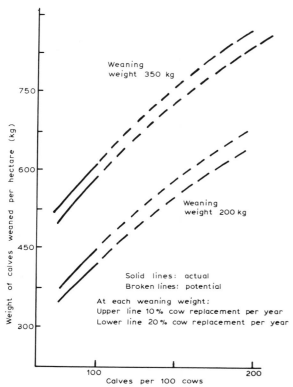

Fig. 9.4. The effect of the number of calves per cow, the weaning weight, and the cow replacement rate on the influencing weight of calves weaned per hectare (when 7.5 t DM ha$^{-1}$ are utilized).

reared for veal, reaching a slaughter weight of 180 kg in four months, or raised extensively, taking four years to reach slaughter condition at weights exceeding 700 kg. However, high interest charges on capital invested have led to the widespread adoption of the more intensive pasture-based systems referred to below.

**18–24-month grass–cereal beef.** Autumn-born calves can be raised on milk and milk substitutes and, if development of the ruminant capacity is encouraged by provision of dry feeds, they can be turned out on pasture at three to six months of age. If high-quality conserved forage is fed in the subsequent winter, they can successfully be brought to slaughter condition at eighteen months (Glover, 1982), but some supplementation, usually with cereal, is normal.

**20–30 month grass-fed beef.** Where the quality of the pasture or conserved forage is lower, cattle grow more slowly, but can still reach slaughter weight in twenty to thirty months, depending on their season of birth and breed. Autumn-born calves of earlier-maturing breeds could finish on grass in their second summer, and spring-born calves and late-maturing breeds could be finished on grass in their third summer. Some typical data for these beef production systems in British conditions are shown in Table 9.1.

TABLE 9.1

The contribution of grass and of concentrate feeds in various beef production systems (based on MLC, 1978)

| | Concentrates consumed (kg per animal) | Proportion of metabolizable energy from grass and forage (%) | Slaughter weight (kg) | Period to slaughter (days) |
|---|---|---|---|---|
| *Dairy-bred beef* | | | | |
| 1. Cereal beef | 1800 | 5 | 395 | 365 |
| 2. 15-month grass–cereal beef | 1200 | 50 | 435 | 455 |
| 3. 18-month grass–cereal beef | 1100 | 60 | 470 | 550 |
| 4. 20-month grass beef | 900 | 70 | 480 | 610 |
| 5. 24-month grass beef | 1000 | 75 | 500 | 715 |
| 6. 36-month grass beef (estimate) | 600 | 90 | 700 | 1100 |
| *Pure-bred beef* | | | | |
| Suckler cow, per year | 250 | 90 | — | — |
| 7. 15-month autumn calves | 900 | 60 | 420 | 455 |
| 8. 24-month spring calves | 700 | 80 | 450 | 715 |

## RESPONSES TO MANAGEMENT

Animal output from pastures is highly dependent on management. Some of the most important factors are considered below.

### Stocking rate

Several authors have considered the implications of stocking rate for production per animal and per hectare (see Ch. 24). The main consequences are illustrated in Fig. 9.5, which shows that, while daily gain per animal declines with increased stocking rate, gain per hectare follows a parabolic curve with a maximum when gain per animal is about half of the maximum (Jones and Sandland, 1974). Much of the progress in grass utilization has resulted from an appreciation of these elements, and the development of grazing systems which approximate to the optimal stocking rate, and avoid overstocking or understocking.

If, for example, the daily growth rate of pasture over a period of time were 80 kg $ha^{-1}$ $day^{-1}$, if the animal could harvest 75% of the herbage produced, and if the daily requirement of suckler beef cows were 12 kg $day^{-1}$, then the number of cows that could be carried would be:

$$\frac{80 \times 0.75}{12} = 5$$

If more cows were carried per hectare, they would be under-nourished, and if fewer were carried there would be under-utilization of the pasture. Under-utilization can be accepted for a period, and allows flexibility in the system but, over a prolonged period, it allows the maturation of the pasture, and so to a decline in its feeding value; in the long term, it leads to the dominance of coarse and less valuable species, and even to the ingress of shrubs. Over-utilization may, on the other hand, depress performance per animal, and depress the growth of the sward; it may result in permanent deterioration of the sward. In unfavourable circumstances it can cause soil erosion.

### The efficiency of grass utilization

The overall efficiency of utilization (%) over a grazing season can be estimated as:

$$\frac{100 \times \text{herbage consumed}}{\text{herbage accumulation}} = \frac{I \times 100}{A - B - C - D + G} \quad (1)$$

where per unit area: $I$ = herbage DM intake during season; $A$ = herbage mass at the beginning of the season; $B$ = herbage mass at the end of the season; $C$ = herbage consumed by non-agricultural fauna; $D$ = losses by decay; and $G$ = the sum of gross herbage accumulation during season.

Leaver (1976) has estimated that, over a grazing season, maximum animal production is achieved if the efficiency of daily utilization is 30 to 40% (Fig. 9.6). Nevertheless, over the grazing season there could be six to twelve defoliations and the overall efficiency of utilization could be as high as 95% of

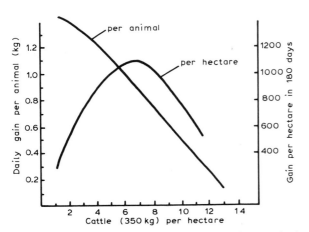

Fig. 9.5. The influence of stocking rate on daily gain per animal and gain per hectare in 180 days (Holmes, 1982).

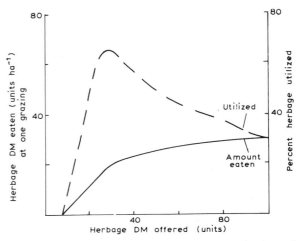

Fig. 9.6. The influence of herbage allowance on intake and percentage utilization of herbage (Leaver, 1976).

the herbage accumulation. Typical efficiencies of overall utilization are from 50 to 75% of herbage growth.

## Grazing systems

Various grazing management systems are practised, ranging from continuous stocking to rotational strip grazing. In strip grazing, cattle are allotted a limited area of pasture, by means of moveable fences, so that the daily herbage allowance is about twice the probable daily herbage consumption. So called "zero grazing" systems are also used where grass is cut daily and fed to housed animals. These systems have been described and their advantages and disadvantages discussed by Holmes (1982).

Many experimental comparisons have now shown that stocking rate is the overriding factor and that, at a given stocking rate, the difference in production per hectare from different grazing systems is small. However, Marsh (1975) concluded that there was an average advantage of 5% in favour of rotational grazing of beef cattle, compared with continuous stocking; the advantage increased at high stocking rates. Rotational systems also offer a degree of flexibility in management not available in continuous grazing systems, unless an additional reserve or buffer area of pasture is provided.

## The response to fertilizers on beef pastures

There has been a growing appreciation of the influence of fertilizers on pasture productivity during the past fifty years. If drainage is satisfactory, and soil acidity has been remedied, applications of nitrogen (Chapter 6), potassium and phosphorus can greatly increase herbage production. Potassium and phosphate are particularly important for grass–legume pastures.

Many experiments on cut swards have shown that grass–clover swards yield 6 to 8 t DM ha$^{-1}$, whereas nitrogen fertilizer may increase yields to 10 to 15 t DM ha$^{-1}$ (Morrison et al., 1980). There have been few comparisons of the response to nitrogen fertilizer when the pasture is grazed by beef cattle. However, Marsh (1975) reported linear responses to fertilizer nitrogen up to an annual application of 600 kg N ha$^{-1}$ and Holmes (1974), reviewing grazing experiments with growing beef cattle, found that

$$G = 740 + 1.807N - 0.00141N^2 \qquad (2)$$

where $G$ = liveweight gain in kg ha$^{-1}$; and $N$ = annual supply of nitrogen in kg ha$^{-1}$; and $r^2 = 0.43$.

## The response of beef cattle on pasture to additional feed

Whereas pasture is the major source of feed for grazing cattle, it is not uncommon to provide additional food supplements. Mineral supplements containing sodium chloride, phosphate, copper sulphate, etc., may improve animal performances, if any of these elements are deficient (Chapter 15). Dramatic improvements as a result of copper supplementation are often recorded.

Cereal supplements are sometimes given to young cattle for a few weeks after they are turned out to grass in spring, and before they are removed from pasture at the end of the grazing season. However, because of high labour costs and the difficulties of individual allocation of feed, these practices are not widespread. Supplements are more often given before slaughter, though experimental studies indicate that such supplementation usually results in only a small increase in liveweight gain, if adequate herbage is available (Table 9.2). In these conditions, supplements reduce herbage intake and grazing time without affecting bite size. However, when pasture is scarce, supplementation results in a greater net effect (Table 10.2). Judicious supplementation might allow a combination of high-level production per animal and per hectare (Beranger and Petit, 1971), but this is difficult to achieve in practice. Systems in which concentrated feeds contribute more than half the total daily feed allowance are feasible, but are only economically viable when such feed is cheap relative to the price of beef, and the cost of calves is low.

It is rare for managed pastures to be deficient in protein content, so provision of protein-rich supplements is seldom beneficial for beef cattle.

## THE BIOLOGICAL EFFICIENCY OF BEEF PRODUCTION SYSTEMS

A consideration of biological efficiency must include two major components: (1) efficiency of

TABLE 9.2

The effect of available herbage (kg DM ha$^{-1}$) and supplements on feed intake and performance in beef production (after Gomez, 1975)

|  | Herbage mass[1] 2900 kg DM ha$^{-1}$ | | | Herbage mass[1] 620 kg DM ha$^{-1}$ | | |
|---|---|---|---|---|---|---|
| Supplement offered (kg day$^{-1}$): | 0.3 | 1.9 | 3.3 | 0.3 | 1.7 | 3.1 |
| Herbage intake (kg DM day$^{-1}$) | 6.1 | 5.5 | 4.2 | 4.9 | 4.2 | 3.4 |
| Total intake (kg DM day$^{-1}$) | 6.4 | 7.4 | 7.5 | 5.2 | 5.9 | 6.5 |
| Daily liveweight gain (kg) | 1.1 | 1.3 | 1.2 | 0.25 | 0.58 | 0.89 |

[1]Livestock were managed so as to maintain herbage mass constant.

utilization — that is, the proportion of the primary production which is consumed by the animal; and (2) efficiency of conversion — that is, the efficiency with which the animal, or animal system, processes the feed consumed into a product useful to man.

**Efficiency of utilization**

The mean utilization of lowland grassland in Britain is about 60% (Lazenby, 1981). Forbes et al. (1980) compared beef- and milk-producing farms in Britain and showed that the average utilization of metabolizable energy from herbage on beef farms (39 GJ ha$^{-1}$, range 15–69) was less than on dairy farms (44 GJ ha$^{-1}$, range 21–82). This was attributed to a combination of lower herbage yields, less efficient harvesting, poorer conservation and less efficient utilization within the animal on the beef farms. However, Hopkins (1983) concluded from a survey of individual fields that herbage utilization was higher (75%) on beef farms than on dairy farms (65%), because beef pastures were grazed longer during the winter period and because dairy cows often received supplementary feed which reduced consumption of herbage. Targets have been proposed, by Holmes (1968, 1982) and Kilkenny et al. (1978), to encourage greater productivity from beef pastures.

**Efficiency of conversion**

The efficiency of conversion depends upon the system of production that is used and the standard of management. Assessments have been made by Leitch and Godden (1953), Large (1970), Holmes (1970, 1977, 1980), Fitzhugh (1978), Bywater and Baldwin (1980) and Spedding et al. (1981), and a reassessment is given in Table 9.3.

The most outstanding features of these studies are: (1) the relative inefficiency of the cow–calf systems, irrespective of the performance of the calf, mainly because of the high cost in animal feed of maintaining the cow; (2) the importance of twinning or fostering additional calves, which increases efficiency by about 50%; (3) systems in which females breed once and are then slaughtered as young cow beef are more efficient than normal cow–calf systems; (4) in Britain and much of Europe, most beef cattle are derived as the by-products of dairy herds, where the cost of the dam is largely borne by milk production. The biological efficiency of such milk/beef systems is much higher than that of cow–calf beef systems.

MODELS OF BEEF SYSTEMS

Quantitative studies of beef production and increasing understanding of the factors determining output have led to greater interest in modelling beef systems (see also Chapter 5). Fitzhugh (1978) compared cow–calf systems with dairying systems, and Glover (1982) has developed a model to deal with the growing beef animal. It is now possible to describe in some detail the many components of beef production systems and to predict, both in biological and in economic terms, the probable outcome of management strategies. However, because variations in the weather from season to season leads to large variations in pasture production, beef producers tend to take a rather conservative attitude to attainment of optimal stocking rate

TABLE 9.3

A comparison of systems of beef producing, showing efficiency of feed conversion and edible output per hectare

|  | Total feed input | | Total edible output | | Efficiency of conversion (%) | | | Edible output from 10 t grass DM ha$^{-1}$ | |
| --- | --- | --- | --- | --- | --- | --- | --- | --- | --- |
|  | metabolic energy (GJ) | crude protein (kg) | energy (GJ) | protein (kg) | metabolizable energy | gross energy | crude protein | energy (GJ) | protein (kg) |
| *Suckler cow systems* | | | | | | | | | |
| 1. Rear beef heifer to 2 years and slaughter | 32 | 320 | 2.7 | 33 | 8.0 | 4.4 | 10.3 | 8.1 | 103 |
| 2. Rear beef heifer, calve at 2 years and slaughter cow at 3 years and calf at 15 months | 86 | 860 | 5.1 | 63 | 6.0 | 3.3 | 7.4 | 6.1 | 74 |
| 3. Rear beef cow, rear 4 calves in 4 years to 15 months slaughter cow | 246 | 2460 | 12.3 | 153 | 5.0 | 2.7 | 6.2 | 5.0 | 62 |
| 4. Rear beef cow, rear 8 calves in 9 years to 15 months slaughter cow | 587 | 5875 | 21.9 | 273 | 3.7 | 2.0 | 4.6 | 3.7 | 46 |
| 5. Rear beef cow, rear 8 calves in 9 years to 24 months slaughter cow | 616 | 6160 | 26.7 | 313 | 4.3 | 2.3 | 5.1 | 4.2 | 51 |
| *Dairy beef systems* | | | | | | | | | |
| 6. 15-month grass cereal | 26 | 260 | 2.4 | 31 | 9.2 | 5.6 | 11.7 | 10.4 | 117 |
| 7. 18-month grass cereal | 33 | 330 | 2.8 | 35 | 8.5 | 5.0 | 10.6 | 9.2 | 106 |
| 8. 24-month grass | 42 | 420 | 3.0 | 35 | 7.1 | 4.1 | 8.3 | 7.6 | 83 |
| 9. 36-month grass | 72 | 720 | 4.7 | 51 | 6.6 | 3.5 | 7.1 | 6.5 | 71 |
| 10. Milk cow, 4 lactations of 5000 kg | 252 | 2520 | 55 | 673 | 22.0 | 13.2 | 27.0 | 24.4 | 270 |

It is assumed that the whole diet contains 100 g crude protein per kg DM, a value which is adequate for beef cattle and likely to be exceeded by forage diets. Diets are calculated primarily to meet energy requirements (Ministry of Agriculture, Fisheries and Food, 1975).

(Chapter 24) and to pasture utilization, so predicted optima are seldom achieved in practice.

**REFERENCES**

Allen, D.M., Bougler, J., Christensen, L.G., Jongeling, C., Petersen, P.H. and Servanti, P., 1982. Cattle. *Livestock Prod. Sci.*, 9: 89–126.

Anslow, R.C., 1967. Frequency of cutting and sward production. *J. Agric. Sci., Cambridge*, 68: 377–384.

Beranger, C. and Petit, M., 1971. Production des jeunes bovins de boucherie à partir de l'herbe. In: *Production de viande par les jeunes bovins. SEI Etude*, No. 46: 279–292.

Bircham, J.S., 1981. *Herbage Growth and Utilization under Continuous Stocking Management*. Thesis, University of Edinburgh, Edinburgh, 250 pp.

Bircham, J.S. and Hodgson, J., 1983. The influence of sward condition on rates of herbage growth and senescence in mixed swards under continuous stocking management. *Grass Forage Sci.*, 38: 323–331.

Bywater, A.C. and Baldwin, R.L., 1980. Alternative strategies in food production. In: R.L. Baldwin (Editor), *Animals, Food and People*, AAAS Selected Symposium, Westview Press, Boulder, Colo., pp. 1–30.

Fitzhugh, H.A., 1978. Bioeconomic analyses of ruminant production systems. *J. Anim. Sci.*, 46: 797–806.

Forbes, T.J., Dibb, C., Green, J.O., Hopkins, A. and Peel, S., 1980. *Factors Affecting the Productivity of Permanent Grassland*. Joint Permanent Pasture Group, Hurley, 141 pp.

Glover, E.H., 1982. *Some Practical and Theoretical Aspects of Intensive Dairy Beef Systems*. Thesis, University of London, London, 286 pp.

Glover, E.H., 1984. All grass beef. In: W. Holmes (Editor), *Grassland Beef Production*. Martinus Nijhof, The Hague, pp. 82–96.

Gomez, P.O., 1975. *The Effects of Pasture Availability and Supplementary Feeding on Beef Production from Pastures*. Thesis, University of London, London, 318 pp.

Hodgson, J., Bircham, J.S., Grant Sheila, A. and King, J., 1981. The influence of cutting and grazing management on herbage growth and utilization. In: C.E Wright (Editor), *Plant Physiology and Herbage Production. Occasional Symposium No. 13*. British Grassland Society, Hurley, pp. 51–62.

Hodgson, J., 1982. Ingestive behaviour. In: J.D. Leaver (Editor), *Herbage Intake Handbook*. British Grassland Society, Hurley, pp. 113–138.

Holmes, W., 1968. The use of nitrogen in the management of pasture for cattle. *Herbage Abstr.*, 38: 265–277.

Holmes, W., 1970. Animals for food. *Proc. Nutrit. Soc.*, 29: 237–244.

Holmes, W., 1974. The role of nitrogen fertilizer in the production of beef from grass. In: *Proceedings No. 142*, The Fertilizer Society, London, pp. 57–69.

Holmes, W., 1977. Choosing between animals. *Philos. Trans. R. Soc. Lond., B*, 281: 121–137.

Holmes, W., 1980. Secondary production from land. In: K.L. Blaxter (Editor), *Food Chains and Human Nutrition*. Applied Science, London, pp. 109–134.

Holmes, W., 1981. Cattle. *Biologist*, 28: 273–279.

Holmes, W. (Editor), 1982. *Grass — Its Production and Utilisation*. Blackwell, Oxford, 295 pp.

Hopkins, A., 1983. Efficiency of utilization: an analysis of grazing data from recorded farms in England and Wales. In: A.J. Corrall (Editor), *Efficient Grassland Farming. Occasional Symposium No. 14*. British Grassland Society, Hurley, pp. 161–164.

Hungate, R.E., 1966. *The Rumen and Its Microbes*. Academic Press, New York, N.Y., 533 pp.

Jones, R.J. and Sandland, R.L., 1974. The relation between animal gain and stocking rate. *J. Agric. Sci., Cambridge*, 83: 335–342.

Kilkenny, J.B., Holmes, W., Baker, R.D., Walsh, A. and Shaw, P.G., 1978. *Grazing Management*. Meat and Livestock Commission, Milton Keynes, 44 pp.

Large, R.V., 1970. The biological efficiency of meat production in sheep. *Anim. Prod.*, 12: 393–401.

Lazenby, A. 1981. British Grasslands, Past, Present and Future. *Grass Forage Sci.*, 36: 243–266.

Leaver, J.D., 1976. Utilisation of grassland by dairy cows. In: H. Swan and W.H. Broster (Editors), *Principles of Cattle Production*. Butterworths, London, pp. 307–327.

Leitch, I. and Godden, W., 1953. *The Efficiency of Farm Animals in the Conversion of Feeding Stuffs to Food for Man*. Commonwealth Bureau of Animal Nutrition, Tech. Comm. No. 14. Commonwealth Agricultural Bureaux, Farnham Royal, 77 pp.

MAFF (Ministry of Agriculture, Fisheries and Food), 1975. *Energy Allowances and Feeding Systems for Ruminants*. MAFF Technical Bulletin, No. 33. HMSO, London, 79 pp.

Marsh, R., 1975. Systems of grazing management for beef cattle. In: J. Hodgson and D.K. Jackson (Editors), *Pasture Utilisation by the Grazing Animal. Occasional Symposium No. 8*, British Grassland Society, Hurley, pp. 119–128.

Micol, D. and Beranger, C., 1984. French beef production systems from grassland. In: W. Holmes (Editor), *Grassland Beef Production*. Martinus Nijhof, The Hague, pp. 11–22.

MLC (Meat and Livestock Commission), 1978. *Beef Improvement Services. Data Summaries on Beef Production and Breeding*. Revised April 1978. MLC, Milton Keynes, 90 pp.

MLC (Meat and Livestock Commission), 1981. *Commercial Beef Production Yearbook 1980–81*. MLC, Milton Keynes, 99 pp.

MLC (Meat and Livestock Commission), 1983. *Beef Yearbook*. MLC, Milton Keynes, 106 pp.

Morrison, J., Jackson, M.V. and Sparrow, P.E., 1980. *The Response of Perennial Ryegrass to Fertiliser Nitrogen in Relation to Climate and Soil*. Tech. Report No. 27. Grassland Research Institute, Hurley, Maidenhead, 71 pp.

Osbourn, D.F., 1982. In: W. Holmes (Editor), *Grass — Its Production and Utilisation*. Blackwell, Oxford, pp. 70–124.

Spedding, C.R.W., Walsingham, J.M. and Hoxey, A.M., 1981. *Biological Efficiency in Agriculture*. Academic Press, London, 383 pp.

Vicente-Chandler, J., Caro-Costas, R., Pearson, R.W., Abruna, F., Figarella, J. and Silva, S., 1964. *The Intensive Management of Tropical Forages in Puerto Rico. Univ. Puerto Rico Agric. Exper. Stn. Bull.*, 187: 152 pp.

Wilkins, R.J. (Editor), 1977. *Green Crop Fractionation. Occasional Symposium No. 9*. British Grassland Society, Hurley, 189 pp.

Winrock International, 1978. *The Role of Ruminants in Support of Man*. Winrock International Livestock Research and Training Center, Petit Jean Mountain, Ark., 136 pp.

Zoby, J.L.F., 1981. *Factors Affecting Herbage Intake and Grazing Behaviour in Cattle*. Thesis, University of London, London, 306 pp.

Zoby, J.L.F. and Holmes, W., 1983. The influence of size of animal and stocking rate on the herbage intake and grazing behaviour of cattle. *J. Agric. Sci., Cambridge*, 100: 139–148.

Chapter 10

# MILK PRODUCTION FROM MANAGED GRASSLANDS

C.W. HOLMES

## INTRODUCTION

Milk production from grassland is mainly determined by three factors: (1) the quantity, quality and seasonal distribution of the feed grown; (2) the proportion of the feed grown which is actually consumed by the cattle; and (3) the efficiency with which the cattle convert the consumed feed into milk (McMeekan, 1960).

The present review discusses these essential components and the factors which influence them. Examples will be taken from New Zealand, Australia, Britain and Ireland, where grazed pasture is either the main, and sometimes the only component of the cattle's diet. Milk production under more intensive conditions is considered in Chapter 9 of Volume 21.

A simplistic framework for some of the discussion to follow is provided by comparing the annual and monthly herbage production with the annual and monthly feed requirements of a cow (Fig. 10.1). From these data it can be calculated that 3.3 cows ha$^{-1}$ can be grazed (if no herbage is wasted), producing 560 kg milkfat ha$^{-1}$ and about 12 000 l milk ha$^{-1}$ annually. Although the total quantity of feed grown annually is sufficient to feed 3.3 cows ha$^{-1}$, this conceals temporary deficiencies and surpluses of feed (Fig. 10.1). The size and the timing of the deficits and surpluses are affected by the date of calving and the date of drying-off, and by stocking rate (Fig. 10.1). Management practices on grassland dairy farms must equate feed required with feed available, and cope with feed surpluses and deficits in practical and profitable ways.

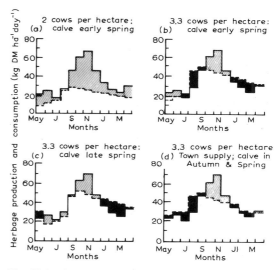

Fig. 10.1. The quantities of herbage grown (——) and eaten (------) in each month of the year; surpluses (hatched) and deficits (black) of herbage are shown for two stocking rates and different calving dates. The data assume 13 t DM ha$^{-1}$ grown annually; 3.9 t DM cow$^{-1}$ eaten annually. (From Massey University and Holmes et al., 1981.)

## PRODUCTIVITY OF GRASSLAND DAIRY FARMS

### Herbage production

#### Annual production

Dairy pastures containing ryegrass (*Lolium perenne*) and clover (*Trifolium repens*), and receiving no nitrogenous fertilizer, can produce about 12 to 16 t DM ha$^{-1}$ in New Zealand (Holmes, 1982; Bryant et al., 1982) and 6 to 7 t DM ha$^{-1}$ in Britain and Ireland (Gordon, 1979; MacCarthy, 1980; Williams, 1980). Nitrogenous fertilizers can increase herbage production of grass–legume pastures by 10 to 15 kg DM kg N$^{-1}$, and are used intensively in Britain and Ireland to achieve greater herbage production (Chapter 6).

The amounts of herbage produced under experimental conditions are well below the potential yields of 23 and 27 t DM ha$^{-1}$, estimated by Brougham (1959) and Williams (1980) for New Zealand and England respectively, for pastures unrestricted by lack of nutrients or water. On the other hand, they are usually greater than the yields actually produced on many dairy farms, because of variations in factors such as climate, soils, fertilizer use and grazing management.

### Seasonal distribution

In many areas of New Zealand, considerable herbage production occurs during winter (Chapter 2). This is in marked contrast to the situation in Britain and Ireland, where virtually no growth occurs from November to February. These differences have important effects on the dairy farming methods which can be adopted.

### Quality

The quality of the herbage has major effects on animal productivity (Hodgson, 1977). Digestibility is usually highest (77%) in spring, but lower in summer (64%) (Hutton, 1962). However, herbage quality is also influenced by grazing management; for example, pastures which had previously been grazed intensely had a digestibility of 73% and contained 45% leaf in late spring, whereas other pastures which had previously been laxly grazed had a digestibility of 67% and contained 29% leaf (C. Hoogendoorn, Massey University, unpubl.). Similar differences in digestibility have also been shown in late winter (Santamaria and McGowan, 1982).

## Milk production

### Theoretical and experimental results

Annual herbage production from good dairy pasture in New Zealand is likely to be about 13 t DM ha$^{-1}$ with the application of very little nitrogen; similar yields can be obtained with applications of 300 to 400 kg N ha$^{-1}$ in Ireland and Britain (see above). If 90% of this herbage is eaten, then the quantities of milk or milkfat (MF) which could theoretically be produced annually range from 7000 to 15 000 l milk ha$^{-1}$ or 400 to 700 kg MF ha$^{-1}$, depending on the feed conversion efficiency of the cows (Hutton, 1963; Leaver, 1976).

Similar levels of milk production have actually been achieved in various experiments in New Zealand (450 to 700 kg MF ha$^{-1}$, at stocking rates of 3.5 to 5.0 cows ha$^{-1}$; Bryant, 1982a; data cited by Holmes and MacMillan, 1982) and in Britain and Ireland (11 000 to 16 500 l milk ha$^{-1}$ at stocking rates of 3.9 to 7.4 cows ha$^{-1}$; McPheely et al., 1975; Castle and Watson, 1975; Gordon, 1979). They have also been achieved by intensively managed herds at

TABLE 10.1

Annual milk production achieved by intensive spring-calving herds at research institutes in three countries

|  | New Zealand[1] | Northern Ireland[2] | Eire[3] |
|---|---|---|---|
| Stocking rate (cows ha$^{-1}$) | 3.5[4] | 2.9 | 3.4 |
| Milk produced (l) |  |  |  |
|   per cow | (3000)[5] | 5500 | 4120 |
|   per hectare | (10 400)[5] | 15 960 | 13 990 |
| Milkfat produced (kg) |  |  |  |
|   per cow | 157 | (220)[5] | (160)[5] |
|   per hectare | 543 | (620)[5] | (540)[5] |
| Concentrates fed (t cow$^{-1}$) | 0 | 0.50 | 0.58 |
| Metabolizable energy derived from forage (% of total) | 100 | 89 | 86 |
| Nitrogen applied (kg ha$^{-1}$) | 0 | 390 | 270 |

Annual herbage production approximately 12 to 13 t DM ha$^{-1}$.
[1]Campbell et al. (1977). [2]Gordon (1979, 1981). [3]MacCarthy (1980). [4]Replacement heifers were also carried. [5]Calculated.

TABLE 10.2

Annual milk production achieved by commercial dairy herds in four countries

|  | New Zealand[1] | Eire[2] | Northern Ireland[3] | England[2] |
|---|---|---|---|---|
| Stocking rate (cows ha$^{-1}$) | 2.3 | 2.37 | 2.35 | 2.0 |
| Milk produced (l) |  |  |  |  |
| per cow | (3190)[4] | 4430 | 4910 | 5150 |
| per hectare | (7340)[4] | 10 538 | 11 530 | 10 250 |
| Milkfat produced (kg) |  |  |  |  |
| per cow | 150 | (175)[4] | (196)[4] | (200)[4] |
| per hectare | 345 | (411)[4] | (450)[4] | 400 |
| Concentrates fed (t cow$^{-1}$) | 0 | 0.43 | 1.1 | 1.7 |
| Metabolizable energy derived from forage (% of total) | 100 | 78 | 76 | 63 |
| Nitrogen applied (kg ha$^{-1}$) | 19 | 170 | not given | 250 |

[1]Anonymous (1981–1982); Crabbe (1983). [2]Hughes (1979). [3]Gordon (1981). [4]Calculated.

research institutes in Ireland and New Zealand; no concentrates or nitrogenous fertilizer were used on the New Zealand farm (Table 10.1).

### Commercial farms

Data about farm productivity in four countries are shown in Table 10.2. When allowance is made for concentrates fed in each country, the data indicate that the New Zealand and Irish farms produced larger quantities of milkfat from forage than the English farms.

The differences in production between research farms and commercial farms (Tables 10.1 and 10.2) suggest that there is considerable scope for increases in commercial farm productivity. This is also shown by the fact that the top 25% of farms in the surveys in Northern Ireland and New Zealand achieved levels of production which were 26% higher than the average for all farms (Gordon, 1981; Crabbe, 1983). However, the effects of variations between farms in soils and climate are not considered in these simple comparisons of farm productivity.

## FACTORS AFFECTING PRODUCTIVITY

### Stocking rate

"... no more powerful force for good and for evil exists than control of the stocking rate in grassland farming. Properly used, it can influence productive efficiency for good more than any other single controllable factor" (McMeekan, 1960). This conclusion was supported by the results of six earlier experiments which showed increases in milk production per hectare of between 17 and 43% caused by increases in stocking rate. Many recent experiments and surveys of commercial farms have also confirmed this statement (Holmes and MacMillan, 1982; and Chapter 24).

The stocking rates for grazing experiments in New Zealand refer to the number of cows grazed, lactating or dry, throughout the year, whereas corresponding data for Britain and Ireland usually refer to a shorter grazing period of 150 to 180 days with lactating cows. In most experiments only cows are grazed, whereas many commercial farms also carry replacement heifers, a practice which reduces milk production per hectare and makes it more difficult to calculate an "effective" stocking rate. For these reasons and others, such as differences between breed of cow, a particular value cited for stocking rate is not necessarily equivalent in all circumstances.

### Growth and composition of pastures

Very low and very high stocking rates can almost certainly depress herbage production, because of the effects of undergrazing and overgrazing on rates of photosynthesis and rates of senescence (Hodgson and Maxwell, 1981). The depression caused by very high stocking rates is shown by a number of experiments with dairy cows (Hutton, 1975; McPheely et al., 1977; Stockdale and King, 1980). High stocking rates can also cause changes in pasture composition, such as a decreased proportion of dead material, an increased concentration of crude protein, an increased digestibility, an increased proportion of clover, and a decreased proportion of erect grasses (such as cocksfoot, *Dactylis glomerata*); these changes generally increase the feeding value of the herbage (Holmes and MacMillan, 1982).

### Annual harvesting efficiency (herbage utilization)

This may be defined as the ratio:

$$\left[\frac{\text{pasture consumed per hectare annually}}{\text{pasture grown per hectare annually}}\right]$$

The most important single effect of increased stocking rate is to increase this ratio, with a resultant decrease in the wastage of pasture through death and decay. Campbell (1966b) calculated herbage utilizations of 0.89 to 1.00, but concluded that these estimates were unreliable because of technical difficulties with the methods. Consequently, although intensively stocked farms may achieve high values for this ratio, claims that it is often as high as 0.95 (Leaver, 1976) must be treated with caution. Values of 0.4 to 0.7 were estimated for three British farms, and 0.8 to 0.9 for intensively stocked New Zealand farms (C.W. Holmes, unpubl.).

### Production per cow and per hectare

An increase in stocking rate is likely to cause a decrease in food intake and milk production per cow, and a decrease in food conversion efficiency by the cow. However, it is also likely to cause an increase in the quantity of feed harvested per hectare, and thus to increase both the total quantity of milk produced per hectare and the total efficiency with which the available herbage is converted into milk (Wallace, 1959).

Many experiments in different countries have shown increases in productivity per hectare with increases in stocking rate (Browne, 1972; Gordon, 1973; McPheely et al., 1977; Holmes and MacMillan, 1982; Journet and Demarquilly, 1979). New Zealand data indicate that an increase of one cow per hectare (over the range 2 to 5 cows ha$^{-1}$) caused a decrease of 18 kg milkfat produced per cow, but an increase of 70 kg milkfat produced per hectare (Fig. 10.2). From this it could be calculated that the maximum level of milkfat production per

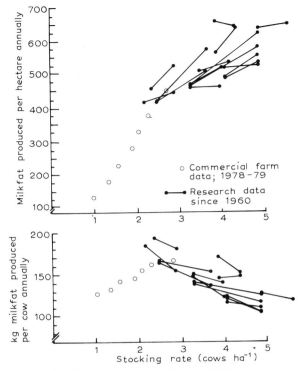

Fig. 10.2. Stocking rates and quantities of milkfat produced per cow and per hectare on research farmlets and commercial farms in New Zealand (Holmes and MacMillan, 1982). Each line joins data from one experiment.

hectare would be achieved at a stocking rate of 5.4 cows ha$^{-1}$, when the level of production per cow would be depressed by about 50 kg milkfat, or 30% (Holmes and MacMillan, 1982); if replacements were also carried, the stocking rate of milking cows would be 4.5 ha$^{-1}$ (Wright and Pringle, 1983).

McMeekan and Walshe (1963) suggested that milk production per hectare would be maximized at a stocking rate which depressed production per cow by 11%. On the other hand, Gordon (1979) concluded that a decrease in annual milk production per cow of 280 l (or 11%) was too great to be practically acceptable in Northern Ireland. The conflict between production per cow and per hectare remains a dilemma associated with the choice of stocking rate for a dairy farm.

High stocking rates often cause decreases in the liveweight of cows (McMeekan and Walshe, 1963). These effects may not be sustainable in the long term, particularly if they also occur in replacement heifers.

## Farm profitability

Surveys of farms in New Zealand (Holmes and MacMillan, 1982) and England (Craven and Kilkenny, 1979) show that farms with higher stocking rates obtain higher gross margins per hectare. However, in New Zealand the farms with higher stocking rates produced more milkfat per cow as well as per hectare (Fig. 10.2), which suggests that the pastures on these farms were probably more productive than those on farms with lower stocking rates.

## Replacement heifers

If replacement heifers are grazed on the dairy farm, they are likely to compete for feed with the milking herd. For example, at an annual replacement rate of 20%, the feed required by the replacement heifers is equivalent to about 13% of that required by the cows. In view of this it may be profitable for a dairy farmer to graze the heifers elsewhere, and so carry more cows on the main farm (Gartner, 1982).

## Calving date

If cows are to rely on grazed pasture, then lactation must be made to coincide as closely as possible with the period of rapid pasture growth (Fig. 10.1). For example, more than 90% of cows in New Zealand calve in early spring and are usually dried off about nine months later, although they may be dried off prematurely in dry summers.

If the period of rapid pasture growth is only six months, then the cows may calve two or three months before pasture becomes available, and be fed indoors on silage, hay and concentrates, until they can be turned out to graze (Gordon, 1979). In England, cows which calved in January produced 10 to 14% more than those which calved in March to July, and 3 to 5% more than those which calved in autumn (Wood, 1970).

The effect of calving date on milk production in New Zealand's seasonal dairy industry is illustrated by data from the southern part of the North Island (Table 10.3). The cows which calved too early, before the start of rapid pasture growth in spring, had low total milk yields per lactation, because of low average daily yields, despite longer lactations. On the other hand, cows which calved too late had low total yields, because of their shorter lactations, despite relatively high average daily yields. All cows were dried off in autumn prior to the slower pasture growth of winter.

In addition to the effect of pasture growth in spring, the choice of calving date is also likely to be influenced by the reliability of pasture growth in summer. For example, in areas where summer rainfall and pasture growth is variable between years, it may be profitable for the cows to calve relatively early in spring. Fortunately, some flexibility about the choice of calving date is possible; for example Bryant (1982a) found only small differences in total productivity between spring-calving herds with mean calving dates of 21 July or 14 August, despite marked differences in daily milk production in early lactation.

The requirement for fresh milk is almost constant throughout the year in most countries. Approximately 5% of cows in New Zealand, and a much larger proportion of cows in Britain, calve in

TABLE 10.3

Effects of calving date on milk production per cow (Anonymous, 1951)

| Month of calving | Percentage of cows calving (%) | Milkfat produced in lactation (kg cow$^{-1}$) | Duration of lactation (day) | Average daily yield of milkfat (kg cow$^{-1}$) |
|---|---|---|---|---|
| June | 3 | 137 | 268 | 0.51 |
| July | 17 | 146 | 273 | 0.53 |
| August | 43 | 152 | 260 | 0.58 |
| September | 18 | 143 | 237 | 0.60 |
| October | 8 | 129 | 213 | 0.61 |
| November | 3 | 114 | 192 | 0.59 |
| December | 1 | 97 | 169 | 0.57 |

autumn in order to take advantage of the higher milk price during winter. However, there is probably little difference in profitability between spring and autumn calving herds (Gleeson, 1973; Nix, 1980). Even when most of the milk is used in manufacturing, the maintenance of a constant daily supply of milk throughout the year would still confer advantages, because of the more efficient use of manufacturing facilities. It will be necessary to offer financial incentives to farmers in order that the cows should calve at times other than spring (Paul and Benseman, 1983). The profitability of all grassland dairy farms is therefore very dependent on effective mating management and cow fertility (MacMillan and Moller, 1976; Esslemont, 1979).

**Drying-off date**

Bryant (1980) showed that cows which were dried off five weeks early produced 10 kg less milkfat; gained 26 kg more live weight and grazed less intensely than cows dried off later.

The immediate decrease in milk production must be weighed against the probable future advantages due to savings in live weight and feed. For example, heifers dried off four weeks early produced 320 l milk less in their first lactation, but 270 l milk more in their second lactation than heifers dried off four weeks later (Gordon, 1979). However, there is no information about the effects of changes in drying-off date on the long-term productivity of grassland dairy farms.

**Grazing management**

Effective grazing management on the dairy farm must ensure that large quantities of high-quality herbage are grown and consumed each year, and that the herbage is rationed and allocated throughout the year so that the stock are fed most efficiently.

Interactions between grazing animals and grazed pastures are complex and important (Hodgson and Maxwell, 1981); in particular, the interaction between the animals' level of feeding and the intensity of grazing must be considered (Hodgson, 1977; Le Du et al., 1979, 1981; data cited by Holmes and MacMillan, 1982).

Many different systems of grazing management have been studied over grazing seasons of 150 to 180 days in Britain. These include variations in rotational or continuous grazing, integration of conservation and grazing, leader and follower cows, and zero grazing, but the effects have usually been small or inconsistent (Leaver, 1976; Journet and Demarquilly, 1979). However, larger effects have been measured in New Zealand and Ireland where cows can be grazed for most, if not all, of the year (McPheely et al., 1977; Bryant, 1981).

**Grazing intensity**

An important dilemma of grazing management is associated with the need to feed cows generously in spring, without allowing the feeding value and growth rate of pastures to deteriorate as a result of the accumulation of reproductive growth and dead material (Hodgson and Maxwell, 1981). Grazing the pastures intensely on every second or third grazing cycle did not increase milk production consistently (Gordon, 1979), whereas mechanical "topping" of the pastures (mowing after grazing) did increase milk production per cow (Bryant, 1982b). However, Bryant's data suggested that the treatment which combined a higher stocking rate with conservation of surplus pasture as silage and hay might have been the most productive in the long term.

In New Zealand, non-lactating cows are grazed on pasture throughout the winter, and are often restricted to a level of feeding close to their maintenance requirements, in order to save herbage until springtime. Their feed intake is usually limited by grazing at high stocking densities (300 to 500 cows ha$^{-1}$ for 24 h) and grazing very intensely (residual herbage mass, 400 to 600 kg DM ha$^{-1}$), which can cause treading damage to the soil or plants (Brown and Evans, 1973) and overgrazing of the plants. The feed intake of cows can also be restricted during winter by limiting the amount of time for which they are allowed to graze, and then moving them onto a "loafing pad". This management increased pasture regrowth rates (Holmes, unpubl. data) and also increased milk production and fertility of the herd in the subsequent spring (McQueen, 1970).

**Grazing rotation**

Several experiments have shown increased milk production from rotational grazing, compared with set-stocking (McMeekan, 1960; McMeekan and Walshe, 1963; Walshe, 1971; Castle and Watson, 1975); when the same stocking rate was used for both treatments, the difference ranged from 8 to 13% in favour of rotational grazing. However, it appeared that rotational grazing enabled stocking rate to be increased, and a combination of rotational grazing with a higher stocking rate produced 16 to 20% more milk than set-stocking (McMeekan and Walshe, 1963; Walshe, 1971). Although there may be no consistent increases in herbage production due to rotational grazing, it allows available herbage to be utilized more effectively in spring (Campbell, 1966a), and it may allow surplus herbage to be recognized more easily (Leaver, 1976). At least one experiment showed that rotational grazing did not increase milk production (Hood, 1974), and some high-producing farms in New Zealand are set-stocked during spring and summer, although it is grazed rotationally for the rest of the year (Simmonds, 1978). Set-stocking during winter, even in New Zealand, is usually considered an invitation to disaster (Hutton, 1966), though use of a long rotation during winter can increase milk production in the subsequent spring, apparently because more herbage is available in early lactation (Bryant and Cook, 1980). However, variations in rotation length during spring and summer have little effect on milk production (Bryant and Parker, 1971; McPheely et al., 1975, 1977).

### Increased herbage production

Milk production per hectare can potentially be increased by increasing herbage production per hectare, but the increase in milk production will be achieved only if at least some of the extra herbage grown is also eaten and converted into milk; this usually requires an associated increase in stocking rate. For example, although nitrogenous fertilizers usually increase herbage production (Chapter 6), nevertheless, marked increases in milk production have been measured only at higher stocking rates (Table 10.4).

The gross financial margin of dairy farms in England is greater with greater use of nitrogen, up to 178 kg N ha$^{-1}$, but with little further increase in gross margin up to 330 kg N ha$^{-1}$ (Aimes, 1980). This is in general agreement with the data in Table 9.4, in which only the response reported by Browne (1972) to 240 kg N ha$^{-1}$ was clearly profitable whereas Gordon (1973) reported that the response to the highest level of nitrogen which he used decreased progressively over a three-year period. Many dairy farmers in New Zealand apply nitrogen in autumn and early spring (average 19 kg N ha$^{-1}$ annually; Anonymous, 1981–82) despite the lack of experimental evidence that nitrogen can produce profitable responses in that country (Bryant et al., 1982; Holmes, 1982).

The application of phosphorus, potassium or lime to volcanic soils in New Zealand caused increases in milk production at higher stocking rates but not at lower ones (Smith, 1964; Anonymous, 1971; Thompson, 1982).

The use of irrigation in New Zealand increases annual herbage production by about 2 to 4 t DM ha$^{-1}$ and milk production by about 80 kg milkfat ha$^{-1}$, at a stocking rate of 4.9 cows ha$^{-1}$ (Hutton, 1975; Holmes and Halford, 1976).

Plant species or cultivars differ in annual and

TABLE 10.4

Effects of nitrogen fertilizer on milk production

| Source | Levels of nitrogen compared (kg N ha$^{-1}$) | Approximate increase in annual milk production per hectare due to the higher level of N, at high stocking rates |
|---|---|---|
| Holmes (1982) | 0, 400 | 100 kg milkfat |
| Bryant et al. (1982) | 0, 110 | 25 kg milkfat |
| Gordon (1973) | 400, 700 | 1160 l milk |
| Browne (1972) | 0, 240 | 2500 l milk |
| McPheely et al. (1977) | 250, 400 | 1000 l milk |
| King and Stockdale (1980) | 0, 224 | 0 |

seasonal herbage production (Chapter 8) and it may be possible to exploit these differences to increase milk production (Mace, 1979; MacCarthy, 1980). In particular, the feeding value of *Trifolium repens* is higher than that of *Lolium perenne* (Rogers et al., 1982), although it may cause bloat in some conditions. Some newer cultivars of grass also have higher feeding values than older cultivars (Wilson and McDowell, 1966; Brookes, 1983).

**Supplementary feeds**

Variations in herbage production between seasons and between years give rise to periods when herbage production does not satisfy feed requirements (Fig. 10.1). The effects of these deficits can be offset, to some extent, by the use of body reserves, and by the use of herbage "stored" *in situ*. However, it may also be necessary to offer feeds other than fresh pasture to supplement the herd's diet, and some of these other feeds, such as silage and hay, are made from surplus herbage at other times of the year (Chapter 26). It may not always be profitable to feed supplements during periods of herbage deficit, even if milk production is reduced by the deficit (e.g., during dry summers: Fig. 10.3).

Based on data cited by Holmes et al. (1981), the consumption of an additional one kilogram of dry matter (11 MJ of metabolizable energy) from a supplementary feed should *theoretically* cause: (a) the production of an extra 2 l of milk, or 90 g milkfat, or (b) an increase of live weight by 300 g, or (c) a decrease of 11 MJ of metabolizable energy in the cow's intake of grazed herbage, or (d) some combination of a, b and c. Evidence reviewed by Bryant and Trigg (1982) and Wilson and Davey (1982) showed that one kilogram of extra dry matter, fed as concentrates, silage or hay in early or late lactation *actually* produced: (a) an extra 0.5 l milk, or 25 g milkfat, plus (b) an extra 150 g live weight, plus (c) an increase in residual herbage mass, due to a decrease in grazed herbage intake. The substitution of supplementary feed for grazed herbage by grazing cows is well documented; for example, an extra one kilogram of concentrate eaten by a cow will increase its total intake only by between 0.3 and 0.5 kg DM (Bines, 1979).

In New Zealand, supplements are not usually given to lactating cows, unless the supply of grazable herbage is severely limited (e.g. during winter), because the responses (see above) are rarely profitable, at least in the short term. However, supplements, mainly hay and silage, are commonly fed to non-lactating cows during winter, and the effects have been reviewed by Grainger and McGowan (1982). By contrast, in Britain concentrates are commonly fed, even when the supply of grazable herbage is plentiful, although the response is usually only about 0.3 kg extra milk per kilogram of concentrate fed (Leaver, 1976; Gordon, 1980). Larger responses have been recorded when the supplies of herbage are severely limited, for example in summer and autumn (Castle and Watson, 1978; Gordon, 1979; Gleeson, 1980) or in spring (Le Du and Newberry, 1981).

Despite these unfavourable responses to feeds given as supplements to animals at pasture, it may be profitable, in some conditions, to feed cows entirely on feeds other than grazed pasture for short periods. For example, an early spring-calving herd can be fed on silage *ad libitum* plus some concentrates for three months until it can be turned out to graze in April (Gordon, 1979).

For maximum profitability in Britain about 0.6 t concentrates should be fed annually per cow in spring-calving herds, but with none fed during the grazing season (Gordon, 1980); the corresponding

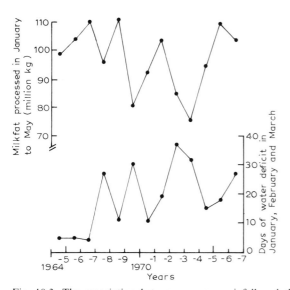

Fig. 10.3. The association between summer rainfall and the production of milkfat in New Zealand during summer and autumn (from Maunder, 1977).

quantity for the autumn-calving herd, is about 0.9 t (Gordon, 1983). Surveys of English farms indicate that the gross financial margin per hectare increases with the annual use of concentrates up to 1.4 t per cow (Aimes, 1980).

The conservation and feeding of hay and silage can have many effects on the dairy farm's productivity. These have not been studied experimentally in the long term, although mathematical models have been constructed (e.g. Miller, 1980).

### Weather and production

Climatic factors including temperature, radiation and rainfall influence herbage production (Chapters 2 and 3), and consequently affect animal production from grassland farms. For example, the monthly milkfat production in New Zealand during the four summer months from January to April was inversely related to the monthly water deficit during the previous month or months (Maunder, 1977; Fig. 10.3). Maunder also reported that the year-to-year variability in milkfat production, over the period 1964 to 1977, was much smaller in the period June to December (coefficient of variation 1%) than in the period January to May (c.v. 3%). The greater variability in summer milk production is associated with the greater variability in summer pasture growth (Baars, 1976; Piggot et al., 1978; see also Chapter 2).

### Efficiency of cows

The efficiency with which feed is converted into milk is influenced by many factors, including feeding, age, health, milking methods and genetic merit of the cows. In particular, cows which have been selected for high yields of milk and milkfat can convert feed into milk more efficiently, and can also produce more milk per hectare, than other cows (McPheely et al., 1977; Bryant and Trigg, 1981; Davey et al., 1983; A.M. Bryant, unpubl. data).

### Quantitative grazing management

Rapid and reasonably accurate assessment of herbage mass on large areas of pasture is now possible, using any one of a variety of different methods (Frame, 1981). Consequently, it will become possible to provide quantitative guidelines for practical management of feeding and grazing. For example, herbage intake by grazing cows can be assessed from the daily herbage allowance or the residual herbage mass (Hodgson, 1977; Le Du et al., 1979, 1981; Holmes and MacMillan, 1982). Thus, these measurements can be used to control the feeding of grazing cows. This approach may also be useful in making rational decisions about the feeding of supplements (Leaver, 1982), and in controlling the intensity of grazing in order to maintain vigorous pasture growth (Hodgson and Maxwell, 1981).

Long-term planning can be assisted by regular assessment of the total herbage mass on the whole farm (the immediate feed supply), particularly in New Zealand over the period from autumn to spring (Holmes, 1974). Feed planning or feed budgeting is now carried out regularly, in one form or another, on a number of dairy farms in New Zealand.

### CONCLUSIONS

The production of milk from grassland dairy farms depends on the growth, consumption and conversion into milk of pasture herbage. Improvements in any of these three components will lead to increased milk production, but further increases in production from the most efficient dairy farms probably depend mainly on increases in herbage production. Intensively managed spring-calving dairy herds in Ireland and New Zealand produce annualy 500 to 600 kg milkfat per hectare from pastures which produce 12 to 13 t dry matter per hectare.

Stocking rate has a dominant effect on milk production per cow and per hectare, and on profitability. Higher stocking rates can be carried if pastures are rotationally grazed, rather than set-stocked; this can give increases in milk production of 10 to 20% by rationing feed in winter and early spring, and by improving herbage quality. Quantitative assessments of herbage mass are useful in grazing management, and in planning the feed supply. The dates of calving and drying-off are also important features of herd management that affect productivity.

Milk production is very dependent on herbage supply, which in turn is dependent on weather. For example, during the summer months in New Zealand milk production is very dependent on summer rainfall. Supplementary feeds given to cows grazing on pastures usually have little effect on milk production, except perhaps when pasture herbage is in very short supply. An excellent and extensive review of milk production from temperate pastures has been published by Leaver (1985) since this chapter was completed.

**REFERENCES**

Aimes, S.J., 1980. Economics of Low and High cost dairying systems. *Ir. Grassland Anim. Prod. Assoc. J.*, 15: 75–78.

Anonymous, 1951. *N.Z. Dairy Board, 27th Annu. Rep.*, p. 78.

Anonymous, 1971. Rep. for Stratford Demonstration Farm, Taranaki.

Anonymous, 1977. Rep. for Waimate West Demonstration Farm, Taranaki.

Anonymous, 1981–82. *58th Farm Prod. Rep. N.Z. Dairy Board*, p. 59.

Baars, J.A., 1976. Seasonal distribution of pasture production in New Zealand in Hamilton, N.Z. *N.Z. J. Exper. Agric.*, 4: 157–161.

Bines, J.A., 1979. Voluntary food intake. In: W.H. Broster and H. Swan (Editors), *Feeding Strategy for the High Yielding Dairy Cow*. Granada, London, pp. 23–48.

Brookes, I.M., 1983. New pasture species for the dairy farm. *Dairyfarming Annu., Massey Univ.*, pp. 91–96.

Brougham, R.W., 1959. The effects of season and weather on the growth rate of a ryegrass and clover pasture. *N.Z. J. Agric. Res.*, 5: 283–296.

Brown, K.R. and Evans, P.S., 1973. Animal treading; a review of the work of the late D.B. Edmond. *N.Z. J. Exper. Agric.*, 1: 217–226.

Browne, D., 1972. Study of nitrogen use for milk production on free draining land. *Anim. Prod. Res. Rep. An foras taluntais, Dublin*, 90.

Bryant, A.M., 1980. *Once-Daily Milking, Effects on Production, Live Weight and Grazing*. Min. Agric. and Fisheries, N.Z. Ag Link, FPP, Bull. 158.

Bryant, A.M., 1981. Maximising milk production from pasture. *Proc. N.Z. Grassland Assoc.*, 42: 82–91.

Bryant, A.M., 1982a. Developments in dairy cow feeding and pasture management. In: *Proc. Ruakura Farmers' Conference*, pp. 75–82.

Bryant, A.M., 1982b. Effects of mowing before or after grazing on milk production. In: K.L. MacMillan and V.K. Taufa (Editors), *Dairy Production from Pasture*. N.Z. Soc. Anim. Prod., Hamilton, pp. 381–382.

Bryant, A.M. and Parker, O.F., 1971. Optimum grazing interval at high stocking rates. In: *Proc. Ruakura Farmers' Conference*, pp. 110–119.

Bryant, A.M. and Cook, M.A.S., 1980. A comparison of three systems of wintering dairy cattle. In: *Proc. Ruakura Farmers' Conference*, pp. 181–188.

Bryant, A.M. and Trigg, T.E., 1981. Progress report on the performance of Jersey cows differing in breeding index. *Proc. N.Z. Soc. Anim. Prod.*, 41: 39–47.

Bryant, A.M. and Trigg, T.E., 1982. The nutrition of the grazing dairy cow during early lactation. In: K.L. MacMillan and V.K. Taufa (Editors), *Dairy Production from Pasture*. N.Z. Soc. Anim. Prod., Hamilton, pp. 185–210.

Bryant, A.M., MacDonald, R.A. and Clayton, D.G., 1982. Effects of nitrogen fertiliser on production of milk solids from grazed pasture. *Proc. N.Z. Grassland Assoc.*, 43: 58–63.

Campbell, A.G., 1966a. Grazed pasture parameters 1. *J. Agric. Sci.*, 67: 199–210.

Campbell, A.G., 1966b. Grazed pasture parameters 2. *J. Agric. Sci.*, 67: 211–216.

Campbell, A.G., Clayton, D.C. and Bell, B.A., 1977. Milkfat production from No. 2 Dairy Ruakura. *N.Z. Agric. Sci.*, 11: 73–86.

Castle, M.E. and Watson, J.N., 1975. Further comparisons between a rigid rotational "Wye College" system and other systems of grazing for milk production. *J. Br. Grassland Soc.*, 30: 1–6.

Castle, M.E. and Watson, J.N., 1978. A comparison of continuous grazing systems for milk production. *J. Br. Grassland Assoc.*, 33: 123–129.

Crabbe, C.L., 1983. Milkfat production per hectare. *Dairyfarming Annu., Massey Univ.*, pp. 19–23.

Craven, J.A. and Kilkenny, J.B., 1979. The structure of the British cattle industry. In: H. Swan and W.H. Broster (Editors), *Principles of Cattle Production*. Butterworths, London, pp. 1–44.

Davey, A.W.F., Grainger, C., Mackenzie, D.D.S., Flux, D.S.F., Wilson, G.F., Brookes, I.M. and Holmes, C.W., 1983. Nutritional and physiological studies of differences between Friesian cows of high or low genetic merit. *Proc. N.Z. Soc. Anim. Prod.*, 43: 67–70.

Esslemont, R.J., 1979. Management with special reference to fertility. In: W.H. Broster and H. Swan (Editors), *Feeding Strategy for the High Yielding Dairy Cow*. Granada, London, pp. 258–259.

Frame, J., 1981. Herbage mass. In: J. Hodgson, R.D. Baker, A. Davies, A.S. Laidlaw and J.D. Leaver (Editors), *Sward Measurement*. Br. Grassland Soc., Hurley, pp. 39–70.

Gartner, J.A., 1982. Replacement policy in dairy herds on farms where heifers compete with cows for grassland. 3. *Agric. Systems*, 8: 249–272.

Gleeson, P., 1973. Pre- and post-calving feeding of dairy cattle. *Ir. Grassland Anim. Prod. Assoc. J.*, 8: 68–78.

Gleeson, P., 1980. Concentrates for cows at Grass. In: *Moorepark Farmers' Conference*. An foras taluntais, Dublin, pp. 26–31.

Gordon, F.J., 1973. The effect of high nitrogen levels and stocking rates on milk output from pasture. *J. Br. Grassland Soc.*, 28: 193–201.

Gordon, F.J., 1979. Some aspects of recent research in dairying. *Ir. Grassland Anim. Prod. Assoc. J.*, 14: 85–109.

Gordon, F.J., 1980. The role of supplements in the spring-calving dairy herd. In: *Supplementation and Effective Use of Grassland for Dairying*. Br. Grassland Soc., Hurley, pp. 1.1–1.17.

Gordon, F.J., 1981. Potential for change in the output of milk from grassland. In: J.L. Jollans (Editor), *Grassland in the British Economy*. Centre for Agricultural Strategy, Reading.

Gordon, F.J., 1983. Feeding concentrates to the autumn-calving cow. *Ann. Rep. Agric. Res. Inst. Northern Irel.*, in press.

Grainger, C. and McGowan, A.A., 1982. The significance of precalving nutrition of the dairy cow. In: K.L. MacMillan and V.K. Taufa (Editors), *Dairy Production from Pasture*. N.Z. Soc. Anim. Prod., pp. 134–171.

Hodgson, J., 1977. Factors limiting herbage intake by the grazing animal. In: *Proc. Int. Meet. on Animal Production from Temperate Grassland*. An foras taluntais, Dublin, pp. 70–75.

Hodgson, J.L. and Maxwell, T.J., 1981. Grazing research and management. In: *Biennal Rep. Hill Farming Res. Org. Scotland*, 1979–80: 169–187.

Holmes, C.W., 1974. The Massey grass meter. *Dairyfarming Annu., Massey Univ.*, pp. 26–30.

Holmes, C.W., 1982. The effect of fertiliser nitrogen on the production of pasture and milk on dairy farmlets. *Proc. N.Z. Grassland Assoc.*, 43: 53–57.

Holmes, C.W. and Halford, R.E., 1976. *Some Effects of Nitrogenous Fertiliser and Irrigation on the Production of Pasture on a Town-Supply Dairy Farm*. Dairy Husbandry Department, Massey University, Palmerston North.

Holmes, C.W. and MacMillan, K.L., 1982. Nutritional management of the dairy herd grazing on pasture. In: K.L. MacMillan and V.K. Taufa (Editors), *Dairy Production from Pasture*. N.Z. Soc. Anim. Prod., Hamilton, pp. 244–274.

Holmes, C.W., Davey, A.W.F. and Grainger, C., 1981. The efficiency with which feed is utilised by the dairy cow. *Proc. N.Z. Soc. Anim. Prod.*, 41: 16–27.

Hood, A.E.M., 1974. Intensive set stocking of dairy cows. *J. Br. Grassland Soc.*, 29: 63–67.

Hughes, H.C., 1979. A comparison of the economics of milk production in Ireland and Great Britain. *Ir. Grassland Anim. Prod. Assoc. J.*, 14: 67–75.

Hutton, J.B., 1962. Studies of the nutritive value of New Zealand dairy pasture. *N.Z. J. Agric. Res.*, 8: 409–424.

Hutton, J.B., 1963. Variations in dairy cow efficiency. In: *Proc. Ruakura Farmers' Conference*, pp. 194–208.

Hutton, J.B., 1966. Preliminary report on farming two cows per acre. In: *Proc. Ruakura Farmers' Conference*, pp. 168–180.

Hutton, J.B., 1975. The effects of irrigation and high stocking rates on seasonal dairy production from grasslands. *Annu. Rep. Res. Div. Min. Agric. Fish., N.Z.*, 1974–75: 58–59.

Journet, M. and Demarquilly, C., 1979. Grazing. In: W.H. Broster and H. Swan (Editors), *Feeding Strategy for the High Yielding Dairy Cow*. Granada, London, pp. 295–321.

King, K.R. and Stockdale, C.R., 1980. The effects of stocking rate and nitrogenous fertiliser on the productivity of irrigated perennial pasture grazed by dairy cows. 2. *Aust. J. Exper. Agric. Anim. Husb.*, 20: 537–542.

Leaver, J.D., 1976. Utilisation of grassland by dairy cows. In: H. Swan and W.H. Broster (Editors), *Principles of Cattle Production*. Butterworths, London, pp. 307–328.

Leaver, J.D., 1982. Grass height as an indicator for supplementary feeding of continuously stocked dairy cows. *Grass Forage Sci.*, 37: 285–290.

Leaver, J.D., 1985. Milk production from grazed temperate pasture. *J. Dairy Res.*, 52: 313–344.

Le Du, Y.L.P. and Newberry, R.D., 1981. The milk production of grazing dairy cows. *Annu. Rep. Grassland Res. Inst. Hurley*, pp. 84–85.

Le Du, Y.L.P., Combellas, J., Hodgson, J. and Baker, R.D., 1979. Herbage intake and milk production by grazing dairy cows. 2. *Grass Forage Sci.*, 34: 249–260.

Le Du, Y.L.P., Baker, R.D. and Newberry, R.D., 1981. Herbage intake and milk production by grazing dairy cows. 3. *Grass Forage Sci.*, 36: 307–318.

MacCarthy, D., 1980. Grassland management for milk production. In: *Moorepark Farmers Conference*. An foras taluntais, Dublin, pp. 13–17.

McMeekan, C.P., 1960. *Grass to Milk*. The N.Z. Dairy Exporter, Wellington.

McMeekan, C.P. and Walshe, M.J., 1963. The interrelationships of grazing method and stocking rate in the efficiency of pasture utilisation by dairy cattle. *J. Agric. Sci.*, 61: 147–166.

MacMillan, K.L. and Moller, K., 1976. Aspects of reproduction in New Zealand dairy herds. *N.Z. Vet. J.*, 25: 220–224.

McPheely, P.C., Browne, D. and Carty, O., 1975. Effect of grazing interval and stocking rate on milk production and pasture yield. *Irish J. Agric. Res.*, 14: 309–319.

McPheely, P.C., Butler, T.M. and Gleeson, P.A., 1977. Potential of Irish grassland for dairy production. In: *Proc. Int. Meet. on Animal Production from Temperate Grassland*. An foras taluntais, Dublin, pp. 5–11.

MacQueen, I.P.M., 1970. The effects of winter management on farm production. *Dairyfarming Annu. Massey Univ.*, pp. 169–180.

Mace, M.J., 1979. Management changes with the intensification of farming the pumice lands over the last ten years. *Proc. N.Z. Grassland Assoc.*, 41: 11–19.

Maunder, W.J., 1977. Weather and climate as factors in forecasting national dairy production. In: *Management of Dynamic Systems in New Zealand Agriculture*. D.S.I.R., Wellington, pp. 1101–1125.

Miller, C.P., 1980. Modelling the contribution of forage crops to production profitability and stability of North Island dairy systems. *Proc. N.Z. Soc. Anim. Prod.*, 40: 64–67.

Nix, J.S., 1980. Economic aspects of grass production and utilisation. In: W. Holmes (Editor), *Grass, Its Production and Utilisation*. Blackwell, London, pp. 216–238.

Paul, K.J. and Benseman, B.R., 1983. Factors affecting flushmilk production in the Waikato. *N.Z. J. Exper. Agric.*, 11: 127–130.

Piggot, G.H., Baars, J.A., Cumberland, G.L.B. and Honore, E.N., 1978. Seasonal distribution of pasture production in New Zealand. XIII South Kaipara, Northland. *N.Z. J. Exper. Agric.*, 6: 43–46.

Rogers, G.L., Porter, R.H.D. and Robinson, I., 1982. Comparison of perennial ryegrass and white clover for milk production. In: K.L. MacMillan and V.K. Taufa (Editors), *Dairy Production from Pasture*. N.Z. Soc. Anim. Prod., Hamilton, pp. 213–214.

Santamaria, A. and McGowan, A.A., 1982. The effect of contrasting winter grazing management on current and subsequent pasture production and quality. In: K.L. MacMillan and V.K. Taufa (Editors), *Dairy Production from Pasture*. N.Z. Soc. Anim. Prod., Hamilton, pp. 359–360.

Simmonds, J., 1978. Set stocking for high production. *Dairyfarming Annu., Massey Univ.*, pp. 108–110.

Smith, B.A.J., 1964. Potash and production. In: *Proc. Ruakura Farmers' Conference*, pp. 204–213.

Stockdale, C.R. and King, K.R., 1980. The effects of stocking rate and nitrogen fertiliser on the productivity of irrigated pastures and grazing dairy cows. 1. *Aust. J. Exper. Agric. Anim. Husb.*, 20: 529–536.

Thompson, N.A., 1982. Lime and dairy production. *Proc. N.Z. Grassland Soc.*, 43: 93–103.

Wallace, L.R., 1959. Grazing management and dairy production. *Proc. N.Z. Inst. Agric. Sci.*, 131.

Walshe, M.J., 1971. Research on intensive use of grassland for dairying in Ireland. *Rev. Cubana Cienc. Agric.*, 5: 143–153 (English ed.).

Williams, T.E., 1980. Herbage production; grass and leguminous forage crops. In: W. Holmes (Editor), *Grass; Its Production and Utilisation*. Blackwell, London, pp. 6–69.

Wilson, G.F. and McDowell, F.H., 1966. The influence of ryegrass varieties on milk yield and composition. *N.Z. J. Agric. Res.*, 9: 1042–1052.

Wilson, G.F. and Davey, A.W.F., 1982. The nutrition of the grazing cow; mid and late lactation. In: K.L. MacMillan and V.K. Taufa (Editors), *Dairy Production from Pasture*. N.Z. Soc. Anim. Prod., Hamilton, pp. 219–235.

Wood, P.D.P., 1970. The relationship between the month of calving and milk production. *Anim. Prod.*, 12: 253–259.

Wright, D.F. and Pringle, R.M., 1983. Stocking rate effects in dairying. *Proc. N.Z. Soc. Anim. Prod.*, 43: 97–100.

Chapter 11

# SHEEP PRODUCTION FROM MANAGED GRASSLANDS

P.V. RATTRAY

## INTRODUCTION

The sheep is a highly versatile species of domestic animal. Throughout the world the various breeds and types occupy a wide range of habitats and environments from the subarctic to the tropics, from arid to high rainfall, and from sea-level to alpine regions; they graze on a range of grasslands from sparse native bunch grass and scrub to productive improved pastures.

Owen (1976) has provided a comprehensive summary of the wide variety of production systems used throughout the world. The major types are: (1) the transhumanic and nomadic, where the flocks follow the feed supply and avoid climatic extremes of cold or drought; (2) the settled village systems where sheep herding and some form of night housing or corralling is usually involved; (3) extensive systems with sheep lightly stocked over large areas of largely unimproved grassland; and (4) intensive systems which are confined more to developed countries.

Sheep produce a range of products useful to man, including meat, fibre, milk, skins, dung (for fertilizer and fuel); they even provide entertainment (ram fighting, performing sheep). Usually there is one major product from a particular system or region, with a number of minor products, these include:

(a) Fine wool production, with meat as a by-product, e.g. Merino sheep in the semi-arid areas of Australia, South Africa, South America and Asia.

(b) Milk production with meat, wool, hides and pelts as by-products, e.g. small-scale peasant farming or communal tribal ownership in southern and eastern Europe, North Africa and Asia.

(c) Lamb meat production, with ewe mutton and wool as subsidiary products, e.g. intensive farming in humid temperate regions of New Zealand, Australia, northern Europe and North America.

(d) On agriculturally marginal areas, such as rangeland or hill country, dual production of wool and meat is common, often as part of a two-tiered industry where breeding and finishing stock are provided to lowland areas or feedlots. The two major products are lamb meat and ewe wool, with ewe mutton, lamb wool, skins and offal as subsidiary products.

## ANNUAL PRODUCTION

The total amount of herbage grown annually influences the stocking rate that can be carried (Chapter 24), and hence the output of meat and wool per hectare. Table 11.1 shows the potential carrying capacity of various regions of New Zealand, assuming a standard feed requirement of 630 kg DM ewe$^{-1}$ yr$^{-1}$ and 90% utilization of pasture grown.

TABLE 11.1

Pasture production and potential carrying capacities in various regions of New Zealand (from Rattray, 1978)

| Region | Average annual pasture production (kg DM ha$^{-1}$ yr$^{-1}$) | Average potential carrying capacity (ewes ha$^{-1}$) |
|---|---|---|
| Waikato | 16 760 | 23.9 |
| Southland | 12 010 | 17.1 |
| Wairarapa | 10 880 | 15.5 |
| Hawke's Bay | 6 740 | 9.6 |
| Central Otago | 2 800 | 4.0 |

Variability of herbage production from year to year often constrains the stocking rate adopted (Chapter 24). Where herbage production is reliable and consistent between years, a carrying capacity close to the potential can be carried with little risk of major shortages in any one year. Where the variation is marked, a much more conservative approach is usually adopted, stocking for the poor years rather than the average years. This is most common in areas prone to drought, where animals may die in very bad years if no supplementary feed is provided. To some extent, the year-to-year variation can be overcome by conserving surplus herbage as hay or silage in good years and using it in poor years (Chapter 26).

Increasing the annual feed intake of a ewe increases the animal's performance in respect of liveweight, lambing and wool production (Table 11.2). Annual feed intake is determined by both herbage production and stocking rate. Food supply at certain times of the year is critical in determining animal performance. These critical periods of feed supply are: (1) pre-mating, which determines the number of lambs born; (2) late pregnancy, which determines the birth weight and survival of lambs; and (3) lactation, which determines lamb growth rates and hence meat output. These critical periods will be considered later in more detail. Wool production by the ewe is also coincidentally influenced most by feed supply at these times, in a parallel manner to other aspects of production.

Maximum production per hectare depends on carrying a sufficiently large number of ewes per hectare to utilize the herbage grown, without too seriously reducing production per ewe (Chapter 24). Typical effects of stocking rate on production, per head and per hectare, are illustrated in Table 11.3. As stocking rate increases, performance per head declines because individual intakes decline, but output per hectare increases because more herbage is utilized (Suckling, 1975; Young and Newton, 1975; Brown, 1977; Egan et al., 1977a, b; Rattray et al., 1978; Reeve and Sharkey, 1980; White et al., 1980; Sharrow et al., 1981). At extremely high stocking rates, performance per head declines so much that production per hectare also declines (Joyce, 1971; Robinson and Simpson, 1975).

## MATCHING FEED REQUIREMENTS AND PASTURE GROWTH

### Seasonal pasture growth and feed requirements

In many areas, 60 to 80% of pasture growth occurs during a three- to four-month period, usually in the spring and early summer (Chapters 2 and 3). The major periods of shortage occur during winter and sometimes in summer or autumn. The extent and duration of these shortages varies from region to region and from year to year.

The feed requirement of the breeding ewe varies several-fold during the year, depending on her physiological status. After weaning and in early-mid pregnancy she can be fed so as to just maintain weight or lose weight slightly. During late pregnancy, requirements rise rapidly to reach a peak in early lactation, two to three times the maintenance level of feeding, depending on the number of lambs suckled (Jagusch and Coop, 1971; Rattray, 1978a).

TABLE 11.2

The effect of annual feed intake on the performance of Coopworth and Romney ewes of average prolificacy (from Rattray et al., 1978)

|  | Feed intake (kg DM ewe$^{-1}$ yr$^{-1}$) | | | | | |
| --- | --- | --- | --- | --- | --- | --- |
|  | 515 | 580 | 645 | 710 | 775 | 840 |
| Mating LW (kg) | 45 | 50 | 55 | 60 | 65 | 70 |
| Pre-mating LW (kg) | 53 | 58 | 63 | 68 | 73 | 78 |
| Lambs weaned per 100 ewes mated | 83 | 93 | 103 | 113 | 123 | 133 |
| Weight of weaned lambs per ewe (kg) | 13 | 17 | 20 | 24 | 27 | 31 |
| Fleece weight (kg) | 4.0 | 4.2 | 4.4 | 4.6 | 4.8 | 5.0 |

# SHEEP PRODUCTION

TABLE 11.3

The effect of stocking rate on animal production and pasture utilization (Rattray et al., 1978)

|  | Stocking rate (ewes ha$^{-1}$) | | |
| --- | --- | --- | --- |
|  | 26 | 21 | 16 |
| *Production per head* | | | |
| Ewe mating weight (kg) | 56 | 60 | 65 |
| Ewe fleece weight (kg) | 4.4 | 4.4 | 4.8 |
| Total wool per ewe (kg) | 5.4 | 5.3 | 6.0 |
| Lambs born per lambing ewe | 1.49 | 1.57 | 1.71 |
| Lambs weaned per lambing ewe | 1.08 | 1.12 | 1.22 |
| Lamb weaning weight (kg) | 20.7 | 21.4 | 23.5 |
| Lamb carcass weight (kg) | 12.3 | 13.3 | 14.3 |
| *Production per hectare* | | | |
| Wool (kg) | 140 | 111 | 96 |
| Weaned lambs (kg) | 582 | 504 | 459 |
| Lamb meat (kg) | 346 | 312 | 280 |
| *Herbage utilized* | | | |
| DM ewe$^{-1}$ yr$^{-1}$ | 620 | 710 | 820 |
| DM ha$^{-1}$ yr$^{-1}$ | 16 120 | 14 910 | 13 120 |

Matching the curves of feed supply and demand usually involves the synchronization of the increasing feed requirements of late pregnancy and lactation with the onset of pasture growth. Unfortunately, herbage production and feed requirements seldom equate perfectly, and periods of surplus and/or deficit occur. Depending on the pattern of seasonal herbage production, different lambing patterns can be used, such as: spring only, autumn only, spring and autumn, or even lambing throughout the year. In areas where variations in seasonal production are most extreme, sheep may be removed from the pastures and fed supplements. Another strategy, in marginal areas, is to use castrated male sheep (wethers) which have no seasonal peak of feed requirement related to reproduction, and which can more easily withstand feed deficits.

In the main, this chapter will consider intensively managed, temperate grasslands with a spring–summer peak in herbage production, that are grazed with dual-purpose sheep. Such pastures are often dominated by perennial ryegrass (*Lolium perenne*) and white clover (*Trifolium repens*) (Chapter 8).

The changing physiological status of the breeding ewe during the year, with its effect on seasonal variability in the demand for herbage, allows some

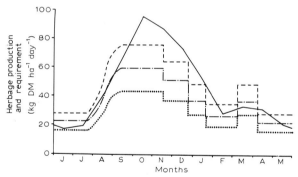

Fig. 11.1. The seasonal pattern of pasture production in the Waikato region of New Zealand in relation to the feed requirements of spring-lambing (August 31) ewes and lambs stocked at 16 (······), 22 (—·—) and 28 (-----) ewes ha$^{-1}$. Note that spring is August to October in the Southern Hemisphere (Rattray, 1978a).

manipulation of the size and duration of the surpluses and deficits by: (1) altering stocking rate (Chapter 24), (2) altering the lambing date, or (3) conserving herbage (Chapter 26).

## Stocking rate

Figure 11.1 illustrates the influence of three stocking rates on the patterns of surplus or deficit from a pasture producing 16.8 t DM ha$^{-1}$ yr$^{-1}$. At the lowest stocking rate (16 ewes ha$^{-1}$) there is never a feed deficit and, on an annual basis, there is a feed surplus of 6.8 t DM ha$^{-1}$. At the intermediate stocking rate (22 ewes ha$^{-1}$) there is a small autumn–winter deficit and a relatively large spring–autumn surplus, resulting in an annual net surplus of 2.9 t DM ha$^{-1}$. The highest stocking rate (28 ewes ha$^{-1}$), in spite of a small spring–summer surplus, results in an overall feed deficit of 0.9 t DM ha$^{-1}$, mainly from late summer to early spring.

### Lambing date

Early lambing dates can result in large feed deficits in late winter and early spring; on the other hand, late lambing increases the summer–autumn deficits and makes it difficult to finish lambs and "flush" ewes before mating. This is illustrated in Fig. 11.2, where the seasonal pattern of herbage production (10.9 t DM ha$^{-1}$ yr$^{-1}$) in a dry summer environment is compared with the pattern of feed requirement for 15.5 ewes ha$^{-1}$ with three different mean lambing dates.

### Conserved feed

Surpluses of herbage can be conserved as high-quality hay or silage for feeding during deficit periods (Rattray, 1983) or, to a limited extent, can be carried forward *in situ* for grazing at a later date.

## BREED AND REPRODUCTIVE RATE

Sheep breeds vary considerably in average liveweight, prolificacy and wool production, and these differences affect production per hectare (Rattray et al., 1978). Since feed requirement is closely related to body weight (Table 11.4), fewer sheep of a larger breed can be carried per hectare. The general relationship between maintenance feed requirement ($M$) and body weight ($W$) is $M = kW^{0.7}$, where $k$ is a constant (Blaxter, 1967), though the total annual feed requirement appears to be linearly related to body weight (Rattray et al., 1978). This is presumably due to the increased food required during pregnancy and lactations, because fecundity in most breeds increases with ewe body weight (Morley et al., 1978). This improved fecundity usually results in better feed conversion. Table 11.2, for example, compares two breeds of average prolificacy with mating weights of 70 kg and 50 kg. The heavier ewe produced 82% more lamb meat, resulting in an improvement of feed conversion efficiency [(kg meat)/(kg intake)] of 26%. However, lighter breeds can be carried at higher stocking rates, which may cancel out such advantages on a per-hectare basis (Rattray et al., 1978). Because

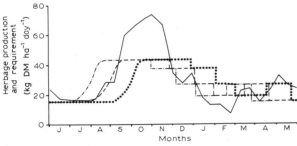

Fig. 11.2. The seasonal pattern of pasture production in the Wairarapa region of New Zealand in relation to the feed requirements of ewes lambing at August 7 (—·—), September 7 (-----) and October 7 (······) and stocked at 15.5 ewes ha$^{-1}$ (Rattray, 1978a).

TABLE 11.4

Average annual intakes, ewe liveweights and production per ewe of various sheep breeds over a five-year period (Rattray et al., 1978)

| Breed | Pasture intake (kg DM ewe$^{-1}$ yr$^{-1}$) | Liveweight (kg) | | Prolificacy (lambs born per ewe lambing) | Fleece weight (kg per ewe) |
|---|---|---|---|---|---|
| | | premating | prelambing | | |
| Control fertility Romney | 635 (100) | 54 (100) | 60 (100) | 1.19 (100) | 4.1 (100) |
| High fertility Romney | 643 (101) | 52 (97) | 61 (103) | 1.69 (142) | 3.6 (88) |
| Coopworth | 716 (113) | 60 (112) | 70 (117) | 1.53 (129) | 4.3 (105) |
| Perendale | 663 (104) | 58 (107) | 65 (108) | 1.42 (119) | 3.8 (93) |

# SHEEP PRODUCTION

TABLE 11.5

The effect of stocking rate and breed on energy conversion by sheep grazing pastures

| Genotype: | Control fertility Romney | | | High fertility Romney | | | Coopworth | | | Perendale | | |
|---|---|---|---|---|---|---|---|---|---|---|---|---|
| Stocking rate (ewes ha$^{-1}$): | 26 | 21 | 16 | 26 | 21 | 16 | 26 | 21 | 16 | 26 | 21 | 16 |
| Annual intake (MJ ewe$^{-1}$) | 10 802 | 11 435 | 12 063 | 10 802 | 10 983 | 12 063 | 11 163 | 12 783 | 14 763 | 11 343 | 11 703 | 12 783 |
| Annual wool production (MJ ewe$^{-1}$) | 106 | 111 | 117 | 96 | 96 | 113 | 117 | 115 | 130 | 98 | 104 | 113 |
| Annual lamb production (MJ ewe$^{-1}$) | 169 | 190 | 227 | 211 | 251 | 317 | 284 | 319 | 383 | 241 | 286 | 347 |
| Conversion efficiency (%) | 2.55 | 2.63 | 2.85 | 2.84 | 3.16 | 3.56 | 3.59 | 3.40 | 3.47 | 2.99 | 3.33 | 3.60 |

body weight is the major determinant of total feed requirements, and the latter is not altered greatly by fecundity (Rattray et al., 1978), comparisons of productivity in terms of liveweight, rather than per animal, appear valid.

Highly fecund breeds and crosses, at similar liveweights, are the most efficient converters of feed to product (Clarke and Rattray, 1983). For example, because of its 42% greater prolificacy (Table 11.4), the High Fertility Romney is 25 to 40% more efficient at conversion of herbage to lamb meat than the Control Romney (Table 11.5). Fecundity is considered further in Chapter 16.

## PASTURE CONDITIONS

Animal performance is determined by both the quantity and the quality of herbage available.

## Herbage quality

Herbage quality is most limiting after flowering, especially if a feed surplus has been left *in situ* and dies or dries off. There is a close relationship between herbage digestibility and the proportion of green material in the sward (Fig. 11.3), since dead material is usually less than 40% digestible. The accumulation of dead material leads to a seasonal trend in herbage digestibility (Fig. 11.4). The low digestibility in late summer can lead to a reduction in the liveweight gains of ewes and lambs (Rattray, 1978b).

Sheep demonstrate a marked preference for green herbage, and actively select against the dead component of the sward (Rattray et al., 1983a). If the proportion of green material in the sward is small, because of the accumulation of dead indigestible material, intake and performance will be severely restricted (Barthram, 1981; Smeaton et al., 1981).

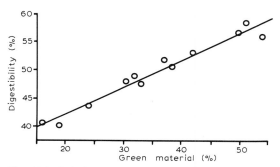

Fig. 11.3. Relationship between sward digestibility and the content of green herbage in the pasture (Rattray, 1978a).

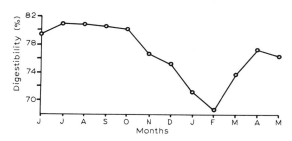

Fig. 11.4. The annual pattern of pasture digestibility in the Waikato region of New Zealand (Rattray, 1978b).

Large proportions of legumes in the sward increase the proportion of green material, intake, digestibility and the efficiency of conversion of digested nutrients, and so increase the liveweight gains of both ewes and lambs (Rattray and Joyce, 1974; Jagusch et al., 1979; Rattray et al., 1983a).

**Quantity of herbage**

As the amount of herbage allocated (kg DM head$^{-1}$ day$^{-1}$) increases, intake and performance increase asymptotically (Fig. 11.5). In addition, herbage mass *per se* also influences performance at a given herbage allowance (Fig. 11.5). This occurs because of the greater ease of grazing when large amounts of herbage are present (Barthram, 1981; Brown, 1977). There is some restriction of intake, if the amount of green herbage on offer is less than 1000 to 1200 kg DM ha$^{-1}$, but only small increases in intake occur if it is more than 2500 kg ha$^{-1}$ (Arnold and Dudzinski, 1967; During et al., 1980; Rattray et al., 1983a).

## GRAZING MANAGEMENT

Continuous and rotational grazing have given similar levels of animal output (Wheeler, 1962; Hill and Saville, 1976; Sharrow and Krueger, 1979), though at very high stocking rates rotational grazing is often superior (Lambourne, 1956; Robinson and Simpson, 1975).

Currently it is common practice to combine the two grazing practices, using rotational grazing at some times of the year and set-stocking at others (Sharrow and Krueger, 1979; Smith et al., 1979; Smeaton, 1983; Sheath et al., 1984). Set-stocking is adopted at times of the year when stock disturbance should be minimized (e.g. with lambing ewes or ewes with newborn lambs at foot). Set stocking is also adopted when feed requirements are highest (e.g. lactating ewes), and when pasture growth is greatest and control of reproductive tiller growth is required (late spring and early summer). Rotational grazing is the preferred method when feed is in short supply and there is a need to build up pasture mass, or when feed requirements are low and some restriction of intake is desirable or essential. Alteration of the length of rotation can be used to adjust herbage allowance and hence intake. It is common practice to speed up the rotation when requirements are high, such as late pregnancy, lactation or pre-mating. The rotation can be slowed when requirements are low, for instance in mid-winter, to allow a build-up of feed reserves for the higher requirements of the lambing ewes in spring; this method therefore improves the matching of feed requirements and pasture growth. Smeaton (1983) has discussed these aspects in detail.

## CRITICAL PERIODS OF THE REPRODUCTIVE CYCLE

**Pre-mating**

Feeding during the pre-mating period can markedly influence the reproductive rate, and so affect the annual output of lambs. Low ovulation rates are a major factor in limiting lambing rate. Heavy ewes have a higher ovulation rate and lambing performance than light ewes or ewes in poor condition (Wallace, 1961; Coop, 1962; Morley et al., 1978; Gunn and Doney, 1979). Mating weight is not the only factor (Gunn et al., 1979), however, and the pattern of liveweight change before mating also markedly influences ovulation rate. Light ewes fed well before mating can ovulate

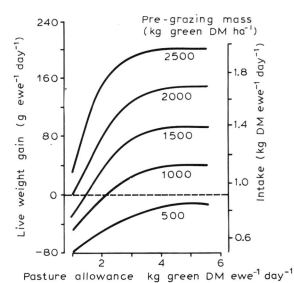

Fig. 11.5. The effect on the liveweight gain and feed intake of ewes of feed allowance and the amount of green feed per hectare initially offered (Rattray et al., 1983a).

# SHEEP PRODUCTION

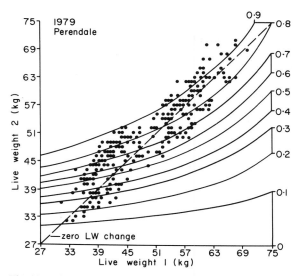

Fig. 11.6. The effect of liveweight before "flushing" (Live weight 1) and before mating (Live weight 2) on the proportion of multiple ovulations in Perendale ewes. The values on the right and top perimeters and associated contours are the proportion of multiple ovulations (Rattray et al., 1983a).

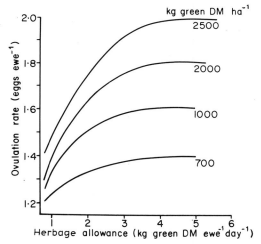

Fig. 11.7. The effect of herbage allowance and the total amount of green herbage on offer on the ovulation rate of ewes (Rattray et al., 1983a).

as much as, or more than, heavy ewes fed poorly before mating (Fig. 11.6). Improved feeding or "flushing" before mating is recommended if conditions allow (Coop, 1967); the ovulation responses to increasing herbage allowance and pre-grazing herbage mass are shown in Fig. 11.7. If pasture conditions do not allow flushing, some form of supplement may be required (Rattray et al., 1983b).

## Pregnancy

Low levels of nutrition, and substantial weight loss, can be tolerated by the breeding ewe in early-to mid-pregnancy without reducing lambing performance, though wool production may be reduced (Monteath, 1971; Rattray and Jagusch, 1978). Ewes in reasonable condition can withstand moderate feed deficiency in late pregnancy too, provided feeding is consistent (Clark, 1978; Rattray et al., 1982a), though inadequate feeding of ewes already in poor condition depresses birth weights and leads to pregnancy toxaemia (Rattray and Trigg, 1979).

## Lactation

Level of feeding during lactation has a much greater effect on total lamb production than feeding during pregnancy. Thus, strict rationing throughout pregnancy may be desirable, in order to carry some winter and spring herbage reserves *in situ* for grazing after lambing. The weights of both the lamb and the ewe are greatly affected by feeding during lactation (Coop, 1950; Rattray et al., 1982b). This affects current lamb output, but can also affect the subsequent reproductive rate of the ewe, and hence future lamb production.

## Growth of young stock

There is a dichotomy of opinion in the literature as to the extent of any carry-over effects of early growth restriction. Allden (1970) reported some studies in which very severe nutritional deprivation, either pre- or post-natal, appeared to have some long-term effect on adult size, wool production and longevity, in addition to the effects on reproduction and milk production during the first lactation. Restrictions until twelve months of age can lower lifetime reproductive rates (Gunn, 1977). According to Moore and Hockey (1982), even short periods of moderate restriction after weaning can delay the onset of puberty and oestrus, and result in reduced reproductive rates at two and three years of age, even though the initial liveweight differences no longer existed. These authors suggested a target weight of 30 kg at six months of age in Romney sheep, which should result in most hoggets showing oestrus in their first breeding season. Recommended feeding levels and grazing practices that allow such growth rates are reported by Jagusch et al. (1979) and During et al. (1980).

It is customary to mate ewes in their second or even third breeding season at about eighteen or thirty months of age, depending on conditions, but well-grown hoggets can be mated at around eight to nine months of age in their first breeding season (Baker et al., 1981), resulting in a shortened generation interval and enhanced lifetime production. If they are preferentially fed, subsequent reproductive rates are not affected, and survival of lambs to weaning after later lambings may be improved, compared with controls that did not lamb as hoggets.

### SHEEP AS CONVERTERS

Sheep are relatively inefficient converters of feed (Owen, 1976). This is partly because ruminants are less efficient converters of highly digestible feed than are monogastric animals, though they are able to convert the structural carbohydrates of plants, which are otherwise indigestible, into useful products. In addition, they can graze grassland on terrain which is unsuitable for cropping.

The efficiency of conversion of food energy to product energy depends on whether or not the overhead feed costs of the breeding ewe are included, since the ewe spends over half the year at maintenance feeding levels. Realistically these costs must be considered in the total cost, in which case the prolificacy of the ewe, as well as the rate of gain of the progeny, becomes an important factor influencing conversion efficiency (Chapter 16). The effect of these factors on conversion efficiency can be seen in Table 11.5, which is based on the results of a farmlet trial (Rattray et al., 1978). For example, High Fertility Romneys were substantially more efficient in lamb-meat production than standard Romneys, especially at high stocking rates. It is also noticeable that Coopworth ewes were equally efficient at all stocking rates, whereas other breeds were more efficient at low levels of stocking.

Estimates of protein conversion are usually meaningless for high quality pastures, because these pastures, in the vegetative state, often contain protein greatly in excess of animal requirements.

### CONCLUSIONS

This chapter briefly summarizes the various systems of sheep production used throughout the world, and the factors influencing the types and levels of production possible. Total feed supply is the major factor influencing output, but this is modified by stocking rate and by the seasonal pattern of herbage production. Adult size appears to be the major factor affecting total feed requirement and, for this reason, more prolific breeds are more efficient producers and converters. Herbage quality, herbage allowance and herbage mass all influence feed intake and sheep production; grazing management can be used, to some extent, to manipulate these factors so as to optimize feeding during critical periods of the reproductive cycle of the breeding ewe. Periods when aspects of production (prolificacy, lamb growth rates, wool growth) can be most affected are pre-mating, pre-lambing and lactation. Management during rearing may also have carry-over effects on adult production.

Sheep producers are faced with complex and very different situations in their various environments; they usually operate systems that have been proven over time to be the most suitable for their particular conditions.

### REFERENCES

Allden, W.G., 1970. The effects of nutritional deprivation on subsequent productivity of sheep and cattle. *Nutrit. Abstr. Rev.*, 40: 1167–1184.

Arnold, G.W. and Dudzinski, M.L., 1967. Studies on the diet of the grazing animal. III. The effect of pasture species and pasture structure on herbage intake of sheep. *Aust. J. Agric. Res.*, 18: 657–666.

Baker, R.L., Clarke, J.N. and Diprose, G.D., 1981. Effect of mating Romney ewe hoggets on lifetime production. *Proc. N.Z. Soc. Anim. Prod.*, 41: 198–203.

Barthram, G.T., 1981. Sward structure and the depth of the grazed horizon. *Grass Forage Sci.*, 36: 130–131.

Blaxter, K.L., 1967. *Energy Metabolism of Ruminants*. Hutchinson and Co., London, 332 pp.

Brown, T.H., 1977. A comparison of continuous grazing and deferred autumn grazing of Merino ewes and lambs at 13 stocking rates. *Aust. J. Agric. Res.*, 28: 947–961.

Clark, D.A., 1978. Effect of pasture reserves and stocking rate on ewe and lamb performance from mid-pregnancy to weaning. *Proc. N.Z. Grassland Assoc.*, 40: 81–88.

Clarke, J.N. and Rattray, P.V., 1983. The scope of dual purpose breeds. In: *Proc. Ruakura Farmers' Conf.*, 35: 15–19.

Coop, I.E., 1950. The effect of level of nutrition during pregnancy and lactation on lamb and wool production of grazing sheep. *J. Agric. Sci., Cambridge*, 40: 311–340.

Coop, I.E., 1962. Liveweight–productivity relationships in sheep. I. Liveweight and reproduction. *N.Z. J. Agric. Res.*, 5: 249–264.

Coop, I.E., 1967. Effect of flushing on reproductive performance of ewes. *J. Agric. Sci., Cambridge*, 67: 305–323.

During, C., Dyson, C.B. and Webby, R.W., 1980. The relationship of pasture parameters to liveweight gain of hoggets on North Island hill country. *Proc. N.Z. Soc. Anim. Prod.*, 40: 98–105.

Egan, J.K., Thompson, R.L. and McIntyre, J.S., 1977a. Stocking rate, joining time, fodder conservation and the productivity of Merino ewes. 1. Liveweights, joining and lambs born. *Aust. J. Exper. Agric. Anim. Husb.*, 17: 566–573.

Egan, J.K., Thompson, R.L. and McIntyre, J.S., 1977b. Stocking rate, joining time, fodder conservation and the productivity of Merino ewes. 2. Birthweight, survival and growth of lambs. *Aust. J. Exper. Agric. Anim. Husb.*, 17: 909–914.

Gunn, R.G., 1977. The effects of two nutritional environments from 6 weeks pre-partum to 12 months of age on lifetime performance and reproductive potential of Scottish Blackface ewes in two adult environments. *Anim. Prod.*, 25: 155–164.

Gunn, R.G. and Doney, J.M., 1979. Fertility in Cheviot ewes. 1. The effect of body condition at mating on ovulation rate and early embryo mortality in North and South Country Cheviot ewes. *Anim. Prod.*, 29: 11–16.

Gunn, R.G., Doney, J.M. and Smith, W.F., 1979. Fertility in Cheviot ewes. 2. The effect of level of pre-mating nutrition on ovulation rate and early embryo mortality in North and South Country Cheviot ewes in moderately-good condition at mating. *Anim. Prod.*, 29: 17–23.

Hill, B.D. and Saville, D.G., 1976. Effect of management system and stocking rate on the performance of ewes and lambs. *Aust. J. Exper. Agric. Anim. Husb.*, 16: 810–817.

Jagusch, K.T. and Coop, I.E., 1971. The nutritional requirements of grazing sheep. *Proc. N.Z. Soc. Anim. Prod.*, 31: 224–234.

Jagusch, K.T., Rattray, P.V. and Winn, G.W., 1979. An evaluation of lucerne and clover crops for finishing weaned lambs. *Proc. Agron. Soc. N.Z.*, 9: 7–10.

Joyce, J.P., 1971. Sheep stocking rates and productivity. In: *Proc. Ruakura Farmers' Conf.*, 23: 13–24.

Lambourne, L.J., 1956. A comparison between rotational grazing and set stocking for fat-lamb production. *N.Z. J. Sci. Tech.*, 37A: 555–568.

Monteath, M.A., 1971. The effect of sub-maintenance feeding of ewes during mid-pregnancy on lamb and wool production. *Proc. N.Z. Soc. Anim. Prod.*, 31: 105–113.

Moore, R.W. and Hockey, H.-U.P., 1982. The effect of early nutrition and hogget oestrus on subsequent reproduction. *Proc. N.Z. Soc. Anim. Prod.*, 42: 41–43.

Morley, F.H.W., White, D.H., Kenney, P.A. and Davis, I.F., 1978. Predicting ovulation rate from liveweight in ewes. *Agric. Systems*, 3: 27–45.

Owen, J.B., 1976. *Sheep Production*. Baillière Tindall, London, 436 pp.

Rattray, P.V., 1978a. Pasture constraints to sheep production. *Proc. Agron. Soc. N.Z.*, 8: 103–108.

Rattray, P.V., 1978b. Effect of lambing date on production from breeding ewes and on pasture allowance and intake. *Proc. N.Z. Grassland Assoc.*, 39: 98–107.

Rattray, P.V., 1983. Use of pasture silage as a summer supplement for ewes. *Proc. N.Z. Grassland Assoc.*, 44: 188–195.

Rattray, P.V. and Jagusch, K.T., 1978. Pasture allowances for the breeding ewe. *Proc. N.Z. Soc. Anim. Prod.*, 38: 121–126.

Rattray, P.V. and Joyce, J.P., 1979. Nutritive value of white clover and perennial ryegrass. IV. Utilisation of dietary energy. *N.Z. J. Agric. Res.*, 17: 401–406.

Rattray, P.V. and Trigg, T.E., 1979. Minimal feeding of pregnant ewes. *Proc. N.Z. Soc. Anim. Prod.*, 39: 242–250.

Rattray, P.V., Garrett, W.N., Hinman, N. and East, N.E., 1974. Effects of level of nutrition, pregnancy and age on the composition of the wool-free ingesta-free body and carcass of sheep. *J. Anim. Sci.*, 39: 687–693.

Rattray, P.V., Jagusch, K.T., Clarke, J.N. and Maclean, K.S., 1978. Romneys, Coopworths and Perendales — optimum feeding levels compared. In: *Proc. Ruakura Farmers' Conf.*, 30: Aglink FPP 164.

Rattray, P.V., Jagusch, K.T., Duganzich, D.M., Maclean, K.S. and Lynch, R.J., 1982a. Influence of pasture allowance and mass during late pregnancy on ewe and lamb performance. *Proc. N.Z. Grassland Assoc.*, 43: 223–229.

Rattray, P.V., Jagusch, K.T., Duganzich, D.M., Maclean, K.S. and Lynch, R.J., 1982b. Influence of feeding post-lambing on ewe and lamb performance at grazing. *Proc. N.Z. Soc. Anim. Prod.*, 42: 179–182.

Rattray, P.V., Jagusch, K.T. and Smeaton, D.C., 1983a. Interaction between feed quality, feed quantity, body weight and flushing. *Proc. Sheep Beef Cattle Soc. N.Z. Vet. Assoc.*, 13: 21–34.

Rattray, P.V., Smeaton, D.C. and Jagusch, K.T., 1983b. When all is said and done feed them. *Proc. Sheep Beef Cattle Soc. N.Z. Vet. Assoc.*, 13: 64–75.

Reeve, J.L. and Sharkey, M.J., 1980. Effect of stocking rate, time of lambing and inclusion of lucerne on prime lamb production in north-east Victoria. *Aust. J. Exper. Agric. Anim. Husb.*, 20: 637–653.

Robinson, G.G. and Simpson, I.H., 1975. The effect of stocking rate on animal production from continuous and rotational grazing systems. *J. Br. Grassland Soc.*, 30: 327–332.

Sharrow, S.H. and Krueger, W.C., 1979. Performance of sheep under rotational and continuous grazing on hill pastures. *J. Anim. Sci.*, 49: 893–899.

Sharrow, S.H., Krueger, W.C. and Thetford, F.O. Jr., 1981. Effects of stocking rate on sheep and hill pasture performance (grazing). *J. Anim. Sci.*, 52: 210–217.

Sheath, G.W., Webby, R.W. and Pengelly, W.J., 1984. Management of late spring–early summer pasture surpluses in hill country. *Proc. N.Z. Grassland Assoc.*, 45: 199–206.

Smeaton, D.C., 1983. Sheep management on hill country. In: *Proc. Ruakura Farmers' Conf.*, 35: 47–53.

Smeaton, D.C., Knight, T.W. and Winn, G.W., 1981. Problems in flushing ewes on North Island hill country. *Proc. N.Z. Soc. Anim. Prod.*, 41: 183–189.

Smith, M.E., McLaren, P.N. and Hopkins, D.R., 1979. Farm production gains following adoption of a hill country grazing management system. *Proc. N.Z. Soc. Anim. Prod.*, 39: 251–253.

Suckling, F.E.T., 1975. Pasture management trials on unploughable hill country at Te Awa. III. Results for 1959–69. *N.Z. J. Exper. Agric.*, 3: 351–436.

Wallace, L.R., 1961. Influence of liveweight and condition on ewe fertility. In: *Proc. Ruakura Farmers' Conf.*, 13: 14–25.

Wheeler, J.L., 1962. Experimentation in grazing management. *Herbage Abst*. 32: 1–7.

White, D.H., McConchie, B.J., Curnow, B.C. and Ternouth, A.H., 1980. A comparison of levels of production and profit from grazing Merino ewes and wethers at various stocking rates in northern Victoria. *Aust. J. Exper. Agric. Anim. Husb.*, 20: 296–307.

Young, N.E. and Newton, J.E., 1975. A comparison of rotational grazing and set stocking with ewes and lambs at three stocking rates. *Anim. Prod.*, 21: 303–311.

Chapter 12

# OTHER DOMESTICATED ANIMALS

J.M. WALSINGHAM

## INTRODUCTION

The utilization of grassland by herbivores is dominated, throughout the world, by cattle and sheep (Chapters 9–11). However, other ruminants are also important; for example, goats and buffalo constitute 16% and 4% respectively of the world's ruminant animals (F.A.O., 1982), and make substantial contributions to supplies of meat and milk for human consumption (Table 12.1), whereas deer have recently increased in importance. In addition, various non-ruminants, such as rabbits, pigs, poultry and horses, derive at least a proportion of their diet from managed grasslands. This chapter is concerned with the general pattern of productivity of these various species in a grassland context; production under more intensive conditions is considered in Volume 21.

## RUMINANTS

### Goats

Goats (*Capra hircus*) probably originated in Asia and are still commoner in the Asian and Africa continents than elsewhere. They thrive in most climates but are commoner in dry zones, where they can survive on the meagre vegetation. In some situations, a relatively small proportion of the goats' diet may be derived from pasture, most coming from scrub, trees and hedgerows.

The size of goat herds varies from one or two goats, serving the milk requirements of a family, to herds of up to 1000, which are kept for meat production and may migrate in search of food supplies. Many breeds of goats exist throughout the world (Mason, 1981), and some are specially suitable for meat, milk and fibre production (Table 12.2). There are large differences between breeds, especially in milk production (Table 12.3).

There have been few investigations into the digestive efficiency of goats, compared with other ruminants. There is no evidence that goats differ from sheep in their ability to digest high-quality herbage (Ndosa, 1980), but in the tropics they are able to utilize roughage better than sheep (Devendra, 1978). The limited amount of data available (Table 12.4) suggest that young goats are slightly less efficient in converting feed into meat than young sheep, which feed on high-quality herbage. However, it should be borne in mind that the sheep, but not the goats, have been highly selected

TABLE 12.1

World production of milk and meat from ruminants in 1981 (data from F.A.O., 1982)

|         | Meat ($t \times 10^3$) | Milk ($t \times 10^3$) |
|---------|------------------------|------------------------|
| Cattle  | 45 548                 | 428 213                |
| Sheep   | 5984                   | 7910                   |
| Goats   | 2049                   | 7559                   |
| Buffalo | 1304                   | 27 943                 |

TABLE 12.2

Some breeds of goats kept for different purposes

| Intensive dairying | Meat | Fibre |
|--------------------|------|-------|
| Anglo-Nubian       | Boer | Angora |
| Saanen             | Cashmere | Cashmere |
| Toggenburg         |      | Pashmina (India) |
| British Alpine     |      | Don (U.S.S.R.) |

TABLE 12.3

Performance of various breeds of dairy goats

| Goat breed | Milk yield per year (kg) | Lactation length[1] (days) | Source |
|---|---|---|---|
| Saanen | 684 | 273 | French (1970) |
| "Dairy goats" | 563[2] | 240 | Morand-Fehr and Sauvant (1978) |
| Indigenous (in the tropics) | 60–500[2] | 126–283 | Devendra (1975) |
| Exotic (in the tropics) | 119–886[2] | 106–344 | Devendra (1975) |
| Exotic in temperate environments | 989–2707[2] | 356–730 | Devendra (1975) |
| Anglo-Nubian | 989 | up to 365 | Devendra and Burns (1970) |
| British Saanen | 1227 | up to 365 | Devendra and Burns (1970) |

[1] Lactation length varies considerably with management and genotype.
[2] Diet not specified, but forage based.

for the purpose. There are few reported comparisons of dairy goats with dairy cows, though average performances in France, where grass is the main dietary component, suggest that goats have considerable potential as converters of herbage into milk (Table 12.5). There is evidence that goats selectively graze scrub and weed species (Lambert et al., 1982; Rolston et al., 1982), but discriminate against legumes (Wilkinson and Stark, 1982). As a result, mixed grazing or alternate grazing may be a useful management tool for improving pasture, especially when scrub (e.g. gorse, *Ulex europaeus*) invades pastures (Rolston et al., 1982).

## Water buffalo

The water buffalo (*Bubalus bubalis*) is, as yet, of minor importance in the world as a meat producer but contributes significant quantities of milk (Table 12.6), as well as being the main draught animal in most of Asia. In temperate regions, the water buffalo, when fed adequately, can compete economically with the dairy cow in production of milkfat (Chalmers, 1974). This is partly due to the high fat content of buffalo milk (Table 12.6). In most areas of the world, grass is the principal

TABLE 12.5

A comparison of the average milk production of cows and goats in France (after Morand-Fehr and Sauvant, 1978)

|  | Cow | Goat |
|---|---|---|
| Lactation length (days) | 280 | 240 |
| Milk yield per year (kg) | 3815 | 563 |
| Milk yield per year per kg liveweight (kg) | 6.36 | 9.38 |
| Average milk yield per day of lactation per kg liveweight (g) | 22.7 | 39.1 |
| Fat corrected milk yield per year (kg) | 3694 | 512 |
| Average fat corrected milk yield per day per kg liveweight (g) | 22.0 | 35.6 |

TABLE 12.4

A comparison of the performance of lambs and kids reared for meat on dried lucerne (after Walsingham, 1981)

|  | Lambs (Suffolk × Kerry Hill) | Kids (British Saanen) |
|---|---|---|
| Age at weaning (days) | 42 | 35 |
| Age at slaughter (days) | 130 | 204 |
| Growth rate, weaning to slaughter (g day$^{-1}$) | 248 | 187 |
| Feed intake (kg DM) | 127.0 | 176.8 |
| Carcass weight (kg) | 16.5 | 17.8 |
| Yield of carcass per unit of feed (kg kg$^{-1}$) | 0.13 | 0.10 |

## TABLE 12.6

The percentage composition of milk of various animal species

|  | Total solids | Fat | Casein | Whey protein | Lactose | Ash |
|---|---|---|---|---|---|---|
| Cow |  |  |  |  |  |  |
| B. taurus | 12.7 | 3.7 | 2.8 | 0.6 | 4.8 | 0.7 |
| B. indicus | 13.5 | 4.7 | 2.6 | 0.6 | 4.9 | 0.7 |
| Water buffalo | 17.2 | 7.4 | 3.2 | 0.6 | 4.8 | 0.8 |
| Horse | 11.2 | 1.9 | 1.3 | 1.2 | 6.2 | 0.5 |
| Goat | 13.2 | 4.5 | 2.5 | 0.4 | 4.1 | 0.8 |
| Deer | >20.0 | 10.0 | n.a. | n.a. | n.a. | n.a. |
| Rabbit | 25.7 | 12.3 | n.a. | n.a. | 1.12 | n.a. |

Data from Coates et al. (1964), Jenness and Sloan (1970), and North of Scotland College of Agriculture (1978).

source of food for buffalo, though this is often supplemented with cereal straws. It is said that water buffalo are able to make more efficient use of fibrous material than cattle, but this has yet to be proved experimentally (Chalmers, 1974).

The predominant use of buffalo as draught animals has overshadowed their potential for meat and milk production and, consequently, there has been little research on their management at pasture.

## Deer

The only type of deer that is employed agriculturally is the red deer (*Cervus elaphus*), but even this has no major importance on a world-wide basis. Its distribution is widespread across the parklands and hills of northern Europe, and it has been successfully introduced into Australasia and South America (Whitehead, 1972). Previously, they were kept, largely for ornament, in parks but were also hunted in the wild state. It is only in recent years that they have begun to be farmed on the lowlands as well as in the uplands of both the United Kingdom and New Zealand (Blaxter et al., 1974). Some analyses have suggested (Bryden, 1978) that it would be more economic to farm deer than sheep on some hill and moorland pastures. Venison has the advantage that it contains less fat than lamb (Table 12.7), and has always been considered a luxury meat in Europe. A large proportion of the income from deer farming in New Zealand, in recent years, has been for the sale of "velvet" (the skin covering growing antlers), used medicinally in the Far East.

Deer thrive on good grassland, but optimum stocking rates are difficult to determine. Most of

## TABLE 12.7

The composition of various types of meat as retailed

|  | Energy (MJ kg$^{-1}$) | Protein (%) | Fat (%) |
|---|---|---|---|
| Lamb | 10.1 | 11.9 | 21.1 |
| Chicken | 5.1 | 12.3 | 7.7 |
| Horse-meat | 3.9 | 15 | 3 |
| Goat-meat | 5.1 | 14 | 7 |
| Rabbit | 4.9 | 17 | 5 |
| Venison | 5.3 | 21.0 | 4 |
| Goose | 8.6 | 10.0 | 18.0 |
| Buffalo | 8.7 | 20.5 | 9.6 |

Data from Chatfield (1954), Watt and Merrill (1963), and Ognjanovic (1974).

## TABLE 12.8

A comparison of the output of carcass meat per hectare of grassland using deer and sheep (after Walsingham, 1981)

|  | Deer | | Sheep | |
|---|---|---|---|---|
|  | low stocking | high stocking | MLC[1] average farm | MLC[1] top third of farms |
| Stocking rate (adult females ha$^{-1}$) | 5 | 12.5 | 10.4 | 12.7 |
| Young reared per female | 0.63 | 0.63 | 1.40 | 1.57 |
| Carcass weight (kg) | 48.5 | 48.5 | 18.0 | 18.0 |
| Carcass output (kg ha$^{-1}$) | 152 | 382 | 262 | 358 |

[1]Meat and Livestock Commission.

the practical experience relates to unfertilized permanent pasture in deer parks, where only about 2.5 hinds per hectare are kept. However, on fertilized pastures in New Zealand, as many as 16 hinds per hectare have been carried (Young, 1978). Any calculation of the comparative efficiency with which deer utilize grassland thus has to make some assumptions about stocking rate. Two examples (Table 12.8) serve to show that, in terms of biological productivity, deer have considerable potential.

The principal difficulty with red deer is that they have nervous dispositions and are difficult to tame, unless hand-reared from birth. They are therefore more difficult to handle than other domestic animals and require expensive fencing. However, experience in the domestic handling of deer, particularly on research farms, is gradually yielding solutions to these problems.

Red deer can convert large quantities of low-quality herbage into good-quality meat, and they are less selective than sheep in their grazing habits (Milne et al., 1976). However, deer lack the capacity of cattle to process fibrous food quickly, since the rumen is small, so that they need alternating periods of feeding and undisturbed rest (Short, 1963).

**NON-RUMINANTS**

Horses (*Equus caballus*) and rabbits (*Oryctolagus cuniculus*) are the most important non-ruminant animals using grassland products. The meat of these two species differs in acceptibility in different parts of the world. The contribution of grass to the diet of the two species also differs throughout the world.

## Horses

Horses consume much the same feeds as cattle and have a similar capacity to digest them (Olsson, 1969). Traditionally, in Europe, horses are kept at pasture and receive supplements of conserved grass (usually hay) and cereals, according to the work they do and their physiological state. Most horses in the United Kingdom are kept for pleasure purposes, and very few are now employed for draught-power or meat production. A similar pattern of decline of working horses and increase in leisure horses has occurred throughout Europe (Staun et al., 1982). However, horse meat is an acceptable component of human diet in some European countries, notably France and Italy. Some of this meat comes from specially bred animals, but the rest is from culled animals of various breeds.

Most horses are kept at pasture during their early growth and development, but mixed grazing with dairy cattle is now much less common than it was once. More information is needed on the nutritional requirement of horses at pasture, and for the rearing of young horses (Cunha, 1980; Staun et al., 1982). Such information could be used to improve the production of horse meat, perhaps from fast-growing breeds, for those countries where horse meat is in demand.

## Rabbits

The other non-ruminants which make considerable use of grassland products are rabbits, though a distinction needs to be made between wild rabbits, an agricultural pest, and domestic rabbits, which are usually primarily housed. Wild rabbits select a variety of plant material from agricultural land, including hedgerow and woodland herbs as well as grasses and cereals (Monk, 1986). The losses in agricultural productivity caused by rabbits are difficult to assess, but rabbits were a major pest of grassland in Australia and New Zealand, and an important pest of crops and pasture in Britain, before the advent of myxomatosis. The severe reduction in rabbit numbers following myxomatosis allowed much semi-natural grassland in England to revert to scrub, including hawthorn, gorse, bracken and blackberry (Wells, 1980).

Rabbit production in most of Europe is now predominantly undertaken in buildings, usually with good temperature and ventilation control systems, and using dried pelleted feeds. This seems unfortunate, in view of the fact that the rabbit's growth on good quality dried grass or legume is as rapid as on the usual diet of only 30% grass (Walsingham and Large, 1977). However, the economics of dried grass and of cereal production have encouraged the development of highly productive and sophisticated housed rabbit systems (Lebas and Matheron, 1982). In the United Kingdom, some 9000 t of grass is utilized in rabbit diets each year. Similarly, small proportions of grassland

TABLE 12.9

Feed intake, meat production and the conversion ratio for various meat-producing animals fed on herbage diets (data from Treacher and Hoxey, 1969; Tayler 1970; and Grassland Research Institute, 1974)

|         | Feed intake[1] by individual (kg DM) | Carcass output (kg) | $E^2$ |
|---------|--------------------------------------|---------------------|-------|
| Cattle  | 3005                                 | 257                 | 8.6   |
| Sheep   | 108–136                              | 19–18               | 17.6–13.2 |
| Rabbits | 3.086                                | 1.028               | 33    |

[1]Other than milk.

[2] $E = \dfrac{\text{carcass output (kg)}}{\text{feed intake (kg DM)}} \times 100.$

production are used in the rest of Europe. However, successful attempts are being made in New Zealand to farm rabbits in outdoor enclosures. This has the advantage that the rapid reproductive rate of rabbits might be used to develop grazing systems which closely match the seasonal pattern of grass growth. Rabbits might then be competitive with cattle and sheep in the conversion of herbage to meat (Table 12.9).

### Geese

Geese (*Anser anser*) are one of the few domestic poultry species which obtain a significant proportion of their diet from grassland. They are good grazers and, provided the grass is short, most of their nutrient requirements can be satisfied without recourse to cereal supplements during the grazing season. Cereal pellets are normally used for about the first month after the young hatch, and diminishing quantities are used in the subsequent four weeks as the birds adapt to grassland. Some cereal supplementation may also be provided in the final fattening period. There have been virtually no attempts to measure the food conversion efficiency of geese at pasture. In addition to the grazing of domesticated geese on grassland, wild geese can be a major pest of managed grassland (Newton and Campbell, 1973).

### Invertebrates

There is great consumer resistance to the meat of invertebrate animals, some of which utilize grassland as their principal feed. The only truly farmed grassland invertebrate is the snail *Helix pomatia*. This snail is an important component of the gastronomy of France and of a number of other countries. Traditionally, in France, snails were collected from the wild and then fattened in timber-edged grassland paddocks with the addition of material like cabbages (Darbishire, 1897). *Helix pomatia* is still difficult to breed in captivity, and the current production methods still rely on collection from the wild, which can affect the viability of natural populations (Welch and Pollard, 1975), followed by a period of fattening. Fattening can occur on various types of waste biomass, though other species prefer dead or decaying material (Williamson and Cameron, 1976). Grassland is likely to be relatively unimportant in snail farming but, conversely, snails may in some years achieve numbers large enough to reduce the productivity of grassland.

### CONCLUSION

It is clear that a variety of other animals, besides cattle and sheep, utilize grassland and convert it into a range of products. Often they are less desirable than the products of cattle and sheep, so that they command a lower market price. However, in some cases the products are more valuable than those of cattle and sheep, are highly acceptable and represent good use of grassland resources. Whilst the acceptability of dietary components can change, and may yet embrace currently unacceptable meats, it seems more likely that developments in food-processing techniques will enable food manufacturers to utilize the products of a variety of animals as protein sources from which to manufacture acceptable foodstuffs.

### REFERENCES

Blaxter, K.L., Kay, R.N.B., Sharman, G.A.M, Cunningham, J.M.M. and Hamilton, W.J., 1974. *Farming the Red Deer*. First report of investigations by the Rowett Research Institute and the Hill Farming Research Organisation, H.M.S.O., Edinburgh.

Bryden, J.M., 1978. Deer versus sheep: a model for analysing the comparative value of deer and sheep farming in private and social terms. *J. Agric. Econ.*, 29(1): 23–28.

Chalmers, M.I., 1974. Nutrition. In: W.R. Cockrill (Editor), *The*

*Husbandry and Health of the Domestic Buffalo*. F.A.O., Rome, pp. 167–194.

Chatfield, C., 1954. *Food Composition Tables*. F.A.O. Nutritional Studies No. 11. F.A.O., Rome.

Coates, M.E., Gregory, M.E. and Thompson, S.Y., 1964. The composition of rabbits milk. *Br. J. Nutrit.*, 18: 583–586.

Cunha, T.J., 1980. *Horse Feeding and Nutrition*. Academic Press, New York, N.Y.

Darbishire, R.D., 1897. A visit to a snail farm. *J. Conch.*, 8: 374.

Devendra, C. and Burns, M., 1970. *Goat production in the Tropics*. Technical Communication No. 19. Commonwealth Bureau of Animal Breeding and Genetics. C.A.B., Farnham Royal.

Devendra, C., 1975. Biological efficiency of milk production in dairy goats. *World Rev. Anim. Prod.*, 11(1): 46–53.

Devendra, C., 1978. The digestive efficiency of goats. *World Rev. Anim. Prod.*, 14(1): 9–22.

F.A.O., 1982. *F.A.O. 1981 Production Yearbook, 35*. F.A.O., Rome.

French, M.H., 1970. *Observations on the Goat*. F.A.O., Rome.

Hunter, P.J., 1969. Slugs and their control. In: *Proc. 5th Br. Insecticide Fungicide Conf.*, pp. 715–719.

Jenness, R. and Sloan, R.E., 1970. The composition of milks of various species: a review. Review Article No. 158. *Dairy Sci. Abstr.*, 32(10): 599–612.

Lambert, M.G., Luscombe, P.C. and Clark, D.A., 1982. Soil fertility and hill country production. *Proc. N.Z. Grassland Assoc.*, 43: 153–160.

Lebas, F. and Matheron, C., 1982. Rabbits. *Livestock Prod. Sci.*, 9: 235–250.

Mason, I.L., 1981. Breeds. In: C. Gall (Editor), *Goat Production*. Academic Press, London, pp. 57–110.

Milne, J.A., MacRae, J.C., Spence, A.M. and Wilson, S., 1976. Intake and digestion of hill land vegetation by the red deer and the sheep. *Nature*, 263: 763–764.

Monk, K., 1986. Feeding strategy of farmland rabbits. Poster Paper. In: *Third European Ecological Symposium: Plant–Animal Interactions, Lund, Sweden*.

Morand-Fehr, P. and Sauvant, D., 1978. Nutrition and optimum performance of dairy goats. *Livestock Prod. Sci.*, 5: 203–213.

Ndosa, J.E.M., 1980. *A comparative Study of Roughage Utilisation by Sheep and Goats*. Thesis, Department of Agriculture and Horticulture, University of Reading, Reading.

Newton, I. and Campbell, C.R.G., 1973. Feeding of geese on farmland in east-central Scotland. *J. Appl. Ecol.*, 10: 781–801.

North of Scotland College of Agriculture, 1978. Red Deer Farming. *College Bull.*, No. 14.

Ognjanovic, A., 1974. Meat and meat production. In: W.R. Cockrill (Editor), *The Husbandry and Health of the Domestic Buffalo*. F.A.O., Rome, pp. 912–960.

Olsson, N.O., 1969. The nutrition of the horse. In: D. Cuthbertson (Editor), *Assessment of and Factors Affecting Requirements of Farm Livestock. Nutrition of Animals of Agricultural Importance, 2*. Pergamon Press, London.

Rolston, M.P., Lambert, M.G. and Clark, D.A., 1982. Weed control options in hill country. *Proc. N.Z. Grassland Assoc.*, 43: 196–203.

Short, H.L., 1963. Rumen fermentations and energy relationships in white-tailed deer. *J. Wildl. Manage.*, 27: 184–195.

Staun, H., Bruns, E., Forde, D.J., Haring, H., Langlois, B. and Minkema, D., 1982. Horses. *Livestock Prod. Sci.*, 9: 217–234.

Tayler, J.C., 1970. Dried forages and beef production. *J. Br. Grassland Soc.*, 25(2): 180–190.

Treacher, T.T. and Hoxey, A.M., 1969. *Internal Report No. 170*. Grassland Research Institute, Hurley.

Walsingham, J.M., 1981. Potential change in animal output from grassland: other animals. In: J.L. Jollans (Editor), *Grassland in The British Economy*. C.A.S. Paper 10. Centre for Agricultural Strategy, Reading, pp. 483–495.

Walsingham, J.M. and Large, R.V., 1977. The performance of New Zealand White rabbits on dried green feeds. Poster paper. In: *British Society of Animal Production Winter Meeting, 69*.

Watt, B.K. and Merrill, A.L., 1963. *Composition of Foods: Raw, Processed, Prepared*. U.S. Dep. Agric. Handbook No. 8. Agricultural Research Service, U.S. Dep. Agric., Washington, D.C.

Welch, J.M. and Pollard, E., 1975. The exploitation of *Helix pomatia*. *Biol. Conserv.*, 8: 155–160.

Wells, T.C.E., 1980. Management options for lowland grassland. In: T.H. Rorison and R. Hunt (Editors), *Amenity Grassland: an Ecological Perspective*. John Wiley and Sons, London, pp. 175–195.

Whitehead, G.K., 1972. *Deer of the World*. Constable, London.

Wilkinson, J.M. and Stark, B.A., 1982. *Goat production*. Chief Scientist's Group, Ministry of Agriculture, Fisheries and Food, London.

Williamson, P. and Cameron, R.A.D., 1976. Natural diet of the landsnail *Cepaea nemoralis*. *OIKOS*, 27: 493–500.

Young, C.D., 1978. *Intensive Deer Management*. North of Scotland College of Agriculture, Aberdeen.

Chapter 13

# GRAZING BEHAVIOUR

G.W. ARNOLD

## INTRODUCTION

The behaviour of grazing animals is, to a large extend, determined by their requirements for food. The behavioural patterns involved include the search for, and location and ingestion of food. An understanding of this behaviour is important in developing successful animal husbandry practices. Husbandry practices range from intensive housing, where man controls most of the factors influencing ingestive behaviour (see Volume 21 of this series), to free-range conditions where there is little or no control. At one extreme, the food available, the time span during which it is available, and the social context of its consumption are closely controlled. At the other extreme, complex interactions of animal and environmental factors determine diet (Fig. 13.1); a detailed review of this subject has been published by Arnold and Dudzinski (1978).

## THE CIRCADIAN PATTERNS OF BEHAVIOUR

The daily routine of the grazing animal is largely determined by times of feeding; other activities accommodate to this.

Sheep and cattle, in temperate regions, usually graze around dawn and again in the afternoon (Fig. 13.2). When daylength is short, these two periods merge; they do little grazing at night compared to horses. The times at which sheep start and stop grazing in the morning, and stop grazing after dark, is influenced by the times of sunrise and sunset, but grazing starts and stops earlier in the morning on hot days; similarly, afternoon grazing is affected by both temperature and humidity (Dudzinski and Arnold, 1979).

In tropical climates, when daytime conditions are hot, cattle may do much of their grazing at night (Payne et al., 1951), though daytime grazing also occurs (Harker et al., 1954, 1961). Above 25°C, night grazing varies from 0% to 70% of total grazing time. This variation may well be influenced by differences in humidity, but too little information is available to substantiate this.

Cattle and sheep seem to sense how hot a day will be. For example, Low et al. (1981) found a correlation between the time that cattle started grazing in the morning and the subsequent maximum temperature. There is also a correlation between temperature and the time at which morning grazing ceases; for example, cattle arrive at water when the temperature is 29°C (Dwyer, 1961).

Night temperatures between −7°C and 9°C have little effect on sheep grazing at night or in the early morning (Arnold, 1982). Little is known about the effects of low temperatures on cattle grazing.

Cloud has the effect of shortening the days (Gonzalez, 1964). Rain has little effect *per se* but, when accompanied by wind, affects the direction of grazing by cattle and sheep, but not horses; however, animals cease grazing in prolonged driving rain.

Patterns of grazing, and the effects of climatic factors, are greatly changed when animals are disturbed, for instance by twice daily milking, or by being moved to new areas of feed, as with rotational grazing.

## TIME SPENT FEEDING

Most wild animals, such as birds, make best use of feeding time through the choice of types of food,

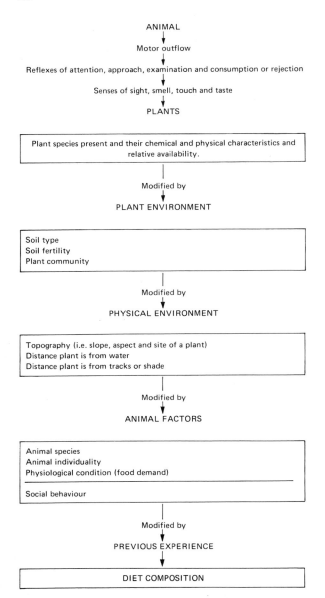

Fig. 13.1. Model of an animal's diet selection.

selection of the patches containing that food, and by optimizing the pattern and speed of movement between patches. There is a strong tendency to maximize the ingestion of energy and other essential nutrients per unit time (or energy) spent in feeding. The control imposed on domesticated animals may not allow this degree of choice, though there is selection between types of food (see below).

Sheep and cattle graze for a similar period daily (4.5—14.5 h), but horses graze for longer (Table 13.1). The time spent grazing varies, depending on the amount of feed available, the animal's feed requirement, and the climate. Feed requirement is, in part, dependent on age. The time that young animals spend grazing is influenced by their dam's milk supply. Grazing time for calves increases rapidly as the milk supply decreases, and by six months of age calves will graze for about 80% of the time their dams spend grazing (Chambers, 1959; Hutchison et al., 1962). Lambs feed independently at an earlier age.

Intake is determined not only by the time spent grazing but also by the number of bites taken per unit time, and by weight taken per bite. For example, when feed requirement is increased during late pregnancy, intake increases by increasing the rate of eating by 27%, though grazing time is similar to that of non-pregnant sheep. However, during early lactation, sheep graze for 7 to 12% longer (Arnold and Dudzinski, 1967a; Arnold, 1975), and also eat 20% faster. Similarly, thin sheep eat more than fat sheep, both by grazing for a slightly longer period and by eating much faster (Arnold and Birrell, 1977). Shorn sheep, subject to cold stress, graze for less time than woolled sheep, but eat faster.

Groups of animals whose grazing times differ may interact socially if grazed together, tending to reduce the difference in their grazing times: this interaction is termed social facilitation. It appears to occur consistently with sheep (Tribe, 1950; Holder, 1962), but not always in cattle (Hodgson and Wilkinson, 1967; Bailey et al., 1974).

## GRAZING TIME AND MANAGEMENT

Various management factors affect the time spent grazing. For example, feeding supplements to grazing animals reduces their grazing time, by between 11 and 28 min per kilogram of concentrate (Stobbs, 1970; Sarker and Holmes, 1974; Combellas et al., 1979).

Livestock are sometimes removed from the pasture or range at night but, even with a 7-h grazing day, are able to maintain their food intake by eating faster (Smith, 1961). Full compensation, however, may not occur under humid conditions, or when feed availability is low.

# GRAZING BEHAVIOUR

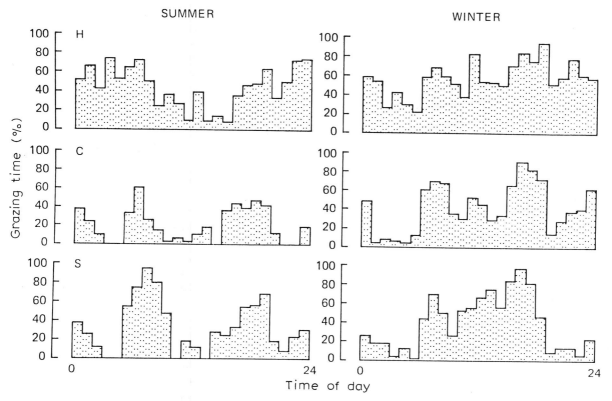

Fig. 13.2. The grazing activity (% of time) of horses (*H*), cattle (*C*) and sheep (*S*) during summer and winter in a temperate environment.

The grazing behaviour of both sheep and cattle is different under strip-grazing, where a daily allowance of herbage is given, from that under continuous grazing, where the structure and mass of the pasture is changing only slowly. Under continuous grazing, grazing time and bites per minute increase as pasture availability decreases. However, Baker et al. (1981a) found that the grazing time of both cows and calves decreased as daily pasture allowance decreased under strip-grazing. The animals appear to "set" their behaviour to the initial conditions of the sward and, once they have eaten the herbage to a certain height, they do not adjust by grazing for longer periods or taking smaller bites, even if the allowance is inadequate to meet their needs. This has important practical implications, and Gibb and Treacher (1976) showed that, if herbage allowance is not more than three times the daily intake, the intake by lambs may be restricted.

Several attempts have been made to relate grazing behaviour to herbage height, density and mass of the pasture and to animal liveweight (e.g. Arnold and Dudzinski, 1967a, b; Arnold, 1975). In one study (Arnold and Dudzinski, 1967b), involving six types of mixed pasture, virtually none of the variation in either grazing time or rate of eating

TABLE 13.1

Grazing times of sheep, cattle and horses grazing together (h day$^{-1}$)

|  | No supplements available | Supplements available |
|---|---|---|
| *Winter — green feed* | | |
| Sheep | 10.4 | 9.6 |
| Cattle | 9.5 | 5.1 |
| Horses | 11.5 | 5.2 |
| *Summer — dry feed* | | |
| Sheep | 9.5 | 6.7 |
| Cattle | 7.7 | 3.0 |
| Horses | 14.2 | 7.2 |

could be explained. In another two studies (Arnold and Dudzinski, 1967a; Arnold, 1975), only between 47 and 66% of the variation in rate of eating and 4 to 32% of the variation in grazing time could be accounted for by pasture yield and height, herbage density and herbage digestibility.

Hodgson (1981), using strip-grazed pastures of ryegrass (*Lolium perenne*), showed that both intake per bite and rate of herbage intake by calves and lambs were affected by pasture height but not by herbage density. The rate of biting was also less sensitive to these factors under strip-grazing, where sward changes are rapid, than under continuous grazing, where changes are slow. However, neither Hodgson (1981) nor Baker et al. (1981a, b) could produce a model that satisfactorily integrated the results of different experiments.

## THE CHOICE OF DIET

### Mechanics of grazing

The ability of animals to graze selectively depends on the structure and size of their jaws and teeth, and on their basic method of grazing. Sheep graze by gripping the herbage between the incisors and the dental pad and either biting or breaking it by jerking the head forwards or backwards. Cattle usually gather vegetation into the mouth with the tongue before biting and tearing it off, unless the vegetation is too short, when they bite off the herbage. Both species move their muzzles in a horizontal plane, when grazing all but very long herbage, and select in a vertical one. Because sheep have smaller mouths, they take smaller bites and can graze closer and more selectively than cattle. Cattle and sheep tend to nibble the leaves from long herbage, such as lucerne and tropical pastures.

Because of the way they graze, taking bites from the canopy surface downwards, the material eaten will often be different from that on offer. However, this does not mean that the animals have been deliberately selective. It may simply mean that it is physically impossible to be non-selective.

### Selective grazing

The composition of the diet is often very different from that on offer. Selection is most intense when food is abundant and varied (Leigh and Mulham, 1966a, b), and may only be slight when swards are simple (Davis, 1964; Arnold et al., 1966; Milne et al., 1982a). Detailed reviews of selection, and factors affecting it, will be found elsewhere (Arnold, 1964a, b; Heady, 1964; Arnold and Hill, 1972).

In general, leaf is eaten in preference to stem, and green (or young) material in preference to dry (or old) material. The selected material usually contains more nitrogen, phosphate and gross energy, but less "fibre" (Chapter 14). Differences in other mineral constituents and soluble carbohydrates are less consistent.

Selection might occur because animals preferred certain material or merely because that material was more easily grazed. It is therefore important to examine the basis of selection. Similarly, it is important to know whether the selected diet is nutritionally more valuable and whether animals exhibit "nutritional wisdom" in diet selection.

Grazing animals use the senses of sight, touch (lips and mouth), taste, and smell in selecting their diet. Sight is used primarily to orient the grazing animal to other animals and to its environment. Sheep and cattle can recognize conspicuous food plants by sight but use sight very little in selection of grazing (Arnold, 1966a).

Both Arnold (1966b) and Kreuger et al. (1974) have shown that touch, taste and smell are used in selective grazing, but that taste is the most important sense. Smell is less important; for example, Milne et al. (1982b) found that only in one of five cases did anosmic sheep select a diet different from that chosen by normal sheep.

The animal's response to sense stimuli is moderated by its current nutritional state. For example, hungry animals usually have lower thresholds of rejection for taste (Goatcher and Church, 1970) and smell.

Presumably, the concentration of specific molecules are the stimuli, but these must be detected from a background of chemical "noise". Differences probably occur between and within animal species, in sensitivity to the stimuli or to the balance between stimuli, such as bitter to sweet taste. As a result, species and individual animals differ in their diet selection.

There is no doubt that failure to understand how ruminants select their food has led to much wasted effort in seeking "palatability factors" in pasture plant species. There have been numerous attempts to relate preferences to the content of such things as "soluble carbohydrates", "nitrogen", "crude fibre", "energy", or "ash", which animals cannot recognize. In cases where correlations are found between such characteristics and preferences, there must be specific compounds, or some physical property of the plant, which determines choice. In the case of ash, it could be related to the content of sodium or potassium salts; with fibre, the ease of harvesting could be the significant factor (Evans, 1964).

A simple correlation between preference rating and the content of a particular chemical compound does not necessarily imply a causal relationship; for example, another causal factor might be correlated with the character measured. Such correlations may be used to construct hypotheses, but these should then be tested experimentally.

Another problem in evaluating relationships between preference and plant attributes is the fact that most analyses are based on the overall concentration of a component, or group of components, frequently determined on oven-dried material. This does not take account of the fact that the components may be concentrated in one part of a plant, or may be modified during the drying process. In addition, maceration and autolysis during chewing could change the chemical forms presented to the taste-buds.

## ANIMAL FACTORS INFLUENCING PREFERENCES

### Physiological status

Generally, only small differences have been found in the diets selected by lambs and older sheep (Hodge and Doyle, 1967; Langlands, 1969; Jamieson and Hodgson, 1979), or between calves and cows (Le Du and Baker, 1981), though Arnold et al. (1982) found that five-months-old sheep selected a diet higher in digestibility and in nitrogen content, and lower in fibre, than did older sheep.

Breeds of sheep differ in the diet they select (Arnold et al., 1982). These differences are at least partly due to differences in taste preferences (Arnold and Hill, 1972). Fewer studies have been made with cattle, but no differences in preferences were found between Hereford and Santa Gertrudes cows grazing on a New Mexico range (Herbel and Nelson, 1966).

Optimal foraging theory would predict that food preferences might differ for animals of different physiological status, or with different food requirements. There is no evidence of this in either sheep or cattle grazing on sown pastures (Arnold, 1967; Le Du and Baker, 1981).

### Animal species

Sheep, cattle and goats differ in food preferences when grazing species–rich swards (Dudzinski and Arnold, 1973; Langlands and Sanson, 1976), but not when grazing simple swards (Le Du and Baker, 1981). Differences in the mechanics of grazing may cause some of these differences, but in general they may be due mainly to differences in feeding strategies leading to optimal foraging. Schwartz and Ellis (1981) found that selectivity in bison (*Bison bison*), cattle, pronghorn (*Antilocapra americana*) and sheep on shortgrass prairie was particularly sensitive to changes in forage quality. Selective feeding occurred when forage conditions permitted, but feed requirements seemed to determine the switch from selective to non-selective grazing. The switch occurs at different points in forage quality for different species.

### Grazing experience

The grazing experience of sheep in early life can influence their later food preferences (Arnold and Maller, 1977; Key and MacIver, 1980). An example is given in Table 13.2.

### Nutritional wisdom

There has been considerable discussion about whether large herbivores exhibit nutritional wisdom. Zahorik and Houpt (1977) have argued that the "learned aversion" model of nutritional wisdom, found in species that eat discrete meals of a single food, cannot apply to large herbivores, where a "meal" can last for hours, and includes a large number of foods that are regurgitated and chewed again. Under these circumstances it is unlikely that animals can relate illness or benefit to eating a particular component.

TABLE 13.2

Differences in scores for utilization of two unpalatable pasture species for ewes and yearlings with different grazing experience in the previous twelve months (Arnold and Maller, 1977)

| | Previous twelve months spent grazing | | | |
|---|---|---|---|---|
| | *Phalaris aquatica* pasture at Canberra, A.C.T. | irrigated pasture at Deniliquin, N.S.W. | rangeland at Deniliquin, N.S.W. | rangeland at Trangie, N.S.W. |
| Yearlings | | | | |
| *Stipa hyalina* | 4.2 | 4.2 | 5.0 | 6.4 |
| *Eragrostis curvula* | 2.4 | 4.8 | 4.6 | 4.8 |
| Ewes | | | | |
| *Stipa hyalina* | 1.0 | 4.0 | 1.0 | 6.0 |
| *Eragrostis curvula* | 4.6 | 4.0 | 5.4 | 6.0 |

A score of 0 means no herbage eaten; 5 denotes >80% of herbage eaten.

Sensory responses have to assure adequate nutrition, but the ruminant, like all other animals, is hedyphagic — that is, food selection is directed at minimizing unpleasant and/or maximizing pleasant olfactory and other sensations (McClymont, 1968). As a result, animals may often select plants not for their nutritional advantages but for their flavour. It would seem that domestic ruminants do not show specific euphagia (food selection directed toward optimal nutrition and avoiding intoxication), and do not preferentially select plants to meet specific nutrient needs.

Evidence of beneficial selection of diet is very limited. Sheep select herbage with a higher phosphate content, even when animals are not phosphate-deficient (Ozanne and Howes, 1971); indeed, they may not be selecting for greater phosphorus content but for a lower content of free phenols, since these two constituents are negatively correlated. Conversely, Gordon et al. (1954) found that grazing sheep and cattle failed to correct a phosphate deficiency when offered a phosphate supplement. Under grazing conditions, sodium-deficient sheep may preferentially select herbage species rich in sodium (Arnold, 1964a) but, since they cannot detect less than 0.10% sodium in herbage, they may be unable to select for sodium in the most critical condition (Arnold, unpubl.).

The selection of green herbage, in preference to dry, is nutritionally advantageous, because green herbage is more nutritious and digestible. However, on luxuriant improved grass pastures, this selection leads to a diet that is richer in protein and lower in readily fermentable carbohydrates than is nutritionally desirable (Arnold et al., 1966). Sheep grazing mixed grass–legume pastures do not always select the legume component, even where it is nutritionally advantageous, because sometimes it is in the least accessible position in the plant canopy (Arnold, 1964c; Mufandaedza, 1981). Whereas there is ample evidence of diet selection among animals grazing managed pastures, there is still relatively little evidence that this selection optimizes the dietary composition. Many more critical studies are needed with domestic grazing animals under conditions of feed abundance to assess the extent to which they exhibit nutritional wisdom.

## REFERENCES

Arnold, G.W., 1964a. Some principles in the investigation of selective grazing. *Proc. Aust. Soc. Anim. Prod.*, 5: 258–271.

Arnold, G.W., 1964b. Factors within plant associations affecting the behaviour and performance of grazing animals. In: D.J. Crisp (Editor), *Grazing in Terrestrial and Marine Environments*. Blackwell, Oxford, pp. 133–154.

Arnold, G.W., 1964c. Supplementation of *Phalaris tuberosa–Trifolium subterraneum* pasture in summer with urea and molasses. *Field Stn. Rec. Div. Pl. Ind. CSIRO (Aust.)*, 3: 37–44.

Arnold, G.W., 1966a. The special senses in grazing animals. I.

# FACTORS AFFECTING INTAKE

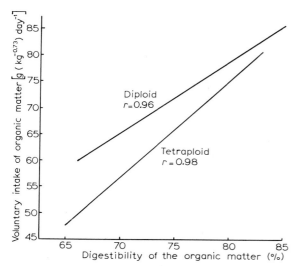

Fig. 14.2. Relation between voluntary intake and digestibility of two Italian ryegrasses (adapted from Osbourn et al., 1966).

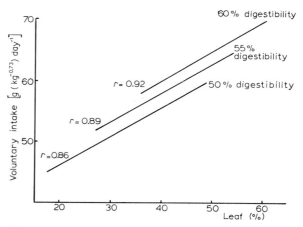

Fig. 14.4. Relation between voluntary intake and leaf percentage of six cultivars of *Panicum maximum* when dry-matter digestibility is held constant (Minson and Laredo, 1972).

digestibility (70%). This difference may have been caused by the later developmental stage of the tetraploid.

Even greater differences in intake have been reported between different species and cultivars of *Panicum* (Minson, 1971; Fig. 14.3). When compared at the same digestibility (55%), the intake of the cultivar Hamil was 27% greater than that of cultivar Kabulabula. This difference was apparently associated with the greater proportion of leaf material in Hamil (47% versus 24%) since, when the different cultivars were compared at the same digestibility, intake was closely correlated with leafiness (Fig. 14.4).

## Leaf versus stem

The greater intake of the leaf material has been confirmed in other studies, where leaf and stem fractions have been fed separately to animals (Fig. 14.5). The intake of leaf was 59% greater than that of the stem at a given digestibility (Fig. 14.5). Similar results have been obtained with cattle eating tropical grasses and legumes (Hendricksen et al., 1981; Poppi et al., 1981a) and sheep eating a

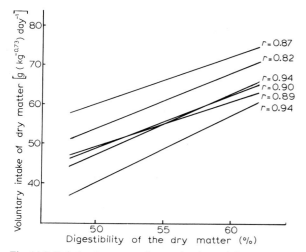

Fig. 14.3. Relation between voluntary intake and digestibility of the dry matter in six *Panicum maximum* cultivars (Minson, 1971).

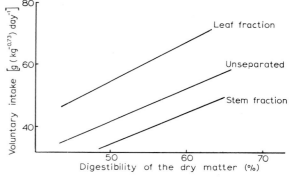

Fig. 14.5. Relation between intake and digestibility of leaf fraction, stem fraction and unseparated forage (data of Minson, 1972; Laredo and Minson, 1973).

temperate grass (Laredo and Minson, 1975). Although the intake of leaf was greater than stem, rumen fill was similar (Table 14.1), so leaf must be retained in the rumen for a shorter time; this is probably the result of more rapid physical breakdown of the leaf fraction (Poppi et al., 1981b).

**Particle size**

Another way of increasing the rate of physical breakdown might be by reducing particle size, for instance by grinding. However, finely-ground feed is dusty so it must be pelleted. Many studies have shown that intake is increased by grinding and pelleting, and that ground and pelleted forages pass more rapidly through the rumen than chopped forage (Campling and Freer, 1966; Laredo and Minson, 1975). This increase cannot be explained by an increase in digestibility, since pelleting invariably depresses digestibility (Minson, 1963).

The effect of grinding and pelleting on intake is greater with forages of low digestibility than high digestibility (Fig. 14.6). This is to be expected, since the fibre in mature feed of low digestibility is coarser and more difficult to subdivide by chewing than the fibre in young forage.

**Legumes versus grasses**

Legumes contain less fibre than grasses of similar digestibility (Minson, 1982b); as a consequence, the intake of legumes is 10 to 50% greater than that of grasses of similar digestibility (Crampton, 1957; Crampton et al., 1960; Heaney et al., 1963; Troelsen and Campbell, 1969; Demarquilly and Weiss, 1970; Milford and Minson, 1966). This difference is associated with the shorter time the legumes are retained in the rumen (17%) and larger quantity of organic matter (14%) in the rumen of sheep fed the legume diets (Thornton and Minson, 1973), though the weight of wet digesta in the rumen is less (7%) for sheep fed legumes; this indicates that legumes, with their lower fibre content, pack more densely in the rumen than do grasses (9.8 versus 8.1 g organic matter per 100 g digesta).

**Other factors limiting intake**

**Water content.** It is often suggested that the intake of succulent forages is depressed by their greater water content. However, most studies have shown no difference in dry-matter intake of grass fed fresh or after partial drying (Heaney et al., 1966; Minson, 1966; Demarquilly, 1970), though intake is depressed if forage is very wet (>82–86% water; Davies, 1962; Verite and Journet, 1970; Wilson, 1978). It has been found with dairy cows that, at water contents above 81.9%, each 1% increase depressed daily intake by 0.34 kg dry matter (Verite and Journet, 1970).

**Protein and mineral deficiency.** Intake is usually depressed when the feed is deficient in any essential nutrient. For crude protein this depression occurs when the content falls below 6 to 8% (Blaxter and Wilson, 1963; Milford and Minson, 1966; Minson and Milford, 1967). Intake of forage is also depressed by deficiencies of sulphur (Playne, 1969a; Kennedy and Siebert, 1972, 1973; Rees et al., 1974; Rees and Minson, 1978), sodium (Joyce and Brunswick, 1975; Minson, 1980), phosphorus (Little, 1968; Playne, 1969b), cobalt (Marston et al., 1938), and selenium (McLean et al., 1962). It seems probable that other deficiencies affect intake, but no examples could be found in the literature. The extent of the depression in forage intake caused by deficiencies varies from slight to a complete cessation of eating, and death in extreme cases.

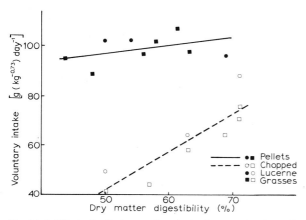

Fig. 14.6. Effect of pelleting on the relation between intake and digestibility (Minson, 1982b).

# FACTORS AFFECTING INTAKE

**Toxic factors.** Where the herbage contains toxic levels of any element (e.g. selenium) or organic compound (e.g. cyanide) the animals become ill, intake is reduced and, in extreme cases, animals die.

## GRAZING FACTORS CONTROLLING INTAKE

Most of the available information concerning the many plant factors that control forage intake has been obtained from pen-feeding studies. There is no reason to believe that the principles so defined do not apply to forage grazed *in situ*. Recently Chenost and Demarquilly (1982) concluded that "when conditions of pasture utilization are such that grazing pressure is relatively high but intake is not restricted, the intake in grazing conditions seems to be similar to that observed indoors". For example, Hodgson (1977) found that the intake of grazing animals was related to the digestibility of the forage eaten.

Despite these similarities between intake of forage in pens and under grazing, there are additional plant factors that affect the intake and productivity of animals grazing pastures (Chapter 13). The most important of these are herbage availability and variation in herbage composition.

There are four reasons for the difference in intake by animals fed cut forage as opposed to grazing: (1) herbage availability; (2) herbage variation; (3) pasture management; and (4) climate, though these various factors interact. Climate, such as extremes of temperature and wind, which depress intake, will not be considered further.

### Herbage availability

When there is little herbage available, intake is restricted. As a result, there is usually an asymptotic relationship between intake and pasture availability (Fig. 14.7). If the pasture is young and relatively homogeneous, feed intake by both sheep and cattle is not restricted as long as available dry matter exceeds 1000 to 1500 kg ha$^{-1}$ (Woodward, 1936; Johnstone-Wallace and Kennedy, 1944; Waite et al., 1950; Arnold and Dudzinski, 1966). One of the main reasons for reduced intake, where there is little available herbage, is that the amount that can be taken with each bite is low (Stobbs,

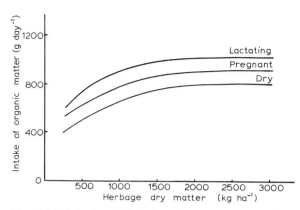

Fig. 14.7. Effect of available herbage dry matter on the intake of herbage by ewes in different physiological conditions (Arnold and Dudzinski, 1966).

1973). This can partly be offset by the animal increasing the number of harvesting bites, but there is an upper limit to the total number of bites animals will take each day (Stobbs, 1973). If the distribution of forage is uneven, as occurs with forage sown in wide rows, the maximum intake is achieved at a lower herbage availability than for a uniform pasture, since there is a smaller effect on bite size.

### Herbage variation

Pastures can be heterogeneous in terms of their leafiness, maturity, species composition, and soiling by excreta. Animals select within this variation and as a result eat more forage, and forage of a higher quality, than might be expected on the basis of the average forage on offer (Arnold and Dudzinski, 1978). Although selective grazing can have nutritional advantages, there can also be disadvantages. For example, animals may continue to search for a component that is present in small proportions, and so intake is reduced. This happened with cattle grazing *Setaria sphacelata* and *Lablab purpureus* (Chacon and Stobbs, 1976; Hendricksen and Minson, 1980), when leaf was selectively grazed. Intake was reduced by more than 50% when selective grazing reduced available leaf from 1700–2400 kg DM ha$^{-1}$ to <400 kg ha$^{-1}$. The main cause of the lower intake was a decrease in bite size from 350 mg per bite to 100 mg per bite.

Similarly, when pastures contain both green and dead material, intake and liveweight gain of both sheep and cattle are more closely related to the quantity of green dry matter per unit area than to the total forage present (Willoughby, 1958; Yates et al., 1964; Arnold and Dudzinski, 1966; 't Mannetje, 1974), since animals eat little dead material if green feed is available ('t Mannetje, 1974; Chacon and Stobbs, 1976; Hendricksen and Minson, 1980).

Animals also selectively graze particular plant species, although the preference can change with time of the year. For example, where pastures contain the legume siratro (*Macroptilium atropurpureum*), cattle may select a diet containing only 2 to 10% legume in the spring, but 62 to 73% legume during the autumn (Stobbs, 1977). Preferences may also be affected by soil fertility; for example, selective grazing of legume was greatly increased when superphosphate was applied to grass–legume pastures (McLean et al., 1981).

## Pasture management

Graziers can manipulate intake by varying stocking rate (Chapter 24) and grazing management. The total amount of forage consumed will be greater if the forage allowance per animal is reduced, but this advantage is usually offset by a reduction in the intake of forage by each animal (Raymond et al., 1956; Tayler and Rudman, 1965; Greenhalgh, 1966, 1970). This reduction can be caused by a low availability of the herbage, once dry matter falls below 1000 to 1500 kg ha$^{-1}$ (see above, p. 141). However, the most likely reason for the lower intake and animal production, when pastures are grazed intensely, is the change in the physical composition of the pasture eaten. At low stocking pressure, the animals have ample opportunity to select leaf but, as grazing intensity is increased, the proportion of leaf in the diet decreases and the proportion of stem increases (Chacon and Stobbs, 1976; Stobbs, 1978; Hendricksen and Minson, 1980); this reduces the animal's appetite.

## CONCLUSION

Forage intake can be limited by a wide range of plant factors, including chemical and physical composition, and by pasture management. All these independent variables must be considered when studying the intake and productivity of grazing animals.

## REFERENCES

Arnold, G.W. and Dudzinski, M.L., 1966. The behavioural response controlling the food intake of grazing sheep. In: *Proc. Tenth Int. Grassland Congr., Helsinki*, pp. 367–370.

Arnold, G.W. and Dudzinski, M.L., 1978. *Ethology of Free-Ranging Domestic Animals*. Elsevier, Amsterdam, 198 pp.

Blaxter, K.L., 1961. The utilization of the energy of grassland products. In: *Proc. Eighth Int. Grassland Congr., Reading*, pp. 479–484.

Blaxter, K.L. and Wilson, R.S., 1963. The assessment of a crop husbandry technique in terms of animal production. *Anim. Prod.*, 5: 27–42.

Blaxter, K.L., Wainman, F.W. and Wilson, R.S., 1961. The regulation of food intake by sheep. *Anim. Prod.*, 3: 51–61.

Blaxter, K.L., Wainman, F.W. and Davidson, J.L., 1966. The voluntary intake of food by sheep and cattle in relation to their energy requirements for maintenance. *Anim. Prod.*, 8: 75–83.

Campling, R.C. and Freer, M., 1966. Factors affecting the voluntary intake of food by cows. 6. A preliminary experiment with ground, pelleted hay. *Br. J. Nutrit.*, 17: 263–272.

Chacon, E. and Stobbs, T.H., 1976. Influence of progressive defoliation of a grass sward on the eating behaviour of cattle. *Aust. J. Agric. Res.*, 27: 709–727.

Chenost, M. and Demarquilly, C., 1982. Measurement of herbage intake by housed animals. In: J.D. Leaver, *Herbage Intake Handbook*. British Grassland Society, Hurley, pp. 95–112.

Crampton, E.W., 1957. Interrelations between digestible nutrient and energy content, voluntary dry matter intake, and the overall feeding value of forages. *J. Anim. Sci.*, 16: 546–552.

Crampton, E.W., Donefer, E. and Lloyd, L.E., 1960. A nutritive value index for forages. *J. Anim. Sci.*, 19: 538–544.

Davies, H.L., 1962. Intake studies in sheep involving high fluid intake. *Proc. Aust. Soc. Anim. Prod.*, 4: 167–171.

Demarquilly, C., 1970. Influence de la déshydratation à basse température sur la valeur alimentaire des fourrages. *Ann. Zootechn.*, 19: 45–51.

Demarquilly, C. and Weiss, Ph., 1970. *Tableaux de la valeur alimentaires des fourrages*. Ministère de l'Agriculture, Institute National de la Recherche Agronomique, Service d'Experimentation et d'Information, I.N.R.A., Versailles, 65 pp.

Demarquilly, C., Boissau, J.M. and Cuylle, G., 1966. Factors affecting the voluntary intake of green forage by sheep. In: *Proc. Ninth Int. Grassland Congr., São Paulo*, pp. 877–885.

Greenhalgh, J.F.D., 1966. Studies of herbage consumption and milk production in grazing dairy cows. In: *Proc. Tenth Int. Grassland Congress, Helsinki*, pp. 351–355.

Greenhalgh, J.F.D., 1970. The effects of grazing intensity on herbage production and consumption and on milk production in strip-grazed dairy cows. In: *Proc. Eleventh Int. Grassland Congr., Surfers Paradise, Qld.*, pp. 856–860.

Heaney, D.P., Pigden, W.J., Minson, D.J. and Pritchard, G.I., 1963. Effect of pelleting on energy intake of sheep from forages cut at three stages of maturity. *J. Anim. Sci.*, 22: 752–757.

Heaney, D.P., Pigden, W.J. and Pritchard, G.I., 1966. The effect of freezing or drying pasture herbage on digestibility and voluntary intake assays with sheep. In: *Proc. Tenth Int. Grassland Congress, Helsinki*, pp. 379–384.

Hendricksen, R.E. and Minson, D.J., 1980. The feed intake and grazing behaviour of cattle grazing a crop of *Lablab purpureus* cv. Rongai. *J. Agric. Sci., Cambridge*, 95: 547–554.

Hendricksen, R.E., Poppi, D.P. and Minson, D.J., 1981. The voluntary intake, digestibility and retention time by cattle and sheep of stem and leaf fractions of a tropical legume (*Lablab purpureus*). *Aust. J. Agric. Res.*, 32: 389–398.

Hodgson, J., 1977. Factors limiting herbage intake by the grazing animal. In: *Proc. Int. Meet. Animal Production from Temperate Grassland, Dublin*, pp. 70–75.

Johnstone-Wallace, D.B. and Kennedy, K., 1944. Grazing management and behaviour and grazing habits of cattle. *J. Agric. Sci., Cambridge*, 34: 190–197.

Joyce, J.P. and Brunswick, L.C.F., 1975. Sodium supplementation of sheep and cattle fed lucerne. *N.Z. J. Exper. Agric.*, 3: 299–304.

Kennedy, P.M. and Siebert, B.D., 1972. The utilization of spear grass (*Heteropogon contortus*). 2. The influence of sulphur on energy intake and rumen and blood parameters in cattle and sheep. *Aust. J. Agric. Res.*, 23: 45–56.

Kennedy, P.M. and Siebert, B.D., 1973. The utilization of spear grass (*Heteropogon contortus*). 3. The effect of the level of dietary sulphur on the utilization of spear grass by sheep. *Aust. J. Agric. Res.*, 24: 143–152.

Laredo, M.A. and Minson, D.J., 1973. The voluntary intake, digestibility, and retention time by sheep of leaf and stem fractions of five grasses. *Aust. J. Agric. Res.*, 24: 875–888.

Laredo, M.A. and Minson, D.J., 1975. The voluntary intake and digestibility by sheep of leaf and stem of *Lolium perenne*. *J. Br. Grassland Soc.*, 30: 73–77.

Little, D.A., 1968. Effect of dietary phosphorus on the voluntary consumption of Townsville lucerne (*Stylosanthes humilis*) by cattle. *Proc. Aust. Soc. Anim. Prod.*, 7: 376–380.

Mannetje, L.'t., 1974. Relations between pasture attributes and liveweight gains on a subtropical pasture. In: *Proc. Twelfth Int. Grassland Congr., Moscow*, 3: 299–304.

Marston, H.R., Thomas, R.G., Murnare, D., Lines, E.W.L., McDonald, I.W., Moore, H.O. and Bull, L.B., 1938. Studies on coast disease of sheep in South Australia. *Counc. Sci. Ind. Res., Bull.*, 113.

McLean, J.W., Thompson, G.G., Iverson, C.E., Jagusch, K.T. and Lawson, B.M., 1962. Sheep production and health on pure species pasture. *Proc. N.Z. Grassland Assoc.*, 24: 57–70.

McLean, R.W., Winter, W.H., Mott, J.J. and Little, D.A., 1981. The influence of superphosphate on the legume content of the diet selected by cattle grazing *Stylosanthes*-native grass pastures. *J. Agric. Sci., Cambridge*, 96: 247–250.

Milford, R. and Minson, D.J., 1966. Intake of tropical pasture species. In: *Proc. Ninth Int. Grassland Congr., São Paulo*, pp. 815–822.

Minson, D.J., 1963. The effect of pelleting and wafering on the feeding value of roughages — a review. *J. Br. Grassland Soc.*, 18: 39–44.

Minson, D.J., 1966. The intake and nutritive value of fresh, frozen and dried *Sorghum almum*, *Digitaria decumbens* and *Panicum maximum*. *J. Br. Grassland Soc.*, 21: 123–126.

Minson, D.J., 1971. The digestibility and voluntary intake of six varieties of *Panicum*. *Aust. J. Exper. Agric. Anim. Husbandry*, 11: 18–25.

Minson, D.J., 1972. The digestibility and voluntary intake by sheep of six tropical grasses. *Aust. J. Exper. Agric. Anim. Husbandry*, 12: 21–27.

Minson, D.J., 1980. Nutritional differences between tropical and temperate pasture. In: F.H.W. Morley (Editor), *Grazing Animals*. Elsevier, Amsterdam, pp. 143–157.

Minson, D.J., 1982a. Effect of chemical composition on feed digestibility and metabolizable energy. *Nutrit. Abstr. Rev., Ser. B*, 52: 591–615.

Minson, D.J., 1982b. Effect of chemical and physical composition of herbage eaten upon intake. In: J.B. Hacker (Editor), *Nutritional Limits to Animal Production from Pastures*. Commonwealth Agricultural Bureaux, Farnham Royal, pp. 167–182.

Minson, D.J. and Laredo, M.A., 1972. Influence of leafiness on voluntary intake of tropical grasses by sheep. *J. Aust. Inst. Agric. Sci.*, 38: 303–305.

Minson, D.J. and Milford, R., 1967. The voluntary intake and digestibility of diets containing different proportions of legume and mature Pangola grass (*Digitaria decumbens*). *Aust. J. Exper. Agric. Anim. Husbandry*, 7: 546–551.

Minson, D.J., Harris, C.E., Raymond, W.F. and Milford, R., 1964. The digestibility and voluntary intake of S22 and H.I. ryegrass, S170 tall fescue, S48 timothy, S215 meadow fescue and Germinal cocksfoot. *J. Br. Grassland Soc.*, 19: 298–305.

Osbourn, D.F., Thomson, D.J. and Terry, R.A., 1966. The relationship between voluntary intake and digestibility of forage crops, using sheep. In: *Proc. Tenth Int. Grassland Congr., Helsinki*, pp. 363–366.

Playne, M.J., 1969a. Effect of sodium sulphate and gluten supplements on the intake and digestibility of a mixture of spear grass and Townsville lucerne hay by sheep. *Aust. J. Exper. Agric. Anim. Husbandry*, 9: 393–399.

Playne, M.J., 1969b. The effect of dicalcium phosphate supplements on the intake and digestibility of Townsville lucerne and spear grass by sheep. *Aust. J. Exper. Agric. Anim. Husbandry*, 9: 192–195.

Poppi, D.P., Minson, D.J. and Ternouth, A.H., 1981a. Studies of cattle and sheep eating leaf and stem fractions of grasses. 1. The voluntary intake, digestibility and retention time in the reticulorumen. *Aust. J. Agric. Res.*, 32: 99–108.

Poppi, D.P., Minson, D.J. and Ternouth, A.H., 1981b. Studies of cattle and sheep eating leaf and stem fractions of grasses. 3. The retention time in the rumen of large feed particles. *Aust. J. Agric. Res.*, 32: 123–137.

Raymond, W.F., Minson, D.J. and Harris, C.E., 1956. The effect of management on herbage consumption and selective grazing. In: *Proc. Seventh Int. Grassland Congr., Palmerston North*, pp. 123–132.

Rees, M.C. and Minson, D.J., 1978. Fertilizer sulphur as a

factor affecting voluntary intake, digestibility and retention time of pangola grass (*Digitaria decumbens*) by sheep. *Br. J. Nutrit.*, 39: 5–11.

Rees, M.C., Minson, D.J. and Smith, F.W., 1974. The effect of supplementary and fertilizer sulphur on voluntary intake, digestibility, retention time in the rumen, and site of digestion of pangola grass in sheep. *J. Agric. Sci., Cambridge*, 82: 419–422.

Stobbs, T.H., 1973. The effect of plant structure on the intake of tropical pastures. 1. Variation in the bite size of grazing cattle. *Aust. J. Agric. Sci.*, 24: 809–819.

Stobbs, T.H., 1977. Seasonal changes in the preference by cattle for *Macroptilium atropurpureum* cv. Siratro. *Trop. Grasslands*, 11: 87–91.

Stobbs, T.H., 1978. Milk production, milk composition, rate of milking and grazing behavior of dairy cows grazing two tropical grass pastures under a leader and follower system. *Aust. J. Exper. Agric. Anim. Husbandry*, 18: 5–11.

Tayler, J.C. and Rudman, J.E., 1965. Height and method of cutting or grazing in relation to herbage consumption and liveweight gain. In: *Proc. Ninth Int. Grassland Congr., São Paulo*, pp. 1639–1644.

Thornton, R.F. and Minson, D.J., 1973. The relationship between apparent retention time in the rumen, voluntary intake and apparent digestibility of legume and grass diets in sheep. *Aust. J. Agric. Res.*, 24: 889–898.

Troelsen, J.E. and Campbell, J.B., 1969. The effect of maturity and leafiness on the intake and digestibility of alfalfa and grasses fed to sheep. *J. Agric. Sci., Cambridge*, 73: 145–154.

Verite, R. and Journet, M., 1970. Influence, de la tenear en eau et de la déhydration de l'herbage sur sa valeur alimentaire pour les vaches laitières. *Ann. Zootechn.*, 19: 255–268.

Waite, R., Holmes, W., Campbell, J.I. and Ferguson, D.L., 1950. Studies in grazing management. II. The amount and chemical composition of herbage eaten by dairy cattle under close-folding and rotational methods of grazing. *J. Agric. Sci., Cambridge*, 40: 392–402.

Weston, R.H., 1982. Animal factors affecting feed intake. In: J.B. Hacker (Editor), *Nutritional Limits to Animal Production from Pastures*. Commonwealth Agricultural Bureaux, Farnham Royal, pp. 183–198.

Willoughby, W.M., 1958. A relationship between pasture availability and animal production. *Proc. Aust. Soc. Anim. Prod.*, 2: 42–45.

Wilson, G.F., 1978. Effect of water content of Tama ryegrass on voluntary intake of sheep. *N.Z. J. Exper. Agric.*, 6: 53–54.

Woodward, T.E., 1936. Quantities of grass that dairy cows will graze. *J. Dairy Sci.*, 19: 347–357.

Yates, J.J., Edye, L.A., Davies, J.G. and Haydock, K.P., 1964. Animal production from a *Sorghum almum* pasture in south-east Queensland. *Aust. J. Exper. Agric. Anim. Husbandry*, 4: 325–335.

# Chapter 15

# MINERALS IN PASTURES AND SUPPLEMENTS

D.I.H. JONES and T.A. THOMAS

## INTRODUCTION

High levels of animal production can now be achieved and sustained on managed pastures, due to improvements in pasture and stock management, increased fertilizer usage and the use of improved forage varieties. Whereas production is predominantly influenced by an adequate supply of forage of suitable digestibility, the efficient utilization of this forage relies on an adequate content of protein and essential mineral elements.

Present evidence suggests that over twenty mineral elements are essential for animals. These include the major elements calcium, chlorine, magnesium, phosphorus, potassium, sodium and sulphur, and the minor elements cobalt, copper, iodine, iron, manganese, molybdenum, selenium and zinc. The essentiality of a further seven microelements: arsenic, chromium, fluorine, nickel, silicon, tin and vanadium has been demonstrated under laboratory conditions, but their essentiality has not yet been shown under field conditions (Underwood, 1981).

There are significant differences between the mineral elements required for plant growth and those required by animals. Forage plants, for example, have no demonstrable requirement for selenium, cobalt or iodine and have a very low requirement for elements such as copper. Many mineral deficiencies in grazing animals therefore occur on pastures where plant growth is normal. Other elements, such as potassium and iron, are required in greater amounts by plants than animals, and deficiencies of these elements do not normally occur in grazing animals.

Most mineral deficiencies in grazing animals tend to occur on infertile soils under extensive systems of management, such as hill land or unimproved range-land; they are often accompanied by deficiencies of energy and protein. However, some mineral deficiencies, such as magnesium, occur under intensive systems of stock and pasture management, when the composition of herbage fails to meet the very high requirements of stock. This is often the result of the availability of one element being reduced by application of another in fertilizer — for instance, the influence of potash fertilizer on the availability of magnesium.

This chapter assesses the occurrence of mineral deficiencies in managed pasture, the factors which influence the mineral status of pastures, and the methods of alleviating mineral deficiency under pasture conditions.

## MINERAL DEFICIENCY IN PASTURE

Mineral deficiencies are widespread throughout the world among animals grazing managed grasslands. The subject has been reviewed by Russell and Duncan (1956) and, more recently, by Underwood (1977, 1981). Economically the most important deficiencies are those attributed to cobalt, copper, iodine, magnesium, phosphorus and selenium.

### Phosphorus

Deficiency of phosphorus, causing bone-chewing and wasting diseases, have been identified in many countries, and was first recognized in South Africa (Theiler and Green, 1932). It is especially prevalent on unfertilized pastures, in climates with a pronounced dry season, where the phosphorus content of herbage falls to as low as 0.05%. The deficiency is often exacerbated by an inadequate supply of digestible feed and a deficiency of protein. The condition does not appear to have been reported in pastures under more intensive systems of management.

## Magnesium

A deficiency of magnesium, often called hypomagnesaemia, tetany or grass staggers, occurs widely in Europe, North America and Australasia, particularly under intensive systems of pasture and stock management. The condition is economically important, with an estimated mortality rate of 1 to 3% in dairy herds in Britain and The Netherlands; individual herds can have mortality as high as 20% (Underwood, 1981). In the United States tetany is a major problem in intensive beef systems; overall mortality has been estimated as 1 to 2%, but losses are as high as 30% in certain areas (Grunes et al., 1970; Reid and Horvath, 1980). Although the disorder is clearly associated with a deficiency of magnesium, the precise relationship between tetany and the chemical composition of the diet is not established. Various environmental and physiological factors influence the incidence of tetany (Allcroft and Burns, 1968; Grunes et al., 1970). Tetany is more prevalent in cold, wet weather (Kemp, 1960), and when nitrogen and potassium fertilizers are applied. Kemp and 't Hart (1957) have shown that it is associated with a high ratio of monovalent to divalent cations in the herbage. Other important factors influencing the availability of magnesium have been reviewed by Grunes et al. (1970), Mayland and Grunes (1979), and Reid and Horvath (1980).

## Cobalt and copper

Cobalt deficiency, variously called bush sickness, wasting disease or pine, occurs in most countries of the world (Russell and Duncan, 1956). It occurs when the cobalt content of pastures falls below 0.07 p.p.m. It is caused by an insufficiency of cobalt in the rumen, which prevents the microorganisms from synthesizing adequate amounts of vitamin $B_{12}$ for the ruminant's needs; animals of all ages are affected. Cobalt deficiency generally occurs on infertile soils, under extensive grazing management, but sub-clinical deficiency may occur on other pastures; this can only be detected by a response to cobalt supplementation or vitamin $B_{12}$ therapy (Underwood, 1981).

Cobalt deficiency often occurs with copper deficiency, and together affect both cattle and sheep in Australia; simple copper deficiency also occurs on soils inherently low in copper (Underwood, 1981). A simple deficiency of copper is, however, rare and is more often induced by an excess of other elements, such as molybdenum and sulphur. High molybdenum content in pasture causes deficiency of copper — for instance on the "teart" soils of Somerset (England) and peat soils in New Zealand. High sulphur content in pastures can also reduce copper utilization, as can iron and zinc (Mills and Williams, 1971). Other unknown factors can induce copper deficiency in sheep on hill pasture in Britain (Russell and Duncan, 1956).

## Iodine

Goitre, due to iodine deficiency, is one of the most widespread of all mineral deficiency diseases. Incidence is related to the iodine content of the soils and the botanical composition of the pastures. The presence of goitrogens may also induce deficiency; clovers contain cyanogenic glucosides, which are goitrogenic through formation of thiocyanate in the rumen, whereas, cruciferous plants also contain goitrogens. Sub-clinical deficiencies of iodine may be widespread, and Underwood (1981) suggests that iodine supplementation would result in significant responses in marginally deficient areas.

## Selenium

Deficiencies of selenium occur in animals throughout the world, and are caused by soils inherently low in selenium (Alloway, 1968). Deficiency occurs if herbage contains less than 0.05 p.p.m. selenium. The commonest disorders associated with selenium deficiency are white muscle disease, ill thrift and various reproductive disorders. Deficiency of selenium is complicated by the influence of vitamin E, by differences in the form of selenium that occur in different soils, and by the interaction with other elements, particularly sulphur (Judson and Obst, 1975).

# MINERALS IN PASTURES

## PASTURE MINERALS AND NUTRITIONAL REQUIREMENT

### Plant and animal requirement

Some indication of which elements are likely to be deficient for animals may be obtained by comparing the mineral content of pasture herbage with those recommended for productive stock (National Research Council, 1975, 1978; Agricultural Research Council, 1980). Table 15.1 compares the requirement of dairy cows and fattening sheep with the values found for the various mineral elements in typical managed pastures. The calculated requirements can only be regarded as an indication, because of differences in the availability and endogenous loss of minerals, which are difficult to determine with precision. The differences in the mineral requirements of plants, compared with animals, is apparent in Table 15.1. This is especially apparent for cobalt, iodine and selenium, which are not required for pasture growth. The animal's requirement for calcium, sodium, zinc and copper also exceeds the minimum requirement for pasture growth and, as a result, deficiencies of most of these elements frequently occur in ruminants fed predominantly on herbage. Conversely, the requirements for potassium, sulphur, iron and manganese by plants clearly exceed those of the ruminant and, as a result, deficiency would not be expected; this is borne out in practice.

### Estimated and observed requirements

Under certain circumstances, previous estimates of the dietary requirements for minerals (Table 15.1) are known to be insufficient. One reason for the disparity is the interaction that occurs between certain elements. Interactions between magnesium and potassium and between copper, molybdenum and sulphur have already been referred to. A better quantitative understanding of these interactions,

TABLE 15.1

Pasture mineral content in relation to ruminant requirements

|  | Pasture content | | Ruminant requirement | |
|---|---|---|---|---|
|  | range[1] | requirement[2] | milking cow[3] | sheep[4] |
| *All values % of DM* | | | | |
| Calcium | 0.2–1.0 | 0.2–0.3 | 0.34 | 0.32 |
| Chlorine | 0.1–2.0 | n.d. | 0.33 | 0.05 |
| Magnesium | 0.1–0.4 | 0.1–0.13 | 0.17 | 0.10 |
| Phosphorus | 0.15–0.5 | 0.3–0.4 | 0.31 | 0.18 |
| Potassium | 1.0–4.5 | 2.0–2.5 | 0.80 | 0.30 |
| Sodium | 0.05–1.0 | 0.2–0.5[5] | 0.11 | 0.11 |
| Sulphur | 0.1–0.4 | 0.25–0.3 | 0.14 | 0.14 |
| *All values p.p.m. of DM* | | | | |
| Cobalt | 0.05–0.3 | n.r. | 0.11 | 0.11 |
| Copper | 2–15 | 4–6 | 10 | 4.5 |
| Iodine | 0.2–0.5 | n.r. | 0.5 | 0.5 |
| Iron | 50–300 | 50–70 | 30 | 30 |
| Manganese | 25–300 | 25–35 | 25 | 25 |
| Molybdenum | 0.5–5.0 | 0.1–0.3 | – | – |
| Selenium | 0.02–0.15 | n.r. | 0.05 | 0.05 |
| Zinc | 15–60 | 10–16 | 23 | 39 |

[1] Represent normal range found in pasture grasses, based on Whitehead (1966) and Butler and Jones (1973).
[2] Minimum concentration required for maximum yield (McNaught, 1970).
[3] Requirement of 500 kg cow yielding 20 kg milk per day (ARC, 1980).
[4] Requirement of 40 kg castrate lamb gaining 200 g per day (ARC, 1980).
[5] Requirement for maintaining K status in normal range, not regarded as essential.
(n.d. = not determined. n.r. = not required.)

and of other factors influencing mineral availability, is needed before more precise estimates of requirement can be made. In other instances, the published values appear to overestimate requirements in comparison with observed responses. Despite a long history of salt supplementation to sheep and cattle, few critical field experiments have demonstrated the benefit of sodium when the sodium content of the diet was more than 0.06% (Morris, 1980). This is considerably lower than the recommended level of sodium (Table 15.1). Similarly, feeding experiments with beef cattle (Call et al., 1978) showed no benefit of phosphorus supplements unless the diet contain less than 0.14% phosphorus; this is again much less than the estimated requirement (Table 15.1).

**Soil ingestion**

Ingestion of soil is an important factor which may influence the supply and availability of minerals to grazing animals (Healy, 1973). Considerable quantities of soil are ingested when little herbage is available and, under such conditions, significant quantities of trace elements such as cobalt, iodine and selenium may be added to the diet. Soil may also have adverse effects, for example, Suttle et al. (1975) found that soil ingestion inhibited copper absorption, presumably due to the ingestion of increased amounts of elements such as molybdenum.

## THE DIAGNOSIS OF MINERAL DEFICIENCIES

If the animals show no visual symptoms, mineral deficiencies may be diagnosed by chemical analysis of soils or pasture, by analysis of animal fluids and tissues, or by observing the effects of supplementation. None of these methods are usually successful alone, since simple deficiencies are rare. It is usually necessary, therefore, to adopt a comprehensive approach, which may involve sampling soil, pasture and certain animal components, together with supplement feeding, before a definite conclusion can be reached (Reuter, 1975; Underwood, 1981). Table 15.2 summarizes the suggested procedures for diagnosing deficiencies of the more important elements.

Soil analysis is generally of very limited value, since the availability of soil minerals to the plant is influenced markedly by such factors as soil pH (Reid and Horvath, 1980). However, deficiency and toxicity of selenium in grazing animals can be related to the selenium content of soil (Kubota et al., 1967). Similarly, cobalt deficiency can be related to available cobalt content of the soil (Mitchell et al., 1957).

Analysis of pasture herbage is undoubtedly of value for most mineral elements, although there are problems of sampling techniques and of determining the availability of minerals to the animal. Sampling difficulties occur because animals graze selectively (Jones, 1981; and Chapter 13), and because grazed herbage is contaminated with soil (Healey, 1973). The availability of minerals to animals depends on the form in which the element is present in the plant. Many elements are found in organic complexes in plants, and the availability to the animal of many of these constituents is not known (Butler and Jones, 1973). The availability of elements is also influenced by the presence of other

TABLE 15.2

Suggested diagnostic analysis for mineral deficiencies

| Element | Analysis |
|---|---|
| Calcium | pasture, blood and bone calcium |
| Phosphorus | pasture, blood and bone phosphorus |
| Magnesium | pasture and blood magnesium and pasture N, K |
| Sodium | pasture, blood and urine sodium and Na:K in saliva |
| Copper | pasture, blood and liver copper |
| Iodine | pasture, milk and urine iodine and blood thyroxine |
| Zinc | pasture, blood and saliva zinc |
| Cobalt | soil, pasture and liver cobalt, blood vitamin $B_{12}$ |
| Selenium | pasture selenium and blood glutathione peroxidase |

Based on Underwood (1981) and Reid and Horvath (1980).

constituents of the diet (see above). The total content of a mineral in pasture therefore only gives an indication of likely deficiency.

Clinical and pathological observations are often the best indications of mineral deficiency (Underwood, 1981), but are rarely specific for one element. For example, anaemia is characteristic of deficiencies of cobalt, copper and iron whereas reproductive disturbances occur with deficiencies of copper, iodine, manganese, phosphorus, selenium and zinc.

Whereas a combination of soil, pasture and clinical tests may indicate the probability of a certain deficiency, the final diagnosis rests on obtaining a positive response to supplementation in controlled experiment. The use of an integrated approach of chemical and biochemical tests, in combination with supplementation experiments, has allowed the mapping of mineral deficiencies for large areas in Australia and the United States (Reid and Horvath, 1980).

## FACTORS INFLUENCING THE MINERAL CONTENT OF PASTURES

Whitehead (1966, 1972), Fleming (1973) and Reid and Horvath (1980) have reviewed the factors influencing the mineral content of pasture herbage. The present discussion will be confined to a consideration of the main principles governing mineral uptake, with particular regard to those elements limiting animal production.

The main factors influencing the mineral content of pasture herbage are: (1) the content and availability of soil minerals; (2) the species and cultivar of the plants present in the sward; (3) the growth-stage of the sward components and the management regime; and (4) fertilizer usage.

### Soil factors

Fleming (1973) points out that, whereas the total content of an element in soil is largely dependent on the parent material, the availability of the element to the plant is mainly determined by other factors, such as soil pH, water content, soil texture and organic matter content. However, there is a close relationship between the availability of some elements, such as cobalt and selenium, and the composition of the parent material. In managed swards, the content of elements such as calcium, nitrogen, phosphorus and potassium is more determined by the application of fertilizers than by inherent soil characteristics. Other elements, such as sulphur and some trace elements, may also be applied as constituents or impurities of fertilizers.

Soil reaction has a major effect on the availability of mineral elements, particularly trace elements. The uptake of all trace elements, other than molybdenum, decreases with increasing pH; manganese is the most affected (Fleming, 1973).

### Plant species and variety

Pasture plants, growing on the same soil and under the same environmental conditions, differ significantly in mineral content. For example, Whitehead (1972) concluded that legumes usually contain more calcium, magnesium, cobalt and iron than grasses, but contain less manganese. Others have reported that the content of all major minerals apart from sodium, and of most trace elements, is greater in legumes than grasses (Fleming, 1973; Reid and Horvath, 1980). The inclusion of a significant proportion of clover in a sward will therefore improve the mineral status of pasture herbage. The presence of various herb species, such as *Cichorium intybus* and *Plantago lanceolata* will also enhance mineral content (Thomas and Thompson, 1948).

Grass species also differ in mineral content (Whitehead, 1966, 1972; Fleming, 1973). Differences in sodium content between grass species and varieties are particularly marked and consistent (ap Griffith and Walters, 1966), whereas some species, such as *Phleum pratense*, appear to be generally poor in mineral content (Fig. 15.1). There are also differences in mineral content between cultivars and genotypes within a species (Cooper, 1973). Consistent differences in sodium (ap Griffith et al., 1965) and iodine content (Alderman and Jones, 1967) have been reported between varieties of the major temperate pasture species. Cooper (1973) found significant differences in the content of most major elements in contrasting cultivars of Italian and perennial ryegrass (*Lolium multiflorum* and *L. perenne*), whereas Davies et al. (1968) reported similar variation in clover. Similarly, Snaydon and Bradshaw (1962, 1969) found

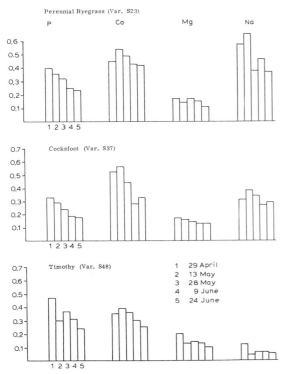

Fig. 15.1. The phosphorus, calcium, magnesium and sodium content of three grass species, perennial ryegrass (*Lolium perenne*), cocksfoot (*Dactylis glomerata*) and timothy (*Phleum pratense*), on five dates from April to June.

large differences in the major mineral content of locally adapted ecotypes of white clover (*Trifolium repens*). Cooper (1973) pointed out that, because of this genetic variation, it would be feasible to improve the mineral content of pasture species by plant breeding. Such an approach has been illustrated by selection for increased magnesium content in Italian ryegrass (Hides and Thomas, 1981).

## Growth stage

Plants take up minerals rapidly during early growth but, as growth accelerates, dry matter accumulates faster than mineral uptake, with a consequent decrease in mineral content. However, there are many exceptions to this, caused by changes in climatic conditions and proportion of leaf to stem during the season. As a result, the effects of maturity on mineral content vary widely (Whitehead, 1966, 1972; Fleming, 1973). There is general agreement that potassium, phosphorus and copper content decline, but results for other elements have been conflicting. Whitehead and Jones (1969) found that the content of manganese, phosphorus and potassium declines but that of zinc increases in legumes, whereas Fleming and Murphy (1968) showed that the content of copper, iron, phosphorus, potassium, sodium and zinc decreases with age in perennial ryegrass. All species do not behave in the same way; Fleming (1973) stated that zinc content increases with age in red and white clover (*Trifolium pratense* and *T. repens*) but decreases in brome grass (*Bromus inermis*) and alfalfa (*Medicago sativa*). Some indication of the variation between species is shown in Fig. 15.1.

## Plant fractions

It has long been recognized that leaves are a richer source of minerals than stems. Fleming (1965), for example, showed that the leaf of cocksfoot (*Dactylis glomerata*), perennial ryegrass and red clover contained more calcium, copper, iron, magnesium, molybdenum, phosphorus, potassium and zinc than the stem. The content of manganese was greater in the leaf than the stem in clover but not in grasses, whereas cobalt levels were similar in leaf and stem for both grasses and clover. Other work on the distribution of mineral elements within pasture plants has been reviewed by Fleming (1973).

## Cutting vs. grazing

The effects of advancing maturity in reducing mineral content can be avoided or reduced by frequent defoliation. However, the regrowth characteristics of different species vary widely, and large differences in the proportion of leaf to stem can occur, with a consequent effect on mineral content. The growth stage when grazed can therefore influence mineral content to a varying extent. Herbage cut for conservation differs in mineral content from grazed herbage, due to differences in growth stage and the effects of animal excreta. The potassium content is particularly affected, and most of herbage potassium is restored to the soil under all grazing systems (Whitehead, 1966). The effects of excreta on other nutrients appears to be more variable; Herriott and Wells (1963), for example, found that calcium content was reduced by excreta, whereas McNaught (1961) found a similar reduction in magnesium. In mixed swards, grazing management can also influence mineral content by changing the botanical composition of swards (Chapter 7), especially the clover content.

**Fertilizer usage**

Lime and fertilizer applications can greatly influence the mineral content of managed pasture. The effects of lime on soil reaction and mineral uptake have already been briefly considered, the effects of nitrogen and other fertilizer nutrients on mineral composition have been reviewed by Whitehead (1970), Fleming (1973), Reid and Jung (1974) and Reid and Horvath (1980).

The effects of added fertilizers on mineral content are often difficult to interpret. The application of a highly deficient nutrient increases growth and mineral uptake, but may not necessarily increase the concentration in the plant; further application usually increases both yield and content, whereas still higher applications increase nutrient concentration without increasing yield (Reid and Horvath, 1980). It is not surprising therefore that the literature on the effects of fertilizer on mineral content should be contradictory.

Whitehead (1970) concluded that applications of nitrogen fertilizer generally increased the content of phosphorus and potassium, and possibly other minerals, when these elements were present in abundant supply, but decreased their content when they were deficient. However, different forms of nitrogen fertilizer, — ammonium or nitrate — affect soil reaction and thus nutrient uptake differently. One important effect of high levels of nitrogen fertilizer is the accumulation of high levels of nitrate (ap Griffith, 1960). Although nitrate can sometimes be toxic to ruminants (Wright and Davison, 1964), there is also evidence that in other cases high levels do not impair health or performance under intensively managed pasture conditions (Hodgson and Spedding, 1966). Fertilizer nitrogen may also have indirect effects on mineral content in mixed swards, by encouraging grass growth relative to clover.

The effect of phosphatic fertilizers on mineral content is generally less than that of nitrogen, and depends on soil conditions. When phosphorus is deficient, application can dramatically increase the phosphorus content of pastures (Reid and Horvath, 1980). The application of basic slag may influence the concentration of trace elements due to its high content of these elements (Fleming, 1973).

Potassium applications, either with or without nitrogen, can have a marked effect on the uptake of other ions. For example, potassium can reduce the content of calcium, magnesium and sodium in herbage (Reid and Horvath, 1980). Repeated applications of potassium, over many years, can result in a steady decline in the sodium content of pastures, though additional application of nitrogen appears to offset the depressing effect on sodium and magnesium.

## METHODS OF MINERAL SUPPLEMENTATION

Underwood (1981) considered that successful procedures now exist for the prevention and control of all mineral deficiencies in farm animals. These include the use of fertilizers, and management to influence the mineral content of pastures, in addition to the direct administration of minerals to the animals in the form of supplements, such as licks, pellets, drenches and injections.

In general, the indirect methods are uncertain methods of improving pasture mineral content. Although successful in some instances, such as cobalt and selenium, interactions with other soil factors often prevent them from being successful elsewhere. Applications of trace elements, for example, are unlikely to be successful if uptake is restricted by soil reaction (p. 149). Similarly, applications of nutrients that are deficient for plant growth are unlikely to increase plant mineral content unless very high, and probably uneconomic, levels are applied. Further complications arise from interactions between elements; applications of copper, for example, will not be effective in counteracting hypocuprosis if soil molybdenum is abundant. Thus, whereas it is important that management, the choice of genus, species and variety, and the use of fertilizer should aim at maximizing mineral content, these measures are not generally fully effective in preventing and controlling deficiency.

For animals receiving a proportion of their diet in the form of concentrates, the addition of supplementary minerals presents little difficulty. Such supplements can usually be commercially formulated to suit local requirements, but often contain minerals not required. Under more extensive pasture conditions, where supplementary feeding is not practised, supplementary minerals can be supplied in the form of salt-based licks, heavy pellets (bullets) or needles lodged in the digestive tract, or injection of slowly absorbed organic complexes. The advantages and disadvantages of these different procedures have been discussed in detail by Underwood (1981). Salt-based licks are considered to be the easiest, cheapest and most common means of supplementation for cobalt, copper, iodine, phosphorus and sodium in grazing stock. Their use is, however, uncertain since there is no guarantee that an individual animal will consume an adequate amount or, indeed, may not consume an excess. Oral drenching has been successfully used for treating copper and selenium deficiency, but has the obvious disadvantage of a large labour requirement. Cobalt deficiency is best treated by the use of heavy pellets or bullets, which lodge in the rumen; copper needles, lodged in the abomasum, have similarly been effective in copper deficiency. Organic complexes of iodine and copper, injected intramuscularly, provide lengthy protection against deficiency of these elements.

## CONCLUSION

It appears that, in many situations, animal production from managed pasture systems will be limited by mineral deficiencies. The incidence of some deficiencies, such as cobalt, will be restricted to certain areas where the soil is inherently deficient in certain minerals; other deficiencies, such as magnesium, are more related to intensive systems of animal production. Underwood (1981) pointed out that effective methods of control are now available for all known deficiencies but, at the same time, he emphasized the probable widespread incidence of sub-clinical deficiencies. The incidence of such deficiencies in managed pastures is presently either not recognized, or is masked by the practice of indiscriminate supplementation with mineral mixes or licks. There is clearly a need for experiments aiming at a more precise definition of the requirement of minerals for optimum production under pasture conditions.

A better understanding is needed of several aspects of pasture minerals and their requirements by stock. For elements such as cobalt or iodine, research is still restricted by the lack of precise and convenient analytical techniques. There is little information available of the state of combination of most mineral elements in plants, and analyses are based on total content after ashing. A more precise definition of the amounts of different elements which exist in ionic or complex forms, and of their availability, would enable requirements to be assessed more precisely.

It may, at present, be economically viable to augment pasture minerals by supplementation where needed. A requirement for reducing inputs into pasture systems would, however, prompt a need to increase pasture mineral content. This could be achieved to a large extent by the inclusion of increased amounts of clover and possibly by breeding forage cultivars inherently higher in mineral content.

## REFERENCES

Agricultural Research Council (ARC), 1980. *The Nutrient Requirements of Ruminant Livestock*. Commonwealth Agricultural Bureaux, Farnham Royal, 351 pp.

Alderman, G. and Jones, D.I.H., 1967. The iodine content of pastures. *J. Sci., Food Agric.*, 18: 197–199.

Allcroft, R. and Burns, K.N., 1968. Hypomagnesaemia in cattle. *N.Z. Vet. J.*, 16: 109–128.

Alloway, W.H., 1968. Agronomic controls over the environmental cycling of trace elements. *Adv. Agron.*, 20: 235–274.

ap Griffith, G., 1960. Nitrate content of herbage at different manurial levels. *Nature, London*, 185: 627–628.

ap Griffith, G., Jones, D.I.H. and Walters, R.J.K., 1965. Specific and varietal differences in sodium and potassium content in grasses. *J. Sci., Food Agric.*, 16: 94–98.

ap Griffith, G. and Walters, R.J.K., 1966. The sodium and potassium content of some grass genera, species and varieties. *J. Agric. Sci., Cambridge*, 67: 81–89.

Butler, G.W. and Jones, D.I.H., 1973. Mineral biochemistry of herbage. In: G.W. Butler and R.W. Bailey (Editors), *Chemistry and Biochemistry of Herbage, 2*. Academic Press, London, pp. 127–162.

Call, J.W., Butcher, E.J., Blake, J.T., Sucont, R.A. and Sharpe, J.L., 1978. Phosphorus influence on growth and production of beef cattle. *J. Anim. Sci.*, 47: 216–225.

Cooper, J.P., 1973. Genetic variation in herbage constituents. In: G.W. Butler and R.W. Bailey (Editors), *Chemistry and Biochemistry of Herbage, 2*. Academic Press, London, pp. 379–417.

Davies, W.E., Thomas, T.A. and Young, N.R., 1968. The assessment of herbage legume varieties. III. Annual variation in chemical composition of eight varieties. *J. Agric. Sci., Cambridge*, 71: 233–241.

Fleming, G.A., 1965. Trace elements in plants with particular reference to pasture species. *Outlook Agric.*, 4: 270–285.

Fleming, G.A., 1973. Mineral composition of herbage. In: G.W. Butler and R.W. Bailey (Editors), *Chemistry and Biochemistry of Herbage, 1*. Academic Press, London, pp. 549–566.

Fleming, G.A. and Murphy, W.E., 1968. The uptake of some major and trace elements by grasses as affected by season and maturity. *J. Br. Grassland Soc.*, 23: 174–184.

Grunes, D.L., Stant, P.R. and Brownell, J.R., 1970. Grass tetany of ruminants. *Adv. Agron.*, 22: 332–374.

Healey, W.B., 1973. Nutritional aspects of soil ingestion by grazing animals. In: G.W. Butler and R.W. Bailey (Editors), *Chemistry and Biochemistry of Herbage, 1*. Academic Press, London, pp. 567–588.

Herriott, J.B.D. and Wells, D.A., 1963. The grazing animal and sward productivity. *J. Agric. Sci., Cambridge*, 61: 89–99.

Hides, D.H. and Thomas, T.A., 1981. Variation in the magnesium content of grasses and its improvement by selection. *J. Sci. Food Agric.*, 32: 990–991.

Hodgson, J. and Spedding, C.R.W., 1966. The health and performance of the grazing animal in relation to fertilizer nitrogen usage. *J. Agric. Sci., Cambridge*, 67: 155–167.

Jones, D.I.H., 1981. Chemical composition and nutritive value. In: *Sward Measurement Handbook*. The British Grassland Society, Hurley, pp. 243–265.

Judson, G.J. and Obst, J.M., 1975. Diagnosis and treatment of selenium inadequacies in the grazing ruminant. In: D.J.D. Nichols and A.R. Egan (Editors), *Trace Elements in Soil Plant Animal Systems*. Academic Press, New York, N.Y., pp. 385–405.

Kemp, A., 1960. Hypomagnesaemia in milking cows: the response of serum magnesium to alterations in herbage composition resulting in potash and nitrogen dressings on pasture. *Neth. J. Agric. Sci.*, 8: 281–304.

Kemp, A. and 't Hart, M.L., 1957. Grass tetany in milking cows. *Neth. J. Agric. Sci.*, 5: 4–17.

Kubota, J., Alloway, W.H., Carter, D.L., Cary, E.E. and Lazor, V.A., 1967. Selenium in crops in the United States in relation to selenium-responsive diseases of animals. *J. Agric. Food Chem.*, 15: 448–453.

Mayland, H.F. and Grunes, D.L., 1979. Soil–climate–plant relationship in the etiology of grass tetany. In: V.V. Rendig and D.L. Grunes (Editors), *Grass Tetany*. American Society of Agronomy, Madison, Wisc., pp. 123–175.

McNaught, K.J., 1961. Some problems in plant analysis as an index of nutrient status. *J. N.Z. Inst. Chem.*, 25: 7–15.

McNaught, K.J., 1970. Diagnosis of mineral deficiencies in grass-legume pastures by plant analysis. In: *Proc. XI Int. Grassland Congr., Surfers Paradise, Qld.*, pp. 334–338.

Mills, C.F. and Williams, R.B., 1971. Problems in the determination of the trace element requirements of animals. *Proc. Nutrit. Soc.*, 30: 82–91.

Mitchell, R.L., Reith, J.W.S. and Johnston, I.M., 1957. Trace-element uptake in relation to soil content. *J. Sci., Food Agric.*, 8: S51–59.

Morris, J.G., 1980. Assessment of sodium requirements of grazing cattle: a review. *J. Anim. Sci.*, 50: 145–152.

National Research Council, 1975. *Nutrient Requirements of Domestic Animals. 5. Nutrient Requirements of Sheep*. NRC, Washington, D.C., 5th ed., 72 pp.

National Research Council, 1978. *Nutrient Requirements of Dairy Cattle, No. 3*. NRC, Washington, D.C., 5th revised ed.

Reid, R.L. and Horvath, D.J., 1980. Soil chemistry and mineral problems in farm livestock. A review. *Anim. Feed Sci. Technol.*, 5: 95–167.

Reid, R.L. and Jung, G.A., 1974. Effects of elements other than nitrogen on the nutritive value of forage. In: D. Mays (Editor), *Forage Fertilization*. American Society of Agronomy, Madison, Wisc.

Reuter, D.J., 1975. The recognition and correction of trace element deficiencies. In: D.J.D. Nicholas and A.R. Egan (Editors), *Trace Elements in Soil Plant Animal Systems*. Academic Press, New York, N.Y., pp. 291–324.

Russell, F.C. and Duncan, D.L., 1956. *Minerals in Pastures: Deficiencies and Excesses in Relation to Animal Health*. Commonwealth Bureaux Animal Nutrition Tech. Commun. No. 15, Commonwealth Agricultural Bureaux, Farnham Royal, 170 pp.

Snaydon, R.W. and Bradshaw, A.D., 1962. Differences between natural populations of *Trifolium repens* L. in response to mineral nutrient. I. Phosphate. *J. Exper. Bot.*, 13: 422–434.

Snaydon, R.W. and Bradshaw, A.D., 1969. Differences between natural population of *Trifolium repens* L. in response to mineral nutrient. II. Calcium, magnesium and potassium. *J. Appl. Ecol.*, 6: 185–202.

Suttle, N.F., Alloway, J. and Thornton, I., 1975. An effect of soil ingestion on the utilisation of dietary copper by sheep. *J. Agric. Sci., Cambridge*, 84: 249–254.

Theiler, A. and Green, H.H., 1932. Aphosphorosis in ruminants. *Nutrit. Abstr. Rev.*, 1: 359–382.

Thomas, B. and Thompson, A., 1948. The cell constituents of some grasses and herbs on the Palace Leas plots at Cockle Park. *Empire J. Exper. Agric.*, 16: 221–230.

Underwood, E.J., 1977. *Trace Elements in Human and Animal Nutrition*. Academic Press, New York, N.Y., 4th ed., 545 pp.

Underwood, E.J., 1981. *The Mineral Nutrition of Livestock*. Commonwealth Agricultural Bureaux, Farnham Royal, 2nd ed., 180 pp.

Whitehead, D.C., 1966. *Nutrient Minerals in Grassland Herbage*. Mimco Publ. 1/1966, Commonwealth Agricultural Bureaux, Farnham Royal, 83 pp.

Whitehead, D.C., 1970. *The Role of Nitrogen in Grassland*. Productivity Bulletin 48, Commonwealth Agricultural Bureaux, Farnham Royal, 202 pp.

Whitehead, D.C., 1972. Chemical composition of grasses and legumes. In: C.R.W. Spedding and E.C. Diekmahns (Editors), *Grasses and Legumes in British Agriculture*. Commonwealth Agricultural Bureaux, Farnham Royal, 49 pp.

Wright, M.J. and Davison, K.L., 1964. Nitrate accumulation in crops and nitrate poisoning in animals. *Adv. Agron.*, 16: 197–247.

Chapter 16

# REPRODUCTION, LIFESPAN AND EFFICIENCY OF PRODUCTION

J.M. WALSINGHAM

## INTRODUCTION

The energetic efficiency of secondary productivity in grassland is very dependent on the extent to which the animal populations can be managed, so as to optimize conversion of resources into harvestable animal products.

The efficiency of conversion appears high when the individual animal is considered (Table 16.1), but this is an inadequate measure of efficiency, since it does not include the overhead costs of the remainder of the animal population — that is, the parents, and the siblings which fail to survive. Clearly an individual animal belongs to a population which has to be sustained for the agricultural process to be maintained. It is with the management of such populations that this chapter is concerned.

There are remarkably few measurements of feed conversion efficiency at pasture, whether of individuals or populations, because of the difficulties of measuring feed intake (Commonwealth Bureau of Pastures and Field Crops, 1961). However, recent developments in the field of transducers do make possible the measurement of the feeding behaviour of individuals at pasture (Penning, 1983).

## REPRODUCTION

Increasing the reproductive rate of any population of animals substantially increases the overall conversion efficiency of the population, largely because the feed costs of the adult female, and a share of those for the adult male, are spread across a larger number of offspring. Variation in reproductive rate, both within and between species, gives rise to considerable variation in the annual output of meat per unit of female body size (Table 16.2) and so to variation in efficiency of conversion (Fig. 16.1).

### Frequency of reproduction

The scope for management or manipulation of the reproductive rate varies greatly and is, for

TABLE 16.1

The percentage efficiency of individual animals

|  | Carcass out (kg) Feed in (kg DM) | Carcass energy Gross feed energy | Muscle protein gain Protein intake | |
|---|---|---|---|---|
| Cattle[1] | 8.6 | 5.2–7.8 | veal | 13.3 |
|  |  |  | beef | 7.7 |
| Sheep[1] | 17.6–13.2 | 11.0–14.6 | lamb | 13.3 |
|  |  |  | hogget | 10.0 |
| Rabbits[1] | 42.0 | 12.5–17.5 |  |  |
| Goats[2] | 5.8 | 3.9 |  |  |

[1]Spedding et al. (1981); [2]Mohammed (1982).

TABLE 16.2

Annual output of meat per unit of female body-size

|  | Liveweight of breeding female (kg) | Pregnancies per year | Average litter size | Carcass weight of progeny (kg) | Carcass output (kg yr$^{-1}$) | Carcass output per year as % of adult female weight (%) |
|---|---|---|---|---|---|---|
| Cattle[1] | 500 | 1 | 1 | 200–300 | 200–300 | 40–60 |
| Sheep[1] | 70 | 1–1.5 | 1–3 | 18–24 | 42–63 | 60–90 |
| Goats[2] | 60 | 1 | 2 | 15 | 30 | 50 |
| Rabbits[1] | 4.5 | 7 | 8 | 1–2 | 56–112 | 1244–2488 |

[1]Spedding et al. (1981); [2]Mohammed (1982), Amoah and Bryant (1983).

example, much greater for rabbits than for cattle. A variety of techniques are used to improve reproductive rate. One objective is to minimize the period when the female is neither suckling nor pregnant, provided this is consistent with the welfare of the animals. In cattle, the ability to detect oestrus, and thus inseminate at the appropriate time, is a means of minimizing the period between calvings (Esslemont, 1974). However, in recent years the national average calving index — that is, the period between successive calvings — has changed little from 390 days. Sheep can normally only be mated between autumn and early spring (Lees, 1969), although some breeds (e.g. Dorset Horn) have a longer breeding season (Newton and Betts, 1967). In most cases it is still only possible to have one litter per year, although three or four litters in two years have been achieved with hormone treatment (Newton and Betts, 1967). One of the major advantages of the rabbit for meat production is its ability to breed frequently throughout the year. On average, six litters per year are born in housed populations, but as many as eleven litters per doe are possible (Walsingham, 1972), if a doe is remated soon after giving birth (Foxcroft and Hasnain, 1973).

**Litter size**

The other main method of improving reproductive rate is to increase the number of young born alive and surviving to maturity. This is much more amenable to management, since considerable genetic variation is known to exist, and hormonal control is also possible now.

With cattle, twinning seems to be of the order of 2 to 3% of calvings (Gordon, 1983). Breed, age, environment and nutrition affect the proportion of twins, but management and feeding offer little opportunity of increasing twinning percentages. Research continues into the possibility of inducing twin pregnancies. Induced twin pregnancies have the advantage that cows can be prepared with appropriate feeding, and assistance can be provided for difficult births if necessary (Gordon, 1983). One of the disadvantages of twinning is the high incidence (90%) of freemartins — that is, infertile female calves (Marcum, 1974). However, this need not matter if the female calves are not for breeding; indeed, freemartins tend to be good meat producers.

Multiple births are more common in other species. For example, the litter size in sheep is often two or three, and can be as great as five or six. However, careful management is necessary if all the offspring are to survive and grow to an acceptable market weight. It is possible to rear young lambs on milk substitute, made from reconstituted cows' milk, if the ewe is unable to cope adequately (Large, 1965). Most ewes suckle twin lambs satisfactorily and some can manage triplets, although the third lamb is often, in practice, removed for artificial rearing.

Rabbits commonly have ten to eleven offspring and can easily suckle eight young. Artificial rearing is more difficult with rabbits; although it has been done experimentally (Barnard, 1962), it is difficult to imagine it being carried out as a commercial enterprise.

**Reproductive span**

Another way in which more progeny can be produced for a relatively small additional input is

# REPRODUCTION, LIFESPAN AND EFFICIENCY OF PRODUCTION

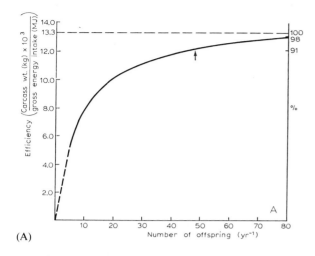

(A)

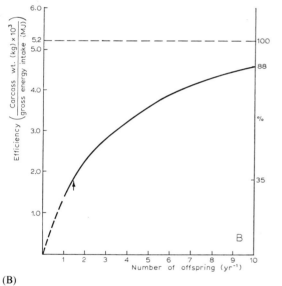

(B)

Fig. 16.1. The effect of reproductive rate on the efficiency of meat production for (A) rabbits and (B) sheep (from Large, 1976). The horizontal broken line at the top of the figures indicates the efficiency of an individual — that is, no allowances are made for the rest of the population. The point on the curve marked by the arrow indicates current levels of achievement.

for young female stock to be mated at the earliest possible opportunity (Table 16.3). Maximum efficiency, at the population level, will then be achieved if such females are able to sustain their reproductive performance over a long period. However, this may not result in the best economic performance, since the carcass of the breeding female will then be of limited value. Consequently, many farming systems sell breeding stock for meat

TABLE 16.3

Age and weight of animals at puberty

|  | Age (months) | Weight (kg) |
| --- | --- | --- |
| Cow[1] | 4–14 | 159–272 |
| Ewe[1] | 7–10 | 27–34 |
| Horse[1] | 15–24 | depends on mature size |
| Goat[2] | 4–8 | depends on mature size |
| Rabbit[3] | 4–7 | depends on mature size |
| Red deer[4] | 17–24 | 60 |

[1]Hafez (1962); [2]Fraser (1971); [3]Sandford (1966); [4]Hall and Henshaw (1983).

before they have reached the end of their full reproductive life-span.

An alternative method of management is for a proportion of breeding females to have offspring only once (i.e. uniparous) and for them to be slaughtered after a minimum period of suckling. Thus, two carcasses are produced in a relatively short period of time.

If the efficiency of energy conversion of individuals is considered, the uniparous female and progeny is almost as efficient as a single individual (Table 16.4). The efficiency of energy conversion of the whole population will depend upon the reproductive span and fecundity of the breed, which will determine the proportion of the females to be bred once before slaughtering. There are considerable practical problems to be overcome before such potential can be realized (T.R. Morris et al., pers. commun., 1983).

## LIFE-SPAN

The long life-span of female animals under natural conditions is probably a function of relatively low reproductive rates, compared with the levels achieved in modern agricultural systems.

Frequent breeding, and the production of large litters, whilst desirable from the point of view of productivity, seems likely to have an effect on the long-term performance of the adult females. The use of management techniques to increase the frequency and size of litters may reduce the fertile life-span of adult females to the extent that they are culled much earlier than previously. Evidence to support this suggestion is difficult to obtain, since

TABLE 16.4

The energetic efficiency $(E_F)$[1] of uniparous and nulliparous female animals (after Spedding et al., 1976)

|  | Nulliparous | | Uniparous females and progeny | | | |
|---|---|---|---|---|---|---|
|  | age at slaughter (days) | $E_F$ | age at slaughter (days) | | litter size | $E_F$ |
|  |  |  | females | progeny |  |  |
| Cattle | 550 | 4.5 | 900 | 550 | 1.0 | 3.6 |
| Pigs | 150 | 33.5 | 325 | 150 | 8.2 | 30.9 |

[1] $E_F = \dfrac{\text{Gross energy in carcass output} \times 100}{\text{Gross energy of feed (dam + progeny)}}$

populations of farm animals consist of highly selected, relatively fertile individuals. Those which exhibit declining fertility are likely to be culled at the earliest opportunity. There is evidence to show that fertility declines with age in laboratory animals, including the rabbit (Adams, 1970), but whether increasing age is also associated with reduced intervals between parities and larger litter sizes remains to be established.

In many agricultural systems, it is desirable, from an economic viewpoint, to cull breeding females either before their carcass value declines or before their reproductive output falls, and if possible before both events. If an earlier culling policy were necessary, it would not of itself necessarily reduce the biological efficiency of populations; the once-bred heifer system mentioned earlier represents an extreme case of early culling. If the heifer were to produce twins, then the level of population efficiency would be further enhanced. This would be equally true of cows and ewes that were culled relatively early; it is quite clear that, from the point of view of theoretical efficiency, it would be better to have a given number of young in the shortest possible number of years rather than spread them out over a long period (Table 16.5). This is primarily because the annual feed costs for maintenance of the adult female are a very high proportion of the total feed costs of the meat-producing system.

## CONCLUSION

The management of animal populations, so as to achieve the optimum output of product per unit of resources employed, depends heavily on obtaining large numbers of viable offspring in a given time and over the life of the dam. Management of such populations is therefore dependent on the achievement of: (1) first parity at an early age; (2) minimal

TABLE 16.5

An example of the effect of feed requirements of the population on the energetic efficiency of sheep production

|  | Energy in feed (MJ) | Energy in carcass (MJ) | $E_F$ |
|---|---|---|---|
| Individual lamb | 2143 | 225 | 10.5 |
| Population |  |  |  |
| ewe and 1 lamb | 12 913 | 281 | 2.17 |
| ewe and 2 lambs | 16 761 | 504 | 3.0 |
| ewe and 3 lambs | 21 035 | 791 | 3.7 |
| ewe and 3 successive single lambs | 38 739 | 843 | 2.18 |

Data from Young and Newton (1975), Penning et al. (1977) and M.J. Gibb (pers. commun., 1979).

intervals between parities; (3) large litter sizes and high survival rate; (4) sustained lifetime reproductive performance, or effective culling policies; and (5) healthy offspring, capable of good food-conversion efficiency. The extent to which these factors can be controlled has been reviewed both for cattle (Balch, 1974) and sheep (Robinson, 1974). In the context of grassland, there can be no substitute for optimum grassland management, which determines the amount and quality of the animals' diet throughout the year. Highly digestible forage is important in maintaining the health and capability of the breeding population and of ensuring rapid growth in the offspring. Traditionally, grass has been supplemented with cereals for many livestock at critical periods (conception, late pregnancy, and late lactation). The increasing cost of supplements, due to increasing fuel costs (Spedding, 1982), has placed greater emphasis on the better management of grassland for livestock populations, and on the introduction of palatable forages, such as clover, which minimize the need for nitrogen fertilizer (Wilkins and Bather, 1981). Since there are health problems associated with some legumes, which can affect rumen function (Clark and Reid, 1974) as well as the ability of livestock to conceive (Newton and Betts, 1968), such developments require even greater management and husbandry skills.

## REFERENCES

Adams, C.E., 1970. Ageing and reproduction in the female mammal with particular reference to the rabbit. *J. Reprod. Fertil. Suppl.*, 12: 1–16.

Amoah, E.A. and Bryant, M.J., 1983. Gestation period, litter size and birth weight in the goat. *Anim. Prod.*, 36(1): 105–110.

Balch, C.C., 1974. Modern developments in productivity of dairy cows. *Proc. Br. Soc. Anim. Prod.*, 3: 27–29.

Barnard, E., 1962. Methods and problems concerned with hand-rearing rabbits. *J. Anim. Tech. Assoc.*, 13: 35–40.

Clark, R.J. and Reid, C.S.W., 1974. Foamy bloat of cattle: a review. *J. Dairy Sci.*, 57: 753–785.

Commonwealth Bureau of Pastures and Field Crops, 1961. *Research Techniques in Use at the Grassland Research Institute, Hurley.* CAB Bulletin 45, Farnham Royal.

Esslemont, R.J., 1974. Economic and husbandry aspects of the manifestation of oestrus in cows. 1. Economic aspects. *A.D.A.S. Q. Rev.*, No. 12: 175–183.

Foxcroft, G.R. and Hasnain, H., 1973. Effects of suckling and time to mating after parturition on reproduction in the domestic rabbit. *J. Reprod. Fertil.*, 33: 367–377.

Fraser, A.F., 1971. *Animal Reproduction. Tabulated Data.* Baillière, Tindall and Cox, London, 28 pp.

Gordon, I., 1983. *Controlled Breeding in Farm Animals.* Pergamon Press, Oxford, 436 pp.

Hafez, E.S.E. (Editor), 1962. *Reproduction in Farm Animals.* Baillière, Tindall and Cox, London, 367 pp.

Hall, M.J. and Henshaw, J., 1983. Red Deer. *Biologist*, 30(1): 4–10.

Large, R.V., 1965. The artificial rearing of lambs. *J. Agric. Sci. Cambridge*, 65: 101–108.

Large, R.V., 1976. The influence of reproductive rate on the efficiency of meat production in animal populations. In: D. Lister, D.N. Rhodes, V.R. Fowler and M.F. Fuller (Editors), *Meat Animals: Growth and Productivity.* Plenum Press, New York, N.Y., pp. 43–55

Lees, J.L., 1969. The reproductive pattern and performance of sheep. *Outlook Agric.*, 6(2): 82–88.

Marcum, J.B., 1974. The freemartin syndrome. *Anim. Breed. Abstr.*, 42: 227–242.

Mohammed, H.H., 1982. *Energy Requirements for Maintenance and Growth: Comparison of Goats and Sheep.* Thesis, University of Reading, Reading, 186 pp.

Newton, J.E. and Betts, J.E., 1967. Breeding performance of Dorset Horn ewes augmented by hormonal treatment. *Expl. Agric.*, 3: 307–313.

Newton, J.E. and Betts, J.E., 1968. Seasonal oestrogenic activity of various legumes. *J. Agric. Sci. Cambridge*, 70: 77–82.

Penning, P.D., 1983. A technique to record automatically some aspects of grazing and ruminating behaviour in sheep. *Grass Forage Sci.*, 38(2): 89–96.

Penning, P.D., Penning, I.M. and Treacher, T.T., 1977. The effect of temperature and method of feeding on the digestibility of two milk substitutes and on the performance of lambs. *J. Agric. Sci. Cambridge*, 88: 579–589.

Robinson, J.J., 1974. Intensifying ewe productivity. *Proc. Br. Soc. Anim. Prod.*, 3: 31–40.

Sandford, J.C., 1966. *The Domestic Rabbit.* Crosby Lockwood, London, 3rd ed., 258 pp.

Spedding, C.R.W., 1970. *Sheep Production and Grazing Management.* Baillière, Tindall and Cassell, London, 435 pp.

Spedding, C.R.W., 1982. Energy use in livestock production. In: D.W. Robinson and R.C. Mollan (Editors), *Energy Management and Agriculture. Proc. First Int. Summer School in Agriculture, Royal Dublin Society*, pp. 337–349.

Spedding, C.R.W., Walsingham, J.M. and Large, R.V., 1976. The effect of reproductive rate on the feed conversion efficiency of meat-producing animals. *World Rev. Anim. Prod.*, 12(4): 43–49.

Spedding, C.R.W., Walsingham, J.M. and Hoxey, A.M., 1981. *Biological Efficiency in Agriculture.* Academic Press, London, 383 pp.

Walsingham, J.M., 1972. *Meat Production from Rabbits.* Technical Report No. 12, Grassland Research Institute, Hurley, 21 pp.

Wilkins, R.J. and Bather, M., 1981. Potential for changes in support energy use in animal output from grassland. In: J.L. Jollans (Editor), *Grassland in the British Economy. CAS Paper 10.* Centre of Agricultural Strategy, University of Reading, Reading, pp. 511–520.

Young, N.E. and Newton, J.E., 1975. *Grasslambs.* Farmer's Booklet No. 1, Grassland Research Institute, Hurley, pp.

Section III

# NUTRIENT CYCLING

Although the principles that govern nutrient cycling are clearly the same in natural and managed grasslands, the rate of cycling and the relative importance of various pathways are often very different. The main factors which give rise to these differences are: (1) inputs of nutrients, especially as fertilizers, and (2) the much greater stocking rates, and hence greater utilization of herbage, in managed grasslands than in natural grasslands.

Many aspects of nutrient cycling in managed grasslands have been studied in detail for several decades. In particular, topics such as nitrogen fixation by legumes, nutrient uptake by swards under various conditions, and the amount, composition and distribution of animal excreta have received considerable attention. However, nutrient cycling in managed grasslands has only quite recently been viewed *in toto*, and considered in a systems context. This approach has brought a more coherent view of the processes involved and their inter-relationships.

Chapters 17 and 18 introduce a systems approach to the cycling of N and P respectively. The following chapters in the section then consider some of the more important factors and processes in more detail. Chapter 19 deals with nutrient returns in excreta, which constitutes a far more important pathway in managed grasslands than in natural grasslands; this has important consequences for speed of cycling, for spatial heterogeneity of cycling, and for nutrient losses from the system. As a result, animal excreta have received much greater attention in managed grasslands than in natural ecosystems. Similarly, studies of nitrogen fixation (Chapter 20) and nitrogen losses (Chapter 21) have received more attention in managed grasslands. In each of these cases, the techniques developed in studies of managed grasslands could with benefit be applied to natural grasslands. Finally, run-off and drainage from managed grasslands are considered in Chapter 22. These processes are important in so far as they constitute important losses of nutrients, and water from grasslands, but also because the intensive management of grasslands affects, usually detrimentally, the quantity and quality of water harvested for catchments.

Chapter 17

# NITROGEN CYCLING IN MANAGED GRASSLANDS

M.J.S. FLOATE

## INTRODUCTION

The supply of nitrogen from the soil is commonly the major factor controlling herbage production from grass-dominant pastures (Chapter 6), particularly when water supply and temperature are not limiting (Chapters 2 and 3). Even production from legume-dominant pastures can be limited by nitrogen supply, though legume growth is more normally limited by the supply of other nutrients from soil.

The atmosphere is the ultimate source of nitrogen for all grassland ecosystems, whether natural or managed, and the primary or secondary productivity of these ecosystems is largely dependent upon the efficiency with which nitrogen is captured from the atmosphere by fixation (Chapter 21), and then cycled through the ecosystem.

The terrestrial part of the cycle is characterized by uptake of nitrogen from soil by plants, and its return to the soil in plant litter and animal excreta; the ends of the food chain are subsequently joined by the processes of decomposition and mineralization (Fig. 17.1). Most of the nitrogen ingested by grazing animals completes the cycle in this way; this feature of grazed ecosystems contrasts with arable cropping systems, in which a large part of the nitrogen taken up by plants is removed from the ecosystem at harvest.

Figure 17.1 shows that there is not a single nitrogen cycle but several partial cycles, which are interlinked. Before discussing these cycles in detail, it is necessary to consider how the generalized nitrogen cycle relates to grassland ecosystems. The generalized cycle (Fig. 17.1) has no boundaries, since the system includes both atmosphere and hydrosphere. As a result no gains or losses occur outside the system. In practice it is more realistic to

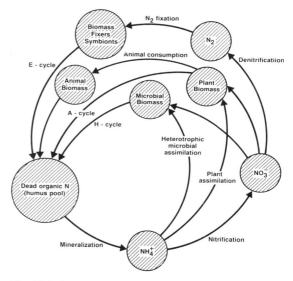

Fig. 17.1. A generalized nitrogen cycle divided into its three partial cycles (after Jansson, 1981).

consider the terrestrial part of the grassland ecosystem, which is mainly concerned with soil, plant and animal components, and to exclude the atmospheric and hydrologic parts of the cycle. It is also more realistic to define the ecosystem geographically, for example as a field or a farm.

The grassland nitrogen cycle illustrated in Fig. 17.2 differs from that in Fig. 17.1 in that some components and transfers have been omitted or combined and certain flows have been subdivided to allow discussion of processes which are especially important in grassland ecosystems. In particular, the soil component has been subdivided and inputs to the defined system, and losses from it, are shown. Fig. 17.2 will form the basis for consideration of the nitrogen cycle in grassland

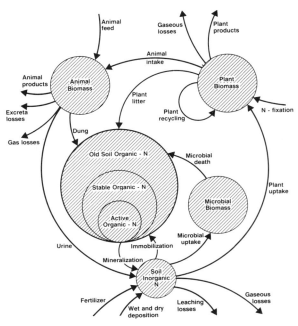

Fig. 17.2. A generalized nitrogen cycle for managed grassland ecosystems, showing the main components and pathways.

ecosystems, with special emphasis on: (1) the nature and size of components; (2) the distinction between inorganic and organic components; (3) gaseous, liquid and solid phases in the cycle; (4) transfer processes between components; (5) factors which control transfer processes; (6) amounts and rates of nitrogen flow; (7) spatial dimensions; (8) time scales within ecosystems; and (9) equilibrium within ecosystems. These features will be illustrated by reference to three typical ecosystems.

## ECOSYSTEM COMPONENTS

Unlike most other elements in the biosphere, nitrogen can occur in solid, liquid and gaseous phases; it can occur as simple inorganic substances or complex organic compounds. Gaseous nitrogen ($N_2$) is by far the largest component, world-wide (Fig. 17.3). The soil pool of organic nitrogen is the next largest component (Fig. 17.3), typically ranging from 1000 to 10 000 kg ha$^{-1}$. The "old" soil organic nitrogen (i.e. that which has remained in the soil for several centuries) is the largest fraction within the organic pool, while the less stable and active fractions, which are closely related to the microbial biomass are smaller.

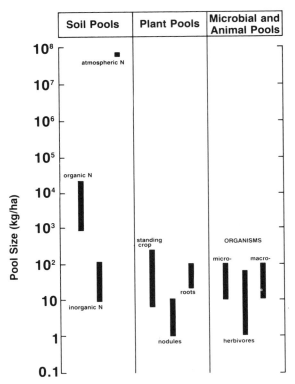

Fig. 17.3. The relative size of nitrogen pools in grassland ecosystems, showing typical ranges of values, kg N ha$^{-1}$ (after Gandar and Ball, 1982).

By comparison, the soil pool of inorganic nitrogen, which consists mainly of ammonium ($NH_4^+$) and nitrate ($NO_3^-$) ions in solution, is small (Fig. 17.3). The inorganic nitrogen pool in the microbial, animal and plant biomass is also small (Fig. 17.3) compared with the organic pool, which consists largely of protein. The plant biomass pool can be separated into roots, stems, leaves and reproductive organs, though probably only separation between root and shoot is practically feasible. In mixed swards, it may be important to separate legume and grass components, because of the special role of the legume–*Rhizobium* symbiosis in nitrogen fixation.

## NITROGEN TRANSFORMATIONS AND FLOWS BETWEEN COMPONENTS

Although inorganic nitrogen in soil solution is the smallest pool in the system, it is the only source

# NITROGEN CYCLING

in the soil from which plants can obtain nitrogen. The large pool of organic nitrogen can only be utilized after mineralization. Mineralization and immobilization are opposing processes within the system, and it is usually only possible to measure their net effect, though isotopic markers can be used to measure the two processes separately. Although micro-organisms are responsible for the release of inorganic nitrogen from organic forms, they also have a requirement for nitrogen, and so can compete with higher plants for inorganic nitrogen.

Nitrogen is taken up by plants from the soil as $NH_4^+$ or $NO_3^-$. After conversion to organic nitrogen, this nitrogen may be recycled within the plant, consumed by herbivores, or pass to the soil organic matter, via plant litter or animal excreta. Recycling within the plant seems to be important in semi-natural and natural grasslands (Clark, 1977; Morton, 1977), but is less important in intensively managed grasslands. In managed grasslands, grazing largely prevents this recycling of nitrogen; this may explain the susceptibility of some species to grazing damage.

Usually less than 15% of the nitrogen ingested by the animal is retained, though more is retained in intensive dairying systems (Henzell and Ross, 1973). The remainder is excreted in dung and urine (Chapter 19), the proportion excreted in urine increasing as the digestibility of the ingested feed increases (Barrow and Lambourne, 1962). Passage through the animal, and especially excretion in urine, speeds the nitrogen cycle, essentially because decomposition of plant material in the rumen is usually more rapid than decomposition of plant litter in the soil (Floate, 1981).

The nitrogen returned in urine is rapidly hydrolyzed to ammonia (Fig. 17.4), and so becomes available to plants. However, this rapid conversion may cause large losses of nitrogen (Ball and Keeney, 1981), because the product is volatile (Chapter 22), especially under the conditions of high soil pH that it induces (Doak, 1952). Recycling of nitrogen via dung is generally much slower, though the speed is very variable (Fig. 17.4), and there is less loss.

There are large differences between ecosystems in the magnitude of inputs, transfers and losses of nitrogen (Fig. 17.5). Some of these differences will be considered more fully in the next section.

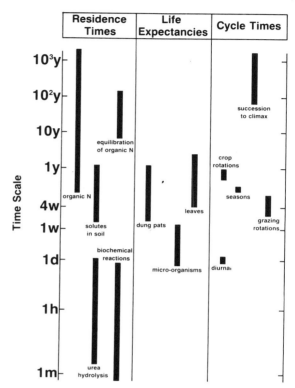

Fig. 17.4. The relative time scales of component processes of the nitrogen cycle in grassland ecosystems (after Gandar and Ball, 1982).

## INPUTS OF NITROGEN

Dry and wet deposition of nitrogen from the atmosphere is usually quite small (Soderlund, 1981), being generally less than 10 kg ha$^{-1}$ yr$^{-1}$, though amounts have increased over the past 20 to 30 years. Non-symbiotic organisms, notably the blue green "algae" (Stewart, 1974), also contribute small amounts of nitrogen. The major biological source of nitrogen is nitrogen fixation by *Rhizobium* in symbiotic association with pasture legumes (Quispel, 1974). Pasture production systems, particularly in New Zealand, have come to rely almost exclusively upon this source of nitrogen, partly because early work in the 1950s (Sears, 1956) showed that clover-based pastures were capable of fixing some 500 to 700 kg N ha$^{-1}$ yr$^{-1}$; however, recent work has demonstrated that such estimates may have been exaggerated (Chapter 20). Modern methods indicate that an average is about 185 kg N

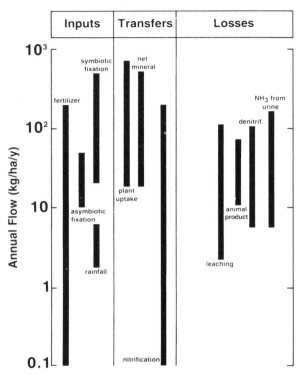

Fig. 17.5. Examples of flows and transformations of nitrogen in grassland ecosystems, showing typical ranges of values, kg N ha$^{-1}$ yr$^{-1}$ (after Gandar and Ball, 1982).

ha$^{-1}$ yr$^{-1}$ (Hoglund et al., 1979). The amounts fixed depend upon the proportion of legume in the sward, the nitrogen status of the soil and additions of nitrogenous fertilizers (Ball and Field, 1982; Keeney and Gregg, 1982). The supply of phosphorus, sulphur, molybdenum and potassium also affects the clover content and so affects nitrogen fixation.

## LOSSES OF NITROGEN

Nitrogen may be lost from grassland ecosystems in solution or in gaseous forms (Chapter 21). Losses are often greater from intensively managed systems of production than from less intensive or natural systems. The inevitable consequence of agricultural production is that nitrogen is removed from the system as meat, milk and wool, or sometimes as conserved forage. These losses are, by definition, greatest in the most intensive systems; but there are other pathways of loss which also seem to be greater in managed than in natural systems. These losses are considered in detail in Chapter 22, so are only briefly outlined here.

Volatile losses of ammonia from plant tissues have been recorded, but are normally only small and may be partially compensated by re-absorption within the sward canopy (Denmead et al., 1974, 1976). More important losses occur through volatilization of ammonia from animal excreta (Ball et al., 1979) and, to a lesser extent, from soils and plants (Freney et al., 1981). There are also large leaching losses from urine patches, where nitrogen is aggregated by animals into zones where the concentration of nitrogen in solution exceeds the capacity of uptake by plants. This loss is much greater from intensively managed systems than from more nearly natural systems, and may explain why the nitrogen cycles of the former are more leaky, whereas the latter are more conservative. Where large amounts of soluble or volatile forms of nitrogen are exposed, and uptake or absorption within the cycle cannot cope in the intensive systems, losses are inevitable but, in more nearly natural systems, uptake by one component is more closely matched to release or return from another, and the opportunities for loss are minimized. In the extreme case, recycling occurs within some plants, when nitrogen supplies are severely limited, and nitrogen is not exposed to pathways which are potentially leaky. Reiners (1981) has suggested that successional changes involving nitrogen fixation by legumes may sometimes be explained by a feedback link with the nitrogen status of the soil, and that this results in a natural nitrogen-conservation strategy. Grass species adapted to low fertility soils are usually slow growing, and have a low nutrient demand so that their growth rate is closely matched to the slow release rate of nitrogen from soil organic matter. By contrast, the more productive species of improved grasslands are able to respond to large applications of nitrogenous fertilizer (Chapters 6 and 8), though uptake seldom exceeds much more than half of that actually applied. In addition to the losses described above, and especially under anaerobic conditions, significant losses may occur through denitrification (Knowles, 1981). If long-term agricultural production is to be sustained, these losses must either be minimized or replaced; in the next section I shall examine how the nitrogen

# NITROGEN CYCLING

cycle is controlled, and later I shall consider the scope for manipulation.

## CONTROL OF NITROGEN CYCLING

The amounts of nitrogen transferred between the various components within grassland ecosystems, their rate of transfer, and the magnitude of gains and losses, are influenced not only by conditions within the system, but also by factors outside the system (Floate, 1977). Chief among these is energy receipt, which provides the driving force for photosynthesis and nitrogen fixation, and which influences temperature-dependent processes.

### Nitrogen inputs

Probably the most important factor controlling the amount and rate of nitrogen flow through grassland ecosystems is the input of fixed nitrogen into the system. This input is largely by fertilizer applications or by symbiotic nitrogen fixation (see above). Both industrial and symbiotic fixation require large amounts of energy ($c.$ 80 MJ kg$^{-1}$ N). The effects of nitrogen supply on nitrogen flow through the ecosystem and on nitrogen losses have already been considered (see above).

### Temperature and water

Temperature has a variety of effects on the nitrogen cycle. One important effect is that the growth rate, and hence nitrogen uptake of plants, increases with temperature. Temperature also affects the fixation of nitrogen by *Rhizobium*, and the mineralization of nitrogen from soil organic matter. In each case the processes reach a maximum rate in the range of 25 to 35°C.

Water supply is the other major factor which affects plant growth and microbial activity; it also affects leaching losses from the soil.

Temperature, water, and supply of raw materials are probably the most important factors which control processes within the ecosystem. These controls are interrelated, and the results of one process may control the supply of nitrogen for the next step in the cycle. In order to manipulate systems, it is of vital importance to be able to identify such rate-determining processes, and to know which factors control those processes.

Soil fertility, especially the supply of phosphorus, potassium and sulphur, may limit the rate of nitrogen cycling if the supply of any one or more nutrients limits plant growth or microbial activity. Soil acidity is also a factor which controls the rate of many biological soil processes, chiefly through restricting the numbers and activity of microbial organisms. This has a profound effect on the relative amounts of old, stable and active organic nitrogen. The control of acidity, by the application of lime to acid soils, offers the manager an opportunity for manipulating the nitrogen cycle, by accelerating the decomposition rate.

## THE DIMENSIONS OF ECOSYSTEMS

One of the major difficulties of dealing with ecosystems is the definition of their limits in space and time. Spatially, ecosystems can be considered at scales ranging from less than 1 m$^2$ to whole catchments, and may even be viewed on a global scale. The time scale is also variable, largely due to the large differences in residence times in some pools (see Fig. 17.4). Gandar and Ball (1982) give excellent illustrations of the range in spatial dimensions involved when nitrogen is studied in ecosystems.

### Spatial dimensions

It is important to realize the heterogeneity that exists in the environment, and the way in which this affects the cycling of nitrogen. For example, soil conditions commonly vary considerably over distances of a few metres (Beckett and Webster, 1971). Grazing animals also create widespread heterogeneity, through the uneven effects of selective grazing (Chapters 2 and 13), trampling and excreta return (Chapter 19). As a result, the input, transfer and loss of nitrogen may be dramatically different within a few hectares (e.g. Hilder, 1966), or even a few centimetres (e.g. Lynch, 1982). Such differences in environmental conditions, on a variety of scales, clearly present difficulties in experimental technique and measurement. For some purposes it may be necessary to study small representative areas, whereas for other purposes it may be permissible to average data over large areas to encompass representative variability. Some appreciation of this variation is obviously necessary for a proper understanding of the errors involved when data are compiled for farm-sized ecosystems.

## Time dimensions

Large differences in residence times, ranging from minutes to thousands of years (Fig. 17.4), obviously pose problems in studying ecosystems. Usually transfer rates are expressed on an annual basis, which encompasses the annual cycles of growth and development in plants and animals, as well as the seasonal patterns of microbial activity. The expression of data on an annual basis, however, usually obscures important changes in certain pool sizes and flow rates during the year. Carran (1982) presented data for nitrogen flows in grazed pasture on four occasions (October, January, March, July), and showed that mineralization of organic nitrogen was greatest in spring, whereas nitrogen fixation and plant death and decay were greatest in summer and autumn; accumulation of nitrogen in soil organic matter dominated the nitrogen flow in winter. The flow of nitrogen from soil to plant, to animal, and back to soil can occur within a few days, so it is therefore possible for several cycles to occur within one year (Henzell, 1968). The participation of the animal in such "multiple recycling" therefore leads to more efficient production per unit of nitrogen in circulation (Floate, 1981), but may also lead to increased losses (Ball et al., 1979; O'Connor, 1981).

## EQUILIBRIUM, STEADY STATE AND BALANCE IN ECOSYSTEMS

One objective of good husbandry in agriculture is to ensure that inputs and outputs of nitrogen should be balanced, so that the system is not depleted. This may be described as a balanced (or stable) system. Losses from the system are undesirable, not only because they represent a loss of valuable resources, but also because such losses may be pollutants in other systems, such as rivers and groundwater.

The maintenance of stability in an ecosystem requires an understanding of the processes involved, and a knowledge of how these can be manipulated. However, one should first distinguish between the terms balance, steady state and equilibrium, because they are often used wrongly and interchangeably.

**Balance** is an accounting term, which refers to the arithmetic sum of gains to and losses from any individual component or the whole system; it is normally calculated at the end of the accounting period. The calculation can only be made if the amounts are expressed in the same units over the same period, but balance may be achieved between a large flow over a short period of time (such as nitrogen uptake by plant growth), and a smaller flow over a longer period of time (such as nitrogen mineralization from organic matter), when both are expressed on an annual basis.

**Steady state** refers to the constancy of the size of any one pool or whole system; this occurs when the rate of loss is matched by the rate of gain. Constancy is therefore maintained at all times, and not just at the end of the accounting period as above.

**Equilibrium** refers to the state of balance between transfers in opposite directions, and is said to be stable when these are equal. The relationship between nitrogen mineralization and immobilization may be said to be in equilibrium when these processes occur at the same rate, or if the amounts transferred by these opposing processes are equal.

Various attempts have been made to calculate nitrogen balances of different ecosystems at different spatial scales. For example, Cooke (1958, 1967) drew up national balance sheets for various nutrients, Henzell and Ross (1973) summarized quantitative data on pool sizes and flow rates in pasture ecosystems. A number of authors have recently provided data for typical ecosystems in a number of countries (Frissel, 1977). Nitrogen balances in grassland and forest ecosystems in New Zealand were the subject of a recent workshop (Gandar and Bertaud, 1982). The objective in Frissel's study (1977) was to reach conclusions on the state of balance in soil, plant and animal pools, using a prescribed format; however, the conclusions were very incomplete because of inadequate data. In addition, the conclusions were based on static balance, whereas the nitrogen cycle is a dynamic concept.

Input–output balances do, however, highlight our lack of knowledge about certain processes in the nitrogen cycle. Failure to account for the fate of nitrogen applied to soil has been an enigma for many years (Allison, 1955, 1966). Only by using radioisotopes in gas-tight lysimeters (Ross et al., 1964) has it been possible to account for all the gains or losses from soils (Martin and Ross, 1968). Such techniques cannot be applied to crops growing in the field, and cannot be used with grazing animals. These are serious deficiencies, since unaccounted losses are usually much greater with animal utilization (Frissel, 1977; Ball and Keeney, 1981). In addition, the technique itself may exaggerate nitrogen losses by creating microclimates which increase ammonia losses.

In order to maintain long-term stability in grassland ecosystems, it is necessary that losses, whether by natural means or those due to farm operations, must be matched by additions, so that the whole system is maintained in a steady state. Certain pathways dominate the system at certain times of the year and, as a result, pool sizes fluctuate within the year. Some of these changes are in response to changes in the natural environment, such as changes in response to temperature and moisture, whereas other changes are in response to deliberate management decisions. The extent to which grassland ecosystems can be manipulated will be examined later.

## CASE STUDIES

Data from three typical, but contrasting, managed grassland ecosystems have been selected to illustrate some of the principles discussed above. The chosen ecosystems are: (1) an unimproved hill pasture at Ballantrae, New Zealand with only 1 to 2% clover in the sward, and a stocking rate of 6 sheep ha$^{-1}$ (Lambert et al., 1982); (2) pasture in the same area as (1), in which the clover content was increased to about 24% by applications of 1.9 t ha$^{-1}$ superphosphate and 1.3 t ha$^{-1}$ lime over a four-year period, and where the stocking rate was 12 sheep ha$^{-1}$ (Lambert et al., 1982); and (3) an intensively managed dairy pasture in New Zealand receiving 448 kg N ha$^{-1}$ yr$^{-1}$ and supporting 4.25 cows ha$^{-1}$ (Ball and Field, 1982). The data (Figs. 17.6, 17.7 and 17.8) conform to the format of the generalized nitrogen cycle illustrated in Fig. 17.2.

Inputs of nitrogen through dry and wet depositions were small (15 kg N ha$^{-1}$ yr$^{-1}$) in all three systems. Non-symbiotic nitrogen fixation was not

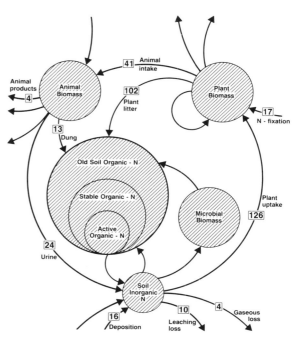

Fig. 17.6. Nitrogen cycle case study for an unimproved hill pasture at Ballantrae, New Zealand (after Lambert et al., 1982).

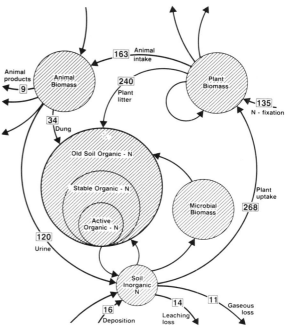

Fig. 17.7. Nitrogen cycle case study for a well fertilized, clover-based pasture at Ballantrae, New Zealand, to which 1.9 t ha$^{-1}$ superphosphate was applied over four years (after Lambert et al., 1982).

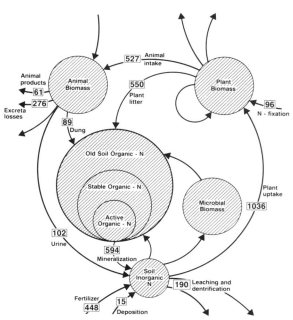

Fig. 17.8. Nitrogen cycle case study for an intensively managed dairy pasture, receiving 448 kg ha$^{-1}$ N, in the Manawatu district, New Zealand (after Ball and Field, 1982).

measured but is also assumed to be small. Nitrogen fixation by legumes was least (17 kg ha$^{-1}$ yr$^{-1}$) in (1) and greatest (135 kg ha$^{-1}$ yr$^{-1}$) in (3), where large amounts of superphosphate had increased the legume content of the sward from about 20% to about 50%. Where large amounts of nitrogen fertilizer were applied in (3), the amount of nitrogen fixation was reduced to 96 kg N ha$^{-1}$ yr$^{-1}$. The increasing input of nitrogen from (1) to (3) increased uptake progressively from 126, through 268 to 1036 kg N ha$^{-1}$ yr$^{-1}$ in systems (1), (2) and (3) respectively; however, it should be noted that (3) includes an estimate of nitrogen content in the roots as well as the shoots. Because of the increasing plant pool of nitrogen through the sequence of the three systems, the amounts returning as plant litter and ingested by grazing animals also increased from (1) to (3). The removal of animal products, and return of dung and urine, also increased from (1) to (2); direct comparison with (3) is not possible, because of the way in which data were presented. Combined nitrogen losses, due to leaching, gaseous, and export of animals increased from 14 kg N ha$^{-1}$ yr$^{-1}$ in (1) to 24 kg N ha$^{-1}$ yr$^{-1}$ in (2) and 466 kg N ha$^{-1}$ yr$^{-1}$ in (3). It is noteworthy that in (3) losses appear to be greater than the nitrogen applied as fertilizer. Urine and mineralization of organic nitrogen are the major pathways by which recycled nitrogen passes via the soil to the plant; Lambert et al. (1982) did not provide data for the amount mineralized. In the intensively managed dairy system (3), it seems that over half of the nitrogen taken up by plant growth may be contributed by mineralization of soil organic nitrogen.

## CONTROL AND MANIPULATION OF GRASSLAND ECOSYSTEMS

Farmers can only control a few of the pools and pathways within grassland ecosystems. In the less intensive systems, where little fertilizer is applied, grazing intensity or grazing management are the major management variables influencing nitrogen cycling. These factors affect the relative amount, pattern, and frequency of nitrogen return to soil in animal excreta and plant litter (Floate, 1981). They also affect the legume component of the sward, and so affect the input of nitrogen by fixation. The beneficial effects of rapid recycling, caused by passage through the animal, are partially offset by increased nitrogen losses (see above). These losses can be reduced by suitable management, such as rotational grazing and feed allocation, by increasing herbage utilization and increasing the legume content of swards.

In more intensive systems using sown pastures, the manager has direct control over both plant and animal components in the system. In particular, he has the option of augmenting nitrogen supply by fertilizers. However, applications in excess of about 100 kg N ha$^{-1}$ yr$^{-1}$ markedly reduce nitrogen fixation by clover (Hoglund and Brock, 1982). Conversely, applications of phosphorus and sulphur, when these are deficient in the soil, increase the clover content and nitrogen fixation.

The plant component of the system can be supplemented by oversowing methods, or completely replaced by cultivation and re-seeding. When such drastic measures are employed, the manager has a wide choice in the composition of the new pasture which is sown (Chapter 8), though limits are set by climatic and soil fertility conditions. Subsequently the allocation of available herbage, its utilization by grazing animals (Chapters 9–11), or its conservation in such forms as hay and silage (Chapter 26), can also be manipulated.

It is important in intensive systems to match the stocking rate to the herbage produced (Chapter 24). This also has the effect of increasing the flow of nitrogen through the animal, rather than via death or decomposition of herbage residues, so increasing the rate of cycling. This occurs because soil organic matter has an extremely long residence time, whereas the nitrogen in urine is transferred or utilized rapidly (Chapters 19 and 21). Grazing intensity and management has effects not only on the turnover rate of nitrogen in the whole system, but also on spatial heterogeneity of the system, through its effects on selective grazing and excretion (Hilder, 1966).

To maximize the efficiency of the system, supply and demand within the system should be closely matched, and losses from the system (Chapter 21) should be minimized. If the manager is to succeed in optimizing his pasture management decisions, he needs a good understanding of the interaction of these processes. Improvements in the future are likely to come from an approach such as modelling, which allows for the use of a range of time bases to accommodate diurnal fluctuations, rotation times and seasonal effects (Chapter 5).

## MODELLING THE NITROGEN CYCLE

It is usually difficult to predict the effects in the whole system of manipulating parts of the system, yet such predictions are obviously important if decisions are to be cost-effective. The need for an accurate prediction is one of the reasons for developing models of nitrogen cycling.

The simplest form of model consists of a diagrammatic description of the nitrogen cycle (e.g. Figs. 17.1, 17.2, 17.6–17.8). The purpose of such models may be simply to describe the system better for purposes of communication. More complex models may be used to investigate the sensitivity of the system to environmental and management variables, to predict likely outcomes of variation in those factors, and to identify an order of priority for research (Van Veen et al., 1981).

The accuracy and reliability of data are not necessarily vitally important for simple descriptive models but, if models are to be used for predictive purposes, it is essential that the model be built on accurate data. Despite the large amount of effort that has been devoted to studying nitrogen in agricultural grassland ecosystems, there are still grave deficiencies in knowledge, especially with respect to transfers and losses in the gas phase, and to losses in solution from the deeper layers of soil. From the farmer's point of view, and from the point of view of environmental quality and water purity, it is highly desirable to quantify and minimize these losses: the farmer's unwanted losses are often another's unwanted gains. At present, the scope for prediction and control is limited by the quality of the data which are available for modelling. Hopefully, increased understanding of mechanisms and processes, and more accurate data, will allow better predictive modelling in the future and give opportunities to conserve, control and optimize the use of nitrogen.

## REFERENCES

Allison, F.E., 1955. The enigma of soil-N balance sheets. *Adv. Agron.*, 7: 213–250.

Allison, F.E., 1966. The fate of nitrogen applied to soils. *Adv. Agron.*, 18: 219–258.

Ball, P.R. and Field, T.R.O., 1982. Responses to nitrogen as affected by pasture characteristics, season and grazing management. In: P.B. Lynch (Editor), *Nitrogen Fertiliser in N.Z. Agriculture*. N.Z. Institute of Agricultural Science, Wellington, pp. 45–64.

Ball, R.P. and Keeney, D.R., 1981. Nitrogen losses from urine-affected areas of a N.Z. Pasture, under contrasting seasonal conditions. In: *Proc. XIV Int. Grassland Congr.*, Lexington, U.S.A., pp. 342–344.

Ball, P.R., Keeney, D.R., Theobald, P.W. and Nes, P., 1979. Nitrogen balance in urine-affected areas of a N.Z. pasture. *Agron. J.*, 71: 309–314.

Barrow, N.J. and Lambourne, L.J., 1962. Partitioning of nitrogen, sulphur and phosphorus between faeces and urine. *Aust. J. Agric. Res.*, 13: 461–471.

Beckett, P.B.H. and Webster, R., 1971. Soil variability, a review. *Soils Fertil.*, 34: 1–15.

Carran, R.A., 1982. Nitrogen flows in a Southland pasture. In: P.W. Gandar and D.S. Bertaud (Editors), *Nitrogen Balances in N.Z. Ecosystems*. DSIR, Palmerston North, pp. 91–94.

Clark, F.E., 1977. Internal cycling of $^{15}$N in short grass prairie. *Ecology*, 58: 1322–1333.

Clark, F.E. and Rosswall, T. (Editors), 1981. *Terrestrial Nitrogen Cycles. Processes, Ecosystem Strategies and Management Impacts. Ecol. Bull. (Stockholm)*, No. 33: 714 pp.

Cooke, G.W., 1958. The nations plant food larder. *J. Sci. Food Agric.*, 9: 761–772.

Cooke, G.W., 1967. *The Control of Soil Fertility.* Crosby Lockwood, London, 526 pp.
Denmead, O.T., Simpson, J.R. and Freney, J.R., 1974. Ammonia flux into the atmosphere from a grazed pasture. *Science*, 185: 609–610.
Denmead, O.T., Freney, J.R. and Simpson, J.R., 1976. A closed ammonia cycle within a plant canopy. *Soil Biol. Biochem.*, 8: 161–164.
Doak, B.W., 1952. Some chemical changes in the nitrogenous constituents of urine when voided on pasture. *J. Agric. Sci. Cambridge*, 42: 162–171.
Floate, M.J.S., 1977. Control of nutrient cycling. In: M.J. Frissel (Editor), *Cycling of Mineral Nutrients in Agricultural Ecosystems. Agro-Ecosystems*, 4: 7–14.
Floate, M.J.S., 1981. Effects of grazing by large herbivores on nitrogen cycling in agricultural ecosystems. In: F.E. Clark and T. Rosswall (Editors), *Terrestrial Nitrogen Cycles. Ecol. Bull. (Stockholm)*, No. 33: 585–601.
Freney, J.R., Simpson, J.R. and Denmead, O.T., 1981. Ammonia volatilisation. In: F.E. Clark and T. Rosswall (Editors), *Terrestrial Nitrogen Cycles. Ecol. Bull. (Stockholm)*, No. 33: 291–302.
Frissel, M.J. (Editor), 1977. *Cycling of Mineral Nutrients in Agricultural Ecosystems. Agro-Ecosystems*, 4: 1–354.
Gandar, P.W. and Ball, P.R., 1982. Nitrogen balances — an overview. In: P.W. Gandar and D.S. Bertaud (Editors), *Nitrogen Balances in N.Z. Ecosystems*. DSIR, Palmerston North, pp. 13–31.
Gandar, P.W. and Bertaud, D.S. (Editors), 1982. *Nitrogen Balances in N.Z. Ecosystems*. DSIR, Palmerston North, 262 pp.
Henzell, E.F., 1968 Sources of nitrogen for grassland pastures. *Trop. Grassland*, 2: 1–17.
Henzell, E.F. and Ross, P.J., 1973. The nitrogen cycles of pasture ecosystems. In: G.W. Butler and R.W. Bailey (Editors), *Chemistry and Biochemistry of Herbage*. Academic Press, New York, N.Y., pp. 227–246.
Hilder, E.J., 1966. Distribution of excreta by sheep at pasture. In: *Xth Int. Grassland Congr., Helsinki*, pp. 977–981.
Hoglund, J.H. and Brock, J.L., 1982. Biological nitrogen inputs in pastures. In: P.W. Gander and D.S. Devland (Editors), *Nitrogen Balances in N.Z. Ecosystems*. DSIR, Palmerston North.
Hoglund, J.H., Crush, J.R., Brock, J.L., Ball, P.R. and Carran, R.A., 1979. Nitrogen fixation in pasture. XII. General discussion. *N.Z. J. Exper. Agric.*, 7: 45–51.
Jansson, S.L., 1981. Rapporteur's comments. In: F.E. Clark and T. Rosswall (Editors), *Terrestrial Nitrogen Cycles. Ecol. Bull. (Stockholm)*, No. 33: 195–199.
Keeney, D.R. and Gregg, P.E.H., 1982. Nitrogen fertilisers and the nitrogen cycle. In: P.B. Lynch (Editor), *Nitrogen Fertiliser in N.Z. Agriculture*, N.Z. Institute of Agricultural Science, Wellington, pp. 19–28.
Knowles, R., 1981. Denitrification. In: F.E. Clark and T. Rosswall (Editors), *Terrestrial Nitrogen Cycles. Ecol. Bull. (Stockholm)*, No. 33: 315–329.
Lambert, M.G., Renton, S.W. and Grant, D.A., 1982. Nitrogen balance studies in some North Island hill pastures. In: P.W. Gandar and D.S. Bertaud (Editors), *Nitrogen Balances in N.Z. Ecosystems*. DSIR, Palmerston North, pp. 35–39.
Lynch, P.B., 1982. *Nitrogen Fertilisers in N.Z. Agriculture*. N.Z. Institute of Agricultural Science, Wellington, 273 pp.
Martin, A.E. and Ross, P.J., 1968. A nitrogen-balance study using labelled fertiliser in a gas Lysimeter. *Plant Soil*, 28: 182–186.
Morton, A.J., 1977. Mineral nutrient pathways in a molinietum in autumn and winter. *J. Ecol.*, 65: 993–999.
O'Connor, K.F., 1981. Comments on Dr Floate's paper. In: F.E. Clark and T. Rosswall (Editors), *Terrestrial Nitrogen Cycles. Ecol. Bull. (Stockholm)*, No. 33: 707–714.
Quispel, A. (Editor), 1974. *The Biology of Nitrogen Fixation*. North-Holland, Amsterdam, 769 pp.
Reiners, W.A., 1981. Nitrogen cycling in relation to ecosystem succession. In: F.E. Clark and T. Rosswall (Editors), *Terrestrial Nitrogen Cycles. Ecol. Bull. (Stockholm)*, No. 33: 507–528.
Ross, P.J., Martin, A.E. and Henzell, E.F., 1964. A gas-tight growth chamber for investigating gaseous nitrogen changes in the soil:plant:atmosphere system. *Nature (Lond.)*, 204: 444–447.
Sears, P.D., 1956. The effect of the grazing animal on pasture. In: *Proc. 7th Int. Grassland Congr.*, Palmerston North, pp. 92–102.
Soderlund, R., 1981. Dry and wet deposition of nitrogen compounds. In: F.E. Clark and T. Rosswall (Editors), *Terrestrial Nitrogen Cycles. Ecol. Bull. (Stockholm)*, No. 33: 123–130.
Stewart, W.D.P., 1974. Blue-gree algae. In: A. Quispel (Editor), *The Biology of Nitrogen Fixation*. North-Holland, Amsterdam, pp. 202–237.
Van Veen, J.A., McGill, W.B., Hung, H.W., Frissel, M. and Cole, C.V., 1981. Simulation models of the terrestrial nitrogen cycle. In: F.E. Clark and T. Rosswall (Editors), *Terrestrial Nitrogen Cycles. Ecol. Bull. (Stockholm)*, No. 33: 25–48.

Chapter 18

# PHOSPHORUS CYCLING IN MANAGED GRASSLANDS

A.G. GILLINGHAM

## INTRODUCTION

There are several important differences between managed grasslands and natural grasslands which affect phosphorus cycling. First, applications of phosphate fertilizer increase the amount of phosphorus in the ecosystem. Secondly, phosphorus and other fertilizers increase pasture growth and so increase the rate of phosphorus cycling. Thirdly, higher stocking rates increase the amount of phosphorus passing through the grazing animal.

The major pathways of phosphorus movement in grazed pasture ecosystems are shown in Fig. 18.1. Each of the components of the pathways will be briefly discussed here, with emphasis on the particular features relating to managed grasslands, as distinct from natural grasslands.

## SOIL PHOSPHORUS

### Forms of phosphorus

Soil phosphorus is considered here as comprising all phosphorus below the soil surface, regardless of its origin (e.g. parent material or fertilizer), and its form (e.g. inorganic or organic). Most of the naturally occurring soil phosphorus is derived from primary minerals of the apatite group; in virgin soils, these provide most of the plant-available phosphorus. In most soils, however, secondary forms of inorganic phosphorus (e.g. calcium, iron and aluminium phosphates) are important regulators of the phosphorus content of the soil solution.

In addition, mineralization of organic matter contributes to the plant-available phosphorus "pool". The significance of organic matter varies with both the total amount and the dynamics of accumulation and mineralization. Organic matter is likely to be more important, therefore, in high-rainfall regions (e.g. New Zealand) where soils may contain 15 to 20% of organic matter in the A horizon (Blakemore and Miller, 1968), than in lower rainfall regions (e.g. parts of the United States and Europe), where soils may contain less than 5% organic matter.

Inorganic soil phosphorus has been classified into four major forms (Ryden et al., 1973): (1) water-soluble compounds, such as monocalcium phosphate; (2) physically adsorbed phosphorus on the amorphous surface coatings of clay minerals and colloids; (3) occluded and chemisorbed phosphorus which, following adsorption, develops a stronger chemical bond; and (4) precipitated phosphorus in discrete, largely insoluble forms. These forms are in dynamic equilibrium with one another, though occluded and precipitated phosphorus apparently contain a proportion of "permanently" fixed phosphorus. Mansell et al. (1981) have argued that any attempt to model the dynamics of transformation and transport of applied phosphorus in soil must recognize this immobilization.

### Phosphorus exchange

Removal of phosphorus from soil solution by plants stimulates the release of phosphorus from inorganic and organic sources, and a new equilibrium concentration of phosphorus in the soil solution. There are complex interactions between the various sources in the formation of the "new" solution phosphorus. As a result, much research

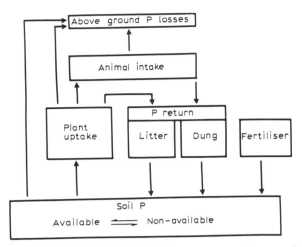

Fig. 18.1. Components of the phosphorus (P) cycle in a grazed and fertilized pasture.

has been conducted to determine the availability of phosphorus in soils, both immediately and in the long term. However, a convenient measure of available soil phosphorus remains an elusive goal in most field situations.

The opposite process, immobilization of phosphorus in soil, either in the short term or the long term, has also attracted much attention. The process has been variously termed phosphate fixation, retention, sorption or reversion (Velayutham, 1980). Two main approaches have been adopted in identifying the major soil components involved in phosphorus sorption or retention. The first method involves removing particular components and measuring the effect on phosphorus sorption (Williams et al., 1958; Saunders, 1965; Syers et al., 1971); the second method involves adding such components to the soil and measuring the effect on phosphorus sorption. Both methods show the importance of amorphous oxides and hydrous oxides of iron and aluminium, which occur as coatings on the surface of clay minerals (Russell and Low, 1954) and calcium carbonate.

### Available phosphorus

The total amount of soil phosphorus in equilibrium with phosphorus in solution represents a measure of that potentially available to plants. This can be estimated by isotopic exchange, using $^{32}P$, a method first used by McAuliffe et al. (1948).

Such estimates of the total exchangeable phosphorus, or "E" value, depends on a true equilibrium being reached between $^{32}P$ and $^{31}P$. Mattingly and Talibudeen (1960) suggested that, for routine work, the amount of phosphorus exchanged in 20 h can be taken as the rapidly exchanging phosphorus, and the amount after 170 h as the total isotopically exchangeable phosphorus. However, these estimates are modified by any factors which change the rate of net phosphorus sorption from the soil solution, such as soil:solution ratio (Mattingly and Talibudeen, 1960), addition of water-soluble inorganic phosphorus (Talibudeen, 1957), and the mineralization of organic phosphorus stimulated by inorganic phosphorus additions (Enwezor, 1966; Smith, 1966) or temperature (Aslyng, 1950).

Two methods have been commonly used to estimate plant-available phosphorus, using isotopically labelled soil. These are based on techniques introduced by Larsen (1950), and by Fried and Dean (1952). Both methods have limitations. The more recently developed techniques of inverse dilution (Mekhael et al., 1965), and double labelling (e.g. Sasaki et al., 1982), may have more general application but, because of their complexity, cannot be used on a routine basis.

Many attempts have been made to correlate plant uptake with various methods of extracting soil phosphorus. Early methods involved mild acids (Truog, 1930) or alkalis (Olsen et al., 1954); these have been widely used for routine analysis in New Zealand (Grigg, 1977) and north-central U.S.A. (Knudsen, 1975). Distilled water alone has also been used as an extractant (Bingham, 1949; Nelson et al., 1953; Martin and Mikkelson, 1960; Hagin et al., 1963; Sissingh, 1969; Ryden and Syers, 1977), and is currently used for determining the phosphorus requirements of arable soils in The Netherlands (Van der Paauw, 1978). Anion exchange resins have also given promising results (DuPlessis and Burger, 1966; Bache and Rogers, 1970; Balleaux and Peaslee, 1975; Brewster et al., 1975).

These various extraction methods have proved only partly successful (Probert and Willett, 1983), and other soil parameters, such as soil buffering capacity (Barrow and Shaw, 1976; Holford and Mattingly, 1976; Helyar and Spencer, 1977), extractable organic phosphorus (Abbott, 1978), or

## PLANT UPTAKE OF PHOSPHORUS

Uptake of phosphorus, a relatively immobile nutrient, occurs largely from the surface soil (Nye and Foster, 1960; Jackman and Mouat, 1972; Gillingham et al., 1980b), and is largely dependent on the root distribution pattern of the plant concerned. Consequently, in mixed grass–legume pastures, the greater competitive ability of many grasses has been attributed to a denser root system exploring a greater volume of soil (Mouat and Walker, 1959; Jackman and Mouat, 1972). The presence of vesicular-arbuscular (VA) mycorrhizal fungi extends the effective size of the root system, and so improves phosphorus uptake and plant growth (Powell, 1977; Powell and Sithamparanathan, 1977; Owusu-Bennoah and Wild, 1980; Abbott and Robson, 1982), though such associations seem to have no ability to extract forms of phosphorus unavailable to non-inoculated plants (Bieleski, 1973). Although VA mycorrhizae can increase phosphorus uptake in soils of low phosphorus status, there is no beneficial effect in well fertilized soils, nor is there any evidence that these mycorrhizae could establish and persist in the soil in competition with indigenous fungal species.

Mass flow, and more particularly diffusion, have been suggested as the major mechanisms by which soil nutrients reach the root surface (Olsen and Watenabe, 1963; Lewis and Quirk, 1967; Omanwar and Robertson, 1970). However, several workers (Bhat et al., 1976; Moghimi and Tate, 1978; Gardner et al., 1983) have suggested that some plants may excrete compounds from their roots which promote the desorption or dissolution of phosphorus, and so increase the availability of soil phosphorus.

Pasture plants which normally grow in soils of low phosphorus status (Asher and Lonergan, 1967; Rorison, 1968) can be said to be more "efficient" in their use of phosphate (Godwin and Wilson, 1976; Blair and Cordero, 1978). Cultivars of white clover (*Trifolium repens*) have also been selected which are better suited to the lower soil fertility and harsher environmental conditions of New Zealand hill country (W.M. Williams et al., 1982).

The concentration of phosphorus in plant material is affected by plant species, type of tissue, age of tissue, and age and physiological condition of the plant. In general, older plant tissue contains less phosphorus than young leaves, largely because of "carbohydrate dilution", but also as a result of some translocation of phosphorus to reproductive or more vigorously growing tissue, especially when plants are in phosphorus-deficient conditions. The proportion of water-soluble inorganic phosphorus in plant leaves increases with increase in total phosphorus content (Bromfield and Jones, 1972); in pasture grown in conditions of high available soil phosphorus, up to 80% of leaf phosphorus may be water soluble (Gillingham et al., 1980a). The minimum phosphorus concentrations in plant tissues for maximum yields are between 2800 and 3600 $\mu g \, P \, g^{-1}$ for ryegrass (*Lolium perenne*) and between 3000 and 4000 $\mu g \, P \, g^{-1}$ for white clover (*Trifolium repens*) (McNaught, 1970). Most grasses have similar phosphate requirements to ryegrass, although *Paspalum dilatatum* appears to tolerate lower concentrations.

The actual phosphorus uptake by pasture is governed largely by growth conditions and therefore varies seasonally. Both growth and phosphorus concentration in the plant fall markedly during periods of moisture stress (Saunders and Metson, 1971).

## ANIMAL INTAKE OF PHOSPHORUS

In an intensively managed pasture, most of the plant material, and hence most phosphorus, is consumed by animals, either by grazing (Chapters 9–11) or as conserved fodder (Chapter 26). Supplementary feeding at certain times of the year provides a further source. In addition, soil ingestion during grazing may contribute up to 12% of the daily intake of phosphorus (Healy, 1967).

Young plants provide a richer source of phosphorus than older plants, and are more palatable and digestible (Minson et al., 1960), so that they are selectively grazed by sheep (Meyer et al., 1957) especially at low grazing pressures (Chapter 14).

## RETURNS OF PHOSPHORUS

Apart from the application of phosphorus fertilizers, phosphorus is mainly returned to the soil either as (1) excreta or (2) dead plant material.

### Excreta

Animal faeces contain almost all the excreted phosphorus. The amount of excretal phosphorus returned to grazed pasture is determined primarily by the amount of herbage consumed and the phosphorus content of that herbage (Chapter 19). To a lesser extent, it is also affected by the physiological state of the grazing animals, which determines the proportion retained in the animal and ultimately exported. The distribution pattern of animal dung is significant to the economy of the phosphorus cycle. More dung is deposited by animals at night than by day (Hancock and McArthur, 1951), resulting in significant phosphorus transfer from established "day" to "night" paddocks on dairy farms (Sears, 1956), and to defined camp-sites on sheep farms (Hilder, 1966), especially in hill country (Gillingham and During, 1973; Gillingham et al., 1980a). The problem can be remedied, on dairy farms, by altering stock management; but the solution is not so simple on hill land where there are a limited number of sites suitable for stock to camp (Gillingham and During, 1973).

### Dead plant material

The amount and type of litter (dead leaves and stem) formed under pasture is influenced by a number of factors. Animals tend to reject herbage low in protein and high in crude fibre (Weir and Torrell, 1959; Arnold, 1960; Blaser et al., 1960; Frame and Hunt, 1971); this herbage also tends to be low in phosphorus. However, prostrate pasture species and material fouled by dung contribute litter of a higher phosphorus content. The phosphorus content of litter also varies during the year, especially with the change in plant growth from vegetative to reproductive condition (Petrie, 1937), and as a result of root demand (R.F. Williams, 1948); this variation is less pronounced in nutrient-deficient plants.

There is relatively little information on the contribution of plant roots to the cycling of phosphorus, largely because of the difficulty of measurement involved. Troughton (1981) showed that, for a number of temperate grasses, the life of roots in a pot trial ranged from about 1.8 to 3.7 years. Compared with the same weight of the above-ground plant material, decomposing roots therefore probably play a much less important role in releasing phosphorus to the soil. However, this return can be accelerated by defoliation, which reduces the life-span of plant roots (Evans, 1973; Troughton, 1981). Little phosphorus is released from litter by direct leaching; the release is predominantly by microbial decomposition (Jones and Bromfield, 1969; Halm et al., 1972). The release of inorganic phosphorus from litter depends on initial attack by fungi, which in turn are attacked by bacteria (Clark and Paul, 1970), which may then be attacked by protozoa (Cole et al., 1977). Partial attack by fungi and bacteria appears necessary before litter becomes acceptable to most soil invertebrates. Because earthworms precondition organic material for subsequent microbial attack (Barley and Jennings, 1959; Edwards and Heath, 1963), they apparently play an important role in the cycling of phosphorus (Sharpley and Syers, 1976; Mansell, 1977).

The nature and rates of mineralization and of immobilization of nutrients are greatly affected by the type of plant material. Young plant material promotes a microbial population explosion, and a rapid exhaustion of the substrate (Birch, 1961); the subsequent death of this population releases inorganic phosphorus. By contrast, the microbial build-up on mature plant material is slower, and a smaller amount of phosphorus is recycled within the decomposer subsystem, with little release of inorganic phosphorus.

The net rate of litter decomposition, and phosphorus mineralization, is greater when the initial organic phosphorus and nitrogen status of the material is high (Kaila, 1954; Floate, 1970a; Singh and Jones, 1976). The rate of decomposition is also greater if the phosphorus status of the soil is high (Blair and Boland, 1978; Batten et al., 1979), and if the temperature is high (Floate, 1970b). An adequate supply of nitrogen, either in the plant material (Sørenson, 1974; Huntjens and Albers, 1978) or by adding nitrogen fertilizer (Bharat and Srivastava, 1982), also increases decomposition rate.

Some detailed measurements have been made of the availability to pasture species of phosphorus released from litter, mainly using isotope techniques (Mansell, 1977; Till and Blair, 1978). For example, the effects of returning clippings to mown grassland have been studied (Lynch, 1947; Elliot and Lynch, 1958). Removal of clippings causes yield reduction, but this was largely attributed to depletion of soil nitrogen and potassium rather than depletion of phosphorus (Wolton, 1963).

Organic phosphorus from litter (and dung) contributes predominantly to soil organic phosphorus in the short term, although some phosphorus is immediately available to plants (Dalal, 1982) as a result of some immediate mineralization in the soil solution or at the root surface. The labile organic-phosphorus fractions, derived from litter or dung, can be more mobile than the inorganic fraction (Read and Campbell, 1981), and may cause some leaching losses (Elliot, 1972).

## ABOVE-GROUND LOSSES OF PHOSPHORUS

The three main routes by which phosphorus is lost above-ground are: export of animal products, transfer in excreta, and soil erosion. These routes offer opportunity for manipulation, so that the "efficiency" of the phosphorus cycle can be varied.

The phosphorus removed in animal products ranges from about 36% of ingested phosphorus, in intensive dairy farming, to about 10% for wool and fat lamb production (Blair et al., 1977), and for sheep rearing and beef production. Phosphorus is also transferred to bare tracks, yards and campsites, when it has little or no effect on pasture production.

The estimated total loss of phosphorus per year by sheep and cattle by export of animal products and by transfer to bare tracks, yards and campsites on flat land is about 0.7 kg per stock unit (During, 1972). On intensively grazed hill land, the comparable loss is 0.9 and 1.1 kg phosphorus per stock unit annually for moderately steep and steep land respectively (Gillingham, 1980). The net loss seems to be little modified by grazing system, and dung distribution patterns can be changed only by very intensive subdivision of paddocks carefully related to topography (Gillingham, 1983).

Most phosphorus losses in surface runoff (Chapter 20), including losses of fertilizer, are in the suspended form (Burwell et al., 1975), although some losses occur in the dissolved inorganic form. Phosphorus losses increase when mineral fertilizers are applied (Sharpley, 1977), or when organic manure is applied, especially to frozen grassland (Kolenbrander, 1977). Johnson et al. (1976) estimated that 20% of the total phosphorus lost from a large (330 $km^2$) catchment was derived from farm sources. However, this represented less than 1% of the total phosphorus applied as fertilizer and manure. Such losses will obviously vary between contrasting conditions. Although these losses are of little importance to farm production, they may have significant effects downstream, especially in ponds or lakes (Syers et al.,1973; Strachan, 1979).

## MODELLING OF THE PHOSPHORUS CYCLE

The amounts of phosphorus present in various compartments of a phosphorus cycle can be determined relatively easily. The transfer rates between compartments, however, are much more difficult to determine, yet are of greater importance. Most studies have therefore been made on systems that are at equilibrium, where components are stable, and where rates of transfer can be more readily determined or estimated — for instance, the ungrazed rangelands studied by Halm et al. (1972) and modelled by Cole et al. (1977). By contrast, the effects of the grazing animal and fertilizer application in intensively managed pastures make measurements or estimates of flux rates of phosphorus much more difficult. The model developed by Blair et al. (1977) represents the most detailed examination to date and has highlighted priority areas for further research.

In New Zealand, most emphasis has been placed on the definition of the size of phosphorus pools and losses (During, 1972; Gillingham and During, 1973; Middleton and Smith, 1978; Parfitt and Lee, 1979; Gillingham et al., 1980a; Parfitt, 1980). Much of this information has been assembled by Cornforth and Sinclair (1982) to provide a basis for phosphorus fertilizer recommendations for grazed pastures. Their scheme estimates losses of phosphorus associated with both "soil" and "animal" factors, and takes into account different soil types, topography and animal stocking rates. This scheme, which places less emphasis on soil phosphorus status than most earlier schemes, provides a rational approach to fertilizer use, which is ultimately the most important reason for studies of the phosphorus cycle in managed grasslands.

## REFERENCES

Abbott, J.L., 1978. Importance of the organic phosphorus fraction in extracts of calcareous soil. *J. Soil Sci. Soc. Am.*, 42: 81–85.

Abbott, L.K. and Robson, A.D., 1982. The role of vesicular arbuscular mycorrhizal fungi in agriculture and the selection of fungi for inoculation. *Aust. J. Agric. Sci.*, 33: 389–408.

Arnold, G.W., 1960. Selective grazing by sheep of two forage species at different stages of growth. *Aust. J. Agric. Res.*, 11: 1026–1033.

Asher, C.J. and Loneragan, J.F., 1967. Response of plants to phosphate concentration in solution culture: I. Growth and phosphorus content. *Soil Sci.*, 103: 225–233.

Aslyng, H.C., 1950. *The Lime and Phosphoric Acid Potentials of Soils, Their Determination and Practical Application.* Thesis, University of London.

Bache, B.W. and Rogers, N.E., 1970. Soil phosphate supply from some Nigerian soils. *J. Agric. Sci., Cambridge*, 74: 383–390.

Balleaux, J.C. and Peaslee, D.E., 1975. Relationships between sorption and desorption of phosphorus by soils. *Proc. Soil Sci. Soc. Am.*, 39: 275–278.

Barley, K.P. and Jennings, A.C., 1959. Earthworms and soil fertility. III. The influence of earthworms on the availability of nitrogen. *Aust. J. Agric. Res.*, 10: 364–370.

Barrow, N.J. and Shaw, T.C., 1976. Sodium bicarbonate as an extractant for soil phosphate. II. Effects of the buffering capacity of a soil for phosphate. *Geoderma*, 16: 273–283.

Batten, G.D., Blair, G.J. and Lill, W.J., 1979. Changes in soil phosphorus and pH in a Red Earth Soil during build-up and residual phases of a wheat-clover Ley farming system. *Aust. J. Soil Res.*, 17: 163–175.

Bharat, R. and Srivastava, A.K., 1982. Microbial decomposition of leaf litter as influenced by fertilisers. *Plant Soil*, 66: 195–204.

Bhat, K.K.S., Nye, P.H. and Baldwin, J.P., 1976. Diffusion of phosphate to plant roots in soil. IV. The concentration distance profile in the rhizosphere of roots with root hairs in a low P soil. *Plant Soil*, 44: 63–72.

Bieleski, R.L., 1973. Phosphate pools, phosphate transport and phosphate availability. *Annu. Rev. Plant Physiol.*, 24: 225–252.

Bingham, F.T., 1949. Soil test for phosphate. *Calif. Agric.*, 3: 11–14.

Birch, H.F., 1961. Phosphorus transformations during plant decomposition. *Plant Soil*, 15: 347–366.

Blair, G.J. and Boland, O.W., 1978. The release of phosphorus from plant material added to soil. *Aust. J. Soil Res.*, 16: 101–111.

Blair, G.J. and Cordero, S., 1978. The phosphorus efficiency of three annual legumes. *Plant Soil*, 50: 387–389.

Blair, G.J., Till, A.R. and Smith, R.C.G., 1977. The phosphorus cycle — what are the sensitive areas? *Rev. Rural Sci.*, III: 9–19.

Blakemore, L.C. and Miller, R.B., 1968. Organic matter in soils. In: *Soils of New Zealand. N.Z. D.S.I.R. Soil Bur. Bull.*, 26(2): 55–67.

Blaser, R.E., Hammes, R.C., Bryant, H.T., Hardison, W.A., Fontenot, J.P. and Engel, R.W., 1960. The effect of selective grazing on animal output. In: *Proc. 8th Int. Grasslands Congr.*, Reading, pp. 601–606.

Brewster, J.L., Gaucheva, A.N. and Nye, P.H., 1975. The determination of desorption isotherms for soil phosphorus using low volumes of solution and an anion exchange resin. *J. Soil Sci.*, 26: 364–377.

Bromfield, S.M. and Jones, O.L., 1972. The initial leaching of hayed-off pasture plants in relation to the recycling of phosphorus. *Aust. J. Agric. Res.*, 23: 811–824.

Burwell, R.E., Timmins, D.R. and Holt, R.F., 1975. Nutrient transport in surface runoff as influenced by soil cover and seasonal periods. *Proc. Soc. Soil Sci. Am.*, 39: 523–528.

Clark, F.E. and Paul, E.A., 1970. The microflora of grassland. *Adv. Agron.*, 22: 375–435.

Cole, C.V., Innis, G.S. and Stewart, J.W.B., 1977. Simulation of phosphorus cycling in semi-arid grasslands. *Ecology*, 58: 1–15.

Cornforth, I.S. and Sinclair, A.G., 1982. Fertiliser and lime recommendations for pastures and crops in New Zealand. *N.Z. Min. Agric. Fish. Occas. Publ.*, 76 pp.

Dalal, R.C., 1982. Effect of plant growth and addition of plant residues on the phosphatase activity in soil. *Plant Soil*, 66: 259–269.

DuPlessis, S.G. and Burger, R. du T., 1966. The availability of different phosphate fractions. *S. Afr. J. Agric. Sci.*, 9: 331–340.

During, C., 1972. Fertilisers and soils in New Zealand farming. *N.Z. Dep. Agric. Bull.*, 409: 312 pp.

Edwards, C.A. and Heath, G.W., 1963. The role of soil animals in breakdown of leaf material. In: J. Doeksen and J. van der Drift (Editors), *Soil Organisms*. North-Holland, Amsterdam.

Elliot, I.L., 1972. Eutrophication and fertilisers. *N.Z. Fertil. Manufactur. Assoc. Press Forum*, Auckland, pp. 22–29.

Elliot, I.L. and Lynch, P.B., 1958. Techniques of measuring pasture production in fertiliser trials. *N.Z. J. Agric. Res.*, 1: 498–521.

Enwezor, W.O., 1966. The biological transformation of phosphorus during incubation of a soil treated with soluble inorganic phosphorus and with fresh and rotted organic materials. *Plant Soil*, 25: 463–466.

Evans, P.S., 1973. The effect of repeated defoliation to three different levels on root growth of five pasture species. *N.Z. J. Agric. Res.*, 16: 31–34.

Floate, M.J.S., 1970a. Decomposition of organic materials from hill soils and pastures. 2. Comparative studies on the mineralisation of C, N and P from plant materials and sheep faeces. *Soil Biol. Biochem.*, 2: 173–185.

Floate, M.J.S., 1970b. Decomposition of organic materials from hill soils and pastures. 3. The effect of temperature on the mineralisation of C, N and P from plant materials and sheep faeces. *Soil Biol. Biochem.*, 2: 187–196.

Frame, J. and Hunt, I.V., 1971. The effects of cutting and grazing systems on herbage production from grass swards. *J. Br. Grassland Soc.*, 26: 163–171.

Fried, M. and Dean, L.A., 1952. A concept concerning the measurement of available soil nutrients. *Soil Sci.*, 73: 263–271.

Gardner, W.K., Barber, D.A. and Parbery, D.G., 1983. The acquisition of phosphorus by *Lupinus albus*. L. *Plant Soil*, 70: 107–124.

Gillingham, A.G., 1980. Hill country topdressing: application variability and fertility transfer. *N.Z. J. Agric.*, 141: 39–43.

Gillingham, A.G., 1983. The role of the grazing animal in nutrient cycling in hill pastures and implications for management. In: *Proc. Hill Land Symp., Corvallis, Ore.*

Gillingham, A.G. and During, C., 1973. Pasture production and transfer of fertility within a long-established hill pasture. *N.Z. J. Exper. Agric.*, 1: 227–232.

Gillingham, A.G., Syers, J.K. and Gregg, P.E.H., 1980a. Phosphorus uptake and return in grazed steep hill pastures. II. Above-ground components of the phosphorus cycle. *N.Z. J. Agric. Res.*, 23: 323–330.

Gillingham, A.G., Tillman, R.W., Gregg, P.E.H. and Syers, J.K., 1980b. Uptake zones for phosphorus in spring by pasture on different strata within a hill paddock. *N.Z. J. Agric. Res.*, 23: 67–74.

Godwin, D.C. and Wilson, E.J., 1976. Prospects for selecting plants with increased P efficiency. *Rev. Rural Sci.*, 3: 131–139.

Grigg, J.L., 1977. Prediction of plant response to fertiliser by means of soil tests. V. Soil tests for phosphorus availability in brown-grey and dry-subhygrous yellow-brown earths. *N.Z. J. Agric. Res.*, 20: 315–326.

Hagin, J., Hillinger, J. and Olmert, A., 1963. Comparison of several ways of measuring soil phosphorus availability. *J. Agric. Sci.*, 60: 245–249.

Halm, B.J., Stewart, J.W.B. and Halstead, R.L., 1972. The phosphorus cycle in a native grassland ecosystem. *Sask. Inst. Pedol. Publ.*, No. R78 (*CCIBP Rep.*, No. 113).

Hancock, J. and McArthur, A.T.G., 1951. Tips on cow management arising from grazing behaviour studies. In: *Proc. Ruakura Farmers' Conf.*, pp. 32–37.

Healy, W.B., 1967. Ingestion of soil by sheep. *Proc. N.Z. Soc. Animal Prod.*, 27: 109–120.

Helyar, K.R. and Spencer, K., 1977. Sodium bicarbonate soil test values and the phosphate buffering capacity of soils. *Aust. J. Soil Res.*, 15: 263–273.

Hilder, E.J., 1966. Distribution of excreta by sheep at pasture. In: *Proc. 10th Int. Grasslands Congr.*, Helsinki, pp. 977–981.

Holford, I.C.R. and Mattingly, G.E.G., 1976. Phosphate adsorption and plant availability of phosphate. *Plant Soil*, 44: 377–389.

Huntjens, J.L.M. and Albers, R.A.J.M., 1978. A model experiment to study the influence of living plants on the accumulation of soil organic matter in pastures. *Plant Soil*, 50: 411–418.

Jackman, R.H. and Mouat, M.C.H., 1972. Competition between grass and clover for phosphate. 2. Effect of root activity, efficiency of response to phosphate and soil moisture. *N.Z. J. Agric. Res.*, 15: 667–675.

Johnson, A.H., Bouldin, D.R. and Goyette, E.A., 1976. Phosphorus loss by stream transport from a rural watershed: quantities, processes and sources. *J. Environ. Quality*, 5: 148–157.

Jones, O.L. and Bromfield, S.M., 1969. Phosphorus changes during the leaching and decomposition of hayed-off pasture plants. *Aust. J. Agric. Res.*, 20: 653–663.

Kaila, A., 1954. Microbial fixation and mineralisation of phosphorus during the decomposition of organic matter. *Soils Fertil.*, 17: 240–243.

Knudson, D., 1975. Recommended phosphorus soil tests. *Bull. N. D. Agric. Exper. Stn.*, No. 499: 16–19.

Kolenbrander, G.J., 1977. Runoff as a factor in eutrophication of surface water in relation to phosphorus manuring. In: J.H. Voorburg (Editor), *Utilization of Manure by Hand Spreading*. Kirchberg, Luxembourg, Comm. Europ. Commun., pp. 181–196.

Larsen, S., 1950. *Studies on the Uptake of Phosphorus in Plants with Radiophosphorus as an Indicator*. K. Veterinaer og Land bohojskole, Copenhagen.

Lewis, D.C. and Quirk, J.P., 1967. Phosphate diffusion in soil and uptake by plants. 1. Self diffusion of phosphate in soils. *Plant Soil*, 26: 99–118.

Lynch, P.B., 1947. Methods of measuring the production from grassland. *N.Z. J. Sci. Technol.*, 28A: 385–405.

Mansell, G.P., 1977. *The Effect of Surface Casting Earthworms on the Plant Availability of Phosphorus in Dead Herbage*. Thesis, Massey University, Palmerston North.

Mansell, G.P., Syers, J.K. and Gregg, P.E.H., 1981. Plant availability of phosphorus in dead herbage ingested by surface-casting earthworms. *Soil Biol. Biochem.*, 13: 163–167.

Martin, W.E. and Mikkelson, D.S., 1960. Grain fertilisation in California. *Calif. Agric. Exper. Stn. Bull.*, 775.

Mattingly, G.E.G. and Talibudeen, O., 1960. Isotopic exchange of phosphates in soil. *Rep. Rothamstead Exper. Stn.*, 248–265.

McAuliffe, C.C., Hall, N.S., Dean, L.A. and Hendricks, S.B., 1948. Exchange reactions between phosphates and soils: Hydroxylic surface of soil minerals. *Proc. Soc. Soil Sci. Am.*, 12: 119–123.

McNaught, K.J., 1970. Diagnosis of mineral deficiencies in grass-legume pastures by plant analysis. In: *Proc. 11th Int. Grasslands Congr.*, Surfers Paradise, Qld., pp. 334–338.

Mekhael, D., Amer, F. and Kadry, L., 1965. Comparison of isotopic dilution methods for estimation of plant available soil phosphorus. In: *Proc. Symposium on Isotopes and Radiation. Soil–Plant Nutrit. Stud., I.A.E.A., Vienna*, pp. 437–438.

Meyer, J.H., Lofgreen, G.P. and Hull, J.L., 1957. Selective grazing by sheep and cattle. *J. Anim. Sci.*, 16: 766–772.

Middleton, K.R. and Smith, G.S., 1978. The concept of a climax in relation to the fertiliser input of a pastoral ecosystem. *Plant Soil*, 50: 595–614.

Minson, D.J., Raymond, W.F. and Harris, C.E., 1960. Studies on the digestibility of herbage. 8. The digestibility of S37 Cocksfoot, S23 Ryegrass and S24 Ryegrass. *J. Br. Grassland Soc.*, 15: 174–180.

Moghimi, A. and Tate, M.E., 1978. Does 2-ketogluconate chelate calcium in the pH range 2.4 to 6.4. *Soil Biol. Biochem.*, 10: 289–292.

Mouat, M.C.H. and Walker, T.W., 1959. Competition for nutrients between grasses and white clover. I. Effect of grass species and nitrogen supply. *Plant Soil*, 11: 30–40.

Nelson, W.L., Meylich, A. and Winters, E., 1953. In: W.H. Pierre and A.G. Normans (Editors), *Soil and Fertiliser Phosphorus*. Agronomy 4. Academic, New York, N.Y.

Nye, P.H. and Foster, W.N.M., 1960. The use of radioisotopes to study plant feeding zones in natural soil. In: *Trans. 7th Int. Congr. Soil Sci.*, 215–221.

Olsen, S.R. and Watenabe, F.S., 1963. Diffusion of phosphorus as related to soil texture and plant uptake. *Proc. Soc. Soil Sci. Am.*, 27: 648–653.

Olsen, S.R., Cole, C.V., Watenabe, F.S. and Dean, L.A., 1954. Estimation of available phosphorus in soils by extraction with sodium bicarbonate. *Circ. U.S. Dep. Agric.*, 939.

Omanwar, P.K. and Robertson, J.A., 1970. Movement of phosphorus to barley roots growing in soil. *Can. J. Soil Sci.*, 50: 57–64.

Owusu-Bennoah, E. and Wild, A., 1980. Effects of vesicular-arbuscular mycorrhiza on the size of the labile pool of soil phosphate. *Plant Soil*, 54: 233–243.

Parfitt, R.L., 1980. A note on the losses from a phosphate cycle under grazed pasture. *N.Z. J. Exper. Agric.*, 8: 215–217.

Parfitt, R.L. and Lee, R., 1979. The efficiency of utilisation of P in superphosphate. *N.Z. J. Exper. Agric.*, 7: 331–336.

Petrie, A.H.K., 1937. Physiological entogeny in plants and its relation to nutrition. 3. The effect of nitrogen supply on the drifting composition of the leaves. *Aust. J. Exper. Biol. Med. Sci.*, 15: 385–404.

Powell, C.Ll., 1977. Mycorrhizas in hill country soils. V. Growth responses in ryegrass. *N.Z. J. Agric. Res.*, 20: 495–502.

Powell, C.Ll. and Sithamparanathan, J., 1977. Mycorrhizas in hill country soils. IV. Infection rate in grass and legume species by indigenous mycorrhizal fungi under field conditions. *N.Z. J. Agric. Res.*, 20: 489–494.

Probert, M.E. and Willet, I.R., 1983. The relationship between labile phosphate and Bray $P_1$ extractable phosphate. *Commun. Soil Sci., Plant Anal.*, 14: 115–120.

Read, D.W.L. and Campbell, C.A., 1981. Biocycling of phosphorus in soil by plant roots. *Can. J. Soil Sci.*, 61: 587–589.

Rorison, I.H., 1968. The response to phosphorus of some ecologically distinct plant species. I. Growth rates and phosphorus absorption. *New Phytol.*, 67: 913–923.

Russell, G.C. and Low, P.F., 1954. Reaction of phosphate with kaolinite in dilute solution. *Proc. Soc. Soil Sci. Am.*, 18: 22–25.

Ryden, J.C. and Syers, J.K., 1977. Origin of the labile phosphorus pool in soils. *Soil Sci.*, 123: 353–361.

Ryden, J.C., Syers, J.K. and Harris, R.F., 1973. Phosphorus in runoff and streams. *Adv. Agron.*, 25: 1–45.

Sasaki, Y., Arima, Y. and Kumazawa, K., 1982. Studies on the radial transport and metabolism of phosphate in corn roots using the $^{32}P$ and $^{33}P$ double-labelling method. *Soil Sci. Plant Nutr.*, 28: 141–145.

Saunders, W.M.H., 1965. Phosphate retention by New Zealand soils and its relationship to free sesquioxides, organic matter and other soil properties. *N.Z. J. Agric. Res.*, 8: 30–57.

Saunders, W.M.H. and Metson, A.J., 1971. Seasonal variation of phosphorus in soil and pasture. *N.Z. J. Agric. Res.*, 14: 307–328.

Sears, P.D., 1956. The effect of the grazing animal on pasture. In: *Proc. 7th Int. Grasslands Congr.*, Palmerston North, pp. 92–101.

Sharpley, A.N., 1977. *Sources and Transport of Phosphorus and Nitrogen in a Stream Draining a Dominantly Pasture Catchment*. Thesis, Massey University, Palmerston North, 291 pp.

Sharpley, A.N. and Syers, J.K., 1976. Seasonal variation in casting activity and in the amounts and release to solution of phosphorus forms in earthworm cases. *Soil Biol. Biochem.*, 9: 227–231.

Singh, B.B. and Jones, J.P., 1976. Phosphorus sorption and desorption characteristics of soil as affected by organic residues. *J. Soil Sci. Soc. Am.*, 40: 389–394.

Sissingh, H.A., 1969. The dissolution of soil phosphoric acid by water extraction in relation to the development of a new P-water method. *Landwirtsch. Forsch.*, 23: 110–120.

Smith, A.N., 1966. The role of inorganic soil phosphates in supplying phosphorus to the wheat plant. *Agrochemica*, XI: 79–81.

Sørenson, L.H., 1974. Rate of decomposition of organic matter in soil as influenced by repeated air drying–rewetting and repeated additions of organic material. *Soil Biol. Biochem.*, 6: 287–292.

Strachan, C., 1979. The Waikato River, a resources study. *N.Z. Water Soil Tech. Publ.*, No. 11: 225 pp.

Syers, J.K., Evans, T.D., Williams, J.D.H. and Murdock, J.T., 1971. Phosphate sorption parameters of representative soils from Rio-Grande do Sul, Brazil. *Soil Sci.*, 112: 267–275.

Syers, J.K., Harris, R.F. and Armstrong, D.E., 1973. Phosphate chemistry in lake sediments. *J. Environ. Quality*, 2: 1–13.

Talibudeen, O., 1957. Isotopically exchangeable phosphorus in soils. 2. Factors influencing the estimation of labile phosphorus. *J. Soil. Sci.*, 8: 86–96.

Till, A.R. and Blair, G.J., 1978. The utilisation by grass of sulphur and phosphorus from clover litter. *Aust. J. Agric. Res.*, 29: 235–242.

Troughton, A., 1981. Length of life of grass roots. *Grass Forage Sci.*, 36: 117–120.

Truog, E., 1930. The determination of readily available phosphorus of soils. *J. Am. Soc. Agron.*, 22: 874–882.

Van der Paauw, F., 1978. The water soluble P method in the assessment of soil P supply. *Landwirtsch. Forsch. Sonderheft (Kongressband 1977)*, 34: 109–120.

Velayutham, M., 1980. The problem of phosphate fixation by minerals and soil colloids. *Phosphorus Agric.*, 77: 1–8.

Weir, W.C. and Torrell, D.T., 1959. Selective grazing by sheep as shown by a comparison of the chemical composition of range and pasture forage obtained by hand clipping and that collected by esophageal fistulated sheep. *J. Anim. Sci.*, 18: 641–649.

Williams, R.F., 1948. The effects of phosphorus supply on the rates of intake of phosphorus and nitrogen and upon certain aspects of phosphorus metabolism in Graminaceous plants. *Aust. J. Sci. Res.*, B. 1: 333–361.

Williams, E.G., Scott, N.M. and McDonald, M.J., 1958. Soil properties and phosphate sorption. *J. Sci. Food Agric.*, 9: 551–559.

Williams, W.M., Lambert, M.G. and Caradus, J.R., 1982. Performance of a hill country white clover selection. *Proc. N.Z. Grassland Assoc.*, 43: 188–195.

Wolton, K.M., 1963. An investigation into the simulation of nutrient returns by the grazing animal in grassland experimentation. *J. Br. Grassland Soc.*, 18: 213–219.

# Chapter 19

# RETURN OF NUTRIENTS BY ANIMALS

N.J. BARROW

## INTRODUCTION

In natural ecosystems most of the nutrients ingested by grazing animals are eventually returned to the soil. In a managed ecosystem, however, the aim of the management is usually to produce a product which can be exported from the system; this leads to the export of nutrients. The proportion of the ingested nutrients which are exported in this manner varies widely, depending on the nutrient concerned and the type of animal product. For example, under intensive dairy production in New Zealand, Middleton and Smith (1978) estimated that 36% of the ingested phosphorus was exported as animal products, though only 16% of the ingested sulphur, and 5% of the ingested potassium was exported. These are probably close to maximum figures for managed ecosystems. For less productive systems, the exports will be smaller and a greater proportion of the ingested nutrients will be recycled within the system; for example, about 20 kg ha$^{-1}$ of phosphorus, 20 kg ha$^{-1}$ of sulfur and 200 kg ha$^{-1}$ of potassium may be recycled (Middleton and Smith, 1978). The amounts of nutrients returned will vary widely with the productivity. Estimates can usually be made from a knowledge of the amounts of herbage consumed and its approximate composition. This chapter is therefore mainly concerned with those aspects of the return of nutrients that affect the future availability of the nutrients. These are: the route followed (faeces or urine); the heterogenous distribution of the nutrients; and the potential availability of nutrients.

## DISTRIBUTION OF NUTRIENTS BETWEEN FAECES AND URINE

Some plant nutrients are excreted largely in the faeces, and others largely in the urine. The individual nutrients must therefore be treated individually (Chapters 17 and 18).

It has long been known that ruminants, such as sheep, usually have only trace amounts of phosphate in their urine (Forbes and Keith, 1914). The amount of phosphate increases slightly when the phosphorus intake increases (Braithwaite, 1976), and may become substantial if the diet is deficient in calcium (Braithwaite, 1976). Urinary excretion of phosphorus also occurs briefly if a large injection of phosphate is given, causing plasma phosphate to reach unusually high levels (Clark et al., 1973). Under most conditions, however, excretion of phosphate is largely in the faeces. Faecal phosphate consists of both inorganic and organic phosphate. Their proportions can be measured, because inorganic phosphate is easily dissolved in dilute acid (Bromfield, 1961; Barrow, 1975). A large proportion is present as inorganic phosphate when the feed contains a high concentration of phosphate (Bromfield, 1961; Barrow and Lambourne, 1962; Floate, 1970). A simple model was proposed to describe this relation (Fig. 19.1). The organic phosphorus content of the faeces was found to average about 0.06 g per 100 g of feed eaten. For feed with a concentration higher than this, the excess would be excreted as inorganic phosphate, mainly dicalcium phosphate (Barrow, 1975). Thus, for feed with 0.24% phosphorus, 75% of the phosphorus in the faeces would be in inorganic form. However, this simple model was found to break down at low concentrations of phosphate,

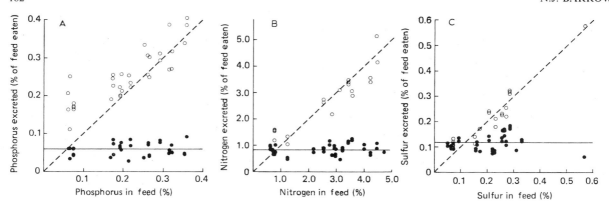

Fig. 19.1 Effect of the phosphorus (A), nitrogen (B), and sulfur (C) concentration in sheep fodder on the excretion of the respective nutrients. In each case the broken diagonal lines indicate balance between intake and excretion. Values above the line indicate net loss by the animal; those below the line, indicate net gain. For phosphorus, open symbols indicate total phosphorus excretion and closed symbols indicate organic phosphorus in the faeces. For nitrogen and sulfur, open symbols indicate total excretion (i.e. faeces plus urine) and solid symbols indicate excretion in the faeces only. (From Barrow and Lambourne, 1962.)

because the sheep were not able to eat enough of the low-quality feed to maintain phosphorus balance (Fig. 19.1).

Nitrogen and sulfur contrast with phosphorus in that the main pathway for excretion is in the urine, though faecal excretion also occurs; faecal excretion is approximately constant, when expressed as a proportion of the feed eaten. For nitrogen (Fig. 19.1B), faecal excretion by sheep is about 0.8 g per 100 g of feed eaten, regardless of the nitrogen content of the feed (Blaxter and Mitchell, 1948; Gallup and Briggs, 1948; Lancaster, 1949a, b; Barrow and Lambourne, 1962). Thus, for an animal in nitrogen balance eating feed with 4% nitrogen, 80% of the ingested nitrogen would be excreted in the urine; the proportion excreted in the urine would decrease as the nitrogen content of the feed decreased. For sulfur, faecal excretion was about 0.11 g per 100 g of feed eaten (Barrow and Lambourne, 1962) (Fig. 19.1C), and the proportion of the sulfur excreted in the urine varied from 90% for a feed of high sulfur content to 6% for a feed with low sulfur (Barrow and Lambourne, 1962).

The ratios of nitrogen and sulfur in the urine and faeces differ from those in the feed. Barrow and Lambourne (1962) found that the N:S ratio in sheep faeces averaged 7:1, but the N:S ratio in herbage was commonly about 15:1; the N:S ratio in urine was greater than that in herbage and is commonly 20:1 (Barrow and Lambourne, 1962).

The cations potassium, calcium and magnesium differ in their pathway of excretion. Potassium is mostly excreted in the urine, with only 10 to 30% in the faeces (Salter and Scholenberger, 1939; L'Estrange and Axford, 1966; Dewhurst et al., 1968). In contrast, faeces are the main excretory pathway for magnesium (L'Estrange and Axford, 1966; Todd, 1969) and for calcium (L'Estrange and Axford, 1966; Braithwaite, 1976). Most of the faecal magnesium is unabsorbed dietary magnesium (Rook and Storry, 1962). Urinary excretion of calcium in ruminants is increased if the animal is deficient in phosphate, or if the diet contains acidic substances (Braithwaite, 1976). The large excretion of calcium and magnesium in faeces means that cation excretion tends to exceed the excretion of nutrient anions. The balance appears to be made up by carbonate. Barrow (1975) showed that calcium carbonate was present in sheep faeces. This is important in determining the form of the other nutrients; for example, because of the alkaline conditions, phosphate is present as dicalcium phosphate (see earlier). These alkaline conditions may also reduce the solubility of nutrients, such as zinc, which form sparingly soluble precipitates of carbonate or hydroxide under alkaline conditions.

Data on the excretion of trace elements have been reviewed by Underwood (1977). According to Underwood, a high proportion of the ingested copper is excreted in the faeces. Most of this is unabsorbed copper, though active excretion of copper also occurs via the bile. Zinc also leaves the body largely in the faeces; this is mostly unabsorbed dietary zinc, though small amounts also enter the faeces via the bile, the pancreatic juice and by direct secretion into the intestine. Manganese is poorly absorbed; only about 1% of dietary manganese is absorbed by cattle, irrespective of dietary concentration. Absorbed manganese is almost totally excreted via the bile, though excretion via the pancreatic juice and directly through the intestinal wall can also occur. Iron, once absorbed, is retained with great tenacity. Iron leaving the body is therefore mostly unabsorbed dietary iron. Earlier work had suggested that cobalt was poorly absorbed, and that most passed out in the faeces. However, more recent work suggests that cobalt is effectively absorbed. The pathway of molybdenum excretion depends on the sulfur content of the feed; on low-sulfur diets, sheep tend to retain molybdenum, and most of the excretion that does occur is in the faeces. On high-sulfur diets, molybdenum is readily excreted in the urine. Selenium is excreted in the faeces, the urine and in the expired air. The amounts and proportions for each pathway depend on the level and form of the intake, the nature of the diet, and the species. Exhalation is only important at high levels of intake. For ruminants, faecal excretion of selenium is generally greater than urinary excretion, and most of the selenium in the faeces consists of unabsorbed selenium from the diet.

## SPATIAL DISTRIBUTION OF THE RETURN

Faeces and urine are not distributed uniformly over a pasture but deposited on discrete areas; this creates small-scale heterogeneity over the grazed area. Large-scale heterogeneity is caused by the tendency for animals to excrete in certain areas (Hilder, 1964).

The small-scale heterogeneity gives rise to fertility mosaics in grazed pastures. The characteristics of these mosaics depend on the area covered by individual excretions, and on the rate of depletion of the increased concentrations (Petersen et al., 1956a, b). However, there is usually a skewed distribution of fertility, with a large number of points with low values and a few with high values (Graley et al., 1960). For cattle, the area covered by one dung pat ranges from 0.05 $m^2$ to 0.13 $m^2$, and the frequency of defaecation varies from 11 to 16 per day (MacLusky, 1960), though the frequency may be much lower under range conditions (Gonzalez, 1964). Sheep faeces tend to be scattered over a greater area, because of the smaller size of faecal pellets, so that, at equal grazing pressure, the heterogeneity is less marked than for cattle.

The volume of urine voided varies with the season and with the kind of pasture grazed. For example, Vercoe (1962) found that, for grazing sheep in Victoria, Australia, the daily urine output ranges from 0.5 l in December to 2.8 l in June. Even larger values were observed by Davies (1965) for sheep fed fresh capeweed (*Arctotheca calendula*) with a high water content, when the urine volume was 13 l per day.

Although the distribution of the affected area will vary with the conditions, the proportion of the total area affected within a short period is always small. For example, MacLusky (1960) showed that the area affected by faeces and urine was, in each case, about 4 $m^2$ per cow per day. Petersen et al. (1956a) concluded that, after ten cow-years, about 6 to 7% of the pasture would have received no faecal return, and about 15% would have received four excretions.

In addition to the small-scale heterogeneity, caused by random return of the excreta, there may also be a large-scale heterogeneity caused by the tendency of animals to rest in certain parts of an area. This can cause a large accumulation of plant nutrients in some parts and a depletion in the others (Hilder and Mottershead, 1963; Hilder, 1964). A similar effect may arise from imposed behaviour, such as collecting dairy cows twice daily in one area for milking. Middleton and Smith (1978) have estimated that 25% of total excretion of nutrients may be ineffective under highly productive New Zealand dairy farming, because of their concentration in farm yards, races and in sheltering and watering areas.

## AVAILABILITY OF PLANT NUTRIENTS IN FAECES

The inorganic phosphate in sheep faeces is readily available to plants if the faeces are ground and mixed with the soil, but not if they remain on the surface (Gunary, 1968). This is consistent with its occurrence as dicalcium phosphate. Fine particles of dicalcium phosphate dissolve readily in soil

(Probert and Larsen, 1970a) and are a good source of available phosphate (Probert and Larsen, 1970b). Hence, when the faeces are incorporated in the soil, the inorganic phosphate is as effective as that in superphosphate (Gunary, 1968). However, if the faeces remain on the surface, the phosphate moves into the soil very slowly, because of its low solubility caused by the high calcium concentration and high pH in faeces. Soil animals that bury dung thus accelerate the return of phosphate to an available form. Where earthworms are scarce, dung beetles may play an important role in this process (Bornemissza, 1960).

The organic phosphorus component in sheep faeces decomposes only slowly. Bromfield (1961) found little increase in inorganic phosphate after eight weeks' incubation at 28°C. However, Floate (1970) measured mineralization that ranged from 3 to 34% after incubation at 30°C for twelve weeks. Mineralization increased with time, and was most marked in material with the highest organic phosphate concentrations. Such results suggest that the organic phosphorus residues from returned faeces would persist in soil for some time, accumulating to a level at which the amount decomposing each year balanced the amount added.

The nitrogen and sulphur contents of faeces are approximately constant per weight of feed eaten (see above), and hence the concentration in the faeces increases as digestibility increases. In turn, the mineralization increases as the concentration increases (Barrow, 1961). Thus, the most rapid mineralization occurs in faeces derived from highly digestible feed. The relationship between concentration and mineralization is similar to that which occurs in plant material, but in faeces the concentration needs to be higher before mineralization occurs (Barrow, 1961). The ratio of nitrogen mineralized to sulfur mineralized was found to be 5.5:1 (Barrow, 1961); this relatively low proportion of nitrogen reflects the small proportion of nitrogen in faeces. Further, much of the nitrogen may be lost as ammonia, because of the alkalinity of the faeces. Large losses of ammonia were noted by Floate (1970), while Gillard (1967) showed that large losses of nitrogen occur from faecal pads that dry out on the soil surface. Burial of the faeces by soil animals may increase the return of nutrients to the available form.

There does not appear to be any information on the availability of the other nutrients in faeces. One may speculate that those nutrients that form sparingly soluble compounds under alkaline conditions, such as zinc, may behave like phosphate and be fully available if the faeces are buried, but not if they remain on the surface.

## AVAILABILITY OF PLANT NUTRIENTS IN URINE

Many of the plant nutrients in urine are in an immediately available form; for example, potassium is present as $K^+$. The remainder are readily converted to an available form; thus, urea is rapidly changed to ammonium. As a result, sheep urine patches are relatively rich in nitrogen and potassium and relatively poor in calcium, magnesium, phosphorus and manganese (Joblin and Keogh, 1979). It does not necessarily follow that the nutrients in urine are fully utilized. Each urine patch contains nutrients collected from a large area but deposited on a small area; it therefore contains a considerable concentration of nutrients, and so are more prone to leaching. In the case of nutrients that react with the soil, there is usually a curved relation between the concentration of the nutrient in the solution and the amount retained by the soil, so that the proportion in the solution increases as the amount present increases. It is the soluble fraction which is prone to loss by leaching, and the high concentration of nutrients in small patches increases the likelihood of leaching loss.

The concentration of nutrients in small patches indirectly increases the probability of other losses. For example, the high concentration of urea causes a rapid release of ammonium in a urine patch which produces a high pH (Doak, 1952). This leads initially to a loss of nitrogen as ammonia (see Chapter 21). Subsequent oxidation of the ammonium would be expected to produce a low pH. This increase and subsequent decrease in pH in a urine patch would change the availability of nutrients such as manganese and molybdenum. Thus, Joblin and Keogh (1979) found that manganese concentration decreased in grass growing in sheep urine patches. The initial high pH would also stimulate breakdown of soil organic matter and so increase the availability of nitrogen and sulfur (Barrow, 1960).

There is also a sequence of change in pastures,

resulting from changes in the acceptability of herbage on the urine patch, and from the faster growth of non-legumes, caused by the high nitrogen levels. For example, Keogh (1973, 1975) reported that sheep grazed urine patches preferentially. If grazing is infrequent, however, the faster growth of non-legumes may compete out the legumes. In sulfur-deficient areas, the growth of non-legumes would be even further stimulated, in relation to legumes, because the high N:S ratio in urine tends to be wide; this favours grass rather than clover (Walker and Adams, 1958). However, where potassium is limiting, the initial grass dominance of a urine patch may be followed by increased legume content, because of the high potassium content of urine.

## REFERENCES

Barrow, N.J., 1960. Stimulated decomposition of soil organic matter during the decomposition of added organic materials. *Aust. J. Agric. Res.*, 11: 331–338.

Barrow, N.J., 1961. Mineralization of nitrogen and sulfur from sheep faeces. *Aust. J. Agric. Res.*, 12: 644–650.

Barrow, N.J., 1975. Chemical form of inorganic phosphate in sheep faeces. *Aust. J. Soil Res.*, 13: 63–67.

Barrow, N.J. and Lambourne, L.J., 1962. Partition of excreted nitrogen sulfur and phosphorus between the faeces and urine of sheep being fed pasture. *Aust. J. Agric. Res.*, 13: 461–471.

Blaxter, K.L. and Mitchell, H.H., 1948. The factorization of the protein requirements of ruminants and of the protein values of feeds, with particular reference to the significance of metabolic fecal nitrogen. *J. Anim. Sci.*, 7: 351–372.

Bornemissza, G.F., 1960. Could dung eating insects improve our pastures? *J. Aust. Inst. Agric. Sci.*, 26: 54–56.

Braithwaite, G.D., 1976. Calcium and phosphorus metabolism in ruminants with special reference to parturient paresis. *J. Dairy Res.*, 43: 501–520.

Bromfield, S.M., 1961. Sheep faeces in relation to the phosphorus cycle under pasture. *Aust. J. Agric. Res.*, 12: 111–123.

Clark, R.C., Budtz-Olsen, O.E., Cross, R.B., Finnamore, P. and Bauert, P.A., 1973. The importance of the salivary glands in the maintenance of phosphorus homeostasis in the sheep. *Aust. J. Agric. Res.*, 24: 913–919.

Davies, H.L., 1965. *Studies in Nutrition and Reproduction in Sheep in South Western Australia.* Thesis, University of Western Australia, Perth, W.A.

Dewhurst, J.K., Harrison, F.A. and Keynes, R.D., 1968. Renal excretion of potassium in the sheep. *J. Physiol.*, 195: 609–621.

Doak, B.W., 1952. Some chemical changes in the nitrogenous constituents of urine when voided on pasture. *J. Agric. Sci.*, 42: 162–171.

Floate, M.J.S., 1970. Decomposition of organic materials from hill soils and pastures. II. Comparative studies in the mineralization of carbon, nitrogen and phosphorus from plant materials and sheep faeces. *Soil Biol. Biochem.*, 2: 173–185.

Forbes, E.B. and Keith, M.H., 1914. Phosphorus compounds in animal metabolism. *Bull. Ohio Agric. Exper. Stn., Tech. Ser.*, No. 5: 212 pp.

Gallup, W.D. and Briggs, H.M., 1948. The apparent digestibility of prairie hay of variable protein content, with some observations of faecal nitrogen excretion by steers in relation to their dry matter intake. *J. Anim. Sci.*, 7: 110–116.

Gillard, P., 1967. Coprophagous beetles in pasture ecosystems. *J. Aust. Inst. Agric. Sci.*, 33: 30–34.

Gonzalez, M.H., 1964. *Patterns of Livestock Behaviour and Forage Utilization as Influenced by Environmental Factors on a Summer Mountain Range.* Thesis, Utah State University, Logan, Utah.

Graley, A., Nicholas, K.D. and Piper, C.S., 1960. Availability of potassium in some Tasmanian soils. 1. The variability of soil potassium in the field and its fractionation. *Aust. J. Agric. Res.*, 11: 750–753.

Gunary, D., 1968. The availability of phosphate in sheep dung. *J. Agric. Sci.*, 67: 295–304.

Hilder, E.J., 1964. The distribution of plant nutrients by sheep at pasture. *Proc. Aust. Soc. Anim. Prod.*, 5: 241–248.

Hilder, E.J. and Mottershead, B.E., 1963. The redistribution of plant nutrients through free-grazing sheep. *Aust. J. Sci.*, 26: 88–89.

Joblin, K.H. and Keogh, R.G., 1979. The element composition of herbage at urine patch sites in a ryegrass pasture. *J. Agric. Sci.*, 92: 571–574.

Keogh, R.G., 1973. *Pithomyces chartarum* spore distribution and sheep grazing patterns in relation to urine patch and inter-excreta sites within ryegrass dominated pastures. *N.Z. J. Agric. Res.*, 16: 353–355.

Keogh, R.G., 1975. Grazing behaviour of sheep during summer and autumn in relation to faecal eczema. *Proc. N.Z. Soc. Anim. Prod.*, 35: 198–203.

Lancaster, R.J., 1949a. The measurement of feed intake by grazing cattle and sheep. *N.Z. J. Sci. Technol.*, (A), 31: 31–38.

Lancaster, R.J., 1949b. Estimation of digestibility of grazed pastures from faeces nitrogen. *Nature*, 163: 330–331.

L'Estrange, J.L. and Axford, R.F.E., 1966. Mineral balance studies on lactating ewes with particular reference to the metabolism of magnesium. *J. Agric. Sci.*, 67: 295–304.

MacLusky, D.S., 1960. Some estimates of the areas of pasture fouled by the excreta of dairy cows. *J. Br. Grassland Soc.*, 15: 181–188.

Middleton, K.R. and Smith, G.S., 1978. The concept of a climax in relation to the fertilizer input of a pastoral ecosystem. *Plant Soil*, 50: 595–614.

Petersen, R.G., Lucas, H.L. and Woodhouse, W.W., 1956a. The distribution of excreta by freely grazing cattle and its effect on pasture fertility. 1. Excretal distribution. *Agron. J.*, 48: 440–444.

Petersen, R.G., Woodhouse, W.W. and Lucas, H.L., 1956b. The distribution of excreta by freely grazing cattle and its effect on pasture fertility. 2. Effect of returned excreta on the residual concentration of some fertilizer elements. *Agron. J.*, 48: 444–449.

Probert, M.E. and Larsen, S., 1970a. The stability of dicalcium phosphate dihydrate in soil. 1. Laboratory studies. *J. Soil Sci.*, 21: 353–358.
Probert, M.E. and Larsen, S., 1970b. The stability of dicalcium phosphate dihydrate in soil. 2. Pot experiments. *J. Soil Sci.*, 21: 359–364.
Rook, J.A.F. and Storry, J.E., 1962. Magnesium in the nutrition of farm animals. *Nutrit. Abst. Rev.*, 32: 1055–1077.
Salter, R.M. and Scholenberger, C.J., 1939. Farm manure. *Ohio Agric. Exper. Stn. Bull.*, 405.
Todd, J.R., 1969. Magnesium metabolism in ruminants. Review of current knowledge. In: *Trace Minerals Study on Isotopes and Domestic Animals, Proc. Panel 1968*, pp. 131–140.
Underwood, E.J. 1977. *Trace Elements in Human and Animal Nutrition*. Academic Press, New York, N.Y., 4th ed., 545 pp.
Vercoe, J.M., 1962. Some observations on the nitrogen and energy losses in the faeces and urine of grazing sheep. *Proc. Aust. Soc. Anim. Prod.*, 4: 160–162.
Walker, T.W. and Adams, A.F.R., 1958. Competition for sulfur in a grass–clover association. *Plant Soil*, 9: 353–366.

Chapter 20

# NITROGEN FIXATION IN MANAGED GRASSLANDS

J.H. HOGLUND and J.L. BROCK

## INTRODUCTION

Nitrogen fixation in grasslands is predominantly due to the symbiotic association between legumes and an appropriate type of *Rhizobium*. In this association, the host legume supplies energy and nutrients for the *Rhizobium* and, in turn, the host benefits from the nitrogen fixed by the *Rhizobium*. Other plants in the community benefit from the association when the legume tissues die, or are eaten by animals, and nitrogen is returned to the soil (Chapters 17 and 19).

The nitrogen fixation rates that have been reported for grasslands range from less than 5 kg N ha$^{-1}$ yr$^{-1}$, in many natural grasslands, to greater than 700 kg N ha$^{-1}$ yr$^{-1}$, in some vigorous legume-dominant pastures. In this chapter we will give examples of nitrogen fixation rates from a range of contrasting ecosystems, and discuss more fully the major factors affecting nitrogen fixation.

In stable ecosystems, nitrogen inputs are balanced by nitrogen losses from the system:

$$N_f + N_i = N_l \qquad (1)$$

or

$$N_f = N_l - N_i \qquad (2)$$

where $N_f$ = nitrogen fixation, $N_i$ = other nitrogen inputs (e.g. fertilizer) and $N_l$ = nitrogen loss. If the system is in balance, nitrogen fixation will be equal to the difference between nitrogen losses and other nitrogen inputs, such as fertilizer nitrogen. In the short term, however, management practices and climate may create a temporary imbalance between nitrogen inputs and losses, mainly through accumulation or depletion of soil organic nitrogen. Such imbalances do not necessarily immediately affect pasture production. Net mineralization of soil organic nitrogen generally has a far greater influence on total plant-available nitrogen than do annual inputs or losses, which may be masked by changes in the turnover rate of nitrogen in the ecosystem. Hence, in the short term, nitrogen fixation may not precisely reflect current net nitrogen losses from the ecosystem.

Nitrogen fixation by legumes is dependent on their growth rate and their ability to obtain fixed nitrogen from the soil. Both these factors are affected by competition from other plant species. Nitrogen losses mainly depend on stocking rate, since passage through grazing animals increases nitrogen mobility (Chapter 19) and hence the risk of loss from the ecosystem (Chapter 22).

## NITROGEN FIXATION IN GRASSLAND ECOSYSTEMS

There is less nitrogen mobility in most "undisturbed" or semi-natural grasslands than in managed grasslands and nitrogen losses are correspondingly less. Nitrogen inputs are usually less than 10 kg N ha$^{-1}$ yr$^{-1}$, and less than half of this is derived from nitrogen fixation (Copeley and Reuss, 1972; Reuss and Cole, 1973; Woodmansee, 1978). Production is almost entirely dependent on the turnover of soil organic nitrogen, and fixation represents only about 2% of the nitrogen flux through the vegetation (Paul, 1976). In these ecosystems, asymbiotic free-living organisms are the predominant nitrogen fixers, and symbiotic nitrogen-fixing associations between legumes and *Rhizobium* species make only a minor contribution.

Agriculturally marginal grasslands usually contain more legumes, but total nitrogen inputs are usually less than 40 kg ha$^{-1}$ yr$^{-1}$. Nitrogen fixation by legumes commonly contributes about half of this total (Haystead and Lowe, 1977; Grant and Lambert, 1979).

Highly productive permanent pastures might also be expected to be in long-term nitrogen balance. Losses of nitrogen from these pastures can be as high as 200 kg N ha$^{-1}$ yr$^{-1}$ (Chapter 21), but nitrogen fixation rates normally approximate these losses. For example, nitrogen fixation in Irish pastures grazed by cattle was within the range 83 to 296 kg N ha$^{-1}$ yr$^{-1}$ (Masterson and Murphy, 1976); the corresponding range in predominantly sheep-grazed pastures in New Zealand was 45 to 390 kg N ha$^{-1}$ yr$^{-1}$ (Edmeades and Goh, 1978; Hoglund et al., 1979). Nitrogen fixation by legumes normally accounted for more than 90% of the nitrogen inputs in these studies, although in tropical pastures significant nitrogen fixation has been reported from the rhizosphere or phyllosphere of tropical grasses (Dobereiner, 1978). Asymbiotic nitrogen fixation is of only minor significance in temperate pastures (Whitehead, 1970; Bergersen, 1973) and, together with nitrogen in rainfall, dust and pollen, usually contributes no more than 15 kg N ha$^{-1}$ yr$^{-1}$ (Sears et al., 1965; Henzell and Ross, 1973). Nitrogen fixation by legumes represents about 40% of measured herbage nitrogen yield in these managed pastures (Hoglund et al., 1979), or about 28% of the total nitrogen flux; this is substantially more than the 2% occurring in undisturbed grassland (see above). The difference mainly reflects the greater losses of nitrogen from intensively grazed pastures.

When conditions allow rapid legume growth, and a rapid build-up of soil organic nitrogen, for example on virgin soil or after cropping, nitrogen fixation by legumes may reach 700 kg N ha$^{-1}$ yr$^{-1}$ in both temperate regions (Sears et al., 1965) and the tropics (Whitney, 1977). Such high rates of nitrogen fixation only occur if available soil nitrogen levels are low, if deficiencies of other nutrients, including phosphorus, sulphur and molybdenum, are eliminated, and if productive legumes are present. Highly effective indigenous strains of *Rhizobium* are often widespread, despite the apparent absence of suitable legume hosts, and rapidly increase following legume introduction (Gibson et al., 1976; Newbould et al., 1982). The main advantage of inoculation with *Rhizobium* spp. is to improve the initial establishment of legume seedlings; but, in some instances, may be necessary for long-term legume survival. Although there may be initially rapid rates of nitrogen fixation, rates subsequently decline as soil nitrogen levels increase, nitrogen uptake replaces nitrogen fixation, and increasing competition suppresses legume growth (Sears, 1960). Where pastures are sown on soils of high organic matter content, accelerated mineralization of nitrogen from the organic matter stimulates the growth of non-legumes, and competition limits initial legume growth and nitrogen fixation.

## NITROGEN FIXATION, LEGUME GROWTH AND SOIL MINERAL NITROGEN

Legumes can use two alternative sources of nitrogen, atmospheric dinitrogen and soil mineral nitrogen. If available, mineral nitrogen is used but, except in soils with a large content of mineral nitrogen, fixation and assimilation of mineral nitrogen usually proceed concurrently. Since nitrogen supply must match nitrogen requirement, in steady state, then:

$$N_a = N_f + N_m \tag{3}$$

or

$$N_f = N_a - N_m \tag{4}$$

where $N_a$ = assimilated nitrogen (i.e. increase in nitrogen content of legume tissue), $N_f$ = nitrogen fixation, and $N_m$ = uptake of mineral nitrogen.

A schematic model of the relationship between legume growth rate, nitrogen fixation, and availability of mineral nitrogen is shown in Fig. 20.1A.

If no mineral nitrogen is available, there is a direct relationship between nitrogen fixation and legume growth (Fig. 20.1A), but increasing the supply of mineral nitrogen reduces nitrogen fixation, causing the relationship between nitrogen fixation and legume growth rate to become non-linear (Fig. 20.1A). The ratio of nitrogen fixation to uptake of mineral nitrogen is not affected until the supply of mineral nitrogen is nearly exhausted by uptake; this indicates a competitive relationship between nitrogen fixation and nitrogen uptake, rather than simple preferential uptake of mineral nitrogen. When the growth rate of the pasture exceeds the capacity of the soil to supply mineral

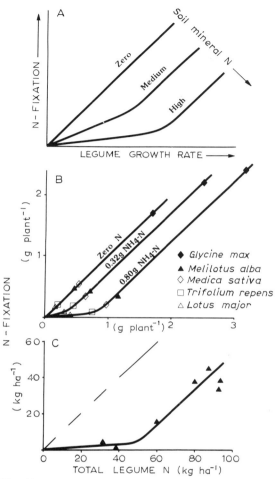

Fig. 20.1. A. The generalized relationship between nitrogen fixation, legume growth rate and the mineral nitrogen content of the soil (Hoglund, 1973). B. Data for five legume species (Allos and Bartholomew, 1959) fitted to the model in A. C. Data for seven genotypes of *Medicago sativa* (Heichel et al., 1978) fitted to the model in A.

nitrogen, a relatively greater increase in the rate of fixation occurs with each increment of legume growth. The data of Allos and Bartholomew (1959), for various legume species (Fig. 20.1B), and those of Heichel et al. (1978) for several lucerne clones (Fig. 20.1C), fit the model well. In particular, genetic variation in growth rate between species (Fig. 20.1B) and between genotypes (Fig. 20.1A) interacts with nitrogen supply in the way predicted.

Considerations of energetics and of nitrogen supply suggest that applications of mineral nitrogen to legumes should increase growth rate (Gutschick, 1981). Certainly the growth rate of various pasture legumes can respond markedly to supplementation with mineral nitrogen in controlled environment (Hoglund, 1973; Gibson, 1976), and in the field (Harris et al., 1973; Hoglund et al., 1974). The growth response of legumes to mineral nitrogen occurs predominantly when climatic conditions are suitable for rapid legume growth, or when nodule development is restricted (Hoglund, 1973; Hoglund et al., 1974). When legume growth is totally dependent on nitrogen fixation, about 15% of assimilated nitrogen is allocated to new nodule development; this restricts the rate of nodule development and legume host growth. Consequently, in the field, legume growth and nitrogen fixation are very dependent on seasonal conditions, which affect not only the potential growth rate of the legume but also availability of soil mineral nitrogen, and competition between grasses and legumes.

The suppression of nitrogen fixation by mineral nitrogen occurs through a number of mechanisms (Eady, 1981; Darrow et al., 1981; Dazzo et al., 1981), involving either a reduction in the supply of metabolites and mineral nutrients to nodules, or the direct effects of nitrogenous compounds on nitrogen fixation enzyme systems. Mineral nitrogen uptake can result in reduced nitrogenase activity, reduced nodule formation, and even loss of existing nodules, though the effects are less marked when legumes are growing rapidly and the demand for nitrogen is great. In the short term, nitrogen fixation may be buffered against fluctuations in supply of mineral nitrogen by the use of high-energy compounds stored in the nodules, and by accumulation of nitrate in vacuoles, so that the nitrate content of legume herbage may exceed 1000 p.p.m. (see Fig. 20.5). Although the response of individual mechanisms to addition of mineral nitrogen may vary, and these responses may vary between different symbiotic associations, the overall net effect of mineral nitrogen on nitrogen fixation is similar for a wide range of species, as shown in Fig. 20.1B. However, if the supply of mineral nitrogen varies with time of year, legumes with a greater proportion of their growth coinciding with periods of low availability of mineral nitrogen will have a greater proportion of their nitrogen supply provided by fixation.

The suppression of nitrogen fixation by mineral

nitrogen seems to occur in two phases (Chen and Phillips, 1977; Carroll et al., 1981). When legumes are exposed to nitrate for seven to nine days, there is a rapid recovery of nitrogenase activity when they are returned to nitrogen-free culture. However, if exposure to nitrate continues beyond this period, recovery is greatly delayed. Control mechanisms involving carbohydrate supply have been suggested, but these are likely to be short-term; longer-term control may be by feedback mechanisms involving nitrogenous compounds. Nitrogen fixation increases following carbon dioxide supplementation, but this is associated with a reduction in nitrate reductase activity (Masterson and Sherwood, 1978), possibly because carbon dioxide enrichment reduces stomatal opening, so reducing transpiration and nitrate uptake. No consistent correlations have been found between apparently energy-wasteful hydrogen evolution by nodules and nitrogen fixation, suggesting that carbohydrate supply may not normally limit nitrogen fixation.

The application of nitrogenous fertilizers to grass–legume pastures reduces nitrogen fixation directly, by the mechanisms already considered. However, in the longer term, nitrogen fixation is reduced more by the effects of grass competition on legume content of the sward (Ledgard et al., 1982). The extent of clover suppression, and thus indirectly nitrogen fixation, can be influenced by competition for nutrients other than nitrogen, such as potassium (Richards, 1976), and by defoliation management.

When applications of nitrogen fertilizer are less than the nitrogen fixing capacity of the sward, reductions in nitrogen fixation are usually slightly less than the amount of nitrogen fertilizer applied (Fig. 20.2). As a result, low rates of nitrogen fertilizer application usually increase the annual herbage yield and the total annual nitrogen harvested from these swards only slightly, though it may shift the seasonal distribution of yield.

## NITROGEN FIXATION AND OTHER EDAPHIC FACTORS

Seasonal variations in soil temperature and moisture considerably affect nitrogen fixation (Masterson and Murphy, 1976; Hoglund et al.,

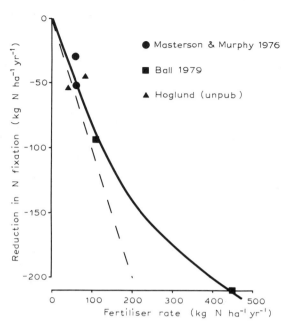

Fig. 20.2. The effect of nitrogen fertilizer on nitrogen fixation in long-term studies of grazed pastures of *Lolium perenne* and *Trifolium repens* (dashed line indicates 1:1 replacement response).

1979). Nitrogen fixation increases with increasing soil temperature in cool, moist environments, but is reduced by increasing soil temperatures when conditions are hot and dry. The infection of roots by *Rhizobium*, and initial nodule development, occur at higher minimum temperatures than those required for nodule growth and nitrogen fixation (Gibson, 1971, 1977). In cooler climates, nitrogen fixation in winter and early spring may depend on the conditions for nodule initiation during the previous autumn. Symbiotic associations in tropical legumes require higher temperatures than those for temperate species, generally in line with the higher temperature requirements of the hosts. Developed nodules can recover from exposure to low temperatures (Bergersen et al., 1963; Ranga Rao, 1977), but high temperatures are usually more damaging (Roughley, 1970; Pankhurst and Gibson, 1973).

Nitrogenase activity per unit weight of nodule is greatest in the range 20 to 30°C, although the optimum depends on previous growing temperatures (Fig. 20.3). Low temperatures reduce activity per unit nitrogenase and thereby increase the time required to develop comparable symbiotic ca-

# NITROGEN FIXATION

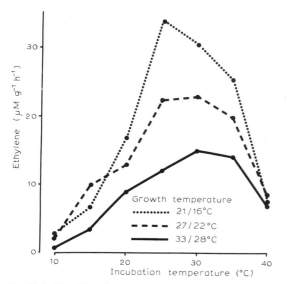

Fig. 20.3. The effect of incubation temperature on the nitrogenase activity of nodules of soybean (*Glycine max*), taken from plants grown at various combinations of day and night temperatures (Gibson, 1976). Nitrogenase activity is measured as ethylene production per gram (fresh weight) of nodule tissue.

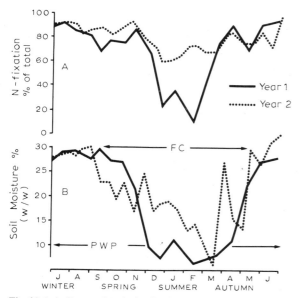

Fig. 20.4. A. Seasonal variation in nitrogen fixation occurring in the surface layer of soil (0–75 mm) beneath a grazed grass–clover pasture as a proportion of total nitrogen fixation. B. Soil moisture in the surface layer of soil (0–75 mm). (*FC* = field capacity, *PWP* = permanent wilting point).

pacity, because of the need for greater nodule development (Hoglund, 1973). In the short term the temperature sensitivity of nitrogenase leads to marked diurnal variation in nitrogen fixation (Halliday and Pate, 1976; Masterson and Murphy, 1976; Murphy, 1983).

Moisture stress also damages nodule structure (Sprent, 1976) but, in general, the effects of temperature and moisture stress on nitrogen fixation in the field are of lower magnitude than those occurring in controlled conditions. This is because nodules can form in deeper soil zones in the field, which are less subject to temperature and moisture stress (Munns et al., 1977; Hoglund and Brock, 1978). When soil moisture is at or near field capacity, over 80% of total nitrogen fixation occurs in the top 75 mm of soil (Hoglund et al., 1979), but under drought conditions a larger proportion occurs at greater depths (Fig. 20.4). Short-term peaks in nitrogen fixation, occurring one to three weeks after rainfall in dry conditions, are usually related to renewal of legume growth (Brock et al., 1983).

## NITROGEN FIXATION AND COMPETITION

In pastures, legumes generally grow with grasses which, in the presence of adequate nitrogen, compete strongly for nutrients, water and light. When the nitrogen supply is inadequate for grass growth, legumes are able to compete more successfully, because of their ability to fix nitrogen. Most reports (e.g. Walker et al., 1956; Whitehead, 1970) show that grasses take up as much nitrogen when growing with legumes as when growing in monocultures. However, other reports show that nitrogen uptake by grass is less in mixtures with legumes than in monoculture (Willoughby, 1954; Davies, 1964; Simpson, 1965), indicating that legumes may be competitive for available mineral nitrogen under some conditions. The presence of nitrate in herbage of *Trifolium repens* collected from soils rich in mineral nitrogen (Fig. 20.5) indicates that this species may be able to compete for mineral nitrogen, especially in dry conditions, when *Lolium perenne* is less vigorous. In this study (Fig. 20.5) only about 50% of the nitrogen harvested from legumes in two years was derived from nitrogen fixation; other studies, on similar pastures, gave estimates ranging from 50 to 100% (Hoglund et al., 1979). The proportion was inversely related to the proportion of legume in the sward and to

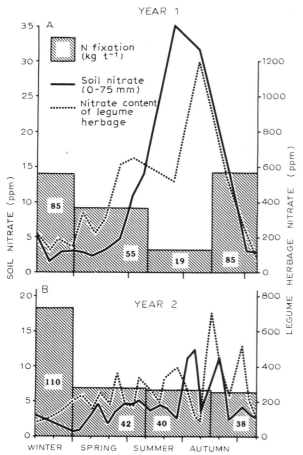

Fig. 20.5. Seasonal variation in the nitrate content of the surface soil and of legume herbage, and nitrogen fixation per unit of growth in legume dry matter in a grazed pasture of *Lolium perenne* and *Trifolium repens* during two years.

soil fertility (Hoglund et al., 1979). West and Wedin (1981) also found that the proportion of nitrogen derived from fixation in *Medicago sativa* varied inversely with the percentage of the legume in the herbage. In mixed swards, negative correlations between the proportion of nitrogen derived from fixation and total legume yield may therefore indicate successful competition by legumes for mineral nitrogen. The competitive ability of legumes is increased when the seasonal growth patterns of the legume and associated grass are complementary rather than synchronous (Chestnutt and Lowe, 1970; Harris and Hoglund, 1980); increased uptake of mineral nitrogen by legumes is then possible. The dominance of *Trifolium subter-raneum* over *Lolium rigidum* in some years (Willoughby, 1954) was attributed to successful competition for mineral nitrogen, as a consequence of earlier germination of the legume. When pastures in summer-dry temperate regions are irrigated, only relatively small increases in nitrogen fixation occur, despite substantial increases in legume growth (Crush, 1979). Under these conditions, grass growth is slight and legumes utilize mineral nitrogen. Similarly, when low temperature restricts the growth of tropical grasses during early spring in subtropical environments, most of the available mineral nitrogen may be taken up by *T. repens* and *Lotononis bainesii* (Vallis et al., 1977). In this instance, however, the proportion of the legume nitrogen derived from fixation was largely independent of legume yield, perhaps indicating that grasses were not active competitors during the period when legumes were obtaining mineral nitrogen.

The net effect on nitrogen fixation of growing a grass in association with a legume depends, on the one hand, on the reduction in mineral nitrogen available to the legume, which will increase nitrogen fixation per unit growth and, on the other hand, on the reduction in legume yield due to grass competition. Multiple regression analysis (Hoglund and Brock, 1978) showed that nitrogen fixation was positively correlated with the growth of both legumes and grasses, and negatively correlated with the nitrate content of legume herbage. Multiple regression analysis of a range of legume–grass mixtures, receiving either 20 kg N ha$^{-1}$ after each grazing or no nitrogen (Harris and Hoglund, 1980) gave the following relationship:

no nitrogen fertilizer: $N_f = 0.77$ (LG) + 0.0016 (GG)

fertilized with nitrogen: $N_f = 0.048$ (LG) + 0.0032 (GG)

where $N_f$ = nitrogen fixation (kg N ha$^{-1}$ day$^{-1}$), LG = legume growth (kg DM ha$^{-1}$ day$^{-1}$) and GG = grass growth (kg DM ha$^{-1}$ day$^{-1}$).

Legume growth had a proportionately greater effect on nitrogen fixation in the absence of nitrogen fertilizer, whereas grass growth had a greater effect when nitrogen fertilizer was supplied. Increased grass growth can increase nitrogen fixation by reducing soil nitrogen, which can completely offset the reduction in legume content of the sward

(Sears et al., 1965). Harris and Hoglund (1980) found that the situation could be aggravated by growing particularly aggressive grasses in association with legumes; the most aggressive grass they tested reduced legume growth and nitrogen fixation to less than half of the corresponding values with the least aggressive grass. Conversely, utilization of mineral nitrogen by legumes reduces the amount available to grasses, and thereby shifts the competitive balance away from grass. This could enhance nitrogen fixation in the longer term, because of the greater proportion of legume in the sward.

The presence of pasture pests and diseases can alter the competitive balance between grass and legume, thereby affecting nitrogen fixation. Controlling soil pests (especially nematodes) can greatly increase nitrogen fixation in pots (Yeates et al., 1977), and in the field (Steele and Shannon, 1982). The weevil *Sitona discoideus* specifically attacks nodules and root hairs of *Medicago sativa* and so reduces nitrogen fixation (Wightman, 1979), but older *M. sativa* plants appear to sustain lower weevil populations because of development of nodules on deeper roots, beyond the range of the *Sitona* larvae.

**EFFECTS OF DEFOLIATION**

Moustafa et al. (1969) found that nitrogen fixation in plots of *Trifolium repens* declined for three days after defoliation, but then rapidly recovered as growth resumed. Other studies have shown slower recovery, depending on the species, environmental conditions and the severity of defoliation. Nitrogen fixation by *Medicago sativa* grown in pots (Vance et al., 1979; Groat and Vance, 1981) declined by 88% within 24 h of defoliation and recovery took 13 to 15 days. Nodules degenerated from their proximal ends after defoliation, but meristem and vascular bundles remained intact. Starch and soluble protein in the nodules was depleted, but there was no massive loss of nodules and the senescent nodules began to regrow and fix nitrogen after shoot growth resumed. Fishbeck and Phillips (1982) demonstrated a 99% reduction in photosynthesis in *M. sativa* following defoliation but only a 75% reduction in nitrogen fixation. They postulated that non-structural carbohydrate reserves in roots and nodules acted as a buffer until photosynthesis was restored.

Defoliation causes a greater reduction in nitrogen fixation than shading (Chu and Robertson, 1974; Halliday and Pate, 1976; Hoglund, unpubl.). Removal of active meristems by defoliation causes a reduced demand for nitrogen, which is probably more rate-limiting than deficient carbohydrate supply. In the field, grazing can affect nitrogen fixation by reducing nitrogen uptake by the sward and also by urine deposition, which increases the mineral-nitrogen available to the legume (Hoglund and Brock, 1978). Nitrogen fixation took 15 to 20 days to recover after grazing; recovery was associated with a decline in both the mineral-nitrogen content of the soil and the nitrate content of the herbage.

In grass-dominant swards, on soils rich in available nitrogen, frequent intensive defoliation is usually advocated to increase the proportion of short and stoloniferous legumes, such as *Trifolium repens* and *T. subterraneum*, whereas infrequent defoliation favours erect species, such as *T. pratense* and *Medicago sativa*. Frequent defoliation during spring, followed by less frequent defoliation during summer, favours growth of *T. repens* (Brougham et al., 1978). Rotational grazing by sheep throughout the year led to more nitrogen fixation in spring, but less during drier conditions in summer and autumn, than set-stocking systems (Brock et al., 1983). Although seasonal variability in nitrogen fixation was closely correlated with legume growth rate under both managements, nitrogen fixation was further negatively correlated with soil mineral nitrogen when under periodic defoliation, but not under continuous grazing.

Sheep selectively graze *T. repens* in mixed swards, causing a reduction in the proportion of legume present and a reduction in nitrogen fixation, compared to pastures grazed with cattle (Lambert et al., 1982). Conversely, goats avoid *T. repens*, resulting in greater legume growth and nitrogen fixation (Clark et al., 1982; P. Rolston, pers. comm.).

**CONCLUSION**

In stable pasture ecosystems, nitrogen fixation is essentially in balance with nitrogen losses, since other nitrogen inputs, e.g. from precipitation, are slight. Hence, although nitrogen fixation rates as high as 700 kg N ha$^{-1}$ yr$^{-1}$ have been reported for systems in which nitrogen is being accumulated, rates of 150 to 200 kg N ha$^{-1}$ yr$^{-1}$ are more common for mature intensively managed pastures. Two main proximal factors control the amount of nitrogen fixation: (1) the amount of legumes present in the sward, and (2) the rate of nitrogen fixation by those legumes. Under conditions of a high content of mineral nitrogen in the soil, vigorous grass growth causes a reduction in legume yields. However, legumes can be active competitors for available mineral nitrogen and, although grasses are generally more competitive, situations occur when legumes obtain the majority of their nitrogen requirements from uptake of mineral nitrogen, thereby causing a reduction in nitrogen fixation.

The factors controlling the amount of legume present and the rate of uptake of mineral nitrogen are considered. These include climatic, soil and management factors which affect the competitive ability of legumes. Any limitation of legume growth by nitrogen fixation is normally only temporary; compensatory responses, involving further growth of existing nodules or the development of new nodules, often in more favourable rooting zones, soon increase nitrogen fixation.

## REFERENCES

Allos, H.F. and Bartholomew, W.V., 1959. Replacement of symbiotic fixation by available nitrogen. *Soil Sci.*, 87: 61–66.

Ball, P.R., 1979. *Nitrogen Relationships in Grazed and Cut Grass-Clover Systems.* Thesis, Massey University, Palmerston North, 217 pp.

Bergersen, F.J., 1973. Symbiotic fixation by legumes. In: G.W. Butler and R.W. Bailey (Editors), *Chemistry and Biochemistry of Herbage, 2.* Academic Press, New York, N.Y., pp. 189–226.

Bergersen, F.J., Hely, F.W. and Costin, A.B., 1963. Overwintering of clover nodules in alpine conditions. *Aust. J. Biol. Sci.*, 16: 920–921.

Brock, J.L., Hoglund, J.H. and Fletcher, R.H., 1983. The effects of grazing management on seasonal variation in nitrogen fixation. In: *Proc. XIV Int. Grassland Congr.*, pp. 339–341.

Brougham, R.W., Ball, P.R. and Williams, W.M., 1978. The ecology and management of white clover-based pastures. In: J.R. Wilson (Editor), *Plant Relations in Pastures.* C.S.I.R.O., East Melbourne, Vic., pp. 309–324.

Carroll, B., Hart, J. and Gresshoff, P.M., 1981. Nitrate inhibition of nodulation and nitrogen fixation in white clover. In: A.H. Gibson and W.E. Newton (Editors), *Current Perspectives in Nitrogen Fixation.* Australian Academy of Science, Canberra, A.C.T., p. 466.

Chen, P. and Phillips, D.A., 1977. Induction of root nodule senescence by combined nitrogen in *Pisum sativum* L. *Plant Physiol.*, 59: 440–442.

Chestnutt, D.M.B. and Lowe, J., 1970. Agronomy of white clover/grass swards — review. *Occas. Symp. Br. Grassland Soc.*, 6: 191–213.

Chu, A.C.P. and Robertson, A.G., 1974. The effects of shading and defoliation on nodulation and nitrogen fixation by white clover. *Plant Soil*, 41: 509–519.

Clark, D.A., Lambert, M.G., Rolston, M.P. and Dymock, N., 1982. Diet selection by goats and sheep on hill country. *Proc. N.Z. Soc. Anim. Prod.*, 42: 155–157.

Copeley, P.W. and Reuss, J.O., 1972. *Evaluation of Biological $N_2$ Fixation in a Grassland Ecosystem.* Technical Rep. No. 152. U.S. I.B.P. Grassland Biome, Fort Collins, Colo.

Crush, J.R., 1979. Nitrogen fixation in pasture IX. Canterbury plains, Kirwee. *N.Z. J. Exper. Agric.*, 7: 35–38.

Darrow, R.A., Crist, D., Evans, W.R., Jones, B.L., Keister, D.L. and Knotts, R.R., 1981. Biochemical and physiological studies on the two glutamine synthetases of rhizobia. In: A.H. Gibson and W.E. Newton (Editors), *Current Perspectives in Nitrogen Fixation.* Australian Academy of Science, Canberra, A.C.T., pp. 182–185.

Davies, W.E., 1964. The yields and competition of lucerne, grass and clover under different systems of management V. Method of growing lucerne and meadow fescue. *J. Br. Grassland Soc.*, 19: 263–270.

Dazzo, F.B., Hrabak, E.M., Urbano, M.R., Sherwood, J.E. and Truchet, G., 1981. Regulation of recognition in the *Rhizobium*-clover symbiosis. In: A.H. Gibson and W.E. Newton (Editors), *Current Perspectives in Nitrogen Fixation.* Australian Academy of Science, Canberra, A.C.T., pp. 292–295.

Dobereiner, J., 1978. Potential for nitrogen fixation in tropical legumes and grasses. In: J. Dobereiner et al. (Editors), *Limitations and Potentials for Biological Nitrogen Fixation in the Tropics.* Plenum, New York, N.Y., pp. 13–24.

Eady, R., 1981. Regulation of nitrogenase activity. In: A.H. Gibson and W.E. Newton (Editors), *Current Perspectives in Nitrogen Fixation.* Australian Academy of Science, Canberra, A.C.T., pp. 172–181.

Edmeades, D.C. and Goh, K.M., 1978. Symbiotic nitrogen-fixation in a sequence of pastures of increasing age measured by a 15N dilution technique. *N.Z. J. Agric. Res.*, 21: 623–628.

Fishbeck, K.A. and Phillips, D.A., 1982. Host plant and *Rhizobium* effects on acetylene reduction in alfalfa during regrowth. *Crop Sci.*, 22: 251–254.

Gibson, A.H., 1971. Factors in the physical and biological environment affecting nodulation and nitrogen fixation by legumes. *Plant Soil, Spec. Vol.*, 1971: 139–152.

Gibson, A.H., 1976. Recovery and compensation by nodulated legumes to environmental stress. In: P.S. Nutman (Editor), *Symbiotic Nitrogen Fixation in Plants.* Cambridge University Press, Cambridge, pp. 385–403.

Gibson, A.H., 1977. The influence of the environment and managerial practices on the legume–*Rhizobium* symbiosis. In: R.W.F. Hardy and A.H. Gibson (Editors), *A Treatise on Dinitrogen Fixation, Section IV: Agronomy and Ecology.* Wiley, New York, N.Y., pp. 393–450.

Gibson, A.H., Date, R.A., Ireland, J.A. and Brockwell, J., 1976. A comparison of competitiveness and persistence amongst five strains of *Rhizobium trifolii. Soil Biol. Biochem.*, 8: 395–401.

Grant, D.A. and Lambert, M.G., 1979. Nitrogen fixation in pasture V. Unimproved North Island hill country. *N.Z. J. Exper. Agric.*, 7: 19–22.

Groat, R.G. and Vance, C.P., 1981. Root nodule enzymes of ammonia assimilation in alfalfa (*Medicago sativa* L.). *Plant Physiol.*, 67: 1198–1203.

Gutschick, V.P., 1981. Evolved strategies in nitrogen acquisition by plants. *Am. Naturalist*, 118: 607–637.

Halliday, J. and Pate, J.S., 1976. The acetylene reduction assay as a means of studying nitrogen fixation in white clover under sward and laboratory conditions. *J. Br. Grassland Soc.*, 31: 29–35.

Harris, A.J., Brown, K.R., Turner, J.D., Johnston, J.M., Ryan, D.L. and Hickey, M.J., 1973. Some factors affecting pasture growth in Southland. *N.Z. J. Exper. Agric.*, 1: 139–163.

Harris, W. and Hoglund, J.H., 1980. Influences of seasonal growth periodicity and N-fixation on competitive combining abilities of grasses and legumes. In: *Proc. XIII Int. Grassland Congr.*, pp. 239–243.

Haystead, A. and Lowe, A.G., 1977. Nitrogen fixation by white clover in hill pasture. *J. Br. Grassland Soc.*, 32: 57–63.

Heichel, G.H., Barnes, D.K. and Vance, C.P., 1978. Dinitrogen fixation of alfalfa (*Medicago sativa* L.) clones and populations: Field measurements with N-15. In: *Proc. of the Steenbock–Kettering Int. Symp. Nitrogen Fixation.*

Henzell, E.F. and Ross, P.J., 1973. The nitrogen cycle of pasture ecosystems. In: G.W. Butler and R.W. Bailey (Editors), *Chemistry and Biochemistry of Herbage, 2.* Academic Press, New York, N.Y., pp. 227–246.

Hoglund, J.H., 1973. Bimodal response by nodulated legumes to combined nitrogen. *Plant Soil*, 39: 533–545.

Hoglund, J.H. and Brock, J.L., 1978. Regulation of nitrogen fixation in a grazed pasture. *N.Z. J. Agric. Res.*, 21: 73–82.

Hoglund, J.H., Dougherty, C.T. and Langer, R.H.M., 1974. Response of irrigated lucerne to defoliation and nitrogen fertiliser. *N.Z. J. Exper. Agric.*, 2: 7–11.

Hoglund, J.H., Crush, J.R., Brock, J.L., Ball, P.R. and Carran, R.A., 1979. Nitrogen fixation in pasture XII. General discussion. *N.Z. J. Exper. Agric.*, 7: 45–51.

Lambert, M.G., Luscombe, P.C. and Clark, D.A., 1982. Soil fertility and hill country production. *Proc. N.Z. Grassland Assoc. Conf.*, 43: 153–160.

Ledgard, S.F., Steele, K.W. and Saunders, W.H.M., 1982. Effects of cow urine and its major constituents on pasture properties. *N.Z. J. Agric. Res.*, 25: 61–68.

Masterson, C.L. and Murphy, P.M., 1976. Application of the acetylene reduction technique to the study of nitrogen fixation by white clover in the field. In: P.S. Nutman (Editor), *Symbiotic Nitrogen Fixation in Plants.* Cambridge University Press, Cambridge, pp. 299–316.

Masterson, C.L. and Sherwood, M.T., 1978. Some effects of increased atmospheric carbon dioxide on white clover (*Trifolium repens*) and pea (*Pisum sativum*). *Plant Soil*, 49: 421–426.

Moustafa, E., Ball, P.R. and Field, T.R.O., 1969. The use of acetylene reduction to study the effect of nitrogen fertiliser and defoliation on nitrogen fixation by field-grown white clover. *N.Z. J. Agric. Res.*, 12: 691–696.

Munns, D.N., Fogle, V.W. and Hallock, B.G., 1977. Alfalfa root nodule distribution and inhibition of nitrogen fixation by heat. *Agron. J.*, 69: 377–380.

Murphy, P.M., 1983. Photosynthate supply and nitrogen fixation in forage legumes. In: *Proc. XIV Int. Grassland Congr.*, pp. 332–335.

Newbould, P., Holding, A.J., Davies, G.J., Rangeley, A., Copeman, G.J.F., Davies, A., Frame, J., Haystead, A., Herriott, J.B.D., Holmes, J.C., Lowe, J.F., Parker, J.W.G., Waterson, H.A., Wildig, J., Wray, J.P. and Younie, D., 1982. The effect of *Rhizobium* inoculation on white clover in improved hill soils in the United Kingdom. *J. Agric. Sci., Cambridge*, 99: 591–610.

Pankhurst, C.E. and Gibson, A.H., 1973. *Rhizobium* strain influence on disruption of clover nodule development at high root temperature. *J. Gen. Microbiol.*, 74: 219–231.

Paul, E.A., 1976. Recent studies using the acetylene-reduction technique as an assay for field nitrogen fixation levels. In: W.D.P. Stewart (Editor), *Nitrogen Fixation by Free Living Micro-organisms.* Cambridge University Press, Cambridge, pp. 259–269.

Ranga Rao, V., 1977. Effect of temperature on the nitrogenase activity of intact and detached nodules in *Lotus* and *Stylosanthes. J. Exper. Bot.*, 28: 261–267.

Reuss, J.O. and Cole, C.V., 1973. Simulation of nitrogen flow in a grassland ecosystem. In: *Proc. U.S. I.B.P. Grassland Biome: Fort Collins, Colo.*

Richards, I.R., 1976. Nitrogen under grazing — responses to fertilizer N and the role of white clover. In: J. Hodgson and D.K. Jackson (Editors), *Pasture Utilization by the Grazing Animal. Br. Grassland Soc. Occas. Symp.*, No. 8: 69–77.

Roughley, R.J., 1970. The influence of root temperature, *Rhizobium* strain and host selection on the structure and nitrogen-fixing efficiency of the root nodules of *Trifolium subterraneum. Ann. Bot.*, 34: 631–646.

Sears, P.D., 1960. Grass/clover relationships in New Zealand. In: *Proc. VIII Int. Grassland Congr.*, pp. 130–133.

Sears, P.D., Goodall, V.C., Jackman, R.H. and Robinson, G.S., 1965. Pasture growth and soil fertility VIII. The influence of grasses, white clover, fertilisers, and the return of herbage clippings on pasture production of an impoverished soil. *N.Z. J. Agric. Res.*, 8: 270–283.

Simpson, J.R., 1965. The transference of nitrogen from pasture legumes to an associated grass under several systems of management in pot culture. *Aust. J. Agric. Res.*, 16: 915–926.

Sprent, J.I., 1976. Water deficits and nitrogen-fixing root nodules. In: T.T. Kozlowski (Editor), *Water Deficits and Plant Growth IV.* Academic Press, New York, N.Y., pp. 291–315.

Steele, K.W. and Shannon, P., 1982. Concepts relating to the nitrogen economy of a Northland intensive beef farm. In:

P.W. Gander (Editor), *Nitrogen Balances in New Zealand Ecosystems.* DSIR, Palmerston North, pp. 85–89.

Vallis, I., Henzell, E.F. and Evans, T.R., 1977. Uptake of soil nitrogen by legumes in mixed swards. *Aust. J. Agric. Res.,* 28: 413–425.

Vance, C.P., Heichel, G.H., Barnes, D.K., Bryan, J.W. and Johnson, L.E., 1979. Nitrogen fixation, nodule development, and vegetative regrowth of alfalfa (*Medicago sativa* L.) following harvest. *Plant Physiol.,* 64: 1–8.

Walker, T.W., Adams, A.F.R. and Orchiston, H.D., 1956. Fate of labeled nitrate and ammonium nitrogen when applied to grass and clover grown separately and together. *Soil Sci.,* 81: 339–351.

West, C.P. and Wedin, W.F., 1981. The contribution of $N_2$ fixation to the nitrogen budget of alfalfa-orchardgrass pastures. In: *Agronomy Abstracts 73rd Annual Meeting of the American Society of Agronomy.* American Society of Agronomy, Madison, Wis.

Whitehead, D.C., 1970. *The Role of Nitrogen in Grassland Productivity.* Commonwealth Agricultural Bureaux Bulletin 48. Farnham Royal, 202 pp.

Whitney, A.S., 1977. Contribution of forage legumes to the nitrogen economy of mixed swards. A review of relevant Hawaiian research. In: A. Ayanaba and P.J. Dart (Editors), *Biological Nitrogen Fixation in Farming Systems of the Tropics.* Wiley, Chichester, pp. 89–96.

Wightman, J.A., 1979. *Sitona humeralis* in the South Island of New Zealand (Coleoptera: Curculionidae). In: T.K. Crosby and R.P. Pottinger (Editors), *Proc. 2nd Australasian Conference on Grassland Invertebrate Ecology.* Government Printer, Wellington, pp. 138–141.

Willoughby, W.M., 1954. Some factors affecting grass–clover relationships. *Aust. J. Agric. Res.,* 5: 157–180.

Woodmansee, R.G., 1978. Factors influencing input and output of nitrogen in grasslands. In: N.R. French (Editor), *Perspectives in Grassland Ecology.* Springer-Verlag, New York, N.Y., pp. 117–134.

Yeates, G.W., Ross, D.J., Bridger, B.A. and Visser, T.A., 1977. Influence of the nematodes *Heterodera trifolii* and *Meloidogyne hapla* on nitrogen fixation by white clover under glasshouse conditions. *N.Z. J. Agric. Res.,* 20: 401–413.

Chapter 21

# NITROGEN LOSSES FROM MANAGED GRASSLAND

K.W. STEELE

## INTRODUCTION

Grasslands occur over a wide range of climates and soil types (Chapters 1–3); as a result, the annual above-ground production of herbage dry matter ranges from less than 1 t ha$^{-1}$ to more than 20 t ha$^{-1}$ (Chapters 2 and 3). Most grasslands have one factor in common, that nitrogen usually limits herbage production. Even for highly productive temperate grasslands, it has been estimated that production could be increased by up to 30% by elimination of nitrogen deficiency (Steele, 1982a).

Losses of nitrogen from grassland ecosystems are important not only because they lead to nitrogen deficiency, and so to reductions in productivity, but also because these losses can cause pollution elsewhere. Nitrogen is lost from grassland ecosystems in various forms including: (a) ammonia ($NH_3$) emission from soil, plant, animal excreta and fertilizer; (b) biological and chemical denitrification; (c) erosion by wind and water; (d) fire; (e) leaching; (f) retention in animals; (g) transfer to unproductive areas in the excreta of animals; and (h) removal in animal products. Nitrate ($NO_3^-$) loss, in waters draining from agricultural lands, can pose a threat to wildlife, and to human health if they contaminate drinking water. Concern has also been expressed that nitrous oxide ($N_2O$), originating from agricultural lands and other sources, may reduce radiation loss from the earth's surface, from nitric acid in the atmosphere thereby contributing to acid rain, and may reach the stratosphere and deplete the ozone layer.

Grassland management practices can have large effects on both the magnitude and the pathways of nitrogen loss. An understanding of these pathways and the factors controlling them provides a rational basis on which to reduce or divert these losses. This chapter examines the various pathways through which nitrogen is lost from grasslands, and discusses management variables which affect the magnitude and pathway of losses.

## GASEOUS LOSSES

The major pathways for gaseous loss of nitrogen are: (a) ammonia emission from soil, fertilizer, animal excreta and plant material; (b) biological and chemical denitrification; and (c) fire. Although not strictly a gaseous loss, nitrogen may also be lost in particulate material carried by wind. Each of these processes will be considered individually.

### Ammonia emission

Most ammonia is lost from animal excreta or fertilizers, though small amounts of ammonia may be released from live plants and decomposing organic matter. Conversely, grasslands can act as a sink for atmospheric ammonia (Catchpoole et al., 1983; Harper et al., 1983), particularly around sunset and sunrise, and when soil concentrations of mineral nitrogen are low (Harper et al., 1983).

### Volatilization from animal excreta

Volatilization loss of ammonia can be large when animal urine is deposited on grassland. Losses of between 25 and 30% have been reported for pastures of *Setaria sphacelata* during late summer in a subtropical region of Australia (Vallis et al., 1982), for pastures of lucerne (*Medicago sativa*), and for mixtures of perennial ryegrass (*Lolium perenne*) with subterranean clover (*Trifo-

*lium subterraneum*) during spring and autumn in southern Australia (Denmead et al., 1974, 1976). Losses of up to 66% have been found during summer for perennial ryegrass–white clover (*Trifolium repens*) pastures in New Zealand (Ball and Keeney, 1983), and losses of up to 80% have been found under dry summer conditions in Western Australia (Watson and Lapins, 1969). In temperate grasslands, average annual losses of urinary nitrogen are probably between 25 and 30% (Ball and Keeney, 1983; Simpson and Steele, 1983).

Factors which increase volatilization of ammonia include: high soil temperature (Martin and Chapman, 1951; Harper et al., 1983), an open plant canopy (Denmead et al., 1976), coarse soil texture (Fenn and Kissel, 1976), high ammonium ($NH_4^+$) concentrations (Martin and Chapman, 1951), high soil pH, low soil cation exchange capacity (Gasser, 1964), and a high wind velocity at the soil surface (Kissel et al., 1977).

The amount of nitrogen excreted in the urine depends on its concentration in the herbage ingested by grazing animals (Chapter 19). Sheep and cattle excrete a relatively constant amount of nitrogen in the faeces (8 g N per kg of dry matter consumed) (Barrow and Lambourne, 1962; Barrow, 1967); the remaining nitrogen is excreted in the urine. When sheep graze temperate grasslands rich in nitrogen, they excrete 70 to 75% of the nitrogen ingested in urine (Sears et al., 1948). Conversely, cattle grazing poor native pasture in South Africa excrete only about 20% of ingested nitrogen in urine (Gillard, 1967). The total amount of nitrogen excreted in urine will depend largely on the stocking rate. Dairy cattle stocked at 4.1 cows $ha^{-1}$, on a pasture producing 16.5 t $ha^{-1}$ of dry matter annually, were found to ingest 530 kg N annually, of which 327 kg was excreted in urine (Steele, 1982b). This is equivalent to an application of nearly 1000 kg N $ha^{-1}$ on the area wetted by urine; the equivalent figure for sheep grazing similar pasture is 500 kg N $ha^{-1}$ (Steele, 1982b).

Approximately 75% of the nitrogen in the urine of ruminants is in the form of urea (Doak, 1952), which rapidly hydrolyses to ammonium carbonate once in contact with the soil (Gasser, 1964). An equilibrium exists between $NH_4^+$ and $NH_3$ in soil solution and between $NH_3$ in soil solution and $NH_3$ in the soil atmosphere. Ammonia is formed above pH 7. Since hydrolysis increases the pH of the soil solution, this favours formation of $NH_3$. The diffusion of ammonia from solution into the atmosphere depends on the difference in partial pressure between air and solution but, since the ammonia content of the atmosphere is usually low (1–10 µg N $m^{-3}$; Simpson and Steele, 1983), the rate of ammonia emission from soils is largely dependent on its concentration in solution.

Losses of gaseous ammonia from dung can be as high as 80% of the nitrogen content, under hot dry conditions where no dung beetles are active (Gillard, 1967), but are normally considerably lower than from urine. For example, losses of less than 10% have been found in laboratory studies (Floate, 1970a, b, c; Floate and Torrance, 1970), and less than 5% in the field (McDiarmid and Watkin, 1972).

**Volatilization from fertilizer**

Surface applications of alkaline fertilizers (such as urea, ammonium hydroxide) lead to larger losses than applications of neutral or acidic fertilizers (such as ammonium sulphate). Losses of about 25% have been reported following application of urea to subtropical pastures (Catchpoole et al., 1983), mostly within fourteen days of application. Such losses are controlled by the same factors which control volatilization from urine and dung.

**Biological denitrification**

The reduction of $NO_2^-$ or $NO_3^-$ to gaseous oxides [nitric oxide (NO) and nitrous oxide ($N_2O$)], and sometimes to gaseous dinitrogen ($N_2$), is brought about by soil micro-organisms, especially under anaerobic conditions. Many types of soil bacteria can carry out the reaction (Knowles, 1981; Firestone, 1982), and populations of denitrifying bacteria in agricultural soils are commonly of the order of 1 to $5 \times 10^6$ per gram of soil (Gamble et al., 1977) — that is, from 1 to 5% of the total heterotrophic population.

Relatively few quantitative estimates have been made of nitrogen loss through denitrification from grasslands, because of difficulties of measurement. Early estimates were generally based on differences; any nitrogen entering a grassland ecosystem which could not be accounted for as soil nitrogen, in leachate, or plant uptake, was considered lost by denitrification. Such estimates are subject to substantial errors. More recently, emission from the soil of nitrous oxide and gaseous nitrogen have been measured.

### Nitrous oxide emission

Measurement of nitrous oxide emission is difficult because it is spatially very variable, with twofold differences occurring over a few metres (Simpson and Steele, 1983).

In mown grasslands in Australia, daily emissions were largest (217 mg N m$^{-2}$) in spring, when the soils were warm and wet, and lowest (0.05 mg N m$^{-2}$) in winter, when the soils were cold and dry (Denmead et al., 1979). Similar rates and patterns of emission have been reported in both Britain (Ryden, 1981; Webster and Dowdell, 1982) and New Zealand (Limmer and Steele, 1982).

In most cases, the loss of nitrogen as nitrous oxide from fertilizer application is relatively small, accounting for less than 2% of that applied (Ryden, 1981; Webster and Dowdell, 1982). However, some exceptionally large losses have been reported (e.g. O'Hara et al., 1984). When the $NO_3^-$ concentration in the soil is less than 1 p.p.m., there may even be a net uptake of $N_2O$ by the soil (Ryden, 1981).

Nitrous oxide can be lost from soils during the oxidation of $NH_4^+$ to $NO_2^-$ and $NO_3^-$ (Blackmer et al., 1980; Goreau et al., 1980). The loss may result from the oxidation of $NH_2OH$, which is an intermediate in the oxidation of $NH_4^+$ to $NO_2^-$ (Hooper and Terry, 1979), or from the reduction of $NO_2^-$ (Ritchie and Nicholas, 1972). It is unlikely to be of practical significance in most managed grasslands.

### Gaseous nitrogen emission

Measurement of the emission of gaseous nitrogen ($N_2$) is very difficult, because of the small quantity produced, compared with the background concentration in the atmosphere. The release of $^{15}N$ from labelled substrates can be measured, but may give artificially high estimates (Focht, 1978). Recently, emission of gaseous nitrogen from soils has been measured by inhibiting $N_2O$-reductase in soils (Klemedtsson et al., 1977; Yoshinari et al., 1977; Ryden et al., 1979), and by using an isotopic dilution technique (Limmer et al., 1982).

Emission of gaseous nitrogen from grasslands may be many times greater than that of nitrous oxide (Ryden, 1981; Limmer and Steele, 1982). The amount of gaseous nitrogen emitted, relative to nitrous oxide, is reduced by low carbon availability (Focht and Verstraete, 1977; Smith and Tiedje, 1979), increasing oxygen partial pressure (Krul and Veeningen, 1977), high $NO_3^-$ concentrations (Blackmer and Bremner, 1978), low soil pH, and low temperature (Nommik, 1956).

### Regulatory factors

The loss of nitrogen through biological denitrification is greatest when soils are nearly neutral in pH (Burford and Bremner, 1975), are above 5°C (Bremner and Shaw, 1958), contain high concentrations of available carbon (Burford and Bremner, 1975) and nitrate (Kohl et al., 1976), and are anaerobic (Cox and Payne, 1973; Greenwood, 1961), though denitrification can occur in aerobic soils if they contain a large population of denitrifying strains of *Rhizobium* (O'Hara et al., 1983).

Most of the losses by biological denitrification occur from the surface layer of mineral soils, but can occur from deeper horizons in organic soils (Limmer and Steele, 1983).

### Non-biological denitrification

Although most losses of nitrogen through denitrification have been attributed to biological activity, losses can also occur through non-biological processes. Under acid conditions (pH < 5), nitrous acid ($HNO_2$) decomposes to nitric oxide (Allison and Doetsch, 1951), though little is likely to escape into the atmosphere if the soil is well aerated (Allison, 1965); measured losses have been in the range 0.2–0.3 kg N ha$^{-1}$ yr$^{-1}$ (Galbally and Roy, 1978).

## Losses of volatile nitrogen compounds from plants

Loss of nitrogen from plants through volatilization of ammonia or amines has only recently been recognized (Wetselaar and Farquhar, 1980), and few data are available.

Ammonia is released and refixed in the photorespiratory cycle (Keys et al., 1978), permitting diffusion of ammonia out of the leaf. Evidence of ammonia emission from Rhodes grass (*Chloris gayana*) was obtained by Martin and Ross (1968), whereas Weiland et al. (1982) found that emission of both reduced and oxidized forms of nitrogen occurred from leaves of sorghum (*Sorghum bicolor*) and Johnson grass (*Sorghum halepense*) and was correlated with transpiration rate, temperature, oxygen concentration, and tissue age. Plants can also absorb volatile nitrogen compounds from the atmosphere. For example, if the concentration of atmospheric ammonia exceeds the critical partial pressure for ammonia in the leaves, absorption will occur (Farquhar et al., 1980).

## WIND EROSION

Wind erosion can remove nitrogen in air-borne particles under dry conditions, especially if the pasture is overgrazed. Losses of soil organic nitrogen during dry seasons, presumably through wind erosion, have been measured in Queensland (Vallis, 1972).

## FIRE

Dry vegetation is often burnt at the end of the dry period to promote regrowth. The combustion gases contain 80 to 90% of the nitrogen in the plant as $NH_3$, $NO_x$ and $N_2O$ (Crutzen et al., 1979; Norman and Wetselaar, 1960).

## SOLUTION LOSSES

Most of the solution losses are by leaching, though some occur by surface runoff (Chapter 22).

### Leaching

Leaching of nitrogen from grasslands occurs mainly as nitrate. In recent years, there has been increasing concern over the leaching of nitrate from agricultural land into streams and groundwaters. Leaching of nitrogen is obviously restricted to areas where, at least seasonally, rainfall exceeds evapotranspiration. Losses are greatest from old grasslands (Harmsen and Kolenbrander, 1965), in high rainfall areas (Owens, 1960), on free-draining soils (Kolenbrander, 1972), at high stocking rates, and with large inputs of fertilizer nitrogen.

Urine is the main source of nitrate in leachate from grazed grasslands, since the amount of nitrogen applied in the urine patch far exceeds the requirements of grassland plants (see above). Under conditions of excess precipitation, much of the excess is leached (Burden, 1982). Leaching losses greater than 100 kg N $ha^{-1}$ $yr^{-1}$ have been measured from intensively grazed grasslands in New Zealand (Steele, 1982c; Steele et al., 1984), and from grasslands heavily fertilized with nitrogen in The Netherlands (Henkens, 1977). As a result, concentrations of up to 58 mg N $l^{-1}$ have been measured in shallow aquifers under intensively grazed grasslands in New Zealand (Baber and Wilson, 1972).

### Surface run-off and erosion

Loss of inorganic nitrogen in surface run-off from most grasslands is usually less than 1 kg N $ha^{-1}$ $yr^{-1}$, though the loss in particulate material is usually in the range from 1 to 5 kg N $ha^{-1}$ $yr^{-1}$ (Vollenweider, 1968; McColl and Gibson, 1979). Losses may be greater if heavy rainfall occurs after overgrazing.

## ANIMAL RETENTION, PRODUCTS AND TRANSFER

The amounts of nitrogen retained in livestock, removed in animal products and transferred to non-productive areas, such as stock-yards and dairy sheds, depends mainly on the number and type of stock.

On an intensive dairy farm with 4.1 cows $ha^{-1}$, consuming 13 t $ha^{-1}$ $yr^{-1}$ of herbage and producing 10 800 kg milk $ha^{-1}$, 66 kg N $ha^{-1}$ will be removed in milk, 46 kg N $ha^{-1}$ will be transferred to non-productive areas, and 8 kg N $ha^{-1}$ will be retained in replacement stock, so that a total of 120 kg N $ha^{-1}$ will be removed annually (Steele, 1982b).

Sheep and beef cattle remove much less nitrogen from grasslands. A liveweight gain of 1000 kg $ha^{-1}$ $yr^{-1}$ would lead to the removal only of 2.4 kg N $ha^{-1}$ in stock, whereas production of 100 kg $ha^{-1}$ of greasy wool would lead to a loss of only 11 kg N $ha^{-1}$ $yr^{-1}$ in wool, though in both cases some nitrogen would be transferred to non-productive areas. These values are for the most productive systems; less productive systems, which are commoner, would have smaller losses.

## MANAGEMENT OF GRASSLANDS TO REDUCE LOSSES OF NITROGEN

Many grassland management practices directly and indirectly affect losses of nitrogen from grassland ecosystems.

### Animal management

Much of the loss of nitrogen occurs as a result of the heavy deposition of excreta around drinking troughs, gateways and camp sites. Fencing and rotational grazing can reduce this transfer and loss.

The intensity of grazing will also influence losses of nitrogen. The total amount of nitrogen in grassland ecosystems in the western United States increases under light or no grazing, is approximately constant at moderate grazing intensity, but decreases at high grazing intensities (Woodmansee, 1979). Net losses of nitrogen may also occur in intensively grazed temperate grasslands, for example in New Zealand (Ball, 1982). Overgrazing should be avoided, particularly during dry periods, since erosion and emission of ammonia are increased.

### Plant factors

Treatments such as phosphorus fertilization, irrigation and drainage, which promote faster growth of pasture, increase the uptake of $NH_4^+$ and $NO_3^-$ from soil solution by plants, and so reduce loss through volatilization or leaching. Combinations of species which grow throughout the year, and are deep-rooted, will also help to prevent loss. Heavy use of nitrogen fertilizers will increase losses of nitrogen.

In many cases, the herbage of grasses and legumes has a nitrogen concentration in excess of animal requirements. This results in high concentrations of nitrogen in urine and greater losses of nitrogen. Excessively high nitrogen concentrations should therefore be avoided.

### Fertilizer management

Loss of nitrogen can be reduced by selecting the most suitable method of application for the fertilizer. For example, alkaline fertilizers or ammonium salts should not be broadcast, because of volatilization of ammonia. In particular, anhydrous ammonia should be injected into the soil, and the soil should not be too dry or too wet at the time of application. Similarly, fertilizer application should be avoided when heavy rainfall is expected, or when the pasture is not growing, since this results in leaching. The loss of ammonia from urea-based fertilizers can be reduced by inclusion of urease inhibitors.

Nitrification inhibitors and slow-release fertilizers can increase the uptake of fertilizer nitrogen in grassland herbage, but the cost of these compounds frequently precludes their use on grasslands.

### Soil factors

Soils with large cation exchange capacities retain more $NH_4^+$, resulting in less volatilization of ammonia. Soils should not be excessively limed, since this increases the rate of ammonia release. Management which improves soil structure will increase the depth to which urine penetrates, and thereby reduce losses of ammonia.

Adequate soil drainage reduces denitrification by preventing waterlogging, while avoiding excessive irrigation reduces leaching losses.

Soil fauna assist the removal of plant residues and animal dung from the soil surface. Dung beetles can substantially reduce gaseous loss of nitrogen from dung. Where dung beetles, earthworms or termites are not present, their introduction should be considered. Besides their influence on losses of nitrogen, the improvement in soil fertility associated with active soil fauna is well documented.

Where legumes are to be introduced to an area previously lacking *Rhizobium* spp., there may be distinct long-term advantages in selecting a non-denitrifying strain of *Rhizobium* as an inoculant, since this will preclude the possibility of loss of nitrogen through rhizobial denitrification.

### Environmental factors

Wind velocity close to the soil surface has a direct effect on gaseous losses of nitrogen by removing gases and so reducing the partial pressure of nitrogen-containing gases and increasing upward diffusion. Wind velocity also has an indirect effect through its effect on evapotranspiration, so perhaps increasing the uptake of mineral nitrogen. Modification of wind velocity, through the planting of shelter belts, can therefore reduce gaseous losses of nitrogen.

## CONCLUDING COMMENTS

Excreta constitute a major pathway by which nitrogen is recycled in grassland ecosystems; in particular, the return of nitrogen in urine greatly speeds the rate of cycling. However, dung and urine patches greatly concentrate the deposition of nitrogen, giving returns of up to 1000 kg N ha$^{-1}$ and causing substantial losses of nitrogen from grassland ecosystems. The recovery of urinary nitrogen by plants in temperate grasslands is generally 30% or less (During and McNaught, 1961; Ball et al., 1979; Ledgard et al., 1982; Ledgard and Saunders, 1982).

Loss of nitrogen from urine occurs by both volatilization of ammonia and leaching, the relative amounts depending on the prevailing climatic conditions. Volatilization of ammonia is favoured by warm–dry conditions, and losses are most severe in arid and semi-arid grasslands; conversely, leaching losses increase as rainfall increases, and are greatest in temperate grasslands. Little of the loss of nitrogen from urine patches is attributed to biological denitrification. Most denitrification occurs close to the soil surface and, once leached from the surface layer, urinary nitrogen is protected from denitrification by the absence of the organisms responsible (Limmer and Steele, 1983). Even in the surface layers of the soil, denitrification is inhibited by the presence of urine.

The amount of nitrogen excreted annualy by grazing animals depends predominantly on the stocking rate. As a result, the potential for absolute loss of nitrogen in the excreta increases as stocking rate increases; but, even at low stocking rates, losses in the excreta may amount to a significant proportion of the annual input of nitrogen.

The total amount of nitrogen lost from grasslands generally increases with increasing animal productivity. It appears that, although the importance of many of the losses of nitrogen depends on the type of grassland management, increased nitrogen loss has to be accepted as a concomitant of increasing grassland productivity. However, a time may be reached when losses of nitrogen from intensively grazed grasslands become unacceptable, because of deleterious environmental effects. Further research to improve understanding of processes which control nitrogen losses should provide a basis on which to improve the efficiency of nitrogen utilization within grassland ecosystems.

## REFERENCES

Allison, F.E., 1965. Evaluation of incoming and outgoing processes that affect soil nitrogen. In: W.V. Bartholomew and F.E. Clark (Editors), *Soil Nitrogen. Agronomy No. 10.* Am. Soc. Agron., Madison, Wis., pp. 573–606.

Allison, F.E. and Doetsch, J.H., 1951. Nitrogen gas production by the reaction of nitrates with amino acids in slightly acidic media. *Soil Sci. Soc. Am. Proc.*, 15: 163–166.

Baber, H.L. and Wilson, A.T., 1972. Nitrite pollution of groundwater in the Waikato region. *Chem. N.Z.*, 36: 179–183.

Ball, P.R., 1982. Nitrogen balances in intensively managed pasture systems. In: *Nitrogen Balances in New Zealand Ecosystems.* DSIR, Palmerston North, New Zealand, pp. 47–66.

Ball, P.R. and Keeney, D.R. 1983. Nitrogen losses from urine-affected areas of a New Zealand pasture under contrasting seasonal conditions. In: *Proc. XIV Int. Grassland Congr. Lexington, Ky, 1981*, pp. 342–344.

Ball, P.R., Keeney, D.R., Theobald, P.W. and Nes, P. 1979. Nitrogen balance in urine-affected areas of a New Zealand pasture. *Agron. J.*, 71: 309–314.

Barrow, N.J., 1967. Some aspects of the effects of grazing on the nutrition of pastures. *J. Aust. Inst. Agric. Sci.*, 33: 254–262.

Barrow, N.J. and Lambourne, L.J., 1962. Partition of excreted nitrogen, sulphur and phosphorus between the faeces and urine of sheep being fed pasture. *Aust. J. Agric. Res.*, 13: 461–471.

Blackmer, A.M. and Bremner, J.M., 1978. Inhibitory effect of nitrate on reduction of $N_2$ by soil microorganisms. *Soil Biol. Biochem.*, 10: 187–191.

Blackmer, A.M., Bremner, J.M. and Schmidt, E.L., 1980. Production of nitrous oxide by ammonia–oxidising chemoautrophic microorganisms in soil. *Appl. Environ. Microbiol.*, 40: 1060–1066.

Bremner, J.M. and Shaw, K., 1958. Denitrification in soil II. Factors affecting denitrification. *J. Agric. Sci.*, 51: 39–52.

Burford, J.R. and Bremner, J.M., 1975. Relationships between the denitrification capacities of soils and total water soluble and readily decomposable soil organic matter. *Soil Biol. Biochem.*, 7: 389–394.

Burden, R.J., 1982. Nitrate contamination of New Zealand aquifers: A review. *N.Z. J. Sci.*, 25: 205–220.

Catchpoole, V.R., Harper, L.A. and Myers, R.J.K., 1983. Annual losses of ammonia from a grazed pasture fertilised with urea. In: *Proc. XIV Int. Grassland Congr., Lexington, Ky., 1981*, pp. 344–346.

Cox, C.D. and Payne, W.J., 1973. Separation of soluble denitrifying enzymes and cytochromes from *Pseudomonas perfectomarinus. Can. J. Microbiol.*, 19: 861–872.

Crutzen, P.J., Heidt, L.E., Krasnec, J.P., Pollock, W.H. and Sieler, W., 1979. Biomass burning as a source of atmosphere gases CO, $H_2$, $N_2O$, NO, $CH_3Cl$ and COS. *Nature*, 282: 253–256.

Denmead, O.T., Simpson, J.R. and Freney, J.R., 1974. Ammonia flux into the atmosphere from a grazed pasture. *Science*, 185: 609–610.

Denmead, O.T., Freney, J.R. and Simpson, J.R., 1976. A closed ammonia cycle within a plant canopy. *Soil Biol. Biochem.*, 8: 161–164.

Denmead, O.T., Freney, J.R. and Simpson, J.R., 1979. Studies on nitrous oxide emission from a grass sward. *Soil Sci. Soc. Am. J.*, 43: 726–728.

Doak, B.W., 1952. Some chemical changes in the nitrogenous constituents of urine when voided on pasture. *J. Agric. Sci.*, 42: 162–171.

During, C. and McNaught, K.J., 1961. Effect of cow urine on growth of pasture and uptake of nutrients. *N.Z. J. Agric. Res.*, 4: 591–605.

Farquhar, G.D., Firth, P.M., Wetselaar, R. and Weir, B., 1980. On the gaseous exchange of ammonia between leaves and the environment. Determination of the ammonia compensation point. *Plant Physiol.*, 66: 710–714.

Fenn, L.B. and Kissel, D.E., 1976. The influence of cation exchange capacity and depth of incorporation of ammonia on ammonia volatilisation from ammonium compounds applied to calcareous soils. *Soil Sci. Soc. Am. J.*, 40: 394–398.

Firestone, M.K., 1982. Biological denitrification. In: J.J. Stevenson (Editor), *Nitrogen in Agricultural Soils. Agronomy No. 22*. Am. Soc. Agron., Madison, Wis., pp. 289–326.

Floate, M.J.S., 1970a. Decomposition of organic materials from hill soils and pastures. 2. Comparative studies on the mineralisation of carbon, nitrogen and phosphorus from plant materials and sheep faeces. *Soil Biol. Biochem.*, 2: 173–185.

Floate, M.J.S., 1970b. Decomposition of organic materials from hill soils and pastures. 3. The effect of temperature on the mineralisation of carbon, nitrogen and phosphorus from plant materials and sheep faeces. *Soil Biol. Biochem.*, 2: 187–196.

Floate, M.J.S., 1970c. Decomposition of organic materials from hill soils and pastures. 4. The effects of moisture content on the mineralisation of carbon, nitrogen and phosphorus from plant materials and sheep faeces. *Soil Biol. Biochem.*, 2: 275–283.

Floate, M.J.S. and Torrance, C.J.W., 1970. Decomposition of the organic materials from hill soils and pastures. 1. Incubation method for studying the mineralisation of carbon, nitrogen and phosphorus. *J. Sci. Food Agric.*, 21: 116–120.

Focht, D.D., 1978. Methods for analysis of denitrification in soils. In: D.R. Nielsen and J.G. MacDonald (Editors), *Nitrogen in the Environment, 2*. Academic Press, New York, N.Y., pp. 432–490.

Focht, D.D. and Verstraete, W., 1977. Biochemical ecology of nitrification and denitrification. In: M. Alexander (Editor), *Advances in Microbial Ecology, 1*. Plenum Press, New York, N.Y., pp. 135–214.

Galbally, I.E. and Roy, C.R., 1978. Loss of fixed nitrogen from soil by nitric oxide exhalation. *Nature*, 275: 734–735.

Gamble, T.N., Betlach, M.R. and Tiedje, J.M., 1977. Numerically dominant denitrifying bacteria from world soils. *Appl. Environ. Microbiol.*, 33: 926–939.

Gasser, J.K.R., 1964. Urea as a fertiliser. *Soils Fertil.*, 27: 175–180.

Gillard, P., 1967. Coprophagous beetles in pasture ecosystems. *J. Aust. Inst. Agric. Sci.*, 33: 30–34.

Goreau, T.J., Kaplan, W.A., Wofsy, S.C., McElroy, M.B., Valois, F.W. and Watson, S.W., 1980. Production of $NO_2^-$ and $N_2O$ by denitrifying bacteria at reduced concentrations of oxygen. *Appl. Environ. Microbiol.*, 40: 526–532.

Greenwood, R.J., 1961. The effect of oxygen concentration on the decomposition of organic materials in soil. *Plant Soil*, 14: 360–376.

Harmsen, G.W. and Kolenbrander, G.J., 1965. Soil inorganic nitrogen. In: W.V. Bartholomew and F.E. Clark (Editors), *Soil Nitrogen Agronomy, 10*. Am. Soc. Agron., Madison, Wis., pp. 43–92.

Harper, L.A., Catchpoole, V.R., Davis, R. and Weir, K.L., 1983. Ammonia volatilisation: Soil, plant and microclimate effects on diurnal and seasonal fluctuations. *Agron. J.*, 75: 212–218.

Henkens, C.H., 1977. Agro-ecosystems in the Netherlands. Part III. In: M.J. Frissel (Editor), *Cycling in Mineral Nutrients in Agricultural Ecosystems. Agro-Ecosystems*, 4: 79–97.

Hooper, A.B. and Terry, K.R., 1979. Hydroxylamine oxidoreductase of *Nitrosomonas* production of nitric oxide from hydroxylomine. *Biophys. Acta*, 571: 12–20.

Keys, A.J., Bird, I.F., Cornelius, M.J., Lea, P.J., Wallsgrove, R.M. and Miffin, B.J., 1978. Photorespiratory nitrogen cycle. *Nature*, 275: 741–743.

Kissel, D.E., Brewer, H.L. and Arkin, G.F., 1977. Design and test of a field sampler for ammonia volitilisation. *Soil Sci. Soc. Am. J.*, 41: 1113–1138.

Klemedtsson, L., Svensson, B.H., Lindberg, T. and Rosswall, T., 1977. The use of acetylene inhibition of nitrous oxide reductase in quantifying denitrification in soils. *Swed. J. Agric. Res.*, 7: 179–185.

Knowles, R., 1981. Denitrification. In: F.E. Clark and T. Rosswall (Editors), *Terrestrial Nitrogen Cycles. Ecol. Bull. (Stockholm)*, 33: 315–319.

Kohl, D.H., Vithayathil, F., Whitlow, P., Shearer, G. and Chien, S.H., 1976. Denitrification kinetics in soil systems: the significance of good fits of data to mathematical forms. *J. Soil Sci. Soc. Am.*, 40: 249–253.

Kolenbrander, G.J., 1972. Eutrophication from agriculture with special reference to fertilisers and animal waste. In: *Effects of Intensive Fertiliser Use on the Human Environment. FAO Soils Bull.*, No. 16: 305–327.

Krul, J.M. and Veeningen, R., 1977. The synthesis of the dissimulatory nitrate reductase under aerobic conditions in a number of denitrifying bacteria, isolated from activated sludge and drinking water. *Water Res.*, 11: 39–43.

Ledgard, S.F. and Saunders, W.M.H., 1982. Effects of nitrogen fertiliser and urine on pasture performance and the influence of soil phosphorus and potassium status. *N.Z. J. Agric. Res.*, 25: 541–547.

Ledgard, S.F., Steele, K.W. and Saunders, W.M.H., 1982. Effect of cow urine and its major constituents on pasture properties. *N.Z. J. Agric. Res.*, 25: 61–68.

Limmer, A.W. and Steele, K.W., 1982. Denitrification potentials: Measurement of seasonal variation using a short-term anaerobic incubation technique. *Soil Biol. Biochem.*, 14: 179–184.

Limmer, A.W. and Steele, K.W., 1983. Effect of cow urine upon denitrification. *Soil Biol. Biochem.*, 15: 409–412.

Limmer, A.W., Steele, K.W. and Wilson, A.T., 1982. Direct field measurement of $N_2$ and $N_2O$ evolution from soil. *J. Soil Sci.*, 33: 499–507.

Martin, J.P. and Chapman, H.P., 1951. Volatilisation of ammonia from surface fertilised soils. *Soil Sci.*, 71: 25–34.

Martin, A.E. and Ross, P.J., 1968. A nitrogen-balance study using labelled fertiliser in a gas lysimeter. *Plant Soil*, 28: 182–186.

McColl, R.H.S. and Gibson, A.R., 1979. Downslope movement of nutrients in hill pasture, Taita, New Zealand, III. Amounts involved and management implications. *N.Z. J. Agric. Res.*, 22: 279–286.

McDiarmid, B.N. and Watkin, B.R., 1972. The cattle dung patch. 2. Effect of a dung patch on the chemical status of the soil and ammonia losses from the patch. *J. Br. Grassland Soc.*, 27: 43–48.

Nommik, H., 1956. Investigations on denitrification in soil. *Acta. Agric. Scand.*, 6: 196–228.

Norman, M.J.T. and Wetselaar, R., 1960. Losses of nitrogen on burning native pasture at Katherine. N.T. *J. Aust. Inst. Agric. Sci.*, 26: 272–273.

O'Hara, G.W., Daniel, R.M. and Steele, K.W., 1983. Effect of oxygen on the synthesis, activity and breakdown of the rhizobium denitrification system. *J. Gen. Microbiol.*, 129: 2405–2412.

O'Hara, G.W., Daniel, R.M., Steele, K.W. and Bonish, P.M., 1984. Nitrogen losses from soils caused by *Rhizobium*-dependent denitrification. *Soil Biol. Biochem.*, 16: 429–431.

Owens, L.D., 1960. Nitrogen movement and transformation in soils as evaluated by a lysimeter study utilising isotopic nitrogen. *Soil Sci. Soc. Am. Proc.*, 24: 372–376.

Ritchie, G.A.F. and Nicholas, D.J.D., 1972. Identification of the sources of nitrous oxide produced by oxidative and reductive processes in *Nitrosomonas europea*. *Biochem. J.*, 126: 1181–1191.

Ryden, J.C., 1981. $N_2$ exchange between a grassland soil and the atmosphere. *Nature*, 296: 235–237.

Ryden, J.C., Lund, L.J., Letey, J. and Focht, D.D., 1979. Direct measurement of denitrification loss from soils. II. Development and application of field methods. *Soil Sci. Soc. Am. J.*, 43: 110–118.

Sears, P.D., Goodall, V.C. and Newbold, R.P., 1948. The effects of sheep droppings on yield, botanical composition and chemical composition of pasture. II. Results for the years 1942–44 and final summary. *N.Z. J. Sci. Tech.*, A30: 231–250.

Simpson, J.R. and Steele, K.W., 1983. Gaseous N exchanges in grazed pastures. In: J.R. Freeney and J.R. Simpson (Editors), *Gaseous Loss of N from Plant–Soil Systems*. Martinus Nijhoff/D W. Junk, The Hague, pp. 215–236.

Smith, M.S. and Tiedje, J.M., 1979. The effect of roots on soil denitrification. *Soil Sci. Soc. Am. J.*, 43: 951–955.

Steele, K.W., 1982a. Nitrogen fixation for pastoral agriculture — biological or industrial? *N.Z. Agric. Sci.*, 16: 118–121.

Steele, K.W., 1982b. Nitrogen in grassland soils. In: P.B. Lynch (Editor), *Nitrogen Fertilisers in New Zealand Agriculture*. N.Z. Institute of Agricultural Science, Wellington, pp. 29–44.

Steele, K.W., 1982c. Concepts relating to the nitrogen economy of a Northland intensive beef farm. In: *Nitrogen Balances in New Zealand Ecosystems*. DSIR, Palmerston North, pp. 85–89.

Steele, K.W., Judd, M.J. and Shannon, P.W., 1984. Leaching of nitrate and other nutrients from a grazed pasture. *N.Z. J. Agric. Res.*, 27: 5–11.

Vallis, I., 1972. Soil nitrogen changes under continuously grazed legume–grass pastures in subtropical coastal Queensland. *Aust. J. Exper. Agric. Anim. Husbandry*, 12: 495–501.

Vallis, I., Harper, L.A., Catchpoole, V.R. and Weir, K.L., 1982. Volatilisation of ammonia from urine patches in a subtropical pasture. *Aust. J. Agric. Res.*, 33: 97–107.

Vollenweider, R.A., 1968. *Scientific Fundamentals of the Eutrophication of Lakes and Flowing Waters with Particular Reference to Nitrogen and Phosphorus as Factors in Eutrophication*. OECD, Directorate of Scientific Affairs, Paris, 220 pp.

Watson, E.R. and Lapins, P., 1969. Losses of nitrogen from urine on soils from South Western Australia. *Aust. J. Exper. Agric. Anim. Husbandry*, 9: 91–97.

Webster, C.P. and Dowdell, R.J., 1982. Nitrous oxide emission from permanent grass swards. *J. Sci. Food Agric.*, 33: 227–230.

Weiland, R.T., Stutte, L.A. and Silva, P.R.F., 1982. Nitrogen volatilisation from plant foliage. *Agric. Exper. Stn., Univ. Ark., Rep. Ser.*, No. 266: 40 pp.

Wetselaar, R. and Farquhar, G.D., 1980. Nitrogen losses from tops of plants. *Adv. Agron.*, 33: 263–302.

Woodmansee, R.G., 1979. Factors influencing input and output of nitrogen in grasslands. In: N. French (Editor), *Perspectives in Grassland Ecology*. Ecological Studies, 32. Springer-Verlag, Heidelberg, pp. 117–134.

Yoshinari, T., Hynes, R. and Knowles, R., 1977. Acetylene inhibition of nitrous oxide reduction and measurement of denitrification and nitrogen fixation in soil. *Soil Biol. Biochem.*, 9: 177–183.

Chapter 22

# RUN-OFF AND DRAINAGE FROM GRASSLAND CATCHMENTS

F.X. DUNIN

## INTRODUCTION

The large amounts of water harvested from grassland ecosystems are gained at some cost, both financially and environmentally. The regulation of outflow from grassland catchments often requires a dense network of small-scale structures, investment in which can be equivalent to the larger-scale undertakings for water conservation and flood mitigation (Cordery and Pilgrim, 1979). Much of the cost in water-quality control has been charged against agriculture, since intensification of grassland management is seen as one of the major causes of erosion, salting and eutrophication. With increasing public pressure for environmental protection and resource conservation, the economic advantages of more intensive grassland management must be debited with any detrimental effects on the availability of fresh-water resources. Consequently, further understanding of the processes and factors governing water transport in grassland systems is of prime importance in formulating policies, both fiscal and physical, to arrest resource degradation. This review focusses on studies which advance this understanding. Case studies, some with unexpected results, are highlighted to illustrate the use of such understanding in providing a rational explanation of apparent anomalies in catchment behaviour.

The management of grassland catchments is primarily designed to optimize the balance between water capture, herbage production and ecological stability (Downes, 1959). A balance between catchment outflow and grassland productivity is central to this aim; management of vegetation is an important determinant of this balance. The role of grassland in river-basin hydrology is briefly reviewed here, and broad comparisons are first made between grassland, forest and bare soil. The hydrologic impact of commonly practised management in grassland is then examined to identify critical factors and processes of water transport in these systems. Finally, the use of models incorporating this understanding is outlined, as a basis for achieving the ecological stability of grasslands.

## A PERSPECTIVE ON GRASSLAND HYDROLOGY

In general, the amount of water harvested from grassland is intermediate between that from forest and bare soil (Rodda, 1976; Bosch and Hewlett, 1982). The greater harvest of water from grassland than from forest can be attributed to lower interception losses, and to reduced uptake of soil water by grassland vegetation. This is illustrated by data, obtained from weighing lysimeters and soil-water measurements, for *Themeda* grassland and young eucalypt forest in New South Wales, Australia (Table 22.1, Fig. 22.1). Although the two sites are separated by 70 km, they are comparable in terms of climate and soil hydraulic properties. Evaporation loss of rain intercepted by the forest canopy accounted for a greater proportion of annual rainfall (15%) than that by the grassland (11%) — a result confirmed by Stewart (1977). The forest lost more water during rainfall, because of greater turbulence in the forest canopy (Pearce et al., 1980), but the grassland lost more water after rainfall.

Soil water is extracted to greater depths by forests than by grassland (Colville and Holmes, 1972). This, together with the greater interception loss of water in forest, is probably sufficient to

TABLE 22.1

Comparison of water losses from grassland and young eucalypt forest in southern New South Wales

|  | Forest | Grassland |
|---|---|---|
| Rainfall (mm yr$^{-1}$) | 587 | 792 |
| Interception loss (mm yr$^{-1}$) | 91 | 91 |
| Percentage loss (%) | 15 | 11 |
| Canopy evaporative loss (mm yr$^{-1}$) (during rainfall) | 43 | 0 |
| Canopy storage loss (mm yr$^{-1}$) (after rainfall) | 48 | 91 |

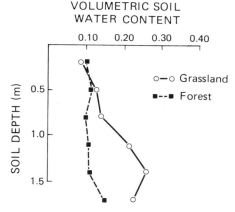

Fig. 22.1. The water content of duplex soils beneath grassland and forest in southern N.S.W., Australia, after a prolonged drought.

account for the harvest of water from grassland being greater, as has generally been observed.

The harvest of water from bare ground exceeds that from vegetated surfaces due, in part, to an absence of interception loss and the limited depth of soil drying (Philip, 1957). The effect of this greater run-off is seen in semi-arid areas of Australia, where run-off from bare areas sustains the growth of mulga (*Acacia aneura*) in lower areas (Slatyer, 1961). In a similar manner, run-off from bare areas has been used for agricultural purposes in arid regions such as in Israel (Shanan et al., 1958).

Although run-off from grassland is intermediate between that from forest and bare soil, in environments which have seasonal water deficits, transpiration losses by grassland can exceed those by forest (Table 22.2) in conditions of adequate precipitation. There may be circumstances, depending on interception, where forests deliver more water than grassland (e.g. Rahkmanov, 1970). Similarly, conversion of grassland to bare catchment conditions, e.g. by burning, does not necessarily result in reduced evapotranspiration or increased run-off. The comparison of observed and computed values (Fig. 22.2) for an experimental grassland catchment indicates little change in evapotranspiration and run-off following burning. Obviously, the ranking of grassland, forest and bare soil for water harvest will depend on the interplay of various site-specific factors. Grassland management can be an important determinant in such rankings (see below).

Vegetation cover of grassland can vary greatly with season, and this can have important hydrological consequences. Firstly, cover greatly affects infiltration rate and so influences surface run-off (Ursic and Thames, 1960; Aston and Dunin, 1979).

TABLE 22.2

Generalized values of transpiration rate ($E_t$), evaporation rate ($E_1$), the ratio of latent heat to available radiant energy ($\lambda E/A$), the Bowen ratio ($\beta$), canopy resistance ($r_c$) and boundary-layer resistance ($r_a$) for wet and dry canopies of temperate forest and of crops (including grassland) during summer (after McNaughton and Jarvis, 1983)

| Measure | Forest | | Crops | |
|---|---|---|---|---|
|  | dry | wet | dry | wet |
| $E_t$ (mm h$^{-1}$) | 0.7 | 0.9 | 1.4 | 0.6 |
| $E_1$ (mm h$^{-1}$) | 0.3 | 0.2 | 0.6 | 0.1 |
| $\lambda E/A$ | 0.2–0.6 | 0.6–4 | 0.7–1.2 | 0.7–1.2 |
| $\beta (H/\lambda E)$ | 0.5–4 | ±0.5 | ±0.5 | ±0.5 |
| minimum $r_c$ (sec m$^{-1}$) | 40–100 | <5 | 20–60 | <5 |
| $r_a$ (sec m$^{-1}$) | 5–10 | 5–10 | 20–200 | 20–200 |

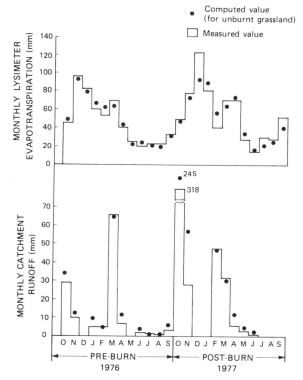

Fig. 22.2. Monthly values of evapotranspiration and run-off from grassland at Krawarree, N.S.W., Australia after burning, compared with predicted values based on unburnt grassland.

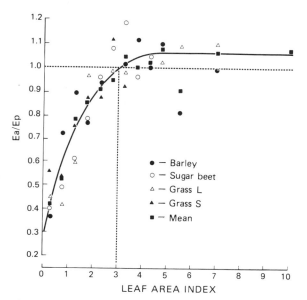

Fig. 22.3. The relation between relative transpiration rate and leaf area index of grassland and crops (after Kristensen, 1974).

Secondly, transpiration increases with leaf area index to a maximum at LAI 3 (Fig. 22.3), as might be expected theoretically (Davidson and Philip, 1956), though an opposite trend in soil evaporation partially offsets this (Ritchie, 1972; Denmead, 1973). Management greatly affects the vegetation cover of grasslands, and consequently has a strong influence on their water balance.

Changes in outflow from grassland catchments may induce very large changes in water quality. Seasonal variation in cover causes erosion rates to vary by at least four orders of magnitude (Moss, 1979); peak rates are associated with sparse cover, rather than bare soil (Schumm, 1977; Singer et al., 1981). Grassland catchments can therefore be major contributors to river sediment. For example, grassland in the Parwan Valley in Victoria, Australia, supplied only 3% of the water in the Werribee River basin, but was the major source of sedimentation (Hexter et al., 1956). Management can greatly affect the export of material, either in suspension or solution, from catchment systems.

## MANAGEMENT INFLUENCES ON GRASSLAND HYDROLOGY

### Agronomic practices

Management practices which increase pasture growth usually increase transpiration (De Wit, 1958; Fischer and Turner, 1978; Tanner and Sinclair, 1983) and also reduce surface run-off (Boughton, 1970); in consequence, water yield is diminished.

Natural grasslands, which are often deficient in nutrients, are conservative in water use (Ripley and Saugier, 1978; Dunin and Reyenga, 1978). The replacement of native grassland by exotic species, usually with applications of fertilizer, increases herbage production and reduces catchment run-off (Dunin, 1970), though withholding fertilizer can increase run-off yield (Hibbert, 1969). If the improved pasture is conservatively stocked, soil may also be conserved.

The widespread use of legumes and phosphatic fertilizer over southern Australia has increased pasture production, resulting in serious reductions of surface run-off for local water supply (De Laine and Vasey, 1961). Preservation of native grassland on marginal hill areas, often with shallow soil, assures water supply which can then be used for supplementary irrigation on more favoured areas (Geddes, 1960).

Deep-rooted grassland species continue to utilize water during drought periods, and so reduce run-off and percolation (Dreibelbis and Amerman, 1964). However, under other conditions, the greater evapotranspiration of shallow-rooted, annual species during the period from autumn until spring reduces annual run-off, compared to that from catchments with deeper-rooted species, and increases the possibilities of water percolation and hence leaching losses, so affecting pollution of the water supply (Dunin and Downes, 1962; Dunin, 1970). Aerial topdressing of hill country with fertilizer, in a mediterranean-type climate of southern Australia, has resulted in the replacement of native perennial grasses with exotic annuals; in this case, the reduced yield of run-off during the growing season is associated with potentially greater erosion during summer, when reduced plant cover causes greater peak flows (Rowe, 1967; Dunin, 1975). Conservative management, with low fertilizer input and low stocking rates, can maintain a balanced sward of native grasses and annual legumes, and so achieve catchment stability under these conditions (Downes, 1958).

## Grazing management practices

Over-grazing of grassland can cause soil erosion (Downes, 1959; Eyles, 1977); several factors contribute to this. Heavy grazing reduces vegetation cover and so reduces infiltration; soil compaction by treading also results in lower infiltration rates (Rhoades et al., 1964; Rauzi and Hanson, 1966; McCarty and Mazurak, 1976; Gifford and Hawkins, 1978). Soil properties are affected more by the number of grazing animals than by the animal species (Hughes et al., 1968). The resulting increase in run-off increases soil erosion, which is accelerated by the depleted cover and increased availability of material, due to trampling (Olive and Walker, 1982). Surface crusting then occurs and reduces still further the capacity of the soil to retain and transmit water (McIntyre, 1958; Hillel and Gardner, 1969, 1970).

A physical interpretation of the hydrological consequences of grazing has proved difficult (Boughton, 1970; Gifford and Hawkins, 1978), but the deleterious consequences of over-grazing can often be halted, and measures taken to restore infiltration characteristics (Gifford and Hawkins, 1978). Cover standards (Costin et al., 1960) have been defined to specify tolerable limits of grazing in fragile systems to avoid irreversible deterioration. These standards are derived empirically, and vary depending on the type of plant cover (Condon et al., 1969) and soil fertility conditions (Toebes et al., 1968). The heavy reliance on empirical methods in combating erosion due to grazing is exemplified in the "universal soil loss equation" (Wischmeier and Smith, 1978), which forms the major basis for specifying conservative grazing practice in U.S.A.

## Soil conservation practices

Mechanical and agronomic conservation practices are designed to maximize retention of precipitation *in situ*, so as to maintain plant growth and prevent erosion. These practices are claimed to benefit small catchments, and to give manageable flows of water of acceptable quality further down the river system (Ogrosky, 1963).

The efficacy of mechanical control measures has been difficult to prove, despite several catchment experiments. Contour banks and terraces have only a minor impact on annual flow from small catchments (Boughton, 1970). Dense networks of banks and terraces have reduced local flooding but have not significantly reduced flow in associated river systems (Sharp et al., 1966). Mechanical control of flood peaks diminishes further downstream in much the same way as flood routing imposed by major regulatory control structures in river systems (Leopold and Maddocks, 1954).

Agronomic practices, in combination with mechanical works, can cause reductions as large as 50% in annual flows from grassland catchments (Sharp et al., 1959), though these effects are slight at the river-basin scale (Leont'yevskiy, 1968; Boughton, 1970). The major impact of this combined approach to erosion control in small catchments is to ensure a hydrologic regime attenuated in both annual yield and in peak flows throughout the year (Adamson, 1974). Large seasonal variation in stream-flow, caused by agronomic measures, can be partially offset by suitable mechanical control.

## Fire and drought

Deliberate burning of grassland is widely practised in an attempt to increase pasture growth (e.g. Groves, 1974), or make new season's growth more accessible to grazing animals (Tothill, 1971), and is prescribed in the western United States to maintain the grazing resource (Wright and Bailey, 1980). Burning has been justified on the basis that it does not affect soil and water loss (Wright et al., 1976); if this is correct, it suggests that evapotranspiration is not affected by burning. This is confirmed by data in Fig. 22.2. However, repeated burning is likely to accelerate erosion (Downes, 1958; Costin et al., 1960), increase nutrient losses (Tiedemann et al., 1979) and reduce vigour of the grassland (Shaw, 1957). In addition, fire may make soils water-repellent, and thus reduce infiltration (De Bano et al., 1970).

Extensive flooding commonly follows drought, as in eastern Australia during the autumn of 1983; this suggests that infiltration is reduced after drought. Furthermore, drought-stricken grassland is commonly overgrazed, which may lead to increased run-off, though other studies (Aston and Dunin, 1980a; Fig. 22.4) indicate that run-off after drought may be little different from predicted values.

## Water quality

Recent concern for the environment had led to increasing interest in the quality, as well as the quantity, of water collected from grassland. Flow rate and availability of erodible material are the major factors controlling the amount of material transported; grassland management can affect both these factors. Loss of plant cover affects flow rate first, but when more cover is lost, sediment transport increases rapidly, and the quality of run-off water deteriorates. Water quality is also affected by leaching. Irrigation, changed land-use and fertilizer use can all increase the solute concentration of drainage water.

Although grassland management at the local level may not have large effects on river flow, it can greatly affect water quality (McCarty, 1967). Simple linear relations between quality and quantity

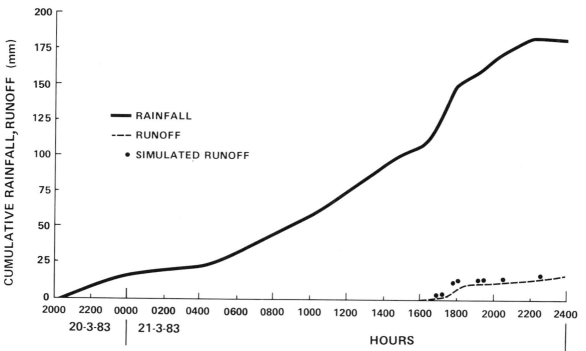

Fig. 22.4. Cumulative rainfall and run-off following drought at Krawarree, N.S.W. Model predictions of run-off (●) are superimposed on the plot.

offer little understanding of contamination in river systems (Olive and Walker, 1982), and more attention needs to be focussed on the few discrete events which affect the export of suspended (Adamson, 1974) and dissolved material (McColl, 1979). Fire and grazing normally affect water quality more than they affect flow rate (McColl, 1979; Tiedemann et al., 1979).

Transport of soluble and insoluble material tends to occur together, because of the plant nutrients adsorbed on the particulate material (Cullen and O'Loughlin, 1982). These losses of plant nutrients in suspension can be of similar magnitude to losses by leaching from the root zone (Sharpley and Syers, 1979). The vertical flow of water through the soil is very slow (Chapman et al., 1982); this poses major problems in identifying potential pollution (Bouwer et al., 1984), and thus making it difficult to plan management so as to protect groundwater systems (Peck and Hurle, 1973). One of the most serious problems of groundwater contamination is salinization, though a wider range of pollutants are increasingly found in groundwater as a result of human activity. Secondary salinization is becoming a serious problem in mediterranean climates, especially in southern Australia (Cope, 1958; Bettenay et al., 1964; Williams, 1982); early agrarian settlement in the Middle East produced similar problems. In southern Australia, the affected area exceeds 500 000 ha (Williams, 1982); it includes some of the more productive land and endangers regional water supply. In Western Australia, the phenomenon is associated with replacement of deep-rooted native woodland (Wood, 1924) by shallow-rooted pasture (Peck, 1975; Sharma et al., 1982). Drainage has increased from near zero to about 50 mm yr$^{-1}$ (Peck, 1975), with the result that high concentrations of solutes in the deep lateritic profiles have been mobilized and appear in streamflow (Williamson and Bettenay, 1979; Stokes and Loh, 1982). Rising water-tables have also led to secondary salinization in outflow areas within five years of the forest replacement (Sharma et al., 1982).

It seems, therefore, that changes in the management of grassland which increase the quantity of water outflow may have a disproportionately large effect on the quality of the outflow. This is true of both surface and groundwater flows. Although there has been considerable success in improving the quality of surface run-off, the problems of polluted groundwater flow generally still await resolution.

## HYDROLOGIC MODELS OF GRASSLAND MANAGEMENT

This chapter has indicated some of the difficulties encountered in predicting changes in catchment outflow associated with grassland management. Even broad generalizations about the direction of change have been questioned in some conditions. Realistic models of water transport in catchments offer a means of extrapolating from past results, often gained from experimental catchments, to more general use. Such models should allow the identification of those factors which can be influenced by management, and the magnitude of their effect if changed. However, doubts have been raised concerning the realism of some models. The conceptual framework of the Stanford model (Crawford and Linsley, 1962) has gained widespread acceptance, and many variants of this prototype have evolved, each with specialized requirements for information.

One of the most critical features of any model is its ability to represent evapotranspiration behaviour (Chapter 5). Several procedures for modelling evapotranspiration of grassland catchments have been proposed and validated over the last decade (e.g. Ritchie et al., 1976). The model by Aston and Dunin (1980b) has been used to predict changes in river flow caused by the replacement of natural grassland with improved pasture in a river basin. The model satisfactorily simulates the behaviour of both small catchments and river basins. The proportional reduction in annual flow, predicted as a result of pasture improvement, progressively diminishes with increased river flow (Fig. 22.5). Thus, the trade-off for greater pastoral productivity appears to be a loss in water yield during critical periods of low flow, but without any benefit in mitigating major floods. A complementary analysis of the effect of pasture improvement on animal output is essential if conflicting objectives are to be reconciled in catchment systems. Complex models may then be needed to determine the effects of management on run-off and erosion (e.g. Foster et al., 1980) and water yield (e.g. Aston and Dunin, 1980b) as well as agricultural output (e.g. Christian et al., 1978).

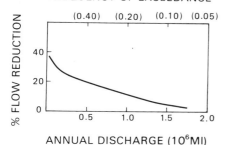

Fig. 22.5. Reduction in run-off, attributable to pasture improvement, in relation to the annual river flow in the Upper Shoalhaven, N.S.W. Flow frequency is also indicated.

Three-dimensional models which represent flow at depth as well as over the surface can provide a rational basis for manipulating both surface flow and percolation (Freeze, 1971). Such models should indicate the effect of surface management on the relationship between surface flow and groundwater recharge. Management strategies can then be planned to minimize pollution of groundwater. This may help to identify critical areas of contamination, and to suggest extraction rates that reduce groundwater pollution.

## CONCLUDING REMARKS

Changes in the management of grassland catchments can change the quantity and quality of outflow, but it is still difficult to predict accurately the effect of specific management practices on the quantity and quality of water flow. Modelling procedures, incorporating interactions between vegetation characteristics and water flow, are needed to predict the outcome of management regimes on plant productivity and water flow. Such models are being evolved (Chapter 5), but the effects of grassland management, especially on the quality of outflow, demand complex formulations. The effect of vegetation on groundwater recharge often extend beyond the root zone. There is now a need to combine both surface and subsurface flows in the description of outflow from catchments. Contemporary modelling is responding to these needs, and there is a prospect that control of the quantity and quality of catchment outflow will be increased by using the results of these studies to improve catchment management.

## REFERENCES

Adamson, C.M., 1974. Effects of soil conservation treatment on runoff and sediment loss from a catchment in southwestern New South Wales, Australia. In: *Proc. Symp. Effects of Man on the Interface of the Hydrological Cycle with the Physical Environment. IASH Publ.*, 113: 1–14.

Aston, A.R. and Dunin, F.X., 1979. Coupled infiltration and surface runoff on a 5 ha experimental catchment, Krawarree, N.S.W. *Aust. J. Soil Res.*, 17: 53–64.

Aston, A.R. and Dunin, F.X., 1980a. The prediction of water yield from a 5 ha experimental catchment, Krawarree, N.S.W. *Aust. J. Soil Res.*, 18: 149–162.

Aston, A.R. and Dunin, F.X., 1980b. Land-use hydrology: Shoalhaven, N.S.W. *J. Hydrol.*, 48: 71–87.

Bettenay, E.A., Blackmore, V. and Hingston, F.J., 1964. Aspects of the hydrologic cycle and related salinity in the Belka Valley, W.A. *Aust. J. Soil Res.*, 2: 187–210.

Bosch, J.M. and Hewlett, J.D., 1982. A review of catchment experiments to determine the effect of vegetation on water yield and evapotranspiration. *J. Hydrol.*, 55: 3–23.

Boughton, W.C., 1970. *Effects of Land Management on Quantity and Quality of Available Water.* Aust. Water Resources Council Research Project 68/2. Univ. of New South Wales, Water Research Laboratory, Manly Vale, N.S.W., Rep. No. 120, 330 pp.

Bouwer, H., Bowman, R.S. and Rice, R.C., 1985. Effect of irrigated agriculture on underlying groundwater. In: F.X. Dunin, G. Matthes and R.A. Grass (Editors), *Relation of Groundwater Quality and Grassland Quality.* Proc. Symp. I.A.H.S., Hamburg, Publ. 146: 13–20.

Chapman, T.G., Bliss, P.J. and Smalls, I.C., 1982. Water quality considerations in the hydrological cycle. In: E.M. O'Loughlin and P. Cullen (Editors), *Prediction in Water Quality.* Aust. Acad. Sci., Canberra, A.C.T., pp. 27–88.

Christian, K.R., Freer, M., Donnelly, J.R. and Armstrong, J.S., 1978. *Simulation of Grazing Systems.* PUDOC, Wageningen, 115 pp.

Colville, J.S. and Holmes, J.W., 1972. Water table fluctuations under forest and pasture in a karstic region of southern Australia. *J. Hydrol.*, 17: 61–80.

Condon, R.W., Newman, J.C. and Cunningham, G.M., 1969. Soil erosion and pasture degeneration in central Australia. *J. Soil Con. Serv. N.S.W.*, 25: 47–92.

Cope, F., 1958. *Catchment Salting in Victoria.* Soil Conservation Authority, Melbourne, Vic., 66 pp.

Cordery, I. and Pilgrim, D.H., 1979. Small catchment design — present problems and guidelines for the future. In: *Hydrology and Water Resources Symposium. Inst. Eng. Aust. Natl. Conf. Publ.*, 79/10: 189–193.

Costin, A.B., 1980. Runoff and nutrient losses from an improved pasture at Ginninderra, Southern Tablelands, New South Wales. *Aust. J. Agric. Res.*, 31: 533–546.

Costin, A.B., Wimbush, D.J. and Kerr, J.D., 1960. Studies in catchment hydrology in the Australian Alps II. Surface runoff and soil loss. *C.S.I.R.O. Div. Plant Ind. Tech. Pap.*, No. 14: 31 pp.

Crawford, N.H. and Linsley, R.K., 1962. The synthesis of continuous stream hydrographs on a digital computer. *Stanford Univ. Dep. Civ. Eng. Tech. Rep.*, No. 12.

Cullen, P. and O'Loughlin, E.M., 1982. Non-point sources of pollution. In: E.M. O'Loughlin and P. Cullen (Editors), *Prediction in Water Quality*. Aust. Acad. Sci., Canberra, A.C.T., pp. 437–453.

Davidson, J.L. and Philip, J.R., 1956. Light and pasture growth. In: *Climatology and Micrometeorology. Proc. Canberra Symp.* UNESCO, Paris, pp. 181–187.

De Bano, L.F., Mann, L.D. and Hamilton, D.A., 1970. Translocation of hydrophobic substances into soil by burning organic litter. *Soil Sci. Soc. Am. Proc.*, 34: 130–133.

De Laine, R.J. and Vasey, G.H., 1961. *A Survey of Farm Water Supplies in Central Parts of Victoria*. University of Melbourne, Melbourne, 66 pp.

Denmead, O.T., 1973. Relative significance of soil and plant evaporation. In: *Plant Response to Climatic Factors. Uppsala Symp., 1970*. UNESCO, Paris, pp. 505–511.

De Wit, C.T., 1958. Transpiration and crop yields. *Versl. Landbouwkd. Onderz.*, 64.6: 88 pp.

Downes, R.G., 1958. Land management problems following disturbance of the hydrologic balance of environments in Victoria, Australia. In: *7th Tech. Meet. International Union for Conservation of Nature and Natural Resources, Athens*, 19 pp.

Downes, R.G., 1959. The ecology and prevention of soil erosion. In: A.L. Keast (Editor), *Biogeography and Ecology in Australia*. Monographiae Biologicae Vol. VIII. Junk, Den Haag, pp. 472–486.

Dreibelbis, F.R. and Amerman, C.R., 1964. Land use, soil type and practice effects on the water budget. *J. Geophys. Res.*, 69: 3387–3393.

Dunin, F.X., 1970. Changes in water balance components with pasture management in southeastern Australia. *J. Hydrol.*, 10: 90–102.

Dunin, F.X., 1975. Changes in the water balance with land modification in southern Australia. In: H.F. Heady, D. Falkenberg and J.P. Riley (Editors), *Watershed Management on Range and Forest Lands*. Utah State University, Logan, Utah, pp. 157–163.

Dunin, F.X. and Downes, R.G., 1962. The effect of subterranean clover and Wimmera Ryegrass in controlling surface runoff from four-acre catchments near Bacchus Marsh, Victoria. *Aust. J. Exper. Agric. Anim. Husb.*, 2: 148–152.

Dunin, F.X. and Reyenga, W., 1978. Evaporation from a *Themeda* grassland I. Controls imposed on the process in a sub-humid environment. *J. Appl. Ecol.*, 7: 317–325.

Eyles, R.L., 1977. Changes in drainage networks since 1820, Southern Tablelands, N.S.W. *Aust. Geogr.*, 13: 377–386.

Fischer, R.A. and Turner, N.C., 1978. Plant productivity in the arid and semiarid zones. *Annu. Rev. Plant Physiol.*, 29: 277–317.

Foster, G.R., Lane, L.J., Nowlin, J.D., Lafler, J.M. and Young, R.A., 1980. A model to estimate sediment yield from field-sized areas development of model. In: *CREAMS — a Field Scale Model for Chemicals, Runoff and Erosion from Agricultural Management Systems, 1. Model Documentation*. Conserv. Research Report No. 26. U.S.D.A. Science and Educ. Admin., Washington, D.C., pp. 36–64.

Freeze, R.A., 1971. Three-dimensional, transient saturated-unsaturated flow in a groundwater basin. *Water Resour. Res.*, 7: 347–366.

Geddes, H.J., 1960. The storage and use of irrigation water on the individual holding. Farrer Memorial Oration. *N.S.W. Agric. Gaz.*, 71: 526–544.

Gifford, G.F. and Hawkins, R.H., 1978. Hydrologic impact of grazing on infiltration. *Water Resour. Res.*, 14: 305–313.

Groves, R.H., 1974. Growth of *Themeda australis* grassland in response to firing and mowing. *Field Stn. Rec. Div. Plant Ind. C.S.I.R.O. (Aust.)*, 13: 1–7.

Hexter, G.W., Leslie, T.I. and Pels, S., 1956. *A Land Survey of the Parwan Valley*, Victoria. Soil Conservation Authority of Victoria Publ., Melbourne, Vic., 32 pp.

Hibbert, A.R., 1969. Water yield changes after converting a forested catchment to grass. *Water Resour. Res.*, 5: 634–640.

Hillel, D. and Gardner, W.R., 1969. Steady infiltration into crust topped profiles. *Soil Sci.*, 108: 137–142.

Hillel, D. and Gardner, W.R., 1970. Transient infiltration into crust topped profiles. *Soil Sci.*, 109: 69–76.

Hughes, J.D., McClatchy, D. and Hayward, J.A., 1968. *Cattle in South Island Hill and High Country*. Tussock Grasslands and Moutain Lands Inst., Lincoln College, Lincoln, 229 pp.

Kristensen, K.J., 1974. Actual evapotranspiration in relation to leaf area. *Nordic Hydrol.*, 5: 173–182.

Leont'yevskiy, B.B., 1968. Problem of allowing for the influence of agricultural melioration practices on flow in the construction of a general plan for the utilization and protection of the water resources of the U.S.S.R. *Sov. Hydrol., Select. Pap.* No. 4: 374–380 (Am. Geophys. Union).

Leopold, L.B. and Maddocks, T., 1954. *The Flood Control Controversy; Big Dams, Little Dams and Land Management*. Ronald Press, New York, N.Y., 278 pp.

McCarty, P.L., 1967. Sources of nitrogen and phosphorus in water supplies. *J. Am. Water Works Assoc.*, 59: 344–363.

McCarty, M.K. and Mazurak, A.P., 1976. Soil compaction in eastern Nebraska after 25 years of cattle grazing management and weed control. *J. Range Manage.*, 29: 384–386.

McColl, R.H.S., 1979. Factors affecting downslope movement of nutrients in hill pasture. *Progr. Water Technol.*, 11: 271–285.

McIntyre, D.S., 1958. Soil splash and formation of surface crusts by raindrops impact. *Soil Sci.*, 85: 261–266.

McNaughton, K.G. and Jarvis, P.G., 1983. Predicting effects of vegetation changes on transpiration and evaporation. In: T.T. Kolowksi (Editor), *Water Deficits and Plant Growth*, 7. Academic Press, London, pp. 1–47.

Moss, A.J., 1979. Thin-flow transportation of solids in arid and non-arid areas. A comparison of processes. In: *The Hydrology of Areas of Low Precipitation. Proc. Canberra Symposium*. IASH Publ., No. 128: pp. 435–446.

Ogrosky, H.O., 1963. Effects of conservation on water yield. *J. Soil Water Conserv.*, 18: 1173–1183.

Olive, L.J. and Walker, P.H., 1982. Processes in overland flow. Erosion and production of suspended material. In: E.M. O'Loughlin and P. Cullen (Editors), *Prediction in Water Quality*. Aust. Acad. Sci., Canberra, A.C.T., pp. 87–119.

Pearce, A.J., Gash, J.H.C. and Stewart, J.B., 1980. Rainfall interception in a forest stand estimated from grassland meteorological data. *J. Hydrol.*, 46: 147–163.

Peck, A.J., 1975. Interactions between vegetation and stream water quality in Australia. In: H.F. Heady, D.M. Falkenberg and J.P. Riley (Editors), *Watershed Management on Range and Forest Lands*. Utah State University, Logan, Utah, pp. 149–155.

Peck, A.J. and Hurle, D.H., 1973. Chloride balance of some farmed and forested catchments in southwestern Australia. *Water Resour. Res.*, 9: 648–657.

Philip, J.R., 1957. Evaporation, and moisture and heat fields in the soil. *J. Meteorol.*, 14: 354–366.

Rahkmanov, V.V., 1970. Dependence of stream flow upon the percentage of forest cover of catchments. In: *Proc. Joint FAO/USSR Int. Symp. Forest Influences and Watershed Management, Moscow*, pp. 55–70.

Rauzi, T. and Hanson, C.L., 1966. Water intake and runoff as affected by intensity of grazing. *J. Range Manage.*, 19: 351–355.

Rhoades, E.D., Locke, L.F., Taylor, H.M. and McIlvain, E.H., 1964. Water intake on a sandy range as affected by 20 years of differential cattle stocking rates. *J. Range Manage.*, 17: 185–189.

Ripley, E.A. and Saugier, B., 1978. Biophysics of a natural grassland: Evaporation. *J. Appl. Ecol.*, 15: 459–479.

Ritchie, J.T., 1972. Model for predicting evaporation from a row crop with incomplete cover. *Water Resour. Res.*, 8: 1204–1213.

Ritchie, J.T., Rhoades, E.D. and Richardson, C.W., 1976. Calculating evaporation from native grassland watersheds. *Trans. Am. Soc. Agric. Eng.*, 19: 1098–1103.

Rodda, J.C., 1976. Basin studies. In: J.C. Rodda (Editor), *Facets of Hydrology*. Wiley, London, pp. 257–297.

Rowe, R.K., 1967. *A Study of the Land in the Victorian Catchment of Lake Hume*. Soil Conservation Authority of Victoria, Melbourne, Vic., 219 pp.

Schumm, S.A., 1977. *The Fluvial System*. Wiley, New York, N.Y., 338 pp.

Shanan, L., Tadmor, N.H. and Evenari, M., 1958. The ancient desert agriculture of the Negev. II. Utilization of the runoff from small watersheds in Abde (Avdat) region. *J. Agric. Res. Stn., Rehovot*, 9: 107–128.

Sharma, M.L., Johnston, C.D. and Barron, R.J.W., 1982. Hydrologic responses to forest clearing in paired catchment study in southwestern Australia. In: E.M. O'Loughlin and L.K. Bren (Editors), *The First National Symposium on Forest Hydrology*. Inst. Eng. Aust. — Aust. For. Council, Canberra, A.C.T., pp. 118–123.

Sharp, A.L., Gibbs, A.E. and Owen, W.J., 1966. *Development of a Procedure for Estimating the Effects of Land and Watershed Treatments on Stream Flow*. Tech. Bull. No. 1352. U.S. Dep. Agric. — U.S. Dept. Int., Washington, D.C., 57 pp.

Sharp, A.L., Owen, W.J. and Gibbs, A.E., 1959. *Two-Year Progress Report: Co-operative Water Yield Procedures Study*. U.S. Dep. Agric., Washington, D.C., 203 pp.

Sharpley, A.N. and Syers, J.K., 1979. Phosphorus inputs into a stream draining an agricultural watershed. II. Amounts contributed and relative significance of runoff types. *Water, Air Soil Pollut.*, 11: 417–428.

Shaw, N.H., 1957. Bunch spear grass dominance in burnt pasture in southeastern Queensland. *Aust. J. Agric. Res.*, 8: 325–334.

Singer, M.J., Walker, P.H., Hutka, J. and Green, P., 1981. Soil erosion under simulated rainfall and runoff at varying cover levels. *C.S.I.R.O. Aust. Div. Soils Rep.*, No. 55.

Slatyer, R.O., 1961. Methodology of a water balance study conducted on a desert woodland (*Acacia anuera* F. Muell.) community in central Australia. *Arid Zone Res.*, 16: 15–24.

Stewart, J.B., 1977. Evaporation from the wet canopy of a pine forest. *Water Resour. Res.*, 13: 915–921.

Stokes, R.A. and Loh, I.C., 1982. Streamflow and solute characteristics of a forested and deforested catchment pair in southwestern Australia. In: E.M. O'Loughlin and L.K. Bren (Editors), *The First National Symposium on Forest Hydrology*. Inst. Eng. Aust. — Aust. For. Council, Canberra, A.C.T., pp. 60–66.

Tanner, C.B. and Sinclair, T.R., 1983. Research or re-search. In: H.M. Taylor, W.R. Jordan and T.R. Sinclair, *Limitations to Efficient Water Use in Crop Production*. Am. Soc. Agron., Madison, Wis., pp. 1–27.

Tiedemann, A.R., Conrad, C.E., Dietrich, J.H., Hornbeck, J.W., Megahan, W.F., Vierick, L.A. and Wade, D.D., 1979. Effects of fire on water — a state-of-knowledge review. *U.S.D.A. For. Serv. Tech. Rep.*, W 0-10: 28 pp.

Toebes, C., Scarf, F. and Yates, M.E., 1968. Effects of cultural changes on Makara Experimental Basin. *Int. Assoc. Sci. Hydrol.*, 13(3): 95–122.

Tothill, J.C., 1971. A review of fire in the management of native pastures with particular reference to northeastern Australia. *Trop. Grasslands*, 5: 1–10.

Ursic, S.L. and Thames, J.L., 1960. Effect of cover types and soils on runoff in Northern Mississippi. *J. Geophys. Res.*, 65: 663–667.

Williams, B.G., 1982. The potential for secondary salinization of surface soils. In: E.M. O'Loughlin and P. Cullen (Editors), *Prediction in Water Quality*. Aust. Acad. Sci., Canberra, A.C.T., pp. 407–415.

Williamson, D.R. and Bettenay, E., 1979. Agricultural land use and its effect on catchment output of salt and water — evidence from southern Australia. *Progr. Water. Technol.*, 11: 463–483.

Wischmeier, W.H. and Smith, D.D., 1978. *Predicting Rainfall Erosion Losses — a Guide to Conservation Planning*. U.S.-D.A. Agric. Handbook No. 537. U.S. Dep. Agric., Washington, D.C., 58 pp.

Wood, W.E., 1924. Increase of salt in soil and streams following the destruction of the native vegetation. *J. R. Soc. W. Aust.*, 10: 35–47.

Wright, H.A. and Bailey, A.W., 1980. *Fire Ecology and Prescribed Burning in the Great Plains — A Research Review*. U.S.D.A. For. Serv. Gen. Tech. Rep. INT-77, 61 pp. Inter. For. Range Exper. Stn., Ogden, Utah.

Wright, H.A., Churchill, F.M. and Stevens, W.C., 1976. Effect of prescribed burning on sediment, water yield and water quality from grazed juniper lands in Central Texas. *J. Range Manage.*, 29: 294–298.

Section IV

# SYSTEMS MANAGEMENT

This final section concentrates on the management factors which determine the economically important output for managed grassland ecosystems: this is usually animal products such as beef, milk and sheepmeat (see Chapters 9–12).

The effectiveness of management regimes for managed grasslands must ultimately be assessed in economic terms, since this is a major (but not sole) criterion for the farmer. Most experimental comparisons of management practices, such as fertilizer application, are assessed in terms of primary productivity (see Chapter 6). Comparisons of animal output are much less frequent, because of the size and cost of grazing experiments, and economic assessments of these results are even less common.

This section begins with some considerations of the economics of production and utilization of herbage (Chapter 23). Economic analyses of grassland productivity proves to be much more complex than those for crop productivity, mainly because herbage usually has no direct economic value, and must first be converted into animal products. In addition, the responses to some important grassland management factors (such as pasture renovation and drainage) can be long-term and subject to considerable year-to-year variation; special economic techniques are needed to cope with this. The three other chapters in this section (Chapters 24–26) deal with the more important management factors determining economic output.

The context of this final section is agricultural, rather than ecological. Although these chapters have little relevance to natural ecosystems, many aspects are relevant to the management of semi-natural grassland ecosystems that are used agriculturally, and some aspects are relevant to such ecosystems managed for wildlife conservation.

Chapter 23

# ECONOMIC CONSIDERATIONS IN THE PRODUCTION AND UTILIZATION OF HERBAGE

C.J. DOYLE

## INTRODUCTION

The production and utilization of herbage on farms involves decisions about both the level of inputs into herbage production and the contribution herbage will make to livestock feed requirements. Whereas an understanding of the biological relationships linking inputs with herbage yields and livestock outputs is necessary for effective decision-making, the decisions themselves must be based on economic considerations. Thus, an increase in herbage yields is only of interest to livestock producers if it leads to higher profits. By the same token, maximizing physical output is only justified if it increases profits. More often than not, the level of inputs that is justified economically is lower than the level necessary to maximize output — that is, the biological optimum. Fig. 23.1 shows the response of herbage yields to fertilizer nitrogen (N) at a site in the United Kingdom, with average herbage-growing conditions (Corrall et al., 1982), and also the associated profits per hectare assuming nitrogen fertilizer costs of £340 per tonne[1] and a value for herbage of £34 per tonne dry matter (t DM). While maximum yield is attained at around 500 kg N ha$^{-1}$, maximum profits are realized at around 350 kg N ha$^{-1}$. The fact that biological and economic optima do not coincide reinforces the need to understand how economic considerations impinge on grassland management decisions.

Economic analysis presents special problems in the grassland context. First, herbage is not a traded commodity and so, in contrast to maize, barley or wheat, it has no actual market price. Accordingly, it is frequently difficult to attach a value to additional grass production. Second, many management decisions, such as whether to resow a pasture or to drain a field, involve investments whose full benefits are only realized over several years. For the investor, the length of time he has to wait to realize the benefits is likely to be as important as the size of the benefits. In other words, for the investor there is likely to be a financial cost attached to time, which should be taken into account in conducting economic evaluations. Third, decisions about grassland management take place in an environment of uncertainty, both as to the yield of herbage and to the prices realized for livestock products. In such an environment, maximizing profits may not be the only consideration in deciding on a particular course of action. Most producers are averse to financial risks, so decisions are often influenced by a desire to minimize them. Thus, the traditional models of economic behaviour, which assume that producers are profit-maximizers, may not be appropriate.

In this review of the economic considerations involved in grassland management, attention has been focused on five aspects: (1) valuing increased herbage production; (2) determining the optimal level of inputs; (3) optimizing the use of herbage as a feed; (4) valuing economic benefits arising at different times; and (5) assessing the impact of risk and uncertainty on decisions.

## VALUING INCREASED GRASS PRODUCTION

Apart from small quantities of hay and artificially dried herbage, very little herbage is sold off

---

[1]Prices throughout the text are based on those ruling in the United Kingdom in 1982.

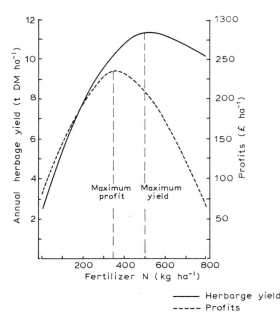

Fig. 23.1 The effect of nitrogen fertilizer on herbage yields and profits (based on data from Corrall et al., 1982).

the farm and so it does not have a directly quotable market price. Instead, the economic value attached to an increase in herbage production must be estimated indirectly, from what is known of the quantities of dry matter needed to produce a given quantity of milk or meat (Middleton, 1973; Doyle, 1982). Estimates of the value of the extra herbage production are arrived at by deducting the costs of all inputs other than herbage from the value of the projected increase in livestock production. However, the resultant value for herbage will depend upon the type of livestock activity being considered, and the current level of returns from that activity.

Let us consider the problem of valuing an increase in herbage production arising from higher usage of fertilizer nitrogen (Chapter 6), and used to increase stocking rates of either dairy cows or beef steers. Fertilizer trials in the United Kingdom have shown that an extra 150 kg N ha$^{-1}$ increases production of herbage/dry matter (DM) by about 3000 kg DM ha$^{-1}$ yr$^{-1}$ (Corrall et al., 1982). Given that the annual herbage requirements of dairy cows and beef steers are typically 3250 and 2220 kg DM head$^{-1}$ yr$^{-1}$ respectively (Doyle and Elliott, 1983), it is possible to calculate the additional number of animals that can be carried, and hence estimate the additional income. Assuming that animals utilize 60% of the extra herbage produced, the potential increases in dairy cow and beef steer numbers would be 0.55 and 0.82 head ha$^{-1}$ respectively. The resultant net increase in profits per hectare in the United Kingdom, at 1982 prices, are shown in Table 23.1. If the whole of these residual profits are attributed to herbage, the derived values for herbage are £59.7 and £34.3 per tonne DM respectively (Table 23.1) — a difference of nearly 75%. A more comprehensive analysis by Doyle and Elliott (1983) shows that the apparent value of herbage may vary by as much as sevenfold, according to the end-use of the herbage.

The financial gains from increasing herbage production depend not only on the use to which it is put, but also on the season of production. For most livestock systems, both the availability of herbage and demands for it are continually fluctuating (Chapters 9, 10 and 11), so its economic value changes over the season (Auld et al., 1979; Doyle and Elliott, 1983). Where herbage production occurs mainly in a period when its availability already exceeds supply, it may be comparatively worthless. In contrast, where the increase in herbage production occurs in a period of shortage, it may allow a significant increase in overall stocking rates and, in consequence, may be extremely valuable. Marked seasonal differences in the value of herbage therefore may be expected.

TABLE 23.1

The calculated values of herbage for dairy and beef enterprises at 1982 U.K. price levels

| | Dairy | Beef |
|---|---|---|
| Financial benefits, £ head$^{-1}$ | | |
| sales of livestock and milk | 960 | 430 |
| less livestock purchases | −150 | −120 |
| less purchased feed | −260 | −130 |
| less miscellaneous costs | −45 | −20 |
| less annual costs of additional investment in housing, etc. | −180 | −35 |
| (1) Net profit | 325 | 125 |
| (2) Increase in stocking rate, head ha$^{-1}$ | 0.55 | 0.82 |
| (3) Increase in net profits, £ ha$^{-1}$ [(1) × (2)] | 179 | 103 |
| (4) Imputed value of herbage in the field, £ (tonne DM)$^{-1}$ [(3) ÷ 3 tonnes] | 59.7 | 34.3 |

# ECONOMY OF GRASS PRODUCTION

Because the value of herbage depends on the season in which it is produced, and the use to which it is put, estimates of its value are only valid for the specific context in which it has been calculated. Accordingly, considerable caution must be exercised in extending the use of any imputed value for herbage from the specific to the general.

## OPTIMAL LEVEL OF INPUTS

A major reason for deriving a value for herbage is to allow calculations of the optimal level of inputs into herbage production. This basically involves finding the level of inputs which will maximize profits. A thorough treatment can be found in several textbooks (Heady, 1952; Dillon, 1968; Doll et al., 1968; Hardaker et al., 1970; Upton, 1976), but the essential principles may be illustrated by the following example. Given a typical relationship between herbage yield per hectare ($Y$, in t DM ha$^{-1}$) and fertilizer nitrogen ($N$, in kg ha$^{-1}$) in the United Kingdom of the type:

$$Y = 2.0 + 0.03N - 0.00003N^2 \qquad (1)$$

then it is possible to derive a corresponding relationship for the profit ($Z$) — that is, revenue less fixed and variable costs — resulting from different amounts of nitrogen fertilizer applied:

$$Z = pY - cN = p(2.0 + 0.03N - 0.00003N^2) - cN \qquad (2)$$

where $p$ is the value of herbage and $c$ is the cost of nitrogen; the cost of application is not often included, but is small. Valuing herbage and fertilizer nitrogen at £34 (t DM)$^{-1}$ and £0.34 kg$^{-1}$ respectively. Fig. 23.2A. shows the profit at various levels of nitrogen input. The corresponding marginal revenue ($MR$) — value of the extra herbage produced by an additional kilogram of nitrogen — is also shown at each nitrogen level (Fig. 23.2B). Mathematically, $MR$ may be expressed as:

$$MR = p(dY/dN) = p(0.03 - 0.00006N) \qquad (3)$$

The marginal cost ($MC$) of each additional kilogram of nitrogen remains constant at £0.34.

It is evident from Fig. 23.2 that the highest profits ($Z$) occur at the point where the marginal revenue equals the marginal cost, namely:

$$p(dY/dN) = c \qquad (4)$$

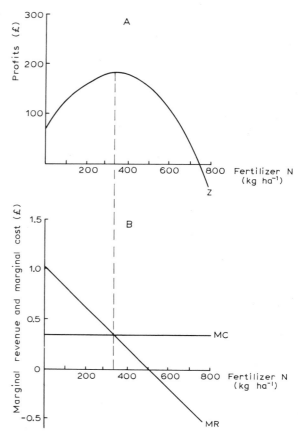

Fig. 23.2. The effect of nitrogen fertilizer on profits ($Z$), marginal revenue ($MR$) and marginal cost ($MC$).

where $c$ is the cost of nitrogen. Rearranging terms this gives:

$$dY/dN = c/p \qquad (5)$$

Thus, the net returns from nitrogen application are maximized at the point where the marginal response of herbage to nitrogen is equal to the ratio of the nitrogen cost to the value of herbage.

### Optimal input of two or more resources

Farmers must usually manipulate several inputs concurrently. In such cases, herbage yield ($Y$) can be expressed as a function of several inputs ($X_1, X_2 \ldots X_n$) and the optimal level of each input $X_i$ is given by:

$$\partial Y / \partial X_i = c_i / p \qquad (6)$$

where $\partial Y / \partial X_i$ is the partial derivative of $Y$ with respect to $X_i$ (all other inputs kept fixed), whereas $c_i$ and $p$ are the respective prices of input $X_i$ and herbage. The response to nitrogen (N), phosphorus (P) and potassium (K) may be described by a quadratic function of the type:

$$Y = b_0 + b_1 N^{0.5} + b_2 P^{0.5} + b_3 K^{0.5}$$
$$+ b_4 (NP)^{0.5} + b_5 (NK)^{0.5} + b_6 (PK)^{0.5}$$
$$+ b_7 N + b_8 P + b_9 K \qquad (7)$$

From equation (6) the conditions for maximizing profits per hectare are then:

$$\frac{\partial Y}{\partial N} = \frac{(b_1 + b_4 P^{0.5} + b_5 K^{0.5})}{2 N^{0.5}} + b_7 = \frac{c_N}{p} \qquad (8a)$$

$$\frac{\partial Y}{\partial P} = \frac{(b_2 + b_4 N^{0.5} + b_6 K^{0.5})}{2 P^{0.5}} + b_8 = \frac{c_P}{p} \qquad (8b)$$

$$\frac{\partial Y}{\partial K} = \frac{(b_3 + b_5 N^{0.5} + b_6 P^{0.5})}{2 K^{0.5}} + b_9 = \frac{c_K}{p} \qquad (8c)$$

where $p$, $c_N$, $c_P$ and $c_K$ are respectively the unit values or costs for herbage, nitrogen, phosphorus and potassium. Using iterative techniques it is possible to solve simultaneously equations 8a–8c and determine the optimal levels of N, P and K.

In fertilizer trials conducted at Koonwaara in Australia (Colwell, 1977), the respective coefficients for the first harvest cut were:

$b_0 = 1021.205 \quad b_1 = -17.145 \quad b_2 = 112.493$
$b_3 = -41.773 \quad b_4 = 15.233 \quad b_5 = 4.151$
$b_6 = -2.868 \quad b_7 = 0 \quad b_8 = -5.431$
$b_9 = 8.089$

Substituting these values into equations 8a–8c and taking values per kg for herbage ($p$), nitrogen ($c_N$), phosphorus ($c_P$) and potassium ($c_K$) of £0.034, £0.34, £0.34 and £0.17 respectively, the optimal levels of N, P and K are about 20, 27.5 and 38.5 kg ha$^{-1}$ respectively.

If the application of potassium (K) is held constant at 38.5 kg ha$^{-1}$, Fig. 23.3 shows how profit (revenue – costs) responds to changes in the applications of nitrogen and phosphorus. It is apparent from this that net returns do not differ significantly over a wide range of fertilizer usage, and that various combinations of nitrogen and phosphorus give similar results. Thus, a producer applying 17.5 kg ha$^{-1}$ of both nitrogen and phos-

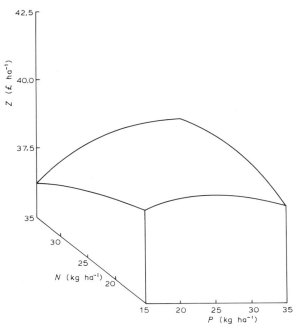

Fig. 23.3. The effect of applications of nitrogen fertilizer and phosphorus fertilizer on the net returns per hectare ($Z$) from grassland (based on data by Colwell, 1977).

phorus will realize the same net profit as one applying 35 kg ha$^{-1}$ of both nitrogen and phosphorus. Given such a situation, two farmers seeking to maximize profits may arrive at very different input levels.

## OPTIMIZING THE USE OF HERBAGE AS A FEED

The management of grassland involves decisions concerning the relative contributions of herbage and purchased feeds to the diet of animals. For a given level of livestock output, this involves identifying the least-cost combination of feed inputs.

Consider the case where the combinations of corn silage ($S$) and concentrates ($C$) needed to increase the liveweight of beef steers from 370 to 400 kg are described by the following relationship (Bhide et al., 1980):

$$C = 202.3428 - 1.27875 S + 0.00615 S^2 \qquad (9)$$

where $C$ and $S$ are the total amounts (kg) of concentrates and silage DM needed to achieve the weight change. Taking the respective unit costs for home-grown concentrates ($p_C$) and silage ($p_S$) as

# ECONOMY OF GRASS PRODUCTION

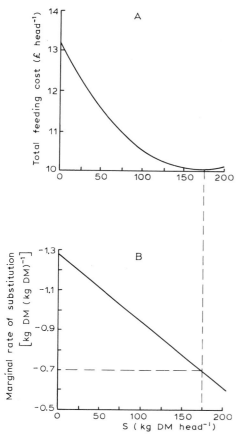

Fig. 23.4. A. The effect of changes in the amounts of silage fed to beef steers to produce a given liveweight change on total feed costs. B. The marginal rate of substitution of silage for concentrates at each level of silage feeding.

£65 and £45.6 (t DM)$^{-1}$, the total costs of achieving the given liveweight change with different levels of silage feeding are shown in Fig. 23.4A. The lowest costs of production occur at a silage input of 175 kg DM head$^{-1}$. Also shown (Fig. 23.4B) is the *marginal rate of substitution* of silage for concentrates at each level of silage feeding, which may be expressed mathematically as:

$$dC/dS = -1.27875 + 0.0033S \quad (10)$$

Thus, at a silage input of 100 kg DM head$^{-1}$, feeding an extra 1.00 kg DM of silage will reduce the requirements for concentrates by 0.95 kg DM. At the optimal feeding level for silage, namely 175 kg DM head$^{-1}$, the marginal rate of substitution is 0.7, which is also the ratio of silage-to-concentrate prices, 45.6/65. This illustrates a basic principle: for a given level of output ($Y$), the least-cost combination of two inputs, $X_1$ and $X_2$, occurs where the marginal rate of substitution equals the inverse ratio of the prices, or

$$dX_1/dX_2 = -p_2/p_1 \quad (11)$$

where $p_1$ and $p_2$ are the respective unit costs of $X_1$ and $X_2$ (Ritson, 1977, pp. 115–119; Nix, 1980).

## Optimizing inputs and output simultaneously

In practice, interest is not so much in the combination of inputs which will minimize the costs of producing a given level of output, but in the combination of inputs which will maximize profits. The following example, based on the work of Østergaard (1979), examines the optimal combination of concentrates and silage for high-yielding dairy cows, during the first twenty-four weeks of lactation. Relationships for total dry-matter intake ($TDMI$: kg DM day$^{-1}$), milk yield ($Y$: kg FCM day$^{-1}$) and liveweight gain ($LWG$: kg day$^{-1}$) in terms of concentrate supply ($C$: kg DM day$^{-1}$) were derived as follows:

$$TDMI = 12.07 + C - 0.034C^2 \quad (12)$$

$$Y = 12.57 + 2.45C - 0.158C^2 \quad (13)$$

$$LWG = -0.098 + 0.006C + 0.008C^2 \quad (14)$$

If it is assumed, for simplicity, that the only feeding stuffs fed to the animals are concentrates ($C$) and silage ($S$), then the daily intake of silage in kg DM day$^{-1}$ may be expressed as

$$S = 12.07 - 0.034C^2 \quad (15)$$

Applying the rule that the optimal level of concentrate feeding per cow is found where the value of the marginal increase in output equals the marginal increase in feeding costs, gives

$$p_m \delta Y + p_g \delta LWG = p_S \delta S + p_C \delta C \quad (16)$$

where $\delta Y$, $\delta LWG$, $\delta S$ and $\delta C$ are small increments in milk, liveweight gain, silage and concentrates and $p_m$, $p_g$, $p_S$ and $p_C$ are the respective unit values. Differentiating equations 12, 13 and 14 and substituting into equation 16, gives the optimal input of concentrates ($C_{opt}$) as

$$C_{opt} = \frac{2.45p_m + 0.006p_g - p_C}{0.316p_m - 0.016p_g - 0.068p_S} \quad (17)$$

TABLE 23.2

Optimal combinations of concentrates (kg DM day$^{-1}$) and silage (kg DM day$^{-1}$) fed per cow at various prices for the two feeds

| Price of concentrates ($p_C$) [£ (kg DM)$^{-1}$] | Price of silage ($p_S$) [£ (kg DM)$^{-1}$] | | | | | |
|---|---|---|---|---|---|---|
| | 0.034 | | 0.051 | | 0.068 | |
| | concentrates fed | silage[1] fed | concentrates fed | silage[1] fed | concentrates fed | silage[1] fed |
| 0.065 | 9.18 | 9.20 | 9.54 | 8.98 | 9.93 | 8.72 |
| 0.098 | 8.11 | 9.83 | 8.43 | 9.65 | 8.77 | 9.45 |
| 0.130 | 7.07 | 10.37 | 7.34 | 10.24 | 7.64 | 10.09 |

[1] Silage intake is derived by substituting the value of $C_{opt}$ from equation 17 into equation 15.

Taking prices for milk ($p_m$) of £0.14 kg$^{-1}$ and for liveweight gain ($p_g$) of £0.70 kg$^{-1}$, Table 23.2 shows how variations in the prices for concentrates ($p_C$) and silage ($p_S$) alter the optimal combination of concentrates and silage.

## VALUING ECONOMIC BENEFITS ARISING AT DIFFERENT TIMES

In the analysis so far considered, the benefits have been assumed to be immediate. However, for most major capital investments, such as land drainage or irrigation, there is likely to be a significant delay between capital expenditure and realization of the benefits. This has important economic implications, since equivalent sums of money received at different points in time need to be valued differently (Mishan, 1971; Organisation of Economic Co-operation and Development, 1972; Ritson, 1977, pp. 286–297). A given sum of money received in a year's time is not seen as the same thing as an equivalent sum received today, so a borrower has to compensate a lender by paying *interest*. The rate of interest ($i$) may be seen as the price paid by the borrower for exchanging the possibility of purchasing a particular good today for the same good in a year's time. If the interest rate ($i$) remains constant, then a unit of money available at once can be exchanged for $(1+i)^n$ units in $n$ years' time. Conversely, one unit of money in $n$ years' time is equivalent to $1/(1+i)^n$ units of money today. In this way a series of annual earnings $E_0$, $E_1$, ... $E_n$ can be reduced to a single value ($V$), termed the *net present value*, as follows:

$$V = E_0 + \frac{E_1}{(1+i)} + \frac{E_2}{(1+i)^2} \cdots + \frac{E_n}{(1+i)^n} \quad (18a)$$

or:

$$V = \sum_{t=0}^{t=n} E_t/(1+i)^t \quad (18b)$$

where $i$, the interest rate, is termed the *discount rate*. In the particular case where the annual earnings are constant over time, so that $E_0 = E_1 = E_n$, then $V$ may be expressed as:

$$V = \frac{1 - \frac{1}{(1+i)^n}}{1 - \frac{1}{1+i}} E_0 \quad (19)$$

where $i$ is greater than zero.

One situation in which the discounting of future earnings might be employed is in deciding whether to drain an area of grassland. On poorly drained land, drainage normally increases both herbage production and the overall utilization of the available herbage. In trials reported by Berryman (1975), the drainage of grassland increased the liveweight gain for a beef enterprise by 120 kg ha$^{-1}$ yr$^{-1}$. At an expected price of £0.9 kg$^{-1}$, this represents an increase in annual revenue of £108 ha$^{-1}$. Against this must be set the costs of drainage, say £750 ha$^{-1}$. Whether drainage is worthwhile depends on whether the long-term benefits ($V$) exceed the costs. Confining attention to a ten-year period, Fig. 23.5 shows the estimated values for $V$ at different discount rates ($i$) using equation 19, and assuming a constant expected annual benefit ($E_0$) of £108 ha$^{-1}$. If annual returns are not

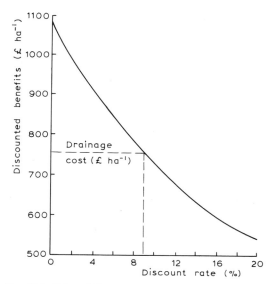

Fig. 23.5. Value of discounted benefits from draining grassland at various discount rates (see text for details of assumptions).

discounted, ($i=0$), drainage is a profitable course of action. However, at discount rates above 9% ($i=0.09$) the opposite appears true. Thus, "discounting" or applying a lower weight to future earnings may alter impressions about the most profitable course of action.

Other instances in which "discounting" might be applied are in the evaluation of the costs and benefits of irrigating fodder crops (Doyle, 1981) and in the assessment of the economic viability of controlling weed species in grassland (Vere and Campbell, 1977).

## RISK AND UNCERTAINTY

So far it has been assumed that the response to various actions, such as drainage or applying fertilizer, is known with certainty. In reality, farmers operate in a world of uncertainty, and this can have a significant influence on their production and investment decisions (Francisco and Anderson, 1972; Bond and Wonder, 1980). Thus, the optimal level of inputs into grassland may fall quite markedly as the effects are considered in terms of profit as opposed to yield, and in terms of financial risks as opposed to profit (Fig. 23.6). As a result, many government policies, such as measures to stabilize prices, are aimed at reducing risks. However, comparatively little is known about farmers' attitudes to risk-taking, mainly because there are significant differences between individual farmers (Young, 1979). Nonetheless, risk is an integral part of all real business decisions, and must be considered alongside profit maximization in evaluating courses of action.

In general, farmers prefer a steady income to one which oscillates between extremes of high and low, even where the latter has a higher mean value. Thus, in deciding to make an investment, the producer will consider both the expected income or profits and the expected variance of the income. Theoretically, it is possible to express the way in which each individual discounts expected profits for the risks involved by means of a *utility function* (Officer and Halter, 1968; Dillon, 1971; Morris, 1974). A popular form of utility function is the quadratic (MacArthur and Dillon, 1971), namely:

$$U(Z) = E(Z) - b\{[E(Z)]^2 + \text{Var}(Z)\} \qquad (20)$$

This basically states that the utility or subjective benefit derived from a risky income, $U(Z)$, is a function of the expected net income, $E(Z)$ and the variance of that income, $\text{Var}(Z)$. Where a farmer is indifferent to risk ($b=0$), equation 20 reduces to $U(Z) = E(Z)$, — that is, maximizing profits is equivalent to maximizing utility. On the other hand, for a producer who is averse to risk, $b$ will be greater than zero and so the strategy chosen will be one giving less than the maximum expected profits.

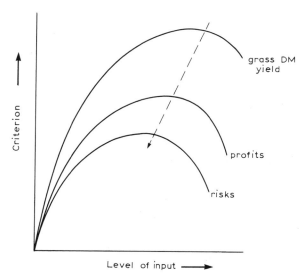

Fig. 23.6. The optimum level of an input as defined by yield, profit and risk criteria.

However, as a measure of the relative satisfaction afforded by incomes with different degrees of riskiness, utility is not an absolute measure of value in the same way that monetary value is. Thus, it is not possible to interpret absolute differences in measured utility between two different utility functions as implying that one individual derives greater absolute satisfaction than another. All that a utility function allows is a comparison for a particular individual of the relative satisfaction afforded by different financial outcomes.

**Impact of risk on the optimal input**

Consider a case where expected herbage yield in kg ha$^{-1}$, $E(G)$, is related to nitrogen usage ($N$) by the function:

$$E(G) = 2000 + 30N - 0.03N^2 \tag{21}$$

At any given level of $N$, herbage yield will vary from year to year, due to climatic variation (Chapters 2 and 6). Let the variance of herbage yield, $Var(G)$, be proportional to the square of the expected yield, so that:

$$Var(G) = 0.02[E(G)]^2 \tag{22}$$

Given values for herbage and nitrogen of £0.034 (kg DM)$^{-1}$ and £0.34 kg$^{-1}$ respectively, expected profits, $E(Z)$, are simply:

$$E(Z) = 0.034 E(G) - 0.34 N \tag{23}$$

whereas the variance of profits, $Var(Z)$, is given by:

$$Var(Z) = (0.034)^2 Var(G) \tag{24}$$

Fig. 23.7 shows the utility values, $U(Z)$, derived from equation 20 for various nitrogen applications ($N$) and different risk aversion coefficients ($b$). If the producer wishes to maximize profits ($b=0$), the optimum nitrogen application is 350 kg ha$^{-1}$. However, if he is highly averse to risk ($b=0.004$), the optimum nitrogen application is 100 kg ha$^{-1}$. Substituting these values for $N$ into equation 23, profits would be £181 and £126 ha$^{-1}$ respectively. Thus, risk considerations lead to lower inputs and lower profits.

**Impact of risk on the optimal combination of inputs**

When several inputs are involved, considerations of risk will result in greater use of those

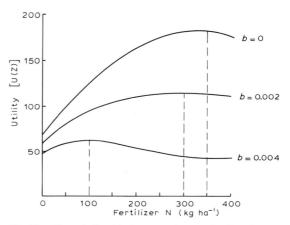

Fig. 23.7. The relationship between nitrogen fertilizer input and utility for various values of the risk-aversion coefficient, $b$.

resources which give a predictable response compared to those giving an unpredictable response. Consider the case of a livestock producer deciding on the amount of concentrate supplement and fertilizer nitrogen that he will use. Let the expected or average livestock output per hectare [$E(Y)$, kg] be a function of expected herbage yields [$E(G)$, kg DM ha$^{-1}$] and the level of concentrate feeding ($C$, kg ha$^{-1}$), so that:

$$E(Y) = C^{0.6}[E(G)]^{0.2} \tag{25}$$

This implies that, for a given expected herbage yield, each 1% increase in concentrates increases livestock output by 0.6%, whereas, for a given concentrate use, each 1% increase in herbage yield increases livestock output by 0.2%. The mean and variance of herbage yields are presumed to be described by the relationships given in equations 20 and 21. Expected profits $E(Z)$, in £ ha$^{-1}$, that is revenue less the costs of concentrates and nitrogen, are then given by:

$$E(Z) = p_Y E(Y) - p_C C - p_N N \tag{26}$$

where $p_Y$, $p_C$ and $p_N$ are the respective unit values of livestock output, concentrates and nitrogen. The variance of profits, $Var(Z)$, may be approximately derived from equations 22, 25 and 26 (Kendall and Stuart, 1958, p. 232) to give:

$$Var(Z) = Var(G)\left[0.2 p_Y \frac{E(Y)}{E(G)}\right]^2 \tag{27}$$

For a given expected output, $E(Y)$, the problem of determining the optimal inputs of concentrates

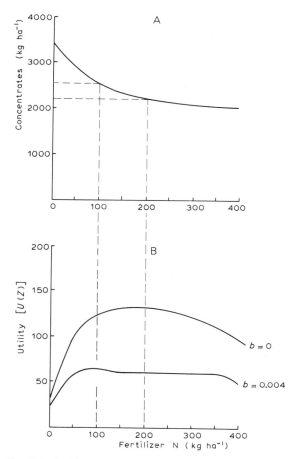

Fig. 23.8. A. The combinations of concentrates and fertilizer nitrogen needed to achieve a livestock output of 600 kg ha$^{-1}$. B. The corresponding utility values for different risk-aversion coefficients, $b$, showing the optimum combinations.

and nitrogen involves determining the combination which maximizes producer's utility, as estimated from equation 20. Thus, Fig. 23.8A shows the combinations of concentrates and nitrogen which will produce an expected livestock output of 600 kg ha$^{-1}$ yr$^{-1}$. Assuming $p_Y = £0.09$ kg$^{-1}$, $p_C = £0.15$ kg$^{-1}$ and $p_N = £0.34$ kg$^{-1}$, the corresponding utility values are shown in Fig. 23.8B. If the producer wishes to maximize profits ($b=0$), the optimum combination is 200 kg N ha$^{-1}$ and 2250 kg ha$^{-1}$ of concentrates. If the producer is averse to risk ($b=0.004$), then the optimum combination is 100 kg N ha$^{-1}$ and 2550 kg ha$^{-1}$ of concentrates. Risk considerations appear to reduce the nitrogen applied and increase the concentrate usage. This may explain, in part, the observation that livestock producers generally apply less nitrogen and use more concentrates than is apparently economically justified.

## CONCLUSIONS

This brief survey indicates some of the complexities of grassland management, when considered from an economic viewpoint. In particular, it illustrates that what is technically possible is not always economic. As a result, producers are unlikely to operate at the point where physical output is maximized, and technical advances which increase output may not translate into proportionate increases in output on farms. For the livestock producer, increased production is not an end *per se*, but only a means to realizing greater profits. However, increased profits are often only a means to another end, namely a higher degree of satisfaction in life. If this is the criterion, then other considerations, such as ease of management and considerations of risk, become important. In a world of uncertainty, both as to forage production and the prices to be realized for livestock products, management decisions may be as much a reflection of risk aversion or ease of management as of profits.

## ACKNOWLEDGEMENTS

I am grateful to John Thornley, John Nix and Roy Snaydon for helpful comments on earlier drafts and to Martin Ridout for statistical advice.

## REFERENCES

Auld, B.A., Menz, K.M. and Medd, R.W., 1979. Bioeconomic model of weeds in pastures. *Agro-Ecosystems*, 5: 69–84.

Berryman, C., 1975. *Improved Production from Drained Grassland*. Agricultural Development and Advisory Service. Field Drainage Experimental Unit Technical Bulletin 75/7. Ministry of Agriculture Fisheries and Food, Pinner, 14 pp.

Bhide, S., Epplin, F., Heady, E.O. and Melton, B.E., 1980. Direct estimation of gain isoquants: An application to beef production. *J. Agric. Econ.*, 31: 29–44.

Bond, G. and Wonder, B., 1980. Risk attitudes amongst Australian farmers. *Aust. J. Agric. Econ.*, 24: 16–34.

Colwell, J.D., 1977. *National Soil Fertility Project, 1*. C.S.I.R.O. Division of Soils, Adelaide, S.A., 195 pp.

Corrall, A.J., Morrison, J. and Young, J.W.O., 1982. Grass production. In: C. Thomas and J.W.O. Young (Editors), *Milk from Grass*. ICI Agricultural Division, Billingham and the Grassland Research Institute, Hurley, 114 pp.

Dillon, J.L., 1968. *The Analysis of Response in Crop and Livestock Production*. Pergamon Press, Oxford, 135 pp.

Dillon, J.L., 1971. Interpreting systems simulation output for management decision-making. In: J.B. Dent and J.R. Anderson (Editors), *Systems Analysis in Agricultural Management*. Wiley and Sons, Sydney, N.S.W., pp. 85–119.

Doll, J.P., Rhodes, V.J. and West, J.G., 1968. *Economics of Agricultural Production, Markets and Policy*. Irwin, Homewood, 557 pp.

Doyle, C.J., 1981. Economics of irrigating grassland in the United Kindgom. *Grass Forage Sci.*, 36: 297–306.

Doyle, C.J., 1982. Economic evaluation of weed control in grassland. In: *Proc. British Crop Protection Conference, Weeds 1982*. British Crop Protection Council, London, pp. 419–427.

Doyle, C.J. and Elliott, J.G., 1983. Putting an economic value on increases in grass production. *Grass Forage Sci.*, 38: 169–177.

Francisco, E.M. and Anderson, J.R., 1972. Chance and choice west of the Darling. *Aust. J. Agric. Econ.*, 16: 82–93.

Hardaker, J.B., Lewis, J.N. and MacFarlane, G.C., 1970. *Farm Management and Agricultural Economics*. Angus and Robertson, Sydney, N.S.W., 201 pp.

Heady, E.O., 1952. *Economics of Agricultural Production and Resource Use*. Prentice Hall, New Jersey, N.J., 850 pp.

Kendall, M.G. and Stuart, A., 1958. *The Advance Theory of Statistics, 1: Distribution Theory*. Butler and Tamer, London, 433 pp.

MacArthur, I.D. and Dillon, J.L., 1971. Risk, utility and the stocking rate. *Aust. J. Agric. Econ.*, 15: 20–35.

Middleton, K.R., 1973. Monetary value of pasture, especially in relation to fertilizer trials. *N.Z. J. Agric. Res.*, 16: 503–507.

Mishan, E.J., 1971. *Cost–Benefit Analysis*. Allen and Unwin, London, 251 pp.

Morris, J., 1974. The utility approach to making decisions under uncertainty. *Oxford Agrar. Aff.*, 3: 15–28.

Nix, J., 1980. Economic aspects of grass production and utilisation. In: W. Holmes (Editor), *Grass: Its Production and Utilization*. Blackwell, Oxford, pp. 216–238.

Officer, R.R. and Halter, A.N., 1968. Utility analysis in a practical setting. *Am. J. Agric. Econ.*, 50: 257–277.

Organisation of Economic Co-operation and Development, 1972. *Manual of Industrial Project Analysis, 1: Methodology and Case Studies*. OECD Development Centre, Paris, 451 pp.

Østergaard, V., 1979. *Strategies for Concentrate Feeding to Attain Optimum Feeding Levels in High Yielding Dairy Cows*. National Institute of Animal Science, Copenhagen, 138 pp.

Ritson, C., 1977. *Agricultural Economics: Principles and Policy*. Granada, London, 409 pp.

Upton, M., 1976. *Farm Production Economics and Resource-Use*. Oxford University Press, Oxford, 357 pp.

Vere, D.T. and Campbell, M.H., 1977. *Investment Considerations in the Control of Serrated Tussock (Nasella trichotoma) on the Central Tablelands of New South Wales*. Technical Bulletin 13, Department of Agriculture, New South Wales, Sydney, 16 pp.

Young, D.L., 1979. Risk preferences of agricultural producers: their use in extension and research. *Am. J. Agric. Econ.*, 61: 1063–1070.

Chapter 24

# STOCKING RATE

DAVID H. WHITE

## INTRODUCTION

The *stocking rate* on grasslands is the number of animals per unit area of land, irrespective of the amount of forage available; the *grazing pressure* or *grazing intensity*, however, refers to the number of animals per unit of available forage (Mott, 1960). "*Carrying capacity*" is the stocking rate at the optimum grazing pressure, but this term can be misleading for reasons to be discussed later.

The stocking rates normally supported on natural or "climax" grassland are generally low. These low stocking rates are largely determined by the low productivity of such grasslands, the need to maintain a stable botanical composition and prevent dominance by unpalatable plants, and the need to prevent erosion. The stocking rate that can be supported by man-made and intensively managed grassland can be several times as great (e.g. Cotsell, 1956). The increases are largely due to greater productivity of the pasture or improved herbage quality, mainly caused by applications of phosphatic or nitrogenous fertilizers, and sometimes by the introduction of more productive or palatable species (Chapter 8).

Stocking rate is now recognized as one of the most powerful management tools available to the producer, allowing him to regulate the amount of herbage available to his animals throughout the year. Increasing the stocking rate reduces the available feed per animal, and so normally reduces production per animal. However, this is usually more than compensated for by the greater number of animals, so that at low or moderate stocking rates the production per unit area of land is substantially increased (McMeekan, 1956; Tribe and Lloyd, 1962), as shown in Fig. 24.1. Before considering this relationship in detail, I shall consider the factors affecting this response.

Stocking rate can affect the botanical composition and productivity of the pastures, the structure and fertility of the soil and, as a result, animal output. It is essential that these effects are considered before examining how stocking rate interacts with other management variables, and how various biological and economic parameters can be used to determine the stocking rate to be adopted in a particular animal-production enterprise.

## STOCKING RATE AND PASTURE

Pasture growth rates are often reduced at high stocking rates, in both set-stocked and rotationally grazed systems (Campbell, 1969; Carter and Day, 1970; Cayley et al., 1980; Reeve and Sharkey, 1980). The reduction in leaf area caused by more intense grazing pressure can result in reduced interception of solar radiation, and hence lower photosynthetic activity in the sward (Davidson and Philip, 1956; Vickery, 1973). The carbohydrate reserves in the roots and crowns of grasses may also be reduced (El Hassan and Krueger, 1980). These aspects are considered in more detail in Chapter 7. At the other extreme, pasture growth rates are reduced at low stocking rates, because of increased senescence and a greater abundance of old leaves that are photosynthetically less efficient (Chapter 7). Pasture growth may also be reduced under these conditions by exhaustion of soil moisture (Noy-Meir, 1978). Between these extremes of herbage availability, pasture growth rates are usually not greatly affected by stocking rate (Langlands and Bennett, 1973a; Baker et al., 1975).

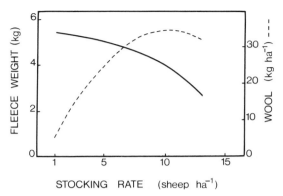

Fig. 24.1. The effect of stocking rate on fleece weight and wool production per hectare (derived from White and McConchie, 1976).

Higher stocking rates often increase the proportion of less palatable species in the pasture (e.g. Davies, 1962; Sharkey et al., 1964; Cameron and Cannon, 1970; Fitzgerald, 1976; Smoliak, 1974). A reduction in the proportion of palatable legume species is likely to reduce nitrogen fixation, and hence the productivity of the sward (Hilder, 1966). Where grazing pressure is excessive, the pasture community becomes unstable and the system "crashes" (Morley, 1966; Noy-Meir, 1975; Anonymous, 1979).

## STOCKING RATE AND SOIL PROPERTIES

High stocking rates usually increase the bulk density of soil and decrease pore space in the surface soil, thus reducing the rate of water infiltration (Rauzi and Hanson, 1966; Langlands and Bennett, 1973a; see also Chapter 20). This compaction occurs as a direct effect of trampling, through a decline in root mass in the surface soil, and through reduced protection of the soil from rainfall (Chapter 20). Changes in the chemical composition of the soil with stocking rate have not been as extensively studied, but appear to be slight (Langlands and Bennett, 1973a).

## STOCKING RATE AND ANIMAL NUTRITION

The direct effect of higher stocking rate is to limit animal intake, especially at those times of the year when herbage is in short supply (Arnold and Dudzinski, 1967; Alden and Whittaker, 1970). This is particularly important when the physiological demands of breeding animals are at their highest, as in late pregnancy and early lactation (Chapters 9–11). Adjustment of lambing or calving dates may be necessary to match nutritive demands with feed supply.

Stocking rate also affects herbage digestibility and botanical composition, and so has a further indirect effect on intake (Arnold et al., 1977; Birrell, 1981).

### Animal liveweight

The effects of stocking rate are greatest when the feed supply is inadequate. For example, in a mediterranean environment the effect is greatest in winter and least in late spring (Fig. 24.2). Adequate feed is almost always available at the lowest stocking rate, so that feed quality accounts for most of the seasonal variation in liveweight. At the high stocking rate, feed is adequate in spring but deficient in winter, leading to variations in liveweight. The variation is not as great as might be expected, because of the effect of *compensatory growth* (Morley et al., 1978a; White et al., 1980).

### Animal reproduction

For reasons outlined above, high stocking rates usually result in lower animal liveweights. This is particularly important at mating (Egan et al., 1977; Reeve and Sharkey, 1980), since ovulation rate is closely associated with maternal liveweight (Allden and Lamming, 1961; Morley et al., 1978b) (Chapter 11). As a result, the number of offspring per female is reduced at high stocking rates of sheep

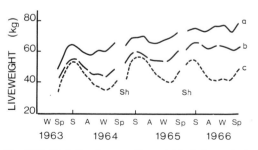

Fig. 24.2. Variation in mean liveweight of sheep stocked at 7.4 (a), 12.4 (b) and 17.4 (c) wethers ha$^{-1}$ at Ruffy in northeastern Victoria, Australia. Data of Cannon (1969). $Sh$ = shearing; $W$, $A$, $Sp$, $S$ = winter, autumn, spring and summer respectively.

(Davies, 1962; Arnold et al., 1971; Egan et al., 1977; Reeve and Sharkey, 1980), and cattle (e.g. Axelsen et al., 1972).

Increases in stocking rate also cause reductions in the milk production of cows (McMeekan and Walshe, 1963; Campbell, 1966). This decrease is associated with a decrease in the ratio of milk produced to feed consumed (Wallace, 1959), and tends to be cumulative from year to year (Bryant and Parker, 1971). Although more milk is usually produced per hectare, proportionately more herbage is consumed by the extra cows, mainly because milk production is reduced by feed deficiency whereas the metabolic requirement of the cows remains reasonably constant. However, the overall efficiency with which pasture is converted into milk is likely to be increased.

### Lamb and calf growth

The growth rates of suckling lambs (Suckling, 1964; Joyce et al., 1976) and calves (Axelsen et al., 1972; Hart, 1972; Leaver, 1974) are usually reduced at high stocking rates. The effect on lambs is greatest during the first six weeks of life (Arnold and Bush, 1962), when the milk yields of the ewes may be affected (Davies, 1963). The effect is greatest if lambing occurs at, or just before, a period of feed deficiency (Davies, 1962; Kenney and Davis, 1974).

A reduction in growth can be critical in meat-production systems, especially if the pastures senesce before the lambs or calves are ready for market. The most important effect on young beef cattle may not be on their growth when feed is plentiful, but whether they gain, maintain or lose weight when the availability of green herbage is limited (Watson et al., 1979). This may delay by months, or even years, the time it takes for animals to reach a marketable weight.

As animals grow, the proportion of fat in their tissues increases, whereas the proportion of bone decreases. Stocking rate appears to have little effect on carcass fat cover, over and above its effect on carcass weight (Arnold et al., 1969; Kirton et al., 1981).

### Fibre production

There is considerable evidence that fleece weight declines with increasing stocking rate, in sheep (Fig. 24.1; Arnold and McManus, 1960; Davies, 1962; Morley et al., 1969) and goats (McGregor, 1984). However, this is partly offset by the higher quality of the wool, with shorter staples, and an increased staple crimp frequency (Sharkey et al., 1962; Cannon, 1972). However, if fibres are too weak, "tender" fleeces will result, particularly in adverse seasons (White and McConchie, 1976; Bircham et al., 1977). These "tender" wools have not suffered a large price penalty in the past (Roberts et al., 1960), but are more likely to do so in future (Rottenbury, 1982).

### Animal health

In addition to greater mortality, due to undernutrition at high stocking rates, wear on incisor teeth is increased, so that the animals have a shorter useful life (Suckling, 1975; Langlands et al., 1984). There is also a greater incidence of pregnancy toxaemia, so that animals must be more closely supervised.

The interactions between livestock numbers, available pasture, parasitic burdens and the environment are exceedingly complex, so that a strong or consistent relationship between stocking rate and parasitological status cannot easily be established (Michel, 1969; Callinan et al., 1982). Nevertheless, close associations between stocking rate and the abundance of the worm *Nematodirus* in lambs (Downey and Conway, 1968; Downey, 1969) and of *Ostertagia* in cattle (Ciordia et al., 1971; Donald et al., 1979) have been found, suggesting that increases in stocking rate may well require increased expenditure on worm control (Morley and Donald, 1980). On the other hand, stocking rate appears to have relatively minor effects on the intensity of infection with the worm *Haemonchus* (Southcott et al., 1970).

## QUANTIFYING ANIMAL OUTPUT IN RELATION TO STOCKING RATE

The quantitative relationship between animal output and stocking rate is still a matter for debate, partly because its shape is obscured by the lack of precision in grazing experiments. As a result, it is difficult to define clear biological and economic optima for stocking rate, since estimation of these

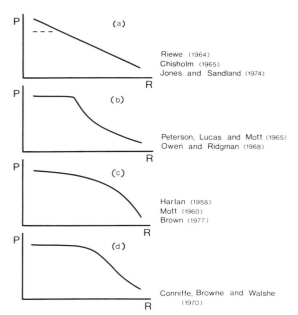

Fig. 24.3. Postulated relationships between production per animal (P) and stocking rate (R).

optima requires rational assumptions concerning the probable shape of the curve. Some postulated relationships between animal production and stocking rate are shown in Fig. 24.3. The various relationships will lead to different estimates of optimum stocking rate (Connolly, 1976; Hodgson, 1971). It is therefore important to investigate how the relationships have been derived and the assumptions on which they are based.

Jones and Sandland (1974) pooled a large amount of experimental data, and found that the ratio (gain per animal/gain at optimum stocking rate) was linearly related to the ratio (stocking rate/optimum stocking rate). They therefore deduced that data for two or three stocking rates were quite sufficient to define the relationship, and so to predict the optimum stocking rate; the stocking rates tested need not even span the optimum. However, Connolly (1976) showed that Jones and Sandland's method would linearize an underlying curvilinear relationship. Such a linear relationship between production per animal and stocking rate (Fig. 24.3a) is computationally appealing and has frequently been suggested (Riewe, 1964; Chisholm, 1965; Cowlishaw, 1969; Langlands and Bennett, 1973b; Jones and Sandland, 1974). This hypothesis assumes that the pasture is homogeneous, and that the quantity of feed available is directly proportional to the area of the pasture (Byrne, 1968). It further assumes that the relationship between production per animal ($P$) and feed available per animal ($F$) approximates a rectangular hyperbola:

$$P = a - b/F \qquad (1)$$

as found by Willoughby (1959) and Arnold and Dudzinski (1967), and that there is an inverse relationship between stocking rate ($R$) and available herbage:

$$R = k/F \qquad (2)$$

It therefore follows that:

$$P = a - b/(k/R) \qquad (3)$$

and thus:

$$P = a - b_1 R \qquad (4)$$

Where $a$, $b$, $b_1$ and $k$ are all constants for a given situation. Linearity also assumes that the large changes in the quantity and quality of herbage available throughout the year (Chapters 2 and 3) do not significantly affect the relationship. This is unlikely to be true in many extensive grazing situations. There are other reasons for doubting the linear relationship. For example, surplus feed is available at low stocking rates; since this feed is less digestible (Birrell, 1981) and there is an upper limit to both intake and daily liveweight gain, it is most unlikely that production per animal will continue to increase linearly at the lowest stocking rates.

The models of Petersen et al. (1965) and Owen and Ridgman (1968) (Fig. 24.3b) assume that at a given stocking rate ($R_0$), feed intake, and hence production, decreases when the total feed available ($TF$) is less than the total voluntary intake of all the animals present. When this stocking rate ($R_0$) is exceeded, production per animal ($P$) is linearly related to the differences between energy intake ($I$) and maintenance requirements ($M$). Thus:

$$P = b(I - M) \qquad (5)$$
$$= b[(TF/R_0) - M]; \; R \leqslant R_0 \qquad (6)$$
$$= b[(TF/R) - M]; \; R > R_0 \qquad (7)$$

Conniffe et al. (1970) extended this concept by dividing the year into periods of restricted and unrestricted feeding. They derived the function:

$$P = A - r\left[R - \left(\frac{R_0}{R}\right)\right]^2; R \geqslant R_0 \quad (8)$$

$$P = A; R < R_0 \quad (9)$$

where $A$ is the maximum possible rate of growth (Fig. 24.3d).

The curvilinear relation between production per animal and stocking rate (Fig. 24.3c) is supported by a considerable number of field studies (e.g. Mott, 1960; White and McConchie, 1976; Brown, 1977; Reeve and Sharkey, 1980; White et al., 1980). Brown (1977) only found a linear relationship (Fig. 24.4a) when stocking rate was confounded with the amount of supplementary feed supplied; he found a curvilinear relationship (Fig. 24.4b) when there was virtually no supplementary feeding. The simple pasture models of Noy-Meir (1978), and more complex simulation models such as those of Arnold and Bennett (1975) and White et al. (1983), which assume sward heterogeneity in both space and time, also support the curvilinear relationship between production and stocking rate.

In conclusion, linearity may be assumed over a small range of stocking rates but, except possibly in environments with both summer and winter rainfall (Langlands and Bennett, 1973b), cannot be presumed over a wide range of stocking rates. If the underlying relationship is curvilinear then the slope of any linear "tangent" will be determined by the range of stocking rates investigated. This can have important consequences, since the estimate of optimum stocking rate will be very sensitive to this slope.

## ECONOMIC RESPONSES TO INCREASING STOCKING RATE

### Estimation of most profitable stocking rates

With the relation between animal production and stocking rate still unresolved, and more difficulties encountered when economics are considered (Chapter 23), it is difficult to define an optimum stocking rate (Blackburn et al., 1973). Tribe and Lloyd (1962) and Chisholm (1965)

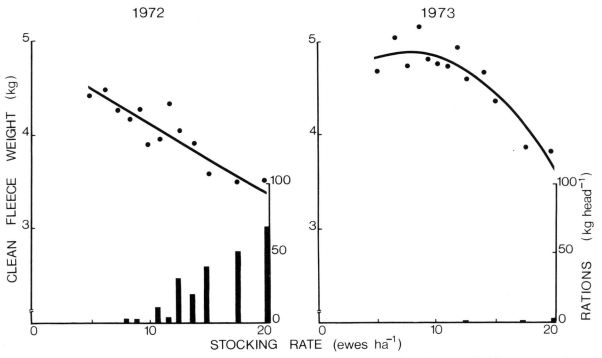

Fig. 24.4. The effect of stocking rate on clean fleece weight in two years. The amounts of supplementary feed (2 parts oats to 1 part hay by weight) used to prevent mortality in each year is also shown. Data of Brown (1977) for ewes with lambs at Kybybolite Research Centre, South Australia.

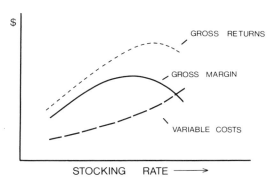

Fig. 24.5. Effect of stocking rate on gross returns, variable costs and gross margin per unit area of land.

pointed out that the stocking rate at which the *gross margin* (i.e. gross returns less *variable costs*) per hectare is maximized is lower than that at which production is maximized (Fig. 24.5), just as in the case for inputs such as fertilizer (Chapter 23). Furthermore, the cost of drought reserves, in areas where year-to-year variations in herbage production are large, forces the long-term stocking rate to a level substantially below the optimum stocking rate for an average season.

It should also be recognized that land is not always the limiting variable, and that expected profit per unit of labour, or per unit of capital invested, can be more decisive criteria in determining whether a stocking rate should be increased (Blackburn, 1969). A lower stocking rate may therefore be adopted, rather than employing extra labour or providing more equipment and facilities.

### Stocking to minimize financial risks

Lloyd (1966) and Jardine (1975) pointed out that, since the top of the gross margin curve is reasonably flat (Fig. 24.5), farmers might choose substantially lower stocking rates to minimize the risk associated with higher stocking rates. McArthur and Dillon (1971) suggested that, in general, farmers seek to maximize the utility or satisfaction of their income, and not their expected net profit (see Chapter 23). Thus risk-averters will use lower stocking rates and accept lower profits for the sake of income stability and a feeling of security. This, of course, does not necessarily imply that such decisions are economically rational. For example, if the stocking rate is too low, the farmer may be unable to increase cash reserves sufficiently in the good seasons to survive the poor seasons (White and Morley, 1977). The choice of stocking rate on predominantly economic grounds should therefore involve a compromise between profit maximization and financial security, and lead to intermediate stocking rates. However, a simulation study by White and Morley (1977) showed that the difference in stocking rate between that which gives most profit, and that which provides most security, is probably not very great (Fig. 24.6).

It is important to consider fixed costs as well as variable costs (White, 1975), and to recognize that the relative prices of inputs and outputs may change with time. In general, the cost of inputs has increased more rapidly than the value of the outputs, so stocking rates have tended to move closer to the most profitable levels.

### Carrying capacity

Mott (1960) defined carrying capacity as the stocking rate at the optimum grazing pressure; it is commonly determined by expressing the number of stock carried during a period of feed shortage in terms of *Ewe Equivalents* (Coop, 1965) or *Dry Sheep Equivalents* (White and Bowman, 1981). Clearly, this is a very subjective estimate which varies according to the severity of the season, the

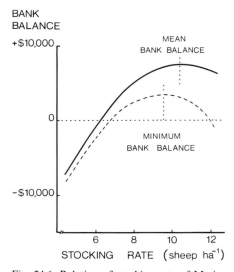

Fig. 24.6. Relation of stocking rate of Merino wethers to the mean and minimum bank balance of a farmer over a number of years. Optima are shown by vertical dotted lines (derived from White and Morley, 1977).

judgement of the farmer, whether or not it is expressed on a monthly or an annual basis, and whether biological or economic criteria are being used to determine the optimum grazing pressure.

## INTERACTION OF STOCKING RATE WITH OTHER MANAGEMENT PRACTICES

Stocking rate cannot be considered in isolation since: (a) other management variables often change concurrently with stocking rate; (b) the stocking rate of one livestock enterprise may affect other enterprises; and (c) other inputs may interact with stocking rate.

### Fertilizer application

Applications of fertilizers supplying nitrogen (Yiakoumettis and Holmes, 1972; Cowan and Stobbs, 1976), phosphate (Carter and Day, 1970) and lime (Bircham et al., 1977) usually increase the amount of herbage available for consumption; the response is typically asymptotic (Chapters 6 and 23). Most of the extra herbage produced must be consumed if an economic benefit is to be gained from the fertilizer (Carter and Day, 1970). The marginal economic returns from increased fertilizer inputs can be substantial if the stocking rate is close to the economic optimum and fertilizer inputs previously small (Morley, 1974), but if the stocking rate is sub-optimal, or the fertilizer inputs already large, little benefit can be expected.

### Lambing and calving dates

The choice of parturition date affects the number and growth rate of the offspring, and the supplementary feed requirements of the flock or herd. The choice of date is affected by seasonal changes in herbage availability (Chapters 9–11), and in the value of products such as prime lamb or veal. Comprehensive field experiments (e.g. Egan et al., 1977; George and Pearse, 1978; Reeve and Sharkey, 1980), backed by simulation studies to cover a wider range of environments and seasons (White et al., 1982, 1983), clearly indicate that these factors can be decisive in determining which date of lambing to adopt, particularly as stocking rates are increased.

### Fodder conservation

The disparity between pasture availability and feed requirements can be reduced by conserving herbage, as hay or silage, in times of plenty for use in times of deficiency (Chapter 26). It is commonly assumed that fodder conservation permits higher stocking rates; however, there is a dilemma in that at high stocking rates, where most conserved fodder is required, there is less surplus to be conserved (Hutchinson, 1971; Bishop and Birrell, 1975). As stocking rate increases, the requirement for supplementary feed increases exponentially (Byrne, 1968).

The benefits of fodder conservation, in terms of animal production, are often negligible and can even be negative (Hamilton, 1974; Egan et al., 1977), mainly because pasture intake declines when supplementary feeding occurs (Allden and Jennings, 1962; Langlands, 1969), and because of the ability of grazing animals to store energy within their body reserves. However, positive responses do occur (Bishop and Birrell, 1975; Birrell et al., 1978), often because of improvements in feed quality caused by cutting, rather than as a direct response to feeding conserved fodder.

Another strategy is to purchase fodder from outside the farm. This may often be preferable to growing it on the farm, particularly given the capital outlay that cropping entails (Morley and Graham, 1971). Whatever feeding strategy is adopted, its value will depend on the stocking rate used.

### Animal genotype

There have been few experiments in which animals of different breeds have been compared over a range of stocking rates, yet measurements of relative intakes, liveweights, efficiency of feed utilization and overall production in the field are essential to identify more desirable breeds for a particular environment. If a farm has to be stocked at a lower rate to support larger or more productive animals, much of the potential economic benefit associated with their use will be lost (Obst, 1986).

Selection for multiple births (Chapter 16) is likely to be associated with greater feed consumption by the dams, and poorer survival and growth rate of the offspring (Joyce et al., 1976; Jones, 1982). These effects are likely to be exacerbated at higher stocking rates (White, 1984).

The effect of selecting particular genotypes on the optimal stocking rates within different environments is poorly understood. The choice will obviously depend upon the criterion of performance used; in general, the relationship between wool production and stocking rate is least likely to be affected by selection, whereas dairy milk production is most likely to be affected and beef production is likely to be intermediate.

## Interaction of stocking rate and grazing management

Systems other than continuous or set-stocking of pastures are advocated in many environments (Chapter 7). These generally involve a substantial amount of paddock subdivision, combined with rotational or strip grazing; animals are moved from one paddock to the next, according to either a predetermined time interval or the availability of pasture feed.

Infrequent defoliation can increase herbage production (Chapter 7), but the effects are very dependent on season (Brougham, 1959). Any advantage in herbage production obtained does not seem to be reflected in animal production, since rotational grazing has rarely shown advantages over continuous grazing in beef production (Chapter 9), in milk production (Chapter 10) or in sheep production (Chapter 11). At higher stocking rates, some increases in production of rotational over continuously grazed systems have been claimed (Conway, 1963; McMeekan and Walshe, 1963; Robinson and Simpson, 1975), but converse results have also been reported (Wheeler, 1962; Boswell et al., 1974). In the various studies, it is important to distinguish between the effects of grazing management *per se* and the effects of stocking rate or more careful animal husbandry (Wheeler, 1962). There are also clear economic disincentives to rotational grazing, because of the additional costs of erecting and maintaining extra fences, and of more frequent supervision and movement of the livestock.

The grazing management that is adopted seems to vary on a broad geographical basis (Booysen et al., 1975). Continuous grazing of both natural ranges and sown pastures, has been widely advocated in Australia and, until recently, North America (Heady, 1961; Wheeler, 1962; Willoughby, 1970), except where lucerne (*Medicago sativa*) pastures are involved (McKinney, 1974), whereas rotational grazing is common in Europe, New Zealand and southern Africa. Booysen et al. concluded that "neither climatic nor biological factors appear to be entirely responsible for these contrasting approaches being so emphatically promoted by researchers and becoming so firmly entrenched in grazing management practice in the different countries". They suggested that the higher the price of land, relative to the cost of the livestock, the higher would be the optimum stocking rate, and the more likely that rotational grazing would be favoured.

## CONCLUSIONS

Choice of stocking rate can profoundly influence the productivity and profitability of a grazing system. Guidelines are therefore required which will assist producers to estimate the optimal level of stocking on their properties, though the objectives of these producers need to be clearly defined.

Stocking rate interacts with many of the physical, biological and economic components of any farming system. In areas where soils are prone to erosion, and the vegetation cannot support high stocking rates, the choice of stocking rate will be determined by the need to maintain a stable ecosystem. However, in other areas, the choice is usually made primarily on economic grounds, though whether this is biased towards maximizing profit or minimizing risks will depend on the financial status and aspirations of the farmer. The advocates of very high stocking rates have misinterpreted the objectives of many farmers.

The lessons of the past help considerably in determining the rate of stocking to adopt. Some farmers have undoubtedly been lured towards financial ruin by using high stocking rates in extremely adverse seasons. At the other extreme, risk-averting farmers, who stock at very low levels, can have such low incomes that they also risk bankruptcy.

Livestock production systems should be analysed in both biological and economic terms, over a range of stocking rates and seasons, before the optimal stocking rate can be estimated with confidence. Preferably stocking rates should be evaluated in terms of the productivity and profitability of the farm as a whole, since farm practices such as stock trading, fodder sales and cropping offer means of varying the enterprise mix and hence complicate the prediction of optimal stocking rates. This has presented immense problems in the past; however, field experiments, backed by simulation studies, can provide the necessary evidence to define more accurately the optimal stocking rate for any particular system.

## REFERENCES

Allden, D.M. and Lamming, G.E., 1961. Nutrition and reproduction in the ewe. *J. Agric. Sci., Cambridge*, 56: 69–79.

Allden, W.G. and Jennings, A.C., 1962. Dietary supplements to sheep grazing mature herbage in relation to herbage intake. *Proc. Aust. Soc. Anim. Prod.*, 4: 145–153.

Allden, W.G. and Whittaker, I.A., McD., 1970. The determinants of herbage intake by grazing sheep: the interrelationship of factors influencing herbage intake and availability. *Aust. J. Agric. Res.*, 21: 755–766.

Anonymous, 1979. Erosion risks with high stocking rates on Kojonup trial. *W. Aust. J. Agric.*, 20: 24.

Arnold, G.W. and Bennett, D., 1975. The problem of finding an optimum solution. In: G.E. Dalton (Editor), *Study of Agricultural Systems*. Applied Science Publishers, London, pp. 129–173.

Arnold, G.W. and Bush, I.G., 1962. The effects of stocking rate and grazing management on fat lamb production. *Proc. Aust. Soc. Anim. Prod.*, 4: 121–129.

Arnold, G.W. and Dudzinski, M.L., 1967. Studies on the diet of the grazing animal. 2. The effect of physiological status in ewes and pasture availability on herbage intake. *Aust. J. Agric. Res.*, 18: 349–359.

Arnold, G.W. and McManus, W.R., 1960. The effect of level of stocking on two pasture types upon wool production and quality. *Proc. Aust. Soc. Anim. Prod.*, 3: 63–68.

Arnold, G.W., Gharaybeh, H.R., Dudzinski, M.L., McManus, W.R. and Axelsen, A., 1969. Body composition of young sheep. II: Effect of stocking rate on body composition of Dorset Horn cross lambs. *J. Agric. Sci., Cambridge*, 72: 77–84.

Arnold, G.W., Axelsen, A., Gharaybeh, H.R. and Chapman, H.W., 1971. Some variables influencing production of prime lamb and wool from pasture. *Aust. J. Exper. Agric. Anim. Husbandry*, 11: 498–507.

Arnold, G.W., Campbell, N.A. and Galbraith, K.A., 1977. Mathematical relationships and computer routines for a model of food intake, liveweight change and wool production in grazing sheep. *Agric. Syst.*, 2: 209–226.

Axelsen, A., Bennett, D., Larkham, P.A. and Coulton, L., 1972. Effect of calving time and stocking rate on production of beef cows. *Proc. Aust. Soc. Anim. Prod.*, 9: 165–170.

Baker, R.D., Hodgson, J., Treacher, T.T. and Rodriguez, J.M., 1975. A research strategy towards the improvement of grazing managements. In: R.L. Reid (Editor), *Proceedings of the Third World Conference on Animal Production*. Sydney University Press, Sydney, N.S.W., pp. 211–216.

Bircham, J.S., Crouchley, G. and Wright, D.F., 1977. Effects of superphosphate, lime and stocking rate on pasture and animal production on the Wairarapa Plains. II. Animal Production. *N.Z. J. Exper. Agric.*, 5: 349–355.

Birrell, H.A., 1981. Some factors which affect the liveweight change and wool growth of adult Corriedale wethers grazed at various stocking rates on perennial pasture in southern Victoria. *Aust. J. Agric. Res.*, 32: 353–370.

Birrell, H.A., Bishop, A.H., Tew, A. and Plowright, R.D., 1978. Effect of stocking rate, fodder conservation and grazing management on the performance of wether sheep and pastures in south-west Victoria. 2: Seasonal wool growth rate, liveweight and herbage availability. *Aust. J. Exper. Agric. Anim. Husbandry*, 18: 41–50.

Bishop, A.H. and Birrell, H.A., 1975. Effect of stocking rate, fodder conservation and grazing management on the performance of wether sheep in south-west Victoria. I: Wool production. *Aust. J. Exper. Agric. Anim. Husbandry* 15: 173–182.

Blackburn, A.G., 1969. Profits from sheep: Which factors are most important. *Wool Techn. Sheep Breed.*, 16: 125–128.

Blackburn, A.G., Frew, M.V. and Mullaney, P.D., 1973. Estimating optimum economic stocking rates for wethers. *J. Aust. Inst. Agric. Sci.*, 39: 18–23.

Booysen, P. deV., Tainton, N.M. and Foran, B.D., 1975. An economic solution to the grazing management dilemma. *Proc. Grassland Soc. S. Afr.*, 10: 77–83.

Boswell, C.C., Monteath, M.A., Round-Turner, N.L., Lewis, K.H.C. and Cullen, N.A., 1974. Intensive lamb production under continuous and rotational grazing systems. *N.Z. J. Exper. Agric.*, 2: 403–408.

Brougham, R.W., 1959. The effects of frequency and intensity of grazing on the productivity of a pasture of short-rotation ryegrass and red and white clover. *N.Z. J. Agric. Res.*, 2: 1232–1248.

Brown, T.H., 1977. A comparison of continuous grazing and deferred Autumn grazing of Merino ewes and lambs at 13 stocking rates. *Aust. J. Agric. Res.*, 28: 947–961.

Bryant, A.M. and Parker, O.F., 1971. Optimum grazing interval at high stocking rate. *Proc. Ruakura Farmers Conf.*, 23: 110–120.

Byrne, P.F., 1968. An evaluation of different stocking rates of Merino sheep. *Wool Economic Res. Rep.*, No. 13 (Australian Bureau of Agricultural Economics), 80 pp.

Callinan, A.P.L., Morley, F.H.W., Arundel, J.H. and White, D.H., 1982. A model of the life cycle of sheep nematodes and the epidemiology of nematodiasis in sheep. *Agric. Syst.*, 9: 199–225.

Cameron, I.H. and Cannon, D.J., 1970. Changes in the botanical composition of pasture in relation to rate of stocking with sheep and consequent effects on wool production. In: *Proc. 11th Int. Grassland Congr.*, pp. 640–643.

Campbell, A.G., 1966. Grazed pasture parameters. III: Relationships of pasture and animal parameters in, and general discussion of, a stocking rate and grazing management experiment with dairy cows. *J. Agric. Sci., Cambridge*, 67: 217–222.

Campbell, A.G., 1969. Grazing interval, stocking rate and pasture production. *N.Z. J. Agric. Res.*, 12: 67–74.

Cannon, D.J., 1969. The wool production and liveweight of wethers in relation to stocking rate and superphosphate application. *Aust. J. Exper. Agric. Anim. Husbandry*, 9: 172–180.

Cannon, D.J., 1972. The influence of rate of stocking and application of superphosphate on the production and quality of wool, and on the gross margin from Merino wethers. *Aust. J. Exper. Agric. Anim. Husbandry*, 12: 348–354.

Carter, E.D. and Day, H.R., 1970. Interrelationships of stocking rate and superphosphate rate on pasture as determinants of animal production. I. Continuously grazed old pasture land. *Aust. J. Agric. Res.*, 21: 473–491.

Cayley, J.W.D., Bird, P.R., Watson, M.J. and Chin, J.F., 1980. Effect of stocking rate of steers on net and true growth rates of perennial pasture. *Proc. Aust. Soc. Anim. Prod.*, 13: 468.

Chisholm, A.H., 1965. Towards the determination of optimum stocking rates in the high rainfall zone. *Rev. Marketing Agric. Econ.*, 33: 5–31.

Ciordia, H., Neville, W.E. Jr., Baird, D.M. and McCampbell, H.C., 1971. Internal parasitism of beef cattle on winter pastures: Level of parasitism as affected by stocking rates. *Am. J. Vet. Res.*, 32: 1353–1358.

Conniffe, D., Browne, E. and Walshe, M.J., 1970. Experimental design for grazing trials. *J. Agric. Sci., Cambridge*, 74: 339–342.

Connolly, J., 1976. Some comments on the shape of the gain-stocking rate curve. *J. Agric. Sci., Cambridge*, 86: 103–109.

Conway, A., 1963. Effect of grazing management on beef production. II. Comparison of three stocking rates under two systems of grazing. *Ir. J. Agric. Res.*, 2: 243–258.

Coop, I.E., 1965. A review of the ewe equivalent system. *N.Z. Agric. Sci.*, 1: 13–18.

Cotsell, J.C., 1956. Sheep investigations at Shannon Vale Nutrition Station with special reference to strategic stocking. *Proc. Aust. Soc. Anim. Prod.*, 1: 24–32.

Cowan, R.T. and Stobbs, T.H., 1976. Effects of nitrogen fertilizer applied in autumn and winter on milk production from a tropical grass–legume pasture grazed at four stocking rates. *Aust. J. Exper. Agric. Anim. Husbandry*, 16: 829–837.

Cowlishaw, S.J., 1969. The carrying capacity of pastures. *Br. Grassland Soc. J.*, 24: 207–214.

Davidson, J.L. and Philip, J.R., 1956. Light and pasture growth. In: *Proc. Canberra Symposium Arid Zone Res.*, 11: 181–187.

Davies, H.L., 1962. Studies on time of lambing in relation to stocking rate in south-western Australia. *Proc. Aust. Soc. Anim. Prod.*, 4: 113–120.

Davies, H.L., 1963. The milk production of Merino ewes at pasture. *Aust. J. Agric. Res.*, 14: 824–838.

Donald, A.D., Axelsen, A., Morley, F.H.W., Waller, P.J. and Donnelly, J.R., 1979. Growth of cattle on phalaris and lucerne pastures. II. Helminth parasite populations and effects of anthelmintic treatment. *Vet. parasitol.*, 5: 205–222.

Downey, N.E., 1969. Grazing management in relation to trichostrongylid infection in lambs. 2. Level of infestation associated with increased stocking rate and its effects on the host. *Ir. J. Agric. Res.*, 8: 375–395.

Downey, N.E. and Conway, A., 1968. Grazing management in relation to trichostrongylid infestation in lambs. 1. Influence of stocking rate on the level of infestation. *Ir. J. Agric. Res.*, 7: 343–362.

Egan, J.K., Thompson, R.L. and McIntyre, J.S., 1977. Stocking rate, joining time, fodder conservation and the productivity of Merino ewes. 1. Liveweights, joining and lambs born. *Aust. J. Exper. Agric. Anim. Husbandry*, 17: 566–573.

El Hassan, B. and Krueger, W.C., 1980. The impact of grazing pressure and season of grazing on carbohydrate reserves of perennial ryegrass. *J. Range Manage.*, 33: 200.

Fitzgerald, R.D., 1976. Effect of stocking rate, lambing time and pasture management on wool and lamb production on annual subterranean clover pasture. *Aust. J. Agric. Res.*, 27: 261–275.

George, J.M. and Pearse, R.A., 1978. An economic analysis of a 10 year rate of stocking trial with fine-wool Merino sheep that lambed in winter, spring or summer. *Aust. J. Exper. Agric. Anim. Husbandry*, 18: 370–380.

Hamilton, D., 1974. Alternatives to dry annual pasture for steers over summer and later effects on liveweight gain during winter. *Proc. Aust. Soc. Anim. Prod.*, 10: 99–102.

Harlan, J.R., 1958. Generalized curves for gain per head and gain per acre in rates of grazing studies. *J. Range Manage.*, 11: 140–147.

Hart, R.H., 1972. Forage yield, stocking rate and beef gains on pasture. *Herb. Abstr.*, 42: 345–353.

Heady, H.F., 1961. Continuous vs. specialised grazing systems: a review and application to the Californian annual type. *J. Range Manage.*, 14: 182–193.

Hilder, E.J., 1966. Rate of turnover of elements in soils: The effects of stocking rate. *Wool Technol. Sheep Breed.*, 13: 11–16.

Hodgson, J., 1971. The influence of grazing pressure and stocking rate on herbage intake and animal performance. In: J. Hodgson and D.D.K. Jackson (Editors), *Pasture Utilization by the Grazing Animal. Br. Grassland Soc. Occas. Symp.*, No. 8: 93–103.

Hutchinson, K.J., 1971. Productivity and energy flow in grazing/fodder conservation systems. *Herb. Abstr.*, 41: 1–10.

Jardine, R., 1975. Two cheers for optimality! *J. Aust. Inst. Agric. Sci.*, 41: 30–34.

Jones, L.P., 1982. Economic aspects of developing breeding objectives. In: J.S.T. Barker, K. Hammond and A.E. McLintock (Editors), *Future Developments in the Genetic Improvement of Animals*. Academic Press, London, pp. 119–136.

Jones, R.J. and Sandland, R.L., 1974. The relation between animal gain and stocking rate: Derivation of the relation from the results of grazing trials. *J. Agric. Sci., Cambridge*, 83: 335–342.

Joyce, J.P., Clarke, J.N., Maclean, K.S. and Cox, E.H., 1976. The effects of stocking rate on the productivity of sheep of different genetic origin. *Proc. Ruakura Farmers Conf.*, 28: 34–38.

Kenney, P.A. and Davis, I.F., 1974. Effect of time of joining and rate of stocking on the production of Corriedale ewes in southern Victoria. 2. Survival and growth of lambs. *Aust. J. Exper. Agric. Anim. Husbandry*, 14: 434–440.

Kirton, A.H., Sinclair, D.P., Chrystall, B.B., Devine, C.E. and Woods, E.G., 1981. Effect of plane of nutrition on carcass composition and the palatability of pasture-fed lamb. *J. Anim. Sci.*, 52: 285–291.

Langlands, J.P., 1969. The feed intake of sheep supplemented with varying quantities of wheat while grazing pastures differing in herbage availability. *Aust. J. Agric. Res.*, 20: 919–924.

Langlands, J.P. and Bennett, I.L., 1973a. Stocking intensity and pastoral production. 1. Changes in the soil and vegetation of a sown pasture grazed by sheep at different stocking rates. *J. Agric. Sci., Cambridge*, 81: 193–204.

Langlands, J.P. and Bennett, I.L., 1973b. Stocking intensity and pastoral production. 3. Wool production, fleece characteristics and the utilization of nutrients for maintenance and wool growth by Merino sheep grazed at different stocking rates. *J. Agric. Sci., Cambridge*, 81: 211–218.

Langlands, J.P., Donald, G.E. and Paull, D.R., 1984. Effects of different stocking intensities in early life on the productivity of Merino ewes grazed as adults at two stocking rates. 3. Survival of ewes and their lambs, and the implications for flock productivity. *Aust. J. Exper. Agric. Anim. Husbandry*, 25: 57–65.

Leaver, J.D., 1974. Rearing of dairy cattle. 5. The effect of stocking rate on animal and herbage production in a grazing system for calves and heifers. *Anim. Prod.*, 18: 273–284.

Lloyd, A.G., 1966. Economic aspects of stocking and feeding policies in the sheep industry in southern Australia. *Proc. Aust. Soc. Anim. Prod.*, 6: 137–147.

McArthur, I.D. and Dillon, J.L., 1971. Risk, utility and stocking rate. *Aust. J. Agric. Econ.*, 15: 20–35.

McGregor, B.A., 1984. Growth and fleece production of Angora wethers grazing annual pasture. *Proc. Aust. Soc. Anim. Prod.*, 15: 715.

McKinney, G.T., 1974. Management of lucerne for sheep grazing on the Southern Tablelands of New South Wales. *Aust. J. Exper. Agric. Anim. Husbandry*, 14: 726–734.

McMeekan, C.P., 1956. Grazing management and animal production. In: *Proc. 7th Int. Grassland Congr.*, pp. 146–156.

McMeekan, C.P. and Walshe, M.J., 1963. The interrelationships of grazing method and stocking rate in the efficiency of pasture utilization by dairy cattle. *J. Agric. Sci., Camb.*, 61: 147–163.

Michel, J.F., 1969. The epidemiology and control of some nematode infections of grazing animals. *Adv. Parasitol.*, 7: 211–282.

Morley, F.H.W., 1966. Stability and productivity of pastures. *Proc. N.Z. Soc. Anim. Prod.*, 26: 8–21.

Morley, F.H.W., 1974. Evaluation by animal production of increases in pasture growth, using computer simulation. In: *Proc. 12th Int. Grassland Congr.*, pp. 320–325.

Morley, F.H.W. and Donald, A.D., 1980. Farm management and systems of helminth control. *Vet. Parasitol.*, 6: 105–134.

Morley, F.H.W. and Graham, G.Y., 1971. Fodder conservation for drought. In: J.B. Dent and J.R. Anderson (Editors), *Systems Analysis in Agricultural Management*. Wiley, Australasia Pty. Ltd., Sydney, N.S.W., pp. 212–236.

Morley, F.H.W., Bennett, D. and McKinney, G.T., 1969. The effect of intensity of rotational grazing with breeding ewes on phalaris–subterranean clover pastures. *Aust. J. Exper. Agric. Anim. Husbandry*, 9: 74–84.

Morley, F.H.W., Axelsen, A., Pullen, K.G. and Nadin, J.B., 1978a. Growth of cattle on phalaris and lucerne pastures: Part 1. Effect of pasture, stocking rate and anthelmintic treatment. *Agric. Syst.*, 3: 123–145.

Morley, F.H.W., White, D.H., Kenney, P.A. and Davis, I.F., 1978b. Predicting ovulation rate from liveweight in ewes. *Agric. Syst.*, 3: 27–45.

Mott, G.O., 1960. Grazing pressure and the measurement of pasture production. In: *Proc. 8th Int. Grassland Congr.*, pp. 606–611.

Noy-Meir, I., 1975. Stability of grazing systems: An application of predator–prey graphs. *J. Ecol.*, 63: 459–481.

Noy-Meir, I., 1978. Grazing and production in seasonal pastures: Analysis of a simple model. *J. Appl. Ecol.*, 15: 809–835.

Obst, J.M., 1986. Sheep production systems. *Proc. Aust. Soc. Anim. Prod.*, 16: 52–64.

Owen, J.B. and Ridgman, W.J., 1968. The design and interpretation of experiments to study animal production from grazed pasture. *J. Agric. Sci., Cambridge*, 71: 327–335.

Petersen, R.G., Lucas, H.L. and Mott, G.O., 1965. Relationship between rate of stocking and per animal and per acre performance on pasture. *Agron. J.*, 57: 27–30.

Rauzi, T. and Hanson, C.L., 1966. Water intake and runoff as affected by intensity of grazing. *J. Range Manage.*, 19: 351–356.

Reeve, J.L. and Sharkey, M.J., 1980. Effect of stocking rate, time of lambing and inclusion of lucerne on prime lamb production in north-east Victoria. *Aust. J. Exper. Agric. Anim. Husbandry*, 20: 637–653.

Riewe, M.E., 1964. An experimental design for grazing trials using the relationship of stocking rate to animal gain. In: *Proc. 9th Int. Grassland Congr.*, pp. 1507–1510.

Roberts, N.F., James, J.F.P. and Burgmann, V.D., 1960. Tenderness in fleece wool. *J. Text. Inst.*, 51: T935–T952.

Robinson, G.G. and Simpson, I.H., 1975. The effect of stocking rate on animal production from continuous and rotational grazing systems. *J. Br. Grassland Soc.*, 30: 327–332.

Rottenbury, R.A., 1982. Towards sale by description. *Proc. Aust. Soc. Anim. Prod.*, 14: 12–15.

Sharkey, M.J., Davis, I.F. and Kenney, P.A., 1962. The effect of previous and current nutrition on wool production in southern Victoria. *Aust. J. Exper. Agric. Anim. Husbandry*, 2: 160–169.

Sharkey, M.J., Davis, I.F. and Kenney, P.A., 1964. The effect of rate of stocking with sheep on the botanical composition of an annual pasture in southern Victoria. *Aust. J. Exper. Agric. Anim. Husbandry*, 4: 34–38.

Smoliak, S., 1974. Range vegetation and sheep production at three stocking rates on *Stipa–Bouteloua* Prairie. *J. Range Manage.*, 27: 23–26.

Southcott, W.H., Langlands, J.P. and Heath, D.D., 1970. Stocking intensity and nematode infection of grazing sheep. In: *Proc. 9th Int. Grassland Congr.*, pp. 888–890.

Suckling, F.E.T., 1964. Stocking rate trials at Te Awa. *Sheep Farming Annual, (Massey University of Manawatu), New Zealand*, pp. 18–32.

Suckling, F.E.T., 1975. Pasture management trials on unploughable hill country at Te Awa. 111. Results for 1959–69. *N.Z. J. Exp. Agric.*, 3: 351–436.

Tribe, D.E. and Lloyd, A.G., 1962. Effect of stocking rate on the efficiency of fat lamb production. *J. Aust. Inst. Agric. Sci.*, 28: 274–278.

Vickery, P.J., 1973. Comparative net primary productivity of grazing systems with different stocking densities of sheep. *J. Appl. Ecol.*, 9: 307–314.

Wallace, L.R., 1959. Grazing management and dairy production. *Proc. N.Z. Inst. Agric. Sci.*, pp. 131–138.

Watson, M.J., Bird, P.R. and Cayley, J.W.D., 1979. Pasture-based feeding systems (perennial pastures). In: G.H. Smith (Editor), *Nutrition of Beef Cattle. Gov. Vic. Agric. Note Ser.*, No. 51: 10–28.

Wheeler, J.L., 1962. Experimentation in grazing management. *Herb. Abstr.*, 32: 1–7.

White, D.H., 1975. The search for an optimal stocking rate: Continuing the saga. *J. Aust. Inst. Agric. Sci.*, 41: 192–194.

White, D.H., 1984. Economic values of changing reproductive rates. In: D.R. Lindsay and D.T. Pearce (Editors), *Reproduction in Sheep*. Australian Academy of Sciences, Canberra, A.C.T., pp. 371–377.

White, D.H. and Bowman, P.J., 1981. *Dry Sheep Equivalents for Comparing Different Classes of Stock*. Agnote No. 1530/81. Government of Victoria, Melbourne, Vic., 4 pp.

White, D.H. and McConchie, B.J., 1976. Effect of stocking rate on fleece measurements and their relationships in Merino sheep. *Aust. J. Agric. Res.*, 27: 163–174.

White, D.H. and Morley, F.H.W., 1977. Estimation of optimal stocking rates of Merino sheep. *Agric. Syst.*, 2: 289–304.

White, D.H., McConchie, B.J., Curnow, B.C. and Ternouth, A.H., 1980. A comparison of levels of production and profit from Merino ewes and wethers grazed at various stocking rates in northern Victoria. *Aust. J. Exper. Agric. Anim. Husbandry*, 20: 296–307.

White, D.H., Bowman, P.J. and Morley, F.H.W., 1982. Choice of date of lambing and stocking rate for Merino wool-producing flocks. *Proc. Aust. Soc. Anim. Prod.*, 14: 38–40.

White, D.H., Bowman, P.J., Morley, F.H.W., McManus, W.R. and Filan, S.J., 1983. A simulation model of a breeding ewe flock. *Agric. Syst.*, 10: 149–189.

Willoughby, W.M., 1959. Limitations to animal production imposed by seasonal fluctuations in pasture and by management procedures. *Aust. J. Agric. Res.*, 10: 248–268.

Willoughby, W.M., 1970. Grassland management. In: R. Milton Moore (Editor), *Australian Grasslands*. Australian National University Press, Canberra, A.C.T., pp. 392–397.

Yiakoumettis, I.M. and Holmes, W., 1972. The effect of nitrogen and stocking rate on the output of pasture grazed by beef cattle. *J. Br. Grassland Soc.*, 27: 183–291.

Chapter 25

# FERTILIZER INPUTS AND BOTANICAL COMPOSITION

R.W. SNAYDON

## INTRODUCTION

At first sight, it may seem strange to consider fertilizers and botanical composition together; however, there are several good reasons for doing so. Firstly, although the botanical composition of sown pastures is initially determined by the species composition of the seed sown, within a few years the botanical composition is largely determined by fertilizer inputs (Chapter 8), though other factors such as climate and grazing management are also important. This effect of fertilizers on botanical compositions is reinforced by the fact that the amount of fertilizer applied varies in relation to the sward type; in particular, leys receive almost twice as much fertilizer as permanent pastures (Chapter 8). Secondly, response to fertilizer is often dependent upon botanical composition; this is particularly true of the effect of legume content on response to nitrogen fertilizer (Chapter 6). Thirdly, the principles underlying the effects of fertilizers and botanical composition on animal output are essentially the same.

Fertilizer inputs are probably the most important management variables determining herbage production. Fertilizer inputs also interact with other important management variables, such as stocking rate, in determining herbage production and herbage quality, and hence animal output (Chapter 24). Potentially, any mineral nutrient which affects either plant growth or animal performance (i.e. about twenty elements) could affect animal production from grazed pastures. However, some of these mineral nutrients (such as copper, boron, and molybdenum) are only rarely deficient in agricultural ecosystems (Chapter 15). Only nutrients affecting pasture production are considered here, since nutrients which affect only animals are usually supplied directly to the animals as mineral supplements, injections, bullets or drenches (Chapter 15). Only rarely are elements (such as cobalt) which affect grazing animals, but do not normally affect plants, applied in fertilizer to the pasture.

The relative importance of various nutrients, in determining pasture production or quality, obviously varies from soil to soil. Deficiencies of nitrogen or phosphate are probably most widespread, though other macronutrients, such as magnesium, potassium and sulphur, can be limiting in many soils. Deficiencies of micronutrients, such as boron, copper and molybdenum, can also occur, especially in agriculturally marginal areas and land recently brought into agricultural production. Apart from applications of specific nutrients, applications of lime [$CaCO_3$ or $Ca(OH)_2$] are often important in modifying soil pH, and so affecting the availability of many mineral nutrients and toxins.

Applications of fertilizer have two important effects on pasture which, in turn, can affect animal performance. The first effect is on herbage growth, and hence on herbage yield. The second effect is on herbage quality, both the organic composition (Chapter 14) and the inorganic composition (Chapter 15). Changes in quality may come about either as a result of changes in the chemical or physical composition of the species already present, or as a result of changes in the botanical composition of the pasture (Chapter 8). It is often difficult to disentangle the effects of herbage yield and herbage quality on animal performance. The difficulty is increased by the interacting effects of stocking rate (Chapter 24).

## INTERACTIONS WITH STOCKING RATE

It is a truism that any increase in herbage production will only increase animal output if the stocking rate is sufficiently great to limit food intake (Chapter 24). There is some doubt about the form of the relationship between animal output and stocking rate at low stocking rates (Chapter 24), so that the effects of increasing herbage production on animal production is difficult to quantify. However, the response of animal production to increased herbage production, at low stocking rates, is likely to be slight (Fig. 25.1A). As a result, if the stocking rate is initially low, benefit from increased herbage production would only be achieved if the stocking rate is increased. The greatest response to additional herbage production should occur at, or above, the stocking rate that gives the highest output (i.e. the biological optimum); however, this is above the economic optimum stocking rate (Chapter 24).

If some management practice, such as the use of a fertilizer or sowing another species, improved the quality of herbage available, this would normally increase herbage intake (Chapter 14), and so increase animal output, regardless of the stocking rate used (Fig. 25.1B). As a result, the benefit of improved quality is gained regardless of stocking rate or feed availability. Since animals eat more of the higher quality feed, the optimum stocking rate is lower than that for the lower quality feed (Fig. 25.1B), and the greatest advantage is achieved slightly below the biological optimum stocking rate and probably close to the economic optimum.

There are several apparent advantages in improving herbage quality, as opposed to herbage quantity. Firstly, increased animal output can be achieved without the additional costs of increasing stocking rate (Chapter 24). Secondly, the advantage can be gained regardless of the current stocking rate. However, it should also be recognized that the situation may be more complex than has been indicated so far. Firstly, herbage quality and herbage quantity are not independent of one another. For example, the accumulation of herbage mass is usually associated with a higher proportion of lignified and senescing material of lower quality (Chapters 7 and 14). On the other hand, a large herbage mass gives animals a greater opportunity for selective grazing (Chapter 13), so that they can obtain a diet of higher quality than the average present. Secondly, the relationship between stocking rate and availability of herbage is complex, being highly dependent upon seasonal changes in pasture growth and the animals' feed requirement (Chapters 9–11). As a result, herbage may be in limiting supply at some times of year (e.g. winter), even at quite low stocking rates, and may be in excess at other times (e.g. spring) even at very high stocking rates. The relationship between stocking rate and animal performance therefore probably varies with season, though this has not been carefully studied.

## FERTILIZER USE

Nitrogen is the fertilizer used in greatest quantity in world agriculture; phosphorus and then potassium are used in the next largest quantities (Stangel, 1976). Much of the nitrogen fertilizer is used on arable crops, and its use on grassland, even when intensively managed, varies widely (Chapter 6). This wide variation reflects different degrees of dependence on nitrogen fixation by legumes (Chapter 21), rather than differences in the inherent nitrogen deficiency of soils. The main effect of nitrogen deficiency is on herbage production (Chapter 6); this can affect animal performance, though the effect depends on stocking rate (Fig. 25.1A). Nitrogen supply can also affect herbage quality, directly and indirectly (see below). Differences in the protein content of herbage, in particular, might affect animal performance.

Phosphorus is deficient in many soils throughout the world. It is especially deficient in the lateritic soils of the tropics, but also in many soils of temperate regions, especially those of volcanic

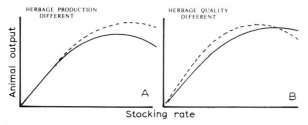

Fig. 25.1. The effect of differences between pastures in (A) herbage production and (B) herbage quality on animal output (ha$^{-1}$) at various stocking rates.

origin, e.g. in New Zealand, those derived from phosphate-deficient parent material, e.g. in Australia (Beadle, 1962), and in acid and podzolized soils, e.g. in upland areas of Britain.

Applications of phosphorus fertilizer to deficient soils increase herbage growth per se (e.g. Ryan and Finn, 1976); they increase the abundance of legumes, at least in the short-term, so increasing nitrogen fixation and hence grass production (e.g. Donald and Williams, 1954; Rossiter, 1964); and they increase the phosphorus content of herbage. The increase in herbage production can increase animal production if the stocking rate is sufficiently high (e.g. Kohn, 1974; Curll, 1977). The increase in phosphorus content of herbage can increase feed intake and so increase animal production (Underwood, 1966). As a result, the use of phosphate fertilizer has played a very important role in increasing the output from grassland, especially in Australia (Donald, 1964) and New Zealand (Levy, 1970).

Although other macronutrients (such as potassium and magnesium) and some micronutrients (such as molybdenum) are also supplied in fertilizers, they will not be considered further, because of their less general use, and the paucity of data concerning their effects on animal production.

## Measurement of response

The ultimate criterion for measuring the agricultural benefits of any management variable, such as fertilizer, is economic (Chapter 23). However, very few such economic analyses have been carried out. Indeed, relatively few studies have been made of whole systems, measuring the effects of fertilizer on animal output; most studies have been of herbage production (Chapter 6). It is difficult, and often impossible, to predict the effects of fertilizers on herbage yield and quality, because of interactions with climate (e.g. season, year and site), soil conditions (e.g. water and other nutrients), grazing/cutting management, and pasture type (Chapter 6). It is even more difficult to predict animal output from data on herbage yield, because of the complicating effects of stocking rate (Chapter 24), herbage quality (Chapter 14) and seasonal patterns of herbage growth and animal demand (Chapters 9–11).

The lack of data on the effects of fertilizer applications on animal output is largely the result of the relatively high cost of such experiments. This is exacerbated by the fact that, if the effect of fertilizer is mainly to increase herbage yield, comparisons need to be made over a wide range of stocking rates (Fig. 25.1A). Some people have attempted to overcome this problem by continuously varying the stocking rate to match the herbage available (so-called "put and take" systems). Not surprisingly, when stocking rate is closely matched to herbage production, as in those studies, any differences in animal output are usually correlated with differences in herbage production. However, there are doubts about the scientific validity of results obtained by the "put and take" method, since it usually takes no account of differences in herbage quality. Similarly, there are doubts about the agricultural validity of the results, since farmers can rarely vary stocking rate continuously in this way, though conservation of excess herbage (Chapter 26) does offer a means of matching feed availability to feed requirement.

## Survey data

In the absence of adequate experimental data, various attempts have been made to use survey data (Minson and Rees, 1976; Morley, 1981) to determine the effects of fertilizer use on animal output (e.g. Rees et al., 1972; Castle et al., 1972; Hawkins and Rose, 1979). The main difficulty is that many other factors (such as soil conditions, climate, stocking rate, other fertilizer inputs, and management) also vary, and may be correlated with the input of the particular factor being investigated. The problem may be reduced by using data from a single farm (e.g. Castle et al., 1972), but this greatly reduces the general applicability of the data. The alternative is to attempt to remove the effects of as many other variables as possible, either by multiple regression analysis (e.g. Hawkins and Rose, 1979) or by the simpler "stabilized group" method (e.g. Minson and Rees, 1976).

Castle et al. (1972), in Scotland, found no correlation between milk production and the application rate of either nitrogen or phosphorus fertilizer on soils with a long history of fertilizer use. Rees et al. (1972) in Queensland, Australia, found a highly significant correlation between butter fat production and phosphorus application, and a smaller but significant correlation with nitrogen application, on soils inherently deficient in phosphorus. Hawkins and Rose (1979), in England and Wales, found a significant correlation between milk production and nitrogen application in 1972, but the correlation declined with time and was not significant in 1977. However, the use of nitrogen fertilizer was correlated with stocking rate, making interpretation of the data uncertain. In the case of beef cattle and sheep (Craven and Kilkenny, 1974), there was little correlation between nitrogen fertilizer use and stocking rate, and no evidence that nitrogen fertilizer application increased profitability.

**Experimental evidence**

Ideally, studies of the effects of fertilizers on animal output should be carried out experimentally, using a range of stocking rates; studies should be long-term, so that they include year-to-year variations and cumulative effects, such as those on reproduction (Reed, 1981). Unfortunately this is rarely, if ever, achieved in studies of fertilizer use.

**Nitrogen fertilizer**

Most studies of the effects of fertilizer on animal output have been concerned with nitrogen fertilizer; these studies have been reviewed by Reed (1981). Since most pastures in temperate regions are composed of mixtures of grasses and legumes, the effects of nitrogen fertilizer on pasture and animal production are complex. Firstly, the effect of nitrogen fertilizer on herbage production is highly dependent upon the legume content of pasture (Whitehead, 1970; Chapter 6). Pastures containing significant amounts of legume are much less responsive to nitrogen fertilizer (see Fig. 6.4); in turn, nitrogen fertilizer rapidly reduces the legume content of the pasture. Secondly, legumes are generally of higher feed quality, and give rise to improved animal performance, compared with grasses (Thomson, 1979). In addition, sheep (Curll, 1982; Milne et al., 1982) and, to a lesser extent, cattle, graze clover selectively; as a result, their diet may differ substantially from that on offer, and their selectivity may rapidly change the composition of pastures. Because of these complex interactions, it is difficult to predict the effect of nitrogen fertilization on animal output, if legumes are present in the pasture. In addition, as Reed (1981) has pointed out, the effect is very dependent on a number of arbitrary decisions in management. The effect of stocking rate has already been briefly discussed, but other factors include the cutting of excess feed for conservation (Chapter 26) and the subsequent use of this feed, the length of the experimental period, and the date on which animals are removed from the experiment at the end of each season and at the end of the experiment (Reed, 1981).

Substantial increases in animal output have been achieved with nitrogen fertilizer, especially in short-term experiments where "put and take" systems of management have been used to match stocking rate to pasture growth (Holmes, 1968). However, few increases have been obtained under set-stocking, especially where the pasture has been legume-rich and the stocking rate within the range of normal practice. For example, in only two studies, of the 25 studies listed by Reed (1981), did grass swards receiving nitrogen fertilizer give greater beef production than grass–legume swards receiving little or no fertilizer.

Few studies have been made of the effect of nitrogen fertilizer on milk production. The most intensive study has been by King and Stockdale (1981), who used two levels of nitrogen fertilizer and five stocking rates; they carried out the study for two years. Unfertilized pasture produced about 4% more milk than fertilized pasture at the two lowest stocking rates, but about 8% less at the two highest stocking rates (Fig. 25.2). This might be expected, on the basis of the effects of nitrogen fertilizer on herbage yield and herbage quality, and the subsequent effect of these changes on animal production at different stocking rates (Fig. 25.1). Indeed, Stockdale and King (1981) showed that nitrogen fertilizer increased herbage production, but reduced herbage quality, presumably because of a reduction in the legume content of the pasture. Since the lowest stocking rate used in their experiment was almost twice the regional average (King and Stockdale, 1981), there would seem to be little benefit from nitrogen fertilizer under present farming conditions in that region. There would be little benefit even at the economic optimum stocking rate (Fig. 25.2), since the economic optimum is somewhat below the biological optimum (Chapter 24).

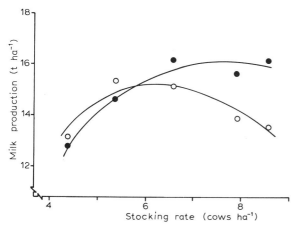

Fig. 25.2. Milk production (t ha$^{-1}$) from grass-clover swards (largely *L. perenne* and *T. repens*), receiving either no nitrogen fertilizer (○) or 224 kg N ha$^{-1}$ yr$^{-1}$ (●) and grazed at various stocking rates. Data are the means of two years. (From King and Stockdale, 1981.)

Since grass-legume swards produce as much herbage as pure grass pastures receiving between 150 and 200 kg N ha$^{-1}$ yr$^{-1}$ (Whitehead, 1970; see Fig. 6.4), and show little response to nitrogen fertilizer over the range 0 to 200 kg N ha$^{-1}$ yr$^{-1}$, applications up to about 150 to 200 kg N ha$^{-1}$ yr$^{-1}$ are unlikely to be economic. This conclusion is supported by a simulation study of a beef-producing system (Doyle and Morrison, 1983). Unfortunately, 80% of pastures in England and Wales, for example, receive less than 200 kg N ha$^{-1}$ yr$^{-1}$ (Leech and Hughes, 1983), the average being 125 kg N ha$^{-1}$ yr$^{-1}$. There is therefore some doubt about the value of much of the nitrogen fertilizer currently used. It seems likely that there may be two economic optima in nitrogen fertilizer use: one based on grass-legume swards and with low or zero nitrogen fertilizer use, and the other based on grass swards and with nitrogen-fertilizer inputs of 300 to 400 kg N ha$^{-1}$ yr$^{-1}$ (e.g. Doyle and Morrison, 1983).

In view of the large expenditure in nitrogen fertilizer for use on grasslands world-wide, but especially in Europe, it is important that the economic optimum for nitrogen fertilizer use should be known under a wide range of soil and climatic conditions, animal enterprises (such as milk, beef or lamb production), stocking rates, and management systems. In view of the cost of long-term grazing experiments, it is unlikely that the necessary information can be obtained experimentally; it will therefore be necessary to construct models to predict likely responses, such as those used by Doyle and Morrison (1983) for nitrogen fertilizer use, or by Curll (1978) for phosphate fertilizer use (see below). These models were both relatively simple, and more complex models will probably be needed.

### Phosphate fertilizer

In spite of the apparent importance of phosphorus fertilizer in the intensification of animal production, especially in Australia and New Zealand (see above), surprisingly few studies have been made of the effect of phosphate fertilizer on animal output. However, Kohn (1974), Ozanne et al. (1976), Southwood et al. (1976), and Curll (1977), amongst others, have shown increases in animal production following applications of phosphate fertilizer. In the absence of comprehensive experimental data, several attempts have been made to model the effects of phosphate on animal production, mainly to determine the economic optimum for phosphate application (Bennett and Ozanne, 1973; Curll, 1978). The models differ markedly in approach, but have proved fairly reliable in predicting responses to phosphate fertilizer.

## BOTANICAL COMPOSITION

As in the case of fertilizer use, there are surprisingly few studies of the effects of botanical composition on animal output. Differences between species in total herbage production, herbage quality, and seasonal patterns of production and quality might affect animal output. Such differences have been considered in Chapter 8.

### Survey data

The use of survey data to determine the effects of the botanical composition of pastures on animal output is fraught with the same dangers as those encountered when investigating the effects of fertilizers (see above). For example, Forbes et al. (1980) and Peel and Green (1984) found that differences in animal output on farms in England were related to the proportions of sown and indigenous species in pastures. However, as Peel and Green (1984) pointed out, the differences in output are more likely to be caused by associated differences in fertilizer input, stocking rate, soil type, climate and management, than by pasture composition *per se*. These other factors are likely to affect both animal output and pasture composition, and so lead to a spurious correlation between botanical composition and animal output. Such surveys are useful for constructing hypotheses, but the hypotheses must then be tested experimentally.

**Experimental evidence**

A considerable number of comparisons have been made of the performance of animals fed on legumes as opposed to grasses (Thomson, 1979). Almost without exception, animals gained more weight and produced more milk when fed on legumes (Thomson, 1979); in some cases animals gained weight twice as rapidly on legume herbage as on grass herbage. The superior performance of animals on legume herbage is largely due to the greater amounts of herbage consumed (e.g. Ulyatt, 1973); this is only partly attributable to the greater digestibility (Chapter 14) of legume herbage. In addition to eating more of the legume herbage, animals also utilize the herbage more efficiently (e.g. Rattray and Joyce, 1974). This may be partly because of the greater protein content of legumes, but other nutritional factors are also important (Reed, 1981).

Numerous comparisons have been made of the performance of animals grazing various grass species, but no comprehensive review has been published. Reed (1972) and Snaydon (1978, 1979, 1981) have reviewed some aspects of the studies. The available evidence is very fragmentary, mainly because studies usually include only two or three species, so that systematic comparison is not possible. In addition, most studies have been carried out at only one stocking rate though, as Fig. 25.1 indicates, differences in animal performance are likely to be very dependent on stocking rate. The various studies have also involved different grazing management systems, which makes comparisons between studies difficult. Some studies have used "put and take" methods (see above) whereas others have used set-stocking.

The most intensive comparison of grass species has been made by Axelsen and Morley (1968); they compared eight species and cultivars under set-stocking over a period of five years. The ranking of species and cultivars differed, depending on the production criterion used, such as wool production or liveweight gain of lambs. They therefore used an "economic index", based on a weighted mean of the criteria. There was a 12% difference in "economic index" between the most productive grass (*Phalaris aquatica*, cv. Australian Commercial) and the least productive grass (*Dactylis glomerata*, cv. Brignoles).

Animal production from various grass species might differ as a result of differences in herbage production, herbage quality, and seasonal patterns in these attributes (Chapter 8). Most of the differences in animal production that have been found, including those found by Axelsen and Morley (1968), seem to be mainly attributable to differences in herbage quality, rather than herbage production. In particular, species with a lower digestibility, such as *Dactylis glomerata, Festuca arundinacea* and *F. rubra*, generally gave lower animal outputs than those with higher digestibility, such as *Festuca pratensis, Lolium perenne* and *Phleum pratense*.

Such comparisons of single species stands are largely academic, since most pastures consist of complex mixtures (Chapter 8), both because of the mixtures sown and because of changes that occur thereafter. Animals are able to selectively graze these mixtures and, as a result, the composition of their diet may be very different from that of the pasture itself (Chapter 13). A number of studies have been made of different mixtures, especially comparisons of permanent pasture versus sown leys. Most of the early studies were invalid, because different stocking rates were used on the different pastures, or because different fertilizer applications were used. One of the most comprehensive studies, though using a "put and take" system of stocking, was made by Mudd and Meadowcroft (1964). Over a total of twelve years of comparison, they found no difference in milk production between permanent pasture and leys, when they received the same fertilizer treatment. Differences in annual milk production followed the usual pattern of differences in herbage production between leys and permanent pasture, discussed in Chapter 8. Herbage production and, as a result of the "put and take" system, milk production was less from the ley in the year of resowing, was greater in the next two years, and was equal or less after that. Gross margins (Chapter 23) were greater from permanent pasture than from leys, because of the greater costs of the latter. Comparisons using set-stocking with sheep (Eyles, 1963, 1964) and dairy cows (Bastiman and Mudd, 1971) showed a slight advantage ($<5\%$) for leys, though this seemed to be mainly due to differences in management, for example the use of extra supplementary feed in the latter study. In both cases, the slight increase in production would not be sufficient to offset higher cost, such as resowing and, in the latter case, supplementary feed and additional veterinary costs (Bastiman and Mudd, 1971).

## CONCLUSIONS

Environmental factors are the driving variables which largely determine the primary and secondary productivity of both natural and managed ecosystems. Except for protected cropping (i.e. glasshouse production) farmers have relatively little control over climatic conditions, but have considerable control over soil conditions, and especially over soil nutrient status, largely by the use of fertilizers. As a result, fertilizer use is probably the most important management variable affecting agricultural output.

The inherent nutrient status of agricultural soils is highly variable, depending on parent material, climate and past history. Most soils, however, are deficient to some degree in one or more macronutrient or micronutrient. Many soils are grossly nutrient deficient when first cleared for agricultural use; phosphate deficiency, in particular, is common. The labile nature of nitrogen in soils (Chapter 22) leads to extensive deficiencies, though this is highly dependent upon the abundance of legumes.

The availability of many nutrients is considerably affected by the nature and speed of nutrient cycling (Chapters 17 and 18); this is greatly affected by grazing management.

The botanical composition of pastures is also greatly affected by soil nutrient status, and therefore by fertilizer use (Chapter 8). Thus, herbage yield and botanical composition are both affected by soil fertility, though there is little evidence that herbage yield is appreciably affected by botanical composition *per se* (Chapter 8).

Animal production is determined by both herbage yield and herbage quality. Since fertilizer applications affect both these attributes, animal output can be greatly affected by fertilizer use. However, the effect is very dependent upon stocking rate, which is one of the most important management variables affecting animal output (Chapter 24).

Although most studies of the effects of phosphate fertilizer have been carried out on soils with a previous history of fertilizer application, most have shown a marked increase in animal production when phosphate was applied.

The effects of nitrogen fertilizer on animal output have been very variable, depending especially on the stocking rate, the abundance of legume in the sward, and the type of animal used. Grass–legume swards, in the absence of nitrogen fertilizer, produce as much herbage as grass swards receiving about 150 kg N ha$^{-1}$ yr$^{-1}$, and show little response to nitrogen fertilizer over the range 0 to 150 kg N ha$^{-1}$ yr$^{-1}$; in addition they produce herbage of higher quality. As a result, little response in animal output to nitrogen might be expected over the range 0 to 200 kg N ha$^{-1}$ yr$^{-1}$. In general, this is confirmed by experimental evidence. Surprisingly, most pastures receiving nitrogen fertilizer receive less than 200 kg N ha$^{-1}$ yr$^{-1}$; the value of these applications are therefore in doubt.

## REFERENCES

Axelsen, A. and Morley, F.H.W., 1968. Evaluation of eight pastures by animal production. *Proc. Aust. Soc. Anim. Prod.*, 7: 92–98.

Bastiman, B. and Mudd, C.H., 1971. A farm scale comparison of permanent and temporary grass. *Exper. Husbandry*, 20: 73–83.

Beadle, N.C.W., 1962. An alternative hypothesis to account for the generally low phosphate content of Australian soils. *Aust. J. Agric. Res.*, 13: 434–445.

Bennett, D. and Ozanne, P.G., 1973. Deciding how much superphosphate to use. *C.S.I.R.O. Div. Plant Ind., Annu. Rep., 1972.* pp. 45–47.

Castle, M.E., MacDaid, E. and Watson, J.N., 1972. Some factors affecting milk production from grassland at the Hannah Institute, 1951–70. *J. Br. Grassland Soc.*, 27: 87–92.

Craven, J.A. and Kilkenny, J.B., 1974. Economics of nitrogen fertilizer on grassland for milk and meat production. *Proc. Fertil. Soc.*, 142: 39–56.

Curll, M.L., 1977. Superphosphate on perennial pastures. 1.

Effects of a pasture response on sheep production. 2. Effects of a pasture response on steer beef production. *Aust. J. Agric. Res.*, 28: 991–1014.

Curll, M.L., 1978. Simulation: an aid to decisions on superphosphate use for beef production. *Agric. Syst.*, 3: 195–204.

Curll, M.L., 1982. The grass and clover content of pastures grazed by sheep. *Herbage Abstr.*, 52: 403–411.

Donald, C.M., 1964. Phosphorus in Australian agriculture. *J. Aust. Inst. Agric. Sci.*, 30: 70–105.

Donald, C.M. and Williams, C.H., 1954. Fertility and productivity of podzolic soil as influenced by subterranean clover and superphosphate. *Aust. J. Agric. Res.*, 5: 664–687.

Doyle, C.J. and Morrison, J., 1983. An economic assessment of the potential benefits of replacing grass by grass–clover mixtures for 18-month beef systems. *Grass Forage Sci.*, 38: 273–282.

Eyles, D.E., 1963. Lamb production from improved permanent and re-seeded pastures at two fixed stocking rates. *Exper. Progr., Grassland Res. Inst.*, 15: 67–69.

Eyles, D.E., 1964. Lamb production from improved permanent and re-seeded pastures at two fixed stocking rates. *Exper. Progr., Grassland Res. Inst.*, 16: 84–86.

Forbes, T.J., Dibb, C., Green, J.O., Hopkins, A. and Peel, S., 1980. *Factors Affecting the Productivity of Permanent Grassland.* Grassland Research Institute, Hurley, 141 pp.

Hawkins, S.W. and Rose, P.H., 1979. The relationship between the rate of fertilizer nitrogen applied to grassland and milk production: an analysis of recorded farm data. *Grass Forage Sci.*, 34: 203–208.

Holmes, W., 1968. The use of nitrogen in the management of pasture for cattle. *Herbage Abstr.*, 38: 265–277.

King, K.R. and Stockdale, C.R., 1981. The effects of stocking rate and nitrogen fertilizer on the productivity of irrigated perennial pasture grazed by dairy cows. 2. Animal Production. *Aust. J. Exper. Agric. Anim. Husbandry*, 20: 537–542.

Kohn, G.D., 1974. Superphosphate utilization in clover ley farming. 1. Effects on pasture and sheep production. *Aust. J. Agric. Res.*, 25: 525–535.

Leech, P.K. and Hughes, A.D., 1984. *Fertilizer Use on Farm Crops in England and Wales.* Ministry of Agriculture, Fisheries and Foods, London, 10 pp.

Levy, E.B., 1970. *Grasslands of New Zealand.* Government Printer, Wellington, 3rd ed., 374 pp.

Milne, J.A., Hodgson, J., Thompson, R.L., Souter, W.G. and Barthram, G.T., 1982. The diet ingested by sheep grazing swards differing in white clover and perennial ryegrass content. *Grass Forage Sci.*, 37: 209–218.

Minson, D.J. and Rees, M.C., 1976. The advantages and disadvantages of using survey data to derive production responses to nutrient inputs by the stabilized group method. *Proc. Aust. Soc. Anim. Prod.*, 11: 289–292.

Morley, F.H.W., 1981. Options in pasture research. *Trop. Grasslands*, 15: 71–84.

Mudd, C.H. and Meadowcroft, S.C., 1964. Comparison between the improvement of pastures by the use of fertilizers and by reseeding. *Exper. Husbandry*, 10: 66–84.

Ozanne, P.G., Purser, D.B., Howes, K.M.W. and Southey, I., 1976. Influence of phosphorus content on feed intake and weight gain in sheep. *Aust. J. Exper. Agric. Anim. Husbandry*, 16: 353–360.

Peel, S. and Green, J.O., 1984. Sward composition and output on grassland farms. *Grass Forage Sci.*, 39: 107–110.

Rattray, P.V. and Joyce, J.P., 1974. Nutritive value of white clover and perennial ryegrass. 4. Utilization of dietary energy. *N.Z. J. Agric. Res.*, 17: 401–408.

Reed, K.F.M., 1972. The performance of sheep grazing different pasture types. In: J.H. Leigh and J.C. Noble (Editors), *Plants for Sheep in Australia.* Angus and Robertson, Sydney, N.S.W., pp. 193–204.

Reed, K.F.M., 1981. A review of legume-based versus nitrogen-fertilized pasture systems for sheep and cattle. In: J.L. Wheeler and R.D. Mochrie (Editors), *Forage Evaluation: Concepts and Techniques.* C.S.I.R.O., Melbourne, Vic., pp. 401–417.

Rees, M.C., Minson, D.J. and Kerr, J.D., 1972. Relation of dairy productivity to feed supply in the Gympie district of south-eastern Queensland. *Aust. J. Exper. Agric. Anim. Husbandry*, 12: 553–560.

Rossiter, R.C., 1964. The effect of phosphate on the growth and botanical composition of annual type pasture. *Aust. J. Agric. Res.*, 15: 61–76.

Ryan, M. and Finn, T., 1976. Grassland productivity. 3. Effects of phosphorus on yield of herbage at 26 sites. *Ir. J. Agric. Res.*, 15: 11–23.

Snaydon, R.W., 1978. Indigenous species in perspective. In: *Proc. 1979 Crop Protection Conf., Weeds*, pp. 905–913.

Snaydon, R.W., 1979. Selecting the most suitable species and cultivars. In: A.H. Charles and R.J. Haggar (Editors), *Changes in Sward Composition and Productivity.* British Grassland Society, Hurley, pp. 178–189.

Snaydon, R.W., 1981. How important is the botanical composition of pastures? *J. Agric. Soc. Univ. College, Wales*, 62: 126–139.

Southwood, O.R., Saville, D.G. and Gilmour, A.R., 1976. The value to Merino ewes and lambs of superphosphate dressing on a subterranean clover ley. *Aust. J. Exper. Agric. Anim. Husbandry*, 16: 197–203.

Stangel, P.J., 1976. World fertilizer reserves in relation to future demand. In: M.J. Wright (Editor), *Plant Adaptation to Mineral Stress in Problem Soils.* Cornell University, Ithaca, N.Y., pp. 3–13.

Stockdale, C.R. and King, K.R., 1981. The effects of stocking rate and nitrogen fertilizer on the productivity of irrigated perennial pasture grazed by dairy cows. 1. Pasture production, utilization and composition. *Aust. J. Exper. Agric. Anim. Husbandry*, 20: 529–536.

Thomson, D.J., 1979. Effect of the proportion of legumes in the sward on animal output. In: A.H. Charles and R.J. Haggar (Editors), *Changes in Sward Composition and Productivity.* British Grassland Society, Hurley, pp. 101–109.

Ulyatt, M.J., 1973. Studies on the causes of differences in pasture quality between perennial ryegrass, short-rotation ryegrass and white clover. *N.Z. J. Agric. Res.*, 14: 352–467.

Underwood, E.J., 1966. *The Mineral Nutrition of Livestock.* Commonwealth Agricultural Bureau, Farnham Royal, 374 pp.

Whitehead, D.C., 1970. *The Role of Nitrogen in Grassland Productivity.* Commonwealth Agricultural Bureau, Farnham Royal, 202 pp.

Chapter 26

# HERBAGE CONSERVATION AND SUPPLEMENTS

H.A. BIRRELL

## INTRODUCTION

The grasslands of the world have been broadly classified into: (1) humid temperate; (2) semi-arid with summer rainfall (continental); (3) semi-arid with winter rainfall (mediterranean); and (4) subtropical, corresponding to the global climatic regions (Moore, 1964). Each of these grassland types has a characteristic seasonal pattern of pasture growth (Fig. 26.1, and Chapters 2 and 3). The large variation in pasture growth during the year presents problems in effectively using herbage, because the seasonal changes in the feed demands of animals, associated with pregnancy, lactation or cold stress, do not always coincide with the pasture growth cycles (Chapters 9–11). The problems can be partly overcome by manipulating the animal's requirements, for example, by varying the type of animal, the rate of stocking or the time of breeding (Chapters 9, 10, 11 and 24). The alternative is to manipulate the feed supply to fit the animal requirement — for example, by using feed supplements or by conserving fodder. Silage and hay can be produced either from surplus herbage or from specifically grown pasture or crops. *Forage crops*, crop residues, grains or browse plants are also used as supplements; these can be produced within the grazing systems, or can be produced outside it and imported. This chapter deals with forage conservation and supplementation in relation to the different pasture types.

## FORAGE CONSERVATION

The practice of conserving is very old. Herbage was conserved as silage in Egypt over three thousand years ago (Jarrige et al., 1982). Recent developments in forage conservation and its utilization have been reviewed by Marsh (1979), Wilkins (1981), Siebert and Hunter (1981), Thomas (1982) and Wilkinson (1983a, b).

### Conservation techniques

One of the most important developments has been the increasing use of silage, although hay is still the main method of conserving forage in many countries, particularly the United States of America and Australia. Jarrige et al. (1982) and Wilkinson (1983a) have shown that well-preserved silage is a better feed for dairy cows than hay made from the same pasture. Cows eat more silage than hay, and produce more milk per unit of feed intake. They also gain more weight, though this tends to be fat rather than muscle, perhaps because less amino acid is absorbed from silage. Improvements in ensiling techniques and in mechanization have increased the quality and consistency of silage, and have given greater confidence in its use. These improvements include: definition of optimum cutting date, precision chopping, wilting, automatic application of additives, and use of polythene seal stacks or bales. Secondary fermentation in silage reduces quality and can cause a variety of metabolic diseases. Secondary fermentation is less likely to occur if the water-soluble carbohydrate content (WSC) is kept above 3% of the total fresh weight. Tropical pasture species, in particular, usually have WSC contents lower than 2%, but the WSC content can be increased by rapid wilting before ensiling (Wilkinson, 1983a). Wilting before ensiling can also reduce the digestibility of silage (Marsh, 1979), but this is u

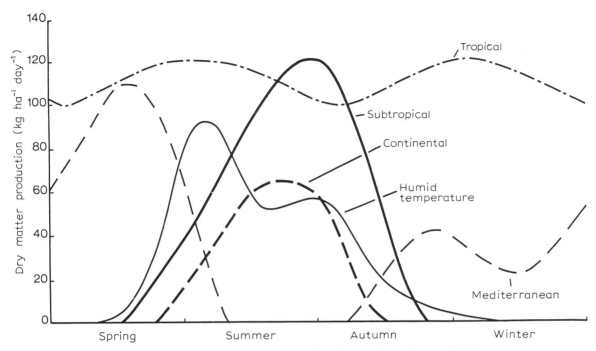

Fig. 26.1. The seasonal patterns of pasture production in several climatic zones (from Snaydon, 1981).

offset by increases in the amount of silage voluntarily eaten so that animal performance is increased.

Additives can be used to improve fermentation and so improve quality and increase animal performance. The most common additives for temperate crops are formic acid and formaldehyde. Additives such as molasses can increase the WSC content, and so reduce the risk of secondary fermentation, whereas urea can improve the nitrogen status of silage made from tropical species. Fine-chopping the silage (less than 7 cm) can increase the amount of silage eaten by both sheep and young cattle (Jarrige et al., 1982).

Fast drying, especially in the early stages, improves the quality of hay. Most of the advances in haymaking have centred on mechanical conditioning to accelerate this initial stage of drying. Conversely, the later stages of drying have presented major problems in haymaking, especially in wetter environments. Rain can often spoil hay before it is sufficiently dry to bale. Barn drying and the use of additives which prevent decomposition have been developed, but are expensive and have had limited application.

Faster and cheaper analytical techniques have improved the chemical evaluation of herbages, and this has led to a grading system for hay in the United States (Rohweder and Baylor, 1980).

**REGIONAL USE OF CONSERVATION**

**Temperate regions**

Severe winters at high latitudes mean that animals must be fed indoors for long periods. Traditionally, large areas of grassland were cut for hay during the short growing season, and the hay was fed during winter. This method is still used extensively in mountain areas of Europe; herbage production in these areas has been increased by chemical fertilization, irrigation, and the introduction of plant species suited to repeated harvesting. In lowland regions, silage has tended to replace hay as winter fodder, and indoor feeding has become commoner as systems have been intensified. For example, in Britain autumn calving is widespread; this causes a peak of feed requirement during autumn and winter, which is met by silage and feed supplements.

## Mediterranean regions

Winters are mild in mediterranean regions and animals can remain at pasture throughout the year, though feed may be deficient in winter. There is an abundance of high-quality forage in spring, followed by an abundance of low-quality dead material during summer (Purser, 1981). The surplus pasture in late spring can be harvested as hay, which should be of higher quality than standing dry feed. This conserved feed can allow higher stocking rates, or improvement in the seasonal nutrition of animals, or can act as a drought reserve (Lloyd, 1959). Unfortunately such benefits may not materialize in practice. In Australia negative correlations have been found between conservation and profitability (Bureau of Agricultural Economics, 1968). One reason for this is that in mediterranean regions, where annual species dominate pastures, spring conservation reduces seed production or removes the seed produced, and so reduces pasture growth in the following winter (Hamilton, 1974). The limitations in these pastures, and problems in improving the nutrition for cattle, have been reviewed by Allden (1982).

## Subtropical and tropical regions

In the developing countries of the tropics, most of the grassland is restricted to areas of poorer soils, areas of temporary fallow or along roadsides; the best land is used for crop production, because of increasing human populations. Social attitudes can also hinder progress in improving livestock nutrition in these countries ('t Mannetje, 1981). Problems in evaluating the nutritive value of the indigenous plant species, their component parts and their byproducts, as well as the difficulty in communicating the applicability of this knowledge to feeding systems, have been discussed elsewhere (e.g. Nitis, 1983).

In subtropical and tropical areas, consistent wet weather during the growth season, plus the heavy lignification of tropical plant species at maturity, hinders the development of effective conservation practices. Silage is probably the most effective method of conserving herbage; however, complete sealing and correct fermentation are major problems. Improved methods of ensiling in tropical areas are under investigation.

A browse plant, *Leucaena leucocephala*, a high-protein legume, has potential in subtropical areas for supplementing low-quality pasture, by increasing nitrogen intake and thus the intake of herbage. The plant has to be prevented from growing beyond browse level, usually by rotational cutting or grazing; this allows a higher carrying capacity and so increased cattle production (Jones and Jones, 1982). Most cultivars of *L. leucocephala* contain a goitrogenic amino acid, mimosine; however, provided the thyroid gland is not overenlarged, animal productivity can be increased by supplements of *L. leucocephala* (Jones and Jones, 1984). Tree lucerne (*Chamaecytisus palmensis*) can also provide a highly nutritious green feed in summer, producing in excess of 11 t of edible dry matter per hectare annually; it has also attracted attention as a possible fodder crop in temperate areas, such as New Zealand (Davies and Macfarlane, 1979) and Australia (Snook, 1982). Usually two to three years are required before the trees are large enough to withstand heavy grazing.

In arid areas of the subtropics, where standing hay is abundant during the dry season, the incorporation of a persistent legume, such as *Stylosanthes* spp., can increase nitrogen intake and hence encourage a greater intake of the dry standing herbage (Romero and Siebert, 1980).

The nutrition of cattle in the dry tropical areas of Australia has recently been reviewed by Winks (1984), who concluded that the many experimental management practices still need to be evaluated in integrated practical situations. Sowing better pasture species in place of native species probably has the greatest potential for improving the seasonal nutrition of grazing animals in these regions (Evans, 1981).

## EFFECTS OF CONSERVATION

Hutchinson (1971) noted that many reports show only small, generally variable, responses to fodder conservation. Using an energy-flow model, he postulated that increasing stock numbers to utilize conserved fodder would reduce the ability of the grazing system to provide fodder. Observations by Bishop and Birrell (1975) on the amounts of hay made and eaten by wether sheep grazed at different stocking rates demonstrate this antagonism (Fig. 26.2).

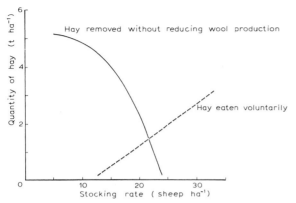

Fig. 26.2. The effect of stocking rate on the quantity of hay which can be removed from a system without incurring a penalty in wool production, and the effect of stocking rate on the quantity of hay which is voluntarily eaten by sheep in a quasi-mediterranean region with year-round grazing.

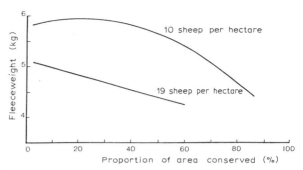

Fig. 26.3. The effect on annual wool production of cutting and removing herbage from different proportions of the pasture during spring, in grazing systems with sheep at different stocking rates. The experiment was carried out in a quasi-mediterranean region with year-round grazing.

In quasi-mediterranean regions, where perennial pasture species can survive the dry summer, it can be advantageous to close off small areas and cut hay for sale, provided the stocking rates of sheep are not too high. The annual wool production may actually increase, because the increased spring grazing pressure (Birrell and Bishop, 1980) removes surplus pasture in spring so allowing sheep (Birrell, 1981; Birrell et al., 1980) and cattle (Watson et al., 1979; Scattini, 1984) to select fresh green pasture which would otherwise be less available beneath an accumulation of dead herbage material, when the pasture growth begins. However, if the grazing pressure is increased too much, either by increasing animal numbers or increasing the area exclosed for conservation, wool production is decreased (Fig. 26.3). Similarly, early conservation in spring can provide a regrowth of nutritious green pasture which lasts well into the summer period. This can be used to advantage for young growing animals (McLaughlin and Bishop, 1969).

## Supplementation in year-long grazing systems

Recent reviews of the behavioural and nutritional factors governing the performance of grazing animals (Allden, 1981; Freer, 1981; Morley, 1981; Saville, 1981) indicate that inability to estimate the amount of pasture material eaten is a major obstacle to improving the nutrition of grazing animals. Partial substitution of pasture by supplements further increases the biological complexity, and makes the measurement of the substitution rate extremely difficult.

Estimates for grazing sheep suggest that, on a dry matter basis, 100 g of grain supplement replace between 30 and 90 g of herbage intake (Allden and Jennings, 1962; Holder, 1962; Langlands, 1969). The lower values relate to sheep supplemented on pasture of low availability, whereas the high values relate to abundant pasture (Langlands, 1969). Similarly, 100 g of hay [60% digestible organic matter (DOM)] can replace 80 g of pasture herbage (Birrell, 1984), though this substitution also depends upon the amount of pasture available and its quality relative to the supplement (Elderidge and Kat, 1980). Sheep eat more high-quality hay (69% DOM) when feeding on dry summer pasture than when feeding on green winter pasture (Fig. 26.4). Hay intake decreases with increasing availability of green feed in winter, but is less affected by the availability of dry feed in summer (Fig. 26.4). Hay intake is also reduced by inclement weather and by a low body weight (Birrell, 1984).

Supplements reduce the time an animal spends grazing pastures, though this reduction is not necessarily proportional to the contribution that the supplement makes to the animal's diet. Sheep receiving supplements may still graze for between $1\frac{1}{2}$ and 3 hours, compared with the normal period of between 9 and 11 hours (Holder, 1962; Birrell, 1984), but the rate of herbage intake is slow because they select the higher-quality components of the sward.

# HERBAGE CONSERVATION AND SUPPLEMENTS

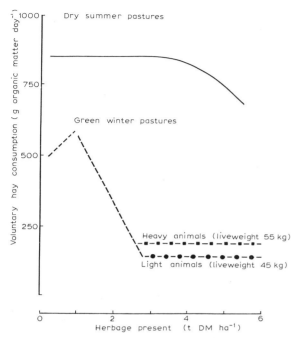

Fig. 26.4. The effect of the amount of herbage present, and its quality, on the amount of hay voluntarily eaten by wethers when fed hay while grazing on dry summer pasture and green winter pasture in a quasi-mediterranean region with year-round grazing (Birrell, 1984).

## FORAGE CROPS

Wheeler (1981) has outlined the advantages and disadvantages of including forage crops in grazing systems. The forage crops used are often physiologically different from the pastures that they supplement. For example, subtropical or drought-resistant species may be used to supplement temperate species during summer, and temperate species may be used to supplement subtropical species during winter.

Table 26.1 shows the yield and quality of some summer forage crops used for dairying under irrigation in a mediterranean-type area of southern Australia (Mason and Pritchard, 1983). In dryland areas, maize has recently received attention, because it tolerates lower soil temperatures at planting than *Sorghum* spp.; this attribute allows earlier growth and a longer growth period before soil moisture in summer is depleted. Lucerne is also a popular special-purpose pasture.

Winter forage cereals (e.g. triticale, oats and

TABLE 26.1

The dry matter yield, digestibility and protein content of irrigated summer grass crops in southern Australia

| | Yield (t ha$^{-1}$) | Digestibility (%) | | Protein (%) | |
|---|---|---|---|---|---|
| | | first cut | mean of 3 cuts | first cut | mean of 3 cuts |
| Shirohie millet | 16.2 | 70.1 | 66.2 | 15.4 | 11.6 |
| Forage sorghum | 17.3 | 66.1 | 63.5 | 11.4 | 9.1 |
| Sweet sorghum | 25–30 | 58–60 | | 5–6 | |
| Maize | 20–26 | 60–64 | | 7–9 | |

Source: Mason and Pritchard (1983).

barley) can produce more dry matter than short-term ryegrasses (*Lolium* spp.) (Craig, 1983), though the value of the practice in mediterranean regions is limited because permanent pastures usually outyield winter forage crops (Shovelton, 1976).

In areas where animals are able to remain outdoors, complementary crops are the commonest method of overcoming seasonal deficiencies in pasture-based systems (Wheeler, 1981). In temperate regions with good summer rains, or where irrigation is available, root crops (e.g. turnips and swedes) grow quickly and store nutrients for later use; they may also complement conserved herbage nutritionally (Nicol and Barry, 1980).

## CONCLUSIONS

In most climatic regions of the world there are large seasonal variations in pasture growth and, in most animal enterprises, there are also large seasonal variations in the feed requirement of animals, as a result of their reproductive cycles or because of climatic factors. One of the most difficult aspects of grassland management is to match feed supply to feed requirement. Conserving herbage in times of feed surplus, to use as a fodder supplement in times of feed deficit, is one method of matching supply to demand. It is most important in areas where winter conditions are severe, but may also be used when summer drought conditions are encountered. Its value is less clear where animals can be pastured throughout the year. When animals are kept indoors under controlled conditions for part of the year, it is easy to determine and formulate rations for conserved herbage and concentrates according to current feeding standards

(National Research Council, 1976; Agricultural Research Council, 1980). It is much more difficult to formulate feeding systems for supplementing animals which are grazing pasture, because the supplements may reduce the animals' intake of pasture herbage. So far only simple models of supplementation have been constructed (e.g. Freer and Christian, 1983); these models largely serve to emphasize the limited conceptual basis of our knowledge of the relationships between animals and plants. Morley (1981) commented that the uncertainties in the biology, the economics, and the managerial decisions in using supplements make it difficult to improve the nutrition of grazing animals. He suggested that perhaps the major benefit from supplementary feeding of grazing animals may come from the psychological comfort obtained by the farmer through knowing his animals are well fed, and that this feeling of security may compensate for any economic losses that occur.

## REFERENCES

Agricultural Research Council, 1980. *The Nutrient Requirements of Ruminant Livestock*. Commonwealth Agricultural Bureaux, Farnham Royal, 314 pp.

Allden, W.G., 1981. Energy and protein supplements for grazing livestock. In: F.H.W. Morley (Editor), *World Animal Science, B1. Grazing Animals*. Elsevier, Amsterdam, pp. 289–308.

Allden, W.G., 1982. Cattle growth in a mediterranean environment. *Aust. Meat Res. Com. Rev.*, No. 42.

Allden, W.G. and Jennings, A.C., 1962. Dietary supplements to sheep grazing mature herbage in relation to herbage intake. *Proc. Aust. Soc. Anim. Prod.*, 4: 145–153.

Birrell, H.A., 1981. Some factors which affect the liveweight change and wool growth of adult Corriedale wethers grazed at various stocking rates on perennial pastures in southern Victoria. *Aust. J. Agric. Res.*, 32: 353–370.

Birrell, H.A., 1984. Effects of pasture availability and liveweight on the feeding of grass hay to sheep. *Aust. J. Exper. Agric. Anim. Husbandry*, 24: 26–33.

Birrell, H.A. and Bishop, A.H., 1980. Effect of feeding strategy and area of pasture conserved on the wool production of wethers. *Aust. J. Exper. Agric. Anim. Husbandry*, 20: 406–412.

Birrell, H.A., Reed, K.F.M. and Bird, P.R., 1980. Seasonal limitations to the nutrition of sheep and beef cattle in the high rainfall areas of south-eastern Australia. *Proc. Aust. Soc. Anim. Prod.*, 13: 32–36.

Bishop, A.H. and Birrell, H.A., 1975. Efficiency of grazing/fodder conservation systems. In: R.L. Reid (Editor), *Proc. 3rd World Conf. Anim. Prod. Melbourne*. Sydney University Press, Sydney, N.S.W., pp. 217–225.

Bishop, A.H., Birrell, H.A. and Tew, A., 1966. The effect of stocking rate during spring on wool production and liveweight gain. *Proc. Aust. Soc. Anim. Prod.*, 6: 161–163.

Bureau of Agricultural Economics, 1968. Wool growing in the Hamilton district of Victoria. *Wool Econ. Res. Rep.*, No. 14: 50 pp.

Craig, A.D., 1983. A review of winter fodder crops in Victoria. In: Fodder Crops for Profit. In: *26th Annu. Conf. Grassland Soc. Vic.*, pp. 38–48.

Davies, D.J.G. and Macfarlane, R.P., 1979. Multiple purpose trees for pastoral farming in New Zealand: with emphasis on tree legumes. *N.Z. J. Agric. Sci.*, 13: 177–186.

Elderidge, G.A. and Kat, C., 1980. The effect of hay supplementation and pasture availability on pasture selection and substitution. *Proc. Aust. Soc. Anim. Prod.*, 13: 245–248.

Evans, T.R., 1981. Overcoming nutritional limitations through pasture management. In: J.B. Hacker (Editor), *Nutritional Limits to Animal Production from Pastures*. Commonwealth Agricultural Bureaux, Farnham Royal, pp. 343–361.

Freer, M., 1981. The control of food intake by grazing animals. In: F.H.W. Morley (Editor), *World Animal Science, B1. Grazing Animals*. Elsevier, Amsterdam, pp. 105–124.

Freer, M. and Christian, K.R., 1983. Application of feeding standards systems to grazing ruminants. In: G.E. Robards and R.G. Packham (Editors), *Feed Information and Animal Production*. Commonwealth Agricultural Bureaux, Farnham Royal, pp. 333–355.

Hamilton, D., 1974. Alternatives to dry annual pasture for steers over summer and late effects on liveweight gain during winter. *Proc. Aust. Soc. Anim. Prod.*, 10: 99–102.

Holder, J.M., 1962. Supplementary feeding of grazing sheep — its effect on pasture intake. *Proc. Aust. Soc. Anim. Prod.*, 4: 154–159.

Hutchinson, K.J., 1971. Productivity and energy flow in grazing/fodder conservation systems. *Herbage Abstr.*, 41: 1–10.

Jarrige, R., 1980. Place of herbivores in the agricultural ecosystems. In: Y. Ruckebusch and P. Thivend (Editors), *Digestive Physiology and Metabolism in Ruminants*. M.T.P. Press, Lancaster, pp. 763–823.

Jarrige, R., Demarquilly, D. and Dulphy, J.P., 1982. Forage conservation. In: J.B. Hacker (Editor), *Nutritional Limits to Animal Production from Pastures*. Commonwealth Agricultural Bureaux, Farnham Royal, pp. 363–387.

Jones, R.J. and Jones, R.M., 1982. Observations on the persistence and potential for beef production of pastures based on *Trifolium semipilosum* and *Leucaena leucocephala* in subtropical coastal Queensland. *Trop. Grasslands*, 16: 24–29.

Jones, R.M. and Jones, R.J., 1984. The effect of *Leucaena leucocephala* on liveweight gain, thyroid size and thyroxine levels of steers in south-eastern Queensland. *Aust. J. Exper. Agric. Anim. Husbandry*, 24: 4–9.

Langlands, J.P., 1969. The feed intake of sheep supplemented with varying quantities of wheat while grazing pastures differing in herbage availability. *Aust. J. Agric. Res.*, 20: 919–924.

Lloyd, A.G., 1959. Fodder conservation in the southern tablelands wool industry. *Rev. Mark. Agric. Econ.*, 27: 5–50.

't Mannetje, L., 1981. Problems of animal production from

tropical pastures. In: J.B. Hacker (Editor), *Nutritional Limits to Animal Production from Pastures.* Commonwealth Agricultural Bureaux, Farnham Royal, pp. 67–85.

McLaughlin, J.W. and Bishop, A.H., 1969. Management of weaner sheep in western Victoria. The influence of time of mowing, addition of oat grain and method of feeding pasture hay upon current and subsequent production. *Aust. J. Exper. Agric. Anim. Husbandry*, 9: 272–277.

Marsh, R., 1979. The effects of wilting on fermentation in the silo and on the nutritive value of silage. *Grass Forage Sci.*, 34: 1–10.

Mason, W.K. and Pritchard, K.E., 1983. Summer fodder crops — Summary for Victoria. In: *Fodder Crops for Profit, 24th Annual Conference of Grassland Society of Victoria*, pp. 68–74.

Moore, C.W.E., 1964. Distribution of grasslands. In: C. Barnard (Editor), *Grasses and Grasslands.* MacMillan, London, pp. 182–205.

Morley, F.H.W., 1981. Management of Grazing Systems. In: F.H.W. Morley (Editor), *World Animal Science, B1. Grazing Animals.* Elsevier, Amsterdam, pp. 379–400.

National Research Council, 1976. Nutrient requirements of beef cattle. In: *Nutrient Requirements of Domestic Animals.* National Academy of Science, Washington, D.C., No. 4, pp. 56.

Nicol, A.M. and Barry, T.N., 1980. The feeding of forage crops. In: *Supplementary Feeding. Occas. Publ. N.Z. Soc. Anim. Prod.*, No. 7: 69–106.

Nitis, I.M., 1983. Feed analyses: the needs of developing countries. In: G.E. Robards and R.G. Packman (Editors), *Feed Information and Animal Production.* Commonwealth Agricultural Bureaux, Farnham Royal, pp. 433–449.

Purser, D.B., 1981. Nutritional value of Mediterranean pastures. In: F.H.W. Morley (Editor), *World Animal Science, B1. Grazing Animals.* Elsevier, Amsterdam, pp. 159–180.

Rohweder, A. and Baylor, J.E., 1980. New forage analyses offer new horizons for hay grading, marketing, evaluating forages. *Forage Grassland Progr.*, 20: 2–4.

Romero, A. and Siebert, B.D., 1980. Seasonal variations of nitrogen and digestible energy intake of cattle on tropical pastures. *Aust. J. Agric. Res.*, 31: 393–400.

Rossiter, R.C., 1966. Ecology of the Mediterranean annual-type pasture. *Adv. Agron.*, 18: 1–56.

Saville, D.G., 1981. Management and feeding of grazing animals during drought. In: F.H.W. Morley (Editor), *World Animal Science, B1. Grazing Animals.* Elsevier, Amsterdam, pp. 335–348.

Scattini, W.J., 1984. Hay conservation for cattle grazing on winter-grazed green panic (*Panicum maximum* var. *trichoglume*) pasture in south-eastern Queensland. *Aust. J. Exper. Agric. Anim. Husbandry*, 24: 20–25.

Shovelton, J., 1976. Pasture beats sod seeding and fodder crops for winter feed. *J. Vic. Dep. Agric.*, 74: 212.

Siebert, R.W. and Hunter, R.A., 1981. Supplementary feeding of grazing animals. In: J.B. Hacker (Editor), *Nutritional Limits to Animal Production from Pastures.* Commonwealth Agricultural Bureaux, Farnham Royal, pp. 409–426.

Snaydon, R.W., 1981. The ecology of grazed pastures. In: F.H.W. Morley (Editor), *World Animal Science, B1. Grazing animals.* Elsevier, Amsterdam, pp. 13–31.

Snook, L.C., 1982. Tagasaste (tree lucerne) *Chamaecytisus palmensis*: A shrub with potential as a productive fodder crop. *J. Aust. Inst. Agric. Sci.*, 46: 209–213.

Thomas, P.C., 1982. Utilization of conserved forages. In: D.J. Thomson, D.E. Beever and R.G. Gunn (Editors), *Forage Protein in Ruminant Animal Production. Occas. Publ. Br. Soc. Anim. Prod.*, No. 6: 67–76.

Watson, M.J., Bird, P.R. and Cayley, J.W.D., 1979. Pasture based feeding systems. In: *Vic. Dep. Agric., Proc. Beef Industry Conference (Dookie)*, pp. 10–35.

Wheeler, J.L., 1981. Complementing grasslands with forage crops. In: F.H.W. Morley (Editor), *World Animal Science, B1. Grazing Animals.* Elsevier, Amsterdam, pp. 239–260.

Wilkins, R.J., 1981. Improving forage quality by processing. In: J.B. Hacker (Editor), *Nutritional Limits to Animal Production from Pastures.* Commonwealth Agricultural Bureaux, Farnham Royal, pp. 389–408.

Wilkinson, J.M., 1983a. Silages made from tropical and temperate crops. I. The ensiling process and its influence on feed value. *World Anim. Rev.*, 45: 36–42.

Wilkinson, J.M., 1983b. Silages made from tropical and temperate crops. II. Techniques for improving the nutritive value of silage. *World Anim. Rev.*, 46: 35–40.

Winks, L.W., 1984. Cattle growth in the dry tropics of Australia. *Aust. Meat Res. Comm. Rev.*, No. 45: 45 pp.

# GLOSSARY

Carrying capacity: the stocking rate at the optimum grazing pressure.

Compensatory growth: the more rapid growth of animals which usually occurs after they have been subjected to a restricted diet and have lost weight.

Concentrate(s): high-quality supplementary feeds, such as cereal grains or special mixtures of feedstuffs.

Conservation: *see* Herbage conservation.

Continuous grazing: management systems in which livestock graze the pasture continuously for long periods, though the stocking rate ($q.v.$) may be varied.

Continuous model: a model which simulates changes by integrating at each moment in time.

Defoliation: partial or complete removal of the above-ground parts of the pasture (i.e. the sward), either by grazing or cutting.

Deterministic model: a model based on fixed mathematical relationships between independent variables.

Digestibility: the proportion (usually %) of the ingested herbage that is digested by the grazing animal, that is:

$$\frac{\text{herbage DM ingested} - \text{DM excreted}}{\text{herbage DM ingested}} \times 100.$$

Usually now measured by *in vitro* methods.

Discount rate: the assumed interest rate in calculating the present value of future income.

Discrete model: a model in which changes occur at fixed time intervals.

Drying-off: the time, at the end of the lactation period, when little milk is produced and milking ceases.

Dry cows: cows not currently being milked.

Dry matter on offer: the amount (usually mass above the ground level) of herbage available to the grazing animal at a given time.

Dry sheep equivalent: a unit of stock number, based on the feed requirement of dry sheep, i.e. non-breeding ewes or wethers ($q.v.$).

Dynamic model: a model in which changes over periods of time are simulated.

Ewe equivalent: a unit of stock number, based on the feed requirement of a breeding ewe.

Finished stock: animals suitable for slaughter.

Fixed costs: those costs which remain constant, regardless of the intensity of management (such as rent, general overheads, regular labour, and machinery), as opposed to variable costs ($q.v.$).

Flush: the practice of giving extra feed to female animals in the period before a critical reproduction phase (e.g. mating or parturition).

Forage conservation: *see* Conservation.

Forage crops: arable crops (e.g. turnips, kale, forage oats or forage maize) grown specifically for grazing, and usually fed at times when the supply of fresh herbage is inadequate.

Grassland: a plant community consisting largely of grasses and legumes, but often containing other dicotyledonous species: usually short ($<1$ m) and normally grazed or cut.

Grazing pressure: the number of animals per unit of feed available.

Gross (financial) margin: a measure of profitability, being the financial returns less the variable costs ($q.v.$) but not the fixed costs ($q.v.$). Usually calculated per unit of land, but can also be calculated per animal.

Heifers: young dairy cows which have not yet produced their first calf.

Herbage: the above-ground parts of a pasture or sward, either *in situ* or after being cut.

Herbage accumulation: usually used in the context of *net* accumulation. The increase in herbage mass over a given period of time, usually between two cuttings or grazings.

Herbage conservation: the preservation of herbage (e.g. as hay or silage) so that it can be fed at times when the supply of fresh herbage is inadequate.

Hogget: a yearling sheep.

Indigenous species: pasture species which occur in grasslands but are not usually sown in leys: species which invade sown grassland.

Leaf area index (LAI): the total leaf area of herbage per unit area of land.

Ley: a recently sown pasture, usually destined to be ploughed out within 1–10 years. For administrative purposes, often defined as pastures less than 5 years (or 7 years) old (*see* Permanent pasture).

Marginal return: *see* Marginal revenue.

Marginal revenue: the value of the extra output obtained as a result of applying an extra unit amount of an input.

Net present value: the discounted value of future income.

Oversowing: methods of sowing desirable species into a pasture without completely destroying the existing pasture. This is often achieved by very heavy grazing followed by seed dispersal, by killing the existing sward by herbicide, or by removing a narrow strip of pasture and "slot-seeding".

Pasture: an area of grassland used for grazing or cutting (*see also* sward).

Permanent pasture: pasture, usually sown many years previously, which has now essentially reached equilibrium with the current environmental and management conditions. In Britain the term is used administratively for pasture more than 5 years old.

Profit: total revenue (income) less total costs.

"Put-and-take": a term usually used for the experimental technique in which the stocking rate is frequently varied to match the amount of herbage available.

Revenue: the income from an enterprise, often also termed "total revenue", equal to the output multiplied by the unit value of the output.

Rotational grazing: grazing management systems in which animals are moved around paddocks in sequence, so that each paddock is grazed off quickly (1–3 days) after a period of rest (3–5 weeks).

Set-stocking: experimental or production systems in which the stocking rate remains constant over long periods.

Sown species: pasture species (usually grasses and legumes) which are commonly sown in leys (*q.v.*).

Static model: a model of an instantaneous state, or which assumes steady-state conditions.

Stochastic model: a model which predicts the expected range of outcomes, based on statistical probability.

Stock unit: a measure of stock numbers based upon feed requirement, and using non-reproductive adult cattle as the basic unit. Thus an adult sheep is usually taken to be 0.15 stock units, though this depends on size and reproductive state (*see also* Ewe equivalent *and* Dry sheep equivalent).

Stocking pressure: *see* Grazing pressure.

Stocking rate: the number of animals per unit area of land, often expressed in stock units (*q.v.*). Note that the stocking rate is calculated on the basis of the total area of land in the grazing system; this is especially relevant when rotational grazing (*q.v.*) is practised. The term "stock density" would be a more accurate term.

Strip grazing: a form of rotational grazing in which an area of pasture is fenced off, sufficient for one day or for half a day of grazing.

Sward: this term has two common meanings: (a) the above-ground portion of a pasture, and (b) an area of grassland (more or less pasture).

Top-dressing: applications of fertilizer.

Total dry matter on offer: *see* Herbage on offer.

Utility function: an expression used to calculate the subjective benefit (or utility) gained from an enterprise, taking into account various goals. The expression normally includes components for net income and risk, but may also include components representing other goals, such as ease of management.

Variable costs: those costs which vary with the intensity of management (e.g. fertilizers, agrochemicals, additional labour), as opposed to fixed costs (*q.v.*).

Wether: a castrated male sheep.

Zero grazing: systems in which herbage is cut and carried to the animals, rather than being grazed *in situ*.

# SYSTEMATIC LIST OF GENERA[1]

## MONERA

### EUBACTERIA

*Rhizobium*

## PLANTS

### PTERIDOPHYTA

**FILICOPSIDA**
Dennstaedtiaceae
*Pteridium*

### MAGNOLIOPHYTA (ANGIOSPERMAE)

**LILIOPSIDA (MONOCOTYLEDONES)**
Poaceae
*Agrostis*
*Alopecurus*
*Anthoxanthum*
*Aristida*
*Arrhenatherum*
*Avena*
*Briza*
*Bromus*
*Chloris*
*Cynosurus*
*Dactylis*
*Digitaria*
*Elymus*
*Festuca*
*Holcus*
*Hordeum*
*Lolium*
*Molinia*
*Nardus*
*Panicum*
*Paspalum*
*Pennisetum*
*Phalaris*
*Phleum*
*Poa*
*Setaria*
*Sorghum*
*Themeda*
*Trisetum*
*Triticum*
*Vulpia*
*Zea*

**MAGNOLIOPSIDA (DICOTYLEDONES)**
Asteraceae (Compositae)
*Arctotheca*
*Bellis*
*Carduus*
*Chrysanthemoides*
*Cichorium*
*Hypochoeris*
*Leucanthemum*
*Taraxacum*
Boraginaceae
*Echium*
Brassicaceae (Cruciferae)
*Brassica*
Caryophyllaceae
*Sagina*
*Stellaria*
Chenopodiaceae
*Atriplex*
*Beta*
Fabaceae
*Chamaecytisus*
*Glycine*
*Lablab*
*Leucaena*
*Lotononis*
*Lotus*
*Medicago*
*Melilotus*
*Macroptilium*
*Stylosanthes*
*Trifolium*
*Ulex*
*Vicia*
Geraniaceae
*Erodium*
Hydrocotylaceae
*Hydrocotyle*
Mimosaceae
*Acacia*
Myrtaceae
*Eucalyptus*
Oxalidaceae
*Oxalis*
Plantaginaceae
*Plantago*
Polygonaceae
*Rumex*
Rosaceae
*Crataegus*
*Rubus*

## ANIMALS

### NEMATODA

*Haemonchus*
*Nematodirus*
*Ostertagia*

### ARTHROPODA

**INSECTA**
Coleoptera
Rhynchophora
*Sitonia*

### MOLLUSCA

**GASTROPODA**
*Helix*

### CHORDATA

**AVES**
Anseriformes
Anatidae
*Anser*

**MAMMALIA**
Artiodactyla
Antilocapridae
*Antilocapra*
Bovidae
*Bison*
*Bubalus*
Cervidae
*Capra*
*Cervus*
Suidae
*Sus*
Lagomorpha
Leporidae
*Oryctolagus*
Perissodactyla
Equidae
*Equus*

---

[1] The taxonomic relationships of plants and animals are shown. Plant genera are listed alphabetically under families, arranged into phyla and subphyla; invertebrates are listed under phyla, vertebrates under orders and families.

# AUTHOR INDEX[1]

Abbott, J.L., 175, *178*
Abbott, L.K., 175, *178*
Abruna, F., *100*
Acock, B., 50, *57*
Acocks, J.P.H., 21, *25*
Adams, A.F.R., 185, *186*, *196*
Adams, C.E., 158, *159*
Adamson, C.M., 208, 210, *211*
Addiscott, T.M., 54, *57*
Adem, L., *27*
Agricultural Research Council, 147, *152*, *252*, *252*
Aikman, D.P., 55, *57*
Aimes, S.J., 107, 109, *110*
Alberda, T., 10, 14, *15*
Albers, R.A.J.M., 176, *179*
Alcock, M.B., 71, *77*
Alderman, G., 149, *152*
Aldrich, D.T.A., 75, *77*
Alexander, K.I., 56, *57*
Allcroft, R., 146, *152*
Allden, D.M., 228, *235*
Allden, W.G., 119, *120*, 228, 233, *235*, 249–250, *252*
Allen, D.M., 91, *99*
Allen, L.H., *59*
Allison, F.E., 169, *171*, 199, *202*
Allos, H.F., 189, *194*
Alloway, J., *153*
Alloway, W.H., 146, *152–153*
Alvarez, F., *135*
Amer, F., *179*
Amerman, C.R., 208, *212*
Ameziane, T.E., *58*
Amoah, E.A., 156, *159*
Anderson, G.W., *25–26*
Anderson, J.R., 223, *226*
Anderson, M.C., 50, *57*
Anderson, W.K., 22, *25*
Angus, J.F., 50–51, *57*
Anonymous, 103, 105, 107, *110*, 228, *235*
Anslow, R.C., 13, *15*, 34, 42, 63, 68, 72, *77*, 85, *86*, 91, *99*
Antonovics, J., 55, *57*
ap Griffith, G., 149, 151, *152*
Appadurai, R.R., 71, *77*

Archer, K.A., 73, *77*
Arima, Y., *180*
Arkell, P.T., 24, *25*
Arkin, G.F., *203*
Armstrong, D.E., *180*
Armstrong, J.S., *211*
Arnold, E., 19–20, *25*
Arnold, G.W., 23–25, *25–26*, 36, 42, 118, *120*, 129–134, *134–135*, 141–142, *142*, 176, *178*, 228–231, *235*
Arundel, J.H., *235*
Asher, C.J., 175, *178*
Ashford, R., 73, *78*
Asiegbu, J.E., 73, *79*
Aslyng, H.C., 174, *178*
Aspinall, D., 37, 39–40, *42*
Aston, A.R., 206, 209–210, *211*
Atkin, R.K., 36, *42*
Atkinson, C.J., 34, *42*
Auld, B.A., 218, *225*
Austenson, H.M., 73–74, *77*
Axelsen, A., 229, *235–237*, 244, *245*
Axford, R.F.E., 182, *185*

Baars, J.A., 10, 12–13, *15*, *16*, 75, *77*, 109, *110–111*
Baber, H.L., 200, *202*
Bache, B.W., 174, *178*
Baghurst, P., *27*
Baier, W., 47–48, *57*
Bailey, A.W., 209, *213*
Bailey, P.J., 130, *135*
Baird, D.M., *236*
Baker, R.D., 66, 68, 81, 87, *100*, *111*, 131–133, *135–136*, 227, *235*
Baker, R.H., *44*
Baker, R.L., 120, *120*
Bakhuis, J.A., 85, *87*
Balch, C.C., 159, *159*
Baldwin, B.A., *136*
Baldwin, J.P., *178*
Baldwin, R.L., 98, *99*
Ball, J., *135*
Ball, P.R., 7, *15*, 164–170, *171–172*, *194–195*, 198, 201–202, *202*
Balleaux, J.C., 174, *178*
Barber, D.A., *179*
Barley, K.P., 176, *178*
Barnard, E., 156, *159*
Barnes, A., 54, *58*
Barnes, D.K., *195–196*

Barnes, R.J., *70*
Barron, R.J.W., *26*, *213*
Barrow, N.J., 23, *25*, 165, *171*, 174, *178*, 181–182, 184, *185*, 198, *202*
Barry, T.N., 251, *253*
Bartholomew, P.W., 74, *77*
Bartholomew, W.V., 189, *194*
Barthram, G.T., 117–118, *120*, *136*, *246*
Barton, G.E., *42*
Bastiman, B., 245, *245*
Bather, M., 159, *159*
Batten, G.D., 176, *178*
Bauert, P.A., *185*
Baylor, J.E., 248, *253*
Bazilevich, N.I., *26*
Beadle, N.C.W., 241, *245*
Bean, E.W., 38–39, *42*
Beckett, P.B.H., 167, *171*
Begg, J.E., 34, 37, *42*, *45*
Behaeghe, T.J., 11, *15*
Bell, B.A., *110*
Benjamin, R.W., *27*
Bennett, D., 23–24, *25*, *42*, 231, *235*, *237*, *243*, *245*
Bennett, I.L., 227–228, 230–231, *237*
Benseman, B.R., 106, *111*
Bentley, J.R., *27*
Beranger, C., 93, 97, *99–100*
Berendse, F., 56, *57*
Bergersen, F.J., 190, *194*
Berrie, A.M.M., 40, *43*
Berry, L.J., 23, *26*
Berryman, C., 222, *225*
Bertaud, D.S., 168, *172*
Betlach, M.R., *203*
Bettenay, E.A., 210, *211*, *213*
Betts, J.E., 156, 159, *159*
Bharat, R., 176, *178*
Bhat, K.K.S., 175, *178*
Bhide, S., 220, *225*
Biddiscombe, E.F., 22–24, *25–27*, 35, *46*
Bieleski, R.L., 175, *178*
Bines, J.A., 108, *110*
Bingham, F.T., 174, *178*
Binnie, R.C., 72–74, *77*
Birch, H.F., 176, *178*
Bircham, J.S., *43*, *78*, 93, *99–100*, 229, 233, *235*
Bird, I.F., *203*
Bird, J.N., 73, *77*
Bird, P.R., *25*, 236, 238, 252–253

---
[1] Page references to text are in roman type, to bibliographical entries in italics.

Birrell, H.A., 22–23, 25, 130, *135*, 228, 230, 233, *235*, 249–251, *252*
Bishop, A.H., *135*, 233, *235*, 249–250, *252–253*
Blackburn, A.G., 231–232, *235*
Blacklow, W.M., 34, *43*
Blackmer, A.M., 199, *202*
Blackmore, V., *211*
Blad, B.L., 49, *58*
Blair, G.J., 175–177, *178*, *180*
Blake, J.T., *152*
Blakemore, L.C., 173, *178*
Blaser, R.E., 72, 75–76, *77*, *79*, 176, *178*
Blaxter, K.L., 116, *120*, 125, *127*, 137–138, 140, *142*, 182, *185*
Bliss, P.J., *211*
Blum, A., 38, *42*
Bogdan, A.V., 61, *68*
Boissau, J.M., *142*
Boland, O.W., 176, *178*
Bond, G., 223, *225*
Bonin, S.G., 73–74, *77*
Boom, R.C., 73, *79*
Boord, C.T., *135*
Booysen, P. de V., 72–73, *77*, *235*
Bornemissza, G.F., 184, *185*
Bosch, J.M., 205, *211*
Boswell, C.C., 234, *235*
Boughton, W.C., 207–208, *211*
Bougler, J., *99*
Bouldin, D.R., *179*
Boundy, C.A.P., *25*
Bouwer, H., 210, *211*
Bowman, P.J., 24, *25*, 232, *238*
Bowman, R.S., *211*
Boyd, J., 33, *46*
Bradshaw, A.D., 83, *86*, 149, *153*
Brady, C.J., 39, *42*
Braithwaite, G.D., 181–182, *185*
Brash, D.W., 14, *16*
Bremner, J.M., 199, *202*
Brenchley, W.E., 4, *5*, 82, *86*
Brewer, D.W., *59*
Brewer, H.L., *203*
Brewster, J.L., 174, *178*
Bridger, B.A., *196*
Briegel, D.J., *25–26*
Briggs, H.M., 182, *185*
Briseno de la Hoz, V.M., 74–76, *77*
Brock, J.L., *15*, 81, *86*, 170, *172*, 191–193, *194–195*
Brockman, J.S., 66, *68*, *70*
Brockwell, J., *195*
Bromfield S.M., 175, 176, *178*, *179*, 181, 184, *185*
Brookes, I.M., 108, *110*
Brougham, R.W., 4, *5*, 7, 10, 13, *15*, 38, *43*, 73, 75, *77*, *78*, 82, 84, *86*, 102, *110*, 193, *194*, 234, *235*
Brown, D.M., 49, 51–52, *59*

Brown, K.R., *16*, 106, *110*, *195*
Brown, L.F., *57*
Brown, R.H., 72, *77*
Brown, S., *43*
Brown, T.H., 21, 24, *25*, 114, 118, *120*, 230–231, *235*
Browne, D., 104, 107, *110–111*
Browne, E., *236*
Brownell, J.R., *153*
Bruns, E., *128*
Brunswick, L.C.F., *15*, 140, *143*
Bryan, J.W., *196*
Bryant, A.M., 101–102, 105–109, *110*, 229, *235*
Bryant, H.T., 75–76, *77*, *178*
Bryant, M.J., 156, *159*
Bryden, J.M., *127*
Budtz-Olsen, O.E., *185*
Bull, L.B., *143*
Burden, R.J., 200, *202*
Bureau of Agricultural Economics, 249, *252*
Burford, J.R., 199, *202*
Burger, R. du T., 174, *178*
Burgmann, V.D., *237*
Burma, G.D., 24, *25*
Burnham, C.P., 14, *16*
Burns, K.N., 146, *152*
Burns, M., *124*, *128*
Burton, G.W., 73–74, *77*, *79*
Burwell, R.E., *59*, 177, *178*
Bush, I.G., *135*, 229, *235*
Butcher, E.J., *152*
Butler, G.W., 147–148, *152*
Butler, T.M., *111*
Byrne, G.F., 56, *57*
Byrne, P.F., 230, 233, *235*
Bywater, A.C., 98, *99*

Calder, F.W., 75, *77*
Caldwell, A.G., 65, *68*
Cale, W.G., 49, 55–56, *57*
Call, J.W., 148, *152*
Callinan, A.P.L., 229, *235*
Cameron, D.R., 52, *58*
Cameron, I.H., 26, 228, *235*
Cameron, R.A.D., 127, *128*
Camlin, M.S., 75, *77*, 84, *86*
Campbell, A.G., 72, *77*, 102, 104, 107, *110*, 227, 229, *235–236*
Campbell, C.A., 177, *180*
Campbell, C.R.G., 127, *128*
Campbell, J.I., *144*
Campbell, J.B., 140, *144*
Campbell, M.H., 219, *226*
Campbell, N.A., *135*, *235*
Campling, R.C., 140, *142*
Cannon, D.J., 228–229, *235–236*
Caradus, J.R., *180*
Carbon, B.A., *26*

Caro-Costas, R., *100*
Carran, R.A., *15*, 168, *171–172*, *195*
Carroll, B., 190, *194*
Carson, R.B., *77*
Carter, D.L., *153*
Carter, E.D., 21, 23, *25*, *27*, 227, 233, *236*
Carty, O., *111*
Cary, E.E., *153*
Castle, M.E., 102, 107–108, *110*, 241, *245*
Catchpoole, V.R., 197–198, *202–204*
Cayley, J.W.D., 21–22, 24, *25*, 227, *236*, *238*, *253*
Chacon, E., 141–142, *142*
Chadwick, M.J., *86*
Chalmers, M.I., 124–125, *127*
Chambers, D.T., 130, *135*
Champness, S.S., 82, *86*
Chancellor, R.J., 82, *87*
Chapman, H.P., 198, *204*
Chapman, H.W., *235*
Chapman, P.E., *17*, *69*
Chapman, T.G., 210, *211*
Charles-Edwards, D.A., 52, *57–58*
Chatfield, C., 125, *127*
Chen, P., 190, *194*
Chenost, M., 141, *142*
Chestnutt, D.M.B., 68, 74, *77*, 192, *194*
Chien, S.H., *203*
Chin, J.F., *25*, *236*
Chisholm, A.H., 230–231, *236*
Christensen, L.G., *99*
Christian, K.R., 48–49, 51, *57*, 210, *211*, *252*, *252*
Christie, A.J.R., *16*
Chrystall, B.B., *237*
Chu, A.C.P., 193, *194*
Church, B.M., 83, *86*
Church, D.C., 132, *135*
Churchill, F.M., *213*
Ciordia, H., 229, *236*
Clark, D.A., 119, *120*, *128*, 193, *194–195*
Clark, F.E., 165, *171*, 176, *178*
Clark, J., 76, *77*
Clark, R.C., 181, *185*
Clark, R.J., 159, *159*
Clarke, J.N., 117, *120–121*, *236*
Clayton, D.C., *110*
Clayton, D.G., *110*
Cleaver, T.J., *58*
Clement, C.R., 83, *86*
Clements, R.J., 56, *60*, 71, 75–76, *79*
Clifford, P.E., 32, 39–40, *42*
Clough, B.F., 37, *42*
Coates, M.E., 125, *128*
Cobby, J.M., *59*
Cole, C.V., 55, *57*, 187, *195*, *172*, 176–177, *178*, *180*
Coleman, D.C., *59*
Collett, B., 43–44, *59*, *79*

# AUTHOR INDEX

Collins, D.P., 85, *87*
Colvill, K.E., 32–34, 39–41, *42*
Colville, J.S., 205, *211*
Colwell, J.D., 220, *225*
Combellas, J., *111*, 130, *135*
Commonwealth Bureau of Pastures and Field Crops, 155, *159*
Condon, R.W., 208, *211*
Conniffe, D., 230, *236*
Connolly, J., 230, *236*
Connor, D.J., 50, *57*
Conrad, C.E., *213*
Conte, P.T., *60*
Conway, A., 229, 234, *236*
Cook, M.A.S., 107, *110*
Cooke, G.W., 168, *171–172*
Coop, I.E., 114, 118–119, *121*, 232, *236*
Cooper, J.P., 14, *16*, 31, 34, 36, *42*, *45*, 149–150, *152*
Cope, F., 210, *211*
Copeley, P.W., 187, *194*
Copeman, G.J.F., *195*
Cordero, S., 175, *178*
Cordery, I., 205, *211*
Cornelius, D.R., 20, 24, *25–26*
Cornelius, M.J., *203*
Cornforth, I.S., 177, *178*
Corral, A.J., 12–13, *16*, 29, *42*, 56, 58, 73, 74, *78*, 217–218, *226*
Cossens, G.G., 8, 12, 14, *16*, *17*
Costin, A.B., *194*, 208–209, *211*
Cotsell, J.C., 227, *236*
Coulter, J.D., 11, *16*
Coulton, L., *235*
Court, M.V., *16*
Cowan, R.T., 233, *236*
Cowling, D.W., 63–64, 66–67, *68*, 84, *86*
Cowlishaw, S.J., *236*
Cox, C.D., 199, *202*
Cox, E.H., *236*
Crabbe, C.L., 103, *110*
Craig, A.D., 251, *252*
Crampton, E.W., 140, *142*
Craven, J.A., 105, *110*, 242, *245*
Crawford, N.H., 210, *211*
Crespo, D.G., 20, 24, *25*
Crist, D., *194*
Cross, R.B., *185*
Crouchley, G., *235*
Crush, J.R., 15, *172*, 192, *194–195*
Crutzen, P.J., 200, *203*
Cullen, N.A., *235*
Cullen, P., 210, *212*
Cumberland, G.L.B., *16*, *111*
Cunningham, G.M., *211*
Cunningham, J.M.M., *127*
Cunningham, R.B., 59, *136*
Cunha, T.J., 126, *128*
Curll, M.L., 66, *68*, 71, 75–76, *78*, 241–243, *245–246*

Curnow, B.C., 27, *122*, *238*
Curry, R.B., 50, *57*
Cuykendall, C.H., 75–76, *78*
Cuylle, G., *142*

Dahl, B.E., 72–73, *78*
Dalal, R.C., 177, *178*
Dale, J.E., 39, *42*, *45*
Dale, W.R., *17*
Daniel, R.M., *204*
Darbisgire, R.D., 127, *128*
Darrow, R.A., 189, *194*
Darwin, C., 3, *5*
Date, R.A., *195*
Davey, A.W.F., 108–109, *110–112*
Davey, J.E., 40, *46*
Davidson, J.L., 7, *16*, *142*, 207, *212*, 227, *236*
Davidson, R.L., 52–53, *57*
Davies, A., 34, *42*, *195*
Davies, D.H., 14, *16*, 63, 65, *68*
Davies, D.J.G., 249, *252*
Davies, G.J., *195*
Davies, H.L., *26*, 140, *142*, 183, *185*, 228–229, *236*
Davies, J.G., *144*
Davies, W.E., VI, *VI*, 4, *5*, 83, *86*, 149, *152*, 191, *194*
Davies, W.J., 37–38, *42*
Davis, I.F., *121*, 132, *135*, 229, *236–237*
Davis, R., 58, *203*
Davison, K.L., 151, *153*
Day, H.R., 21, 23, *25*, 227, 233, *236*
Dayan, E., 47, *57*
Dazzo, F.B., 189, *194*
Deacon, M.J., 38, *45*
Dean, L.A., 174, *179–180*
De Bano, L.F., 209, *212*
De Beyer, J., *16*
De Gooijer, H.H., 65, *68*
De Jong, R., 52, *58*
Deinum, B., 11, *16*, 67, *68*
De Laine, R.J., 207, *212*
Delmas, R.E., *26*
Demarquilly, C., 104, 106, *111*, 137–138, 140–141, *142*, *252*
Denehy, H., *69*
Denmead, O.T., 50–51, *57–58*, 166, *172*, 198–199, *203*, 207, *212*
Detling, J.K., 50, 52, 55, *58*
Devendra, C., 123–124, *128*
Devine, C.E., *237*
De Vries, M., 65, *68*, *87*
Dewhurst, J.K., 182, *185*
De Wit, C.T., 49–51, *58*, 207, *212*
Dibb, C., 83, *86*, 99, *246*
Dickinson, S.E., *16*
Diekmahns, E.G., 67, *69*
Dietrich, J.H., *213*
Diggle, P.J., 55, *58*

Dillon, J.L., 219, 223, *226*, 232, *237*
Dilž, K., *69*
Diprose, G.D., *120*
Dirven, J.G.P., *5*, *86*
Doak, B.W., 165, *172*, 184, *185*, 198, *203*
Dobereiner, J., 188, *194*
Doetsch, J.H., 199, *202*
Doll, J.P., 219, *226*
Donald, A.D., 229, *236–237*
Donald, C.M., 23, *25*, 30, *42*, 71, *78*, 241, *246*
Donald, G.E., *237*
Donefer, E., *142*
Doney, J.M., 118, *121*
Donnelly, J.R., 211, *236*
Dougherty, C.T., *195*
Dovey, M.D., *25*
Dovrat, A., *57*
Dowdell, R.J., 68, *68*, 199, *204*
Downes, R.G., 205, 208, 209, *212*
Downey, N.E., 229, *236*
Doyle, C.J., 218, 223, *226*, 243, *246*
Doyle, J.J., 133, *135*
Drees, E.M., *59*
Dreibelbis, F.R., 208, *212*
Drougsiotis, D., *79*
Dudzinski, M.L., 118, *120*, 129–133, *135*, 141–142, *142*, 228, 230, *235*
Dulphy, J.P., *252*
Duncan, D.A., 23–24, *25*
Duncan, D.L., 145–146, *153*
Dunin, F.X., 206–210, *211–212*
DuPlessis, S.G., 174, *178*
During, C., 118, 120, *121*, 176–177, *178–179*, 202, *203*
Dwyer, D.D., 129, *135*
Dyer, M.I., *58*
Dymock, N., *194*
Dyson, C.B., *121*

Eady, R., 189, *194*
Eagles, C.F., 36, *42*, 44, *46*
East, N.E., *121*
Edelsten, P.R., 56, *58*
Edmeades, D.C., 188, *194*
Edmond, D.E., 76, *78*
Edwards, C.A., 176, *178*
Edwards, C.H.B., *136*
Edye, L.A., *144*
Egan, J.K., 114, *121*, 228–229, 233, *236*
Elberse, W.T., 4, *5*, 81, 82, *86*
Elderidge, G.A., 250, *252*
El Hassan, B., 227, *236*
Ellen, J., 39, 41, *45*
Elliot, C.S., 75, *77*
Elliot, I.L., 75–76, *78*, 177, *178*
Elliott, J.G., 218, *226*
Ellis, J.E., 133, *136*
Engel, R.W., *178*
Enwezor, W.O., 174, *178*

Epplin, F., *225*
Esslemont, R.J., 106, *110*, 156, *159*
Etienne, M., *26*
Evans, L.T., 31, 40, *43*
Evans, P.S., 106, *110*, 133, *135*, 176, *178*
Evans, R.A., 20, *25*
Evans, T.D., *180*
Evans, T.R., *196*, 249, *252*
Evans, W.R., *194*
Evenari, M., *213*
Eyal, E., *27*
Eyles, D.E., 245, *246*
Eyles, R.L., 208, *212*

F.A.O., 123, *128*
Faris, D.G., 39, *43*
Farquhar, G.D., 199–200, *203–204*
Farrar, J.F., 34, *42*
Fenlon, J.S., 29, *42*
Fenn, L.B., 198, *203*
Ferguson, D.L., *144*
Fick, G.W., 49, *58*
Field, T.R.O., 166, 169–170, *171*, *195*
Figarella, J., *70*, *100*
Filan, S.J., *238*
Finn, T., 241, *246*
Finnamore, P., *185*
Firestone, M.K., 198, *203*
Firth, P.M., *203*
Fischer, R.A., 207, *212*
Fiscus, E.L., 51, *58*
Fishbeck, K.A., 193, *194*
Fisher, F.L., 65, *68*
Fisher, M.J., *46*
Fitzgerald, P.D., 11, *16*
Fitzgerald, R.D., 228, *236*
Fitzhugh, H.A., 98, *99*
Fitzpatrick, E.A., 49, *58*
Fitzpatrick, E.N., *26*
Fleming, G.A., 149–151, *153*
Fletcher, G.M., 39, *43*
Fletcher, R.H., *194*
Floate, M.J.S., 165, 168, 170, *172*, 176, *178*, 181, 184, *185*, 198, *203*
Flux, D.S.F., *110*
Focht, D.D., 199, *203–204*
Fogle, V.W., *195*
Fontenot, J.P., *178*
Foran, B.D., *235*
Forbes, E.B., 181, *185*
Forbes, T.J., 92, 98, *99*, 243, *246*
Ford, E.D., 55, *58*
Forde, B.J., 33, *43*
Forde, D.J., *128*
Foster, G.R., 210, *212*
Foster, W.N.M., 175, *180*
Foxcroft, G.R., 156, *159*
Frame, J., 65, *68*, 75–76, *78*, 83, *86*, 109, *110*, 176, *179*, *195*
Francisco, E.M., 223, *226*

Fraser, A.F., 157, *159*
Freer, M., 140, *142*, *211*, 250, 252, *252*
Freeze, R.A., 211, *212*
French, M.H., 124, *128*
Freney, J.R., 166, *172*, *203*
Frew, M.V., *235*
Fried, M., 174, *179*
Frissel, M.J., 168–169, *172*
Fukai, S., 23, *26*

Galbally, I.E., 199, *203*
Galbraith, K.A., 23–25, *25–26*, *235*
Gallimore, J., *44*
Gallup, W.D., 182, *185*
Gamble, T.N., 198, *203*
Gandar, P.W., 164–168, *172*
Gardner, W.K., 175, *179*
Gardner, W.R., 208, *212*
Garrett, M.K., *43*
Garrett, W.N., *121*
Garstang, J.R., 65, *68*
Gartner, J.A., 105, *110*
Garwood, E.A., 40–41, *43*, 63, 68, *69*, 82–83, *86*
Gash, J.H.C., *212*
Gasser, J.K.R., 198, *203*
Gaucheva, A.N., *178*
Geddes, H.J., 208, *212*
Geenty, K.G., 13, *16*
George, J.M., 233, *236*
Gervais, P., 73, *78*
Gerwitz, A., 53, *59*
Gharaybeh, H.R., *235*
Gibb, M.J., 131, *135*, 158
Gibbs, A.E., *213*
Gibson, A.H., 188–191, *194–195*
Gibson, A.R., 200, *204*
Gifford, G.F., 208, *212*
Gifford, R.M., 33, *43*
Gillard, P., 184, *185*, 198, *203*
Gillet, M., 73, *78*
Gillingham, A.G., 15, *16*, 175–177, *179*
Gilmanov, T.G., 53, *58*
Gilmour, A.R., *246*
Gleeson, P., 106, 108, *110*, 111
Glencross, R.N., *25–27*
Glover, E.H., 95, 98, *99*
Gloyne, R.W., 14, *16*
Goatcher, W.D., 132, *135*
Godden, W., 98, *100*
Godwin, D.C., 175, *179*
Goh, K.M., 188, *194*
Gomez, P.O., 98, *99*
Gonzalez, M.H., 129, *135*, 183, *185*
Goodall, V.C., *195*, *204*
Goodwin, P.B., 36, *43*
Goold, G.J., *77*
Gorddard, B.J., *26*
Gordon, F.J., 101–109, *110–111*
Gordon, I., 156, *159*

Gordon, J.G., 134, *135*
Goreau, T.J., 199, *203*
Goudriaan, J., 49–50, *58–59*
Gourlay, R.N., *135*
Goyette, E.A., *179*
Gozlan, G., *42*
Graham, G.Y., 233, *237*
Graham, T.C., *135*
Grainger, C., 108, *110–111*
Graley, A., 183, *185*
Grassland Research Institute, 127
Green, H.H., 145, *153*
Green, J.O., 63, *68*, 81, 85, *86–87*, 99, 243, *246*
Green, L.R., *27*
Green, P., *213*
Greenhalgh, J.F.D., 142, *142*
Greenwood, D.J., 54, *58*
Greenwood, E.A.N., 22–24, *26*
Greenwood, R.J., 199, *203*
Gregg, P.E.H., 166, *172*, *179*
Gregory, M.E., *128*
Gresshoff, P.M., *194*
Griffin, J.L., 74, *78*
Griffis, C.L., *59*
Grigg, J.L., 174, *179*
Grime, J.P., 38, *43*, 81, 83, *87*
Groat, R.G., 193, *195*
Groves, R.H., 209, *212*
Grunes, D.L., 146, *153*
Gunary, D., 183–184, *185*
Gunn, R.G., 118–119, *121*
Gutman, M., 19–20, 23, *26*
Gutschick, V.P., 189, *195*
Grant, D.A., 15, *16*, 81, 85, *86–87*, *172*, 188, *195*
Grant, S.A., 14, *16*, *43*, *78*, *100*

Hacker, J.B., 85, *87*
Hafez, E.S.E., 157, *159*
Haggar, R.J., 64, *69*, 83–85, *86–87*
Hagin, J., 174, *179*
Halford, R.E., 107, *111*
Hall, M.J., 157, *159*
Hall, N.S., *179*
Halliday, J., 193, *195*
Hallock, B.G., *195*
Halm, B.J., 176–177, *179*
Halstead, R.L., *179*
Halter, A.N., 223, *226*
Hamilton, D., 233, *236*, 249, *252*
Hamilton, D.A., *212*
Hamilton, W.J., *127*
Hammes, R.C., *178*
Hancock, J., 176, *179*
Hand, D.W., *57*
Hanks, R.J., 47, *60*
Hanson, A.D., 37, *43*
Hanson, C.L., 208, *213*, 228, *237*
Hardaker, J.B., 219, *226*

# AUTHOR INDEX

Hardison, W.A., *178*
Haring, H., *128*
Harker, K.W., 129, *135*
Harkess, R.D., *78*
Harlan, J.R., 230, *236*
Harmsen, G.W., 200, *203*
Harpaz, Y., 24, *26*
Harper, J.L., 3, *5*, 39, *43*
Harper, L.A., 197–198, *202–204*
Harrington, F.J., 72, 74, 77, *79*
Harris, A.J., 13, *16*, 189, *195*
Harris, C.E., *79*, *143*, *179*
Harris, R.F., *180*
Harris, W., *58*, 71–76, *78*, 192–193, *195*
Harrison, F.A., *185*
Harrison, M.A., 40, *43*
Hart, J., *194*
Hart, M.L., *69*, 146, *153*
Hart, R.H., 77, 229, *236*
Hasnain, H., 156, *159*
Hawkins, R.H., 208, *212*
Hawkins, S.W., 241–242, *246*
Haydock, K.P., *144*
Haystead, A., 188, *195*
Hayward, J.A., *212*
Heady, E.O., 219, *225–226*
Heady, H.F., 19–21, 23–25, *26*, 132, *135*, 234, *236*
Healey, W.B., 148, *153*, 175, *179*
Heaney, D.P., 140, *143*
Heath, D.D., *237*
Heath, G.W., 176, *178*
Heath, S.B., 55, *60*
Hebblethwaite, P.D., 40–41, *43*
Heichel, G.H., 189, *195–196*
Heidt, L.E., *203*
Hely, F.W., *194*
Helyar, K.R., 174, *179*
Henderson, J.D., *78*
Hendricks, S.B., *179*
Hendricksen, R.E., 139, 141, 142, *143*
Henkens, C.H., 200, *203*
Henshaw, J., 157, *159*
Henzell, E.F., *59*, 62, 67, *69*, 165, 168, *172*, 188, *195–196*
Herbel, C.H., 133, *135*
Herriott, J.B.D., 150, *153*, *195*
Hewlett, J.D., 205, *211*
Hexter, G.W., 207, *212*
Hibbert, A.R., 207, *212*
Hickey, M.J., *16*, *195*
Hides, D.H., 150, *153*
Hilbert, D.W., 56, *58*
Hilder, E.J., 167, 171, *172*, 176, *179*, 183, *185*, 228, *236*
Hill, B.D., 118, *121*
Hill, J.L., 132–133, *135*
Hill, M.J., 41, *43*
Hillel, D., 53, *58*, 208, *212*
Hillinger, J., *179*

Hindley, N.L., 24, *27*
Hingston, F.J., *211*
Hinman, N., *121*
Hitz, W.D., 37, *43*
Hockey, H.-U.P., 119, *121*
Hodder, R.M., *136*
Hodge, R.W., 133, *135*
Hodgson, J., 39, *43*, 71–72, 74, *78*, 93, *99–100*, 103, 106, 109, *111*, 130, 132–133, *135–136*, 141, *143*, 151, *153*, *235*, 230, *236*, *246*
Hodgson, J.L., 103, 106, 109, *111*
Hoglund, J.H., *15*, 166, 170, *172*, 188–193, *194–195*
Holder, J.M., 130, *135*, 250, *252*
Holding, A.J., *195*
Holford, I.C.R., 174, *179*
Holmes, C.W., 101–109, *110–111*
Holmes, J.C., *195*
Holmes, J.W., 205, *211*
Holmes, W., 72, 77, 84, *87*, 92–93, 96–98, *100*, 130, *136*, *144*, 233, *238*, 242, *246*
Holt, R.F., *178*
Honore, E.N., *16*, *111*
Hood, A.E.M., 68, 107, *111*
Hoogendoorn, C., 102
Hooper, A.B., 199, *203*
Hooper, J.F., 24, *26*
Hopkins, A., *7*, *16*, 98, *99–100*, *246*
Hopkins, D.R., *121*
Hornbeck, J.W., *213*
Horvath, D.J., 146, 148–149, 151, *153*
Hoste, C.H., 20, 23, *26*
Houpt, K.A., 133, *136*
Howe, C.D., 82, *87*
Howes, K.M.W., 23, *26*, 134, *136*, *246*
Hoxey, A.M., *VI*, *100*, 127, *128*, *159*
Hrabak, E.M., *194*
Hsaio, T.C., 37, *43*
Hughes, A.D., 243, *246*
Hughes, H.C., 103, *111*
Hughes, J.D., 208, *212*
Hull, J.L., *26*, *179*
Humphreys, L.R., 73–74, *78*
Hung, H.W., *172*
Hungate, R.E., 91, *100*
Hunt, H.W., 55, *58*
Hunt, I.V., 73, 75, *78*, 176, *179*
Hunt, L.A., 38–39, *43*
Hunt, R., 38, *43*, 83, *87*
Hunter, P.J., *128*
Hunter, R.A., 247, *253*
Hunter, R.F., 14, *16*
Huntjens, J.L.M., 176, *179*
Hurle, D.H., 210, *213*
Hutchinson, K.J., 55, *58*, 233, *236*, 249, *252*
Hutchison, H.G., 130, *135*
Hutka, J., *213*

Hutton, J.B., 102–103, 107, *111*
Hyder, D.N., 72–73, *78*
Hynes, R., *204*

Idle, A.A., 82, *87*
Innis, G.S., 54–55, *57*, *59*, *178*
Ireland, J.A., *195*
Ive, J.R., *59*
Iverson, C.E., *143*
Ivins, J.D., *79*

Jackman, R.H., 175, *179*, *195*
Jackson, D.K., 74, 76, *78*
Jackson, J.E., *77*
Jackson, M.V., *16*, 66, 68, *69–70*, 75, *78*, *100*
Jagusch, K.T., 114, 118–120, *121*, *143*
James, J.F.P., *237*
Jameson, D.A., 71, *78*
Jamieson, W.S., 133, *135*
Jansson, S.L., 163, *172*
Jardine, R., 232, *236*
Jarrige, R., 247–248, *252*
Jarvis, P.G., 206, *212*
Jenness, R., 125, *128*
Jennings, A.C., 176, *178*, 233, *235*, 250, *252*
Jensen, M.E., 51, *58*
Jensen, S.E., *59*
Jewiss, O.R., 31, 36, 39–41, *43–45*
Jinks, R.L., 39–40, *43*
Joblin, K.H., 184, *185*
Johns, G.G., 52, *59*
Johnson, A.H., 177, *179*
Johnson, H.P., *59*
Johnson, I.R., 57, *58*
Johnson, L.E., *196*
Johnston, C.D., *213*
Johnston, I.M., *153*
Johnston, J.M., *16*, *195*
Johnstone-Wallace, D.B., 141, *143*
Jones, B.L., *194*
Jones, D.I.H., 147–150, *152–153*
Jones, E.L., 84, *87*
Jones, H.G., 37, *45*
Jones, J.G., 33, *46*
Jones, J.P., 176, *180*
Jones, J.W., 52, *59*
Jones, L.P., 234, *236*
Jones, M.B., 29, 31, 37–39, *43*, 54, *58*
Jones, M.G., 4, *5*, 82, *87*
Jones, M.J., 75, *78*
Jones, M.M., 37, *46*
Jones, O.L., 175–176, *178–179*
Jones, R.J., 96, *100*, 230, *236*, 249, *252*
Jones, R.L., 36, *43*
Jones, R.J.A., *16*
Jones, R.M., 249, *252*
Jones, T., 33–34, 36, *45*
Jongeling, C., 99

Joubert, J.G.V., 21, 24, *26*
Journet, M., 104, 106, *111*, 140, *144*
Jowett, D., *86*
Joyce, J.P., 114, 118, *121*, 140, *143*, 229, *236*, 244, *246*
Judd, M.J., *204*
Judson, G.J., 146, *153*
Jung, G.A., 7, *17*, 151, *153*

Kadry, L., *179*
Kaila, A., 176, *179*
Kaplan, W.A., *203*
Kat, C., 77, 250, *252*
Katznelson, J., 55, *58*
Kaufman, P.B., 40, *43*
Kaufmann, M.R., *58*
Kay, B.L., *25*
Kay, R.N.B., *127*
Kays, S., 39, *43*
Keatinge, J.D.H., 35–36, *43*
Keeney, D.R., 165–166, 169, *171–172*, 198, *202*
Keig, G., 51, *58*
Keister, D.L., *194*
Keith, M.H., 181, *185*
Kemp, A., 146, *153*
Kemp, D.R., 34, *43*
Kendall, M.G., 224, *226*
Kennedy, K., 141, *143*
Kennedy, P.M., 140, *143*
Kenney, P.A., *121*, 229, *236–237*
Keogh, R.G., 184–185, *185*
Keoghan, J.M., 72, *78*
Kerr, D., *211*
Kerr, J.D., *246*
Key, C.L., 133, *136*
Keya, N.C.O., *69*
Keynes, R.D., *185*
Keys, A.J., 199, *203*
Kilkenny, J.B., 98, *100*, 105, *110*, 242, *245*
Kimes, D.S., 50, *58*
King, D., 53, *58*
King, J., *43*, 78, 81, *87*, *100*
King, K.R., 103, 107, *111–112*, 242–243, *246*
Kirby, E.J.M., 39, *43*
Kirton, A.H., 229, *237*
Kissel, D.E., 198, *203*
Klemedtsson, L., 199, *203*
Kleter, H.J., 85, *87*
Klute, A., *58*
Knapp, W.W., 51, *59*
Knight, T.W., *121*
Knoerr, K.R., *59*
Knotts, R.R., *194*
Knowles, R., 166, *172*, 198, *203–204*
Knudson, D., 174, *179*
Kohl, D.H., 199, *203*
Kohn, G.D., 241, 243, *246*

Kolenbrander, G.J., 177, *179*, 200, *203*
Koocheki, A., *79*
Kornher, A., *16*
Korte, C.J., 56, *58*, 73–74, *78*
Krasnec, L.E., *203*
Kristensen, K.J., 207, *212*
Krueger, W.C., 118, *121*, 132, *136*, 227, *236*
Krul, J.M., 199, *203*
Kruyne, A.A., 65, *68*, 81, *87*
Kubota, J., 148, *153*
Kumazawa, K., *180*

Lachance, L., 73–74, *78*
Lafler, J.M., *212*
Laidlaw, A.S., 40, *43*
Laing, W.I., *136*
Laissus, R., 67, *69*
Lambert, D.A., 41, *43*
Lambert, J.P., 13, *16*
Lambert, M.G., 14, 15, *16*, 124, *128*, 169–170, *172*, 180, 188, 193, *194–195*
Lambourne, L.J., 118, *121*, 165, *171*, 181, 182, *185*, 198, *202*
Lamming, G.E., 228, *235*
Lancaster, R.J., 182, *185*
Lane, L.J., *212*
Langer, R.H.M., 7, *16*, 34, 38–41, *42–44*, 72, *78*, *195*
Langlands, J.P., 133, *136*, 227–231, 233, *237*, 250, *252*
Langlois, B., *128*
Lapins, P., 198, *204*
Laredo, M.A., 138–140, *143*
Large, R.V., 66, *68*, 98, *100*, 126, *128*, 156–157, *159*
Larkham, P.A., *235*
Larsen, S., 174, *179*, 184, *186*
Laude, H.M., *43*
Lavender, R.H., *78*
Lawlor, D.W., 37, *44*
Lawrence, T., 73, 75, *78*
Lawson, B.M., *143*
Laycock, W.A., *136*
Layzell, D.B., *59*
Lazenby, A., 14, *16*, 63, *69*, 98, *100*
Lazor, V.A., *153*
Lea, P.J., *203*
Leafe, E.L., 13–14, *16*, 29, 31, *43–44*, 46, 59, *79*
Leaver, J.D., 96, *100*, 102, 104, 106–110, *111*, 229, *237*
Lebas, F., 126, *128*
Ledgard, S.F., 190, *195*, 202, *203–204*
Le Du, Y.L.P., 106, 108–109, *111*, 131, 133, *135–136*
Lee, J., 61, *69*
Lee, R., 177, *180*
Leech, P.K., 83, *86*, 243, *246*
Lees, J.L., 156, *159*

Leffelaar, P.A., 55, *58*
Le Houérou, H.N., 20, 23, *26*
Leigh, J.H., 132, *136*
Leitch, I., 98, *100*
Lemeur, R., 49, *58*
Lemon, E.R., *59*
Leont'yevskiy, B.B., 208, *212*
Leopold, L.B., 208, *212*
Leslie, T.I., *212*
L'Estrange, J.L., 182, *185*
Letey, J., *58*, *204*
Levitt, J., 34, *44*
Levy, E.B., VI, *VI*, 4, *5*, 241, *246*
Lewis, D.C., 175, *179*
Lewis, J., *79*
Lewis, J.N., *226*
Lewis, K.H.C., *235*
Lieth, H., 24, *26*
Lightfoot, R.J., *27*
Lill, W.J., *178*
Limmer, A.W., 199, 202, *204*
Lindberg, T., *203*
Lines, E.W.L., *143*
Linsley, R.K., 210, *211*
Little, D.A., 140, *143*
Lloyd, A.G., 227, 231–232, *237–238*, 249, *252*
Lloyd, L.E., *142*
Lloyd, P.S., 81, *87*
Locke, L.F., *213*
Lockyer, D.R., 63–64, 66, *68*, 84, *86*
Lofgreen, G.P., *179*
Loh, I.C., 210, *213*
Loneragan, J.F., 175, *178*
Long, G.A., 20, 24, *26*
Loomis, R.S., 49, *58*
Low, P.F., 174, *180*
Low, W.A., 129, *136*
Lowe, A.G., 188, *195*
Lowe, J., *68*, 192, *194*, *195*
Lucas, H.L., *185*, *237*
Ludlow, M.M., 36–37, *44*, 46, 50, *58*
Lund, L.J., *204*
Luscombe, P.C., 82, *86–87*, *128*, *195*
Luxmoore, R.J., 55, *58*
Lwoga, A.B., *79*
Lynch, P.B., 75–76, *78*, 167, *172*, 177, *178–179*

Mabon, R.M., *135*
MacArthur, I.D., 223, *226*, 232, *237*
MacCarthy, D., 101–102, 108, *111*
MacDaid, D., *245*
MacDonald, R.A., *110*
Mace, M.J., 108, *111*
MacFarlane, G.C., *226*
Macfarlane, R.P., 249, *252*
MacIver, R.M., 133, *136*
Mackenzie, D.D.S., *110*
Maclean, K.S., *121*, *236*

# AUTHOR INDEX

MacLusky, D.S., 183, *185*
MacMillan, K.L., 102–106, 109, *111*
MacQueen, I.P.M., 106, *111*
MacRae, J.C., *128*
Maddocks, T., 208, *212*
M.A.F.F., 67, *69*, 99, *100*
Malafant, K.W.J., *136*
Maller, R.A., 133–134, *135*
Mann, L.D., *212*
Mann, P.P., 22, *26*
Mannetje, L. 't., 142, *143*, 249, *252*
Mansell, G.P., 173, 176–177, *179*
Mansfield, T.A., *42*
Marcum, J.B., 156, *159*
Marsh, R., 97, *100*, 247, *253*
Marshall, C., 32–34, 39–41, *42–44*
Marshall, J.K., *59*
Marston, H.R., 140, *143*
Marten, G.C., 75–76, *78*
Martin, A.E., 169, *172*, 199, *204*
Martin, J.P., 198, *204*
Martin, T.W., *16*
Martin, W.E., 23, *26*, 174, *179*
Mason, I.L., 123, *128*
Mason, W., 73–74, *78*
Mason, W.K., 251, *253*
Masterson, C.L., 188, 190–191, *195*
Matches, A.G., 73, 75–76, *79*
Matheron, C., 126, *128*
Mattingly, G.E.G., 174, *179*
Maunder, W.J., 108–109, *111*
Maunsell, L.A., 12–13, *17*
Maxwell, T.J., 103, 106, 109, *111*
Mayer, J., *42*
Mayland, H.F., 146, *153*
Mazurak, A.P., 208, *212*
McAlpine, J.R., 51, *58*
McArthur, A.T.G., 176, *179*
McAuliffe, C.C., 174, *179*
McCampbell, H.C., *236*
McCarty, M.K., 208, *212*
McCarty, P.L., 209, *212*
McClatchy, D., *212*
McClintock, M.M., *16*
McClymont, G.L., 134, *136*
McColl, R.H.S., 200, *204*, 210, *212*
McConchie, B.J., 27, *122*, 228–229, 231, *238*
McCown, R.L., 52, *58*
McDiarmid, B.N., 198, *204*
McDonald, I.W., *143*
McDonald, M.J., *180*
McDowell, F.H., 108, *112*
McElroy, M.B., *203*
McGill, W.B., 54–55, *58*, *172*
McGowan, A.A., 21, *26*, 102, 108, *111*
McGregor, B.A., 229, *237*
McIlvain, E.H., *213*
McIntyre, D.S., 208, *212*
McIntyre, G.I., 38–39, *44*

McIntyre, J.S., *121*, *236*
McKeown, N.R., 22, *26*
McKinney, G.T., 234, *237*
McLaren, P.N., *121*
McLaughlin, J.W., 250, *253*
McLean, J.W., 140, *143*
McLean, R.W., 142, *143*
McLeod, M.N., 85, *87*
McLusky, D.S., 84, *87*
McManus, W.R., *135*, 229, *235*, *238*
McMeekan, C.P., 101, 103–104, 107, *111*, 227, 229 234, *237*
McNaught, K.J., 147, 151, *153*, 175, *179*, 202, *203*
McNaughton, K.G., 206, *212*
McNeil, D.L., *59*
McNeur, A.J., 76, *79*
McPheely, P.C., 102–104, 106–107, 109, *111*
Meadowcroft, S.C., 85, *87*, 244, *246*
Medd, R.W., 56, *58*, *225*
Mee, S.S., 30, *45*
Megahan, W.F., *213*
Mekhael, D., 174, *179*
Menhenett, R., 36, *44*
Menz, K.M., *225*
Merrill, A.L., 125, *128*
Metson, A.J., 175, *180*
Meyer, J.H., 175, *179*
Meylich, A., *180*
Michel, J.F., 229, *237*
Micol, D., 93, *100*
Middleton, K.R., 177, *179*, 181, 183, *185*, 218, *226*
Miffin, B.J., *203*
Mikkelson, D.S., 174, *179*
Milford, R., 138, 140, *143*
Miller, C.P., 109, *111*
Miller, G.D., *136*
Miller, R.B., 173, *178*
Mills, C.F., 146, *153*
Milne, J.A., 126, *128*, 132, *136*, 242, *246*
Milthorpe, F.L., 31, 37, *42*, *44*, 48–49, 51, *57*
Milton, W.E.J., 82–83, *87*
Minkema, D., *128*
Minson, D.J., 73–74, *79*, 85, *87*, 137–142, *143–144*, 175, *179*, 241, *246*
Mishan, E.J., 222, *226*
Mislevy, P., 73–74, *79*
Mitchell, H.H., 182, *185*
Mitchell, K.J., 11, 13, *16*, 34, 39, *44*
Mitchell, R.L., 148, *153*
MLC (Meat and Livestock Commission), 93–95, *100*
Moghimi, A., 175, *179*
Mohammed, H.H., 155–156, *159*
Moller, K., 106, *111*
Molz, F.J., 51, *58*
Monk, K., 126, *128*

Monson, W.G., 74, *79*
Monteath, M.A., 119, *121*, *235*
Monteith, J.L., 30, 34, *44*, 50, *59*
Mooi, H., *87*
Mooney, H.A., 23, *26*
Moorby, J., 31, *44*
Moore, C.W.E., 4, *5*, 247, *253*
Moore, H.O., *143*
Moore, R.W., 119, *121*
Morand-Fehr, P., 124, *128*
Morley, F.H.W., 25, *26*, 116, 118, *121*, 228–229, 232–233, *235–238*, 241, 244, *245–246*, 250, 252, *253*
Morris, J., 223, *226*
Morris, J.G., 148, *153*
Morris, J.T., 54, *59*
Morris, K., 82, *86*
Morris, R.M., 14, *16*, 83–85, *87*
Morris, T.R., 157
Morrison, J., 9, 11, 14, *16–17*, 62–67, *68–70*, 81–82, *87*, 91, 97, *100*, *226*, 243, *246*
Morton, A.J., 165, *172*
Moss, A.J., 207, *212*
Mott, G.O., 227, 230–232, *237*
Mott, J.J., *143*
Mottershead, B.E., 183, *185*
Mouat, M.C.H., 175, *179*
Moustafa, E., 193, *195*
Mowat, D.N., *79*
Mudd, C.H., 65, *69*, 85, *87*, 244–245, *245–246*
Mufandaedza, O.T., 134, *136*
Mulham, W.E., 132, *136*
Mullaney, P.D., *235*
Munns, D.N., 191, *195*
Munns, R., 37, *44*
Munro, J.M.M., 14, *16*, 63, 65, *68*
Murdoch, J.C., 77
Murdock, J.T., *180*
Murnare, D., *143*
Murphy, C.E., *59*
Murphy, P.M., 188, 190–191, *195*
Murphy, W., 84
Murphy, W.E., 150, *153*
Myers, R.J.K., *202*

Nadin, J.B., *237*
National Research Council, 147, *153*, 252, *253*
Naveh, Z., 20, 23–24, *26*
Ndosa, J.E.M., 123, *128*
Nelson, A.B., 133, *135*
Nelson, C.J., 72, *77*
Nelson, W.L., 174, *180*
Nes, P., *171*, *202*
Neville Jr., W.E., *236*
Newberry, R.D., 106, 108–109, *111*
Newbold, R.P., *204*
Newbould, P., 86, *87*, 188, *195*

Newman, J.C., 211
Newton, I., 127, *128*
Newton, J.E., 114, *122*, 156, 158–159, *159*
New Zealand D.S.I.R., 14, *16*
Ng, E., 58
Ng, T.T., 36–37, *44*
Nicholas, D.A., 22, 24, *25–26*
Nicholas, D.J.D., 199, *204*
Nicholas, K.D., *185*
Nicholls, A.O., 55, *59*
Nicholson, W.G., *77*
Nicol, A.M., 251, *253*
Nitis, I.M., 249, *253*
Nix, H.A., 49, *58*
Nix, J.S., 106, *111*, 221, *226*
Nommik, H., 199, *204*
Nordfeldt, P.H., *16*
Norman, M.J.T., 200, *204*
Norris, I.B., 11, *16*, 34, 36–38, *44–45*
North of Scotland College of Agriculture, *128*
Nowakowski, T.Z., 67, *69*
Nowlin, J.D., *212*
Noy-Meir, I., 19–20, 24, *26*, 227–228, 231, *237*
Nunn, W.R., *135*
Nyahoza, F., 32–33, *44*
Nye, P.H., 54, *59*, 175, *178*, *180*

Obst, J.M., 146, *153*, 233, *237*
O'Connor, K.F., 168, *172*
Officer, R.R., 223, *226*
Ognjanovic, A., 125, *128*
Ogrosky, H.O., 208, *212*
O'Hara, G.W., 199, *204*
Olive, L.J., 208, 210, *212*
Oliver, L.R., *59*
Ollerenshaw, J.H., 36, *44*
Olmert, A., *179*
O'Loughlin, E.M., 210, *212*
Olsen, S.R., 174–175, *180*
Olsson, N.O., 126, *128*
Omaliko, C.P.E., 73, *79*
Omanwar, P.K., 175, *180*
Ong, C.K., 33, 39, 41, *44*
Orchiston, H.D., *196*
Organisation of Economic Co-operation and Development, 222, *226*
O'Rourke, P.A., 50, *59*
Osbourn, D.F., 92, *100*, 138–139, *143*
Østergaard, V., 221, *226*
Østgård, O., *16*, 36, 42, *44*
O'Sullivan, A.M., 81, *87*
Othman, O.B., 36, *42*
Owen, J.B., 113, 120, *121*, 230, *237*
Owen, W.J., *213*
Owens, L.D., 200, *204*
Owusu-Bennoah, E., 175, *180*

Ozanne, P.G., 23, *26*, 134, *136*, 243, *245–246*

Pack, R.J., 19, *26*
Page, E.R., 53, *59*
Pahl, P.J., *135*
Paltridge, G.W., 48, *59*
Pankhurst, C.E., 190, *195*
Papanastasis, V.P., 20, *26*
Parbery, D.G., *179*
Parfitt, R.L., 177, *180*
Parker, J.W.G., *195*
Parker, O.F., 107, *110*, 229, *235*
Parkin, R.J., *25*
Parsons, A.J., 30–31, 33–34, 36, 39, *44–45*, 56, *59*, 72, 74, *79*
Parton, W.J., 51–53, *58–59*
Pate, J.S., 54, *59*, 193, *195*
Paterson, J.G., *25*
Paul, E.A., 176, *178*, 187, *195*
Paul, K.J., 106, *111*
Paull, D.R., *237*
Paustian, K.H., *58*
Payne, J.W.A., 129, *136*
Payne, W.J., 199, *202*
Peacock, J.M., 11, *16*, 31, 35–36, *44–45*, 52, *59*
Pearce, A.J., 205, *212*
Pearse, R.A., 233, *236*
Pearson, R.W., *100*
Peaslee, D.E., 174, *178*
Peck, A.J., 210, *212–213*
Peel, S., *99*, 243, *246*
Pels, S., *212*
Pengelly, W.J., *121*
Penman, H.L., 51, *59*
Penning, I.M., *161*
Penning, P.D., *79*, 155, *159*
Penning de Vries, F.W.T., 50, *59*
Percival, N.S., 13, *16*, *77*
Petersen, R.G., 183, *185*, 230, *237*
Petit, M., 97, *99*
Petrie, A.H.K., 176, *180*
Philip, J.R., 206–207, *212–213*, 227, *236*
Phillips, D.A., 190, 193, *194*
Pierce, M., 37, *45*
Pigden, W.J., *143*
Piggot, G.H., 13, *16*, 109, *111*
Pilgrim, D.H., 205, *211*
Pineiro, J., *78*
Piper, C.S., *185*
Pitt, M.D., 20–21, 23–25, *26*
Playne, M.J., 140, *143*
Plowright, R.D., *235*
Poissonet, P.S., *26*
Pollard, E., 127, *128*
Pollock, C.J., 33–34, 36, *45*
Pollock, W.H., *203*
Poppi, D.P., 130–140, *143*
Porter, R.H.D., *111*

Potter, J.R., 52, *59*
Powell, C.E., 33, *45*
Powell, C.Ll., 175, *180*
Prasad, P.C., *43*
Pratt, B.J., *58*
Price, D.A., *136*
Pringle, R.M., 104, *112*
Pringle, W.L., 20, *26*
Prins, W.H., 62–64, 67, *69*
Pritchard, G.I., *143*
Pritchard, K.E., 251, *253*
Probert, M.E., 174, *180*, 184, *186*
Pullen, K.G., *237*
Purser, D.B., *246*, 249, *253*

Quarrie, S.A., 37, *45*
Quirk, J.P., 175, *179*
Quispel, A., 165, *172*

Rabbinge, R., *58*
Rabotnov, T.A., 82, *87*
Radcliffe, J.E., 8, 10–12, 15, *15–17*
Raguse, C.A., 20, *26*
Raivoka, E.N., *136*
Rakhmanov, V.V., 206, *213*
Ranga Rao, V., 190, *195*
Rangeley, A., *195*
Ranson, K.J., *58*
Raschke, M., 37, *45*
Rattray, P.V., 113–120, *120–121*, 244, *246*
Rauzi, T., 208, *213*, 228, *237*
Raymond, W.F., *79*, 142, *143*, *179*
Read, D.W.L., 177, *180*
Redmann, R.E., *59*
Reed, K.F.M., 24, *25–26*, 242, 244, *246*, *252*
Rees, M.C., 140, *143–144*, 241, *246*
Reeve, J.L., 114, *121*, 227–229, 231, 233, *237*
Reid, C.S.W., 159, *159*
Reid, D., 62, 65–66, *69*, 71, 74, *79*
Reid, R.L., 7, *17*, 146, 148–149, 151, *153*
Reifsnyder, W.E., 51, *60*
Reiners, W.A., 166, *172*
Reith, J.W.S., *153*
Remson, I., 51, *58*
Renton, S.W., *172*
Reuss, J.O., 54–55, *58–59*, 187, *194–195*
Reuter, D.J., 148, *153*
Reyenga, W., 207, *212*
Rhoades, E.D., *59*, 208, *213*
Rhodes, I., 30, *45–46*
Rhodes, V.J., *226*
Rice, R.C., *211*
Richards, I.R., 63, 66, 68, *69*, 75–76, *79*, 190, *195*
Richards, L.A., 51, *59*
Richardson, C.W., *59*, *213*
Richter, J., 51, *59*

# AUTHOR INDEX

Rickard, D.S., 8, 10, 12, *17*
Ridgman, W., 230, *237*
Riewe, M.E., 230, *237*
Ripley, E.A., 207, *213*
Risser, P.G., *59*
Ritchie, G.A.F., 199, *204*
Ritchie, J.T., 52, *59*, 207, 210, *213*
Ritson, C., 221–222, *226*
Robb, J.M., *135*
Roberts, E., 14, *16*
Roberts, N.F., 229, *237*
Robertson, A.G., 193, *194*
Robertson, J.A., 175, *180*
Robertson, N.G., 8, *17*
Robinson, A.R., 73, *78*
Robinson, D.K., *42*
Robinson, G.G., 114, 118, *121*, 234, *237*
Robinson, G.S., *195*
Robinson, I., *111*
Robinson, J.J., 159, *159*
Robson, A.D., 175, *178*
Robson, M.J., 29–31, 33–34, 36, 38–39, 44–45
Rochester, I.J., 73, *77*
Rodda, J.C., 205, *213*
Rodin, L.E., 21, *26*
Rodriguez, J.M., *235*
Rogan, P.G., 32–33, *45*
Rogers, A.L., 22, 24, *25–26*
Rogers, G.L., 108, *111*
Rogers, J.A., 81, *87*
Rogers, N.E., 174, *178*
Rohweder, A., 248, *253*
Rollinson, D.H.L., *135*
Rolston, M.P., 124, *128*, 193, *194*
Romero, A., 249, *253*
Rook, J.A.F., 182, *186*
Rorison, I.H., 175, *180*
Rose, P.H., 241–242, *246*
Ross, D.J., *196*
Ross, D.R., *59*
Ross, P.J., 49–50, 54, *59*, 165, 168–169, *172*, 188, *195*, 199, *204*
Rossiter, R.C., 19, 22–25, *26*, 241, *246*, *253*
Rosswall, T., *171*, *203*
Rottenbury, R.A., 229, *237*
Roughgarden, J., 53, *58*
Roughley, R.J., 190, *195*
Round-Turner, M.A., *235*
Rowe, R.K., 208, *213*
Roy, C.R., 199, *203*
Rozov, N.N., *26*
Rudman, J.E., 72, *79*, 142, *144*
Rumball, P.J., *16*
Russell, F.C., 145–146, *153*
Russell, G.C., 174, *180*
Russell, R.D., 63, *69*
Russo, J.M., 51, *59*
Rutherford, M.C., 21, *26*

Ryan, D.L., *16*, *195*
Ryan, M., 241, *246*
Ryden, J.C., 173–174, *180*, 199, *204*
Ryle, G.J.A., 33–34, 38, *45*, *59*
Ryle, S.M., *44*

Sagar, G.R., 32–33, *42*, *44*
Salehe, I., *135*
Salter, R.M., 182, *186*
Sandford, J.C., 157, *159*
Sandland, R.L., 96, *100*, 230, *236*
Sanson, J., 133, *136*
Santamaria, A., 102, *111*
Santhirasegaram, K., *77*
Sarker, A.B., 130, *136*
Sasalei, Y., 174, *180*
Sastry, P.S.N., *57*
Sato, K., 76, *79*
Saugier, B., 207, *213*
Saunders, W.M.H., 174–175, *180*, 195, 202, *203–204*
Sauvant, D., *128*
Saville, D.G., 118, *121*, 246, 250, *253*
Saxton, K.E., 51, 54, *59*
Scarf, F., *213*
Scattini, W.J., 250, *253*
Scholenberger, C.J., 182, *186*
Schulze, E.-D., *46*
Schuman, G.E., *59*
Schumm, S.A., 207, *213*
Schwartz, C.C., 133, *136*
Scott, D., 12–13, *17*
Scott, H.D., 56, *59*
Scott, J.D., *77*
Scott, N.M., *180*
Scott, R.S., 76, *79*
Sears, P.D., 71, 75–76, *79*, 165, *172*, 176, *180*, 188, 193, *195*, 198, *204*
Seath, D.M., *136*
Seligman, N.G., 19–20, *26*, 54, *59*
Selirio, I.S., 49, 51–52, *59*
Shafie, M.M., *136*
Shanan, L., 206, *213*
Shannon, P.W., 193, *195*, 204
Sharafeldin, M.A., *136*
Sharif, R., 39, *45*
Sharkey, M.J., 114, *121*, 227–229, 231, 233, *237*
Sharma, M.L., 210, *213*
Sharman, G.A.M., *127*
Sharp, A.L., 208, *213*
Sharpe, J.L., *152*
Sharpley, A.N., 176–177, *180*, 210, *213*
Sharrow, S. H., 114, 118, *121*
Shaw, K., 199, *202*
Shaw, N.H., 209, *213*
Shaw, P.G., 70, *100*
Shaw, R.H., 51, *58–59*
Shaw, T.C., 174, *178*
Shawcroft, R.W., 48–49, 51, *59*

Sheard, R.W., 73, *79*
Shearer, G., *203*
Sheath, G.W., 56, *58–59*, 118, *121*
Sheehy, J.E., 31, *45*, 50, 52–56, *59*
Sherwood, J.E., *194*
Sherwood, M.T., 190, *195*
Shim, J.S., *79*
Short, H.L., 126, *128*
Shovelton, J., 251, *253*
Sibma, L., 14, *15*, *17*, 67, *68*
Siebert, B.D., 140, *143*, 249, *253*
Siebert, R.W., 247, *253*
Sieler, W., *203*
Silsbury, J.H., 21, 23, *26–27*, 34, *45*
Silva, P.R.F., *204*
Silva, S., 70, *100*
Simmonds, J., 107, *112*
Simons, R.G., 34, *42*
Simpson, I.H., 114, 118, *121*, 234, *237*
Simpson, J.R., *172*, 191, *195*, 198–199, *203–204*
Sinclair, A.G., 177, *178*
Sinclair, D.P., *237*
Sinclair, T.R., 51, *59*, 207, *213*
Singh, B.B., 176, *180*
Singh, J.S., 47, 50, *59*
Sissingh, H.A., 174, *180*
Sithamparanathan, J., 175, *180*
Slatyer, R.O., 206, *213*
Sloan, R.E., *128*
Smalls, I.C., *211*
Smeaton, D.C., 117–118, *121*
Smith, A., 82–84, *87*
Smith, A.N., 174, *180*
Smith, B.A.J., 107, *112*
Smith, C.A., 130, *136*
Smith, D., 33, 41, *45*
Smith, D.D., 208, *213*
Smith, D.F., 23, *27*
Smith, D.L., 32–33, *45*
Smith, F.W., *144*
Smith, G.S., 177, *179*, 181, 183, *185*
Smith, J.A., *58*
Smith, L.P., 14, *17*
Smith, M.E., 118, *121*
Smith, M.S., 199, *204*
Smith, M.V., 21–22, 25, *27*
Smith, O.L., 55, *59*
Smith, R.C.G., 22, 24–25, *26–27*, 49, 52, 56, *58–59*, *178*
Smith, W.F., *121*
Smoliak, S., 228, *237*
Snaydon, R.W., 3–4, *5*, 81, 83–84, *86–87*, 149, *153*, 244, *246*, 248, *253*
Snook, L.C., 249, *253*
Soderlund, R., 165, *172*
Sørenson, L.H., 176, *180*
Souter, W.G., *136*, *246*
Southcott, W.H., 229, *237*
Southey, I.N., 25, *246*

Southwood, O.R., 243, *246*
Sparrow, P.E., *16*, 62, *69*, *100*
Spedding, C.R.W., VI, *VI*, 67, *69*, 98, *100*, 151, *153*, 155–156, 158, 159, *159*
Spence, A.M., *128*, *136*
Spencer, K., 174, *179*
Spiertz, J.H.J., 39, 41, *45*
Sprent, J.I., 191, *195*
Srivastava, A.K., 176, *178*
Stangel, P.J., 240, *246*
Stant, P.R., *153*
Stapledon, R.G., 73, 79, 83, *87*
Stark, B.A., 124, *128*
Staun, H., 126, *128*
Steele, K.W., 193, *195*, 197–200, 202, *204*
Stephens, M.J., 49, *59*
Stern, W.R., 27, 30, *45*
Stevens, W.C., *213*
Stewart, D.W., *59*
Stewart, J.B., 205, *212–213*
Stewart, J.W.B., 57, *178–179*
Stewart, R.H., 43, 75, 77, 84, *86*
Stewart, W.D.P., 165, *172*
Stewart, W.S., 44
Stiles, W., *16*, 59, 43, 63, *69*
Stirk, G.B., 63, *69*
Stobbs, T.H., 58, 130, *136*, 141–142, *142*, *144*, 233, *236*
Stockdale, C.R., 103, 107, *111–112*, 242–243, *246*
Stoddart, J.L., 39, *45*
Stokes, R.A., 210, *213*
Stolzy, L.H., *58*
Storry, J.E., 182, *186*
St-Pierre, J.C., 73, *78*
Strachan, C., 177, *180*
Stroosnijder, L., 51, *59*
Stuart, A., 224, *226*
Stutte, L.A., *204*
Suckling, F.E.T., 15, *17*, 114, *122*, 229, *238*
Sucont, R.A., *152*
Suttle, N.F., 148, *153*
Svensson, B.H., *203*
Swift, D.M., *58*
Syers, J.K., 174, 176–177, *179–180*, 210, *213*

Tadmor, N.H., 19–20, *27*, *213*
Tainton, N.M., 77, *235*
Talibudeen, O., 174, *179–180*
Talpaz, H., 53, *58*
Tanner, C.B., 207, *213*
Tate, M.E., 175, *179*
Tayler, J.C., 72, 79, 127, *128*, 142, *144*
Taylor, G.B., *26*
Taylor, H.M., *213*
Taylor, J.I., *135*
Taylor, T.H., 75–76, *79*
Tenhunin, J., 50, *59*

Terjung, W.H., 50, *59*
Ternouth, A.H., 27, *122*, *143*, *238*
Terry, C.P., *78*
Terry, K.R., 199, *203*
Terry, R.A., *143*
Tew, A., *235*, *252*
Thairu, D.M., 65, *69*
Thames, J.L., 206, *213*
Theiler, A., 145, *153*
Theobald, P.W., *171*, *202*
Thetford Jr., F.O., *121*
Thiault, M.M., *26*
Thomas, B., 149, *153*
Thomas, C., 63–64, *69*
Thomas, H., 11, *16*, 34, 36–39, 44–45, *69*
Thomas, J.G., 14, *16*, 83–85, *87*
Thomas, P.C., 247, *253*
Thomas, R.G., *143*
Thomas, T.A., 150, *152–153*
Thomas, V.J., *86–87*
Thompson, A., 149, *153*
Thompson, G.G., *143*
Thompson, K., 56, *57*
Thompson, N.A., 107, *112*
Thompson, R., *136*, *246*
Thompson, R.L., *121*, *236*
Thompson, S.Y., *128*
Thomson, D.J., *143*, 242, 244, *246*
Thornley, J.H.M., *57–58*
Thornton, I., *153*
Thornton, R.F., 140, *144*
Tiedemann, A.R., 209–210, *213*
Tiedje, J.M., 199, *203–204*
Till, A.R., 177, *178*, *180*
Tillman, R.W., *179*
Timmins, D.R., *178*
Tinker, P.B., 54, *59*
Tinsley, J., *16*
Todd, J.R., 182, *186*
Toebes, C., 208, *213*
Tomlin, D.C., 73–74, *77*
Torrance, C.J.W., 198, *203*
Torrell, D.T., 176, *180*
Torssell, B.W.R., 55–56, *57*, *59*
Tothill, J.C., 209, *213*
Treacher, T.T., 127, *128*, 131, *135*, *159*, 235
Trenbath, B.R., 84, *87*
Tribe, D.E., 130, *135–136*, 227, 231, *238*
Trigg, T.E., 108–109, *110*, 119, *121*
Trlica, M.J., 57, *59*
Troelsen, J.E., 140, *144*
Troughton, A., 39, *45*, 176, *180*
Truchet, G., *194*
Truog, E., 174, *180*
Turner, J.D., *16*, *195*
Turner, N.C., 36–37, 45–46, 207, *212*
Tweedie, R.L., *136*
Tyson, K.C., 68, *69*, 82, 83, *86*

Underwood, E.J., 145–146, 148–149, 151–152, *153*, 182, *186*, 241, *246*
Ulyatt, M.J., 244, *246*
Upton, M., 219, *226*
Urbano, R.M., *194*
Ursic, S.L., 206, *213*

Vallis, I., 192, *196*, 197, 200, *204*
Valois, F.W., *203*
Van Bogaert, G., *16*
Van Burg, P.F.J., 62–65, 67, *69*
Vance, C.P., 193, *195–196*
Van den Bergh, J.P., *5*, *86*
Van der Meer, H.G., 63, *69*
Van der Paauw, F., 174, *180*
Van Keulen, H., 50–51, *57*, *59*
Van Laar, H.H., 50, *58*
Van Staden, J., 40, *46*
Van Steenbergen, T., 62, 65, *70*
Van Veen, J.A., *171*, *172*
Vasey, G.H., 207, *212*
Veeningen, R., 199, *203*
Velayutham, M., 174, *180*
Vercoe, J.M., 183, *186*
Vere, D.T., 223, *226*
Verite, R., 140, *144*
Verstraete, W., 199, *203*
Vicente-Chandler, J., 91, *100*
Vicente-Chandler, R., 63, 65–67, *70*
Vickery, P.J., 227, *238*
Vieira da Silva, J., 23, *26*
Vierick, L.A., *213*
Viggers, E., *17*
Vine, D.A., 34–35, 38, *46*
Visser, T.A., *196*
Vithayathil, F., *203*
Voightlander, G., 14, *17*
Volk, O.H., 21, *27*
Vollenweider, R.A., 200, *204*
Voss, N., 14, *17*

Wade, D.D., *213*
Wagenet, R.J., 54, *59*
Waggoner, P.E., 51, *60*
Wagnon, K.A., 21, *27*
Waide, J.B., 55, *57*, *57*
Wainman, F.W., *142*
Waite, R., 33, *46*, 141, *144*
Walker, P.H., 208, 210, *212–213*
Walker, T.W., 175, *179*, 185, *186*, 191, *196*
Wallace, L.R., 104, *112*, 118, *122*, 229, *238*
Wallach, D., 49, *60*
Waller, J.E., 10, *15*
Waller, P.J., *236*
Wallsgrove, R.M., *203*
Walsh, A., *100*
Walshe, M.J., 104, 107, *111–112*, 229, 234, *236–237*

# AUTHOR INDEX

Walsingham, J.M., *VI*, *100*, 124–126, *128*, 156, *159*
Walter, H., 21, *27*
Walters, R.J.K., 149, *152*
Wardlaw, I.F., 36, *43*, *46*
Wareing, P.F., 36, *44*
Warington, K., 4, *5*, 82, *86*
Warren Wilson, J., *57*
Washko, J.B., *79*
Wassermann, V.D., 21, 24, *27*
Watenabe, F.S., 175, *180*
Waterson, H.A., *195*
Watkin, B.R., 41, *43*, 56, *58*, *60*, 71, 75–76, *77–78*, 198, *204*
Watkinson, A.R., 55, *57*
Watson, E.R., 24, *26–27*, 83, *87*, 198, *204*
Watson, J.N., 102, 107–108, *110*, 241, *245*
Watson, M.J., 229, 236, *238*, 250, *253*
Watson, S.W., *203*
Watson, V.H., 74, *78*
Watt, B.K., 125, *128*
Webby, R.W., *121*
Webster, C.P., 199, *204*
Webster, R., 167, *171*
Wedin, W.F., 192, *196*
Weeda, W.C., *77*
Weiland, R.T., 199, *204*
Weiner, J., 56, *60*
Weir, B., *203*
Weir, K.L., *203–204*
Weir, R., 37, *44*
Weir, W.C., 176, *180*
Weiss, Ph., 138, 140, *142*
Welch, J.M., 127, *128*
Wellburn, A.R., *42*
Wells, D.A., 150, *153*
Wells, G.J., 84, *87*
Wells, T.C.E., 126, *128*
West, C.P., 192, *196*
West, J.G., *226*
Weston, R.H., 137, *144*
Westrin, S., 50, *59*
Wetselaar, R., 199–200, *203–204*
Wheeler, J.L., 118, *122*, 234, *238*, 251, *253*
White, D.H., 21, *25–27*, 114, *121–122*, 228–229, 231–234, *235*, *237–238*

White, L.D., 22, *27*
Whitehead, D.C., 61–63, 65–68, *70*, 147, 149–151, *153*, 188, 191, *196*, 242–243, *246*
Whitehead, G.K., 125, *128*
Whitlow, P., *203*
Whitney, A.S., 188, *196*
Whittaker, I.A., McD., 228, *235*
Whittaker, R.H., 24, *27*
Wicht, J.E., 21, 24, *27*
Wieling, H., *69*
Wight, J.R., 47, *60*
Wightman, J.A., 193, *196*
Wild, A., 175, *180*
Wildig, J., *195*
Wilkins, R.J., 11, 14, *17*, 75–76, *78*, 91, *100*, 159, *159*, 247, *253*
Wilkinson, J.M., 124, *128*, 130, *135*, 247, *253*
Willet, I.R., 174, *180*
Willey, R.W., 55, *60*
Williams, B.G., 210, *213*
Williams, C.H., 241, *246*
Williams, C.N., 35, *42–43*, *46*
Williams, E.D., 82, *87*
Williams, E.G., 174, *180*
Williams, J.D.H., *180*
Williams, R.B., 146, *153*
Williams, R.D., 32, *46*
Williams, R.F., 176, *180*
Williams, T.E., 63, 66, 68, *69–70*, 75, *78*, 101–102, *112*
Williams, W.A., 24, *27*
Williams, W.M., 175, *180*, 194
Williamson, D.R., 210, *213*
Williamson, P., 38, *46*, 127, *128*
Willoughby, W.M., 142, *144*, 191–192, *196*, 230, 234, *238*
Wilman, D., 67, *70*, 73–76, *77*, *79*
Wilson, A.D., 24, *27*
Wilson, A.T., *202*, *204*
Wilson, D., 30–33, 36–37, *46*
Wilson, E.J., 175, *179*
Wilson, G.F., 108, *110*, *112*, 140, *144*
Wilson, J.H., 50–51, *57*
Wilson, J.R., 37, *46*
Wilson, R.K., 85, *87*
Wilson, R.S., 140, *142*

Wilson, S., *128*
Wimbush, D.J., *211*
Winch, J.E., 73–74, *79*
Winks, L.W., 249, *253*
Winn, G.W., *121*
Winrock International, 91, *100*
Winter, W.H., *143*
Winters, E., *180*
Wischmeier, W.H., 208, *213*
Wofsy, S.C., *203*
Woledge, J., 31, 39, 43, *46*
Wolton, K.M., 64, 66, *68*, *70*, 75–76, *79*, 177, *180*
Wonder, B., 223, *225*
Wood, J.T., *58*
Wood, P.D.P., 105, *112*
Wood, W.E., 210, *213*
Woodhouse, W.W., *185*
Woodmansee, R.G., 23–24, *25*, 54, *58*, 187, *196*, 201, *204*
Woods, E.G., *237*
Woodward, T.E., 141, *144*
Woof, R., *135*
Wray, J.P., *195*
Wright, A.J., 56, *60*
Wright, D.F., 104, *112*, 235
Wright, H.A., 209, *213*
Wright, J.L., *58*
Wright, M.J., 34, *42*, 151, *153*

Yates, J.J., 142, *144*
Yates, M.E., *213*
Yeates, G.W., 193, *196*
Yiakoumettis, I.M., 233, *238*
Yoshida, S., 30, *46*
Yoshinari, T., 199, *204*
Young, C.D., 126, *128*
Young, D.L., 223, *226*
Young, J.A., *25*
Young, J.W.O., 63–64, *69*, 226
Young, N.E., 114, *122*, 158, *159*
Young, N.R., *152*
Young, R.A., *212*
Younie, D., *195*

Zahorik, D.M., 133, *136*
Zoby, 92, *100*

# SYSTEMATIC INDEX[1]

*Acacia*, 257
   *A. aneura* F. Muell. (mulga), 206
*Agrostis* (bent grasses), 82, 83, 85, 257
   *A. capillaris* L. (= *A. tenuis*), 64, 65, 75, 81, 84
   *A. stolonifera* L., 64, 81, 84
   *A. canina* L., 81
*Alopecurus*, 257
   *A. pratensis* L., 84
Anatidae, 257
*Anser*, 257
   *A. anser* L. (goose), 125, 127
Anseriformes, 257
*Anthoxanthum*, 257
   *A. odoratum* L., 84, 85
*Antilocapra*, 257
   *A. americana* (pronghorn), 133
Antilocapridae, 257
*Arctotheca*, 257
   *A. calendula* L. (capeweed), 183
*Aristida*, 24, 257
*Arrhenatherum*, 257
   *A. elatius* (L.) Beauv., 81
Asteraceae (Compositae), 257
Arthropoda, 257
Artiodactyla, 257
*Atriplex*, 24, 257
*Avena*, 257
   *A. sativa* L. (oats), 251
Aves, 257

*Bellis*, 257
   *B. perennis* L., 81
*Beta*, 257
   *B. vulgaris* L. (sugar beet), 207
*Bison*, 257
   *B. bison* L. (bison), 133
Boraginaceae, 257
Bovidae, 257
*Brassica*, 257
   *B. napus* L. (swede), 251
   *B. rapa* L. (turnip), 251
Brassicaceae (Cruciferae), 257
*Briza*, 257
   *B. media* L., 81

*Bromus*, 257
   *B. hordeaceus* L. (= *B. mollis*), 21, 22
   *B. inermis* Leysser, 150
*Bubalus*, 257
   *B. bubalis* L. (water buffalo), 123–125

*Capra* (goat), 257
   *C. hircus* L., 123–125
*Carduus*, 257
   *C. nutans* L., 56
Caryophyllaceae, 257
*Cervus*, 257
   *C. elaphus* L. (red deer), 123, 125, 126
Cervidae, 257
*Chamaecytisus*, 257
   *C. palmensis* Christ (tree lucerne), 249
Chenopodiaceae, 257
*Chloris*, 257
   *C. guyana* Kunth., 65, 138, 199
Chordata, 257
*Chrysanthemoides*, 24, 257
*Cichorium*, 257
   *C. intybus* L. (chickory), 149
Coleoptera, 257
*Crataegus*, 257
   *C. monogyna* Jacq. (hawthorn), 126
*Cynosurus*, 257
   *C. cristatus* L., 81, 85

*Dactylis*, 257
   *D. glomerata* L. (cocksfoot), 22, 36, 38, 64, 81, 83–86, 150, 244
Dennstaedtiaceae, 257
*Digitaria*, 257
   *D. decumbens* Stent, 65, 138

*Echium*, 24, 257
*Elymus*, 257
   *E. repens* (L.) Gould (= *Agropyron repens*), 65
Equidae, 257
*Equus*, 257
   *E. caballus* L. (horse), 123, 125, 126
*Erodium*, 24, 257
   *E. botrys* (Cav.) Bertol., 22, 23
Eubacteria, 257
*Eucalyptus*, 205, 206, 257

Fabaceae, 257
*Festuca* (fescue), 24, 257
   *F. arundinacea* Schreber (tall fescue), 244
   *F. ovina* L. (sheeps fescue), 38, 65, 81, 85

---

[1] In this index, no attempt has been made, for larger taxonomic entities, to list all the pages where subordinate taxa are mentioned. These may be found by use of the Systematic List of Genera (p. 257). For some major groupings, more detailed entries will be found in the General Index.

*Festuca (continued)*

    *F. pratensis* Hudson (meadow fescue), 41, 64, 244
    *F. rubra* L. (red fescue), 64, 81, 82, 84–86, 244
Filicopsida, 257

Gastropoda, 257
Geraniaceae, 257
*Glycine*, 257
    *G. max* (L.) Merr., 189

*Haemonchus*, 229, 257
*Helix*, 127, 257
*Holcus*, 257
    *H. lanatus* L., 64, 81, 82, 84–86
*Hordeum*, 257
    *H. distichon* L. (barley), 207, 217, 251
    *H. murinum* L., 23
Hydrocotylaceae, 257
*Hydrocotyle*, 257
    *H. moschata* G. Forster, 75
*Hypochoeris*, 24, 257

Insecta, 257

*Lablab*, 257
    *L. purpureus* (L.) Sweet, 141
Lagomorpha, 257
Leporidae, 257
*Leucaena*, 257
    *L. leucocephala*, 249
*Leucanthemum*, 257
    *L. vulgare* Lam., 81
Liliopsida (Monocotyledones), 257
*Lolium* (ryegrass), 83, 251, 257
    *L. multiflorum* (Lam) Husnot (Italian ryegrass), 32, 38, 67, 75, 149, 150
    *L. perenne* L. (perennial ryegrass), 7, 9, 13, 14, 21, 29–41, 64–67, 71–75, 81, 83–86, 108, 115, 132, 149, 150, 175, 190–192, 197, 243, 244
    *L. rigidum* Gaudin (Wimmera ryegrass), 192
*Lotononis*, 257
    *L. bainesii* Baker, 192
*Lotus*, 257
    *L. corniculatus* L., 83
    *L. uliginosus* Schkuhr (= *L. major*), 189

*Macroptilium*, 257
    *M. atropurpureum* (DC.) Urban (siratro), 142
Magnoliophyta (Angiospermae), 257
Magnoliopsida (Dicotyledones), 257
Mammalia, 257
*Medicago*, 21, 24, 257
    *M. sativa* L. (lucerne), 21, 83, 140, 150, 189, 192, 193, 197, 234, 251
    *M. truncatula* Gaertner, 21, 23
*Melilotus*, 257
    *M. alba* Medicus, 189
Mimosaceae, 257
*Molinia*, 257
    *M. caerulea* (L.) Moench, 65

Mollusca, 257
Monera, 257
Myrtaceae, 257

*Nardus*, 257
    *N. stricta* L., 81
Nematoda, 257
*Nematodirus*, 229, 257

*Oryctolagus* (rabbit), 257
    *O. cuniculus* L., 123, 125–127, 155–157
*Ostertagia*, 229, 257
Oxalidaceae, 257
*Oxalis*, 257
    *O. corniculatus* L., 75

*Panicum*, 139, 257
    *P. maximum* Jacq., 138, 139
*Paspalum*, 257
    *P. dilatatum* Poir., 13, 175
*Pennisetum*, 257
    *P. clandestinum* Chiov., 13, 138
    *P. purpureum* Schumach. (Kikuyu grass), 65, 66
Perissodactyla, 257
*Phalaris*, 257
    *P. aquatica* L. (= *P. tuberosa*), 134, 244
*Phleum*, 257
    *P. pratense* L. (timothy), 40, 41, 64, 81, 83–85, 138, 149, 150, 244
Plantaginaceae, 257
*Plantago*, 257
    *P. lanceolata* L., 149
*Poa* (meadowgrass), 24, 83, 257
    *P. annua* L. (annual meadowgrass), 65, 75, 81, 82
    *P. pratensis* L. (smooth-stalked meadowgrass), 33
    *P. trivialis* L. (rough-stalked meadowgrass), 75, 81, 82, 85
Poaceae, 257
Polygonaceae, 257
*Pteridium*, 257
    *P. aquilinum* (L.) Kuhu (bracken), 126
Pteridophyta, 257

*Rhizobium*, 164, 165, 167, 187, 188, 190, 201, 257
Rhynchophora, 257
Rosaceae, 257
*Rubus*, 257
    *R. fruticosa* Agg. (blackberry), 126
*Rumex* (dock), 65, 257

*Sagina*, 257
    *S. procumbens* L., 75
*Setaria*, 257
    *S. sphacelata* (Schumach) Stapf & C. E. Hubbard, 138, 141, 197
*Sitonia*, 193, 257
*Sorghum* (sorghum), 251, 257
    *S. bicolor* (L.) Moench, 200, 251
    *S. halepense* (L.) Pers., 200
*Stellaria*, 257
    *S. media* (L.) Vill., 82
*Stylosanthes*, 249, 257

# SYSTEMATIC INDEX

Suidae, 257
*Sus* (pig), 123, 158, 160, 257

*Taraxacum*, 257
   *T. officinale* Weber, 75
*Themeda*, 205, 257
*Trifolium* (clover), 257
   *T. pratense* L. (red clover), 76, 150, 193
   *T. repens* L. (white clover), 7, 13, 66, 67, 72, 75, 81, 83, 84, 108, 115, 149, 150, 175, 189–193, 198, 242
   *T. subterraneum* L. (subterranean clover), 20–23, 192, 193, 197
*Trisetum*, 257

*T. flavescens* (L.) Beauv., 84
*Triticum*, 257
   *T. aestivum* L. (wheat), 217
   *T. aestivum* x *Secale cereale* (triticale), 251

*Ulex*, 257
   *U. europaeus* L. (gorse), 124, 126

*Vicia*, 21, 257
*Vulpia*, 24, 257
   *V. myuros* (L.) C. C. Grelin, 23

*Zea*, 257
   *Z. mays* L. (maize), 217, 251

# GENERAL INDEX

Aberdeen Angus cattle, 93, 94
abscisic acid (ABA) (see also growth regulators), 37, 38
accumulated day degrees ($T$ sum), 64
acid rain, 197
advection, 51
Africa (see also individual countries), 324
altitude, 14
amino acids, 37, 247, 249
ammonia absorption, 200
–, anhydrous, 62, 201
– emission/volatilization, 54, 165, 166, 169, 184, 197–202
ammonium, 54
– nitrate, 62
– sulphate, 198
anaerobic soil (see also waterlogging), 55, 76, 166, 198, 199
Anglo-Nubian goats, 123, 124
angora wool, 123
animal excreta, see dung and urine
– genotype, see breeds
– growth rate, see liveweight gain
– mineral requirement, see individual elements
– production/output, see beef production, milk, sheep
annual dry matter production (see also herbage production), 19–23
aquifers, 200
arsenic, 145
Asia, 124
aspect, 14
assimilate, allocation, 52, 53
– export, 32, 33
–, competition for, 35, 39, 40
–, partitioning, 32, 33, 52, 54
– translocation, 32, 33, 41, 52
– transport/transfer, see translocation
atmosphere, 163
Australia (see also individual states), 21–23, 56, 61, 101, 113, 126, 134, 146, 149, 197–199, 205, 207–210, 220, 234, 241, 247, 249, 251
auxins, see growth regulators

bank balance, 232
barley (*Hordeum distichon*), 207, 217, 251
beef cattle, herbage requirement of, 92
– –, biological efficiency of, 91
– production: contribution to human diet, 91
– –, effect of fertilizers on, 97, 242, 243
– –, effect of seasonal variation on, 91, 92, 98
– – location, 91
– –: stocking rate, 96, 98, 99, 229, 233, 242
– – systems, 93, 94, 95, 99, 221
Belgium, 61

bent grasses (*Agrostis*)
– –, brown (*A. canina*), 81
– –, common (*A. capillaris*), 64, 65, 75, 81, 84
– –, creeping (*A. stolonifera*), 64, 82, 84
biomass, animal, 163, 164, 169, 170
– of microorganisms, 163, 164, 169, 170
–, plant (see also dry matter on offer), 30, 163, 164, 169, 170
bison (*Bison bison*), 133
bite frequency, 92, 130–132, 141
– size, 92, 93, 97, 130, 131, 132, 141
blackberry (*Rubus fruticosa*), 126
bloat, 108
body reserves, 108
– weight, 137
boron, 239
botanical composition, 55, 56, 75, 81–86, 192, 227, 243–245
– –: effect of cages, 19
– –:– of fertilizers, 4, 23, 81, 82, 86, 170, 239, 244, 245
– –:– of grazing/cutting, 4, 74, 75, 76, 77, 81, 84, 86
– –:– of soil conditions, 81, 82, 86, 244
– –:– on animal production, 92, 243, 244
– –:– on herbage production, 75, 81, 92
Bowen ratio, 206
bracken (*Pteridium aquilinum*), 126
breeding time (see also calving and lambing date), 247
breeds (see also individual breeds), 133, 157, 233, 234
–: cattle, 93, 94, 103
–: goats, 123, 124
–: sheep, 114, 116, 117, 119, 120
Britain (see also United Kingdom), beef production systems in, 93, 94, 95
–, milk production in, 103, 107, 108, 146, 218
–, N cycling in, 199
–, N fertilizer use in, 7, 61, 64, 102, 217, 241
–, pasture productivity in, 9–12, 14, 15, 62, 63, 101
–, seasonal pasture production in, 13, 102
–, types of grassland in, 81, 82
British Alpine goats, 123
browse plants, 247, 249
bud(s) inhibition, 40
– regeneration, 74
– tiller, see tiller bud
buffalo, see water buffalo
burning, see fire
bush sickness, see cobalt deficiency

$^{14}$C, 32, 33
$C_4$ species (see also panicoid grasses), 50
caecum, 91
calcium excretion by animals, 182
– in faeces, 182

calcium (*continued*)
– in urine, 184
–, plant content, 147, 149, 150
– requirement by animals, 145, 147
California, 20, 21, 23, 24
calves, diet of, 94, 133
–, grazing behaviour of, 130, 131, 133
–, growth rate of, 229
–, sources of, 93, 94
calving date, 93, 94, 105, 106, 228, 233, 248
canopy photosynthesis, 30, 31
– respiration, 30
– structure, 30, 41, 49, 76
capeweed (*Arctotheca calendula*), 183
capital investment, 222, 232, 233
carbohydrate, non-structural, 193
– reserves, 33, 34, 72, 193, 227
–, soluble, 37, 67
–, structural, 51, 120
– supply, 41, 53, 190
carbon assimilation, 29
– dioxide, 50, 52, 190
carcass-weight, 156
carrying capacity (*see also* stocking rate), 113, 114, 227, 255
cashmere wool, 123
cash reserves, 232
catchment(s), *see* water catchment
cattle (*see also* beef *and* milk production)
– breeds (*see also individual breeds*), 93–95
–, calving date of, 93, 94, 105, 106, 228, 233, 248
–, diet selection of, 132, 133, 242
–, efficiency of feed conversion in, 96–99, 156
–, grazing behaviour of, 92, 93, 129, 130–133
–, herbage intake of, 92, 93, 141, 218
– lactation (*see also* milk production), 102, 105, 106
– production systems, *see* beef, veal *and* milk production
– reproduction (*see also* calving date), 95, 156
–, seasonal feed requirement of, 92, 101, 102
–, stocking rate of, 96, 98, 99, 103, 104, 106, 107, 109, 229, 234
ceiling yield, 30, 38
cell division, 34, 38
– extension/enlargement, 34, 36, 38
– turgor, 37
Charolais cattle, 93, 94
chicken, 125
chlorine, 37, 145
chromium, 145
C:N ratio, 55, 63
clover, red (*Trifolium pratense*), 76, 150, 193
–, subterranean (*T. subterraneum*), 20–23, 192, 193, 197
–, white (*T. repens*), 7, 13, 66, 67, 72, 75, 81, 83, 84, 108, 115, 149, 150, 175, 189–193, 198, 242
cobalt, animal requirement of, 145, 147
– deficiency in animals, 140, 145, 146, 148, 149, 152, 239
– excretion by animals, 183
– in plants, 140, 147
cocksfoot (*Dactylis glomerata*), 22, 36, 38, 64, 81, 83–86, 150, 244
cold tolerance, 34, 36
compensatory growth, 93, 228, 255

competition, 34, 40, 55, 56, 74, 175, 188, 190–194
competitive ability, 55, 56, 175, 192–194
concentrate(s), *see* supplementary feed
conserved fodder, *see* herbage conservation
conservation, *see* herbage conservation
continuous grazing, 29, 40, 71, 93, 97, 106, 118, 131, 132, 234
contour banks, 208
Coopworth sheep, 114, 116, 117, 120
copper, animal requirement of, 145, 147
– deficiency in animals, 145, 146, 149, 239
– excretion by animals, 182
– in plants, 147, 150
– supplements, 97, 148, 152
Corsica, 20
costs, fixed, 219, 255
– of drainage, 217
– of fertilizer, 217, 219, 220
– of resowing, 217
–, marginal, 219
–, variable, 219, 232, 256
crops/cropping, 47, 54, 81, 163, 169, 206, 215, 233, 235, 240, 245, 247, 251
crop growth rate, *see* pasture growth rate
crude protein, *see* protein
culling, 93, 126, 158
culm, 74
cultivar(s), 31, 42, 74, 91, 107, 108, 138, 139, 149, 150, 152, 244
cutting, 4
– compared with grazing, 75, 76, 77
– frequency, 32, 71, 72, 74, 83
– intensity (height), 71, 72, 74, 84
–, timing of, 73
cyanide, 140
cytokynin (*see also* growth regulators), 39, 40

dairy-beef system, 94, 95
dairy cows, 62
– – feed requirement of, 101, 221
– –, lactation period of, 105, 106
– –, milk production of, 101–103, 221
daylength, *see* photoperiod
deer, diet of, 125, 126
–, meat composition of, 125
–, stocking rate of, 125, 126
decomposition, *see* plant litter
defoliation (*see also* grazing *and* cutting)
–: effect on assimilate allocation, 33
–:– on botanical composition, 71, 74, 82, 84
–:– on nutrient content, 150
–:– on nutrient cycling, 176
–:– on pasture growth, 7, 9, 15, 25, 42, 53, 71–73
–:– on pasture composition, 71, 73–75, 77, 81, 86, 190
–:– on pasture quality, 74, 77
–:– on photosynthesis, 30, 32
–:– on seasonal growth, 71, 73
–:– on tillering, 39, 41
– frequency, 10, 71–74, 76, 77, 81
–: grazing and cutting compared, 75–77, 150
–, height/intensity of, 10, 71–73, 76, 77, 81

# GENERAL INDEX

–, timing of, 71, 73, 74
denitrification, 68, 166, 197, 198, 201, 202
density: plant, 39, 55, 56
–, tiller, 39
diagnosis of mineral deficiency, 148
diet selection, see selective grazing
digestibility, 54, 74, 85, 92, 102, 117, 137, 138, 159, 175, 184, 228, 230, 250, 255
–, effect of defoliation on, 74, 102
–, effect of N on, 67
–, leaf, 56, 138–140
– of grasses, 85, 137–139
– of legumes, 137, 140
–, seasonal variation of, 117
discount rate, 222, 223, 255
dock (*Rumex*), 65, 257
Dorset horn sheep, 156
drainage, 51, 97, 201, 210, 215, 217, 222, 223
dried grass, 126, 217
drought (*see also* water stress)
–: effect on animal production, 108, 109, 110, 114
–:– on leaf growth, 36, 37
–:– on N fixation, 191
–:– on pasture production, 9, 11, 12, 36, 37, 52, 72, 73, 232, 251
–:– on photosynthesis, 30, 37
–:– on root growth, 37
–:– on water capture, 206, 208, 209
– reserves, 232
– tolerance, 34, 37
dry matter on offer (DMO), 19–24
dry sheep equivalents, 232, 255
drying off (of cattle), 101, 105, 255
dung, area affected by, 176, 183
– beetles, 184, 198, 201
– breakdown, 165, 184
– burial, 184
– deposition, 54, 183
– distribution, 176, 177, 183
–: effect on herbage production, 75
–, nutrient content of, 75, 165, 169, 170, 177, 181, 184, 198, 202

early bite, 64
ear emergence, 74
earthworms, 184, 201
ecological stability, 205
economic evaluation, 205, 215, 217, 243
– index, 244
– profit, see profit
– optimum, 62, 68, 105, 217, 224, 229, 232, 233, 240, 242, 243
ecosystem balance, 168
– boundaries, 163
– components, 164
– dimensions, 164, 167
– equilibrium, 168, 177
– steady state/stability, 168
ecotypes, 150
efficiency, energetic (*see also* feed conversion), 98, 99, 155, 157, 158, 159

– of herbage conversion, *see* feed conversion
– of herbage utilization, 89, 96, 97, 98, 104, 115, 233, 244
– of protein conversion, 99, 120
Eire (*see also* Ireland), 102
energy flow, 4, 5, 249
England (*see also* Britain *and* United Kingdom), 14, 102, 103, 126, 146
erosion, *see* soil erosion *and* wind
Europe, 13, 14, 61, 113, 125–127, 146, 173, 234
eutrophication, 205
evapotranspiration, 51, 200, 201, 205–210
eutrophication, 205
ewe equivalents, 232, 255
excreta (*see also* dung *and* urine), 47, 54, 61, 63, 66, 71, 150, 151, 197
–: effect on herbage production, 15, 75
– returns to soil, 75, 76, 161, 163–165, 167, 170, 176, 201, 202
–, spatial distribution of, 76, 201

fecundity, 116, 117, 159
feed-back mechanisms, 190
feed conversion of cattle, 89, 93, 99, 101, 102, 109, 125, 127, 155, 229
– – of deer, 126, 127
– – of geese, 127
– – of goats, 155
– – of rabbits, 127, 157, 159
– – of sheep, 116, 118, 127, 155, 157
– – of water buffalo, 125
– deficit, 101, 115, 116, 247, 248, 249, 251
– intake, 54, 71, 72, 89, 96, 97, 114, 118, 228, 233
– –, effect of herbage availability on, 92, 93, 131, 141, 228, 229
– –,– of herbage quality on, 118, 137–141
– –,– of plant cultivar on, 139
– –,– of plant composition on, 137–141
– –,– of plant species on, 137–140
– –,– of sward height on, 93
– requirement (*see also* herbage requirement), 64, 85, 89, 93, 101, 114, 229, 230, 233, 240, 241, 247, 248, 250, 251
– supplements, *see* supplementary feed
fertilizer (*see also* nitrogen, phosphorus *and* potassium)
– application/input, 57, 82, 83, 85, 86, 145, 149, 208, 209, 215, 232, 239, 241, 244, 245
–: economics of use, 215, 232, 241
–: effect on animal output, *see individual elements*
–:– on botanical composition, 82, 239
–:– on pasture productivity, *see* herbage production
– interaction with stocking rate, 227, 233
fescue (*Festuca*), 24, 257
–, meadow (*F. pratensis*), 41, 64, 244
–, red (*F. rubra*), 64, 81, 82, 84–86, 244
–, sheeps (*F. ovina*), 38, 65, 81, 85
–, tall (*F. arundinacea*), 244
field capacity (of soil), 51, 191
finished stock, 92
fire, 197, 200, 206, 207, 209, 210
fleece weight, *see* wool
flood control, 205, 208–210
flowering, 13, 33, 34, 40, 41

fluorine, 145
flushing, 116, 119, 255
fodder, *see* herbage
forage, *see* herbage
– crops, 247, 255
fore-gut, 91
forest, 205, 206
formaldehyde, 248
formic acid, 248
France, 20, 61, 127
freemartins, 158
fructans/fructosans, 33, 34, 67
fungicide, 82

garrigue, 20, 24
gaseous losses of N, *see* ammonia *and* denitrification
genetic merit, 109
genotypes, animal (*see also* breeds), 233, 234
–, plant, 42, 53, 57, 149, 189
Germany, 61
gibberelins (*see also* growth regulators), 36
goat(s) (*Capra hircus*), 123–125
– breeds, 123
–, diet of, 123, 124, 133, 193
–, fibre production of, 123
–, food conversion of, 123
–, meat composition of, 125
–, milk production of, 123, 124
–, reproduction of, 158
goitrogens, 146
goose (*Anser*), 125, 127
gorse (*Ulex europaeus*), 124, 126
grass species, digestibility of, 85, 137–140, 244
– –, dry matter productivity of, 12, 13, 83, 84, 244, 248
– –: effect on animal output, 244, 245
– –: response to water stress, 37, 38
– –: seasonal pattern of production, 13, 85, 244
grass staggers, *see* hypomagnesaemia
grassland, V, 255
– burning, *see* fire
– catchment, 205–207
–, natural, 4, 47, 207, 208, 210, 215, 227, 245
– productivity, *see* herbage production
grazing behaviour, *see also* selective grazing
– –, circadian pattern of, 129, 131
– –, effect of experience on, 133
– –,– of management on, 130, 131
– –,– of pasture on, 131–134, 137
– –,– of physiological state on, 130, 133
– –,– of weather on, 129, 130
– – of cattle, 92, 93, 129–133
– – of horses, 129–131
– – of sheep, 93, 129
– –, social effects of, 129, 130
– compared with cutting, 75–77, 150, 151
–, continuous, 29, 40, 193
–: effect on tillering, 39, 40
– frequency, 32, 40, 47, 56, 61, 71, 84
– intensity, 56, 71, 106, 171, 201, 227
– management, 4, 24, 109, 171, 241

– –: effect on animal output, 97, 106, 107, 109, 118, 120, 234
– –:– on botanical composition, 56, 81, 82, 84, 86, 151, 239
– –:– on feed intake, 141, 142
– –:– on herbage production, 24, 86
– –:– on herbage quality, 102, 141, 142, 151
– –:– on N cycling, 171, 197, 201, 245
– –:– on P cycling, 175–177
– –:– on run off, 206
– –:– on water quality, 209, 210
–, mechanics of, 132
–: overgrazing, 75, 106, 200, 201, 208, 209
– pressure, 227, 228, 255
–, rotational, 71, 75, 97, 106, 107, 118, 165, 170, 193
–, seasonal pattern of, 76
–, selective, *see* selective grazing
– time (*see also* grazing behaviour), 92–94, 106, 129, 130, 131, 250
Greece, 20
grinding (herbage), 140
gross margin, 107, 109, 232, 245, 255
– returns, 232, 255
groundwater, 6
– contamination, 158, 200, 210, 211
– flow, 210
– recharge, 211
growth index, 49
– rate, *see* animal *and* pasture growth rate
– regulators, 34, 36, 40
– substances, *see* growth regulators

hawthorn (*Crataegus monogyna*), 126
hay, 68, 92, 93, 105, 106, 108, 109, 116, 171, 217, 247–250
heat balance, 49
heifers, 103–105, 158, 255
herbage accumulation, 71, 97, 255
– allowance, 118
–, cash value of, 215, 217–219
– composition, *see* herbage quality *and individual elements*
– conservation (*see also* hay, silage *and* supplementary feed), 7, 64, 73, 92–95, 106, 115, 150, 171, 175, 233, 241, 242, 247–252, 255
– consumed, *see* feed intake
–, dead, 72–74, 76, 96, 104, 106, 117, 170, 176, 177
– digestibility, *see* digestibility
– intake, *see* feed intake
– production, economic value of, 215, 217–220
– –, effect of flowering on, 13
– –,– of botanical composition on, 13, 83, 84 86, 227, 239, 243–245, 248
– –,– of cutting on, 66, 83
– –,– of defoliation on (*see also* effect of cutting and grazing), 7, 10, 86
– –,– of drainage on, 222
– –,– of grazing on, 57, 66
– –,– of growing season on, 23, 29
– –,– of nitrogen on, 9, 23. 24, 61–70, 107, 197, 218–220, 227, 233, 239, 243
– –,– of phosphate on, 23, 24, 220, 227, 233, 239
– –,– of potassium on, 23, 24, 220, 239
– –,– of radiation on, 10, 11, 29, 30, 34, 109

# GENERAL INDEX

––,– of rainfall on, 9, 11, 20, 23, 109
––,– of stocking rate on, 76, 103, 104, 227, 228
––,– of sulphur on, 23, 239
––,– of sward age on, 82, 83
––,– of temperature on, 9, 11, 15, 29, 34, 109
–– in Australia, 21–24
–– in California, 20, 21, 23
–– in France, 20
–– in Greece, 20
–– in Israel, 19, 20, 23
–– in mediterranean region, 19–25, 249
–– in New Zealand, 7–12
–– in South Africa, 21
–– in Turkey, 20
–– in United Kingdom, 9–11
–– of grass species, 12, 13, 83, 108
–– of hill pastures, 14, 15
–– of leys, 82, 83, 85
–– of permanent pasture, 85
––, potential, 13, 14, 102
––, seasonal pattern of, 8–10, 12, 13, 29, 57, 63, 64, 73, 85, 86, 89, 93, 98, 101, 102, 113, 114, 120, 218, 219, 228, 230, 233, 240, 247–249, 251
––, year-to-year variation of, 8, 9, 20–22, 25, 91, 105, 109, 114, 215, 217, 224, 232, 242–244
– quality, 85, 86, 89, 227
––, effect of cultivars on, 108
––,– of fertilizer on, 67, 239, 242
––,– effect of species on, 85, 86, 242, 243, 245, 249
––:– on animal performance, 102, 117, 118, 228, 233, 239, 240, 241, 245
––:– on intake, 118, 137–141
––, seasonal variation of, 117, 240
– requirement,
––, critical period of, 114, 118–120
–– of beef cattle, 85, 92, 93
–– of dairy cows, 85, 101, 229
–– of sheep, 85, 114–116, 118–120
–– seasonal pattern of, 218, 240, 247, 251
– utilization (*see also* efficiency), 96, 115, 170
herbicides, 82
Hereford cattle, 93, 94, 133
hill pastures, 14, 15, 82, 145, 169, 175, 208
hogget, 119, 120, 255
Holland, 14
horse(s) (*Equus caballus*), 123, 125, 126
–, feeding systems of, 126
–, grazing behaviour of, 131
–, meat composition of, 125
hydrology, 205, 208
hydrosphere, 163
hypomagnesaemia, 146

indigenous species, 64, 65, 83, 84, 86, 243, 249, 255
indole acetic acid, 40
infiltration rate, *see* water
inoculation with *Rhizobium*, 188
input combinations, 220–222, 224, 225, 241
– costs, 218–223
intake, *see* feed intake

interactions between mineral element, 145–147
– between species and environment, 83, 84, 86
interception loss, *see* rainfall interception
interest, 222, 255
intra-specific variation, *see* genotype
iodine, 145–147, 149, 152
iron, animal requirement of, 145, 147
– deficiency in animals, 145, 149
– excretion by animals, 183
– in plants, 150
Ireland, 101–103, 106
irradiance, *see* radiation *and* light
irrigation, 7, 9, 10, 12, 37, 41, 54, 57, 68, 107, 201, 208, 209, 223, 248, 251
Israel, 19, 20

Kikuyu grass (*Pennisetum purpureum*), 65, 66

labour, 232
lactation, 119
–: effect on lamb weight, 114
–, length of, 105, 106
lambing, date of, 115, 116, 228, 229, 233
– percentage, 115–119, 156, 157
lambs, 114, 124, 125, 157, 229, 244
laterite, 210, 240
leaching (*see also individual elements*), 184, 208, 210
leaf area index (LAI), 30, 31, 37, 38, 53, 55, 71, 72, 93, 207, 255
leaf appearance, rate of, 34, 38–40
– area (see also LAI), 30–32, 38, 72, 227
– digestibility, 56, 138–140
– extension (*see also* growth), 11, 13, 34–38, 56
– growth (*see also* extension), 29, 34, 42
– lifespan, 38, 41
– mortality, 30, 38, 39, 41, 55
– photosynthesis, 31, 33
– primordia, 39
– production, 30
– respiration, 33
– senescence, 30, 50, 55
– sheath, 34, 56
– size, 31
– water potential, 37
legumes, canopy structure of, 49
– content in sward, 66, 67, 169, 170, 185, 187, 188, 190–194, 228, 240, 242, 245
–, digestibility of, 140, 244
–: effect on animal performance, 159, 242, 244
–:– on intake, 137, 140, 142
–:– on live weight gain, 244
–:– on rumen fill, 140
–: nitrogen fixation in, *see* nitrogen fixation
–: response to N, 189
–, tropical, 137, 140, 190, 192
leys, 4, 85, 255
lifespan (of animals), 89, 93, 157, 158
light, diffuse, 50
– interception, 29–31, 37, 41, 49, 50, 56, 72, 227
– intensity, *see* radiation

lignin, 51
lime/liming, 23, 107, 151, 169, 201, 233, 239
Limousin cattle, 93, 94
litter, see plant-litter
– size (animals), 156, 157
liveweight gain, 94, 97, 118, 200–222, 228–230, 244
longevity, see lifespan
lucerne (*Medicago sativa*), 21, 83, 140, 150, 189, 192, 193, 197, 234, 251
lysimeter(s), 205

magnesium, animal requirement of, 145, 147
– excretion by animals, 182, 184
– deficiency in animals, 145, 146, 152
– in faeces, 182,
– in plants, 147, 150, 151, 239, 241
maintenance requirement, 106, 116, 230
maize (*Zea mays*), 217, 251
manganese excretion by animals, 183, 184
– in plants, 147, 149, 150
– in soil, 184
– requirement by animals, 145, 147, 149
maquis, 20
marginal rate of substitution, 221
– revenue, 219, 233, 255
– costs, 219
mating, 120, 228
meadowgrass(es) (*Poa*), 24, 83
–, annual (*P. annua*), 65, 75, 81, 82
–, rough-stalked (*P. trivialis*), 75, 81, 82, 85
–, smooth-stalked (*P. pratensis*), 33
meat composition, 125, 229, 247
mediterranean grasslands, 19–25
Merino sheep, 113, 232
meristem(s), 32, 35, 73, 74, 193
metabolic reserves, 72
micronutrients, see nutrients
milk composition, 125
– fat, 101, 102, 104–109, 124
– production, effect of fertilizer on, 103, 107, 242, 243
– –, factors affecting, 101, 105–110, 229, 241, 242, 244, 247
– – from cattle, 101–103, 123, 124
– – from goats, 123, 124
– – from sheep, 113, 114, 123, 229
– –, location of, 91
– –, theoretical limits of, 101, 102
mineral content of plants (see also individual elements), 149–151
mineralization of nitrogen, 54, 63, 83, 163, 168, 170, 187
– of phosphate, 83
mineral deficiency in animals, 145–148
– – in plants, 145, 147, 149–151
– licks, 152
– nutrients (see also individual elements), 47, 53, 73, 175
– requirement
– – of animals, 145, 147, 148
– – of plants, 145, 147
mineral supplements, see supplements
mimosine, 249
mixed farming, 91

– grazing, 126
model(s), analytical, 47, 48
– of animal production, 98, 99, 109, 233, 235, 243, 252
–, compartmental, 48
–, continuous, 47, 255
–, deterministic, 47, 255
–, discrete, 47, 255
–, dynamic, 47, 255
–, empirical, 47, 48, 54
– of energy flow, 249
–, hydrological, 209–211
–, mechanistic, 48
– of N cycling, 54
–, parameters, 48
– of pasture growth, 10, 24, 25, 47–60, 231, 243
– of P cycling, 177, 178
–, plant-physiological, 47
– of root growth, 53
– of soil water content, 51
–, static, 47, 48, 255
– simulation, 47, 48, 233
–, stochastic, 47, 255
modelling, 1, 5, 24, 47–60, 98, 99, 125
molasses, 248
molybdenum, animal requirement of, 145, 147, 239
– excretion by animals, 183
– in plants, 147, 150
– in soil, 184
– toxicity in animals, 146
mortality, animal, 229
–, leaf, 30, 50, 55
–, plant, 168
–, tiller, 33, 40, 41
mulga (*Acacia aneura*), 206
multivariate analysis, 81, 241
mycorrhiza, 175
myxomatosis, 126

negative feedback, 52
net present value, 222, 255
Netherlands, The, 61, 63, 65, 81, 146, 174, 200
New Mexico, 133
New South Wales, 134, 205, 207, 209
New Zealand, 72, 105, 126, 168, 173, 249
– –, deer production in, 125, 126
– –, grazing management in, 106, 107, 109, 234
– –, hill pastures in, 14, 15, 175
– –, milk production in, 102–110
– –, N cycling in, 168–170, 183, 188, 199–201
– –, P cycling in, 177
– –, pasture productivity in, 7, 8, 13, 15, 101, 113, 165
– –, rabbit production in, 127
– –, seasonal pasture production in, 9–12, 15, 115–117
– –, sheep production in, 113–116
– –, soils in, 173–175
– –, use of N fertilizer in, 61, 68, 165
nickel, 145
nitrate in herbage, 67, 189, 191–193
– in soil, 54, 192, 199
– leaching, 197, 200

GENERAL INDEX

– reductase, 190
– toxicity, 151
nitric acid, 197
– oxide, 198, 199
nitrification, 163, 164, 201
nitrite, 55
nitrogen (N), absorption/assimilation/uptake of, 54, 188
– balance, 168
– cycling, 5, 25, 54, 163–165, 167–171
– deficiency, 7, 54, 197
–, dry deposition of, 163–165, 169, 188
– excretion by animals, 54, 61
– excretion by plants, 54
– export, 181, 197, 200
– fertilizer, 9, 23, 24, 54, 61–70, 151, 159, 167, 170, 176, 190, 192, 199, 200, 201, 217–220, 224, 240, 242, 243
– –: effect on animal production, 97, 101, 102, 107, 224, 225, 233, 240–243, 245
– –:– on botanical composition, 65, 66, 242
– –:– on digestibility, 67, 242
– –:– on herbage production (*see also* herbage production), 9, 23, 62–68, 197, 217, 218, 220, 227, 233, 240, 242, 243, 245
– –:– on leaf growth, 38
– –:– on plant mineral content, 151
– –:– on roots, 54
– –:– on tillering, 39, 41
– – inputs, 163–167, 170, 187, 188, 219, 220
– –, residual effect of, 64
– –, seasonal pattern of application, 64, 67, 85
– fixation, annual rate of, 169, 170, 187, 188, 194
– – by legumes, 7, 54, 56, 57, 84, 161, 163, 165–167, 170, 187, 188, 191, 240, 241
– –, non-symbiotic, 165, 166, 169, 187
– –: response to defoliation, 190, 193
– –: response to grass competition, 84, 190, 194
– –: response to mineral N, 187, 189, 190, 193, 194
– –: response to pests and diseases, 193
– –: response to shading, 193
– –: response to temperature, 190–192
– –: seasonal variation of, 168, 189–192
– flux/flow, 163–166, 168–171, 187, 188
–, gaseous, 163, 164, 188, 198, 199
– immobilisation, 55, 165
– in atmosphere, 163
– in excreta, 182, 194, 197
– in rain, *see* wet deposition
– in soil, 54, 63, 163, 164, 188
– leaching, 54, 64, 68, 197, 200, 202
– losses, 54, 64, 68, 151, 163, 167, 168, 171, 181, 187, 188
– – by export, 166, 169, 170, 200
– –, gaseous, 163–166, 170, 171, 197–202
– – in animal products, 166, 169, 170, 197, 200
– – in solution, 54, 166, 167, 169, 170, 197, 200–202
– metabolism, 54, 56
– mineralization, 54, 63, 83, 163, 164, 168, 170, 184
– movement in soil, 54
–: plant content, 54, 149, 170
–: plant uptake, 38, 163–165
– pool size, 164, 165, 169, 170
– recovery, 67, 68

– recycling in plant, 164–166, 168–170
– response of pastures
– – – –: effect of climate, 63
– – – –:– of defoliation, 65
– – – –:– of grazing, 66
– – – –:– of season, 63, 64
– – – –:– of soil N, 63
– – – –:– of species, 64–67
– – – –:– of temperature, 63
– – – –:– of water supply, 63
–: soil content, 54, 164, 165, 188, 192, 197
–, turnover rate of, 165, 187
–, wet deposition of, 163–166, 169, 170, 188, 194
nitrogenase, 189–191
nitrous oxide, 197–199
nodule(s) development, 189, 190, 193
– formation, 189, 190, 194
– growth, 190, 194
– loss, 189, 193
North Africa, 113
North America (*see also* United States of America), 72, 113, 146, 234
Northern Ireland, 102, 104
N:S ratio, 182, 185
nutrient(s) (*see also individual elements*)
– cycling (*see also* elements), 4, 5, 161, 245
– deficiency in animals (*see also* elements), 145–148
– – in plants (*see also* elements), 145, 147
– losses (*see also* elements), 208–210
–: macronutrients, 1
–: micronutrients, 1, 182
– release (*see also* mineralisation), 83
– returns (*see also* excreta), 75, 76
– translocation, 38
– uptake, 38, 72
nutritional deprivation, 119
– plane, 93
– wisdom, 132–134
nutritive value, *see* digestibility *and* herbage quality

oats (*Avena sativa*), 251
oestrus, 119, 158
ontogeny, 52
optimum foraging strategy, 130, 133
– inputs, 218–224
osmotic regulation (osmoregulation), 37, 38
overgrazing, *see* grassland
ovulation rate, 118, 119
oxygen, 55, 200
ozone, 197

$^{32}$P, 174
$P_{max}$, 50
palatability, 133
panicoid grasses, 85
parasite(s), 229
Park Grass Experiment, 3
parturition, date of (*see also* calving, lambing date), 85
– frequency, 156–158
pasture(s) (*see also* sward)

pastures, annual production of, *see* annual dry matter production *and* herbage production
– composition, *see* botanical composition
– cultivars, 31, 74, 149, 150
– establishment/renovation, 83, 188, 215
– growth rate, 13, 19, 71, 96, 207, 209, 227, 231
– potential production, 13, 14
– productivity, *see* herbage production *and* primary productivity
– regrowth, 29, 31, 42, 53, 72
–, reproductive growth of, 53
–, seasonal production of, *see* herbage production
– species, 31, 53, 55, 61, 72, 74, 81–86, 92, 149, 150
–, vegetative growth of, 52
pathogens, 1, 81
Perendale sheep, 116, 117, 119
permanent pasture, 4, 65, 85, 92, 188, 244, 245, 251, 256
– wilting point, 191
pesticides, 82
pests, 1, 81
phenology, 52, 53, 56
phosphorus (P) adsorbed, 173
–, animal intake of, 174, 175
–, animal requirement of, 97, 140, 145, 147
– cycling, 5, 55, 173, 174, 176–178
– deficiency in animals, 134, 140, 145, 146, 149
– desorption, 175
– effect on animal production, 107, 233, 241–243, 245
– – on herbage production, 220, 227, 233, 241
–, exchangeable, 173, 174
– excretion by animals, 181, 182
– export, 177, 181
– extraction methods, 174, 175
– fertilizer (*see also* superphosphate), 8, 151, 173, 174, 176–178, 201, 220, 233, 240
– fixation, 173, 174
– fluxes, 177
–, forms of, 173
– immobilization, 174, 176
– in diet, 132, 134, 148
– in excreta, 174–177, 181, 182, 184
– in herbage, 134, 140, 147, 149–151, 175, 241
– in litter, 174–177
– in soil, 173, 174, 177, 240
–, inorganic, 173–175, 177, 181, 183, 184
– losses, 174, 176, 177
– mineralization, 173, 176, 177, 184
–, organic, 173, 175–177, 181, 184
–: plant availability, 173, 174, 183, 184
–: plant content, *see* in herbage
– pool size, 173, 177
– recycling, 176, 181
– sorption, *see* fixation
– supplement for animals, 97, 134
– uptake by plants, 38, 174, 175
photoperiod, 10, 11, 53
photorespiration, 50
photosynthesis, 29, 50–52, 54, 71, 72, 193
– in canopy, 30, 31
– in crop, 50, 227

– in leaf, 31, 33
photosynthetic capacity, 30, 39, 52, 53
– efficiency, 30, 56
– potential, 31, 32, 42
phyllosphere, 188
physiological status of animals (*see also* pregnancy, lactation, etc.), 92, 114, 115, 133
pig(s) (*Sus*), 123, 158, 160
plagioclimax, 4, 86
plant analysis, 148, 149
– breeding, 25, 29, 42, 92, 150, 152
– cover, 207–209
– litter, 163, 170
– –, decomposition of, 55, 163, 165, 169–171, 176
plant productivity, *see* herbage production
pooid grasses, 85
Portugal, 20
potassium cycling, 5, 55, 181
– effect on herbage production, 23, 190, 220, 239, 241
– – on milk production, 107
– excretion by animals, 150, 182
– – in urine, 182, 184
– export, 181
–: plant content, 37, 147, 149–151
–: plant requirement, 147, 185
– requirement by animals, 145, 147
poultry, 123
prairie, 52, 133
pregnancy, feed requirement during, 116, 228, 247
– toxaemia, 229
productivity, primary (*see also* herbage production), 5, 19, 23, 24, 89, 91, 98, 245
– secondary, 5, 89, 245
profit, 217–220, 223–225, 232, 234, 242, 249, 256
prolificacy, 95, 114–117, 120, 228, 233, 234
pronghorn (*Antilocapra americana*), 133
protein, 50, 51, 91, 97, 104, 120, 140, 145, 146, 164, 193, 240, 244
puberty, 119, 157
"put and take", 241, 242, 244, 256

Queensland, 200, 241

rabbit(s) (*Oryctolagus*), 123, 125–127, 155–157
– as pests, 126
–, diet of, 126
–, food conversion of, 127
–, meat composition of, 125
–,– production of, 156
– production systems, 126
–, reproductive rate of, 127, 156, 157
radiation (*see also* light)
–: daylength (*see also* photoperiod), 10
–, infra-red, 49
– interception, 30, 49, 227
– receipt, 10, 12, 14, 33, 34, 36, 49, 50
rainfall, effect on animal output, 108–110
–:– on herbage production, 5, 11, 12
–:– on N leaching, 200, 201
–:– on run-off, 200, 228

# GENERAL INDEX

– interception, 51, 205, 206
–, seasonal distribution of, 63
rangelands, 4, 177
red deer (*Cervus elaphus*), 123, 125, 126
regrowth, *see* pasture regrowth
relative growth rate, 38, 49, 83
reproductive rate (*see also* prolificacy), 89, 155, 156, 228, 233, 234, 242
– – of cattle, 95, 156, 229
– – of goats, 158
– – of plants, 56
– – of rabbits, 158, 159
– – of sheep, 114–116, 118, 119 156, 157, 228, 229
– span, 120, 156, 157, 159
– strategy, 53
reseeding, *see* sowing
reserves, body, 73
–, metabolic, 72
residence time (of elements), 167, 168, 171
resistance, aerodynamic, 51
–, boundary layer, 206
–, canopy, 51, 206
–, stomatal, 51
resource allocation (*see also* root/shoot ratio), 84
resowing, *see* sowing
respiration, 30, 31, 33, 36, 50, 51, 55, 72
retention time (in rumen), 140
revenue, 219, 222, 256
rhizobium, 164, 165, 167, 187, 188, 190, 201
rhizome, 74, 84
rhizosphere, 188
risk, 217, 223–225, 232, 234
river sediment, 207
Romney sheep, 114, 116, 117, 119, 120
root(s), activity, 52
– death, 53
– density, 51, 164
– depth, 56, 193, 201, 208, 210
– distribution, 51, 175
– growth, 37, 53, 56
– lifespan, 53, 176
– mass, 52, 176
–, nodal, 32
– penetration, 53
– respiration, 51
– seminal, 32
– zone, 210, 211, 228
root/shoot ratio, 52, 53
rotational grazing, 71, 75, 97, 106, 107, 118, 227, 234, 256
rumen, 91, 138, 140, 146, 159, 165
run-off, *see* water run-off
ryegrass (*Lolium*), 83, 251
–, Italian (*L. multiflorum*), 32, 38, 67, 75, 149, 150
–, perennial (*L. perenne*), 7, 9, 13, 14, 21, 29–41, 64–67, 71–75, 81, 83–86, 108, 115, 132, 149, 150, 175, 190–192, 197, 243, 244
–, Wimmera (*L. rigidum*), 192

Saanen goats, 123, 124
salinization, 205, 210
Santa Gertrudes cattle, 133
savannah, 4
Scotland, 14, 241
seed(s), dispersal, 71
– in soil, 73, 74, 82
– pelleting, 140
– production, 73, 74, 249
– size, 83, 84
selective grazing, 47, 73, 75, 76, 132, 133, 167, 171, 240, 242, 244
– –: age differences, 133, 134
– –: breed differences, 133, 134
– –, effect of experience on, 133, 134
– –,– of season on, 142
– –,– of senses on, 132–134
– –:– on botanical composition, 193, 242
– –:– on mineral intake, 148
– –, factors affecting, 14, 132–134, 141, 142, 185
– –: species differences, 117, 132–134, 193, 242
selenium deficiency in animals, 145, 146, 148, 149, 152
– excretion by animals, 183
– in plants, 140, 147
– toxicity, 141, 148
self-thinning, 55
senescence, leaf, 30, 50, 227
–, plant, 55
–, tiller, 30, 40, 41
set stocking (*see also* continuous grazing), 76, 107, 118, 193, 234, 242, 244, 245, 256
shading, 30–33, 39, 40–42
sheep breeds, 116, 117
–, diet selection of, 117, 132–134, 193, 242
– efficiency of conversion by, 120
–, excretion of nutrients by, 181
–, feed intake by, 115, 116, 120, 137–141, 250
–, grazing behaviour of, 93, 129, 130–134
–, herbage intake of, *see* feed intake
–, lactation of, 114, 118, 119
–, lambing date of, 115–119, 156, 157
–, lamb production of, 113, 120, 156, 244
–, production systems of, 113, 120
–, prolificacy of, 114–117, 158
–, reproductive cycle of, 114–118
–, seasonal feed requirement of, 115, 116
–, stocking rate of, 113–116, 120, 228, 229, 231, 232, 234, 250
shelter belts, 202
shoot apex, 39
– growth, 71
silage, 65, 68, 92, 93, 105, 106, 108, 109, 116, 171, 220, 221, 247
–, economic evaluation of, 220, 221
– intake, 247
–: methods of production, 247, 248
– quality, 247
silicon, 145
Simmental cattle, 93, 94
simulation studies, *see* models and modeling
sink size, 52, 54
siratro (*Macroptilium atropurpureum*), 142
slaughter weight, 95

slope, 14
snails, 127
sodium, animal requirement of, 134, 140, 145, 147, 148
– in plants, 140, 147, 150, 151
soil aeration, 63
– analysis, 148, 174, 175
– CEC, 198, 201
– compaction, 76, 208
– conservation, 208
– erosion, 96, 177, 197, 205–209, 227, 234
– evaporation, 52
– fauna, 201
– fertility, 82, 83, 208
– heterogeneity, 167
– ingestion by animals, 148
– microbial activity, 167, 168, 198
– nutrients, *see individual elements*
– organic matter, 61, 63, 68, 163, 164, 167–171, 173, 176, 184, 187, 188
– pH, 63, 81, 86, 97, 148, 149, 151, 165, 167, 184, 198, 199, 239
– parent material, 149, 245
– pore space, 55
– solution, 164, 177, 184
– structure, 201
– temperature, 63
– water content, 5, 11, 12, 15, 49, 51, 53, 63, 76, 82, 163, 167, 191, 206, 227
– – extraction, 205
– – potential, 51, 54, 55
sorghum (*Sorghum*), 200, 251
South Africa, 21, 24, 113, 198
South America, 113, 125
South Australia, 21
sowing/oversowing (pasture), 81, 86, 170, 217, 239, 245, 249
sown species, 81, 92, 170, 239, 243, 248, 256
species composition, *see* botanical composition
species × environment interaction, 83, 84, 86
species mixtures, 84–86, 191, 192, 244, 245
specific leaf area, 54
starch, 193
stem elongation, 73
steppe, 4
stocking rate, 71, 85, 89, 102, 103, 142, 169, 171, 173, 175, 202, 207, 208, 218, 227–235, 239–245, 249, 250, 256
– – of beef cattle, 96, 98, 99, 229, 233, 242
– –, biological optimum of, 229, 233, 240, 242
– – of dairy cattle, 101, 104, 109, 169, 198, 229, 233, 242
– – of deer, 125, 126
– –, economic optimum of, 105, 229, 232, 233, 240, 242
– –: effect on animal reproduction, 228, 229
– –:– on botanical composition, 104, 227, 228
– –:– on erosion, 200
– –:– on herbage production, 76, 103, 104, 227–229
– –:– on liveweight gain, 96, 228–230
– –:– on milk production, 103, 104, 106, 107, 109, 228
– –:– on nutrient cycling, 169, 171, 173, 175, 178
– –:– on soil conditions, 76, 228
– –, factors affecting response to, 231, 233, 234, 239, 243
– – of goats, 229
– – of sheep, 113–115, 169, 228, 229, 231, 234, 242
stolon, 74, 84
stomata, 37
stream-flow, 208–210
strip grazing, 97, 131, 132, 256
suckler beef system, 93
sugar beet (*Beta vulgaris*), 207
sulphur, animal requirement of, 145, 147
– cycling, 5, 181
– deficiency, 185
–: effect on feed intake, 140
–:– on herbage production, 23, 239
– excretion, 181, 182, 184
– export, 181
– induced Cu deficiency, 146
– in dung, 182, 184
– in plants, 147, 149, 183
– in urine, 182
– mineralization, 184
superphosphate (*see also* phosphate), 142, 169, 170, 184
supplementary feed, 4, 7, 93, 105, 110, 130, 151, 152, 159, 175, 233, 245, 247, 251
– –, economic evaluation of, 220–222, 224, 225, 231, 252
– –: effect on animal output, 97, 98, 108, 109, 233
– –:– on grazing time, 93, 94, 97, 130, 250
– –:– on herbage intake, 93, 97, 108, 233, 250, 252
– –, least cost combination of, 220, 221
supplements, cereal, 95, 97, 126, 127, 159, 247
–, feed, *see* supplementary feed
–, mineral, 97, 146, 148, 149, 151, 152, 239
–, protein, 97
surveys, 81, 241–244
Sussex cattle, 94
sward age, 82
– height, 93
swede (*Brassica napus*), 251
systems management, 215

"teart" pastures (*see also* Mo toxicity), 146
temperate grasslands, botanical composition of, 81, 82
– –, productivity of, 7–17, 65, 91, 92
– –: response to N, 63–66
– –, seasonal production of, 8–12, 91, 92
temperature effect on grazing, 129
– – on leaf death, 55
– – on leaf growth, 34, 35, 36, 41
– – on plant growth, 34, 53, 72, 163
– – on N levels, 54, 169
– – on root growth, 53
– – on tillering, 40, 41
–, soil, 167, 174, 176, 198
termites, 201
tetany, *see* hypomagnesaemia
thistle(s), 56
thyroid, 249
tiller(s) bud, 32, 39
– daughter, 32
– death, 33, 40, 41
– density, 33, 40, 93
– dormancy, 41

# GENERAL INDEX

– emergence, 39, 40
–, primary, 32, 33, 41
– primordia, 39
– production, 29, 30
–, reproductive, 13, 34, 35, 40, 41, 73
–, secondary, 32, 33, 39, 41
–, vegetative, 13, 34, 35, 41, 73
tillering: response to cutting/grazing, 39, 41, 56, 118
–:– to irrigation, 41
–:– to N, 38, 39, 41
–:– to plant density, 39
–:– to shading, 39–41
–:– to temperature, 41
–:– pattern, 40, 41
timothy (*Phleum pratense*), 40, 41, 64, 81, 83–85, 138, 149, 150, 244
tin, 145
Toggenberg goats, 123
tooth wear, 229
top dressing, *see* fertilizer application
"topping", 106
trace elements, *see* micronutrients *and individual elements*
trampling, *see* treading
translocation of nutrients, 38, 39
transpiration (*see also* evapotranspiration), 56, 63, 200
treading, 47, 66, 71, 75, 76, 106, 167, 208
tree lucerne (*Chamaecytisus palmensis*), 249
tri-iodobenzoic acid (TIBA), 40
triticale (*Triticum aestivum* x *Secale cereale*), 251
tropical grasslands, 61, 63, 65, 91, 92, 140, 188, 192, 198, 247–249
Turkey, 20
turnip (*Brassica rapa*), 251
twinning, 94, 95, 98, 156

United Kingdom (U.K.) (*see also* Britain *and separate countries*), 9–15, 61–63, 101, 125, 126, 146, 199, 217, 218, 241
United States of America (U.S.A.) (*see also* North America *and separate states*), 146, 149, 173, 174, 201, 247
uplands (*see also* hill pastures), 82, 241
urea, excretion, 184, 198
– fertilizer, 62, 198, 201
– as silage additive, 248
urine, area affected by, 183, 184
– decomposition, 184, 197
– distribution, 183, 184
– effect on herbage intake, 185
–– on herbage production, 75, 193
–, nutrient content of, 54, 75, 165, 166, 169–171, 181, 184 185, 198, 200–202
utility function, 223, 224, 256
– value, 224, 225

vanadium, 145
veal (*see also* calves), production systems of, 93–95
vegetative propagules, *see* rhizomes *and* stolons
Veld, 21, 24
"velvet", 125
venison, 125
vernalization, 53
Victoria, 21, 207, 228
vitamins, 146
volcanic soils, 7, 12, 240, 241

Wales, 14
water balance, 207
– catchment, 155, 205–211
– conductivity, 53
– content of herbage, 140
– deficiency (*see also* drought), 51, 63, 109
– drainage, *see* drainage
– extraction, 51
– flow in soil, 51, 54, 210, 211
– harvesting, 205, 206
– infiltration, 51, 206, 208, 209
– outflow, 205–211
– percolation, 208, 211
– pollution, 68, 168, 197, 208–210
– potential, *see* soil
– quality, 155, 171, 205, 206, 209–211
– run-off, 51, 161, 177, 206–211
–, soil, *see* soil water
– stress (*see also* drought), 9, 36, 37, 52, 175
– table, 210
– uptake, 72, 73
water buffalo (*Bubalus bubalis*), 123–125
––, diet of, 123, 124
––, meat and milk production of, 123, 124
––, milk quality of, 124, 125
waterlogging (*see also* anaerobic soil), 201
weaning, 93–95, 114, 115, 119
weeds, 56, 75, 76, 81, 223
Western Australia, 22, 23, 198, 210
Western Cape Province, 21
wethers, 115, 149, 256
wheat (*Triticum aestivum*), 217
wilting point, 52
wind, 14, 51, 198, 200–202
white muscle disease, *see* selenium deficiency
wool production, 113–117, 120, 177, 200, 228, 229, 231, 244, 250
worms, *see* earthworms *and* parasites

zero-grazing, 97, 106, 256
zinc deficiency in animals, 145, 147
– excretion by animals, 182–184
– in plants, 147, 150

This document is the property of:-

The Library
Overseas Development Natural Resources
Institute
Tolworth Tower, Surbiton, Surrey
United Kingdom KT6 7DY

# Chirality in Industry II

# CHIRALITY IN INDUSTRY II

Developments in the Commercial Manufacture and Applications of Optically Active Compounds

*Edited by*
**A. N. COLLINS**
*ZENECA Specialities, Manchester, UK*

**G. N. SHELDRAKE**
*The Queen's University of Belfast, UK*

**J. CROSBY**
*ZENECA Pharmaceuticals, Macclesfield, UK*

**JOHN WILEY & SONS**
Chichester · New York · Weinheim · Brisbane · Singapore · Toronto

Copyright © 1997 by John Wiley & sons Ltd,
Baffins Lane, Chichester,
West Sussex    PO19 1UD, England
National    01243 779777
International    (+44) 1243 779777

e-mail (for orders and customer service enquiries): cs-books@wiley.co.uk
Visit our Home Page on http://www.wiley.co.uk
or http://www.wiley.com

All Rights Reserved. No part of this publication may be reproduced, stored in a retrieval system, or transmitted, in any form or by any means, electronic, mechanical, photocopying, recording, scanning or otherwise, except under the terms of the Copyright Designs and Patents Act 1988 or under the terms of a licence issued by the Copyright Licensing Agency, 90 Tottenham Court Road, London, W1P 9HE, UK, without the permission in writing of the Publisher.

*Other Wiley Editorial Offices*

John Wiley & Sons, Inc., 605 Third Avenue,
New York, NY 10158-0012, USA

VCH Verlagsgesellschaft mbH,
Pappelallee 3, D-69469 Weinheim, Germany

Jacaranda Wiley Ltd, 33 Park Road Milton,
Queensland 4064, Australia

John Wiley & Sons (Asia) Pte Ltd, 2 Clementi Loop #02–01,
Jin Xing Distripark, Singapore 0512

John Wiley & Sons (Canada) Ltd, 22 Worcester Road,
Rexdale, Ontario M9W 1L1, Canada

*Library of Congress Cataloguing-in-Publication Data*

Chirality in industry : the commercial manufacture and applications of
  optically active compounds  /  edited by A. N. Collins, G. N. Sheldrake,
  and J. Crosby.
       p.    cm.
  Includes bibliographical references and index.
  ISBN 0 471 93595 6
  1. Chirality.   2. Enantiomers—Separation.   3. Enantiomers–
  Biotechnology.   I. Collins. A. N. (Andrew N.)   II. Sheldrake, G.
  N.   III. Crosby, J.
  QP517.C57C47    1997
  660'.6—dc20                                                     92–16000
                                                                       CIP

*British Library Cataloguing in Publication Data*

A catalogue record for this book is available from the British Library

ISBN 0  471 96680 0

Typeset in 10/12pt Times by Thomson Press (India) Ltd, New Delhi
Printed and bound in Great Britain by Biddles Ltd, Guildford, Surrey
This book is printed on acid-free paper responsibly manufactured from sustainable forestation, for which at least two trees are planted for each one used for paper production.

# Contents

**List of Contributors**  vii

**Foreword**  xi

**Preface**  xiii

**Acknowledgements**  xv

1   **Introduction**  1
    *J. Crosby*

2   **Chiral Drugs: Regulatory Aspects**  11
    *J. J. Blumenstein*

3   **Production Methods for Chiral Non-steroidal Anti-inflammatory Profen Drugs**  19
    *G. P. Stahly and R. M. Starrett*

4   **Synthesis of Enantiomerically Pure Nucleosides**  41
    *B. L. Bray, M. D. Goodyear, J. J. Partridge and D. J. Tapolczay*

**PHYSICAL METHODS AND CLASSICAL RESOLUTION**

5   **Rational Design in Resolutions**  81
    *A. Bruggink*

6   **Resolution Versus Stereoselective Synthesis in Drug Development: Some Case Histories**  99
    *B. A. Astleford and L. O. Weigel*

7   **Crystal Science Techniques in the Manufacture of Chiral Compounds**  119
    *W. M. L. Wood*

8   **Membrane Separations in the Production of Optically Pure Compounds**  157
    *J. T. F. Keurentjes and F. J. M. Voermans*

**BIOLOGICAL METHODS AND CHIRAL POOL SYNTHESES**

9   **Four Case Studies in the Development of Biotransformation-based Processes**  183
    *R. McCague and S. J. C. Taylor*

10  (S)-2-Chloropropanoic Acid: Developments in its
    Industrial Manufacture   207
    S. C. Taylor

11  Development of a Full-scale Process for a Chiral Pyrrolidine   225
    H.-J. Federsel

12  Development of a Multi-stage Chemical and
    Biological Process for an Optically Active Intermediate
    for an Anti-glaucoma Drug   245
    A. J. Blacker and R. A. Holt

13  Synthesis and Applications of Chiral Liquid Crystals   263
    D. Pauluth and A. E. F. Wächtler

14  Bio-transformation in the Production of L-Carnitine   287
    Th. P. Zimmerman, K. T. Robins, J. Werlen and F. W. J. M. M. Hoeks

ASYMMETRIC SYNTHESES BY CHEMICAL METHODS

15  Asymmetric Hydrocyanation of Vinylarenes   309
    A. L. Casalnuovo and T. V. Rajanbabu

16  Enantioselective Protonation in Fragrance Synthesis   335
    C. Fehr

17  Thiamphenicol: a Manufacturing Process Involving a Double Inversion of
    Stereochemistry   353
    L. Coppi, C. Giordano, A. Longoni and S. Panossian

18  Sharpless Asymmetric Epoxidation: Scale-up and
    Industrial Production   363
    W. P. Shum and M. J. Cannarsa

19  Asymmetric Synthesis of Sulphoxides: Two Case Studies   381
    P. Pitchen

20  Asymmetric Reduction of Prochiral Ketones   391
    J.-C. Caille, M. Bulliard and B. Laboue

Index   403

# List of Contributors

**B. A. Astleford**
*Central Process R&D, Lilley Research Laboratories, Lilley Corporate Center 4813, Indianapolis, IN 46285, USA*

**A. J. Blacker**
*ZENECA Process Technology Department, Huddersfield Works, P.O. Box A38, Leeds Road, Huddersfield, UK*

**J. J. Blumenstein**
*Janssen Research Foundation, Technical Regulatory Affairs, Titusville, NJ 08560, USA*

**B. L. Bray**
*Glaxo Wellcome Chemical Development Division, Girolami Research Centre, Research Triangle Park, NC, USA*

**A. Bruggink**
*Chemferm, P.O. Box 3933, 4800 DX Breda, The Netherlands*

**M. Bulliard**
*SIPSY, Route de Beaucouz, B.P. 79, 49242 Avrillé, France*

**J.-C. Caille**
*SIPSY, Route de Beaucouz, B.P., 79, 49242 Avrillé, France*

**M. J. Cannarsa**
*ARCO Chemical Co., 3801 West Chester Pike, Newtown Square, PA 19073, USA*

**A. L. Casalnuovo**
*DuPont Agricultural Products, Stine-Haskell Research Center, S 300/124, Newark, DE 19714, USA*

**L. Coppi**
*Zambon Group SpA, Via Lillo del Duca 10, 20091 Bresso (Milan), Italy*

**J. Crosby**
*ZENECA Pharmaceuticals, Macclesfield, Cheshire SK10 2NA, UK*

**H.-J. Federsel**
*Process Chemistry I, Astra Production Chemicals AB, S-151 85 Sodertalije, Sweden*

**C. Fehr**
*Firmenich SA, Route des Jerne 1, CH-1211, Geneva 8, Swetzerland*

**C. Giordano**
*Zambon Group SpA, Via Lillo del Duca 10, 20091 Bresso (Milan), Italy*

**M. D. Goodyear**
*Glaxo Wellcome Chemical Development Division, Medicines Research Centre, Stevenage, Hertfordshire, UK*

**F. W. J. M. M. Hoeks**
*Department of Biotechnology, R&D, LONZA AG, CH 3930 Visp, Switzerland*

**R. A. Holt**
*ZENECA LifeScience Molecules, P.O. Box 2, Billingham, Cleveland, UK*

**J. T. F. Keurentjes**
*Akzo, International Research Laboratories, Velperweg 76, Postbus 9300, 6800 SB, Arnhem, The Netherlands*

**B. Laboue**
*SIPSY, Route de Beaucouz, B.P. 79, 49242 Avrillé, France*

**A. Longoni**
*Zambon Group SpA, Via Lillo del Duca 10, 20091 Bresso (Milan) Italy*

**R. McCague**
*Chiroscience Ltd, Cambridge Science Park, Milton Road, Cambridge CB4 4WE, UK*

**S. Panossian**
*Zambon Group SpA, Via Lillo del Duca 10, 20091 Bresso (Milan), Italy*

**J. J. Partridge**
*Glaxo Wellcome Chemical Development Division, Girolami Research Centre, Research Triangle Park, NC, USA*

**D. Pauluth**
*Merck Germany/BDH, 64271 Darmstadt, Germany*

**P. Pitchen**
*Rhône-Poulenc Rorer Inc., 500 Arcola Road, P.O. Box 1200, Collegeville, PA 19426-0107, USA*

**T. V. Rajanbabu**
*Department of Chemistry, Ohio State University, Columbus, OH 43210, USA*

**K. T. Robins**
*Department of Biotechnology, R&D, LONZA AG, CH 3930 Visp, Switzerland*

# LIST OF CONTRIBUTORS

**W. P. Shum**
*ARCO Chemical Co., 3801 West Chester Pike, Newtown Square, PA 19073, USA*

**G. P. Stahly**
*Albemarle Corporation, 451 Florida Blvd, Baton Rouge, LA 70801, USA*

**R. M. Starrett**
*Albemarle Corporation, 451 Florida Blvd, Baton Rouge, LA 70801, USA*

**D. J. Tapolczay**
*Glaxo Wellcome Chemical Development Division, Medicines Research Centre, Stevenage, Hertfordshire, UK*

**S. C. Taylor**
*ZENECA Life Science Molecules, P.O. Box 2, Billingham, Cleveland, UK*

**S. J. C. Taylor**
*Chiroscience Ltd, Cambridge Science Park, Milton Road, Cambridge CB4 4WE, UK*

**F. J. M. Voermans**
*Akzo, International Research Laboratories, Velperweg 76, Postbus 9300, 6800 SB, Arnhem, The Netherlands*

**A. E. F. Wächtler**
*E. Merck Germany/BDH, 64271 Darmstadt, Germany*

**L. O. Weigel**
*Central Process R&D, Lilley Research Laboratories, Lilley Corporate Center 4813, Indianapolis, IN46285, USA*

**J. Werlen**
*Department of Biotechnology, R&D, LONZA AG, CH 3930 Visp, Switzerland*

**W. M. L. Wood**
*ZENECA Huddersfield Works, P.O. Box A38, Leeds Road, Huddersfield, West Yorkshire, UK*

**Th. P. Zimmermann**
*Department of Biotechnology, R&D, LONZA AG, CH 3930 Visp, Switzerland*

# Foreword

There is an increasing willingness of industrial organic chemists to publish the results of their process research and development work, much of which is high calibre, and this has been reflected in the number of books and symposia devoted to these topics. This can only be of great benefit to the industry, since we can all learn from each other by examining the details of successes and failures, particularly in conversion of laboratory methods to fully scaled up manufacturing processes. This edition of *Chirality in Industry II* is therefore most welcome.

The editors are to be congratulated, first of all on the successful Volume 1 of this series (now available in paperback—this must reflect its popularity!) but now also on persuading so many industrial chemists to write up their results on methods of making optically active compounds, on large scale as well as in the laboratory. *Chirality in Industry II* continues the high standard set in Volume 1, and also contains new case studies covering the diversity of methodologies—resolutions, asymmetric syntheses, biological methodologies—required to make chiral molecules.

I have enjoyed reading the manuscript prior to publication and I know readers will benefit from reading this text. I sincerely hope the editors will be planning *Chirality in Industry III* in the near future.

Trevor Laird
Scientific Update
(Editor OPRD)

# Preface

The topic of enantioselective synthesis/manufacture continues to be served by a stream of new books[1] and symposia;[2] many of the latter are specifically targeted at industrial aspects of chirality and include individual company symposia. The appearance of the first volumes on stereoselective synthesis in the Houben-Weyl series[3] is particularly timely. Also, we await with keen anticipation the appearance in early 1997 of the new American Chemical Society – Royal Society of Chemistry journal *Organic Process Research and Development*, expecting it to be a rich source of examples of the successful scale-up of methods for making single enantiomer products.

In introducing the second volume of *Chirality in Industry*, we note that once again all the authors are from industry or have strong industrial connections. Whilst it is hoped that Volume II will stand on its own, the intention is that Volumes I and II taken together should present a balanced, up-to-date and comprehensive picture of the technologies required for the production of optically active molecules on the multi-kilogram to high tonnage scales, as well as illustrating the breadth of application of the products of these technologies: pharmaceuticals, agrochemicals, electronics, food, flavours and fragrances. We are gratified to learn that Volume I has found use as a source of case histories for graduate courses and we hope this volume will prove even more useful in that respect.

The present volume is divided into sections on Physical Methods and Classical Resolutions, Biological Methods and Chiral Pool Syntheses and Asymmetric Syntheses. However, these are preceded by a chapter dealing with regulatory issues for chiral drugs and by then by two chapters on chiral drug development which illustrate how widely the industrial practitioner must cast the net in pursuit of practical production methods. Since enantioselective drug manufacture is probably the major driving force for the development of new technologies, regulatory attitudes (Chapter 2) clearly play an important role in dictating the pace and direction of development. Chapter 3 provides an overview of methods for making single enantiomers of one important class of drugs (profens) whilst Chapter 4 presents a more detailed account of the choices made between routes during the development of three specific drug candidates in another very important pharmaceutical class. Here three case histories of the development of processes for anti-viral products are described that embrace enzymatic resolutions, classical resolutions, selective crystallisations of desired diastereoisomers, asymmetric synthesis and the use of membranes.

As with Volume I, we hope that this volume will prove of particular interest to readers who are professionally involved in the scale-up of methods for the production of optically active materials, but also we hope that students and researchers involved in a more academic pursuit of optical activity will benefit from some of the facets of 'large-scale' thinking. An economic solution is most likely to be a simple, elegant solution.

## REFERENCES AND NOTES

1. (a) Sheldon, R. A., *Chirotechnology: Industrial Synthesis of Optically Active Compounds*, Marcel Dekker, New York, 1993; (b) Rahman, A., and Shah, Z., *Stereoselective Synthesis in Organic Chemistry*, Springer, New York, 1993; (c) Jannes, G., and Dubois, V. (Eds), *Chiral Reactions in Heterogeneous Catalysis*, Plenum Press, New York, 1995; *Proceedings of ChiCat, The First European Symposium on Chiral Reactions in Heterogeneous Catalysis*, Brussels, 25–26 October 1993; (d) Seyden-Penne, J., *Chiral Auxiliaries and Ligands in Asymmetric Synthesis*, Wiley–Interscience, New York, 1995; (e) Ojima, I. (Ed.), *Catalytic Asymmetric Synthesis*, VCH, New York, 1993; (f) Noyori, R., *Asymmetric Catalysis in Organic Synthesis*, Wiley, New York, 1994; (g) Eliel, E. L., Wilen, S. H., and Mander, L. N., *Stereochemistry of Organic Compounds*, Wiley–Interscience, New York, 1994; (h) Nogradi, M., *Stereoselective Synthesis: A Practical Approach*, 2nd edn, VCH, Weinheim, 1995; (i) Aitken, R. A., and Kilenyi, S. N. (Eds), *Asymmetric Synthesis*, Blackie, Glasgow, London, 1992; (j) Červinka, O., *Enantioselective Reactions in Organic Chemistry*, Ellis Horwood, London, 1995; (k) Procter, G., *Asymmetric Synthesis*, Oxford Science Publications, Oxford, 1996.
2. The annual Chiral Europe and Chiral USA series of meetings organised by Spring Innovations Ltd, Stockport, UK, have become established features of the 'chiral' calendar. These are supplemented by various fine chemicals conferences and trade shows, such as ChemSpec, which attract many offerings from the chiral manufacturing sector. There are also the International Symposia on Chiral Discrimination, the eighth of which is scheduled to take place in Edinburgh in mid 1996.
3. Helmchen, G., Hoffmann, R. W., Mulzer, J., and Schaumann, E. (Eds), *Stereoselective Synthesis. Houben-Weyl, Methods of Organic Chemistry*, Vols E21a onwward Georg Thieme Verlag, Stuttgart, 1995

March, 1996

A. N. C.
G. N. S.
J. C.

# Acknowledgements

We would like to express our thanks to all of the contributors for their patient cooperation in the preparation of this volume, to ZENECA Specialties and the Queen's University of Belfast for practical assistance, and finally to our families for agreeing to go through the process a second time.

# 1 Introduction

**J. CROSBY**
*ZENECA Pharmaceuticals, Macclesfield, UK*

| | | |
|---|---|---|
| 1.1 | Chiral Pool | 3 |
| 1.2 | Chiral Auxiliaries | 3 |
| 1.3 | Asymmetric Catalysis | 3 |
| 1.4 | Classical Resolution and Crystallisation Methods | 4 |
| 1.5 | Biological Methods | 5 |
| 1.6 | Targets | 6 |
| 1.7 | Enabling Science and Technology | 6 |
| | 1.7.1 Crystallisation | 6 |
| | 1.7.2 Membranes/Separation Technologies | 7 |
| | 1.7.3 Supercritical Fluid Technology | 8 |
| 1.8 | Conclusions and the Future | 8 |
| 1.9 | References and Notes | 9 |

In surveying developments since Volume I was published in 1992, it is clear that the needs of the pharmaceutical industry continue to be a major driving force for development of new and cost-effective methods for manufacturing optically active materials and, inevitably, much of the content of this volume is pharmaceutically oriented. Speed to market remains of paramount importance and can affect the selection of technology, particularly for pharmaceuticals and agrochemicals, and the pressure to achieve rapid commercialisation is even greater than it was four years ago.

The introduction of racemic drugs is becoming increasingly unattractive owing to policy changes made by regulatory agencies. As a result, the practical preparation of non-racemic drugs is a critical issue in the pharmaceutical industry. In 1990, 25% of the 'synthetically' derived drugs on the market were chiral, but only 3% were marketed as pure enantiomers[1] (Figure 1).

The picture is changing very rapidly. Although accurate and up-to-date statistics are difficult to obtain and open to different interpretations, it has been predicted that 75% of man-made pharmaceuticals will be single enantiomers by the year 2000.[2] The corresponding figure for agrochemicals has been estimated at about 20%.[3] The proportion of sales accounted for by resolved chiral agrochemical products is still relatively modest,[4] e. g. *ca.* US$0.7bn for pyrethroid insecticides in 1991[5] out of a $7.7bn insecticide market.[6] The market for enantiopure

*Chirality in Industry II.* Edited by A. N. Collins, G. N. Sheldrake and J. Crosby
© 1997 John Wiley & Sons Ltd

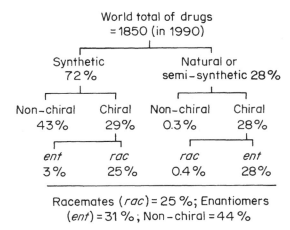

**Figure 1** Distribution of single-enantiomer, chiral and non-chiral drugs

final, formulated pharmaceutical products could be $60–90bn by the year 2000,[7] although the value of resolved chiral intermediates going into these will be much smaller, perhaps $1–2bn;[8] this is nonetheless a big opportunity for the chiral manufacturing sector to address. In a recent report,[9] it was noted that the worldwide market for single enantiomer forms of chiral drugs surged in 1994. The market for dosage forms reached $45.2bn.

The current move to single enantiomer drugs received a push in May 1992 when the FDA issued a policy statement. Although not an edict, and stating that it might be entirely appropriate to develop a chiral drug as a racemate, such official recognition of the chiral drug issue has persuaded drug firms that were not already doing so to start developing single enantiomers. Sales of enantiopure drugs are projected almost to double between 1994 and 1997.[10] This trend is being reinforced by a number of 'racemic switches' in which compounds previously marketed as racemates are relaunched as single isomers. There are only about six racemic switches on the market so far but about 20 compounds are said to be in redevelopment as single isomers, with Merck, Chiroscience and Sepracor being most active in the area. Merck is developing $(S)$-(+)-ibuprofen, Chiroscience compounds include $(S)$-ketoprofen and $(-)$-bupivacaine whilst Sepracor has a large number of candidates including both $(R)$ and $(S)$-ketoprofen for different applications.[9] Switching must be largely a one-off phenomenon, a catching up exercise, but one with the potential to give a significant short to medium-term boost to the demand for chiral intermediates and to provide additional vehicles for technology development.

It is appropriate to include some comments on agrochemicals and on the position of agrochemicals versus pharmaceuticals. There are only two chapters specifically devoted to agrochemical products/intermediates: Chapter 10 in

# INTRODUCTION

Volume II, (S)-chloropropionic acid, and Chapter 4 in Volume I, pyrethroids. This is probably a fair reflection of the less developed application of single enantiomers in this area.[11] The values of bulk actives for agrochemicals fall roughly in the range £10s–£100s kg$^{-1}$; this is, again very approximately, an order of magnitude lower in value than bulk drugs and governs the level of sophistication of the chemistry/technology that can be brought to bear on their production. However, although historically pharmaceuticals have been able to tolerate higher manufacturing costs, this will not necessarily remain the case as we enter the era of health-care reforms and a highly competitive generics market. These changes to the market structure might be expected to stimulate further development of chiral methodologies from within pharmaceuticals that will come closer to meeting the cost constraints of agrochemicals production.

To help set the scene for this volume, the following are presented as examples of developments in the main technology areas which have taken place since the publication of Volume I.

## 1.1 CHIRAL POOL

The 'pool' continues to expand rapidly through the addition of many new single enantiomer products and their intermediates, together with auxiliaries which have been made in response to synthesis needs.

## 1.2 CHIRAL AUXILIARIES

More sophisticated chiral auxiliaries are becoming available in commercial quantities, such as bornane-10,2-sultam and oxazolidinones[12] and (S)-4-(phenylmethyl)-2-oxazolidinone as a 'spin-off' from the manufacture of phenylalanine for the sweetener aspartame.

## 1.3 ASYMMETRIC CATALYSIS

The introduction and scale-up of asymmetric processes continue apace and the practical application of large-scale asymmetric synthesis is certainly not in doubt. The Takasago process for (−)-menthol and related chiral terpenes has, since its inception in 1982, passed the 20 000 tonne production milestone, at an impressive average consumption of only 6 g of the chiral BINAP ligand per tonne of product.[13] Elsewhere, ARCO Chemical Company and SIPSY have developed the Sharpless asymmetric epoxidation (SAE) reaction on an industrial scale.[14] SIPSY has run this process safely at 4000 litre scale to produce multi-tonne quantities of (R)- and (S)-glycidol (Chapter 18). Modified SAE has also been used to produce single-enantiomer sulphoxides on the multi-kilogram scale

(Chapter 19). Seprachem has been particularly active in developing and scaling up several key asymmetric technologies which include the Sharpless asymmetric dihydroxylation and the Jacobsen asymmetric epoxidation;[15] uses of the latter include the production of enantiomers of indene oxide required for HIV protease inhibitors. Another important reaction which is being scaled up is the Corey asymmetric reduction catalysed by oxazaborolidines, as described in Chapter 20.

However, to present a balanced picture, some of the earlier euphoria concerning the prospects for the large-scale use of asymmetric catalysis is now tempered by the realisation that although a reaction may be wonderfully catalytic and selective, in some instances large amounts of other, achiral, reagents may be required to effect the reaction and, overall, create more by-products and result in a higher cost than a superficially less attractive method using auxiliaries/resolving agents. In turn, the use of high molecular weight auxiliaries may be less effective than a simple resolution; hidden steps may also lie in its recovery and re-use. Another aspect which must be considered is selectivity: asymmetric catalysis which only delivers moderate *ee* (or *de*) and moderate yield may not be acceptable. For example, removal of an unwanted enantiomer and chemical purification from an 80% yield reaction with 80% *ee* might sacrifice 25% of the desired enantiomer, netting only 55% of the desired enantiomer. An efficient resolution, on the other hand, might yield an 'uncomplicated' 45% and involve much simpler processing.

Whilst rapid progress is being made in homogeneous asymmetric catalysis, progress in heterogeneous catalysis, which would have obvious processing advantages, lags behind. It has been pointed out[16] that the *ee*s and the range of reaction types obtained by other methods may not be easy to match by heterogeneous catalysis. The field is still largely restricted to hydrogenation and it might be asked whether the way forward here is not to produce heterogeneous catalysts *per se* but rather to focus on methods for rendering proven homogeneous catalysts recoverable.

## 1.4 CLASSICAL RESOLUTION AND CRYSTALLISATION METHODS

Resolution methods employing selective crystallisations are still far from predictable and chirality is still mainly treated as a qualitative property (cf. Chapter 5).[17] Nonetheless, their use is widespread (cf. Chapter 6) and a recent literature survey showed that the number of patents being granted for resolution processes exceeds that for asymmetric synthesis.[18]

A process for diltiazem has been reported[19] which exploits an entrainment resolution. This is said to be very competitive and operate for more than 25 cycles for each enantiomer.

## 1.5 BIOLOGICAL METHODS

At one end of the spectrum, abzymes still seem a long way from providing the basis for industrial processes. Although chiral synthesis is an area where catalytic antibodies should score highly, the industrial potential is still seen as long-term and academic researchers have to resolve some fundamental problems.[20]

At the other end of the spectrum, poly(amino acid) catalysts, which might be regarded as the simplest of enzyme analogues, recently evoked the headline 'The world's cheapest chiral catalyst.'[21] These materials appear to be being considered for development products. For example, poly(L-leucine) has been employed as the chiral epoxidation catalyst in a key step of a reported practical enantioselective synthesis of SK&F 104353, a possible candidate for the treatment of bronchial asthma.[22] Although this asymmetric epoxidation was developed by Julia et al.[23] in the early 1980s, it has not been widely exploited. There are obvious attractions for large-scale application; poly(L-leucine) is stable and inert and could be re-used.

A recent development aimed at improving the robustness of enzymes has been the crosslinking of enzyme microcrystals by bifunctional agents such as glutaraldehyde. The resulting crosslinked enzyme crystals, CLECs®, are reported to have properties superior to both soluble and conventionally immobilised enzymes; they remain active in environments that are otherwise incompatible with enzyme function, including high temperatures and extremes of pH,[24,25] e.g. thermolysin CLECs were fully active for 18 days at 55 °C in ethyl acetate.

A large number of enzymes have become commercially available for organic synthesis. Of about 2500 identified thus far, some 300 are available in a partly purified form. Moreover, because of advances in molecular biology, fermentation and purification techniques, the costs are being reduced to industrially attracive levels.[26] In addition, recognition and use of 'non-nameplate' applications will increase;[27] this is where an enzyme successfully catalyses a reaction other than that for which it is normally employed.

Use of oxynitrilases to produce single-enantiomer cyanohydrins is a process which appears to have good potential (cf. Volume 1, Chapter 14), and has been developed to the semi-technical scale by Peboc, but there does not appear to have been any large-scale use. Another interesting biotransformation which seems to be falling short of its potential is enantioselective nitrile hydrolysis;[28] reasons here are in part related to lack of commercial sources of the enzymes. However, Lonza has recently developed a process for (S)-dimethylcyclopropylcarboxamide.[29]

With respect to bio-production of chiral epoxide targets, there is now more emphasis on enantioselective conversion of racemates versus the earlier monooxygenase approach. Hydrolytic enzymes may offer more promise as they are cofactor-independent and the degradation product may be recovered as optically pure diol.[30]

Searches for extremophiles, organisms obtained from hostile environments, continue[31] and will undoubtedly enrich the armoury of enzymes tolerant of harsh operating conditions.

## 1.6 TARGETS

New general targets include resolved sulphoxides, which are increasingly being sought in the pharmaceuticals area (cf. Chapter 19); they are not amenable to a 'chiral pool' strategy as there are no readily available chiral sulphoxides. Whether accessed by the Kagan modification of the Sharpless epoxidation or enzymically, steric distinction is usually essential for success.

## 1.7 ENABLING SCIENCE AND TECHNOLOGY

There are a number of enabling sciences/technologies which are important to large-scale chiral synthesis, in particular analytical science, crystal science and the use of membranes. It might be asked why neither this volume nor its predecessor has a chapter devoted to analysis, the most important enabling science. Without an ability to analyse, the basis for manufacture, control and legislation all disappear. However, there are to our knowledge no scale-dependent aspects or, at least, none which is peculiar to chiral analysis. For this reason, analytical aspects are omitted but their importance does not go unrecognised.

### 1.7.1 CRYSTALLISATION

Crystallisation science is an enabling science with a high degree of relevance to large-scale manufacture. An understanding of crystallisation principles and techniques is an essential competence for anyone developing large-scale processes for single-enantiomer materials and lies behind many commercially important and successful processes. It was felt appropriate to include a substantial chapter on crystallisation techniques and, although some of it (Sections 7.2 and 7.3) is of general use, it is of particular importance in the production of single-enantiomer products. Examples which may be selected from elsewhere in this volume include the case histories presented in Chapter 6 and the importance of crystal form in processes for naproxen and ibuprofen (Chapter 3). The eutectic composition is a most important parameter to be considered when planning the separation of a racemate. For instance, the eutectic composition dictates that ibuprofen must be at least 90% pure ($S$)-isomer for the crystallisation to result in optical purification. Also, in this case by operating with the sodium salt, a more favourable eutectic is obtained.

# INTRODUCTION

## 1.7.2 MEMBRANES/SEPARATION TECHNOLOGIES

Separation and membrane technologies are developing rapidly. In addition to the Wandrey/Degussa membrane reactors, developed principally for amino acid production and which use ultrafiltration membranes (Volume I, Chapter 20), there is the Bend Research system using selectively permeable membranes[32] and the Sepracor hollow-fibre system,[33] all of which have been well described. To these can now be added the development by Akzo (Chapter 8) of a 'symmetrical' countercurrent extraction/separation system and simulated moving bed technology (SMB), the use of which is being translated from very high tonnage achiral separations to chiral separations.

Chapter 8 describes a broadly applicable membrane technology for the separation of enantiomers: a symmetrical countercurrent extraction system.[34] Two liquids, at least one of which is chiral or contains a chiral adjuvant and which are wholly miscible with each other, flow countercurrently either side of a non-miscible phase, consisting of a liquid supported on a microporous membrane; hollow-fibre modules are used. Enantiomer separation by this method should permit the production of almost any optical isomer. It does not (overtly) require functional handles; clearly interactions are required and direct crystallisation remains the only truly functionality-independent method. Method development should be relatively easy; scale-up is reported to be simple and both enantiomers can be collected. A crucial aspect of the method development is the choice of a chiral selector which will enable sufficient complex formation with the enantiomers, e.g. by H-bonding or electrostatic interaction. In practice, satisfactory results can be obtained with cheap tartaric esters and commercial applications are expected in the next few years.

Although preparative high-performance liquid chromatography (LC) is used for laboratory- and pilot-scale production, chromatographic processes have generally been considered too expensive for large-scale use. However, developments in scale-up continue, with polysaccharide chiral stationary phases proving particularly useful. As an indication of current capabilities of the technology, an example has recently been described in which, using a 50 cm × 10 cm i.d. column, 4 kg of a racemate were processed in 45 h using 510 l of methanol solvent.[35]

UOP has applied the principles of countercurrent adsorption processes, in particular its Sorbex™ SMB process, widely used for the production of commodity chemicals, to the separation of optically pure pharmaceuticals.[36] In the petrochemical sector, the economics of scale reduce separation costs to a few cents/kg; for low-tonnage chiral separations costs are significantly higher owing to high depreciation charges resulting from low throughput rates, high stationary phase costs and high solvent recovery costs. SMB uses an array of, typically, eight or more chromatographic columns linked in a circle. Although solvent consumption is less than for elution chromatography, because SMB takes much more packing to fill the columns the investment is higher, with chiral column packings costing *ca* $8000 kg$^{-1}$.[37] Examples of outputs range up to 25 kg per day. It has been estimated that one of the larger units could produce 10 000 kg per year of ($R$)-3-chloro-1-phenylpropanol, an intermediate for ($S$)-(+)-fluoxetine, at a cost of \$700 kg$^{-1}$. The cost estimate includes racemising/recycling the ($S$)-enantiomer. Since the stationary phase is a significant proportion of the cost, and there is as yet no commercial experience of its lifetime, significant reductions can be expected with future experience and improvements.[38] Projects on the scale of 100 tonnes per annum are said to be under consideration.[39]

### 1.7.3 SUPERCRITICAL FLUID TECHNOLOGY

Is supercritical fluid technology going to emerge as another enabling technology for single-enantiomer production? It could have roles in both enzyme-catalysed reactions and in preparative-scale chiral chromatography. The unique pressure dependence of the physical properties of supercritical fluids presents many opportunities for the control of enzyme activity, specificity and stability in biocatalytic reactors and the rates of reactions involving gaseous reactants such as oxygen should be enhanced.[40] In LC the lower viscosity should lead to greater efficiencies.

## 1.8 CONCLUSIONS AND THE FUTURE

Except perhaps for some areas of amino acid production, pharmaceuticals have been the major user of technologies for the large-scale production of single-enantiomer products, with agrochemicals next, but a long way behind. Single-enantiomer products are also required for flavours and fragrances, in electronics and foodstuffs, e.g. sweeteners (Volume I, Chapter 11) and carnitine (Chapter 14). Chapter 16 illustrates how the needs of the flavour and fragrance industry have stimulated some elegant asymmetric syntheses and Chapter 13 shows how chirality is introduced into liquid crystals. The latter, relatively unsung, corner of chirals manufacture, although small in tonnage and, by pharmaceuticals standards, not very complex, has arguably one of the greatest impacts on our lives through the part it can play in the information technology explosion, for example by enhancing the performance of display devices.

Within the wider field of organic chemistry, the special status of 'chiral' chemistry is sometimes questioned. It is special, witness the activity in symposia and publications devoted to the topic and noted in the Preface, but it cannot stand alone; its successful application usually depends on intelligent integration within a spectrum of supporting technologies. The bio-examples, in particular, reveal the need for a multidisciplinary approach (cf. Chapters 9, 10, 12 and 14). It is of interest to note in this context the changes taking place in the industry after the somewhat enthusiastic spawning of 'chirals' companies in the late 1980s-early 1990s which were often seeking to exploit a single technique or reaction. There has been the realisation that such skills alone are not necessarily sufficient for commercial viability; there have been some departures and some strategic alliances and licensing deals.[41] It is possible to draw parallels with the development of the modern biotechnology industry,[42] which underwent major restructuring in the late 1980s. Companies who have succeeded are those able to integrate their technological capabilities within a broader set of skills, concentrate on selected niches or team up with appropriate partners. Many end users have developed in-house capabilities, reducing the opportunities for independent players. Much of this can be seen to apply to chirals. There is evidence from publications and symposia reports that, in addition to the activities of specialist 'chirals' companies, many of the major chemical companies have active in-house programmes to develop, in particular, catalysts for asymmetric reductions; much of this activity has probably been stimulated by the desire to have proprietary systems which match and extend the capabilities of the Noyori/Takasago technology (cf. Volume I, Chapter 17).

At the research level, all necessary methods exist for making initial quantities of enantiomers for the exploration of property differentiation. There is no longer any question as to whether a desirable isomer can be made, but only if this can be effected economically. Single-isomer development on the large scale had been constrained by the availabilities of techniques, intermediates, catalysts, etc. Now the chiral infrastructure is largely in place, there is no reason why large-scale production of even moderately priced single-enantiomer products should not be contemplated.

We noted in the Epilogue to Volume I that '... the future will see an interesting race between biology, conventional chemistry and separation technology... .' Four years later, all three areas have seen major advances; separation technologies have made a particularly strong showing; abzymes have still to realise their potential; crystallisation techniques continue to be of signal importance.

## 1.9  REFERENCES AND NOTES

1. *Scrip*, Feb. 16 (1993)
2. *Speciality Chem.* Jan./Feb., 6 (1994).
3. Gardner, J. C., and DiCicco, R. L., *Speciality Chem.*, July/Aug., S9 (1994).
4. Silvon, M., *Chem. Mkt. Rep.*, 7 Sept., SR22 (1992).
5. Gupta, N., and Eisberg, N., *Performance Chem.*, Aug./Sept., 19 (1991).
6. Heaton, A., in *The Chemical Industry*, 2nd edn (ed. A. Heaton), Blackie, Glasgow, 1994, p. 218.
7. See, for example, Burke, M., *Chem. Ind. (London)*, 3 Jan., 10 (1994).
8. See, for example, *Performance Chem.*, Feb., 9 (1995).
9. Stinson, S. C., *Chem. Eng. News*, 9 Oct., 44 (1995).
10. Stinson, S. C., *Chem. Eng. News*, 19 Sept., 38 (1994).
11. For a recent review of the manufacture of optically active agrochemicals, see Crosby, J., *Pestic. Sci.*, **46**, 11 (1996).
12. Stinson, S. C., *Chem. Eng. News*, 28 Sept., 76 (1992).
13. Based on figures presented by S. Akutagawa at Chiral '95 USA Symposium, Boston, MA, 15–16 May 1995.
14. Bulliard, M., and Schum, W., in *Proceedings of Chiral '95 USA Symposium*, Boston, MA, 15–16 May 1995, Spring Innovations, 1995, p. 5.

15. Pettman, R., *Speciality Chem.*, July/Aug., S12 (1994).
16. Jannes, G., and Dubois, V. (Eds), *Chiral Reactions in Heterogeneous Catalysis*, Plenum Press, New York, 1995; *Proceedings of ChiCat*, the *First European Symposium on Chiral Reactions in Heterogeneous Catalysis*, Brussels, 25–26 Oct. 1993, p.1.
17. Bruggink, A., and Ariaans, G. J. A., *Pharm. Manuf. Int.*, 85 (1995).
18. Eliel E.L., Wilen S. H., and Mander L. N., *Stereochemistry of Organic Compounds*, Wiley–Interscience, New York, 1994, p. 388.
19. Villa, M., and Pozzoli, C., in *Proceedings of Chiral '95 USA Symposium*, Boston, MA, 15–16 May 1995, Spring Innovations, 1995, p. 13.
20. Johnson, J., *Chem. Ind. (London)*, 20 Feb., 128 (1995).
21. *Chem. Ind (London)*, 4 Sept., 675 (1995).
22. Flisak, J. R., Gombatz, K. J., Holmes, M. M., Jarmas, A. A., Lantos, I., Mendelson, W. L., Novack, V. J., Remich, J. J., and Snyder, L., *J. Org. Chem.*, **58,** 6247 (1993).
23. Julia, S., Masana, J., Vega, J. C., *Angew. Chem., Int. Ed. Engl.*, **19,** 929 (1980).
24. Margolin, A. L., in *Proceedings of Chiral '94 USA Symposium*, Reston, VA, 5–6 May 1994, Spring Innovations, 1994, p. 51.
25. Margolin, A., *Perf. Chem.*, April/May, 23 (1994).
26. See, for example, *Kirk–Othmer's Encyclopedia of Chemical Technology*, 4th edn., Wiley–Interscience, New York, 1993, Vol. 9, p. 672.
27. See, for example, de Zoete, M. C., van Dalen, A. C. K., van Rantwijk, F., and Sheldon, R. A., *J. Chem. Soc., Chem. Commun.*, 1831 (1993); use of lipases to effect aminolysis, a mild procedure for enantioselective synthesis of amides.
28. See, for example, Crosby, J., Moilliet, J., Parratt, J. S., and Turner, N. J., *J. Chem. Soc., Perkin Trans. 1*, 1679 (1994), and references cited therein.
29. Chassin, C., in *Proceedings of InBio'96 Symposium*, 20–21 Feb. 1996, Manchester, Spring Innovations, 1996.
30. See, for example, Weijers, C. A. G. M., and deBont, J. A. M., in *Proceedings of Chiral Europe'95 Symposium*, London, 28–29 Sept. 1995, Spring Innovations, p. 87.
31. (a) Adams, M. W. W., and Kelly, R. M., *Chem. Eng. News*, 18 Dec., 32 (1995);. (b) Govardhan, C. P., and Margolin, A. L., *Chem. Ind. (London)*, 4 Sept., 689 (1995).
32. van Eikeren, P., Brose, D. J., Muchmore, D. C., West, J. B., and Colton, R. H., in *Proceedings of Chiral '92 Symposium*, Manchester, 24–25 March 1992, Spring Innovations, 1992, p. 63.
33. See, for example, Young, J. W., in *Proceedings of the Chiral Synthesis Symposium and Workshop*, 18 April 1989, Stockport, Spring Innovations, 1989, p. 39.
34. Keurentjes, J. T. F., *PCT Int. Pat Appl.* WO 94/07814, 1994.
35. Tachibana, K., in *Proceedings of Chiral Europe '95 Symposium*, 28–29 Sept. 1995, London, Spring Innovations, 1995, p.139.
36. *Performance Chem.*, Feb., 27 (1995).
37. *Chem. Eng. News*, 9 Oct., 52 (1995).
38. Gattuso, M. J., McCulloch, B., House, D. W., and Baumann, W. M., in *Proceedings of Chiral '95 USA Symposium*, 15–16 May 1995, Boston, MA, Spring Innovations, 1995, p. 51.
39. *Chem. Eng. News*, 9 Oct., 58 (1995).
40. Russell, A. J., Beckman, E. J., and Chaudhary, A. K., *Chemtech*, March, 33 (1994).
41. Examples include Eastman/Peboc, *Chem. Eng. News*, Oct. 9, 58 (1995); ARCO/SIPSY, *Chem. Eng. News*, Oct. 9, 44 (1995); Chiroscience's acquisition of rights to DuPont's proprietary asymmetric catalysts for manufacture of pharmaceutical intermediates and actives (Press Release, 6 Nov. 1995).
42. Polastro, E., in *Chiral Reactions in Heterogeneous Catalysis* (ed. Jannes, G., and Dubois, V.), Plenum Press, New York, 1995; *Proceedings of ChiCat*, the *First European Symposium on Chiral Reactions in Heterogeneous Catalysis*, Brussels, 25–26 Oct. 1993, p. 5.

# 2 Chiral Drugs: Regulatory Aspects

**J. J. BLUMENSTEIN**
*Janssen Research Foundation, Titusville, NJ, USA*

| | | |
|---|---|---|
| 2.1 | Introduction | 11 |
| 2.2 | Racemates Versus Single Enantiomers | 12 |
| 2.3 | Pharmacokinetic Evaluation of Enantiomers | 13 |
| 2.4 | FDA Decision Tree for Optically Active Drugs | 14 |
| 2.5 | Differences Between FDA and CPMP Guidelines for Undesired Enantiomers | 16 |
| 2.6 | Drug Product Considerations | 16 |
| 2.7 | Verification of Starting Materials and Intermediates | 17 |
| 2.8 | Conclusion | 17 |
| 2.9 | References and Notes | 17 |

## 2.1 INTRODUCTION

While the importance of drug stereochemistry has been recognised almost since the beginning of pharmacological drug therapy, regulatory consideration of stereochemistry has really only come to the fore in the past decade. Increased regulatory attention to stereoisomeric drugs is due to both technological advances in stereosensitive analytical methods (without which regulation would be impossible) and the increased commercial practicality of stereoselective synthesis. In an era of multinational drug development, the development of policies on stereochemical aspects of drug synthesis now transcends the borders of any one regulatory agency. At the moment, pharmaceutical companies must adapt their strategies to all the regulatory environments where their product will be registered, although attempts to standardise the requirements are being made by bodies such as the International Committee on Harmonisation (ICH). Recently, several regulatory bodies around the world have issued guidelines on the development of drugs with stereogenic centres.[1] The eventual working policy will probably be a conglomeration of the various existing guidelines.

The US Food and Drug Administration (FDA) first raised the topic of stereochemical regulation in its 1987 Drug Substance Guideline,[2] although little emphasis was given to the topic at that time. An internal committee was formed in 1989 to consider further the regulatory concerns related to stereoisomeric drugs and, in response, in 1990 the Pharmaceutical Manufacturers Association

*Chirality in Industry II.* Edited by A. N. Collins, G. N. Sheldrake and J. Crosby
© 1997 John Wiley & Sons Ltd

(PMA) set out its position on the development of drugs with stereogenic elements in a paper published in *Pharmaceutical Technology*.[3] In 1992 a Drug Information Association conference in Paris provided an international forum to examine regulatory requirements for chiral drugs, and these were summarised in a 1993 paper published in *Drug Information Journal*.[4] While international harmonisation may still be a little while in coming, worldwide regulatory bodies have been developing individual positions. The FDA published its stance in 1992, and requested public comment on this. Although individual industries and trade associations provided comments, a revised document has not been issued. More detailed guidelines have been presented in several open forums, but a final guidance document has not become available. The Committee for Proprietary Medicinal Products (CPMP)[5] issued its position in a 1994 policy. Canadian authorities issued draft guidance for comment in 1993, but a final policy has still to be published. This chapter will focus on the FDA policy and discuss recent developments elsewhere which are related to the sections described in that document.

There are two principal scenarios in chiral drug development: the first is the *de novo* development of an optically pure chiral drug, and the second the switch from an existing racemic drug to a single stereoisomer of that drug. Most of the regulatory guidance has focused on the first scenario, although its applicability to switches will also have to be addressed because of the growth of this area.

## 2.2 RACEMATES VERSUS SINGLE ENANTIOMERS

The development of chiral drugs, whether racemates or single enantiomers, raises issues of acceptable control of the manufacture, adequate pharmacological and toxicological assessment, proper characterisation of the metabolism and distribution and appropriate clinical evaluation. The pharmacological activity of the individual enantiomers should be determined, whether developing a single isomer or the racemate. In cases where the pharmacological activities of both enantiomers have been examined, there have been instances (1) in which both enantiomers have similar, desirable activities, (2) where one enantiomer is active and the other lacks pharmacological activity and (3) where each of the enantiomers has a completely different activity (see Chapter 1 in Volume I and Chapter 11 in this volume for examples). There are also intermediate situations, e.g. one isomer has an attenuated activity compared with the other.

Despite the problems identified with some racemates, the hitherto common practice of developing racemates has resulted in few recognised adverse consequences. Although technological advances have, in many cases, made the manufacture of enantiomers more feasible, the development of racemates will still continue to be appropriate for many development programmes. This situation is recognised in all the recent guidelines, although with varying levels of enthusiasm by the different regulators, all of whom have expressed a preference for the development of a single isomer, (even though there are cases where racemates are both appropriate and acceptable). Often that preference is a result of the clarity that is added by only having to study the effects of a single isomer, rather than a history of problems that have resulted from the use of racemates. The decision process when choosing whether to proceed with a racemate or a single-isomer candidate is the same as that for an achiral drug. Factors to be considered include the commercial practicality of the synthesis, the ability to formulate the dosage form, pharmacokinetics and metabolism and the risk–benefit considerations that evolve from safety and efficacy studies. With chiral drugs there is the advantage that these issues may be examined for both the single isomer and the racemate; where the racemate may not be suitable for final development, the single isomer may be, and vice versa. This is recognised by all of the regulators, and the guidelines allow the company to make this decision.

## 2.3 PHARMACOKINETIC EVALUATION OF ENANTIOMERS

Pharmacokinetic evaluations in animals and in clinical studies that do not use a chiral assay may be misleading if the disposition of the enantiomers is different. Therefore, regulators are focusing on techniques to quantify individual stereoisomers in pharmacokinetic studies early in development. The pharmacokinetic profile of each isomer should be characterised in animals and the profile of each enantiomer obtained from Phase I clinical studies. If the pharmacokinetics of the enantiomers are demonstrated to be the same, or exist as a fixed ratio in the target population, an achiral assay or an assay which monitors only one of the enantiomers may be used subsequently. If a racemate is being developed and the pharmacokinetic profiles of the isomers are different, the enantiomers should be monitored individually.

Similarly, in order to evaluate the pharmacokinetics of a single enantiomer, or a mixture of enantiomers, stereoselective assays for *in vivo* samples need to be available early in the development process. This will allow the assessment of the potential for interconversion and the absorption, distribution, metabolism and excretion (ADME) profiles of the individual enantiomers. Unless it proves particularly difficult, the main pharmacological activities of the isomers should be compared *in vitro*, in animals and/or in humans. It is usually sufficient to carry out toxicity studies on the racemate. A relatively benign toxicological profile for the racemate would ordinarily support further development without separate toxicological evaluation of the individual enantiomers. If toxicity is found other than that related to the desired pharmacological effects of the drug, the individual isomers should be evaluated in the study where the toxicity was detected.

Where little difference is observed in the activity and disposition of the enantiomers, racemates may be developed. There are some situations where the development of a single enantiomer is particularly desirable, such as when one enantiomer has a toxicity that the other (pharmacologically desirable) enantiomer does not have. Conversely, when there is rapid interconversion between the isomers *in vivo*, little advantage may be gained by developing a single isomer. In general, it is more important to evaluate both enantiomers clinically when both enantiomers are pharmacologically active but differ significantly in potency, specificity or maximum effect, than when one isomer is essentially inert.

If a racemate has been marketed and the sponsor wishes to develop the single enantiomer (a so-called 'racemic switch'), an appropriate abbreviated pharmacological/toxicological evaluation could be conducted to extend the existing knowledge of the racemate to that of the single isomer. Bridging studies, including a racemate control group, would usually include the longest repeat-dose toxicity study conducted and reproductive toxicity studies in the most sensitive species. Clinical evaluation should include the determination of whether there is a significant conversion to the other isomer, and whether the pharmacokinetics of the single isomer are the same as they were for that isomer as part of the racemate.

## 2.4. FDA DECISION TREE FOR OPTICALLY ACTIVE DRUGS

The FDA has not publicly issued more detailed guidance for the Chemistry, Manufacturing and Controls (CMC) section of applications for new drugs. The other regulatory authorities have expressed only general considerations to be taken into account in the manufacture of chiral drugs, and current regulatory thinking on CMC issues for chiral drugs has been described at recent symposia.[6] The CMC section should contain information about the identity, quality, purity and strength of the drug substance and the drug product, in addition to consideration of the stereogenic characteristics of the drug. Applications for single enantiomer and racemic drugs should include a stereochemically specific identity test and a stereochemically selective assay for both the drug substance (the active ingredient) and the drug product (the final formulated drug). The choice of the controls should be based on the methods of manufacture and the stability characteristics of the drug substance. The approach depends on whether there is one or more than one stereogenic centre present. For compounds with a single chiral centre, inversion may be detected only through the use of stereoselective assay method. However, those with multiple stereogenic centres may be able to be monitored using a standard achiral analytical techniques, since the simultaneous inversion of both centres is unlikely, and inversion of a single chiral centre will afford a diastereoisomer which may be detected through achiral techniques. This concept is illustrated in Figure 1.

The decision tree outlined in Figure 1 breaks down the FDA's requirements for pure enantiomers and racemates. The requirements for racemates are relatively straightforward. The FDA sees the need for an identity test for racemates, whether they have single or multiple chiral centres. Although it may not be intuitively obvious that this is necessary from an analytical standpoint, the

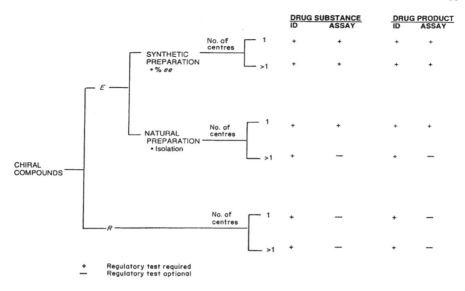

**Figure 1** Enantioselective test requirements for chiral compounds. E, single enantiomer; R, racemate

regulatory stance is that the racemic nature is an identity characteristic that should be included in the specifications. While discussion on this issue has been wide-ranging, the question is often whether the substance exists as a racemic compound or a racemic mixture (conglomerate). Given that some identity tests (e.g. optical rotation) are not sensitive to subtle variations in stereochemical purity, and spontaneous resolution of racemic mixtures to afford an unanticipated, highly stereochemically enriched mixture is very unlikely, the value of this test for racemates should probably be questioned.

As may be expected, the situation with an enantiomerically pure compound is more complex. Single-enantiomer drugs may be further categorised into synthetic preparations and products isolated from natural sources. All synthetic preparations, whether having one or more chiral centres, should have both a stereospecific identity test (to determine the absolute configuration) and a stereoselective assay (to determine the enantiomeric purity). Natural products which have multiple chiral centres will require only an identification method which is stereospecific. However, natural products with only one chiral centre will require a stereoselective assay method in addition to the stereospecific identification method. As with all regulatory guidance, each case should be considered on its own merits. For example, when a stereoselective synthesis of a single isomer with multiple chiral centres is being evaluated, one must consider, *inter alia*, the number of chiral centres, whether it is a convergent or sequential synthesis and the lability of all of the chiral centres.

## 2.5 DIFFERENCES BETWEEN FDA AND CPMP GUIDELINES FOR UNDESIRED ENANTIOMERS

When setting specifications, there is a subtle but important difference in the way the undesired isomer is considered by the FDA and the CPMP. The CPMP specifically regards the undesired isomer as an impurity which, when coupled with recent guidelines on identification and qualification of impurities,[7] provides a clear approach to setting appropriate specification limits. The FDA's comments have suggested that diastereoisomers are regarded as impurities, but the position on the opposing enantiomer is not specified. In the absence of such guidance, it would seem advisable to follow the CPMP position and apply the same criteria to the specification limit as any other impurity.

## 2.6 DRUG PRODUCT CONSIDERATIONS

Stereochemical considerations for the drug *product* (the formulated drug) focus on the specifications and the stability of the drug product. The above discussion (Section 2.4) regarding the decision tree addresses most of the concerns about tests and specifications for both the drug substance and drug product. The FDA has expressed a desire (reflected in the decision tree) to mirror the controls used for the drug substance in the drug product. The CPMP focuses on validating the process to establish that there has been no impact on the stereochemical integrity of the active substance. Given the difficulty of developing validated identification methods for use with a complex pharmaceutical matrix, and the previous discussion regarding identity tests for racemates, the need for stereochemical identity tests in drug products should be re-evaluated. Except where there is a demonstrated need to monitor the level of stereochemical purity (e.g. a single enantiomer with a single chiral centre), the identity test usually provides little additional valuable information beyond the standard chemical identity tests.

The stability considerations for the drug substance and drug product parallel one another. Stressed stability studies at raised temperatures, humidities and light levels on an enantiomerically pure compound should be monitored with a stereochemically selective assay method to identify conditions which may lead to isomerisation. In most cases the products of accelerated stability studies should also be analysed with a stereoselective assay. If isomerisation does not occur, then a stereochemical assay method may not be required for long-term studies, but if isomerisation is observed, a stereochemically selective assay will usually be required. The requirements for long-term studies when isomerisation has been observed under stressed conditions,[8] but not accelerated storage conditions, are difficult to generalise, and must be dealt with on a case-by-case basis following consideration of the storage times and conditions.

## 2.7 VERIFICATION OF STARTING MATERIALS AND INTERMEDIATES

While the decision tree provides an overview of the analytical requirements from a stereochemical standpoint for the final specifications and test methods for a drug substance and drug product, there are many other areas in the CMC section of a registration document which require further attention from a stereochemical viewpoint. In the synthesis of the new drug substance, stereochemical considerations start with controls of the starting materials and reagents. Starting materials with a single chiral centre will require a stereoselective assay method, whereas those with multiple chiral centres may not require a stereochemical assay, since diastereoisomers will usually be detectable by achiral methods and the presence of opposing enantiomers with multiple centres is unlikely.

The issue of the suitability of controls for the stereochemical purity of intermediates is more difficult to handle generically. Stereochemically selective 'in-process' controls will be required when the procedure dictates it is necessary, e.g. to determine when a resolution step is optimal. When a given synthetic step establishes a new chiral centre to give a key intermediate, controls will usually be required to demonstrate that the functionality has been properly introduced. The assay and specifications for a key intermediate should demonstrate that the impurity levels, including the stereochemical impurities, are within established limits. Any subsequent synthetic steps which have the potential to epimerise a chiral centre will require appropriate controls to establish that the optical purity remains within established limits.

## 2.8 CONCLUSION

Although there is currently no harmonised guidance for the development and registration of chiral drugs, regulatory organisations have developed local guidelines. Multinational firms must interpret these local guidelines in developing their approaches to new therapeutic candidates. The local guidelines may allow companies some latitude to make their own development and registration decisions with respect to pure enantiomers and racemates; however, there is a growing consensus on the types of additional information required to support those decisions.

## 2.9 REFERENCES AND NOTES

1. (a) Food and Drug Administration, *FDA's Policy Statement for the Development of New Stereoisomeric Drugs*, FDA, Washington, DC, May 1, 1992; (b) CPMP Working Party on Quality of Medicinal Products, CPMP Working Party on Safety of Medicinal Products, CPMP Working Party on Efficacy of Medicinal Products, *Note for*

*Guidance: Investigation of Chiral Active Substances*; (c) HPB Working Group on Drug Stereochemistry, *Guidelines on the Development of Stereoisomeric Drugs*, Draft, December, 1993.
2. Center for Drugs and Biologics Food and Drug Administration Department of Health and Human Services, *Guideline for Submitting Supporting Documentation in Drug Applications for the Manufacture of Drug Substances*, (FDA, Washington, DC, February, 1987.
3. PMA Ad Hoc Committee on Racemic Mixtures, 'Comments on Enantiomerism in the Drug Development Process', *Pharm. Technol.*, May, 46 (1990).
4. Gross, M., *et al.*, Regulatory Requirements for Chiral Drugs, *Drug Inf. J.* **27**, p. 453 (1993).
5. The CPMP is a scientific committee in the EC comprised of Member States and the Commission which coordinates the registration of applications for new drugs.
6. Jamali, F., *et al.*, Stereochemical Aspects of Drug Development, Drug Information Association Workshop, Rockville, MD, 1994.
7. Food and Drug Administration, HHS. *Guideline on Impurities in New Drug Substances*, *Fed. Regist.*, **61**, 372 (1996).
8. In this context, the term 'stressed stability studies' refers to severe conditions intended to afford visible chemical and/or physical degradation. Accelerated storage conditions, on the other hand, are meant to be a short-term predictor of the behaviour over the full-length storage period.

# 3 Production Methods for Chiral Non-steroidal Anti-inflammatory Profen Drugs

### G. P. STAHLY AND R. M. STARRETT
*Albemarle Corporation, Baton Rouge, LA USA*

| | | |
|---|---|---|
| 3.1 | Introduction | 19 |
| 3.2 | Industrial Processes for (S)-(+)-Naproxen | 22 |
| 3.3 | Industrial Processes for (S)-(+)-Ibuprofen | 27 |
| 3.4 | Other Profens and the Future | 37 |
| 3.5 | References | 38 |

## 3.1 INTRODUCTION

Non-steroidal anti-inflammatory drugs (NSAIDs) comprise a structurally diverse group of compounds that have similar physiological effects.[1] The NSAID designation is somewhat misleading, as these drugs exhibit varying degrees of analgesic and antipyretic, as well as anti-inflammatory, activity. They are widely prescribed to relieve the inflammation encountered with diseases such as arthritis and for relief of low to moderate pain such as headache. In 1986, the 100 most widely prescribed drugs in the USA included seven NSAIDs.[2]

Most NSAIDs are organic acids, aspirin being the prototype. The profens are a subset of the class, each bearing as the common structural feature a 2-aryl-substituted propionic acid. Variation of substituents on the aryl group leads to a variety of different profens, some of which are shown in Figure 1. The most commercially significant are naproxen and ibuprofen.

The mechanism of action of NSAIDs is thought to be inhibition of prostaglandin synthesis.[3] Prostaglandins are not stored in cells but are instead synthesised and released in response to cell damage, appearing in high concentrations in inflammatory exudates. They are also associated with the development of pain resulting from injury or inflammation. The prostaglandin biosynthesis pathway is known, as is the specific step, catalysed by a cyclooxygenase enzyme, which is subject to NSAID interference. All NSAIDs inhibit cyclooxygenase activity, although not all by the same mechanism. Prostaglandin synthesis is believed to occur in all tissues, implying that the effectiveness of an NSAID is primarily influenced by its ability to reach the cyclooxygenase enzyme.

*Chirality in Industry II.* Edited by A. N. Collins, G. N. Sheldrake and J. Crosby
© 1997 John Wiley & Sons Ltd

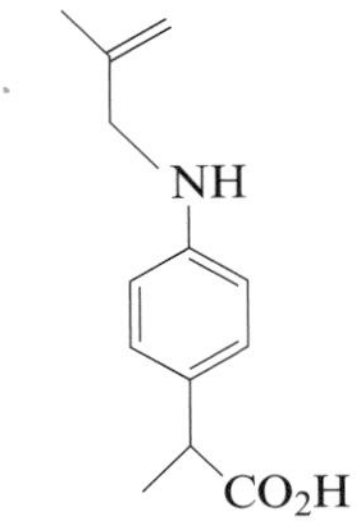

Alminoprofen 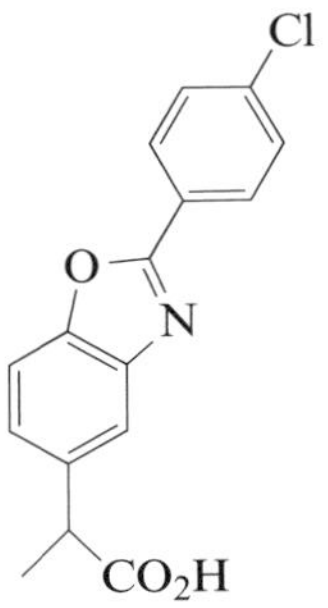   Benoxaprofen 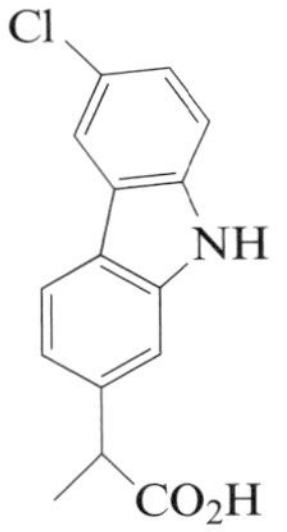   Carprofen 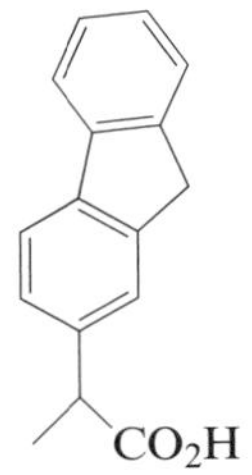   Cicloprofen

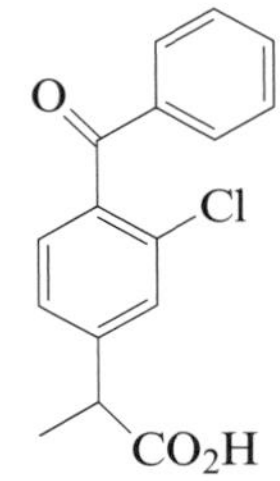

Cliprofen 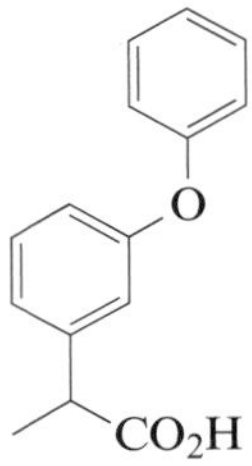   Fenoprofen 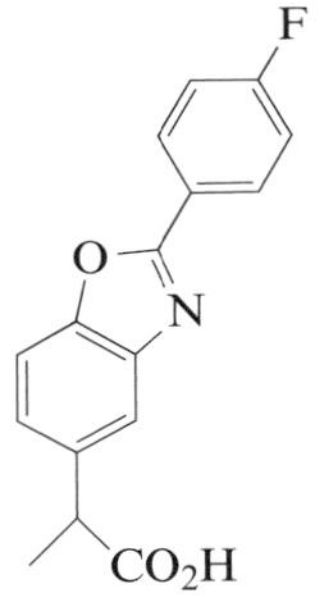   Flunoxaprofen 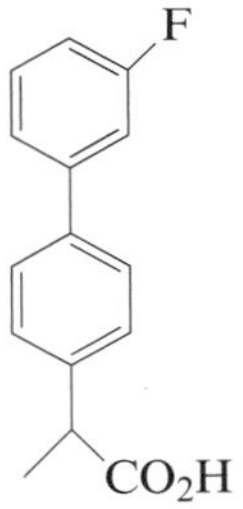   Fluprofen

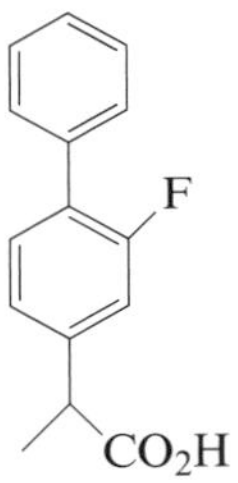

Flurbiprofen 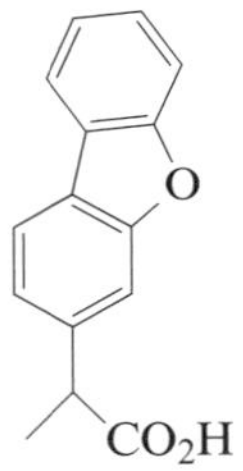   Furcloprofen 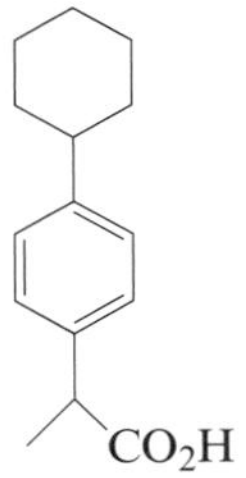   Hexaprofen 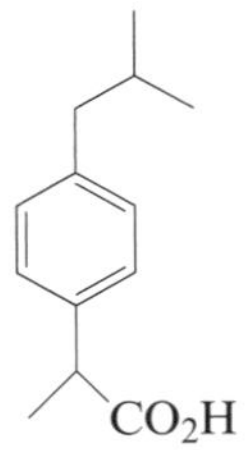   Ibuprofen

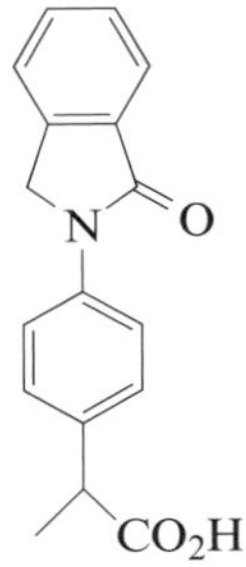

Indoprofen 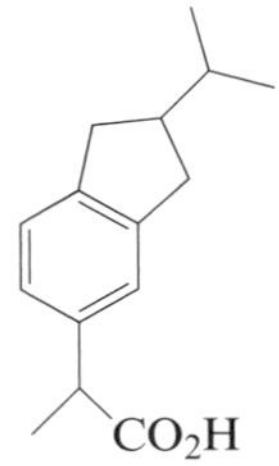   Isoprofen 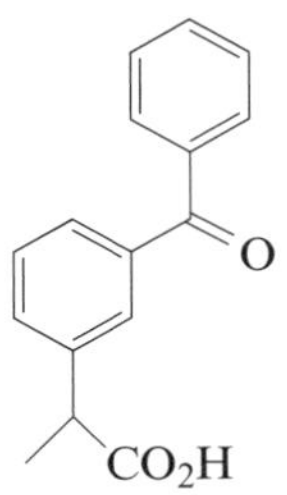   Ketoprofen 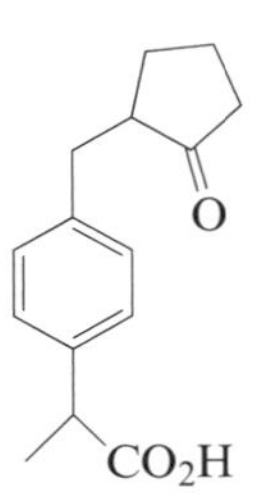   Loxoprofen

**Figure 1** (*continued*)

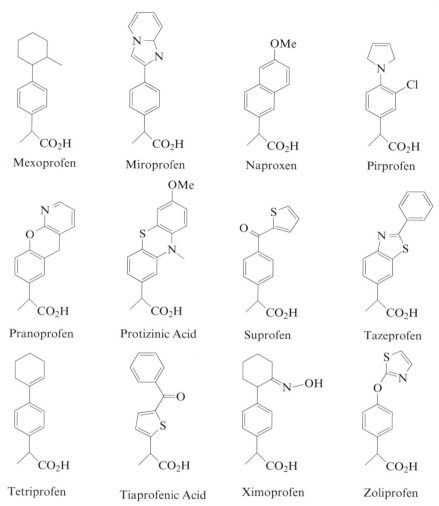

**Figure 1**

Thus, the rate and extent of distribution of NSAIDs in the body may be important in determining the effectiveness of a given NSAID, and the proliferation of different profen structures can probably be attributed to this fact. Delivery of the active inhibiting moiety to the cyclooxygenase enzyme is influenced by the ability of the whole molecule to reach the necessary systems. For example, although both naproxen and ibuprofen have similar analgesic and anti-inflammatory effects, naproxen has a much longer biological half-life.

All of the profen drugs contain a stereogenic centre, which is the carbon atom bearing the carboxylic acid group. In the case of naproxen, early testing clearly showed that the (S)-isomer was more active pharmacologically than the (R)-isomer, although the latter is not thought to be harmful; thus, only the (S)-isomer of naproxen is sold. Ibuprofen, on the other hand, was originally marketed as the racemate. Later, advantages were claimed for the administration of pure (S)-(+)-ibuprofen and development of products containing (S)-(+)-ibuprofen is under way. The development of an optically pure drug where the racemate is already in use has been called a 'racemic switch', and details of the isomer issue with relation to naproxen and ibuprofen are discussed below.

Other profen drugs have also been sold as racemic mixtures. However, the claim that (S)-(+)-ibuprofen may have advantages over the racemate has prompted frantic efforts in the industry to examine all profens for racemic switch potential. Some specific examples are given later in this chapter.

The commercial importance of (S)-(+)-naproxen, and the potential importance of (S)-(+)-ibuprofen and other chiral profens, have resulted in much research into chiral methodologies. Many new ways to synthesise these materials in high optical purity have appeared in the literature.[4] Frequently, new asymmetric reactions are demonstrated using intermediates leading to naproxen or ibuprofen. In this chapter, we review methods for producing chiral profen drugs from an industrial perspective. Specific discussions of naproxen and ibuprofen are followed by a more general section on other profens.

## 3.2 INDUSTRIAL PROCESSES FOR (S)-(+)-NAPROXEN

Naproxen was developed by the Syntex Corporation in the 1960s. Early screening of compounds based on naphthaleneacetic acid showed not only that naproxen is a potent anti-inflammatory agent, but also that the (S)-isomer is much more effective than either the (R)-isomer or the racemate. For example, in one *in vivo* test, (S)-(+)-naproxen was reported to be 28 times more effective than the (R)-isomer, and three times more effective than the racemate.[5] Resolved material for these studies was obtained by crystallisation of the diastereoisomeric salt pair formed from racemic naproxen and (−)-cinchonidine.

Naproxen was commercialised in the UK in 1973 and in the USA in 1977.[6] Manufacturing methods available to Syntex at this time probably did not include catalytic asymmetric synthesis, as the time necessary to develop this or other strategies (e.g. biological processes) would not have allowed a rapid introduction to the market. Consequently, patents indicate that Syntex chose to produce the racemate and develop an efficient resolution.

Scheme 1 shows a likely synthesis of the racemate.[7] Although early production may have relied on (−)-cinchonidine resolution, more efficient technology was developed subsequently and patented. This utilises N-alkyl-D-glucamines as resolving agents,[8] which produce, in the case of N-n-propyl-D-glucamine, (S)-(+)-naproxen (100% *ee*) with only two diastereoisomeric salt crystallisations.[8a]

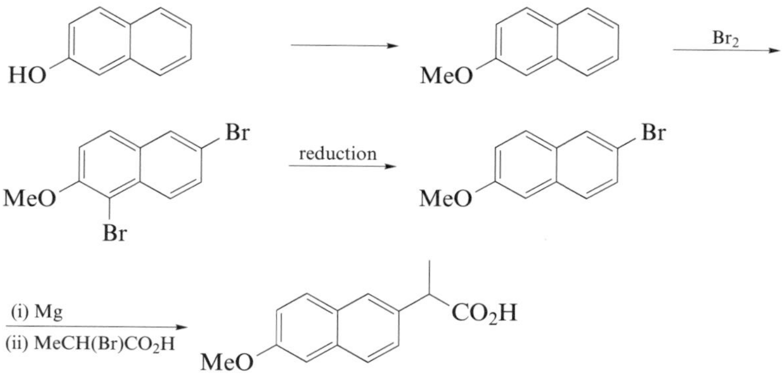

# PRODUCTION METHODS FOR CHIRAL PROFEN DRUGS

**Scheme 1**

The crystalline form of naproxen is important when considering process choices. Naproxen is a racemic compound (see Chapter 7 and Volume I, p. 20). This is determined by examination of a plot of melting point vs composition for a mixture of the (*R*)- and (*S*)-isomers (melting point phase diagram), as shown in Figure 2.[9] A racemic compound has a eutectic (low-melting) composition different from the racemic mixture [50% (*R*)-isomer and 50% (*S*)-isomer]. This is the case for naproxen, whose eutectic composition is about 80% (*S*)-isomer–20% (*R*)-isomer (or, since binary melting point phase diagrams of enantiomeric mixtures are symmetrical, 80% (*R*)-isomer–20% (*S*)-isomer).

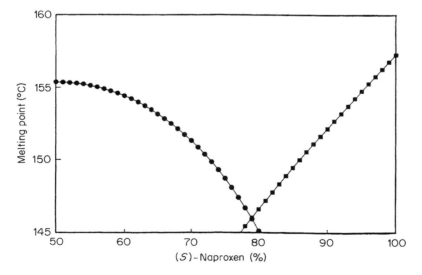

**Figure 2** Calculated melting point phase diagram for naproxen

The eutectic composition is a most important feature to be considered when planning racemate separations. Crystallisation of a given isomer mixture will usually lead to a crystal crop whose composition is 'uphill' on the phase diagram curve compared with the starting composition. This is because the lowest melting (most soluble) composition remains in the mother liquor. Both melting and dissolution processes involve disruption of the crystal lattice, which means that the lowest melting composition is usually the same as the most soluble composition. Also, for a racemic compound, the eutectic composition may not be passed by crystallisation under thermodynamic conditions, so crystallisation of a racemic compound will only yield a crystal crop of increased optical purity if the starting composition lies between the eutectic composition and pure enantiomer.

In the case of naproxen, a mixture only requires about 80% of one isomer in order to allow the optical purity to be upgraded by recrystallisation of the underivatised acid. Thus, any process (resolution or synthesis) that gives >80 wt% (S)-(+)-naproxen can be followed by recrystallisations to yield 100 wt% (S)-(+)- naproxen. The commercial importance of (S)-(+)-naproxen made it a prime target for those seeking new methods for the synthesis of chiral molecules.

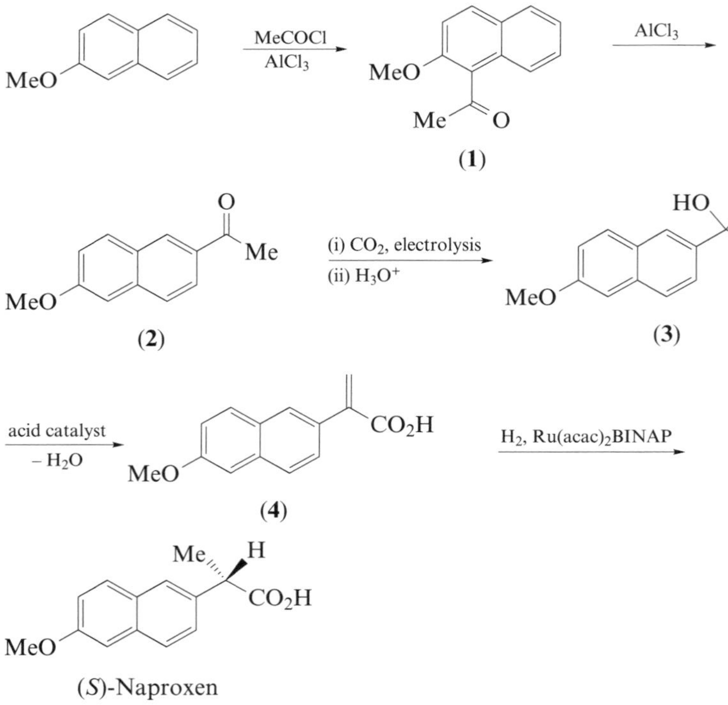

**Scheme 2**

Consequently, many new routes to (S)-(+)-naproxen were reported, the most widely publicised being a number of asymmetric syntheses.

Monsanto chemists developed a route starting with 6-acyl-2-methoxynaphthalene (**2**), as shown in Scheme 2.[9] 2-Methoxynaphthalene, which is the starting material in almost all (S)-(+)-naproxen syntheses, can be acylated with high regioselectivity to give **2**. It is interesting that the kinetic product of acylation, **1**, can be converted to the thermodynamically stable **2** under acylation conditions.[11] This intermediate is then electrochemically carboxylated to give hydroxy acid **3**, the dehydration of which gives the acrylic acid derivative **4**, which is the prochiral starting material for the following asymmetric reduction. This reduction is accomplished using a chiral catalyst similar to that reported by Noyori, and co-workers consisting of Ru(0) bearing a chiral BINAP ligand.[11,12] Monsanto developed a variant that was claimed to show an increased reaction rate compared with the Noyori system[14] and, using this catalyst, the acrylic acid derivative **4** is converted into (S)-(+)-naproxen in 100% yield and up to 98% ee.[10,15] This optical purity is within USP optical purity specifications,[16] but can be improved by recrystallisation if desired. The asymmetric catalyst is critical to this process and, although expensive, it is both robust and recyclable.[10] More recently, a heterogeneous chiral Ru catalyst was developed that converts **4** into (S)-(+)-naproxen with high enantioselectivity.[17]

DuPont researchers developed a process based on the catalytic asymmetric hydrocyanation of 6-methoxy-2-vinylnaphthalene (Scheme 3).[18] The prochiral intermediate **5** can be made most simply by a reduction–dehydration sequence starting with **2**, but also by Ni-catalysed coupling of 6-methoxy-2-bromonaphthalene with vinyl Grignard reagents.[19]

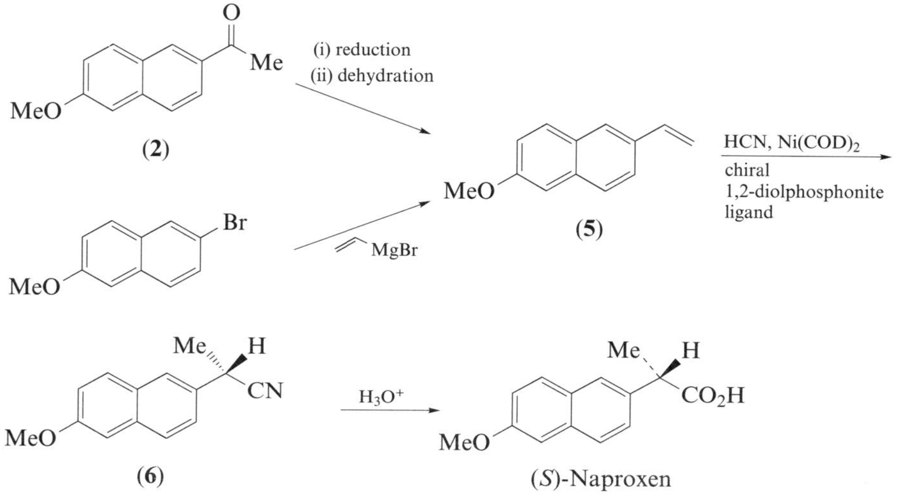

**Scheme 3**

Key to the asymmetric hydrocyanation process was the invention of asymmetric phosphinites, which induce enantioselectivity in the hydrocyanation through ligation of the Ni catalyst (see also Chapter 15). These phosphinites were made by elaboration of mono- and disaccharide hydroxyl groups, affording relatively inexpensive catalysts (based on Ni and sugars) which can be easily modified to optimise the desired activity. Thus, **5** was converted into **6** in >85% yield with reasonable (75–85%) *ee*. The product nitrile exhibits a favourable phase diagram, as does naproxen, and was recrystallised twice to give material in >99% *ee*. Hydrolysis then gave (*S*)-(+)-naproxen.

Another industrial process utilising asymmetric synthesis was developed by Zambon[20] (Scheme 4). An achiral version of the 1,2-aryl shift reaction was invented by Blaschim (Italy), and developed by Zambon into a manufacturing process, the key intermediate being the chiral ketal **7**. The chiral auxiliary, a tartaric acid ester, influences both the bromination and rearrangement reactions so that naproxen of 99% *ee* results. Since the auxiliary is used stoichiometrically, it must be recovered and re-used, as must a resolving agent.

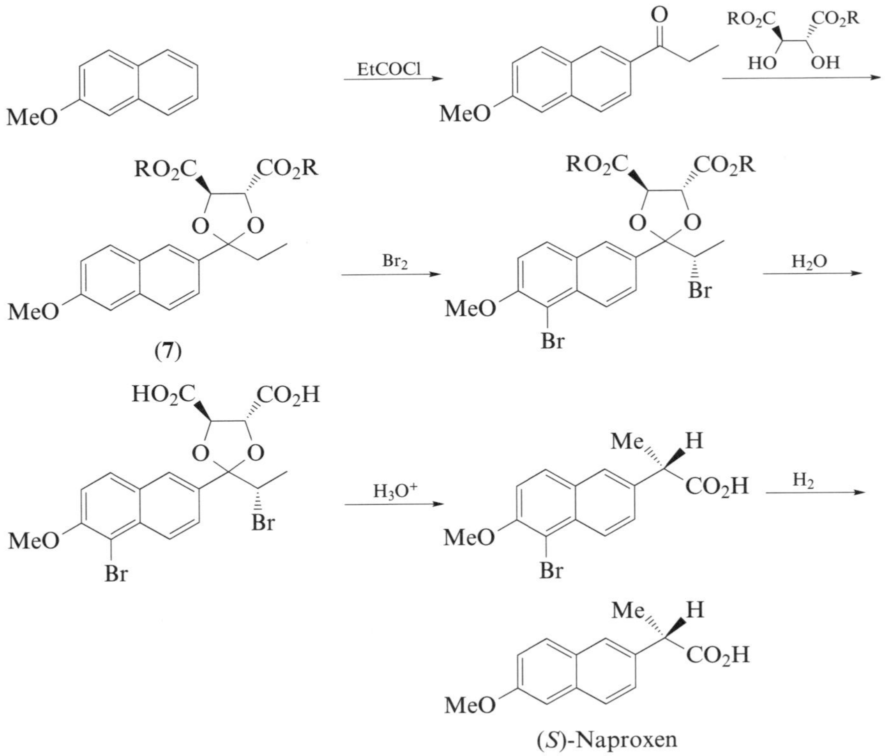

**Scheme 4**

At first glance, some of the newer methods, exemplified by those above, would seem preferable to a classical resolution. For example, the Monsanto asymmetric synthesis utilises selective, high-yielding reactions and produces a product of sufficient optical purity with a chiral catalyst, avoiding the recycle and racemisation steps necessary with resolution. However, other factors must also be considered. Firstly is the price obtainable for the final product, and, in particular, the price stability. It is critical to use the least expensive process available because these costs comprise an increasing percentage of the final bulk drug price. As the price of the bulk drug decreases it becomes more of a commodity chemical, and low-cost production becomes an even more important issue. Secondly, capital expenses are very important. A new process requiring a new plant may not compete economically with an old process in an existing plant, even if the new process has chemical and processing advantages. The cost of building a new plant can override these advantages. Thirdly, the importance of the requirements of governmental regulatory agencies cannot be overlooked (see also Chapter 2). In the USA the Food and Drug Administration (FDA) requires that a significant change to the process to manufacture a drug is preceded by extensive tests, possibly involving clinical trials to show that the resulting product is identical with the original one.[21]

## 3.3 INDUSTRIAL PROCESSES FOR (S)-(+)-IBUPROFEN

Ibuprofen was patented by the Boots Pure Drug Co. in the late 1960s.[22] Efficacy testing at that time revealed that (1) (R)- and (S)-isomers have similar *in vivo* potency[23], (2) only the (S)-isomer inhibits prostaglandin synthetase *in vitro*[23] and (3) after oral administration of the individual enantiomers, metabolites found in the urine are dextrorotatory.[24] Thus, the conclusion was that chiral inversion of the (R)-isomer to the active (S)-isomer occurs *in vivo*,[24] and an enzyme system was proposed to account for this.[25] Subsequent work led to further insights into the inversion process.[26]

All of these studies led to the production and sale of ibuprofen as a racemic mixture. Market introduction in the UK was in 1969 and in the USA in 1974. Numerous research projects looking for new racemate syntheses have been initiated, resulting in a flood of methods, many of which are applicable to a wider range of profen drugs.[27] In 1985 the US patent[22b] on ibuprofen expired, opening the generic prescription market. Even more significant was the FDA decision at that time to allow racemic ibuprofen to be sold over the counter (OTC) in the USA. These events prompted interest in commercial production of racemic ibuprofen throughout the world. Suppliers of bulk active racemate during this period included Boots in the UK and Albemarle Corporation, which was the only US producer; the original route patented by Boots is shown in Scheme 5.[28]

Illustrative of the chemistry developed more recently by others are the routes of the Nippon Petrochemicals Company shown in Scheme 6. Both routes

**Schemes 5**

produce isobutylstyrene (**8**), which is carbonylated in a final step to give ibuprofen. Route A[29] involves acid-catalysed arylation of acetaldehyde, followed by cracking of the resulting adduct to give equal amounts of isobutylstyrene (**8**) and isobutylbenzene, the latter being recycled. In route B,[30] reaction of toluene and ethylene gives perethylated toluenes, which are passed, along with isobutylbenzene, over an acidic transalkylation catalyst. The resulting mixture is fractionally distilled to remove the desired *p*-ethyl(isobutyl)benzene (present at 5–10%). Light fractions, containing lightly ethylated toluenes, are returned to the ethylation reactor and heavy fractions, containing heavily ethylated toluenes and isobutylbenzenes, are fed with fresh isobutylbenzene back to the transalkylation reactor. Dehydrogenation of *p*-ethyl(isobutyl)benzene then gives isobutylstyrene (**8**).

In the USA, a Boots–Hoechst–Celanese joint venture was established which began production in 1993,[31] and the route shown in Scheme 7 is believed to be used in that facility. This chemistry involves three steps, all catalytic, and is a 100% atom-efficient process (i.e. every atom put into the process either becomes part of the final product or is recycled).

In 1989, a US patent assigned to Analgesic Associates was issued claiming that analgesia in animals (including humans) was more rapidly attained and enhanced in effect when (*S*)-(+)-ibuprofen was used compared with the use of the racemate.[32] At about the same time that the Analgesic Associates patent application was undergoing prosecution, a European application was filed by

# PRODUCTION METHODS FOR CHIRAL PROFEN DRUGS

**Schemes 6**

Medice Chem.-Pharm. Fabrik of West Germany.[33] Ultimately, each patent was licensed to a pharmaceutical company: Sterling Drug licensed Analgesic Associate's rights, Merck & Company licensed Medice's US and Canadian rights and Boots Drug Company licensed Medice's UK rights. The claim that (S)-(+)-ibuprofen exhibits superior performance compared with the racemate has led to

**Schemes 7**

recent attempts to market products containing only the (S)-isomer, and the sales potential of (S)-(+)-ibuprofen has prompted research into its synthesis.

Although naproxen was traditionally the molecule of choice for demonstrating new synthetic routes to profen drugs, ibuprofen has received more attention in the last few years, and in some cases these newly developed methods seem to be as effective for ibuprofen as naproxen. For example, Monsanto can apply their electrocarboxylation–asymmetric reduction chemistry to (S)-(+)-ibuprofen synthesis (Scheme 8). The hydrogenation is reported to give (S)-(+)-ibuprofen of 96% ee.[10] A commercially significant variant was reported by Albemarle researchers, who found that an effective chiral catalyst can be generated *in situ* from relatively inexpensive Ru(III) acetylacetonate.[34]

However, there can be differences in the efficiency of a given process when applied to naproxen vs ibuprofen. For example, the DuPont hydrocyanation reaction is reported to hydrocyanate isobutylstyrene (**8**) with much poorer enantioselectivity (50% ee product) than the corresponding reaction for (S)-(+)-naproxen.[18]

Union Carbide Corporation has also developed an asymmetric route to (S)-(+)-ibuprofen (Scheme 9). This process incorporates Carbide's Rh-catalysed hydroformylation (oxo) reaction.[35] The key to the original achiral hydroformylation method was the use of phosphine ligands to modify the Rh catalyst, giving not only the desired reactivity but also control over regioisomer distribution. Thus, olefins can be converted into aldehydes with high linear:branched ratios, the linear isomers being the products of highest value.[36] By using phosphite instead of phosphine ligands, branched rather than linear aldehydes result. In

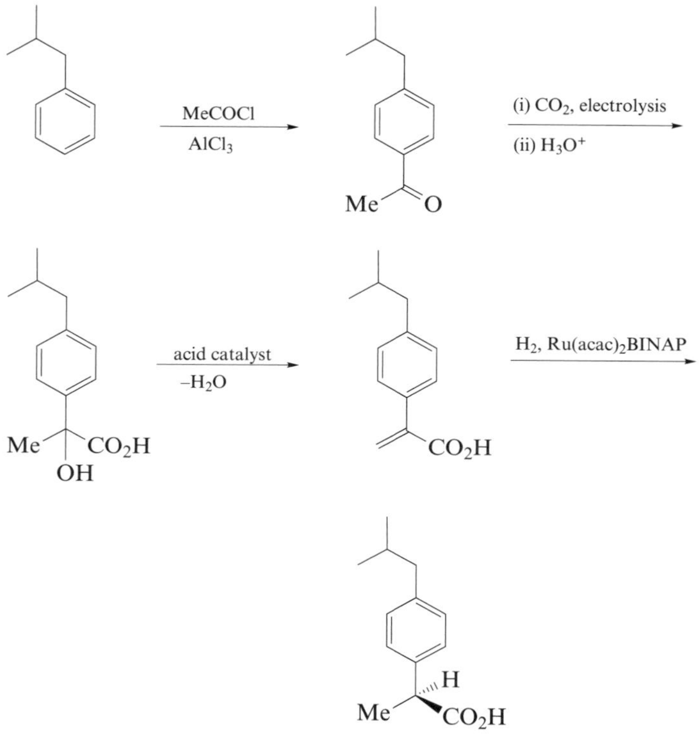

**Scheme 8**

addition, the use of chiral phosphites allowed enantioselective hydroformylation of isobutylstyrene (**8**).[37] The aldehyde product has been obtained in 82% *ee*, and stereospecific oxidation then produces (*S*)-(+)-ibuprofen.

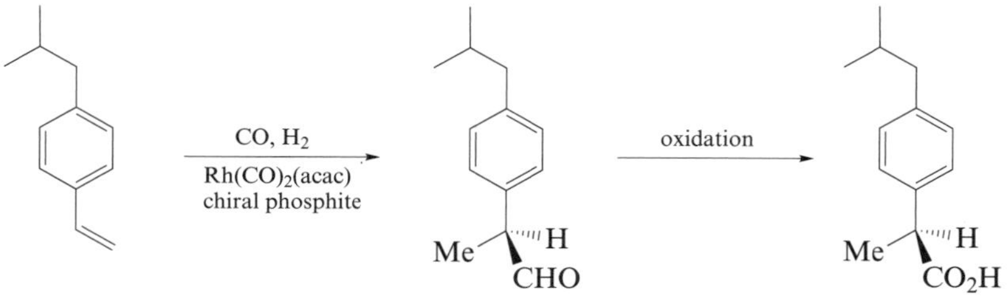

**Scheme 9**

The prochiral precursor used by Union Carbide, isobutylstyrene (**8**), can be made in several ways. The methods of the Nippon Petrochemical Company have already been described (Scheme 6), and other routes are analogous to those used to make methoxyvinylnaphthalene (Scheme 3, compound **5**), including reduction and dehydration of isobutylacetophenone[38] and reaction of isobutylmagnesium chloride with bromostyrene (Scheme 10).[22]

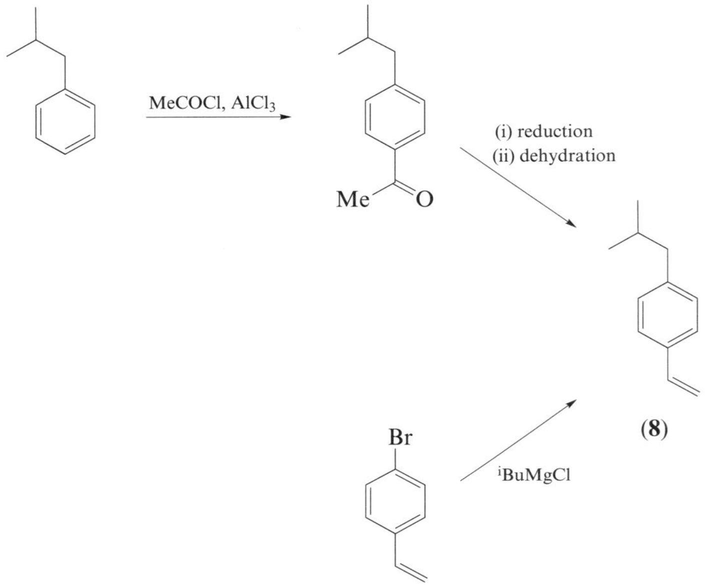

**Scheme 10**

Even more straightforward than hydroformylation would be hydrocarboxylation, in which the acid is produced directly. The achiral version of this reaction works very well on isobutylstyrene (**8**) to give racemic ibuprofen.[29] Alper and Hamel[39] reported such a reaction in 1990.

Merck researchers reported an (*S*)-(+)-ibuprofen synthesis based on the asymmetric addition of an alcohol to a ketene (Scheme 11).[40] This sequence starts with racemic ibuprofen, which is converted into the prochiral ketene. Readily available (lactic acid-derived) chiral alcohols are then used to carry out an asymmetric protonation reaction, affording diastereoisomeric esters with good diastereoselectivity. Thus, all of the racemic mixture can be converted into one enantiomer. When the starting material was ibuprofen, (*S*)-(+)-ibuprofen esters were obtained in 94–99% diastereoisomeric excess (*de*), and an ester of (*S*)-(+)-naproxen was produced in 80% *de*; as with a resolution, the chiral auxiliary must be recovered.

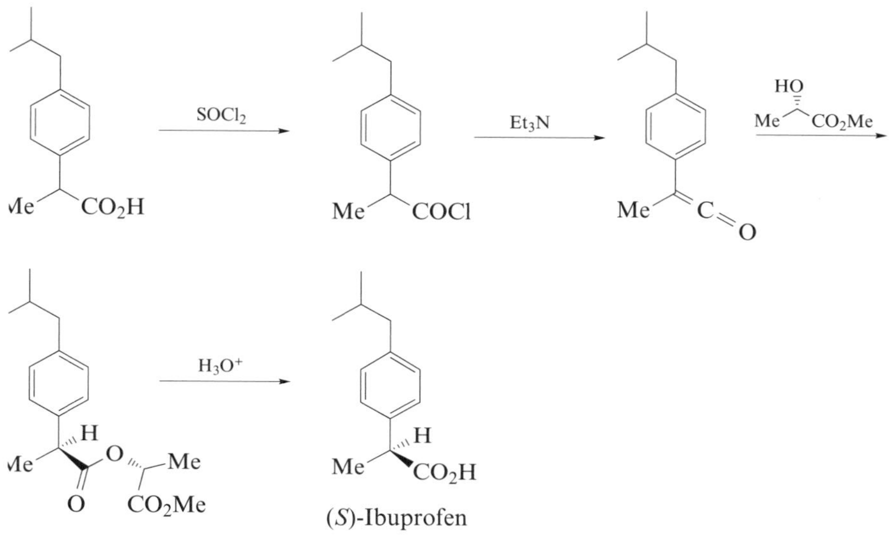

**Scheme 11**

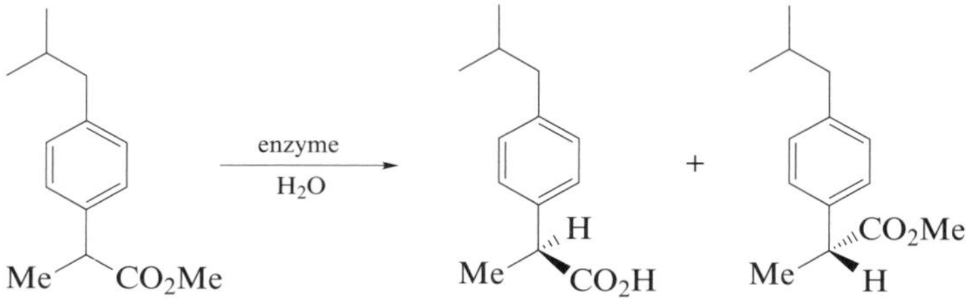

**Scheme 12**

Enzymatic resolution processes based on selective hydrolyses of racemic mixtures of ibuprofen esters have been investigated by several companies. A kinetic resolution of this type is shown in Scheme 12. The Wisconsin Alumni Research Foundation (WARF) made one of the earliest claims to a process of this type,[41] and Sepracor,[42] Gist-Brocades[43] and Rhône-Poulenc[44] have also developed similar chemistry. In each of these cases the processes are reported to have broad substrate specificity, in terms of both the profen moiety and the corresponding starting derivative. The methyl esters of naproxen and ibuprofen are the most common starting materials included in patent examples.

One advantage typically obtained by using enzymes is very high enantioselectivity. For example, Gist-Brocades found that addition of the methyl ester of racemic ibuprofen to growing cultures of *Pseudomonas, Mucor, Arthrobacter* or *Bacillus* species produced (S)-(+)-ibuprofen in 74–98% ee.[43] Gist–Brocades uses a growing microorganism, Sepracor uses isolated enzyme preparations and Rhône-Poulenc uses highly purified enzymes called isozymes. There are advantages and drawbacks to each of the systems, which must be evaluated case by case. The purer the enzyme, the more expensive is its production, but high purity often means high enantioselectivity. Use of a growing microorganism provides a relatively inexpensive enzyme source, but the aqueous systems necessary for fermentation may limit the amount of organic substrate that can be present.

Another important process consideration is the sense of the enantioselectivity. For example, a racemic mixture of the methyl ester of ibuprofen can afford a mixture of (S)-acid and (R)-ester or (R)-acid and (S)-ester, depending on the enzyme used. If the (S)-acid is the desired product, the former situation is better because the product is obtained directly and the unwanted isomer can be racemised and recycled as the methyl ester. If the (R)-acid were to be produced, two additional process steps would be necessary: hydrolysis of the (S)-ester to give product and re-esterification of the (R)-acid for racemisation and recycle.

The use of enzymes depends on their stabilities in non-aqueous solvents. Historically, efforts to overcome poor performance in this area include the membrane reactor system designed by Sepracor, in which the enzyme is immobilised in hollow-fibre reactors, allowing continuous kinetic resolutions.[45] Altus Biologics have developed crosslinked, crystalline enzymes that function in organic solvents, and one example, a lipase from *Candida rugosa*, hydrolyses racemic ibuprofen methyl ester to give 95% ee (S)-(+)-ibuprofen.[46] From a commercial standpoint, the enzyme is viewed simply as a catalyst, whose cost, productivity, selectivity and re-usability have to compete with those of a conventional resolving agent.

Because many companies already have the capacity to produce racemic ibuprofen, they have chosen to investigate new diastereoisomeric salt resolution processes rather than asymmetric routes. Ibuprofen can be resolved by common resolving agents such as α-methylbenzylamine[47] and lysine.[48] Designer amines, **9** and **10**, were developed in Japan specifically for the resolution of ibuprofen.[49] Crystallisation of a mixture of racemic ibuprofen and (−)-**9** from methanol–water, followed by one recrystallisation of the resulting salt, gave a salt containing (S)-(+)-ibuprofen with 100% ee.[49b]

As discussed earlier, the crystalline form of the product needs to be considered when evaluating process choices. Like naproxen, ibuprofen is a racemic compound, this being evident from the calculated binary melting point phase diagram (Figure 3), which shows that the eutectic composition for ibuprofen is about 90% (S)-isomer–10% (R)-isomer [or 90% (R)-isomer–10% (S)-isomer].[9] The calculated diagram agrees reasonably well with results determined experimentally.[9,50]

The eutectic composition dictates that ibuprofen needs to be at least 90% pure (S)-isomer for crystallisation to result in optical purification. However, even

PRODUCTION METHODS FOR CHIRAL PROFEN DRUGS

(9) R = Me
(10) R = Cl

above this composition limit, in practice, purification by crystallisation is rather inefficient. This is generally true when the eutectic is closer to pure isomer. We find that crystallisation of 96% ee (S)-(+)-ibuprofen gives a crystal crop of only about 97–98% ee,[9] which provides much less working margin than is available for naproxen. The result is that any asymmetric process must give nearly specification-grade (S)-(+)-ibuprofen, or multiple recrystallisations will be required to attain the necessary optical purity.

A method to overcome the difficult optical purification of enantiomerically enriched ibuprofen was developed at Albemarle. While ibuprofen itself exhibits an unfavourable phase diagram, that for the sodium salt of ibuprofen is much more favourable. As can be seen in Figure 4, the eutectic composition for sodium ibuprofenate is about 60% (S)-isomer–40% (R)-isomer [or 60% (R)-isomer–40% (S)-isomer], so ibuprofen produced by any method that is at least 60% optically

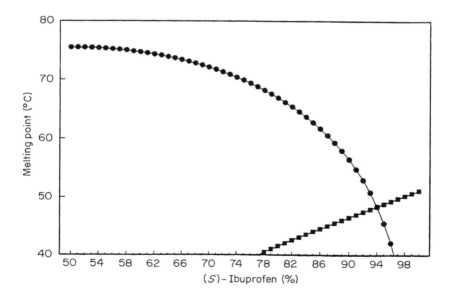

**Figure 3**  Calculated melting point phase diagram for ibuprofen

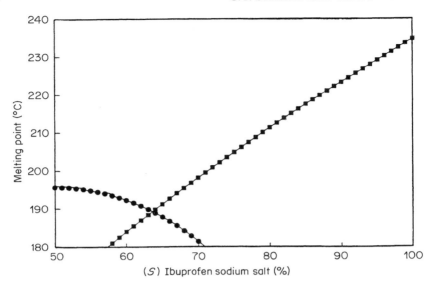

**Figure 4** Calculated melting point phase diagram for sodium ibuproferate

pure can be upgraded via the sodium salt. This method works well, as shown by the one-step upgrade of 76% ee (S)-(+)-ibuprofen to 100% ee (S)-(+)-ibuprofen by treatment with sodium hydroxide in acetone.[9] The method is also of value when coupled with a diastereoisomeric resolution process, reducing the number of steps involving the resolving agent and thereby minimising its recycle and loss.

Which is the best type of process for commercial (S)-(+)-ibuprofen production? As it was with naproxen, the answer is greatly influenced by the plant already in place. Plants assets have been built and are currently producing racemic ibuprofen, so, with an additional capital investment, a resolution process can be added to the racemate production unit; add to this the fact that technology for ibuprofen racemisation is known,[51] the resolution route looks even better. As any unwanted isomer can be easily recycled, the yield of (S)-(+)-ibuprofen from racemic ibuprofen can be nearly 100%.

Another approach, taking advantage of existing racemate production capacity, might be to utilise an intermediate from the racemate process for an asymmetric synthesis. In this case, the intermediate needs to be prochiral (or easily converted into a prochiral intermediate), and any unwanted (R)-isomer resulting from the asymmetric step must be readily convertible back to the prochiral intermediate. This is rarely as simple in practice. Some additional questions can be raised: are the existing plants large enough to produce enough racemic isomer for sale and also for resolution, and how much of the racemic ibuprofen market will (S)-(+)-ibuprofen displace, if any? These questions are difficult to answer until commercial production and sale of (S)-(+)-ibuprofen begin. Even if it is believed that more capacity will be needed, it may be cheaper to expand a current process than to build a new plant with a more efficient process.

One consultant who has closely followed racemic switches has attempted to analyse competitive processes which he believes are being considered for (S)-(+)-ibuprofen production. His conclusion is that the most competitive processes today are likely to be classical resolutions carried out by racemate producers, although asymmetric synthesis and biotransformation routes are not far behind.[52]

## 3.4 OTHER PROFENS AND THE FUTURE

New racemic switch candidates are being aggressively sought by many companies in the pharmaceutical industry. Determination of the physiological behaviours of pure enantiomers derived from drugs currently sold as racemates is widespread.[52] The efforts of Medice and Analgesic Associates to secure and license (S)-(+)-ibuprofen patents have led others to pursue similar strategies for other profen drugs, including flurbiprofen and ketoprofen. Racemic flurbiprofen is sold by Upjohn as Ansaid and racemic ketoprofen by Wyeth-Ayerst as Orudis. Sepracor (Marlborough, MA, USA) has filed or received use patents on (S)-(+)-flurbiprofen and (S)-(+)-ketoprofen claiming that they have enhanced analgesic activity (including faster onset) and reduced side effects (such as gastrointestinal irritation and kidney toxicity).[53] In addition, the same drugs are claimed to inhibit bone loss due to periodontal disease.[54] Analgesic Associates has also remained active in the racemic switch area, concentrating on flurbiprofen and ketoprofen, and a series of US patents have been issued claiming advantages such as faster onset and lower dosage when the (S)-isomers are used.[55]

It is thus likely that the chiral profen drugs with the highest commercialisation potential, besides (S)-(+)-ibuprofen, are (S)-(+)-flurbiprofen and (S)-(+)-ketoprofen. Because the volumes of the racemic forms of these currently sold are relatively small, there has been correspondingly little research aimed at efficient process chemistry. Flurbiprofen and ketoprofen both contain *meta*-substituted aromatic moieties, and the former contains a fluorine atom, making their syntheses inherently more difficult and expensive compared with those of naproxen and ibuprofen.

Here, then, is the immediate future of chiral profen drugs. Methods developed for (S)-(+)-naproxen and/or (S)-(+)-ibuprofen may be readily applicable to (S)-(+)-flurbiprofen and/or, (S)-(+)-ketoprofen, or new methods may be developed. In any case (S)-(+)-flurbiprofen and (S)-(+)-ketoprofen processes based on these methods face less economic competition given the immaturity of the drugs. Of course, marketplace competition will determine whether volumes will grow, but new, efficient chemistry stands a good chance of incorporation into new plants if that growth occurs.

## 3.5 REFERENCES

1. For a review covering the physiology and medicinal chemistry of traditional NSAIDs, see Lombardino, J. G., (Ed.), *Nonsteroidal Antiinflammatory Drugs*, Wiley, New York, 1985.
2. Reuben, B. G., and Wittcoff, H. A., *Pharmaceutical Chemicals in Perspective*, Wiley, New York, 1989, p. 98.
3. Flower, R. J., Moncada, S., and Vane, J. R., in *The Pharmacological Basis of Therapeutics* Gilman, A. G., Goodman, L. S., Rall, T. W., and Murad, F., (Eds.) Macmillan, New York, 1985, Ch. 29.
4. Sonawane, H. R., Bellur, N. S., Ahuja, J. R. and Kulkarni, D. G., *Tetrahedron: Asymmetry*, **3**, 163 (1992).
5. Harrison, I. T., Lewis, B., Nelson, P., Books, W., Roszkowski, A., Tomalonis, A. and Fried, J. H., *J. Med. Chem.*, **13**, 203 (1970).
6. *Chemica Useful Book*, Chemica, Caterham, Surrey, 1991, naproxen profile.
7. Sheldon, R. A., *Chirotechnology*, Marcel Dekker, New York, 1993, p. 349.
8. (a) Holton, P. G., *US. Pat.*, 4246193, 1981; (b) Felder, E., San Vitale, R., Pitre, D., and Zutter, H., *US. Pat.* 4246164, 1981.
9. (a) Manimaran, T. and Stahly, G. P., *Tetrahedron: Asymmetry*, **4**, 1949 (1993); (b) Herndon, R. C., Manimaran, T., and Stahly, G. P., *US Pat.* 5248813, 1993.
10. Chan, A. S. C., *Chemtech*, **23** (3), 46 (1993).
11. Giordano, C., and Villa, M., *Synth. Commun.*, **20**, 383 (1990).
12. Ohta, T., Takaya, H., Kitamura, M., Nagai, K., and Noyori, R., *J. Org. Chem.*, **52**, 3174 (1987).
13. (a) Yoshikawa, S., Saburi, M., Ikariya, T. and Ishii, Y., (Takasago Perfumery) *US Pat.*, 4691037, (1987); (b) Takaya, H., Ohta, T., Noyori, R., Yamada, N., Takesawa, T., Sayo, N., Taketomi, T., Kumobayashi, H. and Akutagawa, S., (Takasago Perfumery) US Pat. 4739084, 1988; (c) Takaya, H., Ohta, T., Noyori, R., Sayo, N., Kumobayashi, H., and Akutagawa, S (Takasago Perfumery) US Pat. 4739085, 1988.
14. Chan, A. S. C., and Laneman, S. A (Monsanto), US Pat. 5144050, 1992.
15. Chan, A. S. C., *US Pat.* 4994607, 1991.
16. Specification is $[\alpha]_D = 63$–$68.5\,°C$ (c = 1, $CHCl_3$), which corresponds to 95–100% ee: *United States Pharmacopeia*, US Pharmacopeial Convention, Rockville, MD, 1994, p. 1053.
17. (a) Wan, K. T., and Davis, M. E., *Nature (London)*, 370, 449 (1994); (b) Wan, K. T., and Davis, M. E., *J. Catal.*, **1**, 148 (1994).
18. RajanBabu, T. V., and Casalnuovo, A. L., *J. Am. Chem. Soc.*, **114**, 6265 (1992).
19. Nugent, W. A. and McKinney, R. J., *J. Org. Chem.*, **50**, 5370 (1985).
20. Collins, A. N., Sheldrake, G. N. and Crosby, J. (Eds), *Chirality in Industry*, Wiley, Chichester, 1992, Ch. 15.
21. *Guideline for Submitting Supporting Documentation in Drug Applications for the Manufacture of Drug Substances*, Food and Drug Administration, Rockville, MD, 1987.
22. (a) Nicholson, J. S. and Adams, S. S., (Boots Pure Drug Co.) *Br. Pat.* 971700, 1964; (b) Nicholson, J. S. and Adams, S. S., (Boots Pure Drug Co.), *US Pat.* 3228831, 1966.
23. Adams, S. S., Bresloff, P., and Mason, G., *Pharm. Pharmacol.*, **28**, 256 (1976).
24. (a) Adams, S. S., Cliffe, E. E., Lessel, B. and Nicholson, and J. S., *J. Pharm. Sci.*, **56**, 1686 (1967); (b) Mills, R. F. N., Adams, S. S., Cliffe, E. E., Dickinson, W., and Nicholson, J. S., *Xenobiotica*, **3**, 589, (1973).
25. Wechter, W. J., Loughhead, D. G., Reischer, R. J., Van Giessen, G. J., and Kaiser, D. G., *Biochem. Biophys. Res. Commun.*, **61**, 833 (1974).
26. (a) Sheldon, R. A., *Chirotechnology*, Marcel Dekker; New York, 1993, p. 57; (b)

Knihinicki, R. D., Day, R. O. and Williams, K. M., *Biochem. Pharmacol.*, **42**, 1905 (1991).
27. Rieu, J.-P., Boucherle, A., Cousse, H. and Mouzin, G., *Tetrahedron*, **42**, 4095 (1986).
28. Boots Pure Drug Co., *Fr. Pat.* 1 545 270, 1968.
29. Shimizu, I., Hirano, R., Matsumura, Y., Nomura, H., and Uchida, K. (Nippon Petrochemicals), *US Pat.*, 4 694 100, 1987.
30. (a) Shimizu, I., Matsumura, Y., Tokumoto, Y., and Uchida, K., (Nippon Petrochemicals) *US Pat.*, 5 097 061, 1992; (b) Tokumoto, Y., Shimizu, I., and Inoue, S. (Nippon Petrochemicals), *US Pat.*, 5 166 419, 1992.
31. (a) *Chem. Mkt. Rep.*, May 17 (243), 20 (1993); (b) *Chem. Mkt. Rep.*, Nov. 22 (244), 41 (1993).
32. Sunshine, A., and Laska, E. M., (Analgesic Associates) *US Pat.*, 4 851 444, 1989.
33. (Medice Chem.-Pharm.), *Eur. Pat. Appl.*, 267 321, 1986.
34. Manimaran, T., Wu, T.-C., Klobucar, W. D., Kolich, C. H., Stahly, G. P., Fronczek, F. R., and Watkins, S. E., *Organometallics*, **12**, 1467 (1993).
35. Fowler, R., Connor, H., and Bachl, R. A., *Chemtech*, 722 (1976).
36. Collman, J. P., Hegedus, L. S., Norton, J. R., and Finke, R. G., *Principles and Applications of Organotransition Metal Chemistry*, University Science Books, Mill Valley, CA, 1987, pp. 625–630.
37. Babin, J. E., and Whiteker, G. T., (Union Carbide) *PCT Int. Pat. Appl.*, WO 93/03839, 1993.
38. Takeda, M., Uchide, M. and Iwane, H., (Mitsibushi Petrochemical) *US. Pat.*, 4 329 507, 1982.
39. Alper, H. and Hamel, N., *J. Am. Chem. Soc.*, **112**, 2803 (1990).
40. (a) Larsen, R. D., Corley, E. G., Davis, P., Reider, P. J. and Grabowski, J. J., *J. Am. Chem. Soc.*, **111**, 7650 (1989); (b) Corley, E. G., Grabowski, J. J., Larsen, R. D., and Reider, P. (Merck), *US Pat.*, 4 940 813, 1990.
41. Sih, C. J., (University of Wisconsin) *Eur. Pat. Appl.*, 227 078, 1986.
42. (a) Matson, S. L. (Sepracor), *US. Pat.*, 4 800 162, 1989; (b) Wald, S. A., Matson, S. L., Zepp, C. M., and Dodds, D. R., (Sepracor) *US Pat.*, 5 057 427, 1991.
43. (a) Bertola, M. A., Marx, A. F., Koger, H. S., Quax, W. J., Van der Laken, C. J., Phillips, G. T., Robertson, B. W., and Watts, P. O., (Gist-Brocades) *US Pat.*, 4 886 750, 1989; (b) Bertola, M. A., De Smet, M. J., Marx, A. F., and Phillips, G. T., (Gist-Brocades) *US Pat.*, 5 108 917, 1992.
44. (a) Cobbs, C. S., Barton, M. J., Peng, L., Goswami, A., Malick, A. P., Hamman, J. P. and Calton, G. J., (Rhône–Poulenc) *US Pat.*, 5 108 916, 1992; (b) Goswami, A., *US Pat.*, 5 175 100, 1992.
45. *Chem. Eng. News*, Sept. 28, 67 (1992).
46. *Chem. Eng. News*, Sept. 19, 72 (1994).
47. (a) Larsen, R. D. and Reider, P. (Merck), *US Pat.*, 4 946 997, 1990; (b) Manimaran, T. and Impastato, F. J. (Ethyl Corp.), *US Pat.*, 5 015 764, 1991; (c) Choudhury, A. A., Kadkhodayan, A. and Patil, D. R. (Ethyl Corp.), *US Pat.*, 5 221 765, 1993; (d) Lin, R. W. (Ethyl Corp.), *US Pat.*, 5 278 334, 1994; (e) Trace, R. L. (Ethyl Corp.), *US Pat.*, 5 278 338, 1994.
48. Tung, H.-H., Waterson, S. and Reynolds, S. D. (Merck), *US Pat.*, 4 994 604, 1991.
49. (a) Nohira, H., *Yuki Gosei Kagaku*, **50**, 14 (1992); (b) Nohira, H. (Nagasi & Co.), *US Pat.*, 5 321 154, 1994.
50. Dwivedi, S. K., Sattari, S., Jamali, F. and Mitchell, A. G., *Int. J. Pharm.*, **87**, 95 (1992).
51. (a) Stinson, S. C., *Chem. Eng. News*, Sept. 28, 46 (1992); (b) Stinson, S. C., *Chem. Eng. News*, Sept. 19, 38 (1994).
52. DiCicco, R. L. in *Proceedings Chiral '93 USA Symposium*, Reston, VA, Spring Innovations, Stockport, 1993, p. 5.

53. (a) Barberich, T. J., and Young, J. W. (Sepracor), *PCT Int. Pat. Appl.*, WO 93 20 809, (1993); (b) Gray, N. M., Wechter, W. J., and Young, J. W (Sepracor), *US Pat.*, 5 331 000, 1994.
54. (a) Wechter, W. J., *PCT Int. Pat. Appl.*, WO 91 02 512, 1991; (b) Wechter, W. J., (Sepracor), *US. Pat.*, 5 190 981, (1993).
55. (a) Laska, E. H. (Miles and Analgesic Associates), *US Pat.*, 4 868 214, 1989; (b) Laska, E. M., Rees, J. A. and Sunshine, A. (Analgesic Associates), *US Pat.*, 4 927 854, 1990; (c) Laska, E. M., Sunshine, A. and Laska, E. (Miles and Analgesic Associates), *US. Pat.*, 4 962 124, 1990; (d) Laska, E. M. and Sunshine, A. (Analgesic Associates), *US Pat.*, 5 286 751, 1994.

# 4 Synthesis of Enantiomerically Pure Nucleosides

**B. L. BRAY[b], M. D. GOODYEAR[a], J. J. PARTRIDGE[b] and D. J. TAPOLCZAY[a]**
*Glaxo Wellcome Chemical Development Division, [a]Medicines Research Centre, Stevenage, Hertfordshire, UK, and [b]Girolami Research Centre, Research Triangle Park, NC, USA*

| 4.1 | Introduction | 41 |
|---|---|---|
| 4.2 | GR85478 | 42 |
| | 4.2.1 Synthesis from (−)-Carbovir | 42 |
| | 4.2.2 Synthesis from (+)-(1R, cis)- Cyclopent-4-ene-1,3-diol Monoacetate | 47 |
| 4.3 | GR95168 | 50 |
| | 4.3.1 Resolution Approaches | 54 |
| | 4.3.2 Use of Palladium-catalysed Functionalisation of (+)-Hydroxyacetate **5** | 60 |
| 4.4 | Lamivudine | 66 |
| | 4.4.1 Synthesis of the Racemate | 67 |
| | 4.4.2 Resolution of Racemic Lamivudine | 69 |
| | 4.4.3 Alternative Strategies | 71 |
| | 4.4.4 A Novel Synthesis of Lamivudine | 72 |
| 4.5 | Conclusions and Summary | 77 |
| 4.6 | References | 77 |

## 4.1 INTRODUCTION

Anti-viral research has been one of the most intensive areas of activity within both the pharmaceutical industry and the academic sector for the past 10 years, and the management of viral diseases still remains a focus for many drug companies. During this time, a wealth of chemistry has been discovered and developed; this chapter charts the development at Glaxo Wellcome of routes to the three compounds **1–3**, each of which has potent activity against a range of viruses.

It is the authors' aim to give the reader the background to the discovery and development of synthetic routes to these compounds, each of which contains at least two stereogenic centres. Each of these drug candidates was required *as a*

*Chirality in Industry II.* Edited by A. N. Collins, G. N. Sheldrake and J. Crosby
© 1997 John Wiley & Sons Ltd

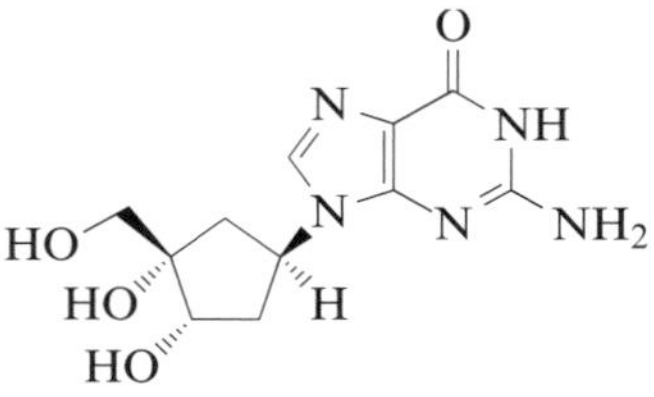

**(1)** GR85478

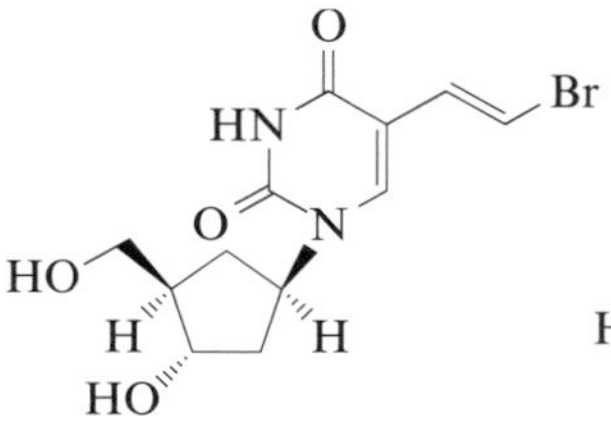

**(2)** GR95168

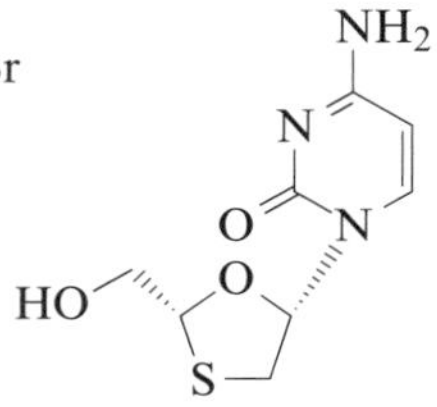

**(3)** Lamivudine

*single enantiomer*, potentially in tonne quantities. Their syntheses are not trivial; the fact that all three are nucleoside analogues prompted some similarities in the synthetic approaches used, but each also posed its own specific challenges. These similarities and differences are explored in this chapter.

## 4.2 GR85478

Herpes simplex viruses exist in two major forms, known as HSV1 and HSV2; infections range from being mildly uncomfortable to extremely painful. The form of infection most people are aware of is HSV1, which manifests itself as cold sores. The compound discussed in this section [GR85478 (**1**)] showed potent activity against both forms of herpes simplex virus;[1] we describe here its synthesis by two routes, the first using a starting material from the chiral pool [the well known anti-viral agent carbovir[2] (**4**)] and the second from the readily available (+)-hydroxyacetate (**5**), obtained from an enzymatic resolution.

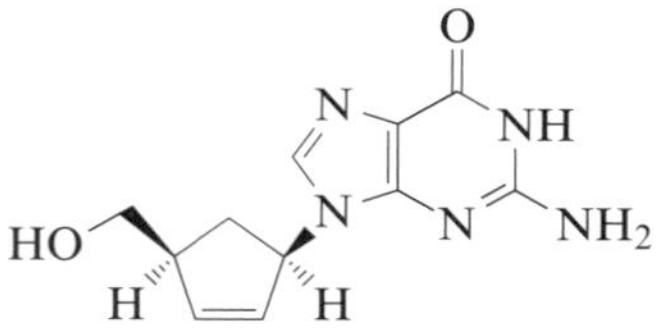

**(4)** (−)-Carbovir

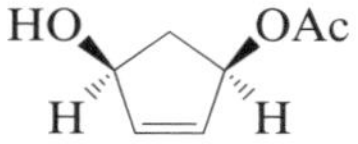

**(5)**

### 4.2.1 SYNTHESIS FROM (−)-CARBOVIR

Conversion of (−)-carbovir (**4**) to GR85478 (**1**), both of which contain a guanine fragment, requires an oxidation at C-4, migration of a double bond, reduction to the corresponding allylic alcohol and selective α-dihydroxylation (Scheme 1).[3]

Careful selection of a protecting group for the guanine moiety of (−)-carbovir (**4**) proved to be critical in this synthesis; the formamidine group seemed the best, and no problems were encountered with oxygen vs nitrogen selectivity (as is

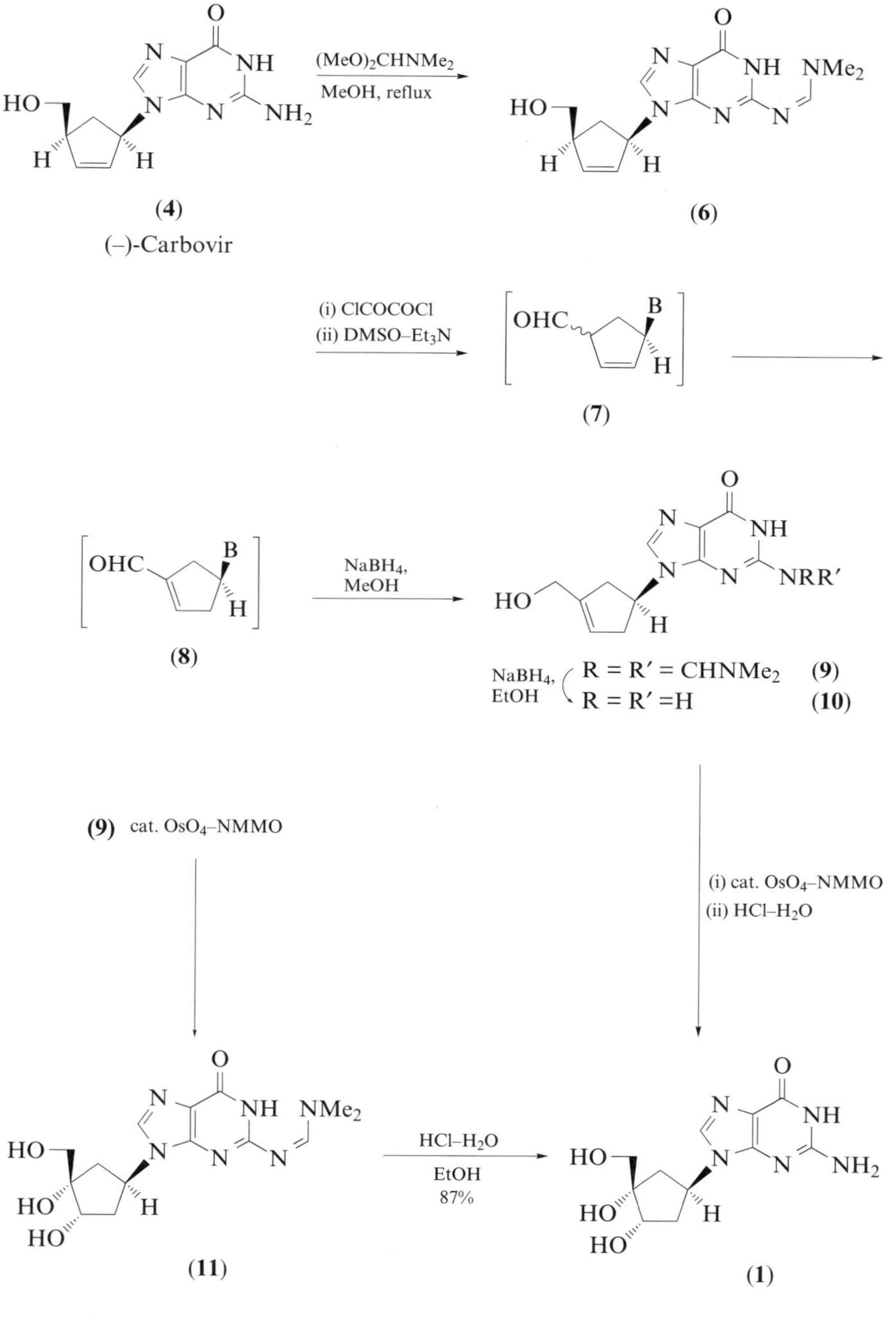

# SYNTHESIS OF ENANTIOMERICALLY PURE NUCLEOSIDES

**(4)** (−)-Carbovir

**Scheme 1**

common with protection of amide, lactam or carbamate groups). Importantly, the organic solubility of the carbocyclic nucleoside was greatly enhanced without significant mass increase using formamidine protection. Treatment of (−)-carbovir with dimethylformamide dimethylacetal in methanol at reflux gave the formamidine **6** in 99% yield. Selective oxidation of the C-4 hydroxymethylene group of **6** using standard Swern conditions[4] provided the $\alpha,\beta$-unsaturated aldehyde **8** in 82% yield. The *cis*-aldehyde initially formed in the oxidation rapidly isomerised to a 1:1 mixture of *cis*- and *trans*-$\beta,\gamma$-unsaturated aldehydes **7** at −50 °C. As the solution warmed to 0 °C, the *cis–trans* mixture (**7**) was completely converted into the desired $\alpha,\beta$-unsaturated aldehyde **8**; the oxidation and isomerisation processes were complete within 1 h. The *cis-trans* mixture of aldehydes (**7**) could be isolated at low temperature, but the mixture was found to be very air sensitive and decomposed at room temperature in a few hours. Oxidation of the C-4 hydroxymethylene group of unprotected (−)-carbovir (**4**) under the same conditions gave a low yield (<15%) of the corresponding unprotected $\alpha,\beta$-unsaturated aldehyde. The low yield was thought to be the result of several factors, such as poor solubility, oxidation at the C-6 amino group and Schiff base formation. Reduction of aldehyde **8** with 0.25 equivalents of sodium borohydride in methanol at 0 °C gave allylic alcohol **9**; the use of a full equivalent of sodium borohydride in ethanol at 25 °C was found to reduce aldehyde **8** directly to the unprotected allylic alcohol **10**.

To complete the synthesis of **1**, a stereofacially-selective *cis*-dihydroxylation of either allylic alcohol **9** or **10** *anti* to the guanine ring was required.[5] Molecular models indicate that osmium tetraoxide-catalysed *cis*-dihydroxylation should occur selectively on the *anti* face (Figure 1); however, from literature precedent, it is clear that a combination of electronic and steric factors generally govern the selectivity of these reactions.

Surprisingly, the first reaction (Scheme 2, **12 → 13**) suggested that the osmium tetraoxide was being attracted to the more sterically congested face of the mole-

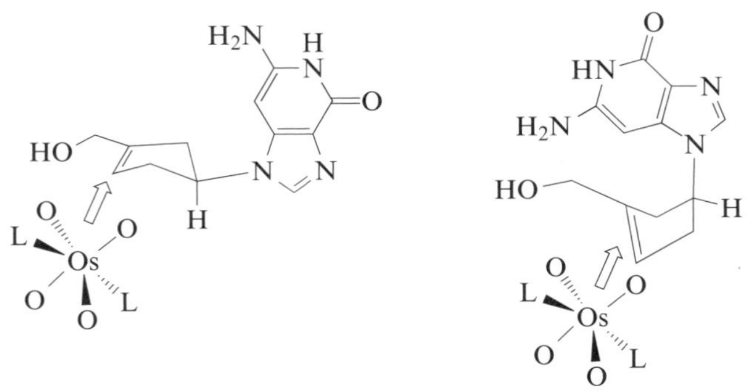

**Figure 1** Osmium tetraoxide-catalysed *cis*-dihydroxylation of allylic alcohol **10**

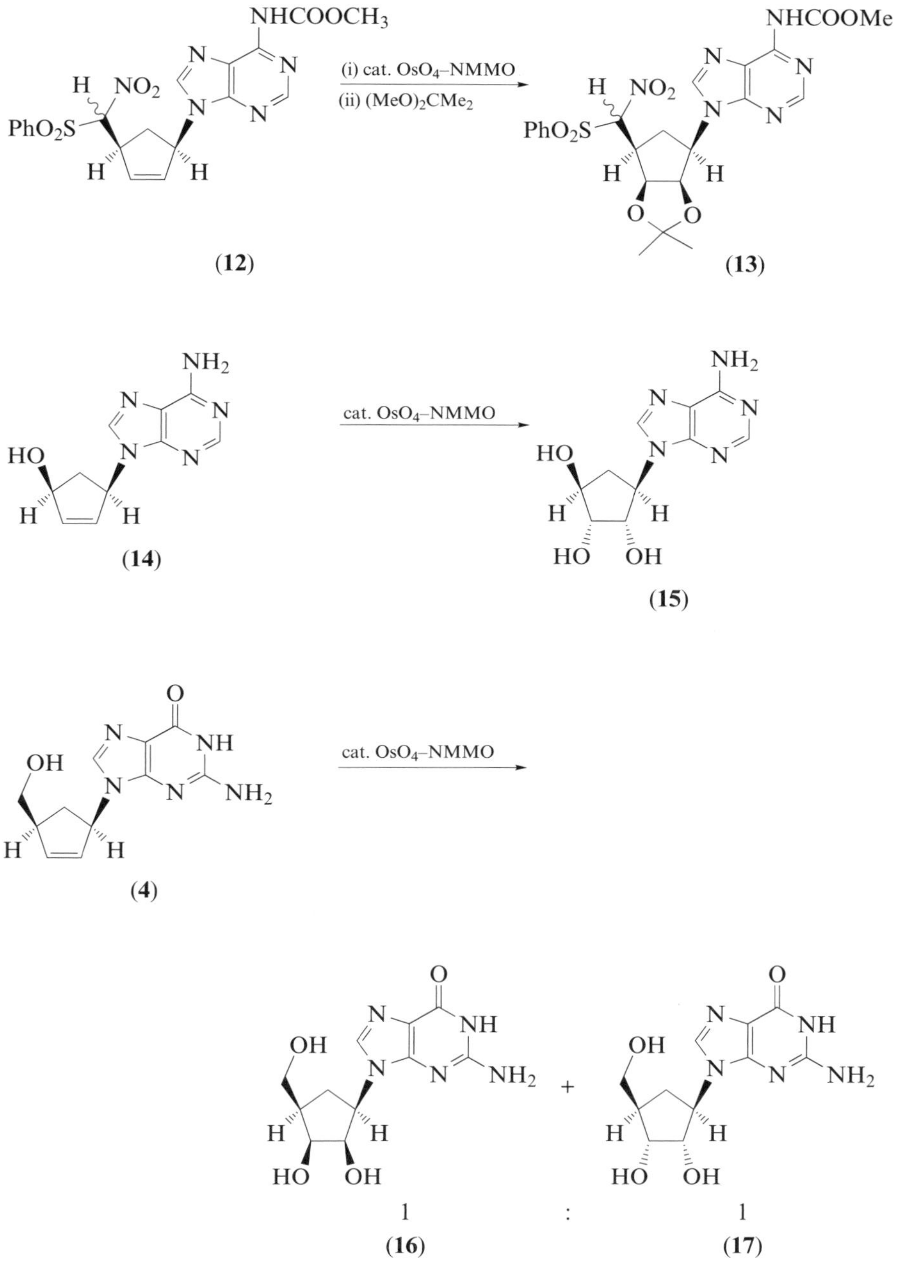

# SYNTHESIS OF ENANTIOMERICALLY PURE NUCLEOSIDES

**Scheme 2** (*continued*)

**Scheme 2** (*continued*)

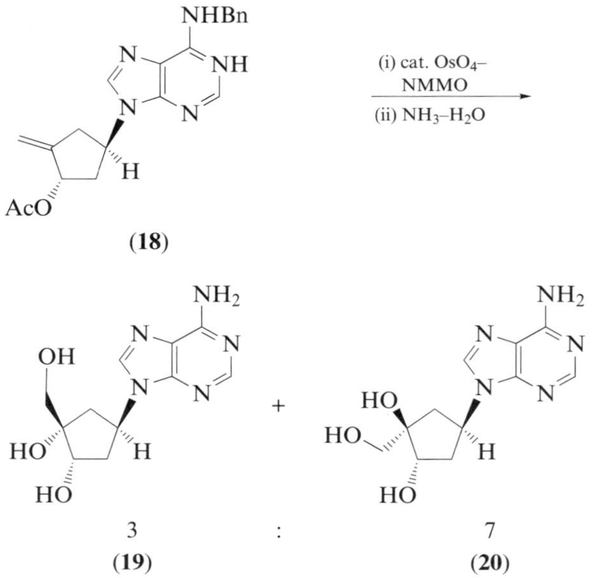

**Scheme 2**

cule by the nitrophenyl sulphone functionality.[6] In the second reaction (**14 → 15**), selective approach of the osmium tetraoxide to the least hindered face of the molecule occurred, as would be predicted,[6] assuming no directing effect by the hydroxyl group. When (−)-carbovir (**4**) was treated with catalytic osmium tetraoxide and *N*-methylmorpholine *N*-oxide in water–acetone at 70 °C, no facial selectivity was obtained (**4 → 16 + 17**)[7]. The final example (**18 → 19 + 20**) illustrates the lack of stereofacial selectivity obtained when both faces of the cyclopentane exocyclic methylene on the cyclopentane ring are hindered.[8]

Treatment of allylic alcohol **9** with 1.5 mol% osmium tetraoxide and *N*-methylmorpholine *N*-oxide in water–acetone at 70 °C for 2 h. provided only the desired *anti*-dihydroxylation product (**11**)[3]. It is unclear why α-dihydroxylation on this allylic alcohol substrate proceeded with total *anti*-selectivity whereas, under identical conditions, the homoallylic alcohol, carbovir (**4**), displayed no selectivity at all. The formamidine protecting group on the guanine base of **9** had no effect on the rate or selectivity of the reaction; gratifyingly, the deprotected allylic alcohol **10** also afforded only the desired *anti*-dihydroxylated product (**1**), under identical conditions.

Although the catalytic *cis*-α-dihydroxylation process worked exceedingly well, removal of all traces of osmium from this water-soluble drug posed a challenge; an acceptable limit for residual osmium in the final drug **1** was not greater than

10 ppm. The guanine ring, like pyridine and other aromatic amines, binds osmium fairly strongly; in fact, osmium tetraoxide is commonly used as a biological staining tool in DNA research. Reduction of the residual osmium level to below 10 ppm required selective and efficient conditions. Pyridine was added to the reaction mixture after hydroxylation of allylic alcohol **10** was complete, then the solution was saturated with hydrogen sulphide and stirred at 70 °C for up to 70 h. Most of the osmium was precipitated as a fine, dark solid, along with sulfur, which was formed as a by-product. The solids were filtered off, the solvent volume was reduced and free base **1** was isolated by direct crystallisation. However, the solid at this stage generally still contained 50–100 ppm of osmium. To remove these final traces, the solid was dissolved in 0.5 M hydrochloric acid, warmed to 70 °C, saturated with hydrogen sulphide and stirred for 24 h. Residual osmium was precipitated and filtered off and the product (**1**) was isolated by crystallisation from the filtrate in 66% overall yield.[9] The residual osmium level after this exhaustive treatment was less than 2 ppm. Interestingly, if either the treatment with hydrochloric acid–hydrogen sulphide or with pyridine–hydrogen sulphide was performed twice, rather than once each in sequence, the residual osmium levels were not reduced below 20 ppm. Although the use of hydrogen sulphide is not desirable on a large scale, this method enabled us to prepare several kilograms of clinical-grade product **1**, essentially free from osmium contamination.

### 4.2.2 SYNTHESIS FROM (+)-(1R, cis)-CYCLOPENT-4-ENE-1,3-DIOL MONOACETATE

The second approach to this molecule used (+)-(1R, cis)-cyclopent-4-ene-1,3-diol monoacetate (**5**) or its enantiomer, which are both commercially available, but expensive.[10] However, either enantiomer can be obtained by an enzyme-catalysed desymmetrisation of cyclopent-4-ene-1,3-diol diacetate (**21**)[11] (Scheme 3).

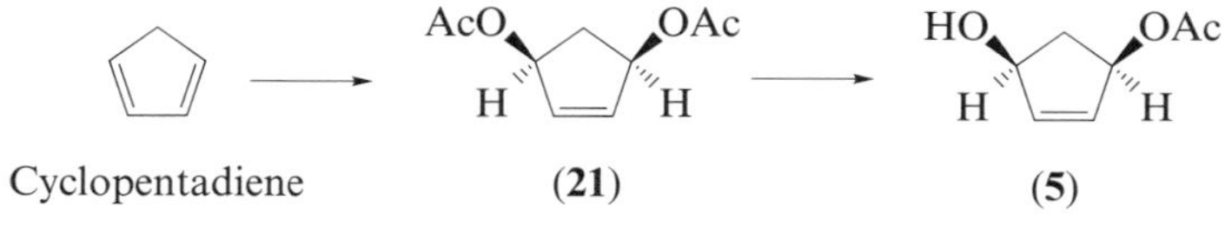

Cyclopentadiene → (**21**) AcO, OAc → (**5**) HO, OAc

**Scheme 3**

In work towards an asymmetric synthesis of (−)-carbovir (**4**), we had found that the pre-formed potassium or sodium salt of 2-amino-6-chloropurine (**23**) can be efficiently alkylated at N-9 with dicarbonate **22** using palladium(0) catalysis to give the *cis*-adduct **24**.[12] Using the dihydroxylation chemistry described above, a strategy for the synthesis of the target compound **1** from (+)-hydroxyacetate **5** would simply require the addition of a one-carbon fragment to **25**, and conversion into **10**, followed by dihydroxylation giving **1** (Scheme 4).

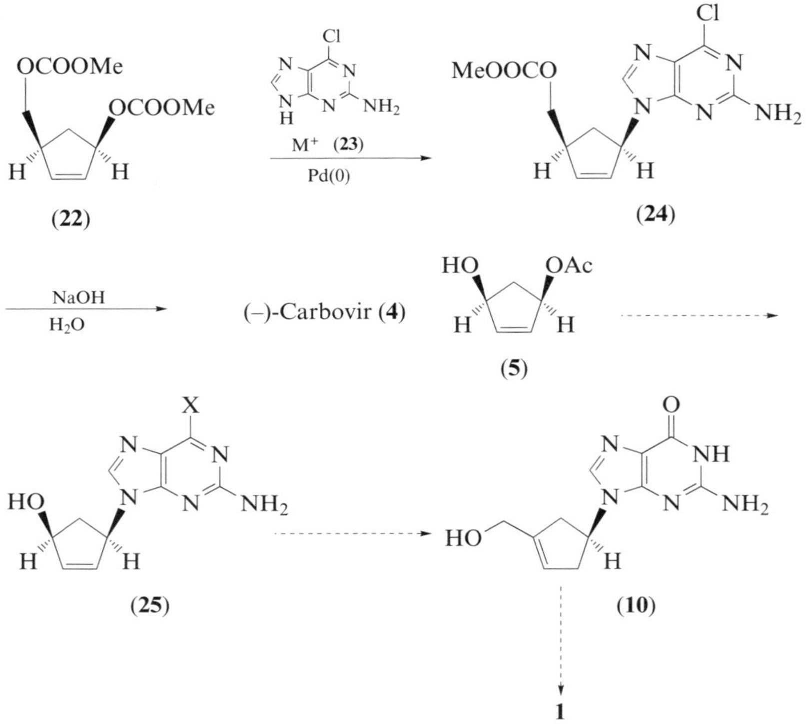

**Scheme 4**

Kilogram quantities of cyclopent-4-ene-1,3-diol diacetate (**21**) were prepared as described in the literature[11d,g] and desymmetrisation was accomplished using a *Pseudomonas* lipase immobilised between membranes in a Sepracor membrane bioreactor (Figure 2).[13] Using this technology, several kilograms of (+)-hydroxyacetate **5** were prepared using less than 0.5g of purified enzyme.[14] The four-stage process, which included a final crystallisation, gave (+)-hydroxyacetate **5** in an overall yield of 30–35%, with an enantiomeric excess of >99% as determined by chiral capillary GC.

An economic plant process requires that palladium(0) is used only in catalytic quantities, and there are two means of regenerating it from the palladium–cyclopentenol π-allyl adduct **26** (Scheme 5).[15] In the presence of a soft nucleophile, palladium(0) is displaced to give the desired alkylation product **27** and palladium(0) (path A). In the presence of a strong base (or no nucleophile at all)

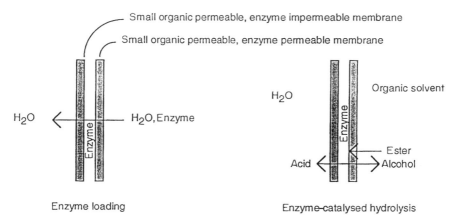

**Figure 2** Membrane bioreactor

elimination is favoured to give, ultimately, the $\beta,\gamma$-cyclopentenone **28**, protonated base or acetic acid and palladium(0) (path B). The sodium salt of 2-amino-6-chloropurine (**23**, M = Na) is not a particularly good nucleophile and competition from path B was difficult to avoid. It was important to add (+)-hydroxyacetate **5** to a solution of pre-formed catalyst and the salt of 2-amino-6-chloropurine. On a large scale, a preformed solution of catalyst and (+)-hydroxyacetate **5** suffered significant degradation by reaction path B during the time it took to add the salt of 2-amino-6-chloropurine. The 1,8-diazabicyclo[5.4.0]undec-7-ene (DBU) salt of compound **23** gave the cleanest reaction.

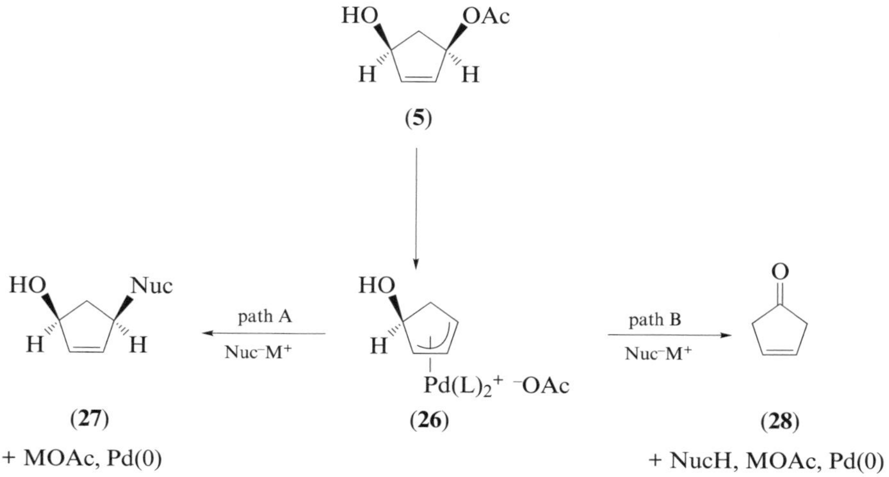

**Scheme 5**

The adduct of chloropurine **23** and (+)-hydroxyacetate **5** was difficult to crystallise (gels were formed), but formamidine **29**, formed by treatment of the crude product with dimethylformamide dimethylacetal, could be isolated as a crystalline solid in 45–51% yield [based on (+)-hydroxyacetate **5**] on a large scale.[16] Treatment of allylic alcohol **29** with methanesulphonic anhydride and pyridine at 0 °C afforded a reactive mesylate **30** which could not be isolated without extensive decomposition. Addition of tetraethylammonium cyanide resulted in cyanide displacement of the mesylate, giving a 1:1 mixture of *cis*- and *trans*-nitriles (**31**). Treatment of this mixture with DBU caused the $\beta,\gamma$-double bonds to migrate into conjugation, affording the desired $\alpha,\beta$-unsaturated nitrile (**32**). Milder bases, such as triethylamine, did not effect this isomerisation process. The mesylate group of **30** was readily displaced with water and, as tetraalkylammonium cyanides are extremely hygroscopic salts and often contain water, the reproducibility of this cyanide displacement was poor. Several cyanide salts were evaluated, but poor solubility often resulted in slow rates of reaction. This problem was overcome by generating an anhydrous tetraalkylammonium cyanide from acetone cyanohydrin in the presence of DBU.[17] Trialkylamines, such as triethylamine, slowly liberate hydrogen cyanide from acetone cyanohydrin, but $^1$H NMR studies showed that the reaction rate using DBU was very fast, and the equilibrium was driven by consumption of cyanide.[17] The salt of DBU and hydrogen cyanide is freely soluble in ethers, acetone and chlorinated hydrocarbons and is a powerful source of nucleophilic cyanide anion.

Treatment of the crude mesylate **30** with six equivalents of acetone cyanohydrin followed by five equivalents of DBU at 0 °C gave a 1:1 mixture of *cis*- and *trans*-nitriles (**31**); on warming to 23 °C, the double bonds spontaneously migrated into conjugation with the nitrile (**31** → **32**), even in the presence of excess acetone cyanohydrin. Treatment of the crude $\alpha,\beta$-unsaturated nitrile **32** with 3M hydrochloric acid removed the formamidine protecting group and hydrolysed the chloropurine to a guanine ring. The resulting $\alpha,\beta$-unsaturated nitrile (**33**) was readily isolated and purified by crystallisation. The complex series of conversions (**29** → **33**) was carried out in two stages in 60% overall yield. Conversion of the $\alpha,\beta$-unsaturated nitrile **33** into the $\alpha,\beta$-unsaturated *n*-propyl ester **34** was carried out in propan-1-ol in the presence of one equivalent of sulphuric acid in 65% yield after crystallisation. Reduction of this ester with DIBAL provided allylic alcohol **10** in high yield without affecting the guanine ring. The desired compound (**1**) was prepared by osmium-catalysed *cis*-dihydroxylation as previously described. This synthesis (see Scheme 6) was superior to the route from carbovir (Scheme 2), and was selected as our method of choice for large-scale manufacture.

## 4.3. GR95168

Varicella zoster (VZV) is the virus responsible for the diseases chickenpox and shingles. People who contracted chickenpox as children have the virus lying dor-

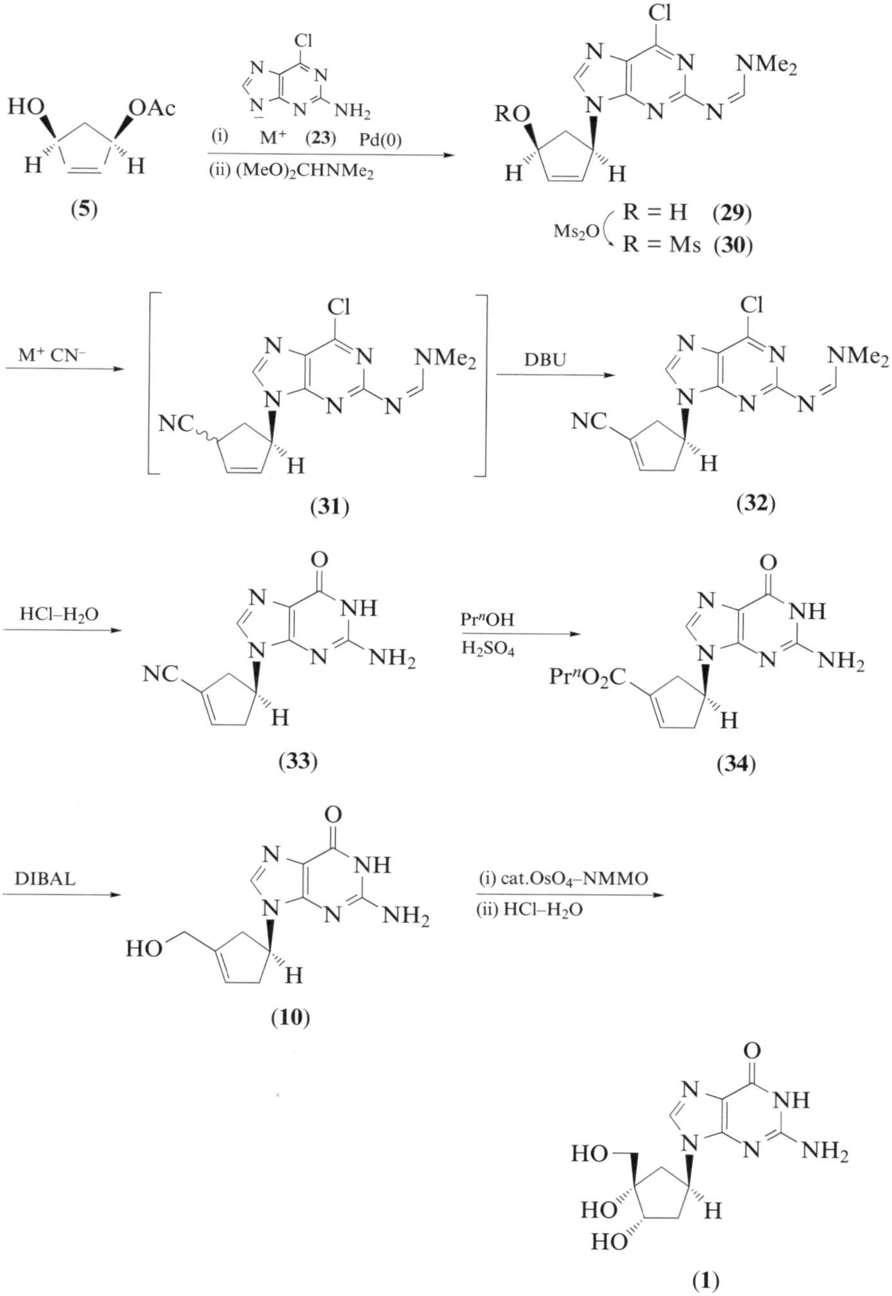

# SYNTHESIS OF ENANTIOMERICALLY PURE NUCLEOSIDES

**Scheme 6**

mant within a large nerve bundle known as the dorsal ganglion. In adults, a mechanism which, as yet, is unknown can lead to the reactivation of this dormant virus, which manifests itself as shingles. This disease is particularly prevalent in the elderly; statistics suggest that 1 in 20 people over the age of 60 will contract it. It is an unpleasant disease that is usually accompanied by skin lesions, nerve damage and an extremely painful neuralgia that may persist long after the visible signs of the disease have disappeared. Shingles affects the quality of life of all the people who contract it; in immunocompromised or very elderly patients, the disease can be fatal. Along with other pharmaceutical companies, Glaxo Wellcome has been actively searching for a treatment for VZV, and compound 2 was shown to have outstanding activity. Two synthetic routes are discussed,[18] the first using a starting material from the chiral pool (lactam 35) and the second from (+)-hydroxyacetate 5 using chemistry developed during work on GR85478 (described in the previous section).

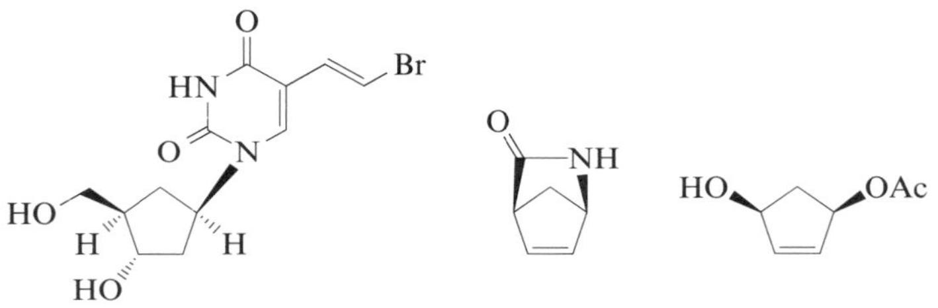

(2)
GR95168

(35)

(5)

The racemate of GR95168, carbocyclic bromovinyldeoxyuridine, was discovered in 1984 as a result of a collaboration between Glaxo Wellcome and Southampton University.[19] It was subsequently shown that the majority of the activity against varicella zoster virus lay in the single enantiomer (2) shown above.[20] This was developed as a potential therapy for shingles and, as a result, large quantities of the enantiopure drug candidate were required for safety studies. The initial synthesis used at Glaxo Wellcome is shown in Scheme 7.[20b]

There are a number of drawbacks to this route as a potential means of synthesis of large quantities of drug substance:

- Column chromatography was required to purify most of the intermediates (which were all non-crystalline).
- The use of diisopinylcamphenylborane gave separation problems during isolation of the reaction product.
- The use of vanadium in stoichiometric quantities would give significant problems in handling waste streams and ensuring that the product was free from contamination.
- The overall yield was low (owing to two low-yielding stages and the large number of steps).

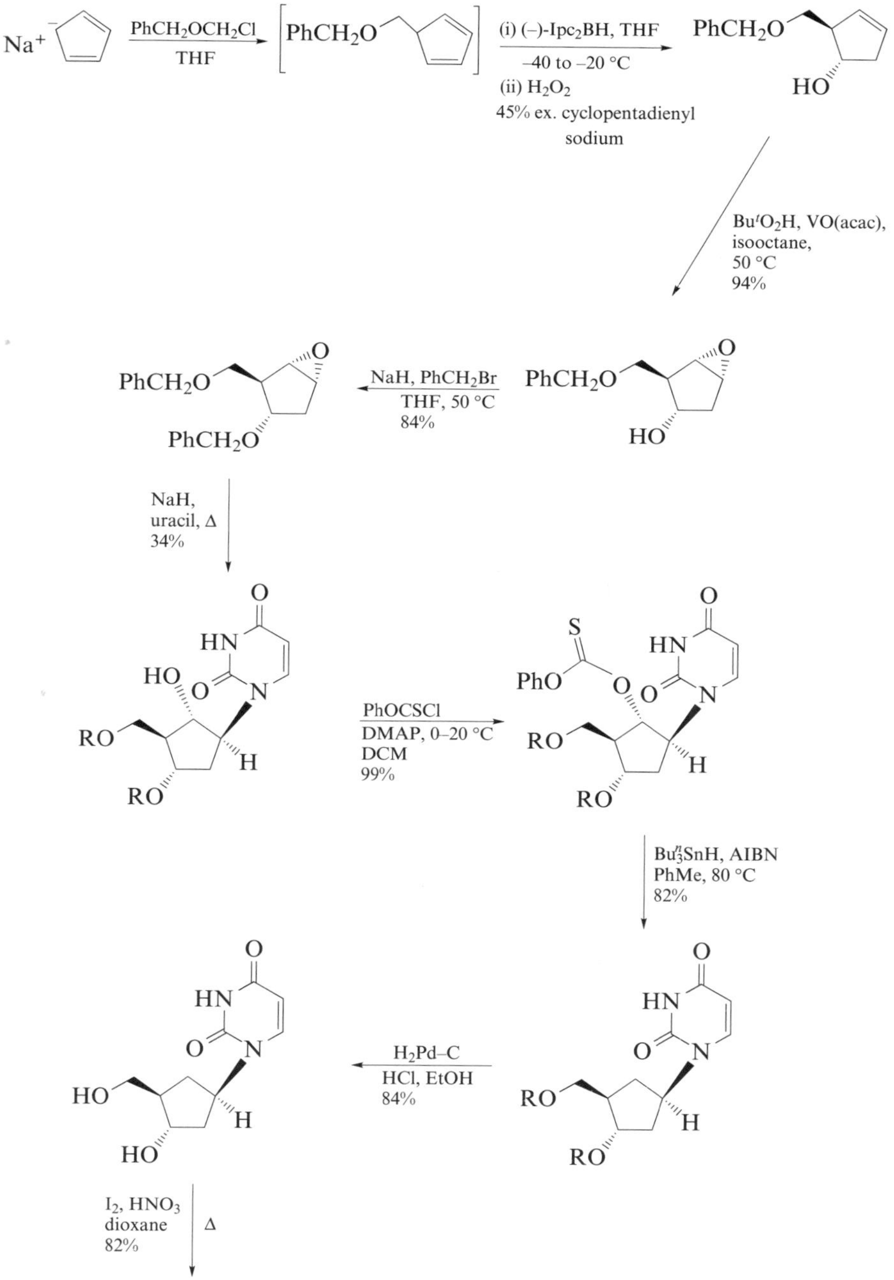

# SYNTHESIS OF ENANTIOMERICALLY PURE NUCLEOSIDES

**Scheme 7** (*continued*)

**Scheme 7** (*continued*)

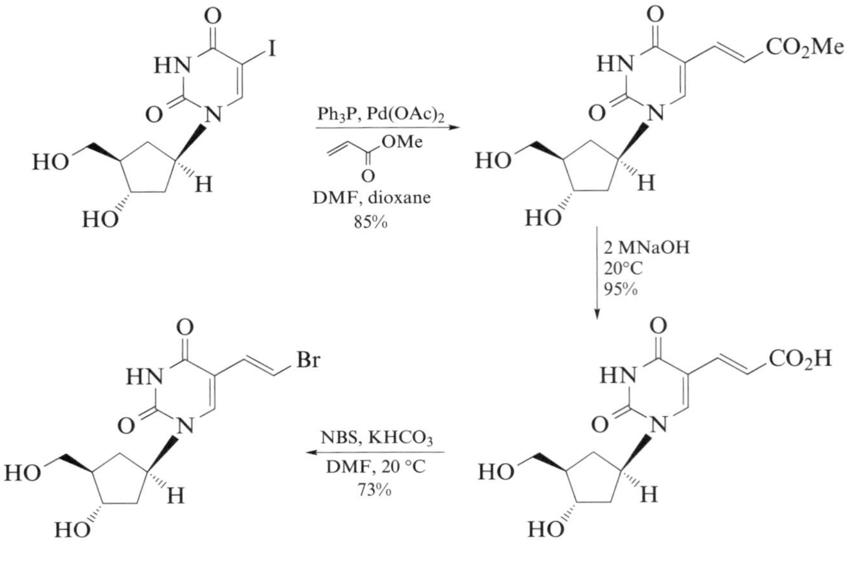

### 4.3.1 RESOLUTION APPROACHES

Our first approach was to use a racemic starting material (**35**),[21] containing the required relative stereochemistry, and to employ a resolution of the ring-opened derivative (**40**) to obtain the desired enantiomer. This approach is shown in Scheme 8.

The secondary hydroxyl group would then be introduced either by hydroboration or by epoxidation and regioselective ring opening. However, all attempts at the selective functionalisation of the 3-position of the cyclopentene (**38 → 37**) were unsuccessful; epoxidation of lactam **35** gave the desired α-epoxide, but all attempts at regioselective ring opening led to inseparable mixtures of isomers. Epoxidation of ring opened derivatives of **35** was less face-selective, and also not suitable. The hydroboration approach also failed, giving mixtures of 3α- and 2β-hydroboration products. Consequently, these approaches were abandoned and the alternative strategy shown in Scheme 9 was proposed.

This approach differs from that shown in Scheme 8 in the oxidation state of the substituent at C-4. A suitably protected racemic lactam is ring opened, resolved and esterified to give the unconjugated ester **43** as a single enantiomer. The double bond would then be moved into conjugation to form the α,β-unsaturated ester (**42**). This contains a trisubstituted alkene within the five-membered ring, and it is well documented that hydroboration of such systems

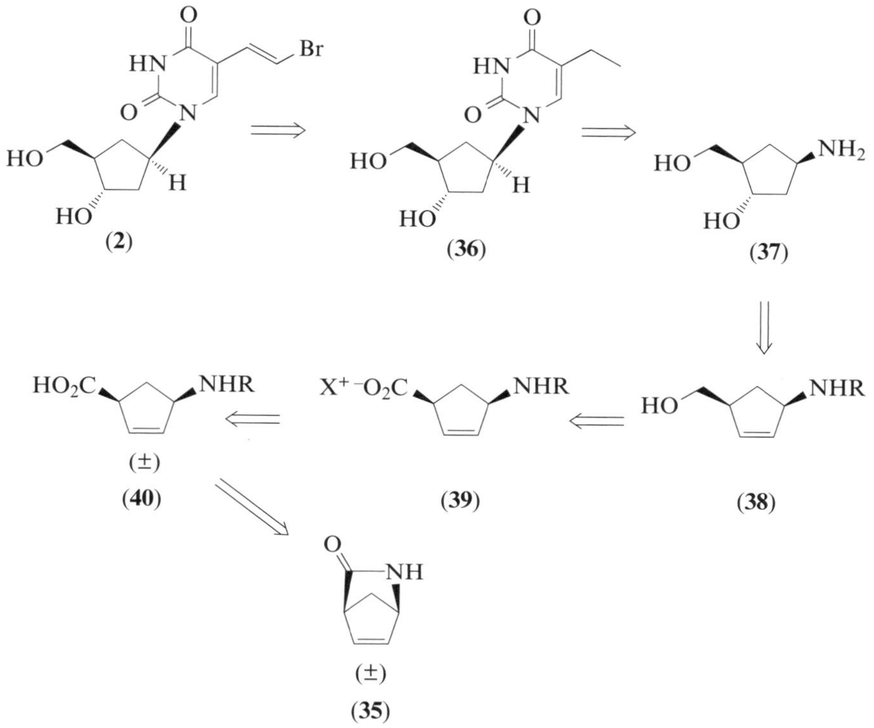

# SYNTHESIS OF ENANTIOMERICALLY PURE NUCLEOSIDES

**Scheme 8**

can be achieved with excellent regiochemical control, although the degree of stereochemical control is less predictable.[22] By using a hindered borane, we envisaged that it might be possible to achieve some degree of selectivity via the approach of the borane from the less hindered α-face; if the amine group, suitably protected, reacted with the hydroborating agent faster than the double bond, then the steric bulk of the top face of the cyclopentane should be increased sufficiently to achieve a very high degree of face-selectivity on addition of a second equivalent of the borane. Brown et al.[23] have reported that tertiary amides are reduced to amines by dialkyl boranes; furthermore, protection of the lactam as an amide was likely to facilitate the ring opening prior to resolution of the free acid. Consequently, we chose to protect lactam **35** as its benzamide, as shown in Scheme 10.

Lactam **35** had previously been prepared by reaction of cyclopentadiene with p-toluenesulphonyl cyanide.[21b] In view of the extremely hazardous nature of the latter, and the need for a synthesis capable of operation on a large scale, we decided to switch to the safer and more readily available chlorosulphonyl isocyanate, which gave racemic lactam **35** as a crystalline solid in 35% yield. The lactam was

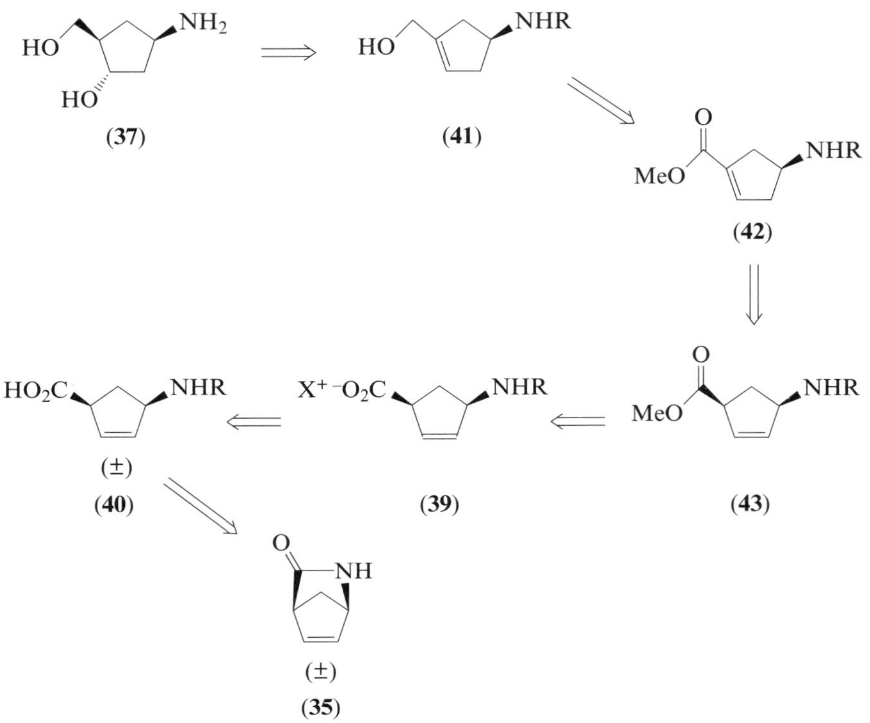

Scheme 9

protected as its benzamide (**44**) and hydrolysed with dilute sulphuric acid giving racemic acid (**40**) as a crystalline solid in 76% overall yield.

Several commercially available chiral amines were screened against the racemic acid in an attempt to obtain a diastereoisomerically pure crystalline salt; fortunately, the cheap and readily available (*R*)-α-methylbenzylamine gave the required diastereoisomer (**39**) as a crystalline salt which could be isolated directly from the reaction mixture in excellent purity (43% yield, 99% *de*, as determined by HPLC). Treatment of this salt with aqueous base followed by methanolic hydrochloric acid gave the enantiopure unconjugated benzamido ester **43** (90% yield, 99% *ee*). The double bond was moved into conjugation to give ester **42** by treatment with DBU in dichloromethane in 90% yield. The use of this solvent in production is not desirable owing to its environmental effects and possible long-term lack of availability. Ethyl acetate proved a suitable replacement.

With the conjugated ester **42** in hand it only remained to reduce the ester to the allylic alcohol and protect in order to evaluate the proposed hydroboration step. This conversion proved far more difficult than first imagined and led to the development of a novel chemoselective reduction.[24] Lithium aluminium hydride gave

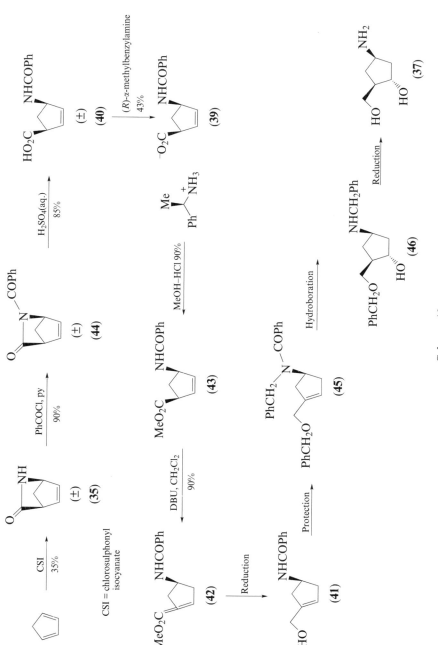

Scheme 10

a lack of selectivity between 1,4- and 1,2-reduction; the same problems occurred with modified borohydride reducing agents. However, DIBAL at $-70\,°C$ gave the required allylic alcohol **41** in reasonable yield (56%). The low yield was due to a side-reaction, in which the benzamide group was also reduced giving amine **47**.

HO⌒⌒NHCH$_2$Ph

(**47**)

Despite numerous changes in reaction conditions, the ratio of the products could not be altered in favour of the desired allylic alcohol **41**. In spite of the low yield, the product could be isolated directly from the reaction mixture, and the process was suitable for scale-up, although it required low temperatures. The main problem was the lack of chemoselectivity and resulting low yield. It was reasoned that the reduction of the amide was occurring as a result of the coordination of DIBAL to the amide as well as to the ester. Based on electronic considerations, the rate of reduction of the amide should be slower than that of the ester; it was hypothesised that selective coordination of the amide carbonyl to some other species, for example a Lewis acid, would prevent subsequent coordination of DIBAL, and thus the reduction of the amide. On adding one equivalent of aluminium chloride, $^1$H NMR spectroscopy showed chelation of the Lewis acid mainly to the amide. On treating this pre-formed complex with DIBAL, we were pleased to find a considerable improvement in yield of **41** to 93%. Furthermore, the reaction could be carried out at $0\,°C$ without a decrease in selectivity. We were now in a position to scale this stage up to pilot plant when required.

Protection of the alcohol and activation of the amide had been carried out using sodium hydride and benzyl bromide in dimethylformamide, the product (**45**) from this stage being isolated as a viscous oil. This procedure was not suitable for scale-up; sodium hydride in DMF is well known as a potentially hazardous mixture and, even if another solvent were used, the oil present in commercial sodium hydride dispersions can be difficult to remove and may contaminate the product. Finally, isolation of the product as a viscous oil is not practical on a large scale. Briefly, the hydroboration of **45** *in situ* was explored, but poor yields were obtained as it was difficult to estimate the correct quantity of borane required. Sodium hydride–DMF was successfully replaced by a phase-transfer system, using toluene and aqueous sodium hydroxide, from which compound **45** could be isolated as a crystalline solid in 90% yield.

We were now in a position to attempt the key stage, for which the reagent diisoamylborane was chosen, as it is known to give good regiocontrol in the hydroboration of trisubstituted alkenes and a reasonable rate of reaction.[25] The reagent is not commercially available but can be readily prepared and used *in situ*. Good control of stereochemistry requires reduction of the amide prior to hydroboration of the alkene. On adding one equivalent of the pre-formed diiso-

# SYNTHESIS OF ENANTIOMERICALLY PURE NUCLEOSIDES

amylborane at 0 °C, we obtained a good yield of amine **48**, indicating a suitable difference in rates of reduction and hydroboration.

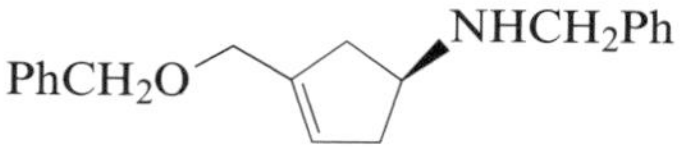

(48) 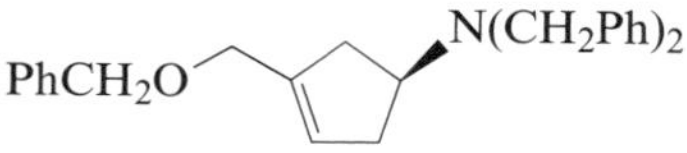   (49)

We repeated the reaction with two equivalents of diisoamylborane at 25 °C, and the required alcohol **46** was isolated directly from the reaction mixture as a crystalline solid in excellent yield (95%). Analysis of product remaining in the mother liquors showed the stereoselectivity of the reaction was 45:1 in favour of 3α- versus 3β-attack. The participation of the reduced benzamide group in affecting the selectivity of the reduction was subsequently demonstrated by carrying out the hydroboration of the structurally and sterically similar tribenzyl compound **49**, which gave a selectivity of only 11:1.

The benzyl groups were readily removed by hydrogenolysis, giving the key intermediate aminodiol **37**. This was successfully converted into the desired drug substance by a three-stage process in which the acryloyl isocyanate **50** was treated with the aminodiol **37** to give the cyclised unprotected uridine **51a**. This was protected as the diacetate **51b** by reaction with acetic anhydride. Conversion to **2** was achieved by using a bromination–dehydrobromination sequence with N-bromosuccinimide and triethylamine, followed by work-up with hydrochloric acid. After extensive development, the entire sequence (from **37** to drug substance **2**) was carried out as a 'one-pot' transformation in excellent overall yield (70%, see Scheme 11).

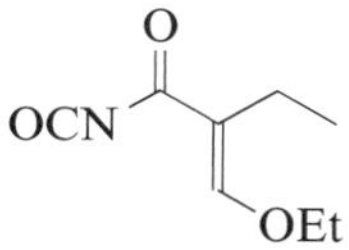

(50)

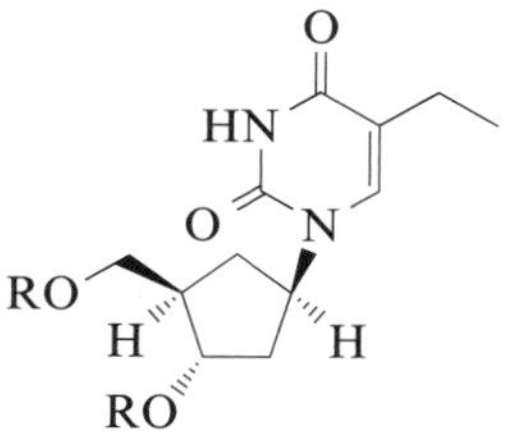

(51a) R = H
(51b) R = Ac

Subsequently, enantiopure benzamido ester **43** became commercially available from two different suppliers and, using this material, we were able to prepare several kilograms of drug substance in our pilot plant using the above route. The final reaction conditions and reagents are outlined in Scheme 11.

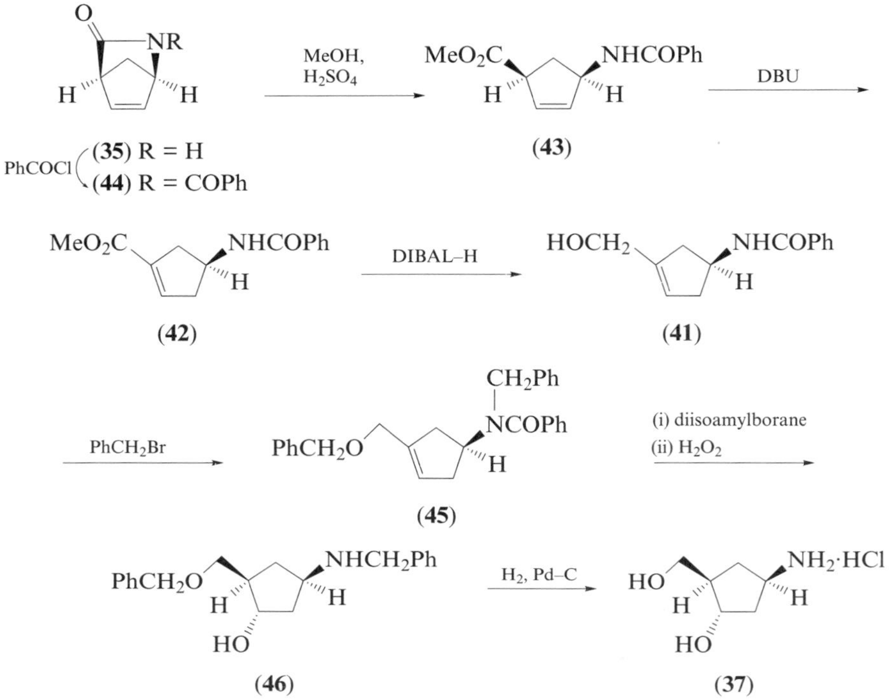

**Scheme 11**

### 4.3.2 USE OF PALLADIUM-CATALYSED FUNCTIONALISATION OF (+)-HYDROXYACETATE 5

Following the success of a palladium-catalysed functionalisation of (+)-hydroxyacetate **5** in the synthesis of GR85478, which was discussed Section 4.2.2, we now applied that chemistry to the synthesis of GR95168. This approach has the advantage of being more convergent than the first method developed.

This strategy requires an N-3-protected 5-substituted uracil (**52**) to be coupled to **5**, followed by addition of a one-carbon fragment with requisite control of the double bond location to yield the allylic alcohol **54**. Hydroboration from the α face, as in the previously discussed route to GR95168, would give the desired *trans*-diol, which could be elaborated subsequently to give the target molecule. This approach is shown in Scheme 12.

We hoped that the palladium-mediated coupling would circumvent the problem of elimination which, in our experience, usually occurred when displacement of a leaving group (such as tosylate or mesylate) on a cyclopentane ring was attempted with a uracil derivative. The first part of the synthesis involved a palladium(0)-catalysed alkylation of a 5-substituted uracil with (+)-hydroxyacetate **5**.

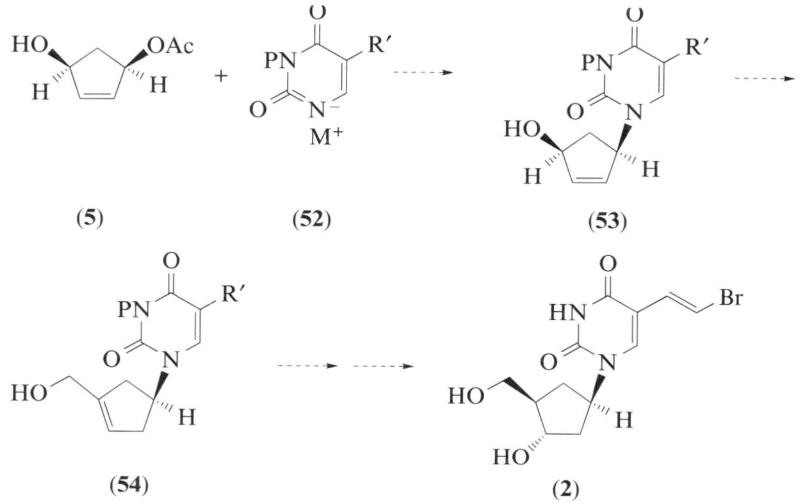

# SYNTHESIS OF ENANTIOMERICALLY PURE NUCLEOSIDES

**Scheme 12**

A series of 5-substituted and N-3-protected, 5-substituted uracil bases was synthesised (see Scheme 13) to evaluate the efficiency of these reactions. The substituent at the 5-position was chosen such that it could be readily converted into the desired vinyl bromide moiety.

Treatment of 5-iodouracil (**55**) with two equivalents of methyl acrylate in the presence of palladium(0) provided the N-1-protected, 5-substituted uracil **56a** in nearly quantitative yield; the Michael addition to N-1 was reversible, and faster than the Heck reaction. Warming the solution of **56a** in dimethyl sulphoxide under vacuum (distilling out the methyl acrylate) removed the N-1 protecting group to give the 5-substituted uracil **57** in 90% yield. Treatment of **56a** with at least three equivalents of methyl acrylate in the presence of palladium(0) gave the N-,N-propyl ester-protected 5-substituted uracil, which, on heating in dimethyl sulphoxide under vacuum, gave **56b**; only the N-1 substituent was removed. Treatment of **56a** with benzoyl chloride afforded **58**, which also, on heating, lost the N-1-propyl ester to give the N-3-protected, 5-substituted uracil **59** in greater than 80% overall yield. N-1-Benzoyl-5-ethyluracil (**62**) was prepared by first treating 5-ethyluracil (**60**) with excess benzoyl chloride to give the $N^1,N^3$ dibenzoyl adduct **61**. Selective removal of the N-1 substituent with potassium carbonate in aqueous dioxane gave N-1-benzoyl-5-ethyluracil (**62**) in >70% overall yield. All of these products were successfully isolated as crystalline solids.

The reactions of these uracil derivatives are summarised in Figure 3. Disappointingly, many of the results suggested that nucleophilic substitution was not competitive with elimination (see Scheme 5, paths A and B). Dialkylation of the uracil to the bis-adduct proved difficult to control. Elimination does not appear

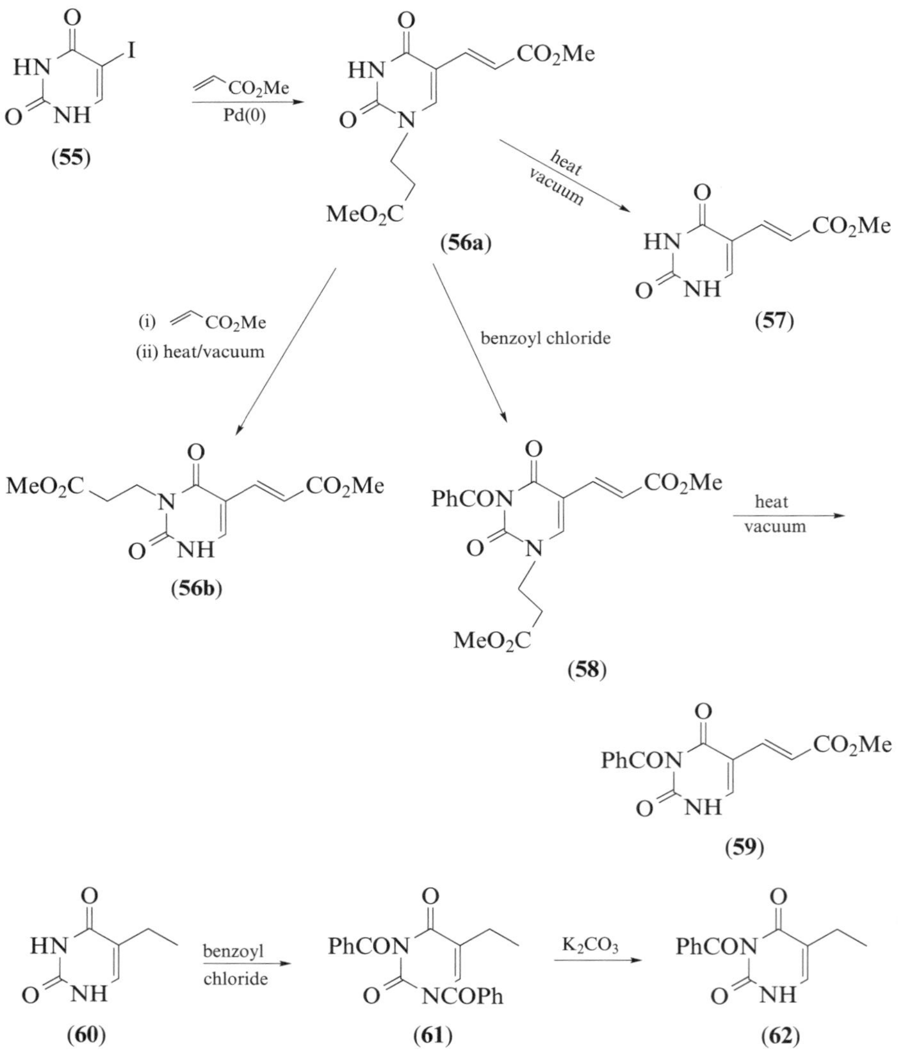

**Scheme 13**

to be competitive with nucleophilic substitution when the uracil base was substituted at both C-5 and N-3 with electron-withdrawing groups (entry 6); this was investigated further.

The synthesis now required a one-carbon homologation and a specific double bond migration (see Scheme 14). Nucleophilic substitution of the mesylate derivatives of allylic alcohols **63a** and **63b** with cyanide anion, as described in the

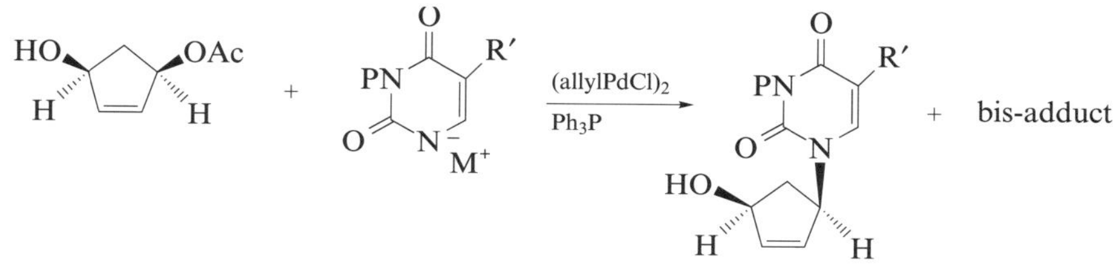

# SYNTHESIS OF ENANTIOMERICALLY PURE NUCLEOSIDES

|   | P | R' | M | Catalyst (mol%) | Yield of 53 | Yield of bis-adduct |
|---|---|----|---|-----------------|-------------|---------------------|
| 1 | H | H | Na, K | 5.0 | 20–40% | 10–15% |
| 2 | H | I | Na, K | 5.0 | 10% | 5% |
| 3 | H | Et | Na, K | 5.0 | 10% | - |
| 4 | H | CHCHCO$_2$Me | Na | 5.0 | 65% | 14% |
| 5 | (CH$_2$)$_2$CO$_2$Me | CHCHCO$_2$Me | Et$_3$NH | 5.0 | 82% | - |
| 6 | COPh | CHCHCO$_2$Me | Et$_3$NH | 0.5 | 96% | - |
| 7 | COPh | Et | DBUH | 5.0 | 72% | - |

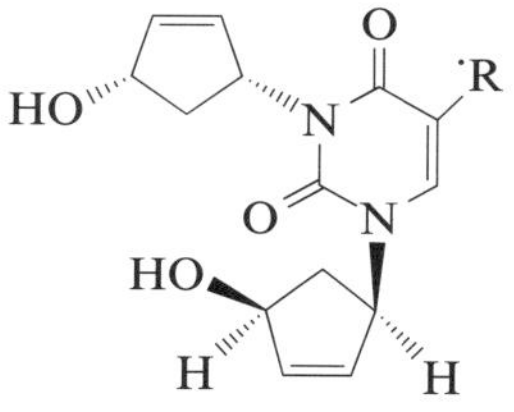

**Figure 3** Reactions of uracil derivatives

synthesis of GR85478 (**1**),[17] proceeded in poor yield (20–30%) and with incomplete isomerisation of the double bond; cyanide is sufficiently basic to promote extensive elimination of the sensitive allylic mesylate derivatives. Work reported by Stille and Scott[26] on the palladium(0)-catalysed carbonylation of vinyl and aryl triflates offered an attractive solution to the conversion of allylic alcohols **63a** and **63b** into α,β-unsaturated carboxylic acids **66a** and **66b**.

Oxidation of allylic alcohols **63a** and **63b** with sulphur trioxide–dimethyl sulphoxide provided the corresponding enones **64a** and **64b** in 87% yield. Selective 1,4-reduction of these enones with KS-Selectride in the presence of *N*-phenyltriflimide provided the desired Δ$^{3,4}$-vinyl triflates **65a** and **65b** in 70–80% yield after chromatography. It was necessary to carry out the reduction in the presence of *N*-phenyltriflimide to avoid extensive elimination of the uracil base, presumably via a rapid enolate isomerisation. The vinyl triflates were converted into the α,β-unsaturated carboxylic acids **66a** and **66b** *in situ* by addition of 5 mol% of palladium(II) acetate, triphenylphosphine, triethylamine and water

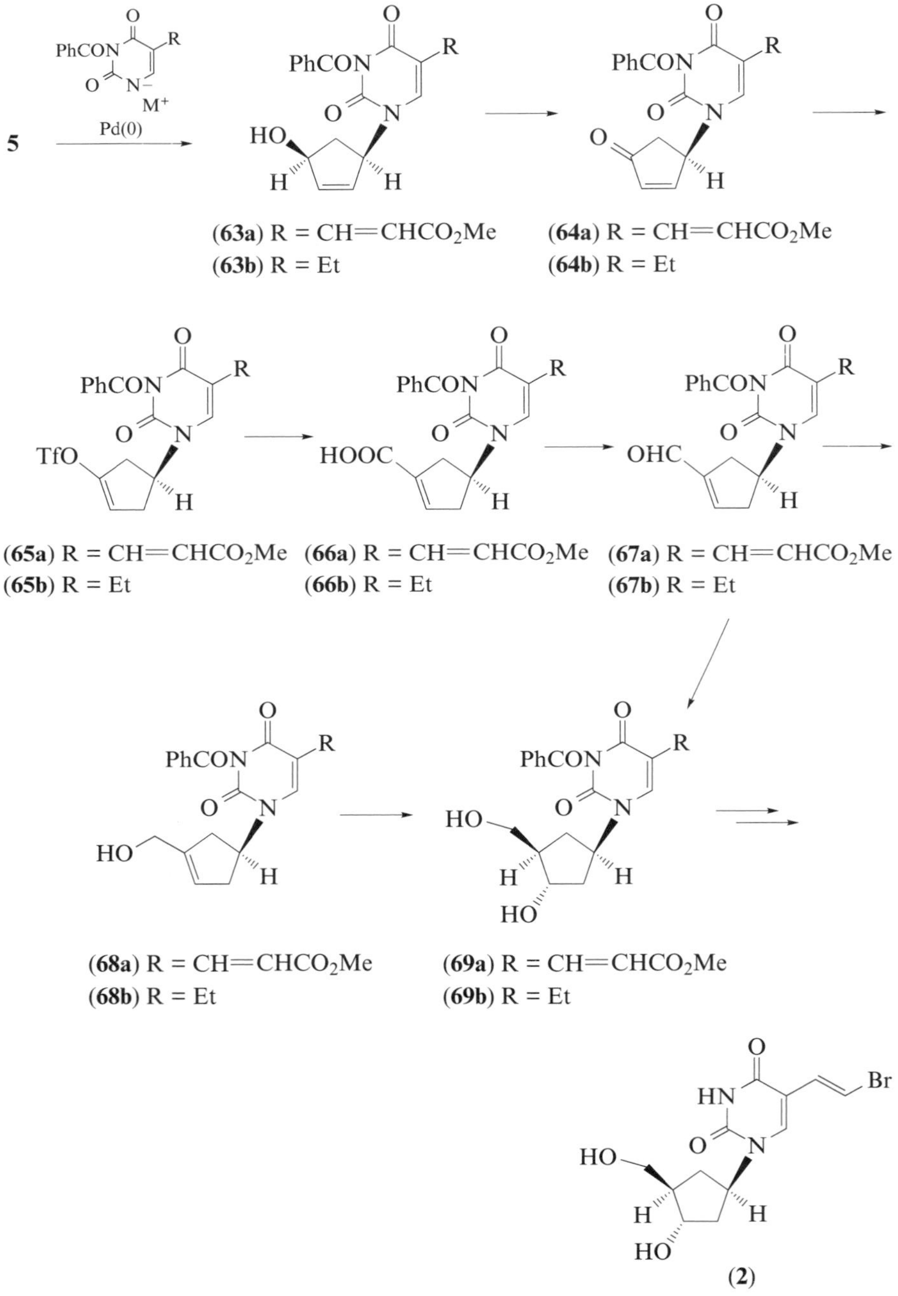
PhCON
O
R
O
N⁻
M⁺
Pd(0)
5
(63a) R = CH=CHCO2Me
(63b) R = Et
(64a) R = CH=CHCO2Me
(64b) R = Et
TfO
HOOC
OHC
(65a) R = CH=CHCO2Me
(65b) R = Et
(66a) R = CH=CHCO2Me
(66b) R = Et
(67a) R = CH=CHCO2Me
(67b) R = Et
(68a) R = CH=CHCO2Me
(68b) R = Et
(69a) R = CH=CHCO2Me
(69b) R = Et
HN
Br
(2)

**Scheme 14**

# SYNTHESIS OF ENANTIOMERICALLY PURE NUCLEOSIDES

pharmacological and safety studies were conducted. Separation of enantiomers on a scale sufficient to determine their individual efficacies and toxicological properties is not usually a difficult problem, but the preparation of several kilograms in order to proceed with clinical studies is often a major challenge to the process research chemist. Lamivudine was no exception to this rule.

As is common with nucleosides and their analogues, it is the *cis* isomers of lamivudine which possess significant anti-viral activity. The two *cis* diastereoisomers were first separated by preparative chiral HPLC using an acetylated β-cyclodextrin column[30] and, although the two isomers were found to have approximately equal activities against HIV, one isomer (the laevorotatory one) was less cytotoxic than the other. When the decision to develop only the single, less toxic isomer was taken, substantial supplies of the racemate had already been prepared. From that point onward, preparation of only the laevorotatory isomer was our goal.

## 4.4.1 SYNTHESIS OF THE RACEMATE

The original synthesis used to prepare the racemate is shown in Scheme 16. The oxathiolane ring was constructed by reaction of hydroxyacetaldehyde [protected as its benzoate ester (**70**)] with mercaptoacetaldehyde, generated by heating its cyclic dimer, dithianediol. The resulting 5-hydroxyoxathiolane **71** was converted into acetate **72** by treatment with acetyl chloride–pyridine, then the acetate group displaced by bis-silylated *N*-acetylcytosine in the presence of titanium(IV) chloride. This coupling reaction gave many problems on scale-up; it was only marginally *cis*-selective, a *cis*:*trans* ratio of about 3:2 was obtained and a substantial amount of *N*-deacetylation occurred. The desired *cis* isomer was partially purified by column chromatography, then further purified by crystallisation. Removal of the acetyl and benzoyl groups was achieved by treatment with sodium methoxide in methanol, and the racemic product was purified by recrystallisation from water.

Although this synthesis was suitable for the preparation of small quantities, it required several modifications to enable manufacture on a large scale to be carried out safely and reliably. The protected hydroxyacetaldehyde **70** was isolated as its hydrogensulphite addition compound (a stable, crystalline solid) which needed no further purification, and distillation of the thermally unstable aldehyde was thus avoided.

The aldehyde **70** was regenerated by treatment with acid, prior to reaction with dithianediol giving hydroxyoxathiolane **71**, which, previously, had not been isolated. This compound exists in solution as a mixture of epimers, caused by rapid ring opening and closing (see Scheme 17). By carefully controlling the crystallisation conditions, the equilibrium between *cis* and *trans* isomers can be continually re-established in solution as crystallisation proceeds, enabling the product to be isolated as the pure, racemic *trans* isomer in about 70% yield.

Control over the relative stereochemistry at C-2 and C-5 of the oxathiolane ring was a major problem. Liotta's group showed[31] that if tin(IV) chloride was

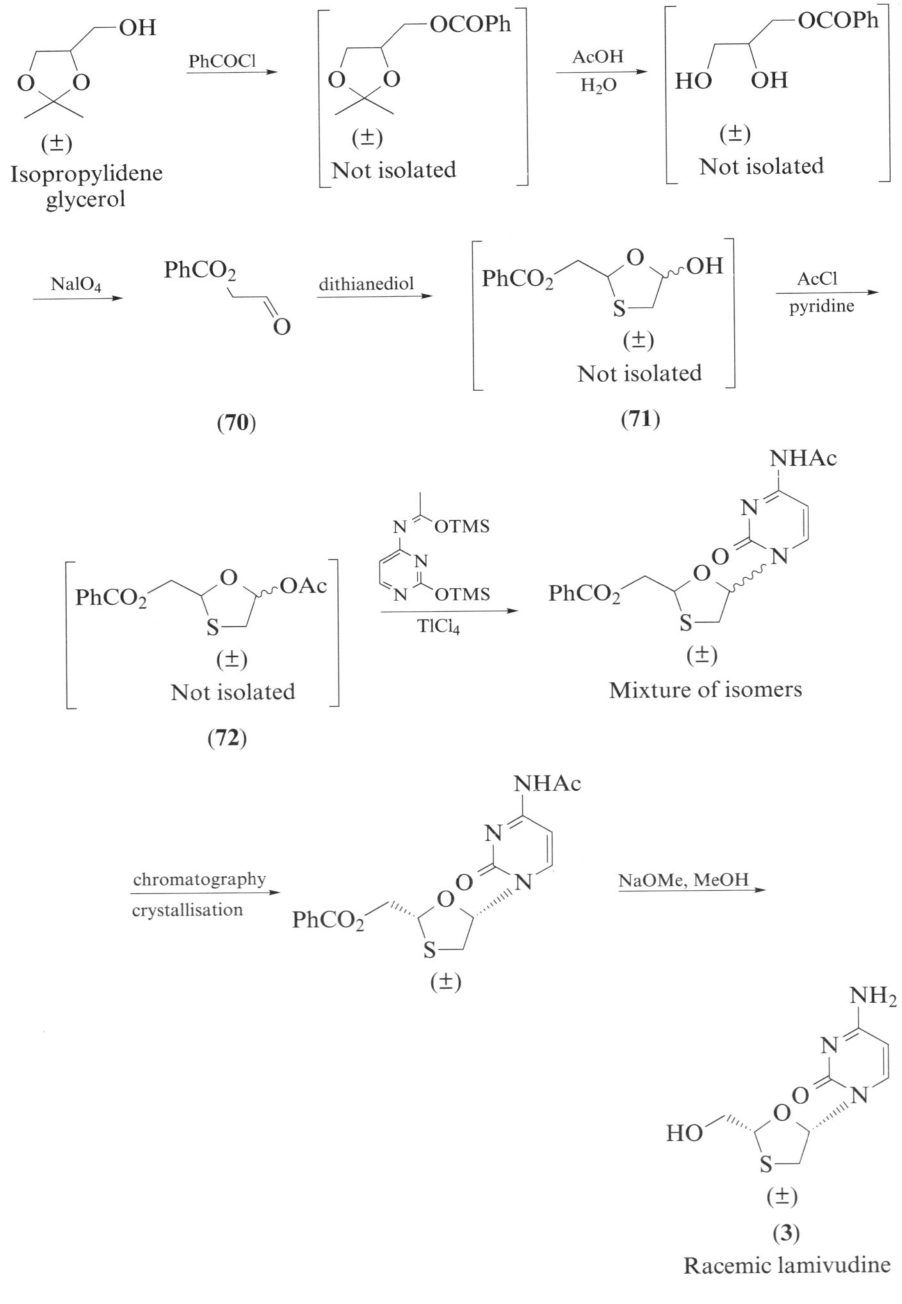
OH
PhCOCl
OCOPh
AcOH
H2O
OCOPh
HO
OH
(±)
Isopropylidene glycerol
(±)
Not isolated
(±)
Not isolated
NaIO4
PhCO2
dithianediol
PhCO2
OH
AcCl
pyridine
(±)
Not isolated
(70)
(71)
NHAc
OTMS
OTMS
TlCl4
PhCO2
OAc
PhCO2
(±)
Not isolated
(±)
Mixture of isomers
(72)
NHAc
chromatography
crystallisation
PhCO2
NaOMe, MeOH
(±)
NH2
HO
(±)
(3)
Racemic lamivudine

**Scheme 16**

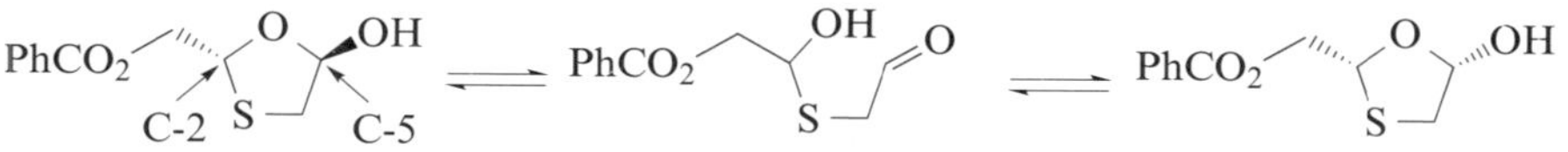

SYNTHESIS OF ENANTIOMERICALLY PURE NUCLEOSIDES 69

Scheme 17

used as a Lewis acid, the *cis* selectivity of the coupling reaction was dramatically increased; no other common Lewis acid had this effect. In our hands, a selectivity of around 10:1 in favour of the *cis*-isomer could be achieved reliably. Work-up of these metal-mediated reactions is often difficult, requiring either extensive chromatography or slow filtrations to remove metal salts. We found that if the reaction mixture was acidified, we could crystallise the desired product as its hydrochloride salt merely by adding a counter solvent. The product was contaminated with only a small quantity of the unwanted *trans*-isomer and required no further purification. On scale-up, it proved necessary to recrystallise the salt to reduce the level of tin-containing impurities, but despite this the new process represented a major improvement. Finally, the benzoyl group was removed by treatment with a basic ion-exchange resin in suspension in ethanol, from which racemic lamivudine was crystallised directly in good yield and purity.

### 4.4.2 RESOLUTION OF RACEMIC LAMIVUDINE

As substantial supplies of the racemate were thus available, it was sensible to attempt resolution of this isomeric mixture. Formation of a diastereoisomeric salt with a chiral acid was tried with some success, and enzymic methods were also investigated. Racemic lamivudine was treated with phosphoryl chloride and trimethyl phosphate to give a mixture of the two phosphates **73a** and **73b**, which were then treated with the *Crotalus atrox* venom 5′-nucleotidase. Only the undesired, 'natural,' isomer (**73b**) was hydrolysed to the alcohol, and was easily separated from the 'unnatural' phosphate; subsequent dephosphorylation, using bacterial alkaline phosphatase, afforded enantiomerically pure lamivudine (**3**).[32] Once separated in quantity, the two isomers were compared, and the absolute configuration of the less toxic enantiomer was established by single-crystal X-ray diffraction (of the 4-bromobenzamide derivative) and found to be the '*unnatural*' (−)-isomer.[32]

Given that this 'unnatural' isomer was the required entity, a much simpler synthetic method could be developed using the enzyme cytidine deaminase.[33,34] This enzyme converts cytidine into uridine, and again only recognises the 'natural' isomer. When racemic lamivudine was treated with cytidine deaminase, the 'natural' isomer was converted selectively into the uridine analogue **74**, leaving the desired enantiomer (**3**) unchanged (see Scheme 18). This substituted uracil could easily be separated from lamivudine by selective adsorption on an ion-exchange column, exploiting the differences in acidity of the hydrogens on the heterocyclic ring. This process (which was subsequently developed into a viable plant process) was used to prepare around 3 tonnes of lamivudine, with an enantiomeric excess[35] typically greater than 99.9%.

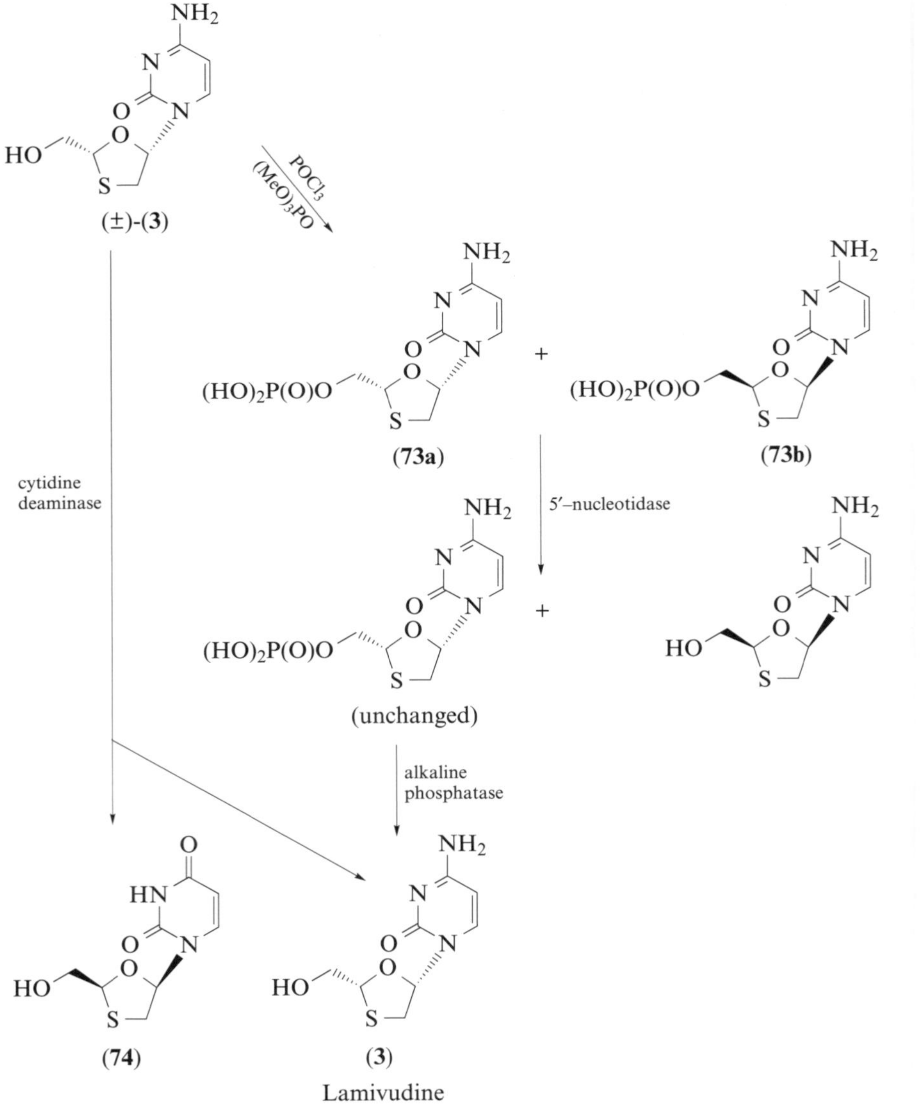

**Scheme 18**

### 4.4.3 ALTERNATIVE STRATEGIES

Although significant advances had been made, we still felt that it should be possible to find a more efficient route. Formation of an enantiopure compound by resolution as the last step of a synthesis is inefficient; at least half of all the input to each stage is lost, giving severe waste-handling problems, and the resulting process is both environmentally and economically unacceptable. We recognised the need for a new approach. An enantioselective synthesis of lamivudine had been published,[36] but this was long and unsuitable for large-scale production. Consequently, we investigated two alternative strategies:

- Synthesis of an intermediate in the current route in chiral form.
- Design of a new synthesis.

To avoid wasting large quantities of the toxic tin(IV) chloride, or the expensive starting material cytosine, the obvious target for a chiral intermediate was the stable, crystalline acetoxyoxathiolane **72**. Successful use of this intermediate requires retention of chirality under the coupling reaction conditions. The stereochemistry at C-5 of the oxathiolane ring is not important; provided that the stereogenic centre at C-2 remains intact and the mechanism of the reaction produces specifically a *cis* arrangement of the two substituents, then the desired enantiomer of lamivudine will be obtained.

Acid **76** was prepared by reaction of aldehyde **70** with mercaptolactic acid (**75**), and resolved via the brucine salt of its *S*-benzyl derivative (Scheme 19). Treatment with lead tetraacetate gave oxathiolane **72**, enantiomerically pure at C-2, although as a mixture of *cis* and *trans* isomers.[37]

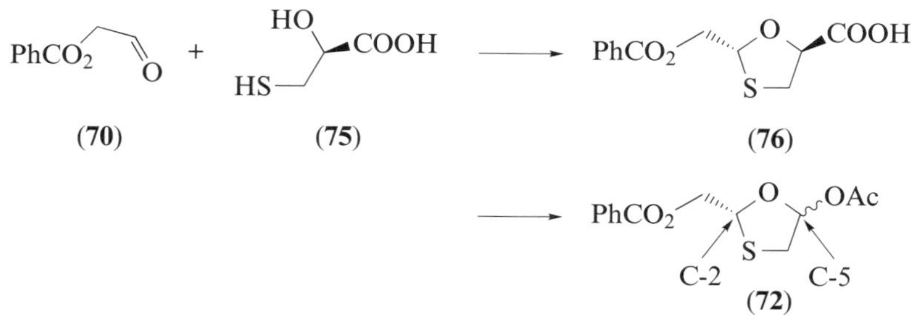

**Scheme 19**

Attempts were made to produce an enantiomerically pure intermediate by enzymic means with the propionate ester **77** as the preferred substrate. On treatment with *Mucor miehei* lipase only the (2*S*)-isomer (**77b**) was hydrolysed, giving a mixture from which the (2*R*)-isomer of the propionate (**77a**) could be isolated, albeit in low yield[38] (Scheme 20).

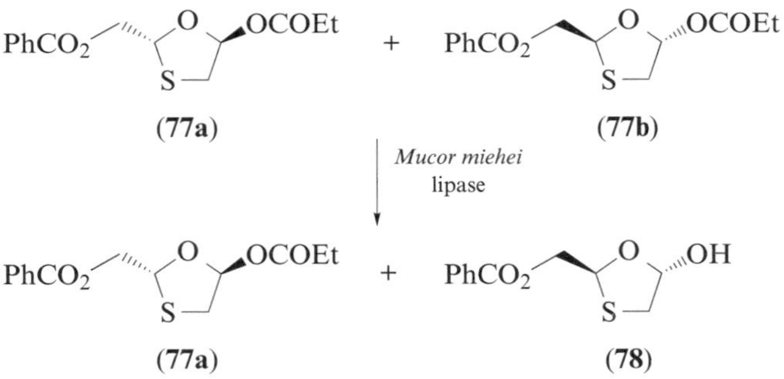

**Scheme 20**

(77a) + (77b) → [Mucor miehei lipase] → (77a) + (78)

Next we investigated the effect of the coupling reaction on the stereogenic centre at C-2. As the reaction tolerates a wide variety of ester and carbonate substituents at C-5, carbonate **79** was selected for investigation. The diastereoisomers were separated chromatographically, and the extra stereogenic centres present in the menthol fragment allowed easy identification of the products of the coupling reaction by either NMR or HPLC.

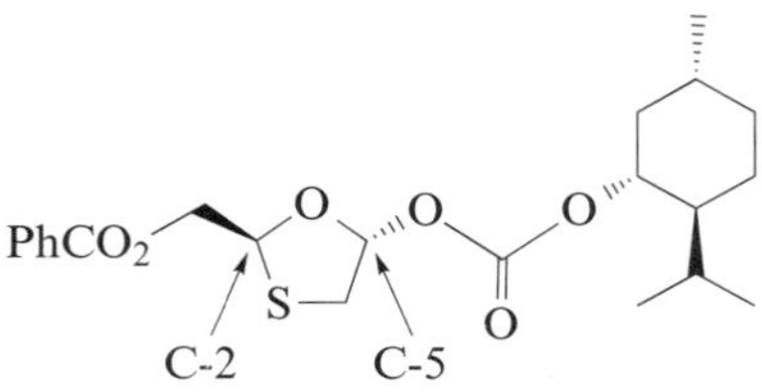

(79)

When compound **79** was treated with bis-silylated cytosine and the resulting benzoate hydrolysed, the lamivudine produced was found (by chiral HPLC) to be *racemic*; all chirality at C-2 had been lost, even though the reaction was *cis*-selective. The stereogenic centre at C-2 could be preserved if iodotrimethylsilane was used as Lewis acid, but this gave a mixture of *cis* and *trans* isomers in a ratio of approximately 1:1. Neither of these methods was attractive for potential scale-up and, with this avenue now seemingly closed, the only option left was to design a new synthesis.

### 4.4.4 A NOVEL SYNTHESIS OF LAMIVUDINE

The chemistry of oxathiolanes with a 2-carboxy substituent derived from glyoxylic acid was being investigated by colleagues at Biochem Pharma. In contrast to those used in the route described in Scheme 16 with a $CH_2$ substituent at C-2, compounds in this series reacted with silylated cytosine in the presence of

# SYNTHESIS OF ENANTIOMERICALLY PURE NUCLEOSIDES

iodotrimethylsilane to give the coupled products with excellent *cis*-selectivity. By making an ester with L-(−)-menthol and separating a single diastereoisomer in an analogous fashion to the menthyl carbonate described above, we showed that the coupling proceeded with *retention* of the stereochemistry at C-2.[39,40]

This chemistry therefore had the potential to form the basis of a more efficient chiral route (see Scheme 21), as the presence of an acid group at C-2 is an obvious handle by which resolution early in the synthesis could be achieved. Condensation of glyoxylic acid with dithianediol gave hydroxy acid **81**. In solution, this compound exists as a mixture of isomers, but by careful control of the crystallisation (as with the hydroxyoxathiolane **71** discussed earlier), it was isolated as the *trans* isomer, although the rate of equilibration in solution was found to be

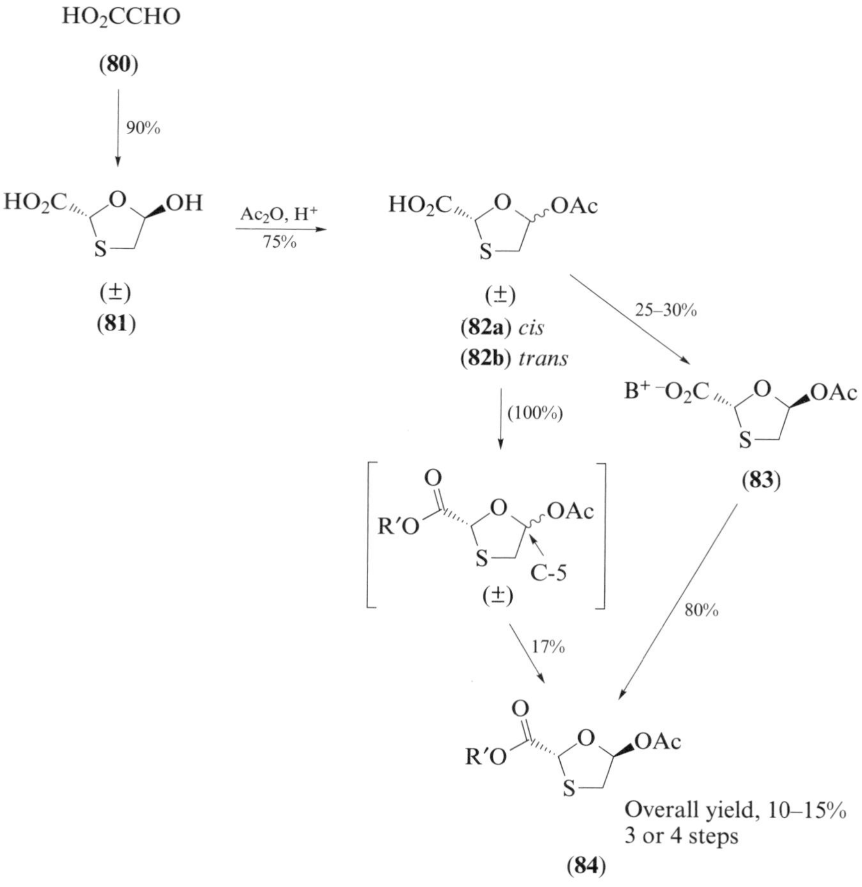

**Scheme 21**

solvent dependent. In DMSO the rate was slow, allowing satisfactory $^1$H NMR spectra to be obtained. In other solvents the rate of equilibration was much faster.

For a satisfactory coupling reaction, the hydroxyl group at C-5 had to be converted into a leaving group, such as an ester or a carbonate. The simplest such group is acetoxy, and this was our first choice (although good results were achieved using other esters and carbonates). In the earlier route, it had been found that addition of acetyl chloride to a solution of hydroxyoxathiolane **71** in dichloromethane and pyridine gave a mixture of *trans* and *cis* isomers of the acetoxy derivative **72** in a ratio of about 70:30. However, if solid hydroxyoxathiolane **71** was added to acetyl chloride–pyridine, a fast acetylation took place with little epimerisation. NMR confirmed that epimerisation of **71** in chlorinated solvents was fast, but clearly not as fast as acetylation. When applied to hydroxy acid **81**, this approach was not successful, however. The acid was much less soluble in chlorinated solvents, reacted slowly and, while the product was of predominately *trans* relative stereochemistry, it was contaminated with many by-products. The major by-product was the dimer **86**, indicating that reaction of intermediate anhydride **85** was of about equal rate to reaction with acetyl chloride–pyridine (see Scheme 22).

Treatment of hydroxy acid **81** with acetic anhydride in the presence of strong acids, such as sulphuric or methanesulphonic acid, gave a clean reaction which

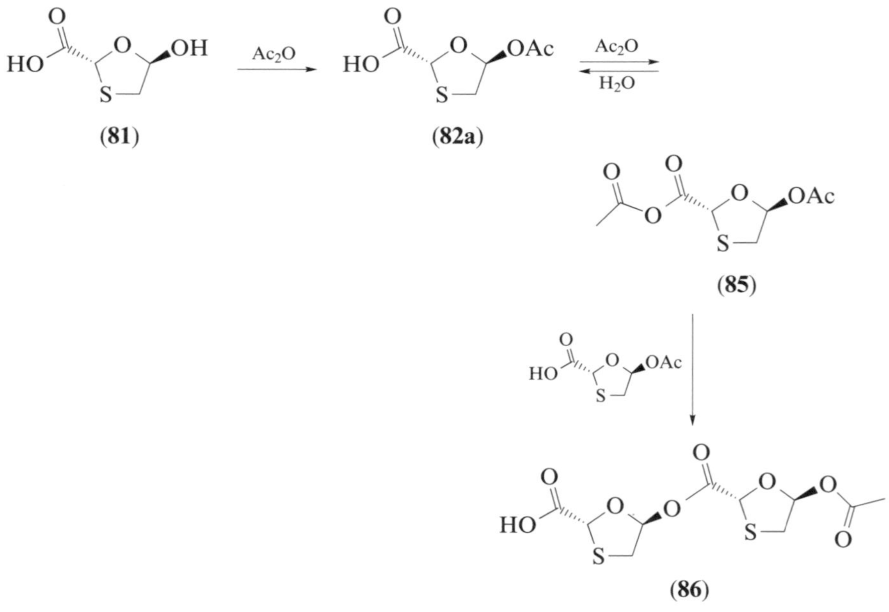

**Scheme 22**

# SYNTHESIS OF ENANTIOMERICALLY PURE NUCLEOSIDES

was rather slow. Under these conditions the compound epimerised again to a 70:30 mixture of *trans* and *cis* isomers, before being cleanly acetylated, giving a mixture of the two acetates **82a** and **82b**. As no reliable method could be developed for the preparation of the pure *trans* isomer (nor, indeed, the *cis* isomer), we investigated methods for the resolution of the mixture which could be reliably reproduced on a large scale in a yield of about 75%.

A common way to resolve racemic acids which was applied to GR95168 (as discussed in Section 4.3) is the formation of diastereoisomeric salts with a chiral base. Several bases were screened for suitability in the resolution of acetoxy acid **82**; the best found was norephedrine. The diastereoisomerically pure salt of *trans*-acid **83** was isolated in 27% yield merely by adding a solution of norephedrine to the racemic mixture of *cis*- and *trans*-acids in isopropyl acetate. Although the yield seems low, this represents a yield of about 80% of the available diastereoisomer present in the four-component mixture.

Concurrently, we investigated the formation of diastereoisomeric esters by reaction of the crude mixture of acids with (−)-menthol. Although this was used initially as a way of both introducing crystallinity to the ester and allowing further purification of the resolved acid, we found that we could isolate a single isomer from the mixture of all four diastereoisomers by careful control of the crystallisation conditions. This represented a yield of 50% of available isomer (although only 17% of all four). Nevertheless, a simple crystallisation of the desired isomer (**84**) from a mixture of four in a ratio of 35:35:15:15 was a significant achievement, as the process could easily be run on a large scale, used no expensive resolving agents and did not require regeneration of the acid.

Following pilot-plant trials, the selective crystallisation of the desired diastereoisomer was chosen in preference to formation of the salt, and the process was developed further before being transferred to full-scale production.

Acetate **84** was treated with silyated cytosine in the presence of iodotrimethylsilane to give ester **87** (Scheme 23). Preliminary investigations showed the reaction gave a *cis*-selectivity of about 93:7, which was comparable to that in the route discussed previously using tin(IV) chloride. On isolation, most of the *trans*

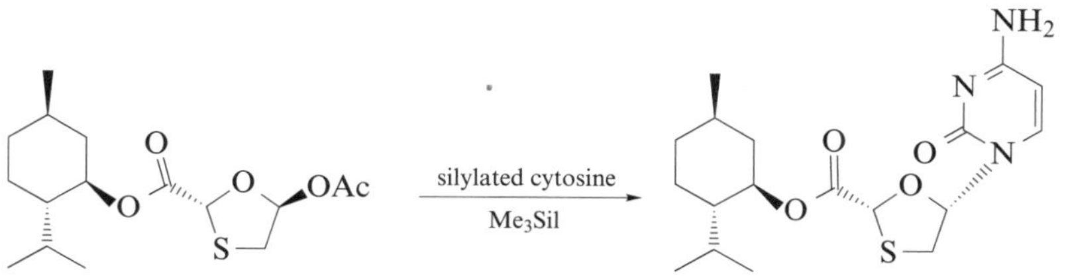

**Scheme 23**

isomer could be removed, giving a product contaminated with only small amounts of unwanted isomers and no racemisation. After optimisation, the work-up was fairly simple: the pure product was isolated directly from the quenched reaction by filtration in a yield of around 75%.

The reason for the high *cis*-selectivity was not immediately apparent. Attempts to isolate intermediates from the corresponding stage of the route shown in Scheme 16 had met with failure; tin(IV) chloride caused decomposition of oxathiolane **72**, and gave a complex mixture with silylated cytosine. However, when ester **84** was treated with iodotrimethylsilane in the *absence* of silyated cytosine, a single compound was formed, which was isolated and identified as iodide **88**. NMR showed a mixture of isomers in solution, with the *trans* predominating, indicating that the reaction was merely an $S_N2$ displacement of iodide by silylated cytosine, and that the *cis*-selectivity resulted from the fact that the major isomer of the iodide was *trans*.

(**88**)

To complete the synthesis, ester **87** was reduced to give the primary alcohol, lamivudine (**3**), using any convenient reducing agent, such as lithium aluminium hydride, but isolation of the drug substance from the reaction mixture was not trivial. In the laboratory, on a small scale, column chromatography was satisfactory, but in the pilot plant a macroreticular resin was preferred.

These resins are non-functionalised and contain pores capable of accepting organic molecules. When an aqueous solution of an organic compound is passed down a column of this resin, the organic components bind to the column and the inorganics do not. On washing the column with water all inorganics are removed, leaving the desired product on the resin. If an organic solvent, such as acetone or an alcohol, is then passed down the column, the resin expands and the bound compound is released. Isolation of the product from an organic phase which is free from inorganic contamination is then possible by concentration and crystallisation. This technique was used successfully for the isolation of lamivudine, passing the solution down a column of Amberlite XAD-16 resin, washing with water, then elution with propan-2-ol. Lamivudine could be crystallised by concentration and cooling, and was obtained in good yield and high purity.

After further development and optimisation, this route was transferred to full-scale production at Glaxo-Wellcome plants, and has been used to produce several tonnes of lamivudine.

# SYNTHESIS OF ENANTIOMERICALLY PURE NUCLEOSIDES

## 4.5 CONCLUSIONS AND SUMMARY

The synthesis on a large scale of the three anti-virals considered here posed many significant problems. The molecules all contain chiral centres which are important with respect to the activity of each drug, and levels of the undesired stereoisomers in the final drug substance had to be minimised.

In the course of developing these products, several chiral strategies have been used:

- Enzymatic resolution.
- Classical resolution with diastereoisomeric salts.
- Selective crystallisation of desired diastereoisomers.
- Asymmetric synthesis.

Practical processes have been developed which we have used to prepare several tonnes of enantiomerically pure intermediates and drugs. New routes to these molecules have also been produced, in some cases requiring the development of new methodology to solve particular problems of selectivity.

## 4.6 REFERENCES

1. Storer, R., Paternoster, I. L., Borthwick, A. D., and Biggadike, K (Glaxo, UK), *US Pat.* 5 155 112, 1992.
2. (a) Vince, R., and Hua, M., *J. Med. Chem.*, **33**, 17 (1990), and references cited therein; (b) Vince, R., and Hua, M., (University of Minnesota), *US Pat.*, 4 950 758, 1990, and 4 931 559, 1990.
3. Bray, B. L., Lichty, M. E., Partridge, J. J., and Turnbull, J. P. (Glaxo, UK), *US Pat.*, 5 233 041, 1993.
4. Mancuso, A. J., and Swern, D., *Synthesis*, 165 (1981).
5. Van Rheenen, V., Kelly, R. C., and Cha, D. Y., *Tetrahedron Lett.*, 1973 (1976).
6. Trost, B. M., Kuo, G. H., and Benneche, T., *J. Am. Chem. Soc.*, **110**, 621 (1988).
7. Bray, B. L., Lichty, M. E., and Partridge, J. J., unpublished results.
8. Sugiyama, H., Sera, T., Dannoue, Y., Marumoto, R., and Saito, I., *J. Am. Chem. Soc.*, **113**, 2290 (1991).
9. Bray, B. L., and Partridge, J. J., (Glaxo, UK), *US Pat.*, 5 438 132, 1995.
10. *Fluka Chemika–BioChemika Analytika 1995–96*, Fluka Chemical, Ronkonkoma, NY, 1995.
11. (a) Wang, Y., Chen, C., Girdaukas, G., and Sih, C., *J. Am. Chem. Soc.*, **106**, 3695 (1984); (b) Laumen, K., Rimerdes, E. M., and Schneider, M., *Tetrahedron Lett.*, **26**, 407 (1985); (c) Deardorff, D. R., Matthew, A. J., McKeekin, S., and Claney, C.L., *Tetrahedron Lett.*, **27**, 1255 (1986); (d) Deardorff, D. R., and Windham, C. Q., *Org. Synth.*, **67**, 114 (1988); (e) Suga, T., and Mori, K., *Synthesis*, 19 (1988); (f) Deardorff, D. R., Windham, C. Q., and Craney, C. L., *Org. Synth.*, **73**, 25 (1995); (g) Bray, B. L., Lackey, J. W., Lovelace, T. C., Mook, R. A., and Partridge, J. J., unpublished results.
12. (a) Deardorff, D. R., Linde, R. G., Martin, A. M., and Shulman, M. J., *J. Org. Chem.*, **54**, 2759 (1989); (b) Lackey, J. W., Mook, R. A., and Partridge, J. J (Glaxo, USA), *US Pat.*, 5 057 630, 1991.
13. Sepracor Inc., Marlborough, MA 01752, USA.

14. Amano International Enzyme Company, Inc., P.O. Box 1000, Troy, VA 22974, USA.
15. Trost, B. M., and Molander, G. A., *J. Am. Chem. Soc.*, **103**, 5969 (1981).
16. Partridge, J. J., and Bray, B. L., (Glaxo, USA), *US Pat.*, 5 329 008, 1994).
17. Partridge, J. J., and Bray, B. L., (Glaxo, USA), *US Pat. Appl.*, 08/271 594, 1994.
18. Coe, D. M., Myers, P. L., Parry, D. M., Roberts, S. M., and Storer, R., *J. Chem. Soc., Chem. Commun.*, 151 (1990).
19. Cookson, R. C., Dudfield, P. J., Newton, R. F., Ravenscroft, P., Scopes, D.I.C., and Cameron, J. M., *Eur. J. Med. Chem.*, **20**, 375 (1985).
20. (a) Cookson, R. C., Dudfield, P. J., and Scopes, D. I. C., *J. Chem. Soc, Perkin Trans. 1*, 399 (1986); (b) Ravenscroft, P., Newton, R. F., and Scopes, D. I. C., *Tetrahedron Lett.*, **27**, 747 (1986); (c) Ravenscroft, P., (Glaxo, UK), *US Pat.*, 4 658 044, 1987.
21. (a) Jagt, J. C., and Van Leusen, A. M., *Recl. Trav. Chim. Pays-Bas*, **92**, 1343 (1973); (b) Jagt, J. C., and Van Leusen, A. M., *J. Org. Chem.*, **39**, 564 (1974); (c) Daluge, S., and Vince, R., *J. Org. Chem.*, **43**, 2311 (1978).
22. (a) Brown, H.C., and Zweifel, G., *J. Am. Chem. Soc.*, **81**, 247 (1959); (b) Brown, H. C., and Daniels, J. J., *Org. Synth.*, 719 (1988); (c) Still, W. C., and Barrish, J. C., *J. Amr. Chem. Soc.*, **105**, 2487 (1983).
23. Brown, H. C., Bigley, D. B., Arora, S. K., and Yoon, N. M., *J. Am. Chem. Soc.*, **92**, 7161 (1970).
24. Bray, B. L., Dolan, S. C., Halter, B., Lackey, J. W., Schilling, M. B., and Tapolczay, D. J., *Tetrahedron Lett.*, **36**, 4483 (1995).
25. Brown, H.C., and Zweifel, G., *J. Am. Chem. Soc.*, **83**, 1241 (1961).
26. (a) Stille, J. K., *Angew. Chem., Int. Ed. Engl.*, **25**, 508 (1986); (b) Scott, W. J., and McMurry, J. E., *Acc. Chem. Res.*, **21**, 47 (1988).
27. Bray, B. L., unpublished results.
28. Fleet, G. W. J., Fuller, C. J., and Harding, P. J. C., *Terahedron Lett.*, 1437 (1978).
29. Soudeyns, H., Yao, X.-J., Belleau, B., Kraus, J.-L., Nguyen-Ba, N., Spira, B., and Wainberg, M. A., *Antimicrob. Agents Chemother.*, **35**, 1386 (1991).
30. Coates, J. A. V., Cammack, N., Jenkinson, H. J., Mutton, I. M., Pearson, B. A., Storer, R., Cameron, J. M., and Penn, C. R., *Antimicrob. Agents Chemother.*, **36**, 202 (1992).
31. Choi, W. B., Wilson, L. J., Yeola, S., and Liotta, D. C., *J. Am. Chem. Soc.*, **113**, 9377 (1991).
32. Storer, R., Clemens, I. R., Lamont, B., Noble, S. A., Williamson, C., and Belleau, B., *Nucleosides Nucleotides*, **12**, 225 (1993).
33. Coates, J. A. V., Mutton, I. M., Penn, C. R., Storer, R., Williamson, C. (IAF International Inc., Canada), *PCT Int. Pat. Appl.*, WO 91/17159, 1991.
34. Mahmoudian, M., Baines, B. S., Drake, C. S., Hale, R. S., Jones, P., Piercey, J. E., Montgomery, D. S., Purvis, I. J., Storer, R., Dawson, M. J., and Lawrence, G. C., *Enzyme Microb. Technol.*, **15**, 749 (1993).
35. Rogan, M. M., Drake, C., Goodall, D. M., and Altria, K. D., *Anal. Biochem.*, **208**, 343 (1993).
36. Beach, J. W., Jeong, L. S., Alves, A. J., Pohl, D., Kim, H. O., Change, C.-N., Doong, S.-L, Schinazi, R. F., Cheng, Y.-C., and Chu, C. K., *J. Org. Chem.*, **57**, 2217 (1992).
37. Humber, D. C., Jones, M. F., Payne, J. J., and Ramsay, M. V. J., *Tetrahedron Lett.*, **33**, 4625 (1992).
38. Cousins, R. P. C., Mahmoudian, M., and Youds, P. M., *Tetrahedron: Asymmetry*, **6**, 393 (1995).
39. Jin, H., Siddiqui, M. A., Evans, C. A., Tse, H. L. A., Mansour,. T. S., Goodyear, M. D., Ravenscroft, P., and Beels, C. D., *J. Org. Chem.*, **60**, 2621 (1995).
40. Mansour, T. S., Jin, H., Tse, A., and Siddiqui, A (Biochem Pharma Inc., Canada), *Eur. Pat.*, EP 0 515 157, 1992.

# PHYSICAL METHODS AND CLASSICAL RESOLUTION

# 5 Rational Design in Resolutions

**A. BRUGGINK**
*Chemferm, Breda, and University of Nijmegen, The Netherlands*

| | | |
|---|---|---|
| 5.1 | Introduction | 81 |
| 5.2 | Background to a Rational Approach | 81 |
| 5.3 | Rational Design and Synthesis | 84 |
| 5.4 | Rationale of a Resolution | 88 |
| 5.5 | Predicting a Resolution? | 92 |
| 5.6 | Asymmetric Transformations, Racemisations and Recycling | 94 |
| 5.7 | Conclusions | 96 |
| 5.8 | Acknowledgements | 97 |
| 5.9 | References | 97 |

## 5.1 INTRODUCTION

In the 100 years since the death of Louis Pasteur, the production of enantiopure compounds by 'classical' resolution has retained a prime position in the chemical industry. In terms of the *number* of products rather than tonnage, it is still the most widely used method, despite the fact that no sound theoretical basis for the method has been available; practical experience and trial and error have been the guidelines for arriving at economic resolution processes. This 'state of the art' has been well presented in two recent books.[1]

## 5.2 BACKGROUND TO A RATIONAL APPROACH

Around 1978, during the development of the enantiopure captopril intermediate D-($-$)-$S$-acetyl-$\beta$-mercaptoisobutyric acid (**1**) (Figure 1), the need for a more rational approach to resolutions through selective crystallisations became apparent to the author and his co-workers at Océ Andeno (now DSM Andeno). Racemic acid **1**, easily obtained by Michael addition of thioacetic acid to methacrylic acid, resisted all attempts at classical resolution. Initially, chemically and optically pure **1** was not available either for reference or for investigation of the preparation of diastereoisomeric salts with resolving agents. Work with the more stable and somewhat less flexible $S$-benzoyl analogue met with more success (confirming our prejudice about the dependence of the resolution result on

*Chirality in Industry II*. Edited by A. N. Collins, G. N. Sheldrake and J. Crosby
© 1997 John Wiley & Sons Ltd

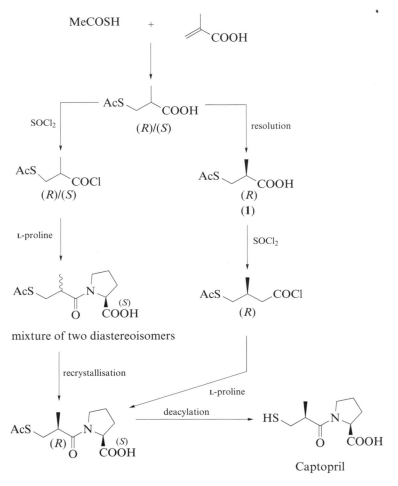

**Figure 1** Captopril syntheses. Left: original Squibb process. Right: DSM Andeno process

conformational rigidity). With the traditional resolving agent cinchonidine an enantiopure product could be obtained,[2] but this did not provide a satisfactory basis for an economic industrial process. The approach suffered from several disadvantages:

- cinchonidine was expensive and difficult to recover;
- thiobenzoic acid was expensive;
- recrystallisations were required to obtain an enantiopure product;
- the unwanted enantiomer cannot be racemised for recycle in the process or recycled for other uses;

# RATIONAL DESIGN IN RESOLUTIONS

- only $S$-acetyl-derived products were registered with the FDA.

There were initial encouraging results using simple amino alcohols as resolving agents; moreover, some of these resolving agents were cheaply available on industrial scale. However, these resolutions were found to be the result of a difference in the *rate* of crystallisation of the two diastereoisomeric salts and *not based on a difference in solubility*. Experiments on a larger scale, which took longer, resulted in both diastereoisomers crystallising.

Eventually, a reliable laboratory-scale resolution was developed using D-(+)-$N$-benzyl-$\alpha$-phenethylamine (not a particularly rigid molecule, given the initial preconception!) as the resolving agent.[2] At this point, only the FDA issue remained to be addressed; resolution of intermediate **1** was still required. As sufficient amounts of enantiopure D-(−)-and L-(+)-$S$-benzoyl-$\beta$-mercaptoisobutyric acid were now available, an efficient debenzoylation–reacylation procedure could, at last, be developed, giving access to the optically pure enantiomers of **1**. Systematic preparation of a range of diastereoisomeric salts and measurement of solubility differences (and other physical parameters; DSC, phase diagrams) indicated some good resolving agents for racemic **1**, leading to the present situation where:

- all captopril is produced via enantiopure **1**;
- the fact that the unwanted enantiomer L-(+)-**1** cannot readily be racemised or recycled does not preclude the economic success of the route while other potentially more economic methods for preparing **1**, e.g. asymmetric addition of thioacetic acid to methacrylic acid (requiring asymmetric protonation) are not available;
- importantly, several of the empirical rules for developing resolutions, such as the need for maximum rigidity in the starting molecules, were found not to apply.

With interest from the Dutch government and universities in the fine chemical industry, it was timely to join forces in a more fundamental research programme on resolutions and chirality. A collaboration between Andeno and the University of Groningen started in 1980 and has evolved into an almost nationwide programme between DSM Andeno, DSM Research, Chemferm and the Universities of Nijmegen and Groningen. An ambitious, long-term programme was set up aimed at:

- rational design and synthesis of new resolving agents (all the commonly used resolving agents are selected on the basis of trial and error);
- revealing, ideally common, key parameters behind successful resolutions;
- defining chirality in quantitative physical parameters and developing methods, ultimately, for predicting good resolving conditions;
- developing asymmetric transformations and more efficient methods for racemisations and recycling.

Interim results from the programme, at approximately its half way point at the time of writing, are presented in this chapter.

## 5.3 RATIONAL DESIGN AND SYNTHESIS

Very few approaches to the rational design of resolutions have been published. After an unsuccessful attempt by Winther in 1895[3] to classify alkaloids as resolving agents, a few publications have appeared in the last two decades. Useful compilations have been given by Wilen, Jacques and Collet,[4-6] and Fogassy and co-workers[7] have found interesting relationships between resolution results and physical parameters of the diastereoisomeric salts. Statistically relevant relationships were found between resolution results in a series of substituted phenylglycines with tartaric acid and parameters such as Taft $\sigma$-values, melting points, heats of fusion, densities and solubilities.

In addition, Fogassy and co-workers[8] have studied solvent effects. Although a lot of data were generated, basic explanations of the resolution results are lacking mainly because crystal structures of the diastereoisomeric salt pairs were not available.[8] The same applies to an extensive study by Arnett and co-workers[9] on the resolution of mandelic acid with $\alpha$-phenethylamine, ephedrine and pseudoephedrine. An attempted crystal study by Gould and co-workers[10] remained incomplete because crystal structures of both partners of the diastereoisomeric pairs were not available. They did indicate, however, that small and subtle differences in interactions within the diastereoisomeric salts could have a profound effect on crystal structure and that the solvent, water, played a decisive role.

Using rationally designed molecular structures for new resolving agents to avoid preconceptions from historical studies, we embarked upon a programme of synthesis. Acidic resolving agents were chosen as the target because only a limited number, particularly of strongly acidic resolving agents, are available for industrial use (tartaric acids, mandelic acids, camphorsulphonic acids). Based on the theoretical and experimental results considered important at that time[1,11] in the design of resolving agents, cyclic phosphoric acids (**2**) were chosen. In making this selection, the following criteria (adapted from ref. 11) were considered:

(i) Strongly acidic and basic resolving agents were preferred over weakly acidic and basic reagents to increase the possibility of salt formation.
(ii) The stereocentre in the resolving agent should be close to the centre of salt formation.
(iii) Rigidity was preferred over flexibility. The incorporation of additional functionality (to increase interactions) should enhance rigidity.
(iv) Both enantiomers should be readily available.
(v) The resolving agent should be chemically and optically stable under the conditions used for the formation, separation and dissociation of diastereoisomeric salts and for recovery of the resolving agent.

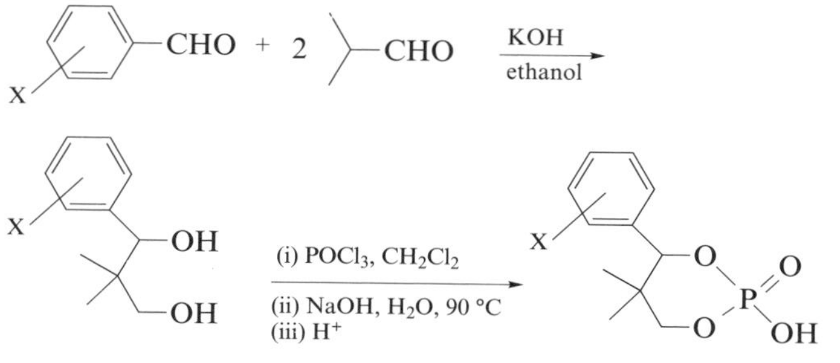

# RATIONAL DESIGN IN RESOLUTIONS

**Figure 2** Synthesis of the cyclic phosphoric acids **2**

| X | X |
|---|---|
| (a) H | (i) 4-OMe |
| (b) 2-F | (j) 3,4-OCH$_2$O |
| (c) 2-Cl | (k) 2-NO$_2$ |
| (d) 2-Br | (l) 3-NO$_2$ |
| (e) 2,4-Cl | (m) 2-thienyl |
| (f) 2,6-Cl | (n) 2-furyl |
| (g) 4-Cl | (o) 2-OEt |
| (h) 2-OMe | (p) 4-Me |

The synthesis of compounds of the general structure **2** is straightforward and an extensive range could be prepared (see Figure 2).[11] Also, resolution was easily accomplished using D-(−)-p-hydroxyphenylglycine, ephedrine or 2-amino-1-phenylpropane-1,3-diol. The enantiopure phosphoric acids proved to be promising resolving agents, and several amines, alcohols and amino acids were resolved through selective diastereoisomeric salt crystallisation, as shown in Table 1. Examples of excellent resolutions were found which afforded diastereopure salts in a single crystallisation in yields of 40% or better.

Recovery of the phosphoric acids is easy owing to their very low solubility in water and common solvents, and their chemical, thermal and optical stability was excellent. Compared with camphorsulphonic acid (until then the best strong acid industrial resolving agent), a much improved strongly acidic resolving agent became available: *p*-chlorophenyl-substituted phosphoric acid. The resolution efficiency of camphorsulphonic acid for 21 bases was compared with that of 4-chlorophenylphosphoric acid, and the results (Table 2) show clearly the better resolving ability of the phosphoric acid.

Before concluding this section, a cautionary remark must be added. Using the same criteria as for the design of the phosphoric acids, the basic analogues **3**, **4** and **5** were prepared (Figure 3). Compounds **3** and **4** were readily resolved by the phosphoric acids **2**, but they did not in themselves perform as good resolving bases. Compound **5** could not even be resolved by **2**. Also, a 'designer' resolving agent for the attractive captopril precursor β-chloroisobutyric acid **6** did not

**Table 1** Resolutions of various amines and amino acids with cyclic phosphoric acids **2**

| Amine /amino acid | Phosphoric acid, **2**. X | Result (small-scale)[a] |
|---|---|---|
| MeO-C$_6$H$_4$-CH$_2$CH(Me)NHEt | H, 4-Cl, 2-OEt, 2,4-Cl$_2$ | − |
|  | 2-OMe | + |
|  | 2-Cl | ++ |
| Cl-C$_6$H$_4$-C(NH$_2$)-C(NH$_2$)-C$_6$H$_4$-Cl | 2-OMe | − |
|  | 2-Cl | + |
|  | H, 2,4-Cl$_2$ | ++ |
| MeO-C$_6$H$_4$-CH$_2$CH(NH$_2$)(Ph) | 4-Cl | − |
|  | H, 2-Cl | + |
|  | 2-OMe, 2-OEt, 2,4-Cl$_2$ | ++ |
| H$_2$N-C(=NH)-SCH$_2$CH(Me)(CO$_2$H) | H, 4-Cl, 2-OEt | − |
|  | 2-OMe, 2-Cl, 2,4-Cl$_2$ | + |
| (CH$_3$)$_2$CH-CH(NH$_2$)(CO$_2$H) | H, 4-Cl, 2-OMe, 2-OEt | − |
|  | 2-Cl | + |
| Ph-CH(NH$_2$)(CO$_2$H) | H, 4-Cl, 2-Cl, 2-OMe | − |
| HO-C$_6$H$_4$-CH(NH$_2$)(CO$_2$H) | H, 4-Cl | − |
|  | 2-OMe, 2-Cl, 2,4-Cl$_2$ | ++ |
| PhCH$_2$CH(NH$_2$)(CO$_2$H) | H, 4-Cl, 2-OEt, 2,4-Cl$_2$ | − |
|  | 2-OMe | + |
|  | 2-Cl | ++ |
| PhCH$_2$CH$_2$CH(NH$_2$)(CO$_2$H) | H, 2-OMe, 2-OEt, 2-Cl | − |
|  | 2,4-Cl$_2$ | + |
| MeSCH$_2$CH$_2$CH(NH$_2$)(CO$_2$H) | 2-OMe | − |
|  | 2-Cl | ++ |

[a] −, No or poor resolution; +, some resolution; ++, excellent resolution.

materialise from this work. Based on model studies and known preferred interactions in crystal structures, resolving agents **7** and **8** were designed and prepared. Resolutions were thwarted either at the stage of preparing enantiopure resolving agent (**8**), or in attempted resolution of acid **6** with amine **7a** or **7b**. Crystalline salts were not obtained even with enantiopure **6** and amine **7a** or **7b**.

# RATIONAL DESIGN IN RESOLUTIONS

**Table 2** Resolution efficiency of 4-chlorophenylphosphoric acid (A) and camphorsulphonic acid (B)

| Base | Resolution efficiency, $S^a$ | |
|---|---|---|
| | A | B |
| D-(−)-2-Aminobutan-1-ol | 0.27 | 0.17 |
| L-(+)-*threo*-2-Amino-1-phenylpropane-1,3-diol | 0.35 | 0.33 |
| (2$S$,3$R$)-(+)-4-Dimethylamino-1,2-diphenyl-3-methylbutan-2-ol | 0.25 | 0.69 |
| (−)-Ephedrine | 0.50 | 0.24 |
| (+)-$N$-Methylephedrine | 0.31 | 0.66 |
| L-(−)-Norephedrine | 0.27 | 0.07 |
| D-(−)-Phenylglycinol | 0.34 | |
| (+)-Pseudoephedrine | | 0.04 |
| D-(+)-α-Amino-ε-caprolactam | 0.18 | |
| D-(+)-α-Dimethylamino-ε-caprolactam | | 0.06 |
| ($R$)-(−)-1-Cyclohexylethylamine | 0.58 | 0.13 |
| L-(−)-α-Methylbenzylamine | | 0.05 |
| (−)-Brucine | 0.13 | 0.54 |
| (+)-Dehydroabietylamine | 0.58 | 0.20 |
| (−)-Cinchonidine | 0.79 | 0.18 |
| (+)-Cinchonine | | 0.06 |
| (+)-Quinidine | 0.39 | 0.05 |
| (−)-Quinine | 0.71 | 0.04 |
| (+)-Hydroquinidine | 0.38 | 0.03 |
| (−)-Hydroquinine | | 0.10 |
| L-(−)-Strychnine | 0.54 | 0.26 |
| Average | 0.41 | 0.20 |
| Number of resolutions with efficiency >0.50 | 6 | 3 |

$^a$ Resolution efficiency $S$ is defined as the difference between solubilities of the separate diastereoisomeric salts divided by the solubility of the most soluble diastereoisomer; see also ref. 14.

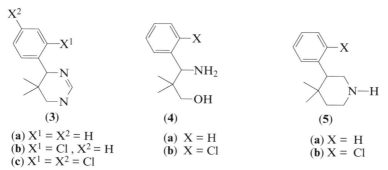

Figure 3 Unsuccessful 'designer' resolving agents

## 5.4 RATIONALE OF A RESOLUTION

One obvious requirement for studying the basis for a resolution, namely that both salts of the diastereoisomeric pair should be available in a pure and well defined form, is not always easy to meet in practice. We found ephedrine and the cyclic phosphoric acids to be ideal pairs for salt preparations and further physical and model studies. Nineteen pairs of diastereoisomeric salts stemming from differently substituted acid–base pairs were prepared[13] and several physical parameters were determined, of which a selection are given in Table 3.

**Table 3** Physical properties of cyclic phosphoric acid/ephedrine salts

| Diastereoisomeric salt (X = arene subst. in acid, Y= arene subst. in ephedrine) | | | Solubility[a] | | | M.p (K) | $\Delta H_{fus}$ (kJ mol$^{-1}$) |
|---|---|---|---|---|---|---|---|
| | | | EtOH | Pr$^i$OH | Pr$^n$OH | | |
| 1 X = H | Y = H | n[b] | 8.8 | 1.37 | 5.33 | 504.1 | 48.5±0.4 |
|  |  | p | 8.4 | 1.30 | 5.22 | 508.3 | 47.6+1.3 |
| 2 X = 2-F | Y = H | n | 22.1 | 3.66 |  | 483.0 | 40.6±0.1 |
|  |  | p | 46.0 | 8.19 |  | 474.2 | 37.1±0.4 |
| 3 X = 2-Cl | Y = H | n | 5.7 | 0.86 | 3.34 | 492.7 | 47.7±0.8 |
|  |  | p | 23.6 | 3.39 | 13.98 | 491.3 | 39.5±1.0 |
| 4 X = 2-Br | Y = H | n | 9.3 | 1.48 |  | 491.6 | 42.6±0.6 |
|  |  | p | 16.4 | 2.56 |  | 482.0 | 40.4±0.8 |
| 5 X = 2,6-Cl$_2$ | Y = H | n | 10.0 | 1.40 |  | 488.4 | 40.6±0.5 |
|  |  | p | 62.6 | 7.56 |  | 462.9$^c$ 476.3$^c$ | 31.8±4.0$^d$ |
| 6 X = 2,4-Cl$_2$ | Y = H | n | 8.1 | 1.29 | 4.9 | 488.5 | 50.4±0.8 |
|  |  | p | 49.5 | 7.62 | 30.9 | 457.4 | 31.9±0.4 |
| 7 X = 4-Cl | Y = H | n | 3.32 | 0.45 |  | 510.7 | 52.0±1.1 |
|  |  | p | 26.7 | 3.99 |  | 478.3 | 41.2±0.5 |
| 8 X = 2-Me | Y = H | n | 5.75 | 0.95 | 3.57 | 487.4 | 54.3±1.0 |
|  |  | p | 17.4 | 2.74 | 11.0 | 498.7 | 46.6+0.3 |
| 9 X = 4-Me | Y = H | n | 17.5 | 2.82 |  | 483.0 | 36.4±0.9 |
|  |  | p | 20.0 | 3.41 |  | 482.9 | 33.6+1.0 |
| 10 X = 2-OMe | Y = H | n | 4.1 | 0.61 | 2.39 | 495.5 | 47.6±1.2 |
|  |  | p | 23.1 | 3.04 | 13.9 | 475.4 | 43.9+0.4 |
| 11 X = 2-NO$_2$ | Y = H | n | 6.70 | 0.79 | 3.10 | 479.5 | 44.0±0.1 |
|  |  | p | 4.36 | 0.44 | 2.07 | 486.6 | 48.6±0.5 |
| 12 X = H | Y =2-Cl | n |  | 1.41 |  | 506.4 | 41.0±0.5 |
|  |  | p |  | 1.28 |  | 499.2 | 47.0±0.2 |
| 13 X =2-Cl | Y = 2-Cl | n |  | 0.21 |  | 534.0 | 60.2±0.7 |
|  |  | p |  | 2.54 |  | 477.8 | 45.6+0.3 |
| 14 X = 2,6-Cl$_2$ | Y = 2-Cl | n |  | 0.99 |  | 507.4 | 45.0±0.6 |
|  |  | p |  | _$^e$ |  | _$^e$ | _$^e$ |
| 15 X = 4-Cl | Y = 2-Cl | n |  | 0.98 |  | 518.8 | 40.8±0.6 |
|  |  | p |  | _$^e$ |  | _$^e$ | _$^e$ |
| 16 X = H | Y = 2,6-Cl$_2$ | n |  | 3.79 |  | 481.5 | _$^d$ |
|  |  | p |  | 0.25 |  | 514.0 | _$^d$ |

(*continued*)

## Table 3 (continued)

| Diastereoisomeric salt (X = arene subst. in acid, Y = arene subst. in ephedrine) | | | Solubility[a] | | | M.p (K) | $\Delta H_{\text{fus}}$ (kJ mol$^{-1}$) |
|---|---|---|---|---|---|---|---|
| | | | EtOH | Pr$^i$OH | Pr$^n$OH | | |
| 17 X = 2-Cl | Y = 2,6-Cl$_2$ | n | | _[e] | | _[e] | _[e] |
| | | p | | 0.13 | | 524.4 | _[d] |
| 18 X = H | Y = 4-Cl | n | | 3.91 | | 452.2 | 43.5 ± 2.0 |
| | | p | | 1.41 | | 492.7 | 45.5 ± 1.1 |
| 19 X = 2-Cl | Y = 4-Cl | n | | _[e] | | _[e] | _[e] |
| | | p | | 0.89 | | 511.3 | 40.5 ± 2.0 |

[a] Solubilities at 25 °C (g/100g).
[b] n Salt = + − or − + ; p salt = + + or − −.
[c] Two points of fusion were observed; the first one is followed by an exothermic crystallisation.
[d] Total change in enthalpy between 460 and 480 K.
[e] Not crystalline.
[f] Reproducible values could not be obtained.

From these data the following conclusions were drawn:

- The efficiency of a resolution may be expressed as $S = (k_p - k_n)/k_p$ [where $k_p$ and $k_n$ are the respective solubility constants for the p and n salts of the diastereoisomeric pair (see refs 14 and 15)] and derived from the differences in solubility within the diastereoisomeric salt pair which correlates fairly well with the experimental results (yield and optical purity) of that resolution (see also Table 4 and 5).
- The resolution efficiency $S$ also correlates fairly well with $\Delta H_f$, the difference in enthalpy of fusion between a diastereoisomeric pair. Using a known thermodynamic model[14] for the crystallising diastereoisomeric pair, it is possible to arrive at a linear relationship between resolution efficiency expressed as $\ln(k_p/k_n)$ and the difference in lattice enthalpy $\Delta H_{\text{solid}}$ as shown in Figure 4.
- Useful correlations of resolution efficiency with other physical parameters could not be obtained, although low solubility and high melting point within a salt pair are generally found (see also the results of Fogassy and co-workers' work[8]).
- Effects of substitution are most profound in the phosphoric acid (see Table 4); substituents in the ephedrine moiety have a smaller influence, although they enhance the effect of the phosphoric acid substituents. In this way, very efficient resolutions for chloro-substituted ephedrines with chloro-substituted phosphoric acids were predicted and found (see Table 4).

In order to find a more definitive explanation of the successful resolutions with these phosphoric acids, the crystal structures of six diastereoisomeric salt pairs were studied,[14–16] and the results are summarised in Table 5. Analysing the crystal structures for an explanation of the efficiency of the resolution revealed no common, single cause.[17]

A subtle interplay of, mainly, hydrogen bonding, van der Waals and electrostatic forces is presumed to be responsible for the resolution results, whereby each

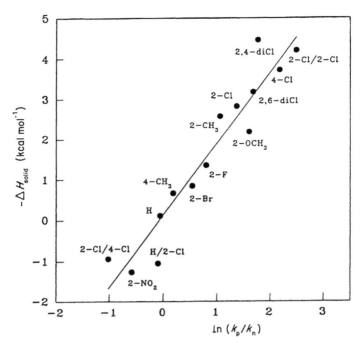

**Figure 4** Plot of $-\Delta H_{solid}$ versus $\ln(k_p/k_n)$. Labels indicate substituents (with unsubstituted ephedrine) or acid–base phenyl substituents

**Table 4** Calculated maximum values of $S$ for the phosphoric acids/ephedrine salts, derived from the solubilities given in Figure 6

| Diastereoisomeric salt | | Calculated value of $S^a$ | | |
|---|---|---|---|---|
| Phosphoric acid | Ephedrine | EtOH(abs.) | $Pr^iOH$ | $Pr^nOH$ |
| 1  X = H | Y = H | −0.05 | −0.05 | −0.02 |
| 2  X = 2-F | Y = H | 0.52 | 0.55 | |
| 3  X = 2-Cl | Y = H | 0.76 | 0.75 | 0.76 |
| 4  X = 2-Br | Y = H | 0.43 | 0.42 | |
| 5  X = 2,6-Cl$_2$ | Y = H | 0.84 | 0.81 | |
| 6  X = 2,4-Cl$_2$ | Y = H | 0.84 | 0.83 | 0.84 |
| 7  X = 4-Cl | Y = H | 0.88 | 0.88 | |
| 8  X = 2-Me | Y = H | 0.67 | 0.65 | 0.68 |
| 9  X = 4-Me | Y = H | 0.13 | 0.17 | |
| 10  X = 2-OMe | Y = H | 0.82 | 0.80 | 0.83 |
| 11  X = 2-NO$_2$ | Y = H | −0.35 | −0.44 | −0.33 |

(*continued*)

# RATIONAL DESIGN IN RESOLUTIONS

**Table 4** (*continued*)

| Phosphoric acid | Ephedrine | | | |
|---|---|---|---|---|
| | Y = H | Y = 2-Cl | Y = 2,6-Cl$_2$ | Y = 4-Cl |
| X = H | −0.05 | −0.09 | —[b] | —[b] |
| X = 2-Cl | 0.76 | 0.92 | −0.93 | −0.64 |
| X = 2,6-Cl$_2$ | 0.81 | + +[c] | + +[c] | + +[c] |
| X = 4-Cl | 0.88 | + +[c] | | |

[a] Positive values of $S$ indicate a more soluble p salt; negative values of $S$ indicate a more soulble n salt.
[b] Not determined; small-scale resolutions indicated poor resolution results.
[c] The better dissolving diastereomeric salt was obtained as an oil; small-scale resolution indicated that very efficient results are obtained.

compound, even within an isostructural series, reaches a unique optimum arrangement for crystal packing. Water or solvent molecules included in the crystals have a stabilising influence only on otherwise unstable crystal packings; inclusion of solvent does not generally lead to much more stable crystals. When inclusion solvents are not available, the non-solvated salt of the crystal pair might become the least soluble, resulting in reversal of the resolution result and, possibly, allowing selection for either enantiomer even though the resolving agent is only available as one enantiomer. Resolution of phosphoric acid **2** (X = o-Cl) with L-(+)-*threo*-2-amino-1-phenylpropane-1,3-diol in methanol vs toluene or chloroform is an example.

**Table 5** Resolution results and crystal structures of six phosphoric acids with ephedrine

| Phosphoric acid | | Resolution efficiency, $S$ | | Space group | Crystal water | Mode of packing | Notes |
|---|---|---|---|---|---|---|---|
| Subst. | Salt | EtOH | Pr$^i$OH | | | | |
| X = H | p | 0.05 | 0.05 | $P2_1$ | No | I$_p$ | Very poor resolution |
| | n | | | $P2_1$ | No | I$_n$ | |
| X = 2-F | p | 0.52 | 0.55 | $P2_1$ | No | I$_p$ | No difference in crystal structure; good resolution. Cause: electrostatic interactions? |
| | n | | | $P2_1$ | No | I$_n$ | |
| X = 2-Cl | p | 0.76 | 0.75 | $P2_1$ | No | I$_p$ | Great difference in crystal packing; very good resolution. Cause: van der Waals interactions |
| | n | | | $P2_1$ | No | III | |
| X = 2-Br | p | 0.43 | 0.42 | C2 | Yes | II | Mediocre resolution. No crystals of *p*-salt without water present |
| | n | | | $P2_1$ | No | I$_n$ | |
| X = 2,6-Cl$_2$ | p | 0.84 | 0.81 | $P2_1$ | Yes | II | Very good resolution. No crystals of *p*-salt without water present |
| | n | | | $P2_1$ | No | III | |
| X = 2,4-Cl$_2$ | p | 0.84 | 0.83 | $P2_12_12_1$ | Yes | II | Very good resolution. No crystals of *p*-salt without water present |
| | n | | | $P2_1$ | No | I$_n$ | |

## 5.5 PREDICTING A RESOLUTION?

In view of the results described above, this would be a brave proposition. Attempts to compute quantitatively the differences in lattice energy of the crystal pairs in Figure 5 were not successful. The results of molecular mechanics calculations supported in only a very qualitative way the correlation of resolution results and lattice energy, and also revealed that different acid–base interactions in the hydrophobic layers of the crystals influence the resolution.[14-16] Although calculating and predicting crystal packings of some rather rigid organic heterocycles are almost within reach,[18] calculation of the far more complicated diastereoisomeric salts is still very difficult (even without addressing the question of the difference in crystal packing with its diastereoisomeric counterpart or the effect of solvent inclusion).

Accepting the severe limitations of the models presented above, a different approach was taken: principal component analysis (PCA). This statistical technique[19] allows the reduction of a large number of physical parameters to a much smaller number of new parameters, the principal components (PCs), which are linear combinations of the original parameters. When a full set of physical data for a series of compounds can be reduced to two or three PCs, useful plots can be obtained of these PCs, the compounds being located on the plots according to their individual properties. Several series of amines, Lewis acids, ketones and monosaccharides have been analysed in this way, and the use of PCA for the selection of the most suitable solvent for a given synthesis has been demonstrated.[19]

A set of 34 chiral amines was characterised using 39 parameters (both measured and calculated data were used). Applying PCA to this data set showed that 90% of the original information could be described with seven PCs, and only two PCs were needed to retain 65% of the initial data. Plotting PC-1 vs PC-2 gave clear groupings of the 34 resolving agents (see ref. 17). However, attempts to correlate the locations on the plot of these resolving amines with the results of the resolving behaviour of the cyclic phosphoric acids have not been successful.

In studies which were limited to a smaller set of 23 amines, characterised by fully reliable physical parameters (26 different properties), an impressive 75% of the original information was retained using only two PCs. A plot of PC-1 vs PC-2 (see Figure 5) shows clustering in two domains: simple amines at the left-hand side and the complex alkaloids on the right. There was, however, still no correlation with the resolution results of the cyclic phosphoric acids. An obvious explanation for this lack of correlation could be the limited contribution of chirality parameters in the data set. Only optical rotations had been used up to this point, and a more structure-related description of chirality was needed. Of the various theoretical models available,[20] we chose the molecular similarity model, which determines the degree of non-superimposability of two enantiomers (i.e. very close to the original definition of chirality).[21,22]

The non-overlap can be computed in terms of molecular properties such as

# RATIONAL DESIGN IN RESOLUTIONS

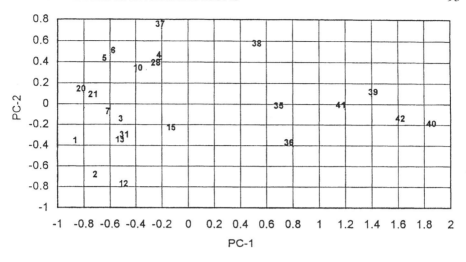

1  (S)-(+)-2-Aminopropanol-1(Alaninol)
2  (R)-(−)-2-aminobutan-1-ol
3  (1S,2S)-(+)-2-amino-1-phenylpropane-1,3-diol
4  (2S, 3R)-(+)-4-Dimethylamino-3-methyl-1,2-diphenylbutan-2-ol (Darvon alcohol)
5  (1R,2S)-(−)-2-Methylamino-1-phenylpropan-1-ol (Ephedrine)
6  (1S,2R)-(+)-(-2-Dimethylamino-1-phenylpropan-1-ol (N-Methylephedrine)
7  (1R, 2S)-(−)-2-Amino-1-phenylpropan-1-ol (Norephedrine)
8  (S)-(−)-2-Amino-3-phenylpropan-1-ol (Phenylalaninol)
9  (R)-(−)-2-Amino-2-phenylethanol (Phenylglycinol)
10  (1S,2S)-(+)-2-Methylamino-1-phenylpropan-1-ol (Pseudoephedrine)
11  (1R, 2R-(−)-2-Amino-1-phenylpropan-1-ol (Norpseudoephedrine)
12  (R)-(+)-3-Aminohexahydroazepin-2-one (α-Amino-ε-caprolactam)
13  (R)-(+)-3-Dimethylaminohexahydroazepin-2-one (α-Dimethylamino-ε-caprolactam)
14  L-(+)-Leucinamide
15  D-(−)-Phenylglyeniamide
16  (+)-3-Aminomethylpinane
17  (R)-(+)-Bornylamine
18  (R)-(−)-Isobornylamine
19  (−)-cis-Myrtanylamine
20  (R)-(−)-1-Cyclohexylethylamine
21  (S)-(–)-1-Phenylethylamine
22  (R)-(+)-1-(1-Naphthyl)ethylamine
23  L-(+)-2,4-Diaminobutyric acid

24  L-(+)-Ornithine
25  (−)-N-Methyl-D-glucamine
26  D-(+)-Glucosamine
27  (S)-(−)-2-Hydroxymethyl-1-methylpyrrolidine (N-methylprolinol)
28  (S)-(−)-3-(1-Methyl-2-pyrrolidinyl)pyridine (Nicotine)
29  (S)-(+)-2-(Hydroxymethyl)pyrrolidine (prolinol)
30  (S)-(+)-2-(methoxymethyl)pyrrolidine (O-methylprolinol)
31  L-(−)-2-Pyrrolidinecarboxylic acid amide (Proliamide)
32  L-(−)-Methyl 2-pyrrolidinecarboxylate (Proline methyl ester)
33  L-(−)-α-(Methylaminomethyl)-3,4-dihydroxybenzyl alcohol (Adrenaline)
34  (R)-(−)-α-(aminomethyl)3,4-dihydroxybenzyl alcohol (Noradrenaline)
35  (−)-Brucine
36  (−)-Strychnine
37  (+)-Dehydroabietylamine
38  (−)-Cinchonidine
39  (+)-Cinchonine
40  (+)-Quinidine
41  (−)-Quinine
42  (+)-Hydroquinidine
43  (−)-Hydroquinine
44  (−)-Sparteine
45  (+)-Calycanthine
46  (−)-Eburnamonine
47  α-Lobeline

**Figure 5** Plot of PC-1 versus PC-2 for 23 resolving bases (out of 47 studied)

electron density, electrostatic potential, electric field and van der Waals volume. An interesting application of this concept has been given by Seri-Levi and Richards,[23] who showed a correlation between the chirality indices of several chiral drugs and their eudismic ratios (i.e. a relationship between drug–receptor interaction and molecular chirality).

Whether chirality indices will turn the PCA approach into a useful tool remains to be seen, although its application in the field of drug–receptor interactions shows promise. Although drug–receptor interactions are extremely complicated, they resemble more closely the interaction of two molecules than the multi-molecule interactions of a crystallisation process. Predicting the latter is still beyond present models.

Models to describe or predict 1:1 interactions of molecules are more readily available. For example, resolution of enantiomers using chiral column chromatography can nowadays be modelled fairly well[24] using techniques which have even been used to develop new or improved homochiral stationary phases for chromatographic resolution of enantiomers. Leusen and co-workers[25] have developed a protocol to improve a chiral stationary phase in a rational way. The method starts with an existing molecular structure from which the geometry of low-energy configurations of the diastereoisomeric complexes of substrate and chiral stationary phase are computed. These calculations are repeated after small, intuitive changes towards lower energy configurations are introduced in the original molecular structure.

## 5.6 ASYMMETRIC TRANSFORMATIONS, RACEMISATIONS AND RECYCLING

There are two prerequisites for an industrially viable resolution: an efficient resolving agent and an efficient recycling procedure. As well as recycling the resolving agent, it is increasingly important for an economical process to recycle the unwanted enantiomer (typically obtained in 60–70% yield and in lower enantiopurity than the desired enantiomer). Other aspects of particular importance to industrial resolution processes are efficient recovery of solvents and the processing and minimisation of inorganic salts. Resolutions through selective crystallisation of diastereoisomeric salts are difficult to imagine without the production of at least one equivalent of an inorganic salt.

Considering next the reuse of the unwanted enantiomer: the principal possibilities are shown in Figure 6. Optimising the yield of the required enantiomer can be considered at three distinct levels:

- *For enantiomers:* racemisation or inversion of configuration in a given process, or use in another synthesis are possible choices.
- *For diastereoisomeric salts:* racemisation or inversion processes again offer the best prospects for industrial viability.

# RATIONAL DESIGN IN RESOLUTIONS

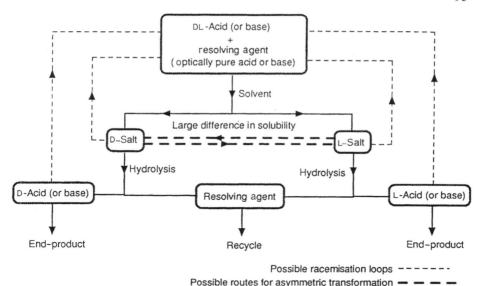

**Figure 6** General scheme for resolution through diastereoisomeric salts

- *In situ racemisation*, i.e. crystallisation-induced asymmetric transformation or 'deracemisation,' is the most desirable situation, and likely to be the basis of the most economic resolution.

Unlike resolutions and resolving agents,[26] no compilation of examples of methods for racemisations and inversions is available, but a useful overview has been given by Eliel and Wilen (see ref. 6, pp. 424–440). In industrial practice, bespoke solutions are usually required.

A few guidelines can be given, of which the most general is to put the resolution step early in a multi-step synthesis. Good racemisation procedures can often be developed when the chiral centre contains a hydrogen, or is the result of a reversible synthesis. Although early resolution is essential when the co-produced enantiomer cannot be re-used, it is not important when efficient *in situ* racemisation is possible. A late resolution without racemisation can easily be 2–4 times more costly than an early resolution with recycling of the racemised co-enantiomer.[26]

*In situ* racemisation and recycling of the unwanted diastereoisomeric salt are preferred to racemisation of the isolated, unwanted enantiomer from an economic point of view. The great difficulty in finding a suitable resolving agent at all makes the task of finding one compatible with racemisation conditions even more difficult. Simple, stable resolving amines or acids, such as camphorsulphonic acid and the cyclic phosphoric acids, are good candidates for surviving

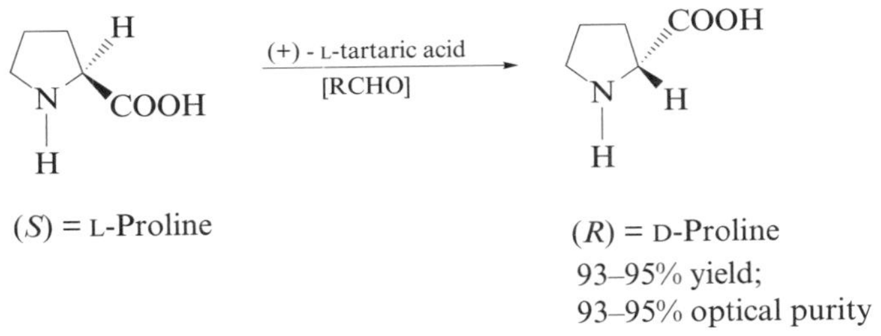

**Figure 7** Transformation of L-proline into D-proline

(S) = L-Proline

(R) = D-Proline
93–95% yield;
93–95% optical purity

demanding racemisation conditions. When resolving agents are stable to the racemisation conditions, the search for a process with asymmetric tranformation is more likely to succeed. This combination of selective crystallisation with racemisation in solution is receiving increasing attention and, where it can be applied, has resulted in favourable process economics. Good examples come from the work of DSM (resolution of aminoamides with mandelic acid using benzaldehyde as the racemisation catalyst[27]), Japanese work (resolution of amino acids with sulphonic acids and salicylaldehyde as racemisation promoter[28]) and the synthesis of D-proline from readily available L-proline[29] (see Figure 7). In the latter example an equilibrium is set up between D- and L-proline through racemisation catalysed by n-butyraldehyde, followed by precipitation of the less soluble D-proline- L-tartrate.

## 5.7 CONCLUSIONS

Industry today can choose from five well established methods for the production of enantiopure compounds: selective crystallisation, asymmetric synthesis, biocatalysis, fermentation and synthesis from chiral pool compounds. Although the level of development and the number of industrial applications vary greatly between methods, there is no single 'best' technology; every product has its own most economic production process governed by volumetric productivity, economy of scale, product life cycle, production location, etc. All techniques are approaching the ideal of '100% yield and 100% enantiopurity,' i.e. of furnishing, in principle, a quantitative yield of a single enantiomer or diastereoisomer.

Predictable process design is still some way off, however. Syntheses from the chiral pool and stoichiometric asymmetric synthesis are the most advanced in this respect, but methods which employ catalysis or selective crystallisations are still far from predictable. At present, chirality is still treated as a qualitative property; a description in terms of quantitative parameters related to fundamental molecular properties could be of great help in the rational design of any process for homochiral products and, in particular, for the design of classical resolutions.

## 5.8 ACKNOWLEDGEMENTS

Major parts of the studies presented in this chapter are the results of the research work of Dr G. J. A. Ariaans, Dr F. J. J. Leusen and Dr A. D. van der Haest. The author also acknowledges the successful design and synthesis of the cyclic phosphoric acids by Dr W. ten Hoeve at an early stage of the project and the inspiring and stimulating contributions of Professor Dr H. Wynberg, without which this project would not have been possible.

## 5.9 REFERENCES

1. (a) Collins, A. N., Sheldrake, G. N., and Crosby J., *Chirality in Industry*, Wiley, Chichester, 1992; see Ch. 2; (b) Sheldon, R. A., *Chirotechnology*, Marcel Dekker, New York, 1993; see Ch. 6.
2. Océ-Anderno BV, *Eur. Pat.*, EP 8.831, 8.833 and 35.811.
3. Winther, C., *Ber. Dtsch. Chem. Ges.*, **28**, 300 (1895).
4. Jacques, J., Collet, A, and Wilen, S. H., *Enantiomers, Racemates and Resolutions*, Wiley, 1981.
5. Collet, A., and Jacques, J., *Bull. Soc. Chim. Fr.*, 3857 (1972); 3330 (1973).
6. Eliel, E. L., and Wilen, S. H., *Stereochemistry of Organic Compounds*, Wiley, Chichester, 1994; see Ch. 7.
7. (a) Fogassy, E., Faigl, F, and Acs, M., *Tetrahedron*, **41**, 2837 (1985); (b) Fogassy, E., Faigl, F., and Acs, M., *Tetrahedron*, **41**, 2841 (1985); (c) Acs, M., in *Problems and Wonders of Chiral Molecules* (ed. M. Simonyi), Akadémiai Kiadó, Budapest, 1990, p.111; (d) Fogassy, E., Acs, M., Faigl, F., Simon, K., Rohonczy, J., and Ecsery, Z., *J. Chem. Soc. Perkin Trans. 2*, 1881 (1986); (e) Fogassy, E, Faigl, F., Acs, M., Simon, K., Kozsda, É., Podányi, B., Czugler, M., and Reck, G., *J. Chem. Soc. Perkin Trans. 2*, 1385 (1988), (f) Czugler, M., Csöregh, I., Kálmánm, A., Faigl, F., and Acs, M., *J. Mol. Struct.*, **196**, 157 (1989); (g) Simon, K., Kozsda, É., Böcskei, Z., Faigl, F., Fogassy, E., and Reck, G., *J. Chem. Soc., Perkin Trans. 2*, 1395 (1990).
8. (a) Kozma, D., Nyéki, A., Acs M., and Fogassy, E., *Tetrahedron: Asymmetry*, **5**, 315 (1994); (b) Kozma, D., Acs, M., and Fogassy, E., *Tetrahedron*, **50**, 6907 (1994).
9. Zingg, S. P., Arnett, E. M., McPhail, A. T., Bothner-By, A. A. and. Gilkerson, W. R., *J. Am. Chem. Soc,*. **110**, 1565 (1988).
10. (a) Gould, R. O., and Walkinshaw, M. D., *J. Am. Chem. Soc.*, **106**, 7840 (1984); (b) Gould, R. O., Gray, A. M., Taylor, P., and Walkinshaw, M. D., *J. Am. Chem. Soc.*, **107**, 5921 (1985); (c) Gould, R. O., Kelly, R., and Walkinshaw, M. D., *J. Chem. Soc., Perkin Trans. 2*, 847 (1985).
11. (a) ten Hoeve, W., and Wynberg, H., *J. Org. Chem.*, **50**, 4508 (1985); (b) Wynberg, H., and ten Hoeve, W., *Eur. Pat. Appl.*, EP 180 276; *Chem. Abstr.*, **105**, 134150u (1989); *US Pat.*, 4 814 477.
12. v. d. Haest, A. D., Wynberg, H., Leusen, F. J. J., and Bruggink, A., *Recl. Trav. Chim. Pays-Bas*, **109**, 523 (1990).
13. v. d. Haest, A. D., Wynberg, H., Leusen, F. J. J., and Bruggink, A., *Recl. Trav. Chim. Pays-Bas*, **112**, 230 (1993).
14. Leusen, F. J. J., Bruins-Slot, H. J., Noordik, J. H., v. d. Haest, A. D., Wynberg, H., and Bruggink, A., *Recl. Trav. Chim. Pays-Bas*, **110**, 13 (1991).
15. Leusen, F. J. J., Bruins-Slot, H. J., Noordik, J. H., v. d. Haest, A. D., Wynberg, H., and Bruggink, A., *Recl. Trav. Chim. Pays-Bas*, **111**, 111 (1992).

16. Leusen, F. J. J., Noordik, J. H., and Karfunkel, H. R., *Tetrahedron*, **49**, 5377 (1993).
17. Ariaans, G. J. A., Leusen, F. J. J., and Bruggink, A., in *Proceedings of the Chiral '92 Symposium*, 1992, pp. 99–106.
18. Karfunkel, H. R., Rohde, B., Leusen, F. J. J., Gdanitz, R. J., and Rihs, G., *J. Comput. Chem.*, **14**, 1125 (1993).
19. Carlson, R., *Design and Optimization in Organic Synthesis*, Elsevier, Amsterdam, 1992.
20. See, for example; Buda, A. B., a. d. Heyde, T., and Mislow, K., *Angew. Chem.*, **104**, 1012 (1992).
21. Kuz'min, V. E., Stel'makh, I. B., Bekker, M. B., and Pozigun, D. V., *J. Phys. Org. Chem.*, **5**, 295 (1992).
22. Gilat, G., *J. Phys. A*, **22**, L545 (1989); *Found. Phys. Lett.*, **3**, 189 (1990).
23. (a) Seri-Levy, A., West, S. and Richards, W. G., *J. Med. Chem.*, **37**, 1727 (1994); (b) Seri-Levy, A. and Richards, W. G., *Tetrahedron: Asymmetry.*, **4**, 1917 (1993).
24. Däppen, R., Karfunkel, H. R., and Leusen, F. J. J., *J. Chromatog.*, **469**, 101 (1989).
25. Däppen, R., Karfunkel, H. R., and Leusen, F. J. J., *J. Comp. Chem.*, **11**, 181 (1990); see also Leusen, F.J.J., PhD Thesis, University of Nijmegen, 1993, Ch. 2.
26. Bruggink, A., Hulshof, L. A., and Sheldon, R.A., *Pharm. Manuf. Int.*, 139 (1990).
27. Boesten, W. H. J., *Neth. Pat.*, 90 00386, 90 00387, 1990.
28. Yoshioka, R., Tohyama, M., Ohtsuki, O., Yamada, S., and Chibata, I., *Bull. Chem. Soc. Jpn.*, **60**, 649 (1987); *Bull. Chem. Soc. Jpn.*, **58**, 433 (1985); *Eur. Pat. Appl.*, 70114, 1983 (to Tanabe).
29. Shiraiwa, T., Shinjo, K., and Kurokawa, H., *Chem. Lett.*, 1413 (1989).

# 6 Resolution Versus Stereoselective Synthesis in Drug Development: Some Case Histories

**B. A. ASTLEFORD and L. O. WEIGEL**
*Eli Lilly & Company, Indianapolis, IN, USA*

| | | |
|---|---|---|
| 6.1 | Introduction | 99 |
| | 6.1.1 Enantiomers and Drug Development | 100 |
| 6.2 | Duloxetine (LY248686) | 101 |
| | 6.2.1 Enantiospecific Ketone Reduction | 101 |
| | 6.2.2 Resolution–Deracemisation | 103 |
| | 6.2.3 Applications to Other Systems | 104 |
| 6.3 | Loracarbef (LY163892) | 105 |
| | 6.3.1 Synthesis from Sodium Erythorbate | 107 |
| | 6.3.2 Resolution via a Phthalimido Group | 110 |
| | 6.3.3 Resolution of other β-Lactam Intermediates | 111 |
| 6.4 | Human Steroid 5-α-Reductase Inhibitor (LY300502) | 111 |
| | 6.4.1 Synthesis via a Chiral Auxiliary | 111 |
| | 6.4.2 Resolution via Methanolysis of Lactams | 112 |
| 6.5 | Conclusion | 114 |
| 6.6 | Acknowledgements | 115 |
| 6.7 | References and Notes | 115 |

## 6.1 INTRODUCTION

In the last decade, many new and exciting techniques for the production of enantiomerically pure compounds have been discovered. These methods include numerous enantiospecific reactions, asymmetric catalytic transformations, deracemisations, the use of chiral pool substrates, enzymatic methods and others.[1] Classical resolution may seem to be a 'lost art' when compared with these newer methodologies. Although apparently lacking elegance, the resolution of enantiomers by the formation of diastereoisomeric derivatives still plays a critical role in the production of non-racemic chemicals.[2] The objective of this chapter is to examine three case histories of drug development which illustrate the utility of resolutions.[3]

*Chirality in Industry II.* Edited by A. N. Collins, G. N. Sheldrake and J. Crosby
© 1997 John Wiley & Sons Ltd

## 6.1.1 ENANTIOMERS AND DRUG DEVELOPMENT

Although numerous stereoselective and asymmetric transformations are available to chemists, relatively few are selected and developed for large scale synthesis. In reality, the manufacture of enantiomerically pure drugs is not entirely driven by overall yield efficiency, but also by the practicality of the synthesis. Issues such as speed-to-market, generally small production requirements, economics and short effective patent life are primary considerations in the route selection and the finalised manufacturing protocol. In early drug development, Chemical Process Research and Development (CPR&D) groups within the industry are invariably charged with two critical functions: rapidly providing kilograms of the target compound and early route selection for the manufacture of the drug. Enantiospecific syntheses are usually tried first because of the potential for a high yield. In many cases, this approach is more time consuming and substrates are often unsuitable. In the end, enantiomers are often obtained by classical resolution owing to time constraints, economics and reliability. However, one drawback is that resolutions are difficult without a useful 'handle' (i.e. covalent or ionic resolving agent). Even with these limitations, classical resolutions are still widely used. In 1993, a survey of the synthetic methods for the preparation of over forty enantiopure drugs in Lilly's CPR&D group showed that nearly two thirds employed ionic or covalent resolutions (Table 1).[4]

In this chapter, three enantiospecific syntheses in an early stage of development will be presented. In each case history the asymmetric strategies were

**Table 1** Syntheses of Enantiopure Drugs at Lilly CPR & D in 1993[a]

| | |
|---|---|
| 'Chiral pool' or semi-synthetic | 5.5 (20%) |
| Asymmetric synthesis | 6.0 (22%) |
| Resolution (ionic or covalent) | 15.5 (57%) |

[a] Data gathered from 40 drug candiates at phase I to phase III of development

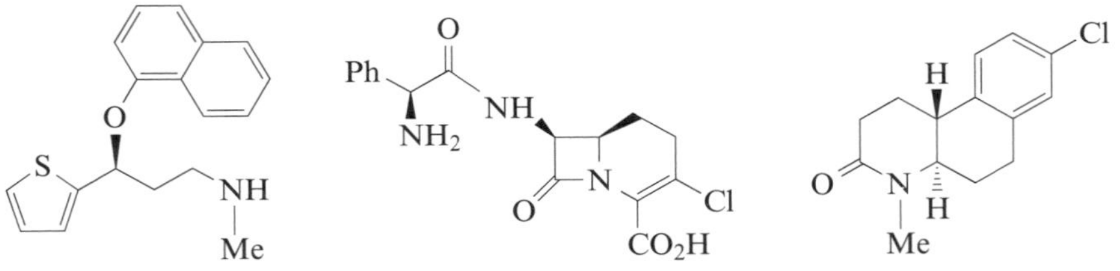

**Figure 1** Duloxetine (**1**), Loracarbef (**2**) and LY300502 (**3**)

altered and 'specialised' resolutions were invented. The three cases are duloxetine (**1**), loracarbef (**2**) and the 5-α-reductase inhibitor LY300502 (**3**) (Figure 1).

## 6.2 DULOXETINE (LY248686)

Duloxetine (LY248686, **1**) is a potent inhibitor of the serotonin (5HT) and norepinephrine (NE)[5] uptake carriers. The drug is in late phase II clinical trials for depression and/or urinary incontinence, and is similar to several other drug candidates possessing the non-racemic [3-(1-aryloxy)-3-arylpropanamine] backbone. Brown, Sharpless and others[6a–c] have published enantioselective syntheses of related drugs with strategies similar to that used for duloxetine (Scheme 1).

### 6.2.1 ENANTIOSPECIFIC KETONE REDUCTION

Initial development efforts centred on a yield-efficient enantioselective synthesis. Duloxetine (**1**) was prepared in kilogramme quantities in about 60% overall yield from 2-acetylthiophene (**4**) *via* a four-stage sequence of Mannich reaction of 2-acetylthiophene; asymmetric reduction of ketone **5**, alkylation of alcohol **6** with 1-fluoronaphthalene and *N*-demethylation of adduct **7** to yield **1** (Scheme 1).[7]

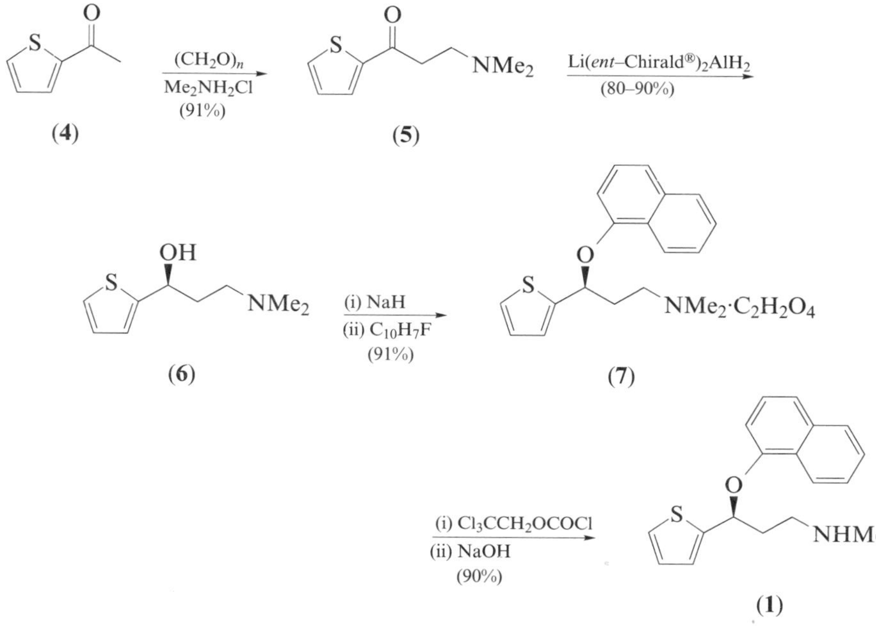

**Scheme 1** Enantiospecific synthesis of duloxetine

During the course of this work, several methods for the enantiospecific reduction of ketone **5**, including reduction with yeast,[8] Corey's oxazaborolidine–borane system[9] and BINAP–LAH[10], were investigated. The reduction of the Mannich base **5**[11] with the kinetically formed Yamaguchi–Mosher–Pohland (YMP)[12] complex derived from *ent*-Chirald® (**8**) was selected for development. Thus, reaction of two equivalents of the ligand **8** with lithium aluminium hydride in toluene at low temperature in a continuous-flow reactor afforded the metastable complex **9** (Scheme 2). Immediate reduction of **5** in the flow reactor provided the amino alcohol **6** in approximately 85–90% ee and 90% yield.

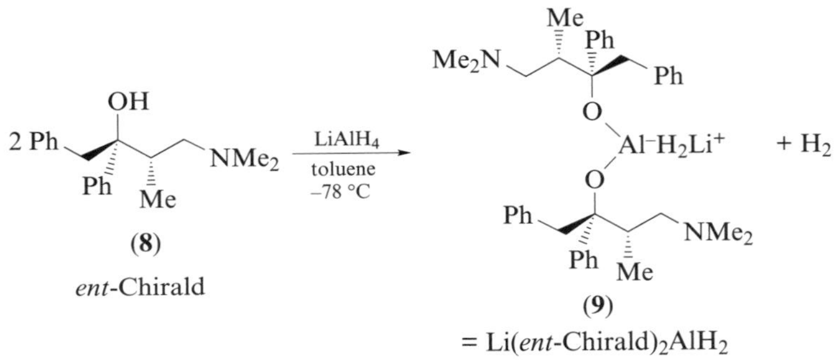

**Scheme 2** Preparation of Li(*ent*-Chirald)$_2$AlH$_2$

Chirald® is a trade name for amino alcohol **12**, which is a resolved intermediate in the Lilly production of Darvon® (Scheme 3). Both Chirald® and *ent*-Chirald are available from the process. Treatment of Mannich base **10** with benzylmagnesium chloride followed by resolution of the diastereoisomeric mixture with 0.5 equivalents of camphorsulphonic acid [(−)-CSA] in ethanol yielded the resolved salt of Chirald (**11**) (Scheme 3).[13]

The CSA salt of *ent*-Chirald (**14**) was obtained from the mother liquors containing Chirald (and a small amount of the diastereoisomers **13**) by crystallisation in acetone. Neutralisation of each isolated salt and recovery of the (−)-CSA afforded Chirald (**12**) and *ent*-Chirald (**8**). Thus, only 0.5 mole equivalents of (−)-CSA was employed in the resolution of both antipodes. Since *ent*-Chirald (**8**) was a waste stream product, the net cost to the duloxetine reduction process involved only the crystallisation of *ent*-Chirald (**8**) from acetone and recovery of the (−)-CSA.[13] The free availability of the auxiliary was the primary reason for selection of this route.

Although large amounts of duloxetine (**1**) were produced, this process was *not* taken to production. A total input of nearly 5 kg of the complex **9** in 100 l of toluene was required to produce 1 kg of alcohol **6** from ketone **5**. Even with the low cost and recycle of *ent*-Chirald, the overall efficiency was, at best, comparable with that of a classical resolution.

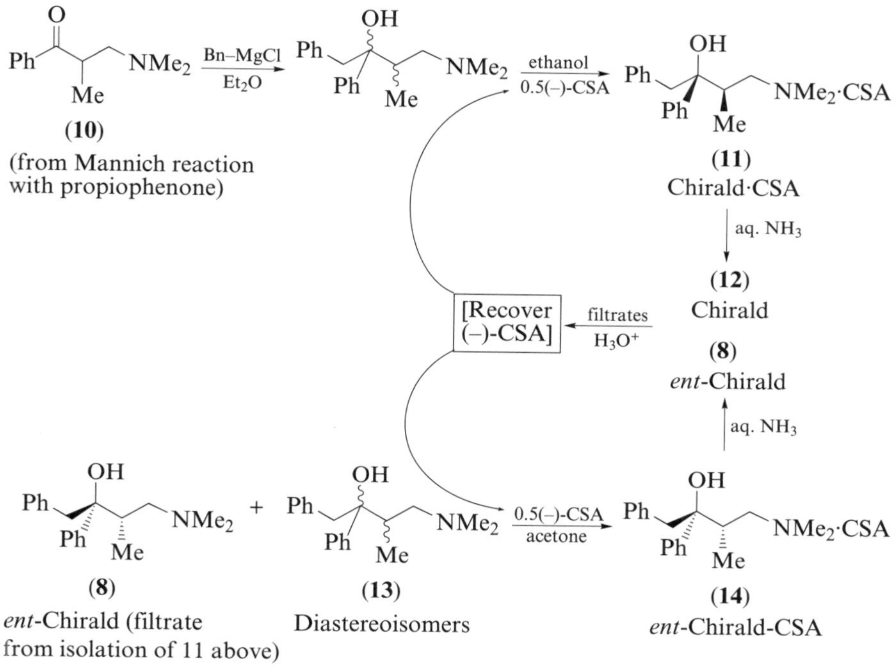

# CASE HISTORIES OF RESOLUTIONS

**Scheme 3** Industrial source of *ent*-Chirald (**8**) and Chirald (**12**)

## 6.2.2 RESOLUTION–DERACEMISATION

Two critical observations were made during the development of the syntheses of duloxetine. First, classical resolution of several intermediates with (*S*)-mandelic acid (MDA) proved highly efficient. Second, racemisation of several synthetic intermediates proved extremely facile in mineral acid at 25 °C.[14] These observations were developed into a new 'net deracemisation' synthesis of duloxetine using a simple iterative sequence of resolution–racemisation–recycle (RRR) (Scheme 4).

Mannich reaction of acetylthiophene **4** (see Scheme 1) and direct sodium borohydride reduction gave the racemic amino alcohol **15**. Treatment of the latter compound with 0.45 equivalents of MDA in methyl *tert*-butyl ether (MTBE) effected immediate precipitation of salt **16** (initially 20–60% *de* at 25 °C). However, on heating this heterogeneous mixture at reflux, pure salt **16** with 95–98% *de* was isolated in 42–43% yield (95% of theory). The MTBE filtrate contained mostly the enantiomer **17** and small amounts of MDA salts. Acidification of this MTBE filtrate with hydrochloric acid effected clean racemisation of the hydrochloride salt of compound **17** in the aqueous layer. Neutralisation with sodium hydroxide allowed re-extraction of racemic amine **15** into the original

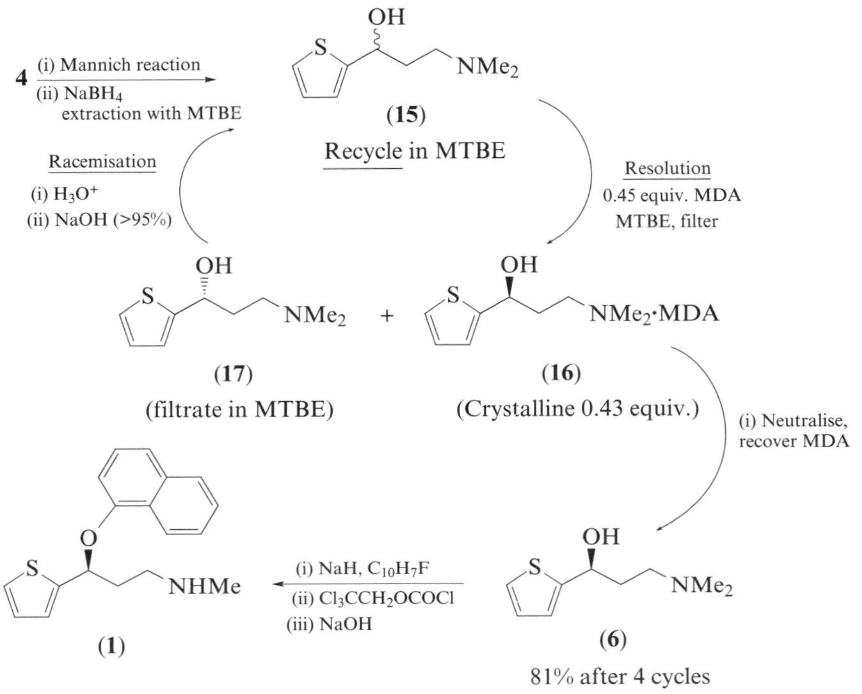

**Scheme 4** Resolution–racemisation–recycle (RRR) synthesis of duloxetine (**1**)

MTBE layer. The dried solution of amine **15** was again subjected to the resolution sequence and, after four cycles, a total of 81% of alcohol **6** was obtained. Completion of the synthesis produced duloxetine (**1**) in the same overall yield as the asymmetric reduction sequence. Thus, a simple acid–base treatment allows recycle of the opposite enantiomer. The robustness and simplicity of the RRR protocol to prepare duloxetine (Scheme 4) proved to be more efficient and cost-effective than the enantioselective route (Scheme 1).

6.2.3 APPLICATIONS TO OTHER SYSTEMS

The enantiomers of 1-phenylpropane-1,3-diol (**18**) and 1-phenyl-3-chloropropan-1-ol (**19**) are starting materials in the synthesis of several drugs related to duloxetine [e.g. nisoxetine (**20**), seproxetine (**21**), tomoxetine (**22**), (*S*)-fluoxetine (**23**) and (*R*)-fluoxetine (**24**)].[6] Like the intermediate in the duloxetine synthesis, both 1-phenylpropane-1,3-diol and 1-phenyl-3-chloropropan-1-ol undergo simple acid-catalysed racemisation. When incorporated into a resolution sequence [e.g. the enzymatic resolution of 1-phenylpropane-1,3-diol (**18**)[6d]], these results provide formal RRR-based syntheses of **20**–**24** (Scheme 5).

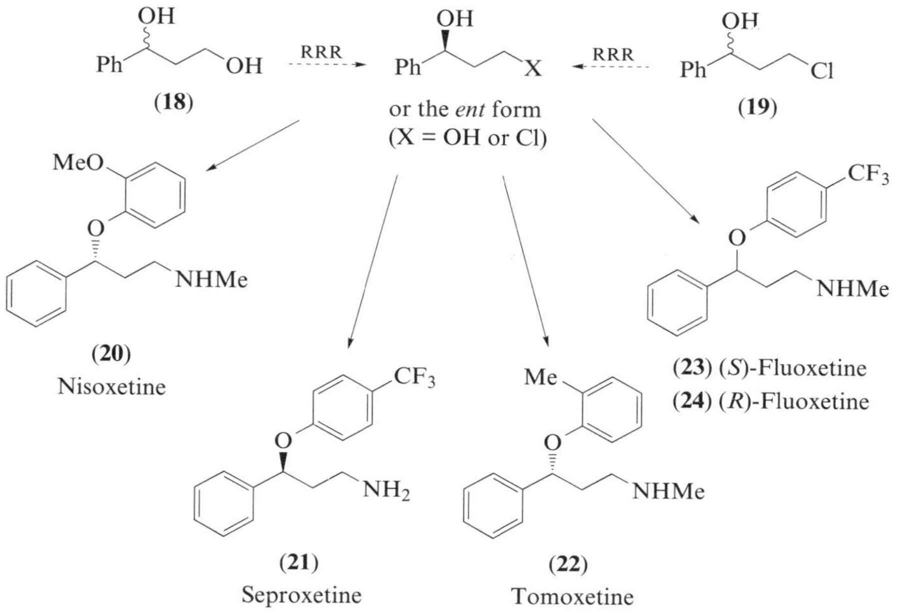

# CASE HISTORIES OF RESOLUTIONS

**Scheme 5** Application of RRR to other drugs

(18), (19): starting diols/chlorohydrins; intermediate (X = OH or Cl) or the *ent* form
(20) Nisoxetine
(21) Seproxetine
(22) Tomoxetine
(23) (*S*)-Fluoxetine
(24) (*R*)-Fluoxetine

## 6.3 LORACARBEF (LY163892)

Loracarbef (**2**) is a carbacephalosporin antibiotic with extended *in vivo* stability.[15a,b] This oral antibiotic is currently on the market and has found specialised use in the treatment of paediatric ear infections. Owing to high doses and intense competition from other β-lactam antibiotics, the cost of manufacturing was a critical issue in the development of loracarbef.

There are many elegant syntheses of loracarbef.[15c] Several critical synthetic intermediates leading to loracarbef (or closely related carbacephalosporins) are shown in Figure 2. The varying synthetic strategies from this group of intermediates are noteworthy:

(i) the racemic, bicyclic β-lactam **25**[16] is resolved via enzymatic acylation of the amino group;
(ii) the monocyclic β-lactam **26**,[17a] which is derived from a chiral auxiliary-based synthesis, is cyclised to a six-membered ring via rhodium-catalysed formation of the carbene and insertion into the N—H bond;
(iii) the amino acid **27**,[18] derived from the serine hydroxymethyl transferase (SHMT)-catalysed reaction of glycine with pentenal, is converted into a β-lactam by a Mitsunobu reaction on a β-hydroxy amide derived from the β-hydroxy acid;

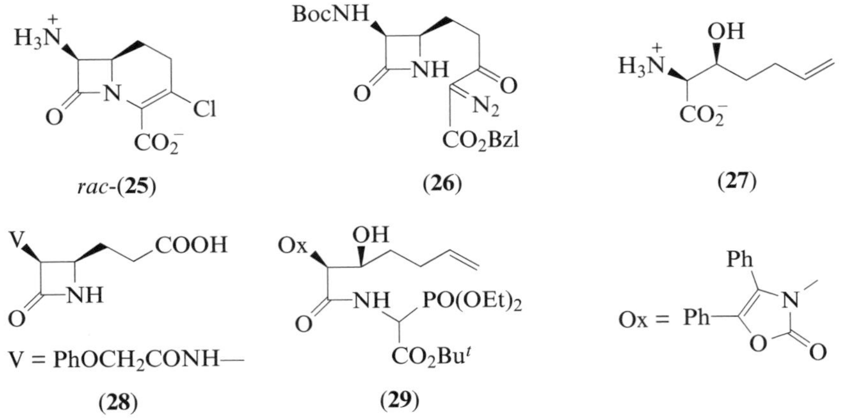

**Figure 2** Selected intermediates for carbacephalosporin synthesis

(iv) the penicillin-derived acid **28**[19] is alkylated on nitrogen with a bromoglycinate, and later closed to the six-membered ring via a Dieckmann reaction;

(v) The racemic aldol-derived intermediate **29**[20] is converted into the β-lactam, subjected to ozonolysis [which converts the alkene into the aldehyde and cleaves the diphenyl oxazolin-2-one (Ox) protecting group] and then cyclised via an internal Wittig reaction.

Many published syntheses of carbacephalosporins utilise the classical [2 + 2] ketene–imine cyclisation (Staüdinger reaction)[15–17,21] to construct the cis-β-lactam. Lilly chemists have utilised a Dieckmann closure of Staüdinger-derived cis-β-lactams for the production process of loracarbef (**2**) (Scheme 6). Thus, the

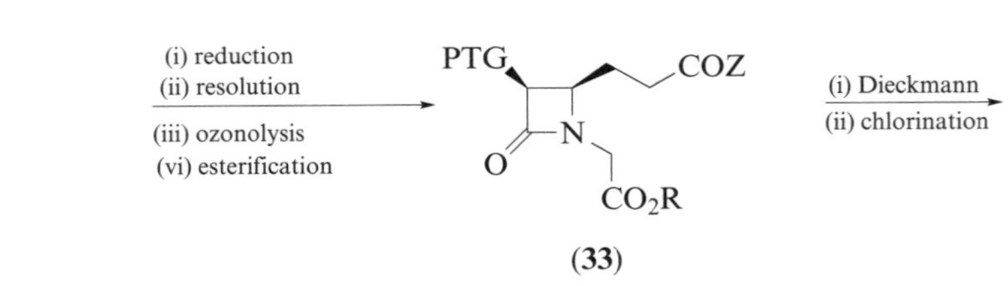

**Scheme 6** Generalised Lilly syntheses of loracarbef (Lorabid)   *(continued)*

# CASE HISTORIES OF RESOLUTIONS

**Scheme 6**  (*continued*)

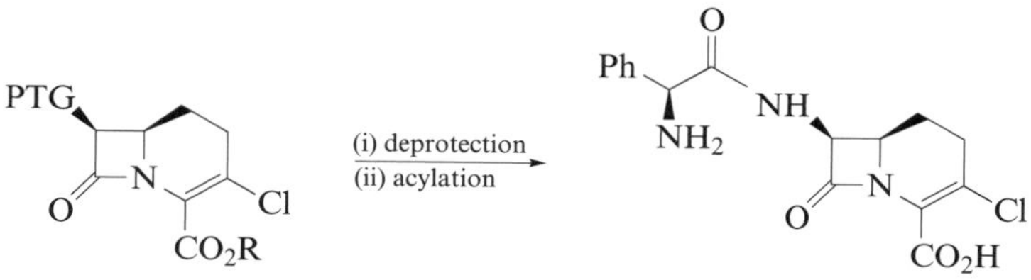

PTG = PhthN— or PhOCH$_2$CONH— or EtOCOCH=C(Me)NH—;
R = Me, p-(O$_2$N)PhCH$_2$, Bu$^t$, PhO; Z = PhO, PhS

Staüdinger reaction of the protected ketene **30** (derived *in situ* from the acid chloride or anhydride and triethylamine) with the imine **31** (derived from 2-furanylacrolein and a glycinate ester) affords the corresponding β-lactam **32**. Generally, high yields are observed with appropriate protecting groups (e.g. 90% using phthalimidoacetyl chloride; PTG = PhthN, R = Me). The generalised sequence depicted in Scheme 6 affords the diester **33**, which, when subjected to a Dieckmann cyclisation and then chlorination, gives **34**. Deprotection of PTG and R followed by acylation with (*R*)-phenylglycine then affords loracarbef (**2**).

## 6.3.1 SYNTHESIS FROM SODIUM ERYTHORBATE

Building upon the [2 + 2] cycloaddition–Dieckmann strategy above, an alternative non-resolution-based synthesis of loracarbef (**2**) from a common food preservative has been developed. A combination of observations by Cohen et al.[22] on the oxidation of isoascorbic acid and the work of Hubschwerlen and Schmid[23] and Evans and Williams[24] was employed for an asymmetric ketene–imine Staüdinger reaction. The synthesis of a β-lactam with the proper framework, stereochemistry and oxidation state was rapidly realised from the very inexpensive ($10 kg$^{-1}$) chiral pool intermediate, sodium erythorbate (**35**)[25] (Scheme 7).

In practice, sodium erythorbate was oxidised (hydrogen peroxide, 80%) and the intermediate lactone diol was monotosylated to afford tosylate **36** (80–88%). Treatment of the latter with sodium ethoxide afforded the crystalline, enantiomerically pure epoxide **37** (91–95%).[26] For larger scale preparations, the conversion of sodium erythorbate into the 4-hydroxyepoxycrotonate **37** was accomplished, without isolation or purification of intermediates, in approximately 75% yield. In one operation, epoxy alcohol **37** was subjected to Swern oxidation and the resulting aldehyde was condensed with *tert*-butyl glycinate. The

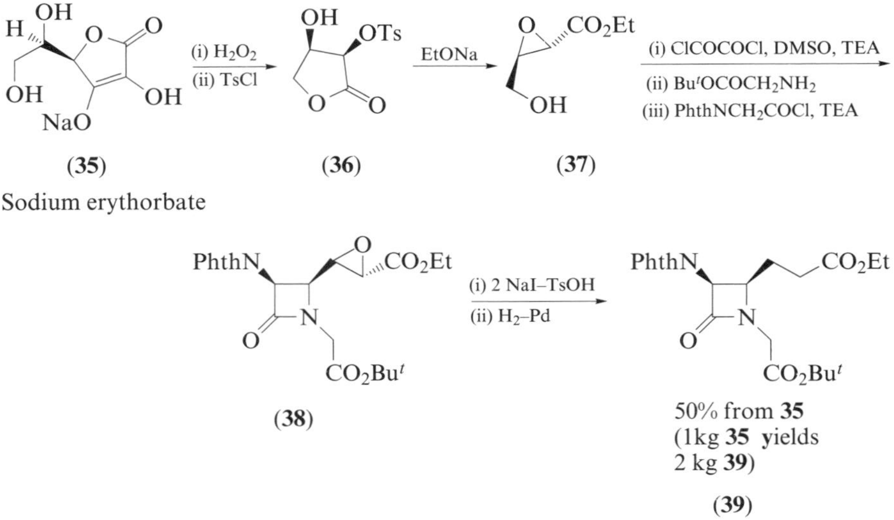

Scheme 7  Synthesis of a loracarbef precursor from sodium erythorbate

resulting unstable imine, whilst still in solution, was then treated with phthalimidoacetyl chloride in the presence of triethylamine to afford the monocyclic β-lactam 38 in 75% yield from 4-hydroxyepoxycrotonate 37 with 92–94% *de*. One recrystallisation of 38 upgraded the *de* to >99%. However, crystallisation of subsequent intermediates increased the *de* to similar levels. The epoxide 38 was easily deoxygenated by treatment with sodium iodide–tosic acid and, without purification of the intermediate alkene, was hydrogenated to afford 39 (87%).[27]

At this stage, the Dieckmann precursor 39 of loracarbef had been generated with high efficiency:

(i) four of the six carbons of sodium erythorbate (35) are incorporated into the nucleus of loracarbef (2);
(ii) two new stereogenic centres were created with high *de*;
(iii) the chiral pool material was inexpensive (<$10 kg$^{-1}$);
(vi) the mass conversion efficiency was very high (1 kg of sodium erythorbate gave 2 kg of 39).

Unfortunately, the precursor 39 was not suited for a Dieckmann condensation[28] owing to unactivated esters and the reactivity of the phthalimido group. The ester and phthalimido protecting groups of the monocyclic lactam 39 were modified (without isolation) by sequential cleavage of the *tert*-butyl ester (trifluoroacetic acid, TFA), conversion of the phthalimido group to the *o*-pyrrolidinocarbonylbenzamide (OPCB)[29] group with pyrrolidine, and ethyl ester hydrolysis (sodium hydroxide, 73% from precursor 39) to yield the diacid 40

(Scheme 8). The diacid **40** was converted into **41** by phenol esterification (dicyclohexylcarbodiimide, phenol, 80%) and then selective transesterification (sodium *p*-nitrobenzylalkoxide, 81%). Dieckmann condensation of **41** (sodium *tert*-butoxide, 70%) and chlorination [(PhO)$_3$PCl$_2$][30] with concurrent reclosure of the OPCB group to PhthN afforded **42**. Removal of the PNB group (H$_2$–Pd, 90%) and the phthalimido group (methylhydrazine, 90%)[18] yielded the enantiopure nucleus **43**. Acylation of **43** with the mixed carboxylic–carbonic anhydride derived from (*R*)-phenylglycine and isobutyl chloroformate (90%) provided loracarbef (**2**).

Although the cost of the starting material is low and the conversion to an advanced intermediate efficient, this sequence was never scaled up. What are the shortcomings of this synthesis? The molecular gymnastics required to convert sodium erythorbate (**35**) into intermediate **41** and the protecting group 'shuffle' resulted in a synthesis which was simply too long (20 functional group interconversions or C—C bond formations). The cumulative cost of performing multiple steps with an overall yield of about 12% quickly outweighs the cost benefit of starting with sodium erythorbate. Even if yields could be dramatically increased, the cumulative cost of processing, isolation, handling and clean-up of numerous intermediates would disfavour the development of this process for manufacture.

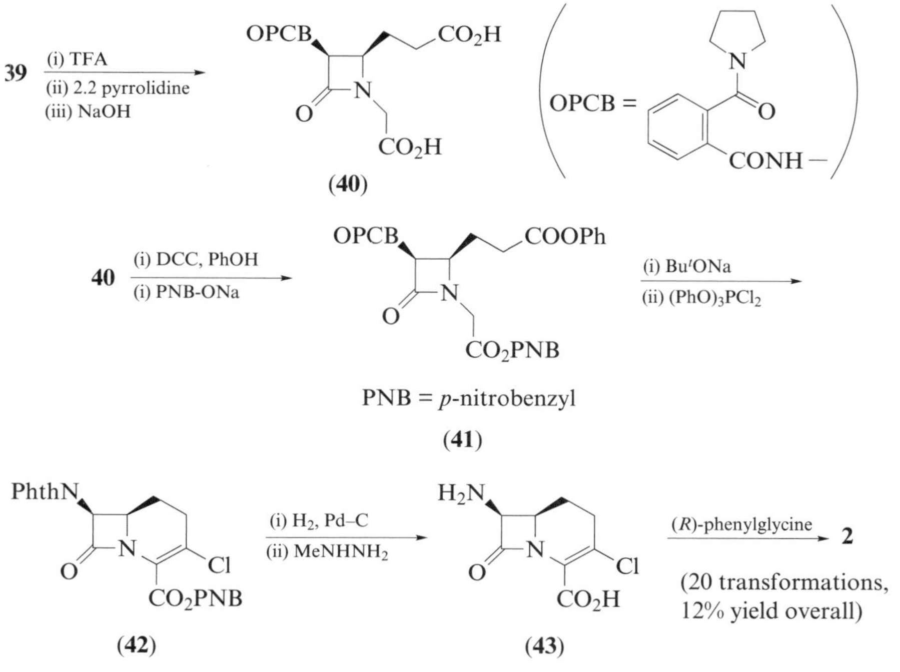

**Scheme 8** Completion of sodium erythorbate route to loracarbef

## 6.3.2 RESOLUTION VIA A PHTHALIMIDO GROUP

Examination of protecting groups in the synthesis of loracarbef has shown that the phthalimido group could be converted into a bisamide in aprotic solvents by treatment with a primary or secondary amine. These intermediates are re-closed to the phthalimido group[29] by treatment with dilute mineral acid or with a combination of boric acid–hydrofluoric acid. Based on the classical resolution of alcohols via reaction with phthalic anhydride,[31] a similar sequence for the resolution of phthalimides with amines was devised. The sequence was demonstrated in a resolution after the first step in the loracarbef synthesis (Scheme 6).[29,31] Thus, when the racemic β-lactam **44** (obtained in 90% yield from a Staüdinger reaction) was treated with (S)-α-methylbenzylamine in hot THF, the pure diastereoisomer **45** crystallised directly from the reaction mixture (95% de). Treatment of amide **45** with mineral acid[29b] gave the resolved intermediate[29c] **46** (Scheme 9).[33] Conversion of the latter to loracarbef (**2**) was then accomplished as depicted in Scheme 6.

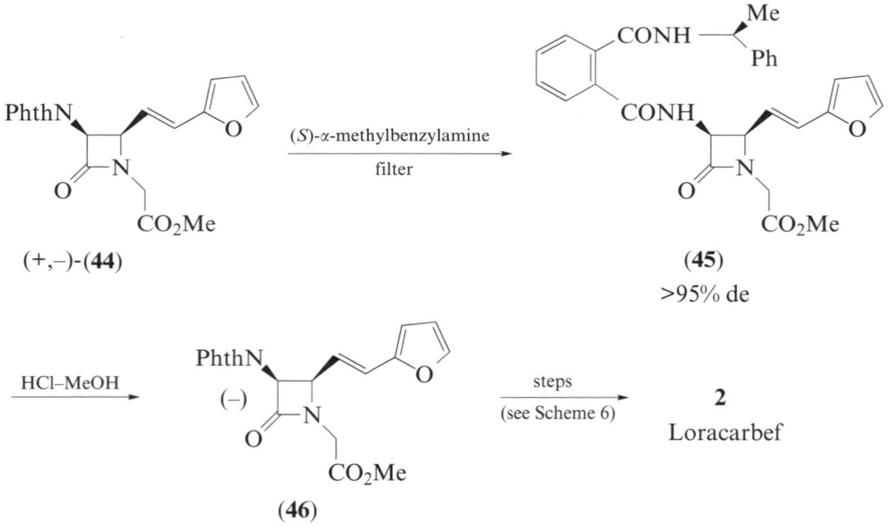

**Scheme 9** Resolution of a loracarbef intermediate via a phthalimido derivative

This methodology represents a simple resolution sequence in which a racemic amine derivative (phthalimido) is resolved with a non-racemic amine. The phthalimido resolution process was replaced by a more efficient enzymatic acylation–resolution[34] sequence utilising $PhOCH_2CO_2Me$ early in the synthesis (see Scheme 6, PTG = $PhOCH_2CONH$ and R = Me).

## CASE HISTORIES OF RESOLUTIONS

### 6.3.3 RESOLUTION OF OTHER β-LACTAM INTERMEDIATES

This new methodology has been applied to the resolution of other β-lactam precursors of loracarbef. The racemic phthalimido intermediates **47, 44** and **48** (Figure 3) have been opened with (R)- or (S)-α-methylbenzylamine or with (S)-2-(methoxymethyl)pyrrolidine (MMP),[29c] and the diastereoisomers separated by crystallisation or chromatography. The separation of amides derived from **47, 44** and **48** represent an early resolution of several carbacephem[28] targets, including loracarbef.[15]

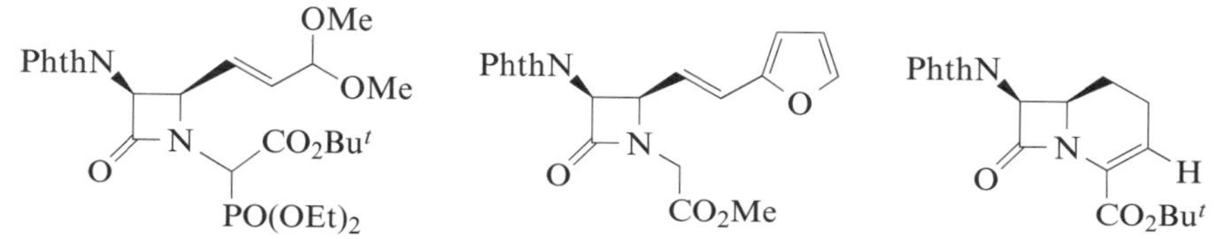

PhthN / OMe / OMe / N / O / CO$_2$Bu$^t$ / PO(OEt)$_2$
with PhCH(Me)NH$_2$
**(47)**

PhthN / N / O / CO$_2$Me / O (furan)
with MMP
**(44)**

PhthN / N / O / H / CO$_2$Bu$^t$
with MMP
**(48)**

**Figure 3** β-Lactams resolved by opening of phthalimido with PhCH(Me)NH$_2$ or MMP

### 6.4 HUMAN STEROID 5-α-REDUCTASE INHIBITOR (LY300502)

The benzoquinolinone LY300502 (**3**) has been synthesised[35] and identified as a potent, non-competitive inhibitor of 5-α-reductase in cell cultures derived from foreskin fibroblasts.[36] This activity may have implications for the treatment of acne, male pattern baldness and benign prostatic hyperplasia.

#### 6.4.1 SYNTHESIS VIA A CHIRAL AUXILIARY

The initial synthesis of **3** was based upon an approach using a covalent chiral auxiliary (Scheme 10).[37] Friedel–Crafts reaction of ethylene and the acid chloride derived from p-chlorophenylacetic acid (**49**) afforded tetralone **50**. Conversion of the tetralone **50** to an enamine with (S)-α-methylbenzylamine and immediate treatment with acryloyl chloride afforded the aza-annulation product **51** (79% from tetralone **50**). The desired diastereoselective reduction of the endocyclic alkene in compound **51** proved difficult, since the carbon–nitrogen bond of the α-methylbenzylamine auxiliary underwent competitive reduction. Reduction of **51** with excess sodium cyanoborohydride in formic acid afforded the *trans* fused tricyclic amide **52** (admixed with the *cis* isomers) in low *de*. Attempts to

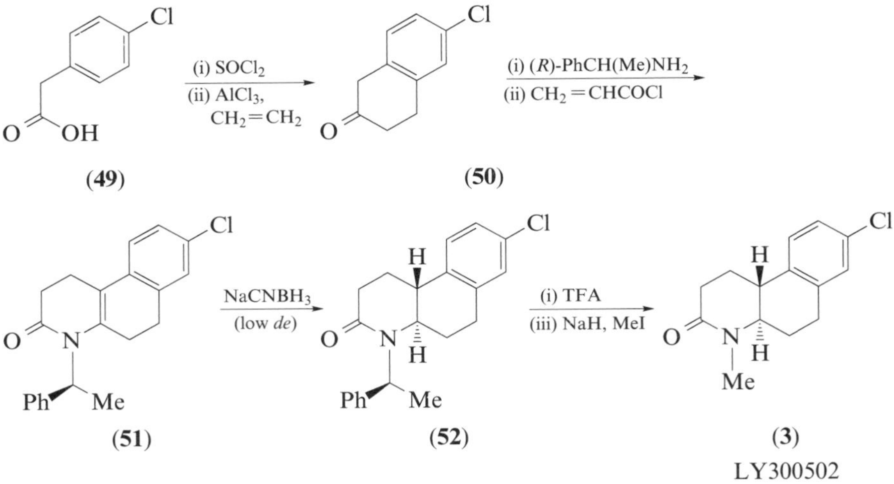

**Scheme 10** Asymmetric synthesis of LY300502 via a chiral auxiliary

optimise the reduction of aza-annulation product **51**, or to obtain pure amide **52** by crystallisation, were not fruitful. Chromatographic separation of the four-component mixture afforded compound **52** in pure form. Removal of the chiral auxiliary with TFA and methylation of the intermediate amide provided LY300502 (**3**) (and its enantiomer) in about 9% overall yield from tetralone **50** (Scheme 10).

The low *de* combined with moderate yields and reliance upon chromatographic separations precluded further development of this sequence. Chiral chromatographic[38] separations of LY300502 (**3**) have been reported but production of kilogram quantities of enantiomerically pure product by this method is not yet practical.

### 6.4.2 RESOLUTION VIA METHANOLYSIS OF LACTAMS

Direct 'chemical' resolutions of δ-lactams similar to LY300502 are not generally viable. One tenable strategy is defined by a simplified sequence of δ-lactam opening (e.g. alcoholysis of the lactam to the ester) followed by an appropriate resolution of the amino ester (chemical, kinetic or enzymatic) and relactamisation.[39] This approach proved very successful in the development for the production of kilogram quantities of LY300502 (Scheme 11). Aza-annulation of 6-chlorotetralone (**50**) under specialised conditions using pyrrolidine enamine and acrylamide gave the ene-lactam **53** (85%). Ionic reduction of the latter with triethylsilane–trifluoroacetic acid and *N*-methylation with methyl chloride (obtained in the next step, see below) in the presence of sodium iodide–sodium

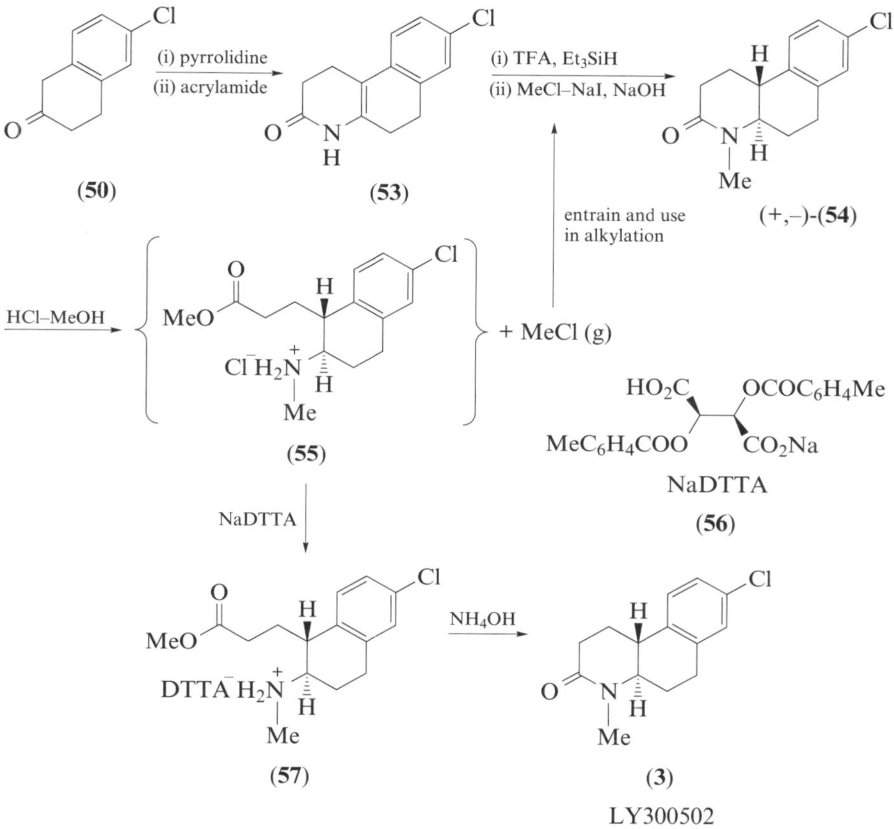

# CASE HISTORIES OF RESOLUTIONS

**Scheme 11** Lactam resolution route to LY300502 which 'breeds' methyl chloride

hydroxide afforded the *trans-* fused bicyclic lactam (±)-**54** (90%). Methanolysis of lactam **54** with hydrochloric acid–methanol afforded hydrochloride **55** and methyl chloride. Although **55** could be isolated and characterised, this was not necessary. Treatment of the solution of **55** with sodium $(R,R)$-di-*p*-toluoyltartaric acid (NaDTTA, **56**) gave **57** in 45% yield (90% of theory, $de > 98\%$). Isolation of the DTTA salt **57** also effected the removal of small amounts of the *cis* isomer of lactam **54,** as well as all other related substances. Treatment of **57** with ammonia solution and recovery of the DTTA afforded pure LY300502 (**3**) in >95% yield and 99% *ee*.

The overall process, although it contains a 'midstream' resolution, is short and efficient. Only 2 kg of 6-chlorotetralone (**50**) are required for the production of 1 kg of the enantiomerically pure product, with four pilot plant operations from **50**, in 30% overall yield. Additionally, by selection of hydrochloric acid and methanol for the alcoholysis sequence, the by-product, methyl chloride, is entrained and utilised in the alkylation sequence, obviating the need to purchase and ship methyl iodide. This new resolution sequence thus 'breeds' the $N$-methylation reagent. As with duloxetine and loracarbef, this resolution methodology is based on simple and robust acid–base chemistry. The protocol has also been utilised in the resolution of derivatives of LY300502 (**3**).[40]

## 6.5 CONCLUSION

These case histories illustrate resolution processes which were developed in preference to stereoselective syntheses. Some of the critical factors from these histories in route selection and development may be summarised as follows:

(i) *External ligands or auxiliaries*. The use of high molecular weight auxiliaries (ligands) is, in some cases, less effective than a simple resolution. Care should be exercised in the use of an auxiliary, and extra 'hidden' steps may also lie in recovery and re-use of the auxiliary.
(ii) *Balance of costs*. Avoid undue focus on the cost of an enantiomerically pure starting material. In the case of loracarbef, the raw material cost was quickly overtaken by cumulative processing expenses.
(iii) *Covalent auxiliaries*. The use of a covalent auxiliary generally involves introduction, transfer of chirality and removal. This approach is best when the *de* and the yields are high and the auxiliary fully complements an efficient synthesis route.
(iv) *Yields and Efficiency*. Although step yields are very important, individual yield considerations become secondary after the execution of a large number of steps.
(v) *Moderate selectivities*. In many cases, moderate *ee* (or *de*) and moderate yields cannot be tolerated (e.g. LY300502). For example, removal of an unwanted enantiomer and chemical purification from a reaction with a yield of 80% and 80% *ee* might sacrifice 25% of the desired enantiomer, netting 55% of the desired enantiomer. An efficient resolution might yield an 'uncomplicated' 45%. The 'value added' by the stereoselective transformation must be weighted against the 10% increase in yield over the resolution sequence.
(vi) '*Molecular gymnastics*'. As in the case of the synthesis of loracarbef, forcing chiral pool starting materials through circuitous routes to meet stereochemical demands of the target diminished the process efficiency.
(vii) *Speed to market* ('penny-wise, pound-foolish'). In many development circumstances, limited human resources and time-frames will not warrant a multi-year, multi-million dollar effort. This aspect of development is more pronounced as drugs evolve to be potent at submilligram doses, and production is based upon kilograms rather than tonnes.

In route selection and development, concerns for the absolute stereochemistry should not entirely outweigh the basic chemical synthesis. From the experiences above, one should consider the best, most efficient synthesis of the racemate and then superimpose the absolute stereochemical demands upon that synthetic route. In many cases, this may result in a resolution-based process. Certainly, bioorganic transformations and catalytic enantioselective reactions will become profoundly important to the single enantiomer drug industry. However, we can expect simple, direct and robust resolution processes to be of industrial importance for some time to come.

## 6.6 ACKNOWLEDGEMENTS

We gratefully acknowledge Professors David Evans and Barry Sharpless for the disclosure of various manuscripts in advance of publication. We also thank Professors Edward C. Taylor, Leo Paquette, William Roush, Andrew Meyers and Paul Wender for fruitful discussions throughout these programmes. We also thank our colleagues and co-authors for providing the experimental development, materials, insight and support throughout the course of this work: Marvin Hansen, Mike Martinelli, Charles Barnett, Jerry Misner, Chris Schmid, Jerry Bryant, Lowell Hatfield, Alyssa Kneisley, Gerald Thompson, Don Brannon, Jack Deeter, Gib Staten, Jeff Frazier, Mike Staszak, Jim Aikins, Jim Dunigan, John Gardner, Perry Heath, Billy Jackson, Steve Pedersen, John Rizzo, Eddie Tao, Jeff Ward, Joe Kennedy, Samantha Janisse, Jim Audia, Ken Hirsch, Dave Jones, Dave Lawhorn, Loretta McQuaid, Tom Kress, Jim Wepsiec and Bob Waggoner. We also thank Jill and Mary for their support and patience.

## 6.7 REFERENCES AND NOTES

1. (a) Helmchen, G., Hoffmann, R. W., Mulzer, J., and Schaumann, E. (Eds), *Stereoselective Synthesis,* Georg Thieme, Stuttgart, 1995, Vols. 1–5; (b) Atkinson, R. S., *Stereoselective Synthesis*, Wiley, Chichester, in press; (c) Norgradi, M., *Stereoselective Synthesis*, VCH, Weinheim, 1987; (d) Ojima, I. (Ed.), *Catalytic Asymmetric Synthesis*, John Wiley, New York, 1993; (e) Noyori, R., 'Enantioselective Catalysis with Metal Complexes—An Overview,' in *Modern Synthetic Methods*, Springer, Berlin, 1989; (f) Bartmann, W., and Sharpless, K. B. (Eds), *Stereochemistry of Organic and Bioorganic Transformations*, VCH, Weinheim, 1987; (g) Stille, J. R., and Barta, N. S., in *Studies in Natural Products Chemistry—Stereoselective Synthesis* (ed. Atta-ur-Rahman), Elsevier, Amsterdam, in press.
2. Collins, A. N., Sheldrake, G. N., and Crosby, J. (Eds), *Chirality in Industry*, Wiley, Chichester, 1992.
3. Weigel, L. O., Seminar presented at the 12th SCI Process Development Symposium entitled 'Racemates, Resolutions, and Results,' Cambridge, 12 December, 1994.
4. Weigel L., unpublished survey from Lilly Chemical PR&D Indianapolis, IN, 3 June, 1993.

5. (a) Robertson, D. W., Wong, D. T., and Krushinski, J. H., Jr, *Eur. Pat. Appl.*, EP 273 658, 1988; *Chem. Abstr.*, **109**, 170224n (1988); (b) Robertson, D. W., Thompson, D. C., Beedle, E. E., Reid, L. R., Bymaster, F. P., and Wong, D. T., presented at the 194th National Meeting of the American Chemical Society, New Orleans, LA, 30 Aug.–4 Sept. 1987, MEDI 49; (c) Wong, D. T., Robertson, D. W., Bymaster, F. P., Krushinski, J. H., and Reid, L. R., *Life Sci.*, **43**, 2049 (1988); (d) *Drugs Future*, **11**, 134 (1986); (e) Ankier, S. I., *Prog. Med. Chem.*, **23**, 121 (1986); (f) Robertson, D. W., Krushinski, J. H., Fuller, R. W., and Leander, J. D., *J. Med. Chem.*, **31**, 1412 (1988); (g) Fuller, R. W., *J. Clin. Psychiatry*, **47** (April Suppl.), 4 (1986).
6. (a) Srebnik, M., Ramachandran, P. V., and Brown, H. C., *J. Org. Chem.*, **53**, 2916 (1988); (b) Gao, Y., and Sharpless, K. B., *J. Org. Chem.*, **53**, 4081 (1988); (c) Mitchell, D., and Koenig, T. M., *Synth. Commun.*, **25**, 1231 (1995). (d) Boaz, N. W., *US Pat.*, 4 921 798, 1990; *Chem. Abstr.*, **116**, 105864b (1992).
7. Weigel, L. O., Deeter, J., Staten, G. S., Frazier, J., and Staszak, M., *Tetrahedron Lett.*, **31**, 7101 (1990).
8. Kumar, A., Ner, D. H., and Dike, S. Y., *Tetrahedron Lett.*, **32**, 1901 (1991).
9. Corey, E. J., and Reichard, G. A. *Tetrahedron Lett.*, **30**, 5207 (1989).
10. Noyori, R., Tomino, I., and Tanimoto, Y., *J. Am. Chem. Soc.*, **101**, 3129 (1979).
11. Weigel, L. O., Deeter, J., Staten, G. S., Frazier, J., and Staszak, M., presented at the Joint 23rd Central/24th Great Lakes Regional Meeting of the American Chemical Society, Indianapolis, IN, 29 May 1991, ORGN349.
12. (a) Yamaguchi, S., Mosher, H. S., and Pohland, A., *J. Am. Chem. Soc.*, **94**, 9254 (1972); (b) Yamaguchi, S., and Mosher, H. S., *J. Org. Chem.*, **38**, 1870 (1973); (c) Brinkmeyer, R. S., and Kapoor, V. M., *J. Am. Chem. Soc.*, **99**, 8339 (1977).
13. Carbinolamine 12 has been referred to as 'Darvon alcohol' at various points in the literature. This use of the Darvon® trademark has not been authorised by Eli Lilly and Company. For this work, non-resolved **8** was obtained from Eli Lilly Industries, Mayaguez, Puerto Rico. Resolution was effected according to Pohland, A., and Sullivan, H. R., *J. Am. Chem. Soc.*, **77**, 3400 (1955).
14. Weigel, L. O., presented at the 205th National Meeting of the American Chemical Society, Denver, CO, 31 March 1993, ORGN261.
15. (a) First public disclosure of loracarbef: Hirata, T., Matsukuma, I., Mochida, K., and Sato, K., presented at the 27th Interscience Conference on Antimicrobial Agents and Chemotherapy, New York, 4–7 October 1987; (b) Sato, K., Okachi, R., Matsukuma, I. Mochida, K., and Hirata, T., *J. Antibiot.*, **42**, 1844 (1989); (c) For a recent review of carbacephalosporins, see Zhang, T. Y., and Hatfield, L. D., in *Comprehensive Heterocyclic Chemistry*, (ed. E. F. V. Scriven and C. W. Rees), 2nd edn, Elsevier, London, 1996, vol. 1, Chap. 20.
16. Matsukuma, I., Yoshiiye, S., Mochida, K., Hashimoto, Y., Sato, K., Okachi, R., and Hirata, T., *Chem. Pharm. Bull.*, **37**, 1239 (1989).
17. (a) Evans, D. A., and Sjogren, E. B., *Tetrahedron Lett.*, **26**, 3787 (1985); (b) for a similar approach, see Bodurow, C. C., Boyer, B. D., Brennan, J., Bunnell, C. A., Burks, J. E., Carr, M. A., Doecke, C. W., Eckrich, T. M., Fisher, J. W., Gardner, J. P., Graves, B. J., Hines, P., Hoying, R. C., Jackson, B. G., Kinnick, M. D., Kochert, C. D., Lewis, J. S., Luke, W. D., Moore, L. L., Morin, J. M. Jr., Nist, R. L., Prather, D. E., Sparks, D. L., and Vladuchick, W. C., *Tetrahedron Lett.*, **30**, 2321 (1989).
18. Jackson, B. G., Doecke, C. W., Farkas, E., Fisher, J. W., Gardner, J. P., Gazak, R. J., Kroeff, E. P., Misner, J. W., Pedersen, S. W., Staszak, M. A., and Vicenzi, J. T., presented at the 'Chiral USA 1994' Conference, Reston, VA, Spring 1994.
19. Deeter, J., Hall, D. D., Jordan, C. L., Justice, R. M., Kinnick, M. D., Morin, J. M., Jr, Paschal, J. W., and Ternansky, R. J., *Tetrahedron Lett.*, **34**, 3051 (1993).
20. Lotz, B. T., and Miller, M. J., *J. Org. Chem.*, **58**, 618 (1993).

# CASE HISTORIES OF RESOLUTIONS

21. (a) Cooper, R. D. G., Daugherty, B. W., and Boyd, D. B., *Pure Appl. Chem.*, **59**, 485 (1987); (b) Hegedus, L. S., Montgomery, J., Narukawa, Y., and Snustad, D. C., *J. Am. Chem. Soc.*, **113**, 5784 (1991).
22. Cohen, N., Banner, B. L., Lopresti, R. J., Wong, F., Rosenberger, M., Liu, Y. Y., Thom, E., and Liebman, A. A., *J. Am. Chem. Soc.*, **105**, 3661 (1983).
23. Hubschwerlen, C., and Schmid, G., *Helv. Chim. Acta*, **66**, 2206 (1983).
24. Evans, D. A., and Williams, J. M., *Tetrahedron Lett.*, **29**, 5065 (1988).
25. Dunigan, J., and Weigel, L. O., *J. Org. Chem.*, **56**, 6225 (1991).
26. (a) Weigel, L. O., *US Pat.*, 5 097 049, 1992; (b) Manchand, P. S., Luk, K. C., Belica, P. S., Choudhry, S. C., Wei, C. C., and Soukup, M., *J. Org. Chem.*, **53**, 5507 (1988).
27. Weigel, L. O., Frazier, J. W., and Staszak, M. A., *Tetrahedron Lett.*, **33**, 857 (1992).
28. Hatanaka, M., and Ishimaru, T., *Tetrahedron Lett.*, **24**, 4837 (1983).
29. (a) *Chem. Eng. News*, 10 September, 26 (1990); (b) Weigel, L. O., and Astleford, B., *Tetrahedron Lett.*, **32**, 3301 (1991); (c) formation of **46** was also accompanied by significant amounts of PhthNCH(Me)Ph. Use of (*S*)-2-(methoxymethyl)pyrrolidine in these resolution sequences suppresses the formation of this undesired product.
30. Hatfield, L. D., Lunn, W. H., Jackson, B. G., Peters, L. R., Blaszczak, L. C., Fisher, J. W., Gardner, J. P., Dunigan, J. M., in *Recent Advances in the Chemistry of β-Lactam Antibiotics* (ed. G. I. Gregory), Royal Society of Chemistry, London, 1980, p. 109.
31. Kenyon, J., *J. Chem. Soc.*, 2540 (1922).
32. Weigel, L.O., *US Pat.*, 5 169 945, 1992.
33. These intermediates can also be converted into the free amine by treatment with methylamine; Wolfe, S., and Hasan, S. K., *Can. J. Chem.*, **48**, 3572 (1970).
34. (a) Zmijewski, M., Briggs, B., and Thompson, A., *New J. Chem.*, **18**, 425 (1994); (b) Zmijewski, M. J., Briggs, B. S., Thompson, A. R., and Wright, I. G., *Tetrahedron Lett.*, **32**, 1621 (1991); (c) Boyer, B. D., and Eckrich, T. M., *Eur. Pat.*, EP 365 212, 1990; (d) Brennan, J., and Eckrich, T. M., *Eur. Pat.*, EP 365 213, 1990.
35. Jones, C. D., Audia, J. E., Lawhorn, D. E., Hirsch, K. S., McQuaid, L. A., Neubauer, B. L., Pike, A. J., Pennington, P. A., Stamm, N. B., and Toomey, R. E., *J. Med. Chem.*, **36**, 421 (1993).
36. Neubauer, B. L., Goode, R. L., Gray, H. M., Hanke, C. W., Hirsch, K. S., Hsiao, K. C., Jones, C. D., Lawhorn, D. E., McQuaid, L., Toomey, R. E., Valia, K., and Audia, J. E., *Drugs Future*, **20**, 144 (1995).
37. Audia, J. E., Lawhorn, D. E., and Deeter, J. B., *Tetrahedron Lett.*, **34**, 7001 (1993).
38. Brandt, M., Mohammed, H., Levey, M. A. and Holt, D. A., *Bioorg. Med. Chem. Lett.*, **4**, 1365, (1994).
39. (a) Astleford, B. A., Janisse, S. K., and Weigel, L. O., presented at the 206th National Meeting of the American Chemical Society, Chicago, IL, 23 August 1993, ORGN076; (b) Astleford, B. A., Deeter, J., Janisse, S. K., and Weigel, L. O., presented at the 206th National Meeting of the American Chemical Society, Chicago, IL, 23 August 1993, ORGN075; (c) Astleford, B. A., Audia, J. E., Deeter, J., Heath, P. C., Janisse, S. K., Kress, T. J., Wepsiec, J. P., Weigel, L. O., *J. Org. Chem.*, **61**, 4450 (1996).
40. Kuo, F., and Wheeler, W. J., *J. Labelled Compd. Radiopharm.*, **34**, 915 (1994).

# 7 Crystal Science Techniques in the Manufacture of Chiral Compounds

## W. M. L. WOOD
*ZENECA Huddersfield Works, Huddersfield, UK*

| | | |
|---|---|---|
| 7.1 | Introduction | 119 |
| 7.2 | Basic Crystal Science | 120 |
| | 7.2.1 Crystals, Polymorphs and Solvates | 120 |
| | 7.2.2 Crystal Habit | 125 |
| | 7.2.3 Habit Modifiers | 126 |
| 7.3 | Development of Crystallisation Processes from Solution | 130 |
| | 7.3.1 Supersaturation—the Driving Force | 130 |
| | 7.3.2 Choice of Solvent | 131 |
| | 7.3.3 Solubility Curves and Their Measurement | 131 |
| | 7.3.4 Nucleation, the Metastable Zone and Control of Crystal Size | 133 |
| | 7.3.4.1 Measurement of Metastable Zone Width | 134 |
| | 7.3.4.2 Control of Crystal Size and Size Distribution | 137 |
| | 7.3.5 Rate of Desupersaturation | 140 |
| | 7.3.6 Isolation of Metastable Phases | 140 |
| 7.4 | Application of Crystal Science to Chiral Systems | 143 |
| | 7.4.1 Separation of Enantiomers by Diastereoisomer Formation | 144 |
| | 7.4.2 Separation of Diastereoisomers | 145 |
| | 7.4.3 Resolution by Direct Crystallisation of Enantiomers | 148 |
| | 7.4.3.1 Seeding Within the Metastable Zone | 152 |
| | 7.4.3.2 Nucleation Inhibition of One Enantiomer | 154 |
| 7.5 | Conclusion | 154 |
| 7.6 | References | 155 |

## 7.1 INTRODUCTION

Crystal science techniques provide powerful aids in the manufacture of chiral compounds. Their applicability depends on the properties of those compounds or their intermediates, and is unpredictable unless those properties have been measured. Synthetic organic chemists often have little understanding of the factors which govern such aspects as crystal size, size distribution, the ease of separation

*Chirality in Industry II.* Edited by A. N. Collins, G. N. Sheldrake and J. Crosby
© 1997 John Wiley & Sons Ltd

of the crystals from the mother liquors and the purity of the isolated crystals, all of which are important in developing a satisfactory crystallisation process.

It is intended that this chapter will help to provide that understanding, and will show how the science may be used in a practical way to achieve, and optimise, the isolation of chiral compounds by a variety of techniques, by reference to some real examples of the application of the science to the manufacture of chiral compounds. The chapter deliberately contains no mathematical equations. For those readers interested in obtaining greater detail about particular areas of the science and the underlying theory, it is suggested that they refer initially to the treatise on crystallisation by Mullin[1] and the references therein.

## 7.2 BASIC CRYSTAL SCIENCE

### 7.2.1 CRYSTALS, POLYMORPHS AND SOLVATES

A crystalline material is one in which the species present (atoms, ions, molecules) are arranged in an ordered three-dimensional structure. Under the appropriate conditions of supercooling or supersaturation, crystals are formed by nucleation followed by diffusion of the species to the surface of the nuclei and then their incorporation into the ordered structure or lattice of the crystal, causing its growth. Often there are several different possible packing arrangements of the species under different crystallisation conditions, resulting in the formation of different crystal lattices. The material is then said to be polymorphic, and the different lattices are polymorphs. A well known example is carbon, which can exist as both graphite and diamond crystalline forms. Figure 1 (from the data of Haisa and co-workers[2,3]) shows the way the molecules are arranged in the unit cell of the crystal lattices of the orthorhombic and monoclinic polymorphs of paracetamol ($p$-hydroxyacetanilide).

The lattice is not limited to only one type of molecule or set of ions. Where crystallisation is from solution, then the lattice may contain solvent molecules bound into the structure, and the resulting crystals are solvates. It is also fairly common (as with chiral racemates) for the lattice to contain more than one type of non-solvent molecule. The species co-crystallise to form a mixed crystal with a well-defined and fixed ratio of the different molecules or ions present in the lattice. Potash alum, $KAl(SO_4)_2 \cdot 12H_2O$, and sodium ammonium tartrate, $NaNH_4C_4H_4O_6 \cdot 4H_2O$, are well known inorganic and organic examples, respectively. Sometimes it is found that crystals of solute grown from solution contain variable amounts of impurities which are also present in the solution, even after washing to remove adhering solution and any material adsorbed on the crystal surface. The impurity is incorporated in the crystal lattice as a 'solid solution' and the amount present is a function of the concentration in solution.

The detailed arrangement of the species in the lattice may be determined by X-ray diffraction (XRD) studies on a single crystal. The resolution given by such

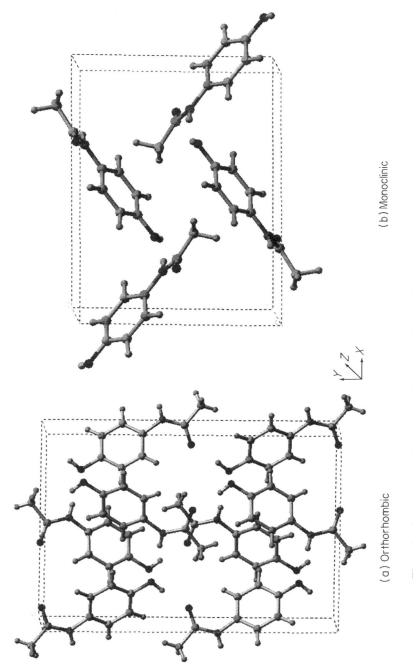

**Figure 1** Arrangement of the molecules in two polymorphs of paracetamol (*p*-hydroxyacetanilide)

(a) Orthorhombic

(b) Monoclinic

studies is sufficiently precise that the technique may be used to determine not only the relative position of the species in the lattice but also, where the species are molecules, the conformation of those molecules and also their molecular structure. This is particularly useful for large molecules such as proteins. Growth of suitable single crystals for these studies can be difficult; it is much simpler to produce a mass of small crystals, and the diffraction pattern produced from a sample of powder or paste is very useful for identification purposes since each polymorph or solvate gives a different pattern. Although, at present, it is not possible to obtain the detailed structural information from such samples, considerable advances are being made toward this end.[4] The use of XRD is described in detail by Glusker et al.[5]

The different solvates or polymorphs are known as crystal forms or phases. Generally, at a given temperature, only one phase will be stable, the others being metastable. A metastable phase can form under certain conditions, particularly in rapid crystallisations, but will tend to transform in time to a more stable phase, usually via solution. The rate-controlling processes in such transformations are discussed by Davey et al.[6] The stable phase at a particular temperature will always have the lowest solubility at that temperature. The exception is if a phase transition temperature exists, below which one phase is stable and above which another phase is stable. At the transition temperature, both phases have equal stability and the same solubility.

The difference in solubility of different polymorphs can alter the effectiveness of a product in end use. For example, with drugs, the lower solubility of a stable polymorph may be desirable to control the rate of release. On the other hand, the stable polymorph of a drug may have such a low solubility under physiological conditions that it is difficult to achieve a therapeutic dose. In these cases, the isolation and formulation of a metastable phase may provide the necessary improvement in solubility.

The different behaviour of polymorphs and solvates on heating provides further analytical techniques for the identification of the crystal form. On heating a solvate crystal, a temperature will be reached beyond which some or all of the solvent molecules will have sufficient energy to break free of the lattice, and the crystals desolvate. Partial desolvation can occur where some solvent molecules are held more tightly in the lattice, but these too will usually break free as the temperature is raised further. The pattern of loss in weight versus temperature is consistent for a particular solvate phase, and may be measured by thermogravimetric analysis (TGA). From a knowledge of the molecular weights of the crystal species and the solvent, and the weight loss, the ratio of solvent molecules to solute in the crystals may be obtained. Figure 2 shows the TGA trace given by gypsum ($CaSO_4 \cdot 2H_2O$). The weight loss caused by dehydration starts at about 40 °C and is complete by 280 °C. The weight loss of 20.95% compares well with the theory weight loss of 20.91%.

The loss of solvent molecules from a solvate is an endothermic process, as is the melting of crystals. Because of their different lattice energies, polymorphs

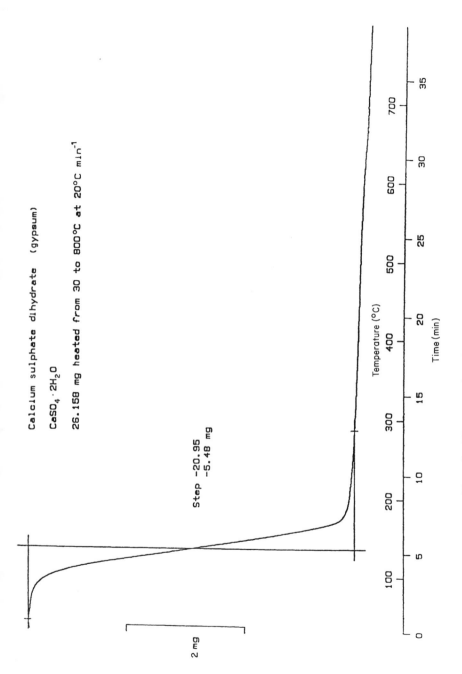

**Figure 2** Thermogravimetric analysis (TGA) trace for gypsum

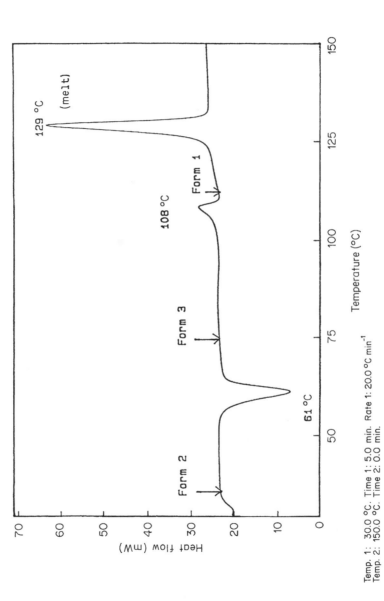

**Figure 3** DSC trace for a red dyestuff (DISPERSOL Scarlet CG)

CRYSTAL SCIENCE IN MANUFACTURE OF CHIRAL COMPOUNDS    125

have different melting points, and therefore another technique often used to identify polymorphs and solvates is differential scanning calorimetry (DSC). A useful introduction to DSC is provided by McNaughton and Mortimer.[7]

Figure 3 shows the output for a sample of a dyestuff undergoing two polymorphic transformations to successively more stable forms before the most stable form melts. Such transformations are exothermic, producing a negative peak on the chart. Melting is endothermic, producing a positive peak. The second transformation, at 108 °C, produces a small positive peak indicating that Form 3 melts and the melt then almost immediately recrystallises as Form 1, the overall heat flux of the two processes being endothermic.

### 7.2.2 CRYSTAL HABIT

The shape (habit, morphology) of crystals and their size distribution are important in that they control the ease and efficiency with which the crystals are separated by filtration or centrifugation from the mother liquor (either melt or solution). Ideally, for fast and efficient separation, the crystals should have a narrow size distribution, be no smaller than about 50 $\mu$m and have a low aspect ratio, with similar length in all directions. Control of the crystal size and size distribution is discussed later in Section 7.3.4.2.

Crystal habit is important in that, while low aspect ratio crystals pack together to form a porous, easily washed filter cake, thin, plate-like crystals quickly form a thin layer with low permeability next to the filter medium, after which filtration is very slow. Their high surface area holds more mother liquor, which can be difficult to remove by washing. Needle-shaped crystals form a porous cake, but tend to block the pores of the filter medium.

The habit of an undamaged crystal is governed by a combination of its lattice symmetry and the rate at which the species is/are incorporated into the different possible sites or growth planes on the nucleus and growing crystal. For example, consider the simple case where growth rates along the $A$, $B$ and $C$ axes of a crystal are the same. The resulting crystal will be a cube (see Figure 4).

If the growth rates along the $A$ and $B$ axes are the same as before, but that along the $C$ axis is only one tenth of that rate, the resulting crystal would be a thin slab. Where growth along both the $B$ and $C$ axes is only one tenth that along the $A$ axis, the result is a flat-sided rod. The growth rate of a particular face is related to the ease with which the species become incorporated into the face, and the distance between successive layers of species perpendicular to the plane of the face.

The habit of an undamaged crystal can vary depending on the environment in which it is grown. This is because other species, such as the solvent in which the crystals form or particular impurities which may be present, can affect the growth rate of one or more faces by preferentially associating with or attaching to those faces and slowing the rate at which the solute species can be incorporated into the growth sites. If the energy with which the impurities or solvent are attached to the growing faces is relatively low, then, as the temperature of the

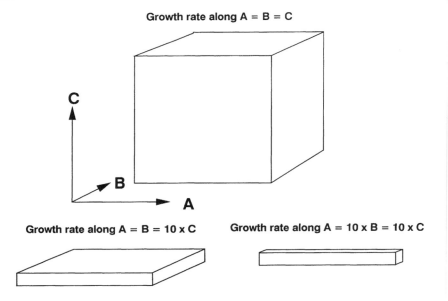

**Figure 4** Effect of growth rate of different faces on crystal habit

system at which crystal growth is occurring is raised, the habit may change as growth sites become unblocked. The micrographs in Figure 5 illustrate the effect on crystal habit of crystal growth at different temperatures for the chlorophenylhydrazone $ClC_6H_4NHN=CHCOOH$. It is seen that a relatively small increase in reaction temperature, from ambient to 40 °C, results in a major change in the crystal habit. XRD showed that the crystals all have the same form.

The effects of different solvents on crystal habit are well known. The section on the physical constants of organic compounds in older editions of the *Handbook of Chemistry and Physics*[8] gives many examples of compounds crystallised from different solvents having different crystal habits; therefore, in the development of any crystallisation process, it should be routine to examine the crystals produced, using an optical microscope. The crystals should, wherever possible, be examined in their mother liquors, prior to any solid–liquid separation step. This will quickly give an idea of the crystal size and size distribution as well as their habit, and will also show whether they have suffered much attrition damage, for example due to agitation (such damage results in fragments which widen the distribution and can cause problems in filtration).

### 7.2.3 HABIT MODIFIERS

Impurities, whether present as by-products or by deliberate addition, can have a profound effect on crystal habit, even at very low concentrations. The presence of

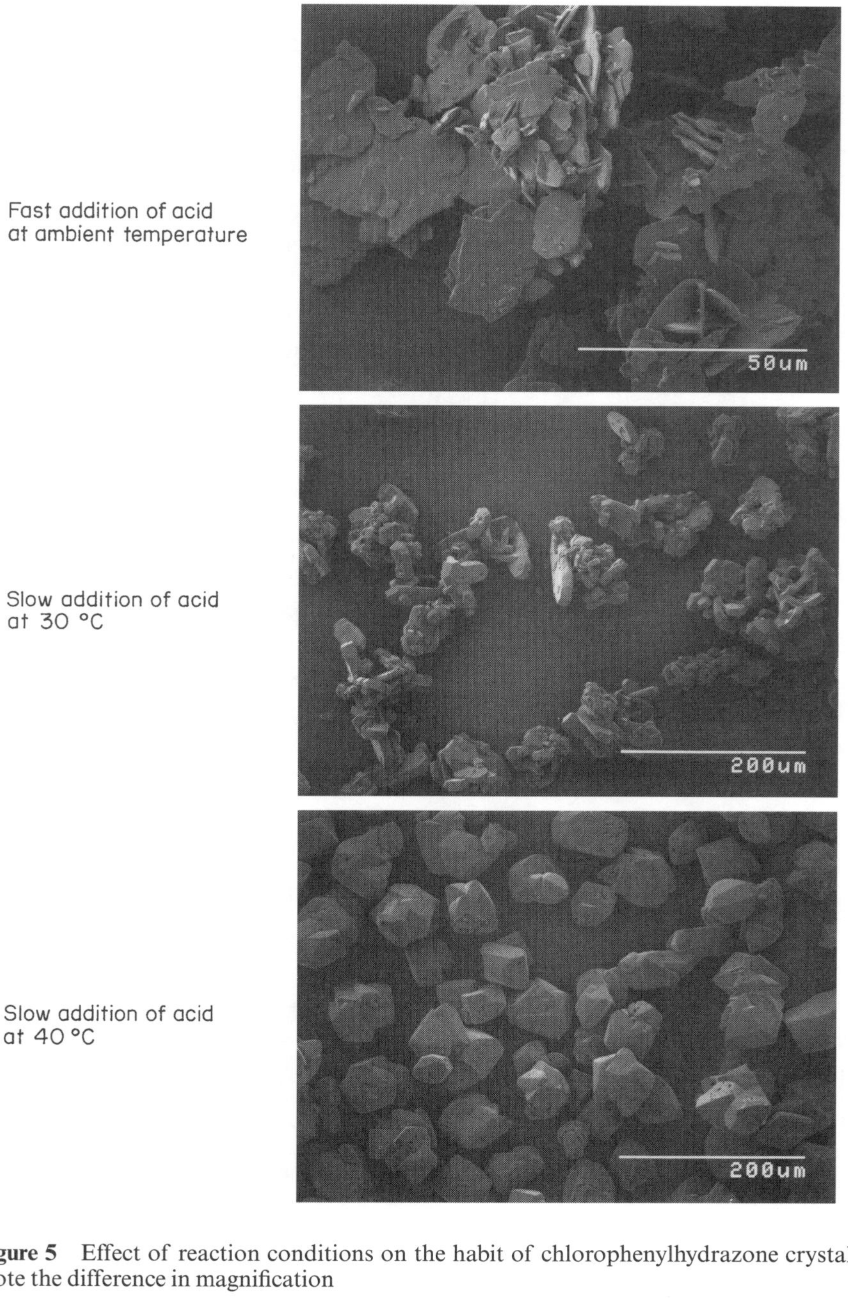

CRYSTAL SCIENCE IN MANUFACTURE OF CHIRAL COMPOUNDS 127

Fast addition of acid at ambient temperature

Slow addition of acid at 30 °C

Slow addition of acid at 40 °C

**Figure 5** Effect of reaction conditions on the habit of chlorophenylhydrazone crystals. Note the difference in magnification

Cubic
(no habit modification)

Octahedral
(modified with urea)

Dendritic
[modified with hexacyanoferrate (II)]

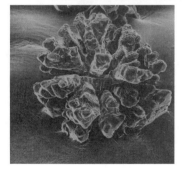

**Figure 6** Sodium chloride crystals of different habits

# CRYSTAL SCIENCE IN MANUFACTURE OF CHIRAL COMPOUNDS 129

even 1 ppm of hexacyanoferrate(II) during the crystallisation of NaCl from aqueous solution is sufficient to change the habit from the normal cubes to dendrites, while higher levels inhibit nucleation. The presence of urea causes salt to grow as octagonal crystals (see Figure 6).

In the past, the experimental approach to finding an effective habit modifier to achieve a desired change in crystal habit was empirical. The effects of large numbers of materials, particularly natural products such as gums and gelatines, were examined in the hope of finding an effective modifier. With the increasing availability of single-crystal XRD and, particularly, the development of software to derive the lattice structure from the diffraction data, the precise arrangement of species in the various faces of a crystal can be determined much more easily and cheaply than before. A more structured approach to the development of effective habit modifiers to control growth at chosen faces has thus become possible. For example, consider a perpendicular slice through a particular crystal face (Figure 7). The molecules in this face are shown in diagrammatic form with an attachment end (the circle) and a receiver end (the triangle). The face grows because of the favourable energetics of fitting the attachment end of the incoming molecule

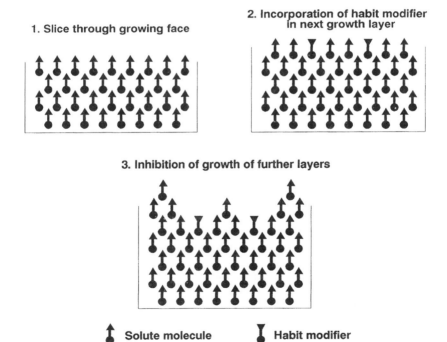

**Figure 7** Effect of a habit modifier on growth rate

to the receiving ends of the molecules in the existing face, together with further interactions with its adjacent neighbours in the new face.

If impurity molecules are present of similar size and structure, with an identical attachment end, but with a different receiving end, then the likelihood of such a molecule becoming incorporated into a growth site in the crystal face may be very similar to that for the normal molecules. If it does become incorporated, then when the next layer begins to grow the energetics of the interaction between the attachment end of a normal incoming molecule and the different receiving end of the impurity will not be favourable and incorporation will not take place. Growth above and adjacent to such sites will be inhibited.

This effect is offset by the fact that the surface of a growing crystal in contact with a supersaturated solution is in a constant state of flux, with molecules continually separating from the surface and becoming incorporated. The rate of growth depends on the relative rates of separation and incorporation. Depending on the interactions with its neighbours in the new face, the impurity may, on average, stay for a longer or shorter time than a normal molecule in the growing face. If shorter, then the impurity will probably have little effect on the growth rate. To be effective, it has to stay at least as long as or preferably longer than a normal molecule, so that it is still present during the formation of the next growth layer.

## 7.3 DEVELOPMENT OF CRYSTALLISATION PROCESSES FROM SOLUTION

Crystallisation may occur from a melt, from solution, from the gaseous phase or from a supercritical fluid. The most common and useful method for optically active compounds is crystallisation from solution, and we shall concentrate on this.

### 7.3.1 SUPERSATURATION—THE DRIVING FORCE

A prime requirement for crystallisation from solution is that the concentration of the solute in the solvent exceeds its equilibrium solubility, to provide the driving force of supersaturation. Supersaturation may be created in a number of ways, for example:

— by cooling a saturated solution;
— by evaporation of solvent from a saturated solution;
— by addition of a miscible non-solvent to a saturated solution;
— for ionic salts, by use of the common ion effect;
— by salting out;
— by reaction to form the solute *in situ*.

In most crystallisations aimed at the isolation of optically active molecules,

CRYSTAL SCIENCE IN MANUFACTURE OF CHIRAL COMPOUNDS 131

one of the first four of these methods are used, particularly the cooling of a saturated solution, and the rest of this section will concentrate on this method.

## 7.3.2 CHOICE OF SOLVENT

The choice of solvent for a crystallisation process from solution is governed by many considerations, depending particularly on the method of creation of supersaturation. For example, for a cooling crystallisation the solvent should provide a high concentration solution at elevated temperature and a relatively steep solubility versus temperature curve. If supersaturation is to be achieved by evaporation, the solvent should not only provide a high concentration solution but also have a relatively low boiling point and latent heat of vaporisation. If addition of a non-solvent to a solution is to be used, then the chosen solvent system should show a relatively steep relationship between solubility and solvent composition and the component solvents should be easily separated from each other after use, e.g. by distillation, to allow their recycle. On the manufacturing scale, the choice of solvent is also often limited on the grounds of the toxicity of any solvent residues which may be retained in the isolated product, and/or fire and explosion hazards.

## 7.3.3 SOLUBILITY CURVES AND THEIR MEASUREMENT

The yield of crystals given by a single crystallisation step not involving a reaction depends on the solubility of the solute at the initial and final conditions of temperature and/or solvent composition in the crystallisation, and the relative volumes of solvent at the start and end of the crystallisation. For example, in a simple cooling crystallisation a high yield depends on a steep relationship between solubility and temperature. In order to be able to design a crystallisation process with high yield, it is important to know how the solubility varies with the changing conditions during crystallisation, and so the solubility curve must be determined.

A further important factor which must be considered is the solubility of impurities or by-products present in the system. Since a prime objective of a crystallisation process is to isolate a pure product from an impure starting material, the choice of final conditions in a crystallisation which would lead to impurities also crystallising should be avoided. Therefore, the solubility of the impurities should also be known. In principle, a solubility curve is measured by holding stirred dispersions of the solute crystals at a number of different conditions (fixed temperatures, fixed solvent compositions) for sufficient time to reach equilibrium, and the concentration of solute in solution is then measured. The equilibrium may be approached either from a higher temperature, by cooling a suspension and crystallising, or by heating a suspension and dissolving. There are a number of practical factors to consider. It is always best to use the representative, impure system if possible, rather than pure solute and solvent, since impurities can sometimes affect the solubility of the solute. If the impurities form solid solutions in the product crystal lattice, then the solubility of the product can be increased significantly above that of pure product; a doubling or more of solubility is not uncommon in such cases. In addition, the presence of impurities may inhibit crystal growth, as discussed in Section 7.2.2, and this effect can be so strong that, if equilibrium is approached by cooling/crystallisation, growth can cease when the solution is still significantly supersaturated. If such behaviour is suspected, it is therefore good practice to measure initially the equilibrium solubility by cooling/crystallisation from solution, and then use the crystals produced to measure their solubility by dissolution. If the result by dissolution is significantly smaller than that by crystallisation, then crystal growth is ceasing when the solution is still supersaturated. Such growth inhibition can have a major effect on the yield from a crystallisation, and further studies may be justified to investigate which species may be the inhibitors. It may be possible to avoid their presence by altering the upstream chemistry or by removing them by a pre-treatment.

There are potential problems which can complicate the solubility measurements, the most common being the crystallisation of the solute as different polymorphs or solvates. If a phase transition temperature (PTT) occurs within the solubility curve temperature range, then the equilibrium solubility curve should display a discontinuity, as shown in Figure 8.

However, there is no guarantee that, as the PTT is crossed, the phase transition actually takes place. What was the stable phase becomes metastable beyond the PTT, but if the new stable phase does not nucleate, the crystals present will

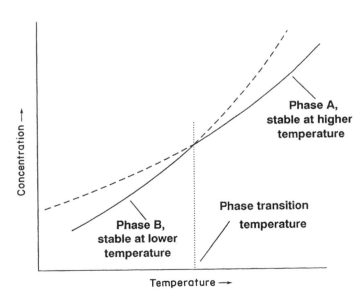

**Figure 8**  Solubility curve showing a phase transition

remain as the metastable phase, and the measured solubility will be of that phase. Therefore, instead of showing the discontinuity, the measured solubility curve will be smooth (dotted line in Figure 8). For this reason, it is important that the nature of the crystalline phase present at equilibrium at the extremes of high and low temperature likely to be used in practice are established by powder XRD. If the XRD patterns show different phases to be present at the different temperatures, then it is good practice to conduct two series of solubility curve measurements, one with rising temperature and the other with cooling. If, as sometimes happens, the phase transition does not occur, the PTT will be where the two solubility curves intersect. There may be more than one PTT within the temperature range.

The solubility curve defines the equilibrium concentration of solute at a given temperature and solvent composition. However, during a crystallisation process the system is not at equilibrium; supersaturation is being created, and crystal nucleation and growth occur towards equilibrium.

### 7.3.4 NUCLEATION, THE METASTABLE ZONE AND CONTROL OF CRYSTAL SIZE

A simple solubility versus temperature curve is shown in Figure 9. Consider a solution at the concentration and temperature represented by point A. This point lies beneath the solubility curve and so represents an undersaturated solution. If this solution is now cooled, the system will travel along a horizontal line on the

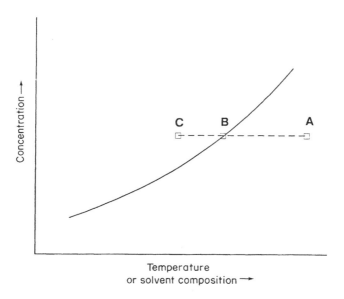

**Figure 9** A simple solubility curve

diagram until, at point B, the system becomes just saturated. Further cooling causes the system to become supersaturated, but nucleation to form crystals does not usually occur immediately the system becomes supersaturated.

Only when cooling is continued to some point C does sufficient supersaturation exist for nucleation to occur. By analogy with reaction kinetics, in which there is an activation energy, there is an 'activation supersaturation' which has to be exceeded before nucleation occurs. The gap between points B and C (the 'metastable zone') may be small, perhaps less than 1 °C in a cooling crystallisation, but may be large, particularly for large molecules. Cases where the gap is many tens of degrees are not uncommon. It may take some hours (the induction time) for nucleation to take place at point C, but as the supersaturation is increased still further, so the induction time normally reduces rapidly. Assuming the solubility curve has been determined, the position of point C relative to B in a cooling crystallisation may be established.

### 7.3.4.1 Measurement of Metastable Zone Width

In the first method, a slightly undersaturated solution is prepared, for example by filtering the equilibrium solution from a dissolution solubility curve experiment. The solution is then re-heated to several degrees above its original equilibrium temperature, and maintained at this temperature for a period of several hours to ensure that any nuclei which may have passed through the filter have redissolved, then cooled to within a few degrees of the saturation temperature. At elevated temperatures or with volatile solvents, great care must be taken to minimise any loss of solvent, for example by using pressure rather than vacuum filtration.

The undersaturated solution is then slowly cooled in a stirred vessel and the temperature at which the first sign of nucleation is observed is noted. This is tedious to do visually, and a preferred method is to use a fibre-optic colorimeter (Figure 10), which detects the onset of nucleation by attenuation of the light beam. If the nucleation temperature is reached quickly, then the experiment should be repeated using a slower cooling rate. The same solution can be re-used, provided that the temperature is raised again and held at several degrees above the saturation temperature to ensure dissolution of all the nuclei. It is generally found that the points at which nucleation occurs lie on a curve which is roughly parallel to the solubility curve (Figure 11), the region between the curves being the metastable zone. Within this zone any crystals present, for example by the addition of seed crystals, will grow but the solution will not self-nucleate.

The second method for measuring the metastable zone width is more time consuming but gives additional information. The same equipment is used, but the temperature of the solution is quickly reduced to a chosen value and then maintained. Again, the time taken for the first nuclei to appear is noted. If no nucleation occurs within several hours, the solution is reheated to above its equilibrium solubility temperature, then re-cooled to a slightly lower temperature. A graph is then plotted of the induction time at each temperature before nucleation

# CRYSTAL SCIENCE IN MANUFACTURE OF CHIRAL COMPOUNDS

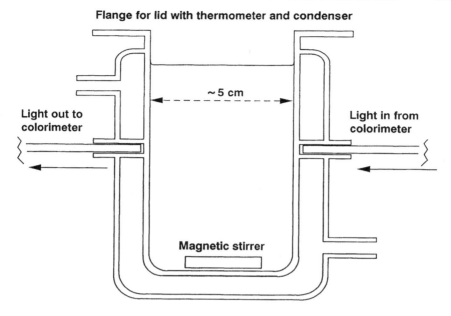

**Figure 10** Jacketed vessel for metastable zone measurement. The side tubes are used for fitting fibre-optic cables from a Brinkmann fibre-optic colorimeter

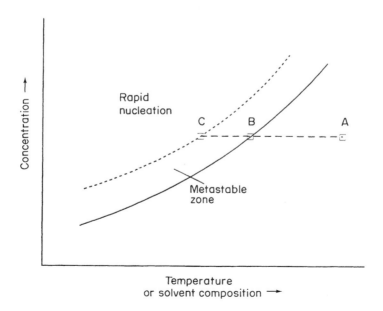

**Figure 11** Solubility curve and metastable zone

occurs versus the degree of supersaturation created at that temperature, which is obtained from the solubility curve. A typical graph is shown in Figure 12.

This graph is particularly useful in that it indicates how long a solution can withstand a given level of supersaturation without nucleating, which is important in designing the combination of the area of heat transfer surface and the temperature difference ($\Delta T$) across that surface to be used in a cooling crystallisation. The solution close to the heat transfer surface will cool quickly to the surface temperature, and then agitation will cause re-heating by mixing with the bulk. A knowledge of the induction time versus supersaturation curve allows the calculation of the minimum surface temperature allowable if crash nucleation is not to occur at, or close to, the heat transfer surface.

In Figure 11, the nucleation point curve is deliberately shown as a dotted line. This curve is in practice rather variable for a particular solvent–solute system, its position depending particularly on the scale of the experiment. Nucleation is a statistical process: the larger the volume of a supersaturated solution, the more likely it is that nucleation will occur and the narrower is the metastable zone.

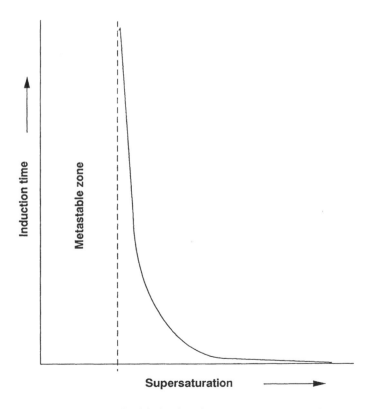

**Figure 12** Graph of induction time versus supersaturation

In the case of crystallisation by addition of a miscible non-solvent, the metastable zone can be measured in similar fashion to that for cooling crystallisation but, in this case, the temperature is held constant and the solvent composition is changed by addition of the non-solvent. It is important that the non-solvent is dispersed as quickly as possible to avoid localised high supersaturations and consequent crash nucleation.

### 7.3.4.2 Control of Crystal Size and Size Distribution

The number of nuclei formed per unit volume of solution during a nucleation event is controlled primarily by the rate at which supersaturation is created and the nucleation point curve is crossed. Crossing it very rapidly, e.g. by very rapid cooling, can increase the number of nuclei formed by up to three orders of magnitude compared with slow cooling. If each nucleus grows to a crystal then, since a fixed amount of the solute will crystallise from the solution, a larger number of nuclei must result in each crystal being smaller.

Figures 13–15 show how the concentration of solute in solution can vary during a cooling crystallisation. Figure 13 shows the case where the cooling rate is

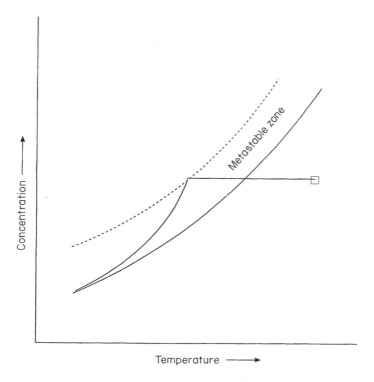

**Figure 13** Supersaturation path for unseeded system with slow cooling. One nucleation event gives small crystals of uniform size

relatively slow. After the nucleation event, the concentration of solute falls as growth occurs on the nuclei. As cooling continues, the concentration remains within the metastable zone, although its position relative to the solubility curve will depend on the relative rates of cooling, and hence creation of supersaturation, and of crystal growth, which discharges that supersaturation. The resulting crystals will be of relatively uniform size which, as described above, will be governed by the number of nuclei initially formed.

If cooling is very rapid, as in Figure 14, the rate at which supersaturation is created may exceed the rate at which it is discharged via crystal growth on the nuclei created at the first nucleation event, particularly if habit modifiers are also present which reduce the crystal growth rate. In this case, further nucleation events occur as shown, the second and subsequent crops of nuclei giving rise to successively smaller final crystals. When crystallisation is complete, the crystal size distribution will be wider than if there had been only one nucleation event, which can result in problems with the subsequent isolation of the crystals from the mother liquor. The wider size distribution gives rise to a less permeable cake

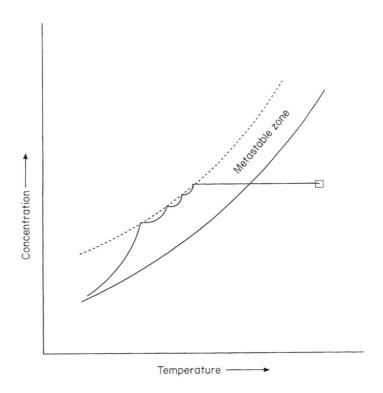

**Figure 14** Supersaturation path for unseeded system with fast cooling. Several nucleation events give a wide distribution of very small crystals

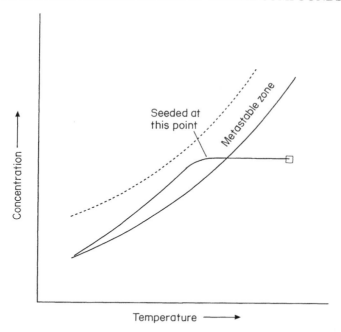

**Figure 15** Supersaturation path for seeded system with slow cooling. No nucleation event; the crystal size depends on the number and size of seeds added

or blocks the filter medium, and such multiple nucleation should generally be avoided.

Figure 15 shows the case where the solution, cooled until it is supersaturated but remaining within the metastable zone, is then seeded with solute crystals and cooled slowly so that the system never self-nucleates. The resulting crystals will be of a uniform size which is dependent on the number of seed crystals added. In this way, much larger crystals may be formed than achievable by self-nucleation. Seeding may also be used to induce crystallisation of a particular metastable polymorph which would not normally nucleate under the crystallisation conditions. For this to be feasible, the metastable zones for both polymorphs must overlap and seeding must be carried out within this overlap region. This is discussed further in Section 7.3.6.

While addition of seeds on the laboratory scale is usually straightforward, on the plant scale this can be difficult. Also, the window of metastable zone where seed must be added is often rather narrow. Where control of crystal size rather than polymorph is desired, an alternative to the addition of seed crystals is the use of ultrasound. This has been shown to induce nucleation at lower supersaturations than necessary for natural nucleation, i.e. it narrows the metastable zone.[9] An example of the practical use of ultrasound in a chiral system is given in ref. 10.

The curves in Figures 13–15 are for a cooling crystallisation, but similar arguments can be made for systems where the supersaturation is created by addition of a non-solvent or by a pH change, for example the crystallisation of a phenol by addition of acid to a solution of the phenolate salt. In such systems it is very important to control the rate of addition of the non-solvent or the acid/alkali used to adjust the pH, and to ensure good agitation to obtain rapid mixing.

Where habit modifiers are being used, their effect on crystal size is not easily predictable. As they act by slowing the growth rate of different faces, after a given growth time the crystals will tend to be smaller than in the absence of modifiers. However, the modifiers may be so efficient at blocking the growth sites that they may inhibit the initial nucleation of the crystals. The resulting increase in supersaturation may then result in crash nucleation, producing very many more nuclei than from a normal nucleation event and smaller final crystals. Alternatively, the modifier may moderate the number of nuclei formed at nucleation which are capable of growth, leading to fewer, larger crystals.

Thus, the nucleation conditions can be used to control the crystal size and size distribution. Such control depends not only on measuring the solubility curve and the width of the metastable zone as described above, but also the rate of desupersaturation via crystal growth.

### 7.3.5 RATE OF DESUPERSATURATION

Desupersaturation of a supersaturated solution by crystal growth takes time, and this has to be taken into account when controlling a crystallisation process. In a given system, the rate of desupersaturation of the solution in a supersaturated dispersion of crystals depends both on the temperature and on the surface area of the crystals present on which growth is occurring. For a given degree of supersaturation, (the percentage by which the concentration exceeds the equilibrium solubility at a given temperature), as the temperature is raised the rate of crystal growth generally increases because the rates both of diffusion of solute species to the crystal surface and their re-orientation to fit into the growth sites increase. In addition, as the surface area of crystals present increases so the mass of solute that can crystallise in unit time also increases. Both of these effects increase the rate of desupersaturation.

Where the crystals have low solubility and the solute is created by reaction, as in the precipitation of a simple ionic salt, the rate of desupersaturation can be very fast. However, with molecular crystals, particularly where the molecules are large, half-lives for desupersaturation of several hours are known, even where there is a large area available for growth.

### 7.3.6 ISOLATION OF METASTABLE PHASES

If, in a polymorphic system, it is desired to isolate a phase which is metastable at ambient temperature, then there are several possible ways of achieving this depending on the system and, in particular, the solubility curves of the phases and their metastable zone widths. Where nucleation occurs via the creation of high supersaturations, it is often found that the predominant nucleating phase is a metastable phase. This empirical observation was first noted by Ostwald and was the basis of his 'Rule of Stages'.[11] In practice, the rule does not always hold,[12] and where it does it is because the kinetics of nucleation and crystal growth of the metastable phase are faster than those of the stable phase.

Even where it does hold, some nuclei of the stable phase are likely to be created. If nuclei of the stable crystal phase exist in a dispersion of a metastable phase, then in time the system will change so that only crystals of the stable phase remain. Therefore, while crystallisation via the creation of very high supersaturation may be worth trying, it does not guarantee lasting success. In addition, the crystals formed are likely to be very small. There is, however, an approach which is likely to be more lastingly successful and give larger crystals.

Consider first a system with a phase transition temperature (PTT) within normal operating temperatures, up to the boiling point of the solvent. Let us refer to the stable phase at lower temperatures as S and the metastable phase as M. The solubility curves and metastable zone widths versus temperature are shown qualitatively in Figure 16.

If controlled nucleation is made to occur above the PTT at temperature $T_1$, then the crystals which form will be the stable phase at that temperature, i.e. M. Phase S is very unlikely to nucleate because the solute concentration stays within

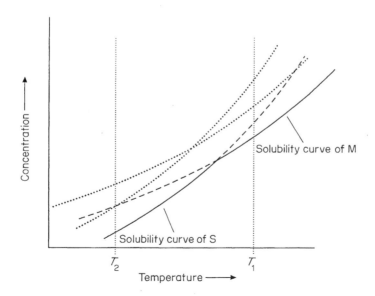

**Figure 16** Solubility curves and metastable zones of a polymorphic system with a phase transition temperature

its metastable zone. If the system is then cooled slowly, at such a rate that the solute concentration remains only slightly above the solubility curve of M, then the temperature could be reduced to $T_2$ before the solution becomes sufficiently supersaturated with respect to S that S nucleates. Therefore, if the crystals are isolated above $T_2$, they should consist of pure M.

Consider now a system which does not have a PTT at a useful temperature. In order to isolate M from such a system, its solubility curve must lie within the metastable zone of S, as shown in Figure 17. So long as this is so, then it should be possible to seed the system with M seeds within the metastable zone of S and within the narrow band of conditions between the solubility curve of M and the metastable zone limit of S, and then apply controlled cooling to achieve growth of M while avoiding nucleation of S. This of course assumes a source of M seeds is available.

If the metastable zone of the stable phase S in a particular solvent does not overlap the solubility curve of M, as in Figure 18, then any saturated solution of M will lie outside the metastable zone of S, and so S will nucleate. Therefore, it is impossible to isolate only phase M from such a system. The only options are to investigate different solvent systems, to try to find one in which the solubility curve of M does lie within the metastable zone of S, or possibly search for, or design, a nucleation inhibitor for phase S, and hence widen its metastable zone.

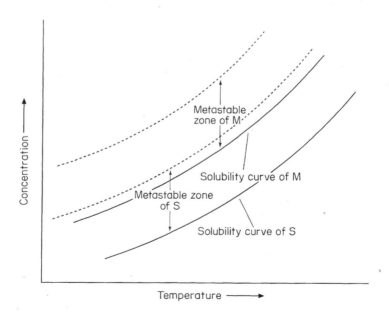

**Figure 17** Case where solubility curve of M lies within the metastable zone of S. M may be isolated by seeding within the narrow region between the solubility curve of M and the metastable zone limit of S

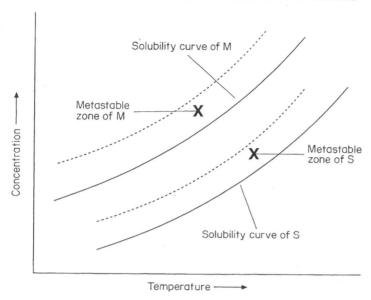

**Figure 18** Case where solubility curve of M lies outside the metastable zone of S. M cannot be isolated by seeding

Occasionally, what was thought to be the stable phase at a given temperature turns out to be metastable with respect to a new phase which spontaneously nucleates for the first time, and thereafter isolation of the original metastable phase may prove difficult if not impossible. This can be a very serious problem, particularly in the pharmaceuticals area, where physical form may dramatically affect the bioavailability of the drug. The phenomenon of 'disappearing polymorphs' is discussed in ref. 13.

## 7.4 APPLICATION OF CRYSTAL SCIENCE TO CHIRAL SYSTEMS

As in the manufacture of non-chiral molecules, crystallisation is used routinely to separate chiral molecules from impurities formed during their synthesis. In addition, crystal science may be applied in the isolation of individual enantiomers from a racemic starting material, and this subject is covered in depth in the excellent book of Jacques et al.[14] This chapter is intended as an introduction, illustrating the various ways in which the science discussed above may be applied to chiral systems, with some specific examples.

A successful crystallisation process to separate a mixture of solutes depends on utilising or controlling one or more of the following properties of those solutes:

— different solubility profiles versus temperature or solvent composition;
— different metastable zone widths;
— their different rates of crystal growth.

For molecules with a single chiral centre, the opportunities to achieve a good separation of the enantiomers from a racemate by application of crystal science appear at first sight to be limited, their individual physical properties (melting points, latent heats of fusion and solubility in non-chiral solvents) being the same, as are their metastable zone widths and rates of crystal growth. If, however, a second chiral centre is introduced to form a diastereoisomer, then the opportunities increase.

### 7.4.1 SEPARATION OF ENANTIOMERS BY DIASTEREOISOMER FORMATION

The basis for the majority of racemate resolutions is fractional crystallisation of diastereoisomers, and the subject is reviewed in depth by Jacques et al.,[14] Bayley and Vaidya[15] and Bruggink in Chapter 5 of this volume.

In favourable cases, where the metastable zone width for the more soluble diastereoisomer is wide, the yield of the less soluble diastereoisomer need not be

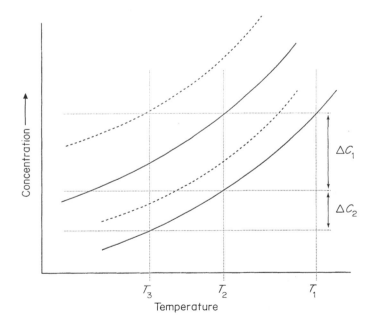

**Figure 19** Solubility curves and metastable zones of two diastereoisomers. On crystallising a 50:50 mixture of diastereoisomers, the yield of the less soluble diastereoisomer can be increased by making use of the wide metastable zone of the more soluble diastereoisomer

CRYSTAL SCIENCE IN MANUFACTURE OF CHIRAL COMPOUNDS   145

limited solely by the difference in the solubilities of the two diastereoisomers. It should be possible to drive the crystallisation further, creating conditions where the more soluble diastereoisomer is supersaturated, but still within the metastable zone, so that it does not nucleate. Figure 19 shows the solubility curves and metastable zones of two diastereoisomers. For a 50:50 mixture of the diastereoisomers, a solution can be made which is just saturated in the less soluble diastereoisomer at temperature $T_1$. Cooling to $T_2$ will cause nucleation and growth of that diastereoisomer and give a crystallised yield equivalent to the concentration difference $\Delta C_1$. It will also make the solution just saturated in the more soluble diastereoisomer. However, the solution can be cooled further to $T_3$, still within the metastable zone of the more soluble diastereoisomer, further increasing the yield of crystals of the other diastereoisomer by an amount equivalent to $\Delta C_2$ and still avoiding nucleation of the more soluble diastereoisomer.

### 7.4.2  SEPARATION OF DIASTEREOISOMERS

An interesting example of this approach from the author's experience concerned the separation of the diastereoisomers of a molecule with two chiral centres. The structure of the molecule cannot be disclosed, for commercial reasons, but nevertheless it provides a good illustration of the application of crystal science in several ways.

The relative solubilities of two diastereoisomers [$(RR')$ and $(RS')$] versus temperature in the chosen solvent are shown in Figure 20. From this solvent the diastereoisomers crystallise as solvates. Given the difference in solubility, by

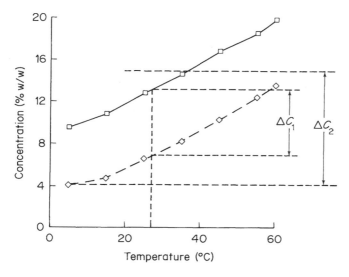

**Figure 20**  Solubility curves for ($\square$)($RR'$) and ($\diamond$)($RS'$) diastereoisomers

**Table 1** Yield and purity of (RR′) and (RS′) diastereoisomers from first crystallisation sequence

| Form | Yield (%) | Purity (%) |
|---|---|---|
| (RR′) | 77 | 98 |
| (RS′) | 49 | 98 |

making a solution at 60 °C containing 26.5% of the 50:50 mixture of the diastereoisomers, that solution will be just saturated in the ($RS'$)-isomer but well undersaturated in the ($RR'$)-isomer. If the solution is then cooled to 27 °C the solution at equilibrium will have become just saturated in ($RR'$), and ($RS'$) will have crystallised to the equivalent of $\Delta C_1$. At temperatures lower than 30 °C, the ($RR'$) form will also be supersaturated.

However, the solution was prepared from non-solvated crystals of the mixed diastereoisomers and, since these are metastable with respect to the solvated crystals, they have a higher solubility. Therefore they continue to dissolve when added to a solution saturated with respect to the solvated ($RS'$) phase, and so a significantly supersaturated solution in ($RS'$) can be prepared containing 30% of mixed diastereoisomers. Also, it was found that the width of the metastable zone for ($RR'$) was so wide that the resulting solution could safely be cooled to 5 °C and held without ($RR'$) nucleating. The yield of ($RS'$) crystals could therefore be increased to the equivalent of $\Delta C_2$. Desupersaturation of ($RS'$) from the solution proceeded with a half-life of 100 min, and so, once nucleated, the system had to be held at 5 °C for about 7–8 h before the concentration of ($RS'$) in solution reached equilibrium solubility.

The crystals of ($RS'$) were then filtered off and the liquors at 5 °C seeded with ($RR'$) seeds. The half-life for desupersaturation of ($RR'$) was found to be 10 h, so the system required holding for about 50 h to reach equilibrium before filtration! The isolated yields and purities of the products from this first crystallisation sequence are shown in Table 1.

If the final filtrates were then re-heated to 60 °C and further mixed diastereoisomers and solvent added to raise the concentration of ($RR'$) to its original value, the whole process could be repeated. Since the concentrations of ($RR'$) and ($RS'$) in the final filtrates will be the same as at the end of the first crystallisation sequence, it follows that, in theory at least, the isolated yields of both isomers should be 100%, based on the weights of isomers added at the start of the second crystallisation sequence. In practice, the yields were found to be ($RR'$) 98% and ($RS'$) 95%, the main causes of yield loss being handling and washing of the filtered crystals.

There is a limit to the number of times that the filtrates may be recycled in such a process before undesirable side-effects are observed. This is because impurities build up in the filtrates, which may begin to interfere with the habit or size of the crystals or their growth rates. Even where these effects are absent, the concentration of the impurities will eventually rise to a point where they themselves crystallise, reducing the purity of the isolated products.

A more sophisticated example of the isolation of a desirable diastereoisomeric pair with biological activity from a pair with no activity is provided by the process for the synthetic pyrethroid cyhalothrin (Figure 21). The synthetic route[16] produces four of the eight possible isomers of this molecule (those with a *cis* configuration at the cyclopropane ring), which comprise two diastereoisomeric pairs known as *cis* A and *cis* B. The insecticidal activity resides only in the *cis* B pair.

During the development of this product, it was found that on cooling a solution of all four isomers, crystals were obtained each of which contained equal amounts of the *cis* B pair of enantiomers (an example of a racemic compound crystal; see below). The crystal growth rate was very slow, and the process was complicated by the fact that the non-crystallising pair of isomers acted as a solvent for the crystallising pair, leading to the observation that the crystallisation rate fell on increasing the initial concentration.

However, the hydrogen atom on the chiral carbon bearing the cyano group is fairly acidic and proton exchange occurs readily in the presence of base. As reprotonation can occur from either side of the planar carbanion, the presence of

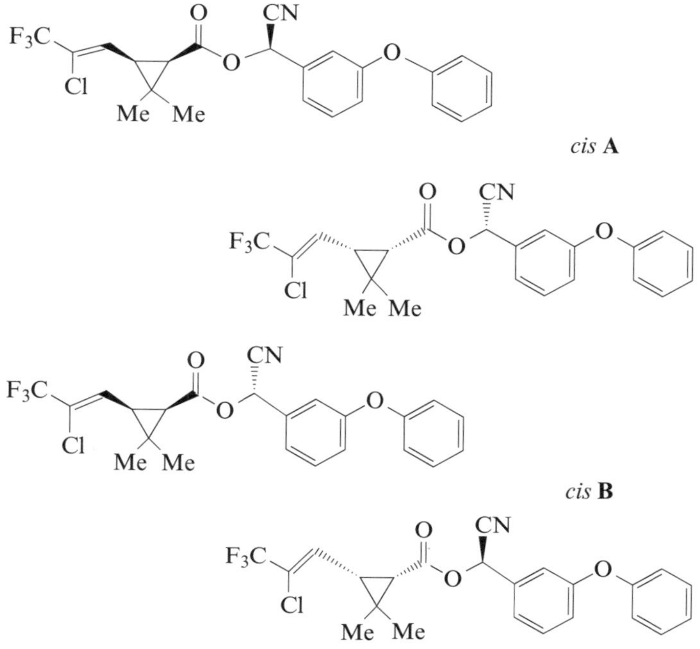

**Figure 21** The *cis* A and *cis* B diastereoisomeric pairs of isomers of cyhalothrin

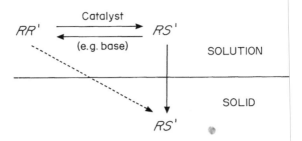

As *RS'* crystallises, it is replaced in solution by epimerisation of *RR'* to maintain the *RR'* ⇌ *RS'* equilibrium

**Figure 22**  Second-order asymmetric transformation

base causes epimerisation of the diastereoisomers at this chiral centre. If crystallisation is carried out in the presence of a base, then, as the active diastereoisomer pair *cis* B crystallises, so the remaining diastereoisomers in solution epimerise to maintain the equilibrium and help to maintain the concentration of *cis* B in solution. The process is known as a second-order asymmetric transformation (see Volume I, p. 26) and is shown diagramatically in Figure 22.

This approach may also be used in resolutions for diastereoisomeric salts, where the chemistry of the target chiral molecule allows *in situ* racemisation during crystallisation. An example is described by Yoshioka *et al.*,[17] involving the resolution of DL-*p*-hydroxyphenylglycine with (+)-1-phenylethanesulphonic acid. Epimerisation was achieved by holding the solution/dispersion of the diastereoisomers at 100 °C in acetic acid containing some salicylaldehyde. After 5 h the slurry was cooled and filtered, giving a yield of 85% of the D-(+)-diastereoisomer, based on the initial weight of the DL-(+)-form.

### 7.4.3 RESOLUTION BY DIRECT CRYSTALLISATION OF ENANTIOMERS

Let us now consider the use of crystal science to obtain single enantiomers from racemates directly. As mentioned above, enantiomers have identical solubilities and metastable zone widths in non-chiral solvents. Therefore, if one is to crystallise one enantiomer but not the other, one has either to work within the limits of the metastable zone, seeding with crystals of an individual enantiomer, or else use chiral nucleation inhibitors to widen the metastable zone of one of the enantiomers.

An additional complication is that when a racemate crystallises, the crystals formed can be of two types (Figure 23). The more common is where the enantiomers co-crystallise to form a mixed crystal containing both enantiomers in equal proportions, known as a racemic compound.

# CRYSTAL SCIENCE IN MANUFACTURE OF CHIRAL COMPOUNDS

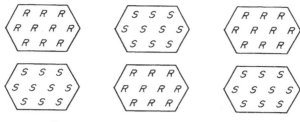

**Racemic mixture (conglomerate)**

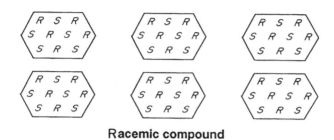

**Racemic compound**

**Figure 23** Crystal forms of a racemate

If a racemate produces only this type of crystal, crystallisation may not be used to separate the enantiomers except by formation of a diastereoisomer. Seeds of pure enantiomer added to a supersaturated solution of such a system within the metastable zone will simply dissolve.

Crystal science may be applied, however, in a minority of systems (perhaps 10%) in which the racemate crystallises to give crystals in which the lattice contains either one or the other enantiomer. These are referred to as racemic conglomerate (or sometimes racemic mixture) crystals. The crystals may have an asymmetric habit (although not always[18]), and where this is the case the (R) and (S) crystals will be mirror images of each other. Louis Pasteur performed the first resolution of a racemate in 1848[19] by manually sorting out the (R) and (S) crystals of sodium ammonium tartrate, although this approach is hardly an industrial procedure!

It is important, therefore, to determine whether a racemate crystallises as racemic compound or conglomerate crystals. Several methods may be used, but all rely on having a sample of one of the enantiomers in a relatively pure crystalline form, ideally recrystallised from the same solvent system and under similar conditions as the racemate. Probably the most widely used method is to compare the infrared spectra of the crystals of the racemate and of a pure enantiomer in the solid state. If the spectra are identical, then the racemate is a con-

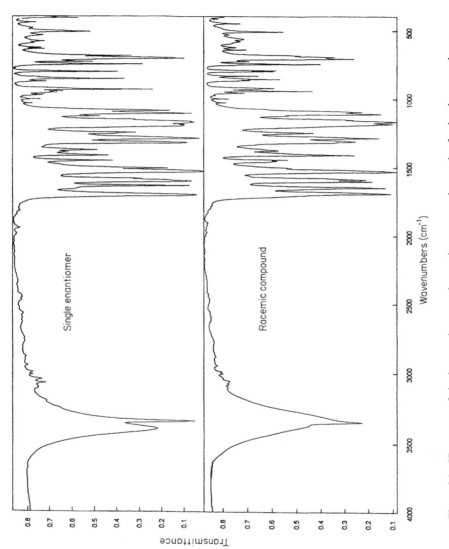

**Figure 24** IR spectra of single enantiomer and racemic compound crystals of a development drug

glomerate. If the spectra are different, then it is most likely that the racemate forms compound crystals (although it is also a possibility that the racemate is a conglomerate, but in a different polymorph or solvate phase from the pure enantiomer). To minimise this possibility it is important that the crystal samples should be isolated under similar conditions. Figure 24 shows the IR spectra of crystals of a single enantiomer and of a racemic compound of a development pharmaceutical. Although the spectra are very similar, there are several significant differences.

Another method is to compare the powder XRD patterns of the racemate and pure enantiomer crystals. If they are identical, then the racemate is a conglomerate. If they are different, then again it is most likely, although not certain, that the racemate gives compound crystals.

If the crystals melt without decomposing, another method which is particularly useful for compounds that are liquid at room temperature, and which has not been reported before now, is to measure the heat of fusion of the racemate and single enantiomer crystals using DSC. For example, the latent heat of fusion of the pure L-enantiomer of 2-chloropropionic acid, MeCHClCOOH, was measured in a sub-ambient DSC and found to be about 84 $J g^{-1}$. The value for the racemate was about 149 $J g^{-1}$. Now, if the racemate was a racemic conglomerate, its latent heat of fusion in $J g^{-1}$ would be expected to be equal to that of the single enantiomer, less some heat of mixing of the enantiomers in the liquid phase. That its heat of fusion is almost double that of the single enantiomer indicates that the racemate cannot be a conglomerate, but rather a racemic compound.

It should be noted that there may be a phase transition temperature on one side of which the racemate crystallises as a conglomerate and on the other as compound crystals. Sodium ammonium tartrate has such a transition temperature at 28 °C. Below this temperature, it forms conglomerate tetrahydrate crystals. Between 28 °C and 35 °C it forms compound monohydrate crystals, while above 35 °C it forms separate anhydrous compound crystals of sodium tartrate and ammonium tartrate. It was fortuitous that Pasteur's final crystallisation temperature was below 28 °C! There are also cases[14] where conglomerate crystals are the stable phase above a transition temperature, specifically, where the conglomerate crystals have a lower degree of solvation than the compound crystals.

As described earlier, the existence of such a phase transition temperature may be detected from discontinuities in the solubility versus temperature curve, and confirmed by XRD, DSC and TGA on the crystals present at equilibrium above and below the transition temperature.

It is good practice to resolve the racemate at as early a stage as possible in the synthesis, assuming the subsequent chemistry is stereospecific. The intermediates formed in the steps after the chiral centre has been introduced into the molecule, as well as the final product, can all be checked to see whether they form racemic conglomerate crystals. Assuming a racemate is found which gives conglomerate crystals, crystallisation techniques may be applied to resolve it.

### 7.4.3.1 Seeding Within the Metastable Zone

The basis of one approach, which has been used commercially by Merck to resolve an α-methyldopa intermediate, was first described by Jungfleisch[20] in 1882. He showed that on careful seeding of a supersaturated solution of sodium ammonium tartrate (at a temperature below 2 °C) within the metastable zone (which is wide for this system) with one seed crystal each of the pure enantiomers, these grew without further nucleation to give two large crystals which were enantiomerically pure. Modern practice is to create a supersaturated solution within the metastable zone and then seed with the pure enantiomers in separate crystallisers using the equipment shown diagramatically in Figure 25.

In each crystalliser the supersaturation is controlled so that it remains within the metastable zone and nucleation of the other enantiomer does not occur. Care must be taken that seeds of the unwanted enantiomer are not inadvertently introduced into the crystallisers via the feed solution. This approach works particularly well for the α-methyldopa intermediate because the enantiomers have a very wide metastable zone. The temperature difference between the dissolver and the crystallisers is quoted as 60 °C.[21] Where the metastable zone is narrow, the control of supersaturation becomes very critical.

While it is not feasible on an industrial scale to sort mirror image racemic conglomerate crystals of the same size, they might be sorted by sieving if the crystals differ significantly in size. This is the basis of a process first described by Dowling,[22] where the crystals of different sizes are obtained by seeding a supersaturated solution simultaneously with large crystals of one enantiomer and very small crystals of the other. Dowling applied the method to glutamic acid salts, while Watanabe and Noyori[23] applied it to acetylglutamic acid.

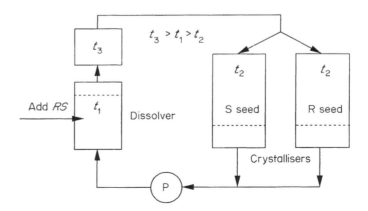

**Figure 25** Process for resolution by seeding with individual enantiomers within metastable zone (after Merck process for an α-methyldopa intermediate)

If the chemistry of the enantiomers permits an easy *in situ* racemisation under crystallisation conditions, then the opportunity exists to use a second-order asymmetric transformation process as described above for diastereoisomers. For example, in the synthesis of the agrochemical paclobutrazol (Figure 26), the penultimate stage involves the formation of a racemic chiral ketone. Since the reduction of this ketone to the final product using borohydride is completely stereoselective, then if the ketone is resolved to obtain the (S) enantiomer, the final product will be the desired (2S,3S)-diastereoisomer.

It was found that crystals grown from aqueous methanol were conglomerate crystals[24] and that the metastable zone width was about 5 °C. In addition, in the presence of base, the ketone is in equilibrium with the enolate form and racemisation is very rapid. Therefore, the system is a good candidate for seeding to produce either desired enantiomer.

Although the ability to form racemic conglomerate crystals is a necessary requirement for successful resolution of a racemate by direct crystallisation, it

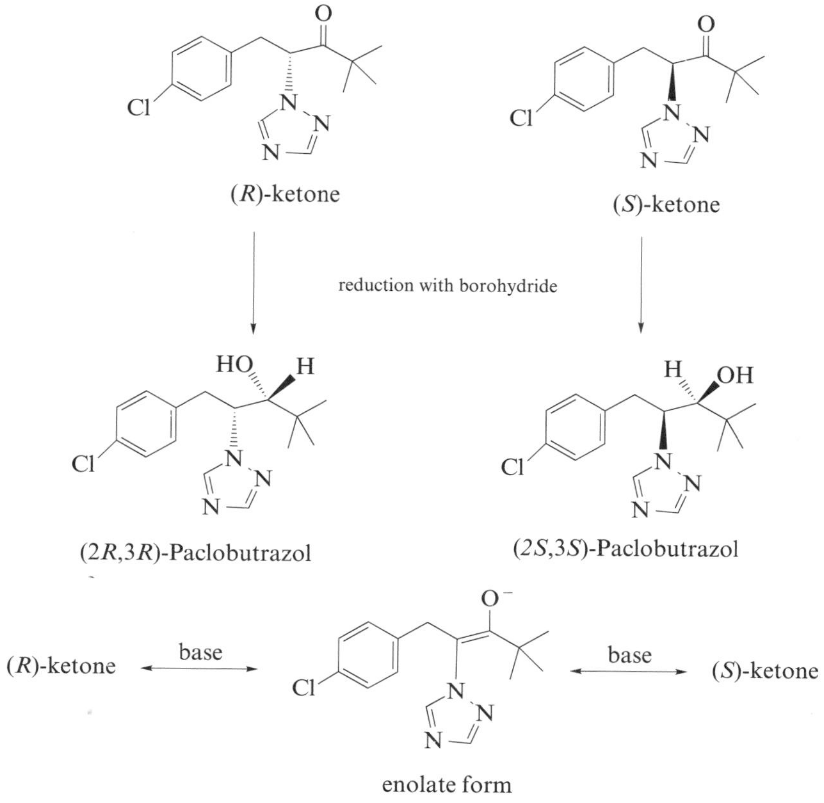

**Figure 26** Paclobutrazol chemistry

does not guarantee it. Some conglomerates, while giving crystal lattices containing only one enantiomer, form crystals which contain domains of both lattices in a single crystal. This was found to be the case with the above ketone, if high supersaturations within the metastable zone were used. Apparently, enantiomerically pure seed was nucleating the other enantiomer at its surface. Only when a low supersaturation, given by 1 °C of undercooling, was used did this effect disappear to give the product with a high enantiomeric purity. This behaviour at higher supersaturation is consistent with observations on $\beta$-phenylglyceric acid.[25]

When this aspect became understood, the process was changed from a cooling crystallisation to an evaporative crystallisation, allowing much closer control of the supersaturation during the process.

### 7.4.3.2 Nucleation Inhibition of One Enantiomer

The last crystallisation technique for resolving a racemate that will be discussed involves the use of additives to inhibit selectively the nucleation and growth of one of the enantiomers. As discussed in Section 7.2.3, such inhibitors or habit modifiers should be similar in structure to the target crystallising molecule. Given that it should affect one enantiomer but not the other, it follows that the agent must itself be chiral.

This approach has been investigated in detail[26] for certain amino acids. Thus, if a racemic solution of DL-asparagine is crystallised in the presence of D-glutamic acid, the first crystals to appear are L-asparagine. The metastable zone for the D-asparagine has effectively been widened due to the nucleation-inhibiting effect of the D-glutamic acid. The design of more potent polymeric chiral inhibitors has been described.[27]

It has been mentioned several times that the solubility of enantiomers is the same in non-chiral solvents. In a single enantiomer chiral solvent, however, there is the possibility for the solubilities to be slightly different. More importantly, chiral solvents can themselves sometimes inhibit nucleation and/or growth of one enantiomer relative to the other, and the successful resolution of enantiomers by direct crystallisation has been reported using isopropyl D-tartrate[28] and $(-)$-$\alpha$-pinene[29] as the chiral solvents.

## 7.5 CONCLUSION

The successful application of crystal science in the isolation of chiral compounds depends very much on the whims of Nature in determining, for example, whether diastereoisomeric salts can be prepared with suitably large differences in solubility and/or metastable zone width, or whether a racemate crystallises as racemic compound or conglomerate crystal forms from a chosen solvent system. The likelihood of success may be improved by judicious choice of the stage in the route at which resolution is attempted, bearing in mind particularly the advantages offered by second-order asymmetric transformations, where they are feasible.

Development of a successful process depends on finding out just how generous Nature has been, by measuring solubility curves in a range of solvents, the metastable zone widths, and the crystal phase(s) present at various temperatures and by checking the crystal habits. Crystal science can then be applied to optimise the process to ensure the isolation of the desired crystal phase as uniform, easily isolated crystals. Where Nature is kind, or accepts a helping hand, crystal science can provide the basis for very effective processes in the manufacture of chiral compounds.

## 7.6 REFERENCES

1. Mullin, J. W., Crystallisation, 3rd edn, Butterworth-Heinemann, Oxford, 1993.
2. Haisa, M., Kashino, S., and Maeda, H., *Acta Crystallogr., Sect. B*, **30**, 2510 (1974).
3. Haisa, M., Kashino, S., Kawai, R., and Maeda, H., *Acta Crystallogr., Sect. B*, **32**, 1283 (1976)
4. Fagan, P. F., Roberts, K. J., Docherty, R., Charlton, A. P., Jones, W., and Potts, G. D., *J. Chem. Soc., Chem. Commun.*, in press.
5. Glusker, J. P., Lewis, M., and Rossi, M., *Crystal Structure Analysis for Chemists and Biologists*, VCH, New York, 1994.
6. Davey, R. J., Cardew, P. T., McEwan, D., and Sadler, D. E., *J. Cryst. Growth*, **79**, 648 (1986).
7. McNaughton, J. L., and Mortimer, C. T., in *IRS; Physical Chemistry Series 2*, Butterworths, London, 1975, Vol. 10; subsequently reprinted by Perkin-Elmer Corpo., Norwalk, CT, USA, from whom it may be ordered, code PE 0993–8669.
8. See, e.g. Weast, R. C. (Ed.), *Handbook of Chemistry and Physics*, 58th edn, CRC Press, Cleveland, OH, 1977.
9. Martin, P. D., Phillips, E. J., and Price, C. J., in *1993 I. Chem. E. Research Event*, University of Birmingham, Birmingham, 1993, pp. 516–518, ISBN 0 85295308 9).
10. Noguchi Research Foundation, *Fr. Pat.*, 1 389 840, 1965. *US Pat.*, 3 365 492, 1968.
11. Ostwald, W., *Z. Phys. Chem.*, **22**, 289 (1897).
12. Cardew, P. T., and Davey, R. J., in *Proceedings of Symposium on Tailoring of Crystal Growth*, (ed. by J. Garside), I. Chem. E. NW Branch, Manchester, 1982, pp. 1.1–1.7.
13. Dunitz, J. D., and Bernstein, J., *Acc. Chem. Res.*, **28**, 193 (1995).
14. Jacques, J., Collet, A., and Wilen, S. H., *Enantiomers, Racemates and Resolutions*, Wiley–Interscience, New York, 1981.
15. Bayley, C. R., and Vaidya, N. A., in *Chirality in Industry* (ed. A. N. Collins, G. N. Sheldrake and J. Crosby), Wiley, Chichester, 1992, pp. 69–77.
16. Robson, M. J., Cheetham, R., Fettes, D.J., and Crosby, J., *1984 British Crop Protection Conference Proceedings*, BCPC Publications, Croydon, 1984, Vol. 3, pp 853–857.
17. Yoshioka, R., Tohyama, M., Ohtsuki, S., Yamada, S. and Chibata, I., *Bull. Chem. Soc. Jpn.*, **60**, 649 (1987).
18. Addadi, L., Berkovitch-Yellin, Z., Weissbuch, I., Lahav, M., and Leiserowitz, L., *Top. Stereochem.*, **16**, 1 (1986).
19. Pasteur, L., *C. R. Acad. Sci.*, **26**, 535 (1848).
20. Jungfleisch, M. E., *J. Pharm. Chim.*, 5ème Ser., **5**, 346 (1882).
21. *Chem. Eng.*, 247 (1965).
22. Dowling, B. B., *US Pat.* 2 898 558, 1959.

23. Watanabe, T., and Noyori, G., *Kogyo Kagaku Zasshi*, **72**, 1083 (1969).
24. Black, S. N., Williams, L. J., Davey, R. J., Moffatt, F., Jones, R. V. H., McEwan, D. M., and Sadler, D. E., *Tetrahedron*, **45**, 2677 (1989).
25. Furberg, S., and Hassel, O., *Acta Chem. Scand.*, **4**, 1020 (1950).
26. Addadi, L., Berkovitch-Yellin, Z., Domb, N., Garti, E., Lahav, M., and Leiserowitz, L., *Nature (London)*, **296**, 21 (1982).
27. Zbaida, D., Weissbuch, I., Shavit-Gati, E., Addadi, L., Leiserowitz, L., and Lahav, M., *React. Polym.*, **6**, 241 (1987).
28. Luttringhaus, A., and Berrer, D., *Tetrahedron Lett.*, 10 (1959).
29. Groen, M. B., Schadenberg, H., and Wynberg, H., *J. Org. Chem.*, **36**, 2797 (1971).

# 8 Membrane Separations in the Production of Optically Pure Compounds

**J. T. F. KEURENTJES and F. J. M. VOERMANS**
*Akzo Nobel Central Research, Arnhem, The Netherlands*

| | | |
|---|---|---|
| 8.1 | Introduction | 157 |
| 8.2 | Enantioselective Membranes | 158 |
| | 8.2.1 Selective Polymers | 158 |
| | 8.2.2 Selective Liquid Membranes | 162 |
| 8.3 | Membrane-assisted Separations | 168 |
| | 8.3.1 Liquid–Liquid Extractions | 168 |
| | 8.3.2 Liquid-membrane Fractionation | 171 |
| | 8.3.3 Micellar-enhanced Ultrafiltration (MEUF) | 174 |
| | 8.3.4 Membrane-assisted Kinetic Resolutions | 175 |
| 8.4 | Conclusion | 178 |
| 8.5 | References | 179 |

## 8.1 INTRODUCTION

In the past two decades, the range of conventional separation techniques, such as distillation, crystallisation and extraction, has been extended by a wide range of membrane separation processes. Traditionally, membranes are found in biomedical applications (e.g. blood dialysis and plasmaphoresis[1]) and reverse osmosis (e.g. desalination[2]). However, material developments over the past 15 years have opened the way to many industrial applications. Membrane separations often provide opportunities as cost-efficient alternatives to separations which are troublesome or impossible using classical methods and, since most membrane processes are performed at ambient temperature, they can offer clear advantages compared with conventional separation processes.[3,4]

There are two general types of membrane processes for enantiomer separations: either direct separation using an enantioselective membrane, or separation in which a non-selective membrane assists an enantioselective process. Obviously, the most direct method is to apply enantioselective membranes which provide a selective barrier, allowing preferential transport of one of the enantiomers of a

*Chirality in Industry II.* Edited by A. N. Collins, G. N. Sheldrake and J. Crosby
© 1997 John Wiley & Sons Ltd

racemic mixture. These membranes can be of either the dense polymer or liquid type. In the latter case a selective carrier is usually added to the liquid membrane. Non-selective membranes can also provide essential non-chiral separation characteristics, thus making a chiral separation based on enantioselectivity outside the membrane technically and economically feasible. Several configurations can be envisaged for this purpose:

(1) hollow-fibre membrane fractionation (liquid–liquid extraction);
(2) liquid-membrane fractionation;
(3) micellar-enhanced ultrafiltration (MEUF);
(4) membrane-assisted kinetic resolution.

In this chapter, an overview is given of the (potential) application of membrane separation processes in the separation of racemic mixtures; although not exhaustive, it should give an impression of the developments in this field. Application of the different process options will be discussed with special regard to scale operations. Many of the systems described in this chapter still are in an early stage of development whereas others are currently being scaled up to pilot plant or are already applied at full scale.

## 8.2 ENANTIOSELECTIVE MEMBRANES

### 8.2.1 SELECTIVE POLYMERS

Enantioselective polymer membranes usually consist of a non-selective porous support coated with a thin layer of an enantioselective polymer. The selective layer is made from a chiral polymer or a polymer containing chiral side groups, such composite materials having the mechanical strength to withstand a pressure gradient over the membrane. The selectivity of such a membrane is determined by enantiospecific interactions between the isomers to be separated and the top-layer polymer matrix, resulting in one enantiomer being selectively dissolved in the membrane. This separation mechanism is termed 'solution/diffusion,' which is similar to the mechanism found in membrane processes like reverse osmosis and pervaporation.[5] In principle, these membranes can be used in a similar process configuration to reverse osmosis (RO) or ultrafiltration (UF) membranes, and a pressure difference across the membrane is the driving force for transport. Scale-up of the system to the desired productivity is relatively easy, as RO and UF are well established technologies with well known guidelines for operation on a large scale. When a membrane does not yield the required optical purity in a single step, a cascade of membrane units can result in a feasible process.[6]

The performance of an enantioselective membrane is dependent on two factors: the permeability $P$ (flux) and enantioselectivity ($\alpha$), the latter being defined as the ratio of the permeabilities of the L- and D-enantiomers:

$$\alpha = \frac{P_\text{L}}{P_\text{D}} \tag{1}$$

The search for enantioselective polymers has been mainly in the context of the separation of racemic amino acids. The polymers commonly used as stationary phases in chromatography can be used as starting points, e.g. polysaccharides (especially cellulose derivatives), acrylic polymers, poly(α-amino acids) and polyacetylene-derived polymers.

Poly-L-glutamate with amphiphilic side-chains (1) forms an α-helix, which is selectively permeable for water-soluble optical isomers, such as α-amino acids.[7-10] Dialysis experiments (with no trans-membrane pressure) showed that the resolution of LD-tryptophan can be almost complete (Figure 1), but that the flux

$$+\text{COCHNH}+_n$$
$$|$$
$$(\text{CH}_2)_2$$
$$|$$
$$\text{COO}+\text{CH}_2\text{CH}_2\text{O}+_m \text{—}\langle\text{—}\rangle\text{—}\text{C}_9\text{H}_{19}$$

(1)

n-Nonylphenoxyhexa(oxyethylene)-1-ol poly(γ-methyl-L-glutamate) (NON6-PLG)

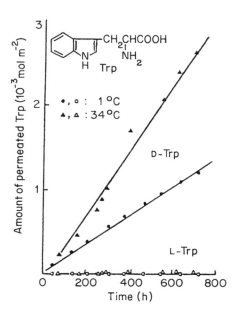

**Figure 1** Permeation behaviour of D- and L-tryptophan from the racemic mixture through poly-L-glutamate-derived membranes (NON-6-PLG) in a dialysis configuration. Reproduced with permission from ref. 9

through these membranes is extremely low (of the order of $3 \times 10^{-6}$ mol m$^{-2}$h$^{-1}$ at 2 mmol l$^{-1}$ in the feed). Similar fluxes, but lower selectivities, were reported for the separation of DL-tyrosine ($\alpha = 6.4^7$) and DL-serine ($\alpha = 3^9$). Polypropene membranes with $(-)$-$\beta$-pinene side groups [PDPSP (**2**)][11] exhibit a higher selectivity ($\alpha = 13.4$) for the resolution of DL-tryptophan than the poly-L-glutamate-derived membranes when a dialysis configuration is used. An increase in the trans-membrane pressure to 8 bar results in a significant increase in the permeation rate (ca. $7 \times 10^{-5}$ mol m$^{-2}$ h$^{-1}$, assuming the membrane is 10 $\mu$m thick), but the selectivity values approach unity. When a membrane is made from a blend of PDPSP and PTMSP {poly[1-(trimethylsilyl)prop-1-yne]} a pressure-resistant membrane results, yielding $\alpha = 2.3$ at 8 bar (flux ca. $5.4 \times 10^{-4}$ mol m$^{-2}$h$^{-1}$, assuming a 10 $\mu$m thick membrane).

(**2**)

Poly{1-[dimethyl(10-pinanyl)silyl]prop-1-yne} (PDPSP)

To improve the permeability of the derivatised poly-L-glutamate membranes[7] and the PDPSP membranes,[11] alternative membranes consisting of a chiral backbone [poly(methyl-L-glutamate)] and flexible short siloxane side-chains have been synthesised by Aoki et al.[12] Permeation rates increase significantly to $5 \times 10^{-4}$–$6 \times 10^{-3}$ mol m$^{-2}$h$^{-1}$ at 10 bar trans-membrane pressure (assuming a 10 $\mu$m thick membrane), although the selectivity values drop to 1.2–1.35.

A different approach is to immobilise the chiral component in a porous membrane. For example, an amino acid can be immobilised in the pores of a polysulphone UF membrane (Figure 2a).[13] Depending on the trans-membrane pressure, selectivities for the separation of DL-phenylalanine were between 1.25 and 4.1, at permeabilities between $10^{-6}$ and $10^{-7}$ m s$^{-1}$, respectively (Figure 2b). A rather dilute feed stream is used (1–4 mol m$^{-3}$), and this results in low solute fluxes. Although with these selectivity values many cycles are required to obtain high enantiomeric ratios, fluxes are at a level commonly found for reverse osmosis.[5] In another case, bovine serum albumin (BSA) has been immobilised on a polysulphone UF membrane which is selective for amino acids, and selectivities of 1.2 have been obtained with a corresponding flux of 0.016 mol m$^{-2}$h$^{-1}$.[14]

Only a few non-amino acids have been resolved using enantioselective polymer membranes. Examples are the $\beta$-blocker oxprenolol (**3**),[15] stilbene oxide (**4**)[16] and butan-2-ol,[11] using cellulose tris(3,5-dimethylphenylcarbamate) (**5**), tri-(p-chlorophenyl)carbamate (**6**) and PDPSP (**2**) membranes, respectively.

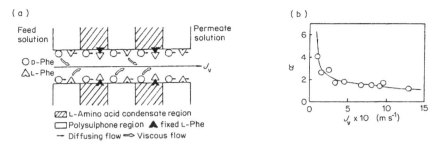

**Figure 2** (a) Transport mechanism for the separation of DL-phenylalanine by an enantioselective UF membrane. (b) Effect of volume flux on the separation factor when DL-phenylalanine is separated by an enantioselective UF membrane ($T = 37\,°C$). $J_v$ = volume flux.

(3) Oxprenolol

(4) Stilbene oxide

(5) Cellulose tris(3,5-dimethylphenylcarbamate)

(6) Cellulose tris($p$-chlorophenylcarbamate)

On examining the data for enantioselective polymer membranes, it is clear that most experiments have been performed on the laboratory scale, using membrane surface areas typically in the range 1–100 cm². Most experiments have been performed using a dialysis type of experiment, and the selectivity thus obtained represents an 'intrinsic' property of the membrane. When the flux is increased by increasing the trans-membrane pressure, this usually results in a sharp decrease in selectivity. Although the selectivities reported so far are encouraging, large-scale application of these membranes does not seem very likely in the short term.

## 8.2.2 SELECTIVE LIQUID MEMBRANES

A liquid film can be immobilised in the pores of a membrane by capillary forces. Two miscible liquids that do not wet the porous membrane can then be kept apart by the immobilised film. This liquid membrane may contain an enantio-selective carrier which selectively forms a complex with one of the enantiomers of a racemic mixture at the feed side, and transports it across the membrane, where it is released into the receptor phase (Figure 3). To avoid leakage of the carrier from the liquid membrane, it must not dissolve in the feed liquid or the receptor phase, and to achieve sufficient selectivity it is necessary to minimise non-selective transport through the bulk of the membrane liquid. Transport of the enantiomers to be separated can be driven by a gradient in concentration or pH.

The simplest way to apply enantioselective liquid membranes is to use a chiral liquid, e.g. chiral alcohols [nopol (7) and (S)-(−)-2-methylbutan-1-ol] immobilised in the pores of a polyethylene film for the separation of amino acid hydrochlorides.[17] Nevertheless, most methods for enantioselective transport across a

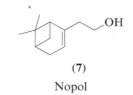

(7)
Nopol

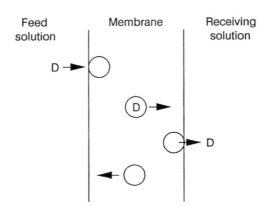

**Figure 3** Schematic representation of a liquid membrane. The carrier selectively forms a complex with one enantiomer at the feed side and releases it in the receiving solution

liquid membrane make use of chiral selector molecules added to an achiral liquid membrane, and an obvious advantage of these systems is the possibility of tailoring the selector molecules to the application.

Despite the fact that cyclodextrins (8) are used extensively in gas chromatography,[18] liquid chromatography[19] and capillary electrophoresis,[20] they have been little used as selective carriers in liquid membranes. Cyclodextrins are expected to be selective in transporting cyclic and heterocyclic compounds,[19] e.g. racemic 1-ferrocenylethylphenylsulphide (9), which has been resolved using β-cyclodextrins at a permeation rate of $10^{-3}$ mol m$^{-2}$h$^{-1}$.[21]

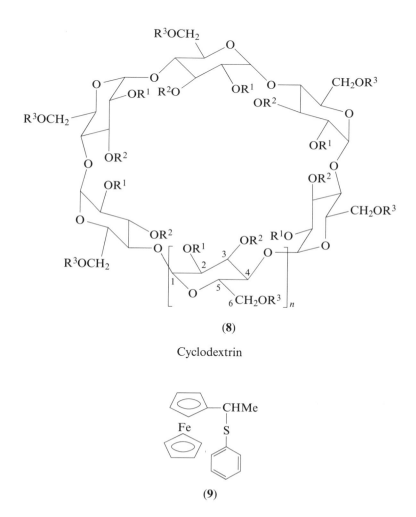

(8)

Cyclodextrin

(9)

1-Ferrocenylethylphenylsulphide

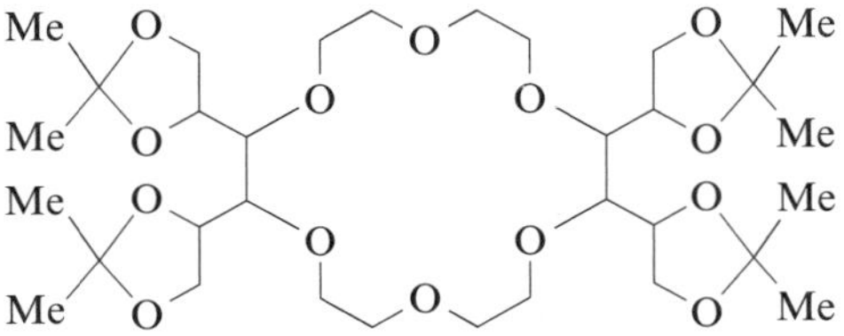

**(10)**

1,2:1′,2′:5,6:5′,6′-Tetra-*O*-isopropylidene-3,3′:4,4′-bis-*O*-oxydiethylenedi-D-mannitol

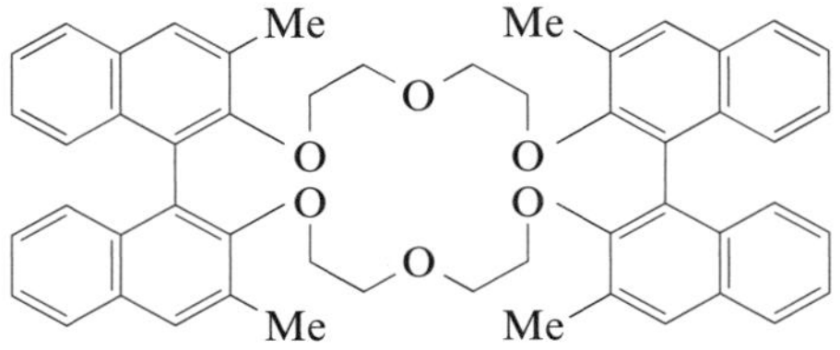

**(11)**

3,3′-Dimethylbis(α,α′-dinaphthyl)-22-crown-6

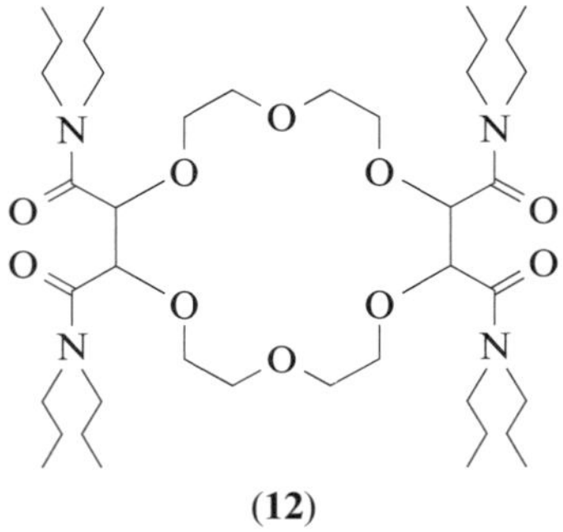

**(12)**

*N,N,N′,N′,N″,N″,N‴,N‴*-Octapropyl-1,4,7,10,13,16-hexaoxacyclooctadecane-2,3,11,12-tetracarboxamide

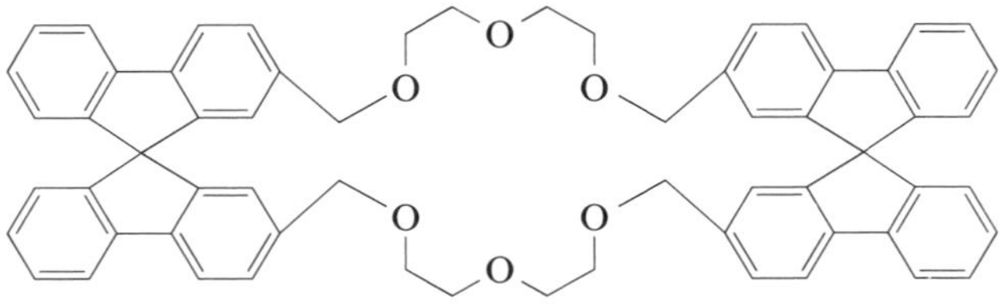

**(13)**

2,2:2′,2′-Bis(2,5,8-trioxanonamethylene)di(9,9′-spirobifluorene)

# MEMBRANE SEPARATIONS FOR OPTICALLY PURE COMPOUNDS 165

Chiral crown ethers are lipophilic compounds, showing selectivity towards chiral amine salts. Chirality is introduced by chiral side groups, such as tartaric acid derivatives (e.g. **10**),[22,23] dinaphthyl compounds (**11**),[24,25] and carboxamides (**12**).[26,27] 9,9'-Spirobifluorene crown ethers (**13**) sometimes exhibit high enantioselectivities for α-amino alcohols;[28] enantioselectivies of up to 9.5 have been reported. Binding of the OH group of the α-amino alcohol is probably the key factor in chiral recognition. Selectivities up to 19.5 have been achieved with binaphthyl crown ethers for the resolution of phenylglycine,[25] in which binding of the aromatic moiety is essential for obtaining selectivity.

Surfactant solutions have also been used for chiral separations in liquid membrane systems. Bile salt micelles, immobilised in a cellulose support, have been used for the enantioselective transport of 1,1'-binaphthol ($\alpha = 1.9$).[29] In a similar experiment, dansylphenylalanine has been separated using ionene (OIM-6) micelles, resulting in enantioselectivities of 2.1, whereas without the micelles no transport occurs.

Amino acid derivatives have been used as carriers in liquid membrane transport[30,31] and in chiral liquid–liquid extractions[32,33] for the separation of amino acids. However, other compounds can also be transported selectively. Examples include the separation of propranolol (**14**) and bupranolol (**15**),[31] using $N$-$n$-alkyl-hydroxyprolines (**16**) as the selective carriers, resulting in enantioselectivities of 1.24 and 1.16, respectively. The transport mechanism is ion pairing (Figure 4). $N$-(3,5-Dinitrobenzoyl)leucine can be transported selectively through a liquid membrane containing fatty esters or amines of ($S$)-$N$-(1-naphthyl)-leucine in hexane.[34] Here, a bulk liquid membrane system was used (Figure 5), in

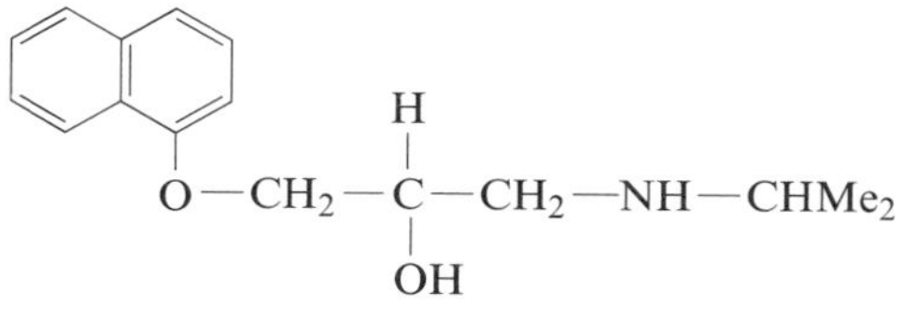

(**14**) Propranolol

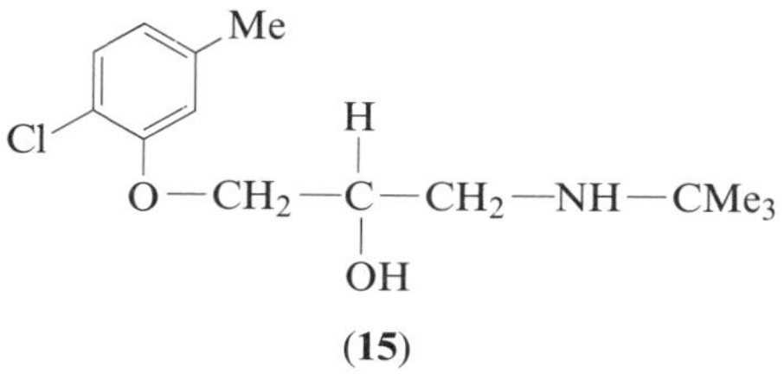

(**15**) Bupranolol

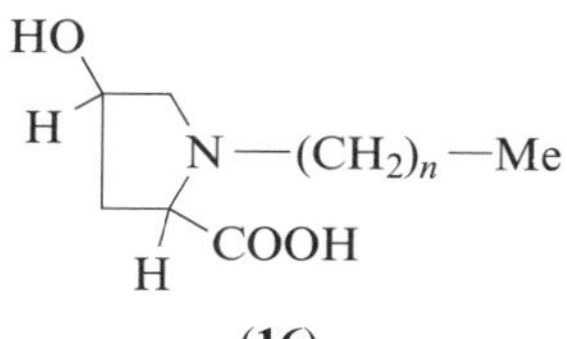

(**16**) $N$-$n$-Alkylhydroxyproline

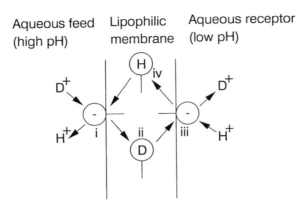

**Figure 4** Principle of ion pairing for selective transport across a liquid membrane. (i) Carrier head group deprotonates at donor/membrane interface and undergoes ion-pair interaction with positively charged drug; (ii) neutral ion pair diffuses through the lipophilic membrane; (iii) at the membrane–receptor interface the carrier re-protonates owing to the lower pH, thereby releasing drug into receptor phase; (iv) carrier diffuses back

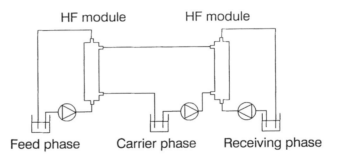

**Figure 5** Principle of a bulk liquid membrane. In the first hollow-fibre membrane module enantioselective extraction from the feed phase occurs to the carrier phase. The carrier phase is passed to the second hollow-fibre unit, where the enantiomer is released into the receiving phase

which the membrane liquid is circulated between an absorption and a desorption hollow-fibre unit. Formation and dissociation of the chiral complex take places at different locations, so that appropriate process conditions for absorption and desorption (such as temperature) can be readily achieved. Using this configuration, *ee* values in excess of 95% were obtained in a single step.

Enantioselective supramolecular transport of complexes of lasolocid A (**17**) with metal ammines and amines resulted in enantioselectivities of 1.22 for 1,2-diaminoethane, 1.74 for diethylenetriamine and 1.38 for the complex of

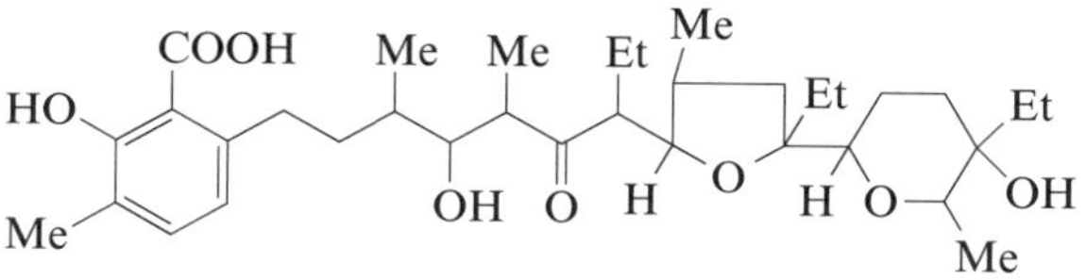

**(17)**

Lasolocid A

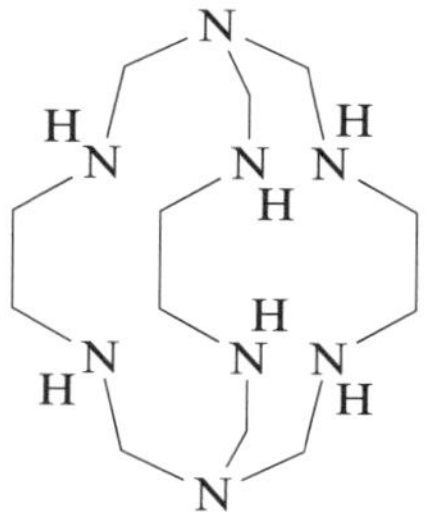

**(18)**

1,3,6,8,10,13,16,19-Octaazabicyclo[6.6.6]eicosane

1,3,6,8,10,13,16,19-octaazabicyclo[6.6.6]eicosane (**18**), using a counter gradient of ammonium ions as the driving force. Permeation rates through this bulk liquid membrane containing 1 mM of **17** in chloroform are of the order of $5 \times 10^{-3}$ mol $m^{-2}h^{-1}$.[35]

Although high selectivities can be achieved in liquid membrane separations, there are potential drawbacks. When the feed stream is depleted in one enantiomer, the enantiomer ratio in the permeate decreases rapidly (Figure 6). This is similar to a kinetic resolution,[36] where the initial *ee* of the desired product can be high, but decreases as the feed mixture is depleted in the more rapidly converted component. To avoid this either requires racemisation of the feed or, alternatively, a system containing the two opposite selectors can be used,[25] so that the feed stream remains virtually racemic. This was illustrated by using binaphthyl crown ethers as carriers in a W-tube configuration for the separation of phenylglycine.

Although selectivities reported for enantioselective transport through liquid membranes can be high, several passes are required to obtain purities in excess of 99%. The problems encountered in this type of multi-stage process are comparable with those in crystallisation processes, which, owing to the many recycles required, lead to a relatively complicated flow scheme and expensive process equipment. In the authors' opinion, the use of supported enantioselective liquid membranes is relatively complicated for large-scale separations unless the product has the required purity after a single step.

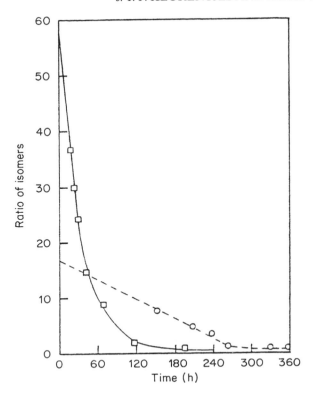

**Figure 6** Enantiomer ratio of isomers penetrating a 0.7 M β-CD liquid membrane *vs* time. The solid line represents the ratio of *p*- to *o*-nitroaniline. The dashed line represents the ratio of (+)-(*R*)- to (−)-(*S*)-1-ferrocenylethylphenylsulphide. Reproduced with permission from ref. 21

## 8.3 MEMBRANE-ASSISTED SEPARATIONS

### 8.3.1 LIQUID–LIQUID EXTRACTIONS

Enantiomers can be separated using conventional liquid–liquid extraction, provided that enantioselectivity is incorporated into one or both of the liquids. Usually, enantioselectivity is achieved by adding a chiral additive (the chiral selector) to just one of the two phases. In principle, the selectors applied in liquid–liquid extraction can be the same as the selectors applied in supported liquid membranes. Examples are the resolutions of DL-valine and DL-isoleucine in an extraction system consisting of *n*-butanol and water containing *N*-*n*-alkyl-L-hydroxyproline and copper(II) ions.[32,33] Chiral ammonium salts, based for example on norephedrine (**19**), can be separated by partitioning between an aqueous

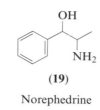

(19)

Norephedrine

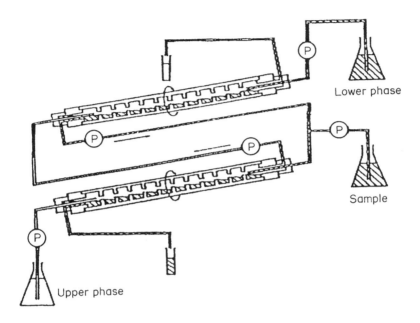

**Figure 7** Schematic representation of the countercurrent extraction apparatus used by Takeuchi et al.[33] for the resolution of amino acids.

and a lipophilic phase containing esters of tartaric acid.[37–39] Nevertheless, despite the relatively high separation factors obtained in these systems (1.25–3.54), a significant number of stages are required to obtain high optical purities, and to achieve this number of stages a good countercurrent flow is essential. This can be achieved relatively easily in an arrangement described by Takeuchi et al.[33] (Figure 7); however, scale-up of this type of equipment is expected to be difficult.

Equipment for liquid–liquid extractions is often described in terms of a required number of transfer units (NTU) for a given degree of separation:[40]

$$\mathrm{NTU} = \frac{\Lambda}{\Lambda-1} \ln\left[\left(1-\frac{1}{\Lambda}\right)\cdot\frac{x_i}{x_o} + \frac{1}{\Lambda}\right] \quad (2)$$

where $x_i/x_o$ is the ratio between the inflow and outflow concentrations in the feed stream and $\Lambda$ is the extraction factor, defined as

$$\Lambda = \frac{mF_e}{F_f} \qquad (3)$$

$F_e$ and $F_f$ are the flow rates of the two liquids, respectively, and $m$ is the distribution ratio over the two phases. The apparatus characteristics are combined in the height of a transfer unit (HTU):

$$\mathrm{HTU} = \frac{v}{K_o A} \qquad (4)$$

where $v$ is the superficial flow velocity, $K_o$ is the overall mass-transfer coefficient and $A$ is the specific surface area of the equipment. The height of the separation apparatus ($H$) now equals

$$H = \mathrm{NTU} \times \mathrm{HTU} \qquad (5)$$

Back-mixing and flooding often limit the performance of conventional extraction equipment, and this can be avoided when the liquid–liquid interface is immobilised in the pores of a membrane (Figure 8). This type of process is called 'hollow-fibre membrane extraction' (and is also referred to as 'pertraction' or 'dispersion-free extraction'). The advantage of hollow-fibre membrane extraction does not result from large mass-transfer coefficients (which are similar to those obtained with conventional equipment), but from the extremely high surface area per unit volume, typically 30 times larger for fibres than for packed towers, and 100 times larger than for conventional extractors,[41] thus reducing HTU accordingly. Extensive literature is available on the (potential) applications of hollow-fibre extractions.[42–45]

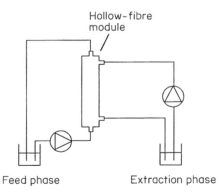

**Figure 8** Schematic representation of hollow-fibre membrane extraction

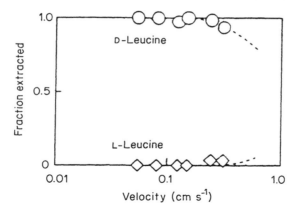

**Figure 9** Separation of DL-leucine in hollow-fibre membrane extraction using a N-n-dodecyl-L-hydroxyproline solution in octanol as the enantioselective extraction liquid. The modules used were 32 cm long and contained 96 Celgard X-20 fibres. Reproduced with permission from ref. 46

As an example, DL-leucine has been separated using N-n-dodecyl-L-hydroxyproline in octanol as the extraction phase flowing countercurrently with an aqueous phase[46]. Polypropylene hollow fibres (Celgard X-20) were used, in which the pores were filled with a cross-linked poly(vinyl alcohol) gel. In experiments with an organic to aqueous flow ratio of 4 (so that $\Lambda_D$ and $\Lambda_L$ were 1.32 and 0.64, respectively), DL-leucine was separated almost completely (Figure 9).

When complete resolution is required, it follows from equation 2 that $\Lambda_D$ and $\Lambda_L$ have to be on the opposite sides of unity. Because the gross distribution over the two phases usually deviates substantially from unity, the flow ratios have to be adjusted accordingly, which may result in excessive phase volumes. In addition to extreme dilution of the separated enantiomers, this may also lead to significant losses of valuable material.

### 8.3.2 LIQUID-MEMBRANE FRACTIONATION

To avoid the extreme flow ratios often encountered in liquid–liquid extractions, a completely symmetrical system has been developed at Akzo Nobel in which two miscible chiral liquids with opposite chiral selectors are used, separated by a non-miscible liquid membrane (Figure 10[47]). The selector has to be chosen to be incapable of passing the liquid membrane, i.e. highly lipophilic in the case of an aqueous liquid membrane and charged or highly polar in the case of a non-polar liquid membrane. Using an aqueous liquid membrane, enantioselectivities have been determined for several racemic drugs using a range of chiral selectors (Table 1).

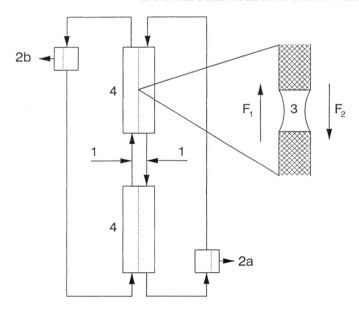

**Figure 10** Schematic representation of the Akzo Nobel enantiomer separation system. Two liquids containing opposing enantiomers of the selector ($F_1$ and $F_2$) flow countercurrently through the column (4). The racemic mixture is added in the middle of the column (1) and the enantiomers are recovered from the outflows (2a and 2b)

**Table 1** Enantioselectivities[47] determined for several drugs in the liquid-membrane fractionation system[a]

| Compound | Selector | Solvent | Anion | α |
|---|---|---|---|---|
| Norephedrine (19) | 0.25 M DHT | Heptane | – | 1.19 |
| | 0.25 M DHT | Heptane | – | 1.50* |
| | 10 wt% PLA | Chlorofom | – | 1.07 |
| | 10 wt% PLA | Chloroform | – | 1.10* |
| Ephedrine (20) | 0.10 M DHT | Heptane | 0.06 M $PF_6^-$ | 1.30 |
| Mirtazapine (21) | 0.10 M DHT | Heptane | – | 1.05 |
| | 0.25 M DHT | Heptane | – | 1.06 |
| | 0.50 M DHT | Heptane | – | 1.08 |
| | 0.25 M DBT | Decanol | – | 1.06 |
| | 10 wt% PLA | Chloroform | – | 1.04 |
| | 0.10 M DHT | Heptane | 0.01 M $PF_6^-$ | 1.07 |
| | 0.50 M DHT | Heptane | 0.03 M $PF_6^-$ | 1.16 |
| Phenylglycine | 0.25 M DHT | Heptane | – | 1.06 |
| Salbutamol (22) | 0.25 M DHT | Heptane | 0.018 M $BPH_4^-$ | 1.06 |
| | 0.25 M DHT | Cyclohexane | 0.0035 M $BPh_4^-$ | 1.04 |
| Terbutaline (23) | 0.43 M DHT | Dichloroethane | 0.5 M $PF_6^-$ | 1.14 |
| | 2.15 M DHT | Heptane | 0.0045 M $BPh_4^-$ | 1.05 |
| Propranolol (14) | 0.25 M DHT | Heptane | – | 1.03 |
| Ibuprofen (24) | 0.10 M DHT | Heptane | – | 1.10 |

[a] All experiments were performed at room temperature, except those marked with asterisks, which were performed at 4°C. In some cases a lipophilic anion was used to increase the solubility of the drug in the organic phases. $PF_6^-$ = hexafluorophosphate; $BPh_4^-$ = tetraphenylborate; DHT = dihexyl tartrate; DBT = dibenzoyl tartrate; PLA = poly (lactic acid)

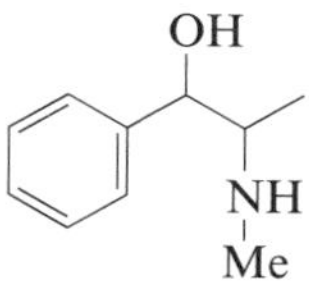

# MEMBRANE SEPARATIONS FOR OPTICALLY PURE COMPOUNDS

(20) Ephedrine

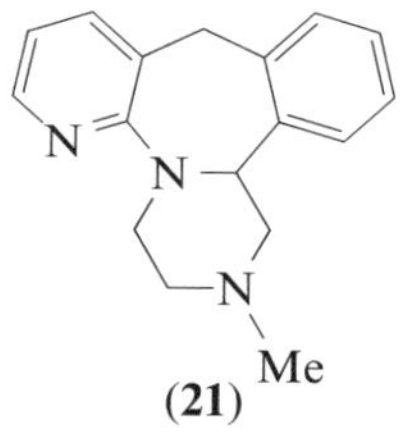

(21) Mirtazapine

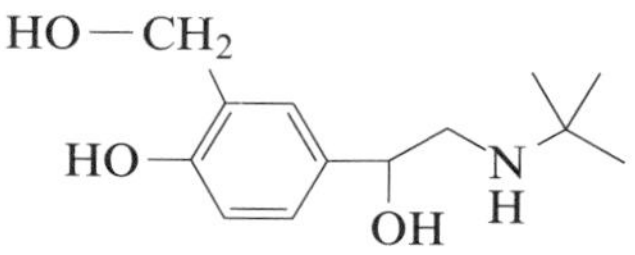

(22) Salbutamol

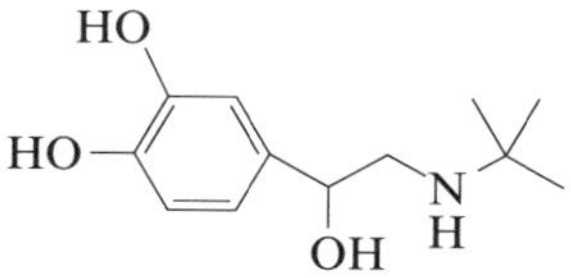

(23) Terbutaline

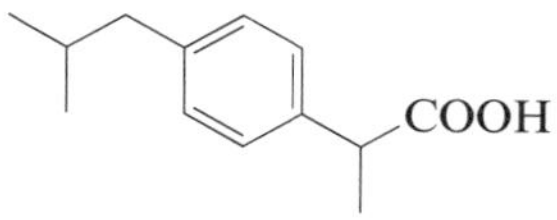

(24) Ibuprofen

Using this configuration, both enantiomers can be recovered completely with an enantiomer ratio, L/D, in the outflow given by

$$\frac{L}{D} = \frac{\frac{\Lambda_D}{\Lambda_D - 1}\left[\exp\left(\frac{\Lambda_D - 1}{2\Lambda_D}\text{NTU}\right) - \frac{1}{\Lambda_D}\right]}{\frac{\Lambda_L}{\Lambda_L - 1}\left[\exp\left(\frac{\Lambda_L - 1}{2\Lambda_L}\text{NTU}\right) - \frac{1}{\Lambda_L}\right]} \quad (6)$$

Owing to the extremely high surface area per unit volume in the hollow-fibre membrane modules, relatively low selectivities are sufficient to separate a racemic mixture into its component enantiomers. Typical values for HTU have been determined experimentally,[48] and range from 2 to 6 cm at flow velocities between 0.01 and 2 mm s$^{-1}$. From equation 6 the required NTU can be calculated for 99% pure products: these values are 190, 100 and 50 for enantioselectivities of 1.05, 1.1 and 1.2, respectively. From equation 5, it then follows that the total length of the apparatus typically is of the order of 2–5 m.

To demonstrate the utility of this method for obtaining both enantiomers with high purity, experiments were performed using racemic norephedrine (**19**). In Figure 11 the enantiomer ratio of norephedrine in the two outflows is given during start-up, and it can be concluded that the system reaches equilibrium within 24 h, both enantiomers being recovered with >99% purity.

The successful application of the separation system described above depends strongly on the ability to find an efficient chiral selector molecule, and molecular mechanics methods have proved a versatile tool for this purpose.[49,50] Currently, this system is being scaled up, and application on a large scale should be possible in the near future.

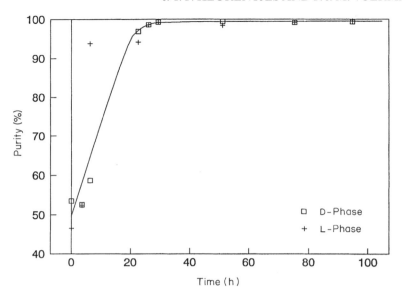

**Figure 11** Enantiomer ratio in the outflows versus time for the separation of racemic norephedrine using the system in Figure 10

### 8.3.3 MICELLAR-ENHANCED ULTRAFILTRATION

Micellar-enhanced ultrafiltration (MEUF), a micelle-based membrane separation system, has been used for the separation of amino acids.[51] An amphiphilic amino acid derivative, L-5-cholesteryl glutamate (**25**), is used as a chiral co-surfactant in micelles made with the non-ionic surfactant Serdox NNP™ 10. Cu(II) ions are added for the formation of ternary complexes between phenylalanine and the amino acid co-surfactant. The basis for the separation is the difference in stability between ternary copper complexes formed with D- or L-phenylalanine

(**25**)

L-5-Cholesteryl glutamate

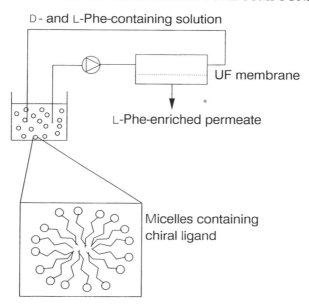

**Figure 12** Principle of micellar-enhanced ultrafiltration (MEUF). The D-enantiomer of a racemic mixture is preferentially bound to the micelles, which are retained by the membrane. The bulk liquid containing the L-enantiomer is separated through the membrane [51]

and L-5-cholesteryl glutamate. The micelles (and the enantiomer preferentially bound to them) are retained by an ultrafiltration membrane which is permeable to the uncomplexed phenylalanine enantiomer (Figure 12).

The 'operational' selectivity of this system reaches values of around 4 when a 2:1 ratio between the chiral co-surfactant and phenylalanine is used. A single-stage MEUF system can, therefore, serve as an effective low-grade separation method, although complete resolution requires a multi-stage MEUF apparatus. The industrial potential of MEUF obviously lies in the possibility of processing large volumes of dilute racemic mixtures inexpensively.

### 8.3.4 MEMBRANE-ASSISTED KINETIC RESOLUTIONS

Until now, the use of membranes to facilitate kinetic resolutions has been limited to enzymatic resolutions, which can be classified into two main types. In one system the enzymes are immobilised inside or on the membrane, and in the other the enzymes are kept in solution and are retained by the membrane. According to Matson and Quinn,[52] the potential benefits of these systems compared with conventional process technology (such as stirred tank reactors) are as follows:

- Membrane bioreactors can improve the efficiency of enzyme-catalysed bioconversions by integrating the conversion and product purification steps and by eliminating certain pieces of process equipment and attendant energy requirements.
- Membrane bioreactors increase yields in bioconversions that are subject to by-product inhibition (for reasons of unfavourable thermodynamics or enzyme inhibition).
- Membrane bioreactors will prove useful in extending the utility of enzyme and whole-cell catalysis to organic chemical syntheses that are not currently amenable to biocatalysis because of various process engineering limitations.

This combination of an enzymatic conversion and product separation using ultrafiltration is being applied to the production of optically pure α-amino acids,[53] and has been scaled up to a level of 100 tonnes annually. For this purpose, a hollow-fibre polysulphone or regenerated cellulose membrane with a molecular weight cut-off of 10 000 is used to retain the enzyme (acylase I). The substrate (N-acetyl-DL-amino acids) and product (L-amino acid) are capable of passing through the membrane for downstream processing. A similar process has been described for the production of menthyl-(−)-laurate from racemic menthol in a non-aqueous solvent using ceramic membranes to retain lipase from *Candida cylindracea*.[54] Other examples of kinetic resolutions using membrane-retained enzymes are the production of N-acetyl-L-phenylalanine[55] and N-acetyl-L-methionine.[56]

It has already been shown for liquid–liquid extractions that hollow-fibre membranes can be used efficiently to create intensive contact between two immiscible phases. Using a membrane in which the enzyme is immobilised, a lipophilic product can effectively be separated from an aqueous-soluble reactant, and vice versa. Examples are the enzymatic resolution of glycidyl butyrate[57] and the enantiospecific hydrolysis of *trans*-3-(4-methoxyphenyl)glycidic acid methyl ester [MPGM (**26**)], a key intermediate in the synthesis of diltiazem[58]. For the hydrolysis of MPGM, lipase from *Serratia marcescens* was immobilised in the pores of an asymmetric polyacrylonitrile UF membrane [59]. (+)-MGPM dissolved in toluene is selectively hydrolysed to (2$S$,3$R$)-(+)-3-(4-methoxyphenyl)glycidic acid and methanol. Enzyme inhibition caused by *p*-methoxyphenylacetaldehyde derived from the glycidic acid could effectively be suppressed by the addition of sodium hydrogensulphite to the aqueous phase (Figure 13).

(**26**)

3-(4-Methoxyphenyl)glycidic acid methyl ester (MPGM)

# MEMBRANE SEPARATIONS FOR OPTICALLY PURE COMPOUNDS

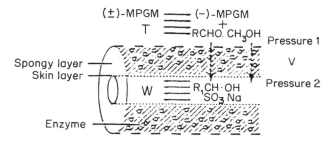

**Figure 13** Behaviour of the products in the membrane bioreactor used for the production of (−)-MPGM. Reproduced with permission from ref. 59

Subsequently, (−)-MPGM is recovered from the toluene solution by crystallisation. The productivity of the membrane reactor is *ca* 40 kg of (−)-MPGM per $m^2$ annually. This system has been developed by Sepracor Inc., and has been in operation as part of the process for the production of diltiazem hydrochloride since 1993.

A similar system has been developed by Matsumae *et al.* for the separation of (*R,S*)-ibuprofen (**24**) using proteases which enantioselectively hydrolyse the (*R*)-ibuprofen esters[60] (Figure 14). A 100% yield of (*S*)-ibuprofen is achieved by racemisation of the (*R*)-ibuprofen and, in principle, this arrangement can be used for the separation of other acids, such as naproxen and 2-chloropropionic acid (see also Chapters 3 and 10). Another process, comprising a membrane extraction and a permselective membrane, has been used by van Eikeren *et al.*[61]

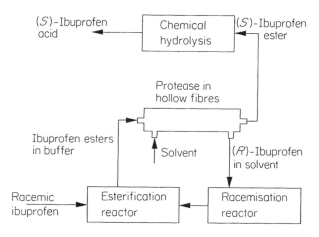

**Figure 14** Membrane bioreactor for the production of (*S*)-ibuprofen[60]

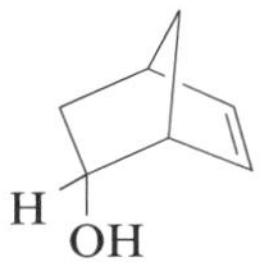

**(27)**

(*R*)-*endo*-Norbornen-2-ol

to produce (*R*)-*endo*-norbornen-2-ol (**27**) from the racemate at a scale of tens of kilograms.

The membrane-assisted enzymatic resolutions are particularly useful technologies which are amenable to large-scale use; the membrane separation required to enhance the enzymatic process is often established technology, which makes the process relatively easy to operate on a multi-kilogram scale. A potential drawback, often encountered when enzymes with a wide range of substrate specificity (such as lipases and esterases) are used, is the fact that the conversion rate can decrease the less the substrate resembles the natural substrate. Moreover, at the moment the use of isolated enantiospecific enzymes on a large scale is operationally easier for those enzymes not requiring a cofactor such as NADH or $FADH_2$. Once simpler cofactor-regeneration technologies have been developed, the large-scale use of redox enzymes will be extended substantially. Although it is sometimes difficult to obtain products in the 99% enantiomeric excess range using conventional process equipment, the combination of the enzymatic conversion with a membrane separation often results in very high optical purities.

## 8.4 CONCLUSION

One of the key advantages of membrane separation processes for the production of optically pure products is their ready applicability to large-scale use. This is particularly important for hybrid processes when the effectiveness of a new enantioselective technique can be enhanced with membrane technology in which scale-up procedures are well established, and can be implemented rapidly and efficiently. For extraction processes, only a moderate enantioselectivity is required to achieve high optical purities, and process configurations are similar to conventional unit operations such as extraction and distillation.

Enantioselective membranes are still at an early stage of development and large-scale implementation does not look likely in the short term. In the authors' opinion, considerable improvements will have to be made in the separation factors before this technology will be viable on a manufacturing scale. Nevertheless, enantioselective polymer membranes have considerable potential for large-scale operation because of the precedence of techniques such as reverse osmosis and ultrafiltration.

## 8.5 REFERENCES

1. Dorson, W. J. and Pierson, J. S., *J. Membr. Sci.*, **44**, 35 (1989).
2. Sourirajan, S. and Matsuura, T., *Reverse Osmosis/Ultrafiltration, Process Principles*, National Research Council Canada, Ottawa, 1985.
3. Strathmann, H., in *Synthetic Membranes: Science, Engineering and Applications* (ed. P. M. Bungay, H. K. Lonsdale and M. N. de Pinho), NATO ASI Series C: Mathematical and Physical Sciences, Vol. 181, Reidel Dordrecht, 1986.
4. Hwang, S. T., and Kammermeyer, K., *Membranes in Separations*, Wiley–Interscience, New York, 1975.
5. Rautenbach, R., and Albrecht, R., *Membrane Processes*, Wiley Chichester, 1989.
6. Keurentjes, J. T. F., Linders, L. J. M., Beverloo, W. A. and van 't Riet, K., *Chem. Eng. Sci.*, **47**, 1561 (1992).
7. Maruyama, A., Adachi, N., Takatsuki, T., Torii, M., Sanui, K., and Ogata, N., *Macromolecules*, **23**, 2748 (1990).
8. Ogata, N., in Proceeding of the 1st International Conference on Frontiers in Polymer Research (ed. P. N. Prasad and J. K. Nigam) Plenum Press, New York, 1991, p. 103.
9. Ogata, N., *Macromol. Symp.*, **77**, 167 (1994).
10. Ogata, N., *Polym. Prepr. Am. Chem. Soc. Div. Polym. Chem.*, **34**, 96 (1993).
11. Aoki, T., Shinohara, K., and Oikawa, E., *Makromol. Chem. Rapid. Commun.*, **13**, 565 (1992).
12. Aoki, T., Tomizawa, S., and Oikawa, E., *J. Membr. Sci.*, **99**, 117 (1995).
13. Masawaki, T., Sasai, M., and Tone, S., *J. Chem. Eng. Jpn.*, **25**, 33 (1992).
14. Higuchi, A., Hara, M., Horiuchi, T., and Nakagawa, T., *J. Membr. Sci.*, **93**, 157 (1994).
15. Yashima, E., Noguchi J., and Okamoto, Y., *J. Appl. Polym. Sci.*, **54**, 1087 (1994).
16. Linder, C., Nemas, M., Perry, M., and Ketraro, R., *Br. Pat. Appl.*, 2 233 248, 1991.
17. Bryjak, M., Kozlowski, J., Wieczorek, P., and Kafarski, P., *J. Membr. Sci.*, **85**, 221 (1993).
18. Venema, A., Henderiks, H. H., and van Geest, R., *J. High Resolut. Chromatogr.*, **14**, 676 (1991).
19. Armstrong, D. W., Ward, T. J., Armstrong, R. D., and Beesley, T. E., *Science*, **232**, 1132 (1986).
20. Nielen, M. W. F., *J. Chromatogr.*, **637**, 81 (1993).
21. Armstrong, D. W., and Jin, H. L., *Anal. Chem.*, **59**, 2237 (1987).
22. Curtis, W. D., Laidler, D. A., Stoddart, J. F., and Jones, G. H., *J. Chem. Soc., Chem. Commun.*, 835 (1975).
23. Behr, J. P., Girodeau, J. M., Hayward. R. C., Lehn J. M., and Sauvage, J. P., *Helv. Chim. Acta*, **63**, 2096 (1980).
24. Yamaguchi, T., Nishimura, K., Shinbo, T., and Sugiura, M., *Chem. Lett.*, 1549 (1985).
25. Newcomb, M., Toner, J. L., Helgeson R. C., and Cram, D. J., *J. Am. Chem. Soc.*, **101**, 4941 (1979).
26. Bussmann, W., Lehn, J.-M., Oesch, U., Plumeré, P., and Simon, W., *Helv. Chim. Acta*, **64**, 657 (1981).
27. Holy, P., Morf, W. E., Seiler, K., Simon, W., and Vigneron, J. P., *Helv. Chim. Acta*, **73**, 1171 (1990).
28. Prelog, V., and Mutak, S., *Helv. Chim. Acta*, **66**, 2274 (1983).
29. Hinze, W. L., Williams, R. W., Fu, Z. S., Suzuki, Y., and Quina, F. N., *Colloids Surf.*, **48**, 79 (1990).
30. Pirkle, W. H., and Bowen, W. E., *Tetrahedron: Asymmetry*, **5**, 773 (1994).
31. Heard, C. M., Hadgraft, J., and Brain, K. R., *Bioseparation*, **4**, 111 (1994).

32. Takeuchi, T., Horikawa, R., and Tanimura, T., *Anal. Chem.*, **56**, 1152 (1984).
33. Takeuchi, T., Horikawa, R., and Tanimura, T., *Sep. Sci. Technol.*, **25**, 941 (1990).
34. Pirkle, W. H., and Doherty, E. M., *J. Am. Chem. Soc.*, **111**, 4113 (1989).
35. Chia, P. S. K., Lindoy, L. F., Walker, G. W., and Everett, G. W., *J. Am. Chem. Soc.*, **113**, 2533 (1991).
36. Sheldon, R. A., *Chirotechnology, Industrial Synthesis of Optically Active Compounds*, Marcel Dekker, New York, 1993.
37. Prelog, V., and Dumic, M., *Helv. Chim. Acta*, **69**, 5 (1986).
38. Prelog, V., Mutak, S., and Kovacevic, K., *Helv. Chim. Acta*, **66**, 2279 (1983).
39. Prelog, V., Stojanac, Z., and Kovacevic, K., *Helv. Chim. Acta*, **65**, 377 (1982)
40. Beek, W. J. and Muttzall, K. M. K., *Transport Phenomena*, Wiley, London, 1975.
41. Dahuron, L., and Cussler, E. L., *AIChE J.*, **34**, 130 (1988).
42. D'Elia, N. A., Dahuron, L., and Cussler, E. L., *J. Membr. Sci.*, **29**, 309 (1986).
43. Yang, M. C., and Cussler, E. L., *AIChE J.*, **32**, 1910 (1986).
44. Prasad, R., and Sirkar, K. K., *AIChE J.*, **34**, 177 (1988)
45. Keurentjes, J. T. F., Sluijs, J. T. M., Franssen, R. J. H., and van 't Riet, K., *Ind. Eng. Chem. Res.*, **31**, 581 (1992).
46. Ding, H. B., Carr P. W., and Cussler, E. L., *AIChE J.*, **38**, 1493 (1992).
47. Keurentjes, J. T. F., *PCT Int. Pat. Appl.*, WO 94/07814, 1994.
48. Keurentjes, J. T. F., Nabuurs, L. J. W. M., and Vegter, E. A., in *Proceedings of Chiral USA '94 Symposium*, Spring Innovations, Stockport, 1994, p. 39.
49. Lipkowitz, K. B., and Baker, B., *Anal. Chem.*, **62**, 770 (1990).
50. Aerts, J., *J. Comput. Chem.*, in press.
51. Creagh, A. L., Hasenack, B. B. E., van der Padt, A., Sudholter, E. J. R., and van 't Riet, K., *Biotechnol. Bioeng.*, **44**, 690 (1994).
52. Matson S. L., and Quinn, J. A., *Ann. N.Y. Acad. Sci.*, **469**, 152 (1986).
53. Bommarius, A. S., Drauz, K., Groeger, U., and C. Wandrey, in *Chirality in Industry* (ed. Collins, A. N., Sheldrake, G. N., and Crosby, J.), Wiley, Chichester, 1992, p. 371.
54. Rios, G. M., Lambert, F., and Jallageas, J. C., *Entropie*, **27**, 31 (1991).
55. Chen, S., *Eur. Pat. Appl.*, EP 273 679, 1988.
56. Schmidt-Kastner, G., and Goelker, C., *GBF Monogr. Ser. 9 (Techn. Membr. Biotechnol.)*, 201 (1986).
57. Wu, D. R., Cramer, S. M., and Belfort, G., *Biotechnol. Bioeng.*, **41**, 979 (1993).
58. Matsumae, H., Furui, M., and Shibatani, T., *J. Ferment. Bioeng.*, **75**, 93 (1993).
59. Matsumae, H., Furui, M., Shibatani, T., and Tosa, T., *J. Ferment. Bioeng.*, **78**, 59 (1994).
60. Matson, S. L., Wald, S. A., Zepp, C. M, and Dodds, D., *PCT Int. Pat. Appl.*, WO 89 09765, 1991.
61. Van Eikeren, P., Muchmore, D. C., Brose, D. J., and Colton, R. H., *Ann. N.Y. Acad. Sci.*, **672**, 539 (1992).

# BIOLOGICAL METHODS AND CHIRAL POOL SYNTHESES

# 9 Four Case Studies in the Development of Biotransformation-based Processes

R. McCAGUE and S. J. C. TAYLOR
*Chiroscience Ltd, Cambridge, U.K.*

9.1 Introduction . . . . . . . . . . . . . . . . . . . . . . . . . . . 184
9.2 Resolution of the Carbocyclic Nucleoside Synthon
    2-Azabicyclo[2.2.1]hept-5-en-3-one with Lactamases . . . . . . . . 184
    9.2.1 Routes to Carbocyclic Nucleosides . . . . . . . . . . . . . 184
    9.2.2 Identification of Microbial Strains for the Resolution of the
          $\gamma$-Lactam 2-Azabicyclo[2.2.1]hept-5-en-3-one . . . . . . . . 186
    9.2.3 Development of a Whole-Cell Bioresolution
          Process for the (−)-$\gamma$-Lactam . . . . . . . . . . . . . . . 186
    9.2.4 Development of an Immobilised Lactamase Resolution
          Process for the (+)-$\gamma$-Lactam and the Ring-opened
          (−)-Amino Acid . . . . . . . . . . . . . . . . . . . . . 187
    9.2.5 Downstream Chemistry from the Resolved $\gamma$-Lactam or
          Amino Acid . . . . . . . . . . . . . . . . . . . . . . . 188
9.3 Resolution of a Versatile Hydroxylactone Synthon
    4-*endo*-Hydroxy-2-oxabicyclo[3.3.0]oct-7-en-3-one
    by Lipase De-esterification . . . . . . . . . . . . . . . . . . . . 190
    9.3.1 Substrate Synthesis . . . . . . . . . . . . . . . . . . . . 190
    9.3.2 Development of a Biotransformation for Resolution
          of the Hydroxylactone . . . . . . . . . . . . . . . . . . . 191
    9.3.3 Esterification versus Hydrolysis . . . . . . . . . . . . . . 192
    9.3.4 Downstream Chemistry from the Resolved
          Hydroxylactone . . . . . . . . . . . . . . . . . . . . . . 192
9.4 Integration of an Acylase Biotransformation with Process
    Chemistry: a One-pot Synthesis of *N*-BOC-L-3-(4-
    thiazolyl)alanine and Related Amino Acids . . . . . . . . . . . . 194
    9.4.1 Uses of Unnatural Amino Acids . . . . . . . . . . . . . . 194
    9.4.2 Overall Synthetic Route to *N*-BOC-3-(4-thiazolyl)alanine . . 195
    9.4.3 An Ester-based Route . . . . . . . . . . . . . . . . . . . 195
    9.4.4 Development of a One-pot Process Based on Aminoacylase
          Resolution . . . . . . . . . . . . . . . . . . . . . . . . 197
    9.4.5 Recycle of the D-Enantiomer . . . . . . . . . . . . . . . . 199

*Chirality in Industry II.* Edited by A. N. Collins, G. N. Sheldrake and J. Crosby
© 1997 John Wiley & Sons Ltd

    9.4.6   Comparison of Different Biotransformation Processes . . . 200
    9.4.7   Application of the Aminoacylase Process to Other
            Unnatural Amino Acids . . . . . . . . . . . . . . . . . . . . 200
9.5   Dynamic Resolution of an Oxazolinone by Lipase Biocatalysis:
      Synthesis of (S)-*tert*-Leucine . . . . . . . . . . . . . . . . . . . . . . 201
    9.5.1   Use of (S)-*tert*-Leucine . . . . . . . . . . . . . . . . . . . . 201
    9.5.2   Azlactones for the Dynamic Resolution of Amino Acids . . 201
    9.5.3   Development of a Lipase-catalysed Dynamic Resolution . . 202
9.6   Conclusions . . . . . . . . . . . . . . . . . . . . . . . . . . . . . . . . 203
9.7   Acknowledgements . . . . . . . . . . . . . . . . . . . . . . . . . . . 205
9.8   References . . . . . . . . . . . . . . . . . . . . . . . . . . . . . . . . 205

## 9.1   INTRODUCTION

Biotransformations are playing an increasing role in the manufacture of single-enantiomer intermediates, especially for pharmaceutical agents. Advantages of biotransformations are (i) they are catalytic and take place under mild, generally near ambient, conditions, (ii) when the biocatalyst is derived from a microbial strain the scaleability is unlimited, as the biocatalyst can be provided by growing the microbial strain to whatever biomass is necessary, and (iii) there is much scope to develop a transformation that is both highly enantioselective and highly economic.

    In this chapter, a selection of biotransformation-based processes are described that afford intermediates for pharmaceutical agents. In each case the optimisation of the biotransformation with respect to enantiospecificity, conversion rate and enzyme utilisation is described. Generally, a biotransformation cannot stand on its own; it needs to be integrated within the total process. Frequently, the biotransformation chosen for development will be governed by the chemical route, so consideration is given here to the integration of the biotransformation within the overall process chemistry, including the development of the product isolation and discussion of means of utilising the unwanted enantiomer. Finally, we draw some broad principles from the specific cases studied that should be of assistance to process scientists in the future development of improved biotransformation-based processes.

## 9.2   RESOLUTION OF THE CARBOCYCLIC NUCLEOSIDE SYNTHON 2-AZABICYCLO[2.2.1]HEPT-5-EN-3-ONE WITH LACTAMASES

### 9.2.1   ROUTES TO CARBOCYCLIC NUCLEOSIDES

Carbocyclic nucleosides are analogues of the natural nucleosides in which the ribosyl oxygen has been replaced by a methylene linkage. They have attracted much interest as therapeutic agents as the absence of the labile ribosidic linkage results in compounds with greatly improved metabolic stability to nucleases and correspondingly greater bioavailability. Two therapeutic applications of carbocyclic nucleosides are (i) as antiviral agents (e.g. carbovir for the treatment of HIV[1] and c-AFG and c-BVDU for Herpes simplex[2,3]) and (ii) as adenosine agonist cardiac vasodilators (e.g. c-NECA[4]) (Scheme 1). For use as therapeutic agents, the carbocyclic nucleosides need to be available as single enantiomers, which, for the above mentioned examples, is the configuration corresponding to that of the natural nucleosides. However, while analogues of natural nucleosides are generally prepared by synthetic manipulation of natural nucleosides or built up from ribose, which is available in single-enantiomer form, there is no such ready access to carbocyclic nucleosides. There is a carbocyclic nucleoside natural product, aristeromycin (Scheme 1), which may be obtained from fermentation of strains of *Streptomyces*;[5] however, the synthetic elaboration into the target compounds might be tortuous. Nevertheless, a clever procedure has been devised by Glaxo to convert the adenine base of aristeromycin into the guanine base required for the antiviral agents.[5] Apart from this, carbocyclic nucleosides are usually constructed by total synthesis (a review of the various approaches has appeared[6]). Of the various options, a particularly versatile approach starts with the bicyclic γ-lactam synthon 2-azabicyclo[2.2.1]hept-5-en-3-one (**1**), first described by Jagt and van Leusen,[7] and prepared by cycloaddition of cyclopentadiene with chlorosulphonyl isocyanate, *p*-toluenesulphonyl cyanide or methanesulphonyl cyanide.[8] The application of the resulting racemic lactam **1** to the

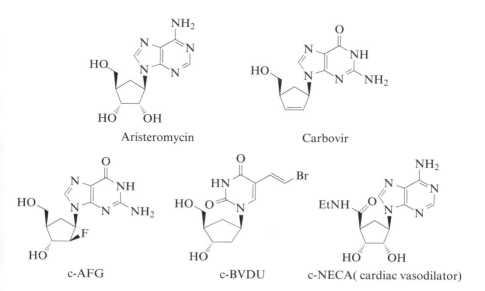

**Scheme 1** Examples of carbocyclic nucleosides

synthesis of carbocyclic nucleosides has been demonstrated by the group of Vince *et al.*; generally, the lactam ring is opened and the heterocyclic base is constructed from the resulting amine.[9] However this route leads to a racemic carbocyclic nucleoside, and, while methods to resolve such products exist, it would be more convenient to have access to the single enantiomer of the lactam. Thus, a synthesis can be envisaged (Scheme 2) whereby lactam **1** is resolved and the appropriate enantiomer is carried forward into the remainder of the synthesis.

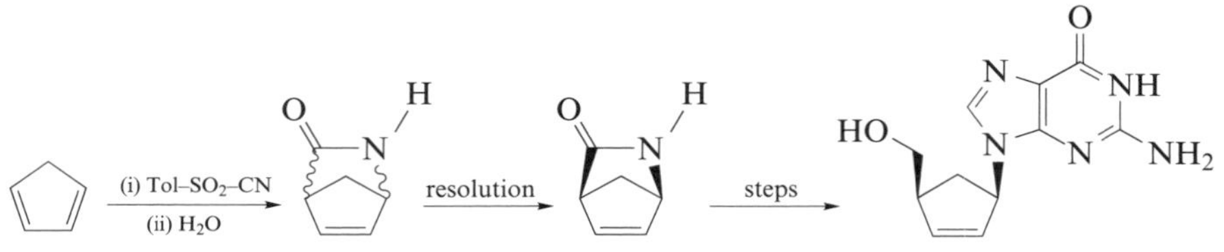

**Scheme 2**  Overall route to a carbocyclic nucleoside

### 9.2.2 IDENTIFICATION OF MICROBIAL STRAINS FOR THE RESOLUTION OF THE γ-LACTAM 2-AZABICYCLO[2.2.1]HEPT-5-EN-3-ONE

With the aim of obtaining a single enantiomer of the lactam, efforts were made to locate a biocatalyst capable of cleaving the amide linkage of the lactam enantiospecifically. While available isolated enzymes (such as proteases) failed to act on the lactam, enzymes were discovered in microbial strains that were screened which did possess such activity. Screening turned out to be relatively straightforward owing to the very high optical rotation of the enantiopure lactam {$[\alpha]_D$ = 557 ($c$ = 1, CH$_2$Cl$_2$)}, and a number of potentially useful strains were identified quickly. Moreover, strains were found with enantiocomplementary lactamases that cleave either the (+)- (**1b**) or the (–)-enantiomer (**1a**) of the lactam[10] (Scheme 3). In the former case, the (–)-lactam **1a**, which has the required configuration for the synthesis of carbocyclic nucleosides of the natural configuration, is not consumed. This was verified by conversion to (–)-carbovir,[11] the absolute configuration of which had already been established by stereochemical correlation to aristeromycin.[5]

### 9.2.3 DEVELOPMENT OF A WHOLE-CELL BIORESOLUTION PROCESS FOR THE (–)-γ-LACTAM

Following an initial laboratory resolution of the lactam **1**, work was undertaken to establish a viable manufacturing process. Cells of suitable strains, for instance *Pseudomonas* spp., were grown in an appropriate medium comprising glucose as carbon source, yeast extract and mineral salts. Cells grown in this way and resuspended in a neutral buffer gave a rapid transformation of racemic lactam at up to

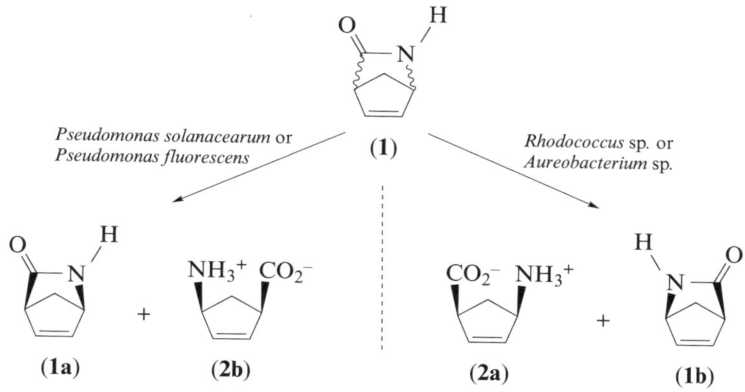

# CASE STUDIES OF BIOTRANSFORMATION PROCESSES

**Scheme 3** Enantiocomplementary lactamase resolutions

$100 \text{ g l}^{-1}$ in aqueous solution such that bioconversion was essentially complete in about 3h at ambient temperature. Whilst the isolated microbial enzyme was found to be of limited stability, it was relatively stable within the cells. Furthermore a convenient procedure has been found in which the cells are grown by fermentation and the resulting cell mass is frozen and stored. The biotransformation can then be conducted at a later time, and addition of the frozen cell mass to an aqueous lactam solution is all that is needed to effect the biotransformation. This procedure has the advantage that the biotransformation can be performed at a different site from the fermentation, thereby improving manufacturing flexibility. Also, the freeze–thaw process causes some degree of cell lysis, which liberates the enzyme and increases the rate of the transformation. A further feature of this biotransformation is that control of the pH is not required, since the amino acid formed acts as its own buffer, maintaining the mixture at around neutral pH. After the biotransformation, the resulting aqueous solution contains the required (–)-lactam (**1a**), the amino acid (**2b**) of opposite configuration and cell debris. The lactam can be selectively extracted into an organic solvent, such as dichloromethane, leaving the amino acid behind. Such a procedure lends itself to scale-up because of its simplicity and the fact that the microbial strain can be grown to the necessary biomass to provide sufficient catalyst. At the time of writing, the process has been used to produce enantiopure (–)-**1a** on a 1 tonne scale.

## 9.2.4 DEVELOPMENT OF AN IMMOBILISED LACTAMASE RESOLUTION PROCESS FOR THE (+)-γ-LACTAM AND THE RING-OPENED (–)-AMINO ACID

As indicated above, the enzymes that provide (–)-lactam **1a** are of limited stability outside of cells, so the transformation is more practicable with whole cells. It was discovered, however, that a lactamase present in an *Aureobacterium* species, which has the opposite stereoselectivity, is remarkably robust.[12] It is stable when isolated and, moreover, it maintains its activity up to 70 °C. These characteristics are ideal for establishing an immobilised enzyme system (see Volume I, Chapter 19). The lactamase was purified by a procedure which included ammonium sulphate fractionation and anion-exchange chromatography, and was immobilised directly on a glutaraldehyde-activated solid support. The biotransformation was then operated simply by cycling an aqueous solution of racemic lactam (up to 200 g l$^{-1}$) through the immobilised enzyme reactor until all the (−)-enantiomer (**1a**) was hydrolysed. At this stage there was no evidence of enantiomer contamination in either the remaining substrate or the product; an enantiospecificity value, $E$, of >7000 was calculated, which is at the limits of sensitivity of the assay methods used. The immobilised lactamase proved to be extremely stable; a continuous transformation using a small reactor revealed little loss of activity during 8 months, after which time the conversion was discontinued owing to mechanical failure of the support.

Clearly, with such a high level of catalyst re-use, the contribution of the cost of the catalyst to the cost of the single enantiomer product becomes minimal. Indeed, the volume productivity (amount of substrate converted by a given amount of biomass in a given time) has been increased some 20 000-fold by the development. A feature of biotransformation processes is that with development, substantial economic improvements can often be realised. Another feature of an immobilised enzyme biotransformation process is that the downstream work-up is greatly simplified; there is no interference from proteinaceous matter during the isolation and large volumes of solvent are not needed. After the transformation the mixture is simply slurried in acetone, which leaves the (+)-lactam (**1b**) in solution, while the (−)-amino acid (**2a**) crystallises, and can be collected by filtration. Noteworthy, also, is that this biotransformation provides the amino acid directly in its neutral (zwitterionic) form, which cannot be achieved by chemical cleavage of the lactam (in which acid hydrolysis would produce the amine salt of the acid).

### 9.2.5 DOWNSTREAM CHEMISTRY FROM THE RESOLVED γ-LACTAM OR AMINO ACID

Given the availability of the single enantiomer of the lactam 2-azabicyclo[2.2.1]hept-5-en-3-one (**1a**), established routes based on the racemate may be used to provide enantiopure carbocyclic nucleosides such as (−)-carbovir. Thus, acid hydrolysis of the lactam (**1a**), esterification, $N$-acetylation and calcium borohydride reduction give the amino alcohol (**3**), which is then converted into the carbocyclic nucleoside carbovir[11,12] (Scheme 4).

Carbocyclic nucleosides in which the ring has all the functionality of the ribose, required for the cardiac vasodilator adenosine agonists (e.g. c-NECA), are accessed through osmium tetraoxide-catalysed dihydroxylation. Here, the rigid bicyclic framework ensures that the dihydroxylation is almost completely

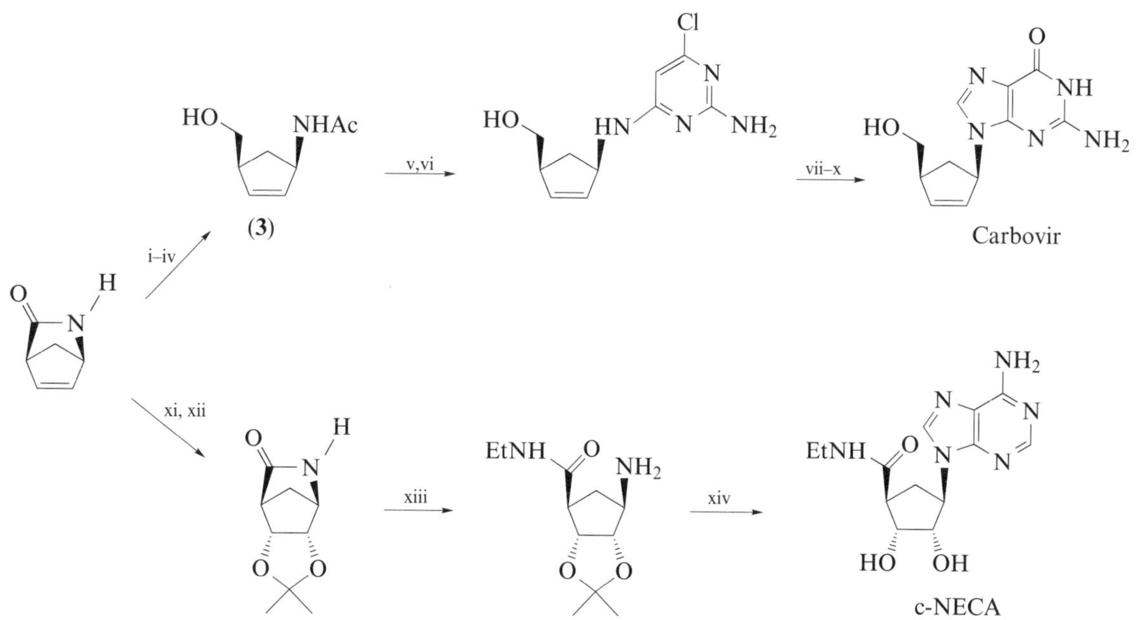

**Scheme 4** Conversion of the single-enantiomer bicyclic lactam into carbocyclic nucleosides

Reagents: i, aq. HCl; ii, (MeO)$_2$CMe$_2$; iii, Ac$_2$O; iv, Ca(BH$_4$)$_2$; v, HCl, aq. EtOH; vi, 2-amino-4,6-dichloropyrimidine; vii, 4-ClC$_6$H$_4$N$_2$$^+$Cl$^-$; viii, Zn, HOAc; ix, (EtO)$_3$CH; x, NaOH; xi, N-methylmorpholine N-oxide, OsO$_4$ (cat.), aq. acetone; xii, (MeO)$_2$CMe$_2$; xiii, EtNH$_2$; xiv, standard steps.

stereoselective for the *exo*-face whereas, when carried out on ring-opened lactam substrate, such alkene functionalisation tends to give isomer mixtures.[13] In the case of the immobilised enzyme biotransformation described in the previous section [which gives the (−)-amino acid **2a** and the (+)-lactam **1b**], it is necessary to use the amino acid product (**2a**) to obtain carbocyclic nucleosides of the natural configuration. The amino acid **2a** is converted easily into its benzoylamide methyl ester derivative which is an established precursor of certain carbocyclic nucleosides. One drawback of any such biocatalytic kinetic resolution is that it leaves an unwanted isomer. The inversion of the (+)-lactam into the (−)-lactam requires a multi-step sequence in which the key step is a skeletal rearrangement;[14] this does not seem an economic option at present, however.

## 9.3 RESOLUTION OF A VERSATILE HYDROXYLACTONE SYNTHON 4-*ENDO*-HYDROXY-2-OXABICYCLO[3.3.0]OCT-7-EN-3-ONE BY LIPASE DE-ESTERIFICATION

### 9.3.1 SUBSTRATE SYNTHESIS

As discussed in the previous section, conversion of the bicyclic lactam **1a** into a carbocyclic nucleoside usually requires the heterocyclic base to be built on to the amine nitrogen (although displacement of the nitrogen, via the *N,N*-ditosylate, has recently been reported[15]). In seeking a route whereby the base could be introduced intact as a nucleophile, we discovered that the bicylic hydroxylactone **4** was suitable for this purpose.

The racemic hydroxylactones **4** and **5** are synthesised simply by the cycloaddition of glyoxylic acid (as a 50% aqueous solution) and cyclopentadiene, which is effected by mixing the neat components at ambient temperature (Scheme 5). The

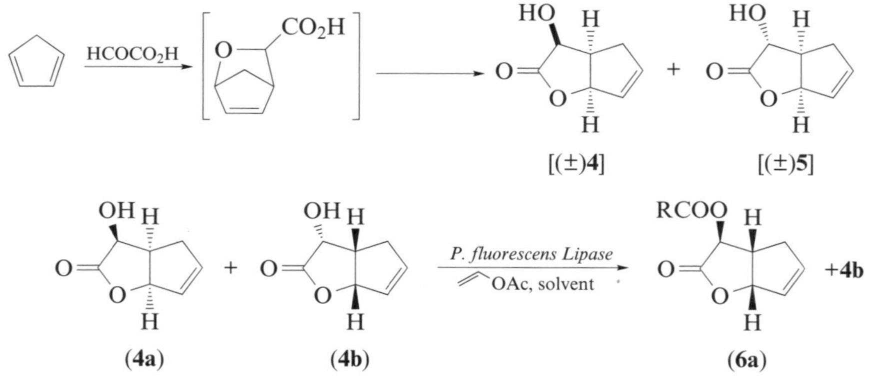

**Scheme 5**  Products from the cycloaddition of cyclopentadiene with glyoxylic acid

reaction was first described by Lubineau et al.[16] as an example of a reaction accelerated by water. The reaction is presumed to proceed initially by a [4 + 2]-cycloaddition of glyoxylic acid and cyclopentadiene in a hetero-Diels–Alder reaction to form a bicyclo[2.2.1] system. This adduct then undergoes rapid electrocyclic rearrangement to the more stable [3.3.0] system, resulting in a mixture of the *endo* (**4**) and *exo* (**5**) isomers in a ratio of about 4:1. On a small scale these diastereoisomers can be separated by column chromatography or by fractional crystallisation; however, the former cannot be scaled up easily and the latter gives poor yields. Hence the challenge was to develop a resolution with selectivity for the *endo*-isomer in addition to enantiocontrol.

### 9.3.2 DEVELOPMENT OF A BIOTRANSFORMATION FOR RESOLUTION OF THE HYDROXYLACTONE

The first resolution[17] identified in the laboratory employed *Pseudomonas fluorescens* lipase to catalyse the acetylation of the hydroxylactone using vinyl acetate in an organic solvent. Here, the lactone isomer (**4a**) with the correct absolute configuration for synthesis of the natural nucleoside was converted into the corresponding acetate product (**6a**; R = Me). The acetate and remaining alcohol enantiomer could be separated by chromatography, but this procedure was not convenient on a large scale. For this reason, we examined the corresponding hydrolytic biotransformation.

Screening for hydrolysis of the acetate, propionate and butyrate esters of a mixture of the diastereoisomers **4** and **5** by commercially available lipases soon revealed hydrolysis of the butyrate ester by *Pseudomonas fluorescens* lipase as a method of choice (Scheme 6).[18] Thus, a crude mixture of the *endo*- and *exo*-butyrate esters (**6** and **7**; R = Pr$^n$) was resolved efficiently in aqueous buffer, with continuous addition of NaOH to maintain neutral pH. The hydrolysis typically gave *endo* product of >92% ee, with minimal hydrolysis of the *exo*-isomer. The enantiomeric ratio (*E* value) for the butyrate ester hydrolysis was calculated as >200, and this process was subsequently scaled for use with multi-kilogram quantities of substrate, with an isolation procedure which avoids chromatography. Initially, solvent extraction studies (heptane–water) showed that if a suffi-

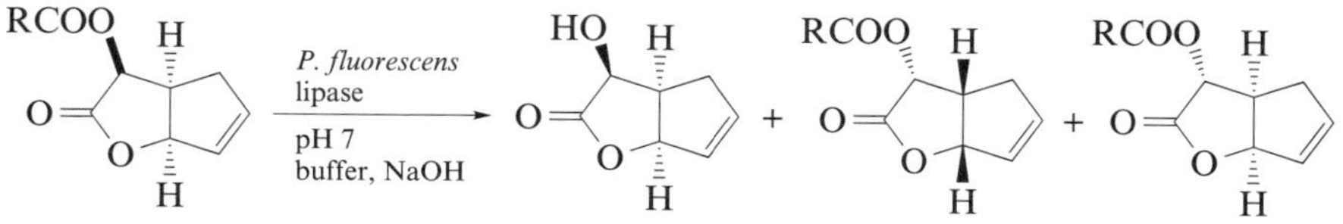

(R = Me, Et, Pr$^n$)

**Scheme 6** Resolution of 4-*endo*-hydroxy-2-oxabicyclo[3.3.0]oct-7-en-3-one [formulae follow the recommendation of Maehr, H., *J. Chem. Educ.*, **62**, 114 (1985)]

ciently hydrophobic ester was used (e.g. the butyrate), the ester could be selectively partitioned into the organic phase while the alcohol formed in the biotransformation was retained in the aqueous phase. Furthermore, the neat butyrate ester was immiscible with the aqueous phase, so centrifugation of the biotransformation mixture separated most of the unwanted ester impurities as an oily phase. The alcohol (**4a**) could then be recovered by simple solvent extraction, followed by a recrystallisation which enhanced the *ee* to 99%.

There was one complication: both acetate and butyrate esters were found to undergo spontaneous pH-dependent ring opening of the lactone to give a carboxylate salt, with faster decomposition at higher pH. This resulted in some loss of product for the butyrate ester (10%), but a much higher loss for the more soluble acetate ester, providing further justification for the choice of butyrate ester as substrate.

### 9.3.3 ESTERIFICATION VERSUS HYDROLYSIS

In view of the loss of some of the product to ring hydrolysis, the enzymatic esterification approach in which, hopefully, such ring opening should not occur, was re-investigated. Furthermore, conducting biotransformations in non-aqueous media frequently offers benefits such as increased enzyme stability or easier work-up. A biocatalytic, irreversible transesterification seemed, at first, to be the preferred route for resolution of the lactone **4**. However, trials showed that acylation of the hydroxylactone mixture by *Pseudomonas fluorescens* lipase, using vinyl acetate or vinyl butyrate in toluene, was slow and not as selective, with *E* values reduced to 30–40; (it is fairly common to find such differences in enantioselectivity between biocatalytic acylations and aqueous hydrolyses). Consequently, the acylation reaction required considerably more enzyme than the hydrolysis, and gave a less pure product after isolation.

### 9.3.4 DOWNSTREAM CHEMISTRY FROM THE RESOLVED HYDROXYLACTONE

The optically pure (–)-hydroxylactone **4a** proved to be a versatile building block for key chiral compounds, as, for example, in an alternative synthesis of the anti-HIV agent carbovir[18] (Scheme 7). This was achieved by reduction of the lactone to a triol, followed by oxidative cleavage of the vicinal diol and another reduction to a diol. Selective protection of the diol, then nucleophilic displacement via a $\pi$-allyl palladium complex intermediate using the intact heterocyclic base 2-amino-6-chloropurine and, finally, a deprotection yielded (–)-carbovir.

The (+)-hydroxylactone **4b**, on the other hand, can be used to provide intermediates for hypocholesteremic reagents,[19] where a key step is a Baeyer–Villiger oxidation of a cyclopentanone to a $\delta$-lactone (Scheme 8). (+)-Hydroxylactone **4b** has also been used as a precursor to the antifungal agent brefeldin A, in what is probably the shortest synthesis of this compound to date (Scheme 9).[20]

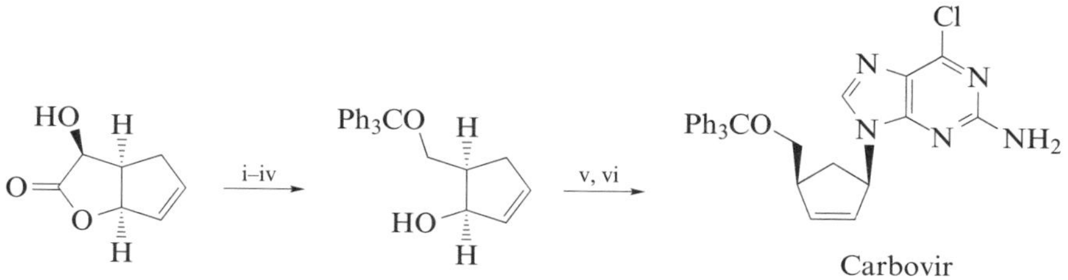

# CASE STUDIES OF BIOTRANSFORMATION PROCESSES 193

**Reagents:** i, LiAlH$_4$; ii, NaIO$_4$; iii, NaBH$_4$; iv, Ph$_3$CCl;
v, Ac$_2$O; vi, 2-amino-6-chloropurine, Pd(PPh$_3$)$_4$ (cat.)

**Scheme 7** Conversion of the resolved hydroxylactone into a carbocyclic nucleoside

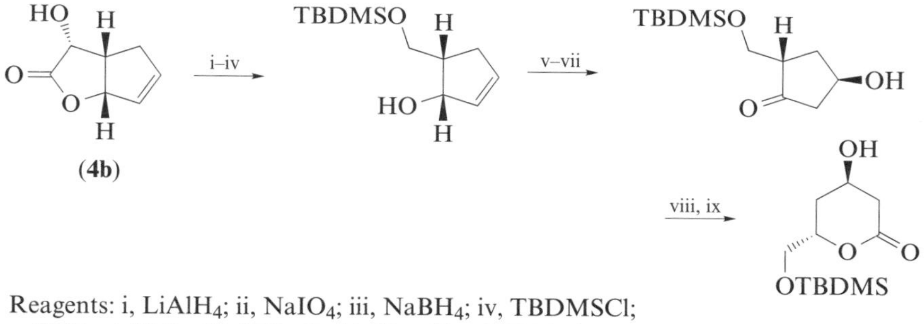

**Reagents:** i, LiAlH$_4$; ii, NaIO$_4$; iii, NaBH$_4$; iv, TBDMSCl;
v, PCC; vi, H$_2$O$_2$, NaOH; vii, Al–Hg; viii, MCPBA; ix, AcOH

**Scheme 8** Conversion of the resolved hydroxylactone in a precursor of hypocholesterolaemic agents

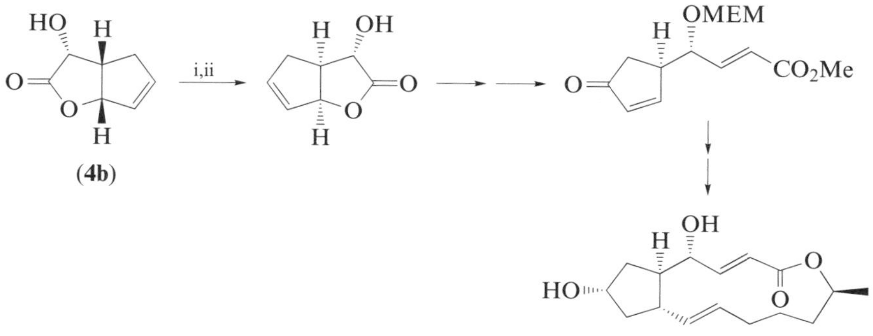

**Reagents:** i, Ph$_3$P, ClCH$_2$CO$_2$H, DEAD; ii, thiourea, EtOH

**Scheme 9** Conversion of resolved hydroxylactone **4b** into brefeldin A

## 9.4 INTEGRATION OF AN ACYLASE BIOTRANSFORMATION WITH PROCESS CHEMISTRY: A ONE-POT SYNTHESIS OF N-BOC-L-3-(4-THIAZOLYL)ALANINE AND RELATED AMINO ACIDS

### 9.4.1 USES OF UNNATURAL AMINO ACIDS

Unnatural amino acids are of growing importance as components of therapeutic oligopeptides, where replacement of a natural amino acid in an enzyme or receptor binding agent may provide a required inhibitory or antagonist activity or, alternatively, may improve properties such as oral bioavailability or resistance to metabolic degradation. In addition, unnatural amino acids may themselves be versatile synthetic intermediates or have potential for use as chiral auxiliaries (see also Volume I, Chapters 8 and 20). One such unnatural amino acid is N-BOC-3-(4-thiazoyl)alanine (BOC-TAZ) (**9**). This has been used as a component of antihypertensive inhibitors of the enzyme renin, where it acts as a mimic of histidine.[21,22] The complete structures of two such renin inhibitors are shown in Scheme 10. The following sections describe the synthesis of the single enan-

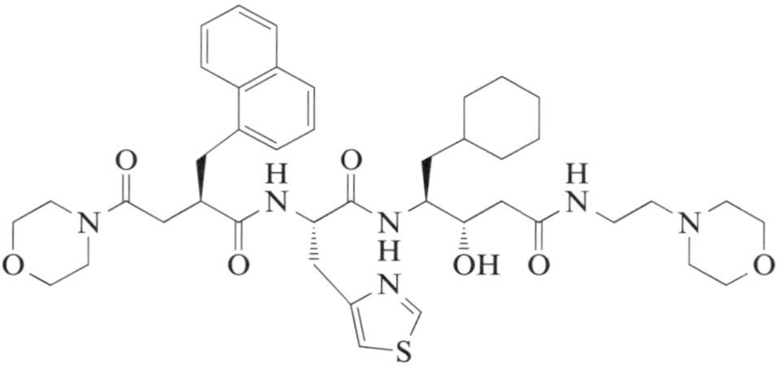

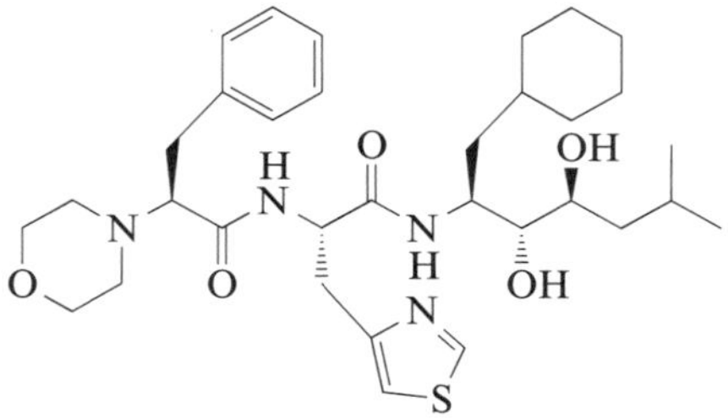

**Scheme 10** Examples of renin inhibitors containing L-thiazolylalanine

tiomer L-BOC-TAZ, although it should be noted that the method is applicable to a range of different amino acids.

## 9.4.2 OVERALL SYNTHETIC ROUTE TO N-BOC-3-(4-THIAZOYL)ALANINE

Today there are a great many synthetic approaches to amino acids.[22] One very effective synthesis of phenylalanine derivatives starts with a condensation reaction of a glycine anion equivalent with the appropriate benzaldehyde derivative; however, this requires the corresponding aldehyde to be available. In the case of L-BOC-TAZ (**9**) the appropriate aldehyde is not readily available, necessitating synthesis from 4-chloromethylthiazole hydrochloride (**8**), which, in turn, is prepared by the condensation of thioformamide with 1,3-dichloroacetone.[24] This approach, which utilises deprotonated diethyl acetamidomalonate as a glycine anion equivalent, is shown in Scheme 11. The process development steps used to obtain the single enantiomer of the protected amino acid from the acetamidomalonate condensation product are described below. The initial target was to produce 20 kg of single enantiomer product, and this was accomplished within a three month timescale.

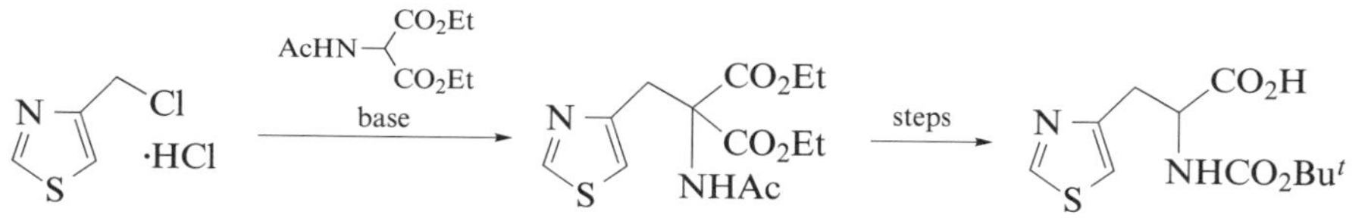

**Scheme 11**  Overall route to N-BOC-3-(4-thiazolyl)-L-alanine

## 9.4.3 AN ESTERASE-BASED ROUTE

Our first experimental approach was based on the use of an esterase[25,26] to effect the key resolution, as outlined in Scheme 12. This initially appears to be an attractive route, since the unwanted D-enantiomer (**10b**), which remains as the ester after the transformation, is readily extracted from the mixture and racemised with ethoxide to recycle to the process.

However, from a process point of view the approach was very cumbersome and inefficient. The condensation step was carried out in ethanol with ethoxide as base, which was problematic because of the need to use ethanol free of other alcohols in order to avoid mixed ester products. In a typical procedure, the anion of the acetamidomalonate was generated by treatment with one equivalent of ethoxide in ethanol at about 60 °C. To this solution was added the chloromethylthiazole, generated as a solution of free base by the prior addition

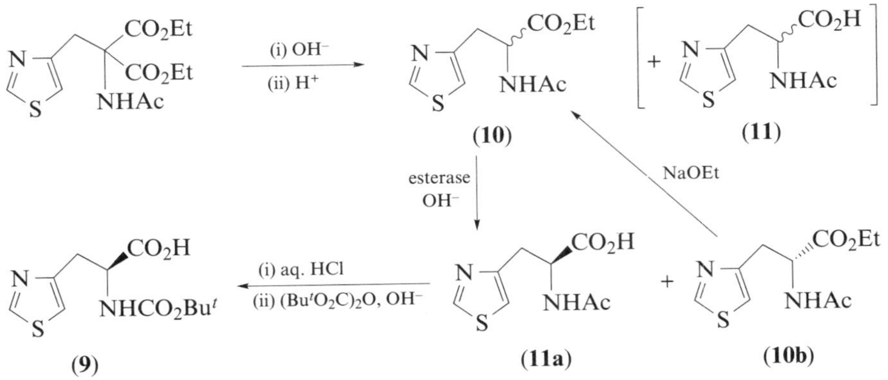

**Scheme 12** Esterase-based route to *N*-BOC-3-(4-thiazoyl)-L-alanine

of 1 equivalent of ethoxide to a suspension of the hydrochloride in ethanol. The chloromethylthiazole, which is a potent vesicant, had to be used quickly since it polymerises readily. After the condensation reaction, which proceeds smoothly at 60–80 °C, the diester condensation product may be isolated by drown-out into cold water, in which it is only slightly soluble. The next step is selective hydrolysis and decarboxylation to give the monoester. On the laboratory scale, this can be effected fairly easily by adding sufficient aqueous lithium hydroxide (ca two equivalents) to a solution of the diester and warming until hydrolysis is complete, then acidifying and heating to promote decarboxylation. On the plant, it would not be safe to add all the lithium hydroxide at once, and it was added portionwise at 55 °C. HPLC analysis mid-way through the addition showed that a substantial amount of monoester **10** had already formed, indicating that decarboxylation was occurring directly under these conditions. Moreover, when a sample of the mixture was treated with more hydroxide, it was evident that the monoester hydrolysed faster than the diester. Whatever the method used to carry out the diester hydrolysis/decarboxylation, at least 10% of the racemic *N*-acetylamino acid **11** was formed.

To avoid the unwanted D-enantiomer being carried through the biotransformation step, it was necessary to remove the racemic acid **11** from the ester **10**. This was done by extraction of the ester **10** into an organic solvent at neutral pH, then solvent removal prior to dissolution in water for the biotransformation. It quickly became apparent that such a process was highly inefficient. The ester **10** was produced in a predominantly aqueous solution (from the hydrolysis), and extracted into organic solvent only to be redissolved in water. Also, a significant loss of yield resulted from chemical 'over-hydrolysis.' Biotransformation of the ester **10** took place readily with proteases such as α-chymotrypsin or subtilisin, to give the L-*N*-acetylamino acid **11a** and residual D-ester **10b**, which could be extracted from the resulting mixture and recycled. Concentrated acid had to be

# CASE STUDIES OF BIOTRANSFORMATION PROCESSES

**Table 1** Steps in the ester resolution process to L-BOC-TAZ

| Step | | Solvent | pH |
|---|---|---|---|
| 1 | Condensation | Ethanol | Basic |
| 2 | Diester crystallisation | None | |
| 3 | Ester hydrolysis | Aq. ethanol | Basic |
| 4 | Decarboxylation | Water | Acidic |
| 5 | Isolation of racemic ester | Dichloromethane | Neutral |
| 6 | Biotransformation | Water | Neutral |
| 7 | Removal of D-ester | | |
| 8 | Amide hydrolysis | | Acid |
| 9 | Butoxycarbonylation | Aq. methanol | Basic |
| 10 | Product extraction | Dichloromethane | Acid |

used for the hydrolysis of the L-acetylamino acid **11a**, and, hence, a lot of base was required for neutralisation prior to the butoxycarbonylation. However, a fundamental problem with the esterase bioresolution was the lack of enantiospecificity for the thiazoylalanine substrate (**10**); the product was only around 92% *ee*, which was not sufficient for its purpose. The *ee* of the BOC-protected thiazoylalanine **9** could be raised by formation of the salt with (*S*)-α-methylbenzylamine, allowing completion of the synthesis (note that this procedure only enhances the *ee* of pre-enriched material, and cannot be employed for *de novo* classical resolution of the racemate).

A complete synthesis of the *N*-BOC-L-thiazoylalanine **9** had thus been achieved, but the process was far from satisfactory. The process steps are summarised in Table 1, and the problems are (i) too many extraction steps, (ii) too many pH changes, (iii) yield loss from the chemical formation of racemic amide acid, (iv) use of pure ethanol as solvent (which requires an excise license) and (v) lack of enantiopurity in the product from the biotransformation, which entails an extra step to improve the *ee*.

### 9.4.4 DEVELOPMENT OF A ONE-POT PROCESS BASED ON AMINOACYLASE RESOLUTION

The problems described above prompted the development of an alternative process. Effective resolution of *N*-acetylamino acids had previously been shown for a wide range of phenylalanine analogues by Whitesides *et al.*[27] and it was discovered that racemic amide acid **11** (a by-product from the esterase-based process) readily underwent stereoselective amide hydrolysis with aminoacylase from pig kidney or *Aspergillus niger*. Moreover, the L-amino acid **12** formed proved to be of high enantiomeric excess (> 98% *ee*). This bioresolution formed the basis of the one-pot process to L-BOC-TAZ (**9**) which is outlined in Scheme 13.

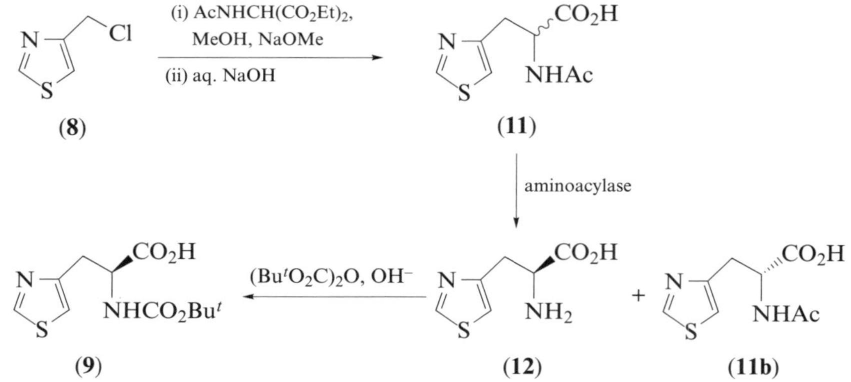

**Scheme 13** Aminoacyclase-based one-pot process to $N$-BOC-3-(4-thiazoyl)-L-alanine

The first advantage in carrying out the resolution at the amide stage is that the initial acetamidomalonate condensation could use methanol instead of ethanol. Thus, while the condensation in methanol at 60 °C resulted in a product mixture of ethyl and methyl esters, these were hydrolysed in the next step. The hydrolysis was performed by slow addition of hydroxide solution, maintaining the mixture at pH 9–10. At this pH, not only does hydrolysis occur at a reasonable rate, but also the initial product undergoes decarboxylation. The decarboxylation causes a reduction in pH and, by careful control of the base addition, the resulting mixture can be kept approximately neutral; in this way a specific acidification step for the decarboxylation is not necessary.

A further feature of this hydrolysis method is that it was carried out with addition of the aqueous base, with concurrent removal of the methanol by distillation. The remaining aqueous solution could be used directly for the biotransformation, for which an acylase had been identified that was sufficiently robust to perform the hydrolysis in what was, essentially, a crude reaction mixture. After the biotransformation the mixture of L-amino acid (12) and D-acetylamino acid 11b was in aqueous solution, thus avoiding the acidic hydrolysis step used in the esterase process. The butoxycarbonylation of the amino acid could be carried out directly with di-*tert*-butyl dicarbonate upon basification and addition of methanol as co-solvent. The BOC-amino acid 9 could then be extracted directly from the aqueous reaction mixture with *tert*-butyl methyl ether (TBME) and was of high enantiomeric excess. Thus, the sequence from chloromethylthiazole (8) to L-BOC-TAZ had been accomplished without removal of the mixture from the reaction vessel. This combines optimum vessel utilisation and minimum solvent waste and provides a demonstration of the effective integration of the biotransformation into the chemical process. The complete process, summarised in Table 2, is much simplified compared with the previous esterase-based process.

# CASE STUDIES OF BIOTRANSFORMATION PROCESSES

**Table 2** Steps in the one-pot aminoacylase-based process

| Step | | Solvent | pH |
|---|---|---|---|
| 1 | Condensation | Methanol | Basic |
| 2 | Hydrolysis/decarboxylation | | Basic |
| 3 | Biotransformation | Water | Neutral |
| 4 | Butoxycarbonylation | | Basic |
| 5 | Product extraction | TBME | Acidic |

## 9.4.5 RECYCLE OF THE D-ENANTIOMER

In the esterase-based process, the unwanted enantiomer is readily extracted as unreacted ester and is racemised with base. It is less straightforward with the shorter, acylase-based route, but the D-acetylamino acid (**11b**) can be recovered after concentration of the aqueous solution and then racemised by heating with acetic anhydride. The mechanism of this racemisation involves activation of the acid function, which causes cyclisation to an oxazolinone (azlactone) (**13**), which tautomerises to the enol **14** (Scheme 14).

Hydrolysis of the azlactone under alkaline conditions then provides the racemised acetylamino acid, which can be recycled to the biotransformation stage. In principle, the azlactone (**13**) could be a substrate for biotransformation; this approach has been used for a synthesis of L-*tert*-leucine, as described in Section 9.5.

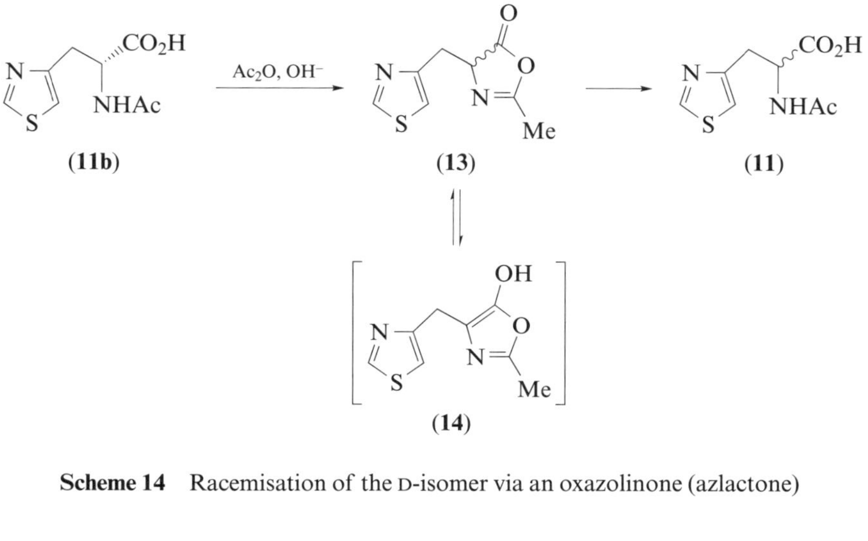

**Scheme 14** Racemisation of the D-isomer via an oxazolinone (azlactone)

### 9.4.6 COMPARISON OF DIFFERENT BIOTRANSFORMATION PROCESSES

Another route worthy of consideration combines the esterase and aminoacylase reactions. Here, the racemic ester **10** is prepared, as described in Section 9.4.3, but is not separated from the racemic amide acid by-product (**11**). The esterase biotransformation reaction is then carried out on this mixture, and the product is subjected to the aminoacylase reaction which provides the L-amino acid **12**. The D-ester (**10b**) left behind from the tandem biotransformations can be selectively extracted and racemised with base for recycle. In terms of the number of solvent changes, a comparison of the three possible routes shows six changes for the esterase-based process, two changes for the aminoacylase-based process and four changes for the combined enzyme process. Owing to its greater simplicity and the higher biocatalyst costs in the combined method, the aminoacylase route is preferred, even though recycling of the unwanted enantiomer is more difficult.

### 9.4.7 APPLICATION OF THE AMINOACYLASE PROCESS TO OTHER UNNATURAL AMINO ACIDS

Initially, work described in Section 9.4.4 used an aminoacylase powder in solution that could not be recovered after use. More recently, an acylase has been identified and immobilised that performs the transformation when the acid amide solution is simply passed through a column of the supported enzyme. This has the double benefit of giving a protein-free solution for the downstream steps and allowing enzyme re-use, which markedly improves the process economics. Additionally, the process has been found to be applicable to other unnatural alanine derivatives such as L-2-naphthylalanine (**15**), L-4-cyanophenylalanine (**16**), L-2-pyridylalanine (**17**) and L-allylglycine (**18**) (Scheme 15), with appropriate process adjustments to account for the different solubilities of these compounds.

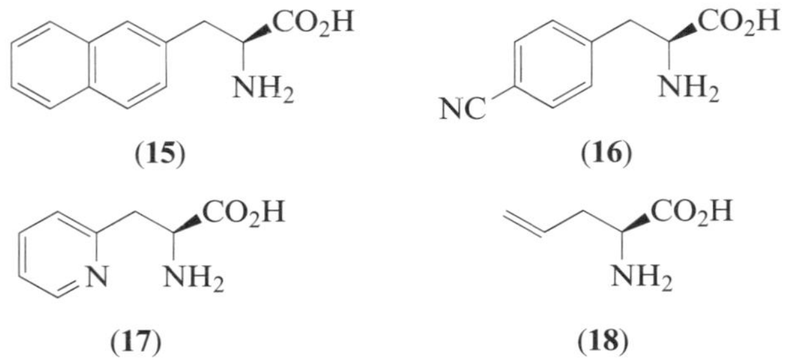

**Scheme 15** Other amino acids prepared by the acylase-based resolution process

# CASE STUDIES OF BIOTRANSFORMATION PROCESSES

## 9.5 DYNAMIC RESOLUTION OF AN OXAZOLINONE BY LIPASE BIOCATALYSIS: SYNTHESIS OF (S)-tert-LEUCINE

### 9.5.1 USES OF (S)-tert-LEUCINE

The unnatural amino acid (S)-tert-leucine (tert-butylglycine) is useful as a lipophilic, hindered component of peptides[28] where, because of its bulky nature, cleavage by peptidases is disfavoured, giving peptides of improved metabolic stability. tert-Leucine is also useful as the building block for a number of chiral auxiliaries and ligands, such as those shown in Scheme 16, where the presence of the bulky tert-butyl group makes these compounds particularly effective for asymmetric synthesis. However, the tert-butyl group also makes the synthesis of the optically active amino acid more challenging.

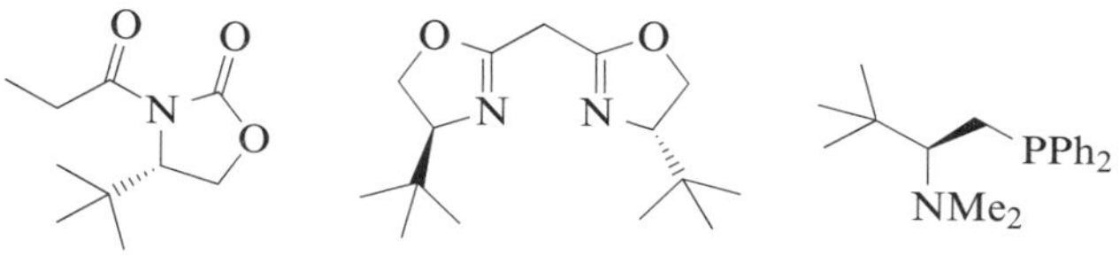

**Scheme 16** Products derived from L-tert-leucine

### 9.5.2 AZLACTONES FOR THE DYNAMIC RESOLUTION OF AMINO ACIDS

One previously reported solution to the synthesis of (S)-tert-leucine involved overall asymmetric reductive amination of the corresponding prochiral keto acid (3,3-dimethyl-2-oxobutanoic acid) with a coupled enzyme system developed by Degussa comprising a dehydrogenase and enzymic cofactor recycling.[29] Chiroscience have developed an alternative approach based on a hydrolase resolution of a 5-(4H)oxazolinone (azlactone) intermediate in which a cofactor is not required. Appropriate azlactones are prepared by cyclodehydration of an N-acylated amino acid with acetic anhydride (cf. Section 9.4.5). Generally, azlactones are readily ring opened by a variety of nucleophiles to amino acid derivatives, and, because of the relatively acidic C-4 hydrogen (pKa 8.8–9), are easily racemised. These compounds have received a lot of attention because of their transient formation during peptide synthesis, leading to partial racemisation. Lipases and proteases have been used to effect asymmetric ring opening of azlactones using either water or alcohols as nucleophiles;[30] facile racemisation of the less reactive enantiomer permits a dynamic resolution and the complete conversion of a racemate to an optically pure amino acid derivative.

## 9.5.3 DEVELOPMENT OF A LIPASE-CATALYSED DYNAMIC RESOLUTION.

An azlactone (**19**) of racemic *tert*-leucine is readily prepared by acetic anhydride treatment of the *N*-benzoyl derivative (Scheme 17). Whilst water can be used as the nucleophile for enzyme-catalysed ring opening, unselective, non-enzymatic hydrolysis of the azlactone reduces the *ee* of the resulting acid. This problem can be overcome by using an alcohol as the nucleophile in a water-free system. Lipozyme (an immobilised *Mucor miehei* enzyme) was found to be the enzyme of choice and, using *n*-butanol as the nucleophile, a highly selective reaction was possible, giving the ester **20** in 97% *ee* and 90% yield (Scheme 17); lower alcohols give lower rates in the resolution reaction and lower enantiomeric excesses.

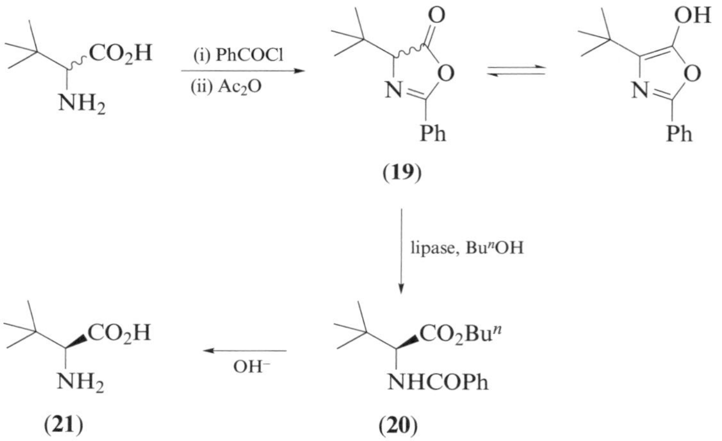

**Scheme 17** Dynamic biocatalytic resolution process for L-*tert*-leucine

The initial laboratory procedure for the resolution[31] was not considered suitable for manufacture since it was performed at a high dilution, with a high loading of enzyme (substrate-to-enzyme ratio 10:1) and still required 5 days for complete conversion. Also, in the early experiments it was necessary to include triethylamine, otherwise the yield and product *ee* were reduced; presumably the triethylamine effected the racemisation of the (*R*)-azlactone. However, during development it was found that the concentration of the substrate could be increased to 20%, and conditions were established whereby the reaction was complete within 24 h; it was also found that the triethylamine was no longer necessary to obtain a high yield and enantiomeric excess. It seems that the higher concentration of the substrate is sufficient to promote racemisation of the (*R*)-azlactone.

# CASE STUDIES OF BIOTRANSFORMATION PROCESSES

Importantly, the enzyme could be reused, although this resulted in decreased yields and *ee*. It is possible that the azlactone somehow modifies the lipase to a form which has reduced enantiospecificity, but which is strongly inhibited by triethylamine. Deprotection of the ester amide **20** to the amino acid **21** is not as straightforward as might be expected. Normally, an amino acid protected in this way would be cleaved by acid hydrolysis but, in this case, such conditions led to partial racemisation caused by transient recyclisation to the stereochemically labile azlactone **19**. This recyclisation probably reduces adverse steric interactions caused by the bulky *tert*-butyl group. A solution to this problem was to perform a sequential deprotection, hydrolysing the ester first using enzymatic (Alcalase) or chemical (NaOH/water/methanol) methods. Debenzoylation, with minimal racemisation, was effected using caesium hydroxide, and it was shown later that the complete hydrolysis could be carried out with potassium hydroxide without racemisation. Note that the *tert*-butyl group probably reduces the risk of base-induced racemisation by hindering abstraction of the α-hydrogen.

Overall, while the process may formally be described as a resolution, by virtue of the *in situ* recycling it is a deracemisation, and is as effective as an asymmetric synthesis from a prochiral starting material. The viability of the process is dependent on the economics of supply of the racemate. However, it is possible to make the azlactone directly from the inexpensive *N*-benzoylglycine (hippuric acid) (**22**) via the glycine azlactone (**23**) as shown in Scheme 18,[32] thus providing a particularly attractive overall route.

Clearly, while the above case history has focussed on L-*tert*-leucine, the methodology could be applied equally to other *tert*-alkyl glycine derivatives. For example, using the route shown in Scheme 18, condensation of the glycine azlactone **23** with different ketones would be followed by introduction of the final alkyl group using an appropriate copper-coupled Grignard reagent.

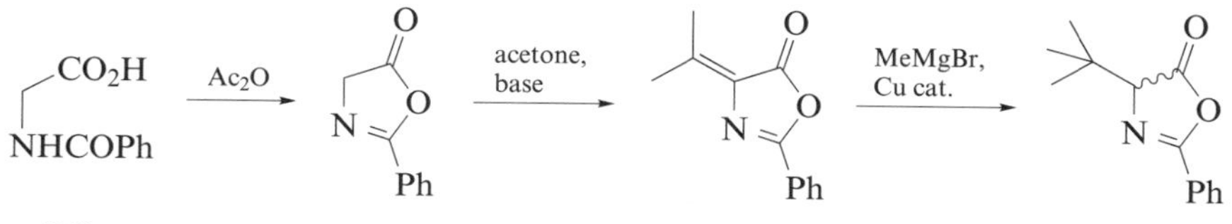

**Scheme 18** Alternative route to the azlactone bioresolution substrate for L-*tert*-leucine

## 9.6 CONCLUSIONS

Four examples have been described of development work carried out on biocatalytic resolution procedures to provide scaleable processes. Combining the learning from these case studies, it is possible to highlight aspects that are of general importance when establishing bioresolutions.

First, it is important to consider the most appropriate substrate, not only for the biotransformation itself (e.g. in respect of enantiospecificity, conversion rate, etc.) but also to facilitate product isolation. Thus, in the case of the bicyclic hydroxylactone (Section 9.3), it was found that the butyrate ester rather than the acetate allowed solvent partitioning as a scaleable means of separation of the biotransformation product. It is better still if the substrate can be chosen such that the product crystallises directly during the work-up. Also, isolation of the product can be facilitated when the biotransformation has been developed, as in the case of the immobilised lactamase (Section 9.2), to operate at a high concentration (200 g $l^{-1}$).

It is important when developing a biotransformation-based step to consider its fit within the overall synthetic pathway, as exemplified by the process development for the $N$-BOC-TAZ (Section 9.4). Here, the preferred biotransformation led to simplification of the overall process, taking into consideration issues such as the number of pH changes and solvent exchange steps. It is also of note here that a bioresolution involving amide hydrolysis has a significant advantage over an esterase resolution in having less, or no, requirement for pH control during the transformation, which facilitates operation in non-specialised equipment. This is illustrated well by the case of the lactamase process (Section 9.2), in which the transformation required only an aqueous solution of the substrate and the biocatalyst. To integrate a biotransformation with the total process, it can be valuable to combine, in the earlier stages of a project, a bioresolution with a method for enhancing the enantiomeric purity of the product, typically a crystallisation technique. This gives an interim process with the capability of providing enantiomerically pure product during the development and optimisation of the biocatalytic resolution (see Section 9.4).

A choice which frequently has to be made in a biocatalytic resolution is whether to work in the hydrolytic direction, in aqueous solution, or in the esterification (or transesterification) mode in a low-water system. The hydrolytic approach frequently has the advantages of higher biotransformation rates and, often, greater enantioselectivity, but can suffer from competing chemical hydrolysis; these were considerations for both the bicyclic hydroxylactone (Section 9.3) and the *tert*-leucine (Section 9.5), which are, preferably, carried out in aqueous and organic solvent systems, respectively.

Lastly, a problem with all resolutions is the utilisation of the unwanted enantiomer—a most important consideration in providing the best economics. In the case of the lactamase work (Section 9.2), and the bicyclic hydroxylactone (Section 9.3), utilisation of the unwanted enantiomer is not practical, but the resolution takes place early in the total synthesis, so lessening the economic impact. In the case of the $N$-BOC-thiazoylalanine acylase resolution approach (Section 9.4), a racemisation procedure for the unwanted D-enantiomer was established via an oxazolinone (azlactone) intermediate. This concept was extended in a biotransformation of the azlactone itself in a process for ($S$)-*tert*-leucine where, owing to *in situ* racemisation of the unwanted enantiomer, the process is as efficient as an asymmetric synthesis.

## 9.7 ACKNOWLEDGEMENTS

We are grateful to colleagues and collaborators at Chiroscience and the University of Exeter whose efforts have provided the results discussed in this chapter.

## 9.8 REFERENCES

1. Coates, J. A. V., Ingall, H. J., Pearson, B. A., Penn, C. R., Storer, R., Williamson C., and Cameron, J. M., *Antiviral Res.*, **15**, 161 (1991).
2. Borthwick, A. D., Butt, S., Biggadike, K., Exall, A. M., Roberts, S. M., Youds, P. M., Kirk, B. E., Booth, B. R., Cameron, J. M., Cox, S. W., Marr C.L.P., and Shill, M., *J. Chem. Soc., Chem. Commun.*, 656 (1988).
3. Scopes, D. I., Newton, R. F., Ravenscroft, P., and Cookson, R.C., *Eur. Pat.*, 0 104 066, 1984.
4. Chen, J., Grim, M., Rock, C., and Chan, K., *Tetrahedron Lett.*, **30**, 5543 (1989).
5. Exall, A. M., Jones, M. F., Mo, C.-L., Myers, P. L., Paternoster, I. L., Singh, H., Storer, R., Weingarten, G. G., Williamson, C., Brodie, A. C., Cook, J., Lake, D. E., Meerholtz, C. A., Turnbull, P .J. and Highcock, R. M., *J. Chem. Soc., Perkin Trans. 1*, 2467 (1991).
6. Borthwick, A.D., and Biggadike, K., *Tetrahedron*, **48**, 571 (1992).
7. Jagt, J. C., and van Leusen, A.M., *J. Org. Chem.*, **39**, 564 (1974).
8. Griffiths, G.J., and Previdoli, F.E., *J. Org. Chem.*, **58**, 6129 (1993).
9. Daluge, S., and Vince, R., *J. Org. Chem.*, **43**, 2311 (1978); Vince, R. and Hua, M., *J. Med. Chem.*, **33**, 17 (1990).
10. Taylor, S. J. C., Sutherland, A. G., Lee, C., Wisdom, R., Thomas, S., Roberts, S.M., and Evans, C., *J. Chem. Soc,. Chem. Commun.*, 1120 (1990).
11. Evans, C. T., Roberts, S.M., Shoberu K.A., and Sutherland, A.G., *J. Chem. Soc., Perkin Trans. 1*, 589 (1992).
12. Taylor, S. J. C., McCague, R., Wisdom, R., Lee, C., Dickson, K., Ruecroft, G., O'Brien, F., Littlechild, J., Bevan, J., Roberts, S. M., and Evans, C.T., *Tetrahedron: Asymmetry*, **4**, 1117 (1993).
13. Vince, R., and Daluge, S., *J. Org. Chem.*, **45**, 531 (1980).
14. Palmer, C., and McCague, R., *J. Chem. Soc., Perkin Trans. 1*, in press.
15. Jung, M. E., and Rhee, H., *J. Org. Chem.*, **59**, 4917 (1994).
16. Lubinau, A., Augé, J,. and Lubin, N., *Tetrahedron Lett.*, **32**, 7529 (1991).
17. MacKeith, R. A., McCague, R., Olivo, H. F., Palmer, C. F., and Roberts, S.M., *J. Chem. Soc., Perkin Trans. 1*, 313 (1993); *ICT Int. Pat. Appl.*, WO 93/00826, 1993.
18. MacKeith, R. A., McCague, R., Olivo, H. F., Roberts, S. M., Taylor, S. J. C., and Xiong, H., *Bioorg. Med. Chem.*, **2**, 387 (1994).
19. McCague, R., Olivo, H. F., and Roberts, S. M., *Tetrahedron Lett.*, **34**, 3785 (1993).
20. Casy, G., Gorins, G., McCague, R., Olivo, H. F., and Roberts, S.M., *J. Chem. Soc., Chem. Commun.*, 1085 (1994); Carnell, A.J., Casy, G., Gorins, G.A., Kompany-Saeid, A., McCague, R., Olivo, H.F., Roberts, S.M., and Willetts, A.J., *J. Chem. Soc., Perkin Trans. 1*, 3431 (1994).
21. Rosenberg, S. H., Spina, K. P., Woods, K.W., Polakowski, J., Martin, D. L., Yao, Z., Stein, H. H., Cohen, J., Barlow, J. L., Egan, D. A., Tricarico, K. A., Baker, W. R., and Kleinert, H. D., *J. Med. Chem.*, **36**, 449 (1993).
22. Nishi, T., Saito, F., Nagahori, H., Kataoka, M., Morisawa, Y., Yabe, Y., Sakurai, M., Higashita, S., Shogi, M., Matsushita, Y., Ijima, Y., Ohizumi, K., and Koike, H., *Chem. Pharm. Bull.*, **38**, 103 (1990).
23. Duthaler, R.O., *Tetrahedron*, **50**, 1539 (1994).

24. Caldwell, W. T., and Fox, S. M., *J. Am. Chem. Soc.*, **73**, 2935 (1951).
25. Miyazawa, T., Iwanaga, H., Ueji, S., Yamada, T., and Kuwata, S., *Chem. Lett.*, 2219 (1989).
26. Hsiao, C.-N., Leanna, M. R., Bhagavatula, L., and De Lara, E., *Synth. Commun.*, **20**, 3507 (1990).
27. Chenault, H. K., Dahmer, J., and Whitesides, G. M., *J. Am. Chem. Soc.*, **111**, 6354 (1989).
28. Fauchere, J.-L., and Petermann, C., *Helv. Chim. Acta*, **63**, 824 (1980).
29. Drauz, K., Bommarius, A, Knaup, G., Kottehahn, M. and Schwarm, M., presented at 'Chiral Europe '94' Symposium 19–20 Sept. 1994, Nice, (Spring Innovations).
30. Crich, J. Z., Brieva, R., Marquart, P., Gu, R.-L., Fleming, S., and Sih., C. J., *J. Org. Chem.*, **58**, 3252 (1993).
31. Turner, N. J., Winterman, J. R., McCague, R., Parratt, J. S. and Taylor, S. J. C., *Tetrahedron Lett.*, **36**, 1113 (1995).
32. Miyazawa, T., Nagai, T., Yamada, T., Kuwata, S. and Watanabe, H., *Mem. Konan Univ., Sci. Ser.*, **23**, 51 (1989).

# 10 (S)-2-Chloropropanoic Acid: Developments in Its Industrial Manufacture

**S. C. TAYLOR**
*ZENECA LifeScience Molecules, Billingham, UK*

| | | |
|---|---|---|
| 10.1 | Introduction | 207 |
| 10.2 | Market Requirements for (S)-2-Chloropropanoic Acid | 208 |
| 10.3 | (S)-2-Chloropropanoic Acid Synthesis Options | 210 |
| 10.4 | The ZENECA Approach to (S)-2-Chloropropanoic Acid | 211 |
| | 10.4.1 Biological Dehalogenation | 211 |
| | 10.4.2 Identification of an (R)-2-Chloropropanoic Acid Dehalogenase | 212 |
| | 10.4.3 Properties of (R)-2-Chloropropanoic Acid Dehalogenase and *Pseudomonas putida* NCIMB 12018 | 213 |
| | 10.4.4 Downstream Processing Requirements | 213 |
| | 10.4.5 Process Development Overview | 214 |
| | 10.4.6 Developments Through Molecular Biology | 219 |
| | 10.4.7 Dehalogenation as a General Route to Chiral Molecules | 220 |
| 10.5 | Conclusions | 221 |
| 10.6 | Acknowledgements | 222 |
| 10.7 | References | 222 |

## 10.1 INTRODUCTION

(S)-2-Chloropropanoic acid [(S)-CPA] was one of the first commercially valuable single-enantiomer materials to be identified and has, as a consequence, received considerable attention from industrial and academic groups exploring synthesis options. Although structurally a relatively simple molecule, its importance as a large-volume homochiral building block and the cost sensitivity associated with its major agrochemical end markets have posed a difficult challenge to businesses and groups seeking to exploit the market opportunity. Numerous options for its production have been proposed and investigated at laboratory and pilot scale. During the 1980s, ZENECA (formerly ICI) invented and developed a novel bio-technological process which, since 1989, has been operated in a full-scale commercial plant.

*Chirality in Industry II.* Edited by A. N. Collins, G. N. Sheldrake and J. Crosby
© 1997 John Wiley & Sons Ltd

It is the purpose of this chapter to explore the background interest in (S)-CPA and its end markets, to review the chemical and biological production options, to present a historical picture of how the current ZENECA bio-process has evolved and, finally, to draw some conclusions as to the application of biological stereoselective dehalogenation to other chiral molecules.

## 10.2 MARKET REQUIREMENTS FOR (S)-2-CHLOROPROPANOIC ACID

The opportunity for (S)-CPA falls into two distinct commercial areas: an undeveloped market as a pharmaceutical intermediate and a growing agrochemical market where it is a key building block for a major group of herbicides.

The importance of chirality to the pharmaceutical industry is now well recognised, reflecting the common situation where the desired biological activity resides only in one of a pair of enantiomers or, as is often the case, an undesirable activity is associated with the other enantiomer. The principal driving forces for making pharmaceutical compounds as individual enantiomers rather than racemates have been product efficacy and regulatory. Cost reduction, through the raw materials savings that can come from moving to a single enantiomer, has generally been of lesser importance. (S)-CPA has not been widely used as a chiral building block in the pharmaceutical sector, partly reflecting its lack of availability in high enantiomeric purity at the time when processes for the production of many of today's active molecules were being established. The non-steroidal anti-inflammatory group of therapeutics can be synthesised from 2-halopropanoates, e.g. 2-bromopropanoic acid or 2-chloropropanoic acid (CPA) and their esters. Products such as naproxen, ibuprofen and flurbiprofen have historically been made as racemates, but it is now clear that single-enantiomer forms will be marketed, creating an opportunity for (S)-CPA and its analogues from which they can be produced.[1]

Chirality is also important in the agrochemical industry, with pesticidal activity often localised in single enantiomers. Here, regulatory issues have been, and remain, less profound than in pharmaceuticals, and the principal reason for making products as single active enantiomers rather than racemates has been to achieve greater cost effectiveness in use. This has not been the sole reason, however, and some companies have recognised the opportunity to gain product differentiation and possibly lead regulatory change in favour of the more environmentally sound single-enantiomer forms. This is well illustrated by the group of herbicides based on the phenoxypropanoic acid structure.

There are many examples of the phenoxypropanoic acid herbicides such as mecoprop and dichlorprop[2] which have been produced and sold since the 1950s. All products are based on the same core structure (Figure 1) but with differing substitution on the aromatic ring. This can range from simple chloro or methyl substitution as in dichlorprop and mecoprop through to complex heteroaromatic systems as in fluazifop.

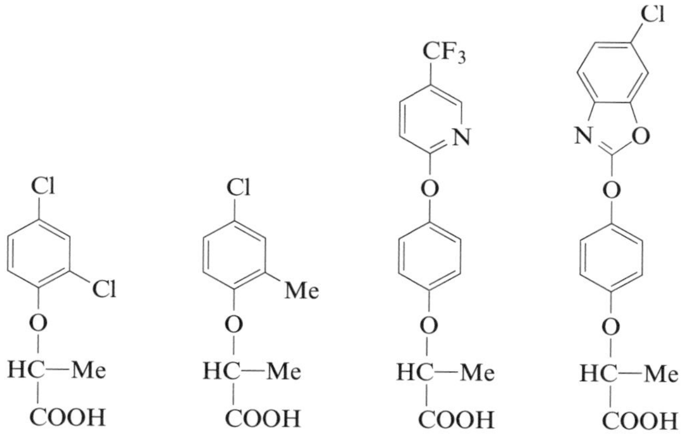

# INDUSTRIAL MANUFACTURE OF (S)-2-CHLOROPROPANOIC ACID

**Figure 1** Examples of phenoxypropanoic herbicides

Products such as fluazifop are particularly valuable herbicides giving excellent post-emergent control of annual and perennial grass weeds in broadleaved crops at low dose rates of <0.5 kg a.i.ha$^{-1}$ even as the racemic form. The older products such as mecoprop are used to control broadleaved weeds in cereals, but at much higher application rates of 1.5–3 kg a.i. ha$^{-1}$ in the racemic form. Thus, very substantial volumes of these herbicides are applied worldwide and, although they are readily biodegradable, the presence of 50% inactive isomer is becoming an environmental issue. All of these products show herbicidal activity in the (R)-enantiomer, with the (S)-enantiomer having much reduced or no activity. Until recently, all major products were made as racemates with, in most cases, the chiral centre being derived from low-cost, racemic CPA, a salt or an ester derivative. In order to produce the single active (R)-enantiomers, the existing chemistry from CPA to final herbicide may be used, as it does not cause significant loss of enantiomer purity through racemisation during processing when used with (S)-CPA with minor proprietary process modifications. (S)-CPA, or in some instances the ester, can thus be substituted as the starting material, allowing the use of existing manufacturing plant and hence minimising new capital requirements.

One of the earliest phenoxypropanoate herbicides to be converted from racemic to single-enantiomer form was the ZENECA product fluazifop. Similar strategies were adopted by A. H. Marks in the UK and BASF in Germany with regard to mecoprop and other older, commodity herbicides, where the lower specific activity of the racemic products has meant a significant environmental load. Conversion to a single-isomer product allows for an effective doubling in manufacturing capacity for the final product and offers cost saving through reduced raw material use. Coupled with the environmental and potential regulatory benefits, this has given a strong case for moving such products to single-enantiomer form. In recent years, most of the major agrochemical companies have either initiated the switch of racemic phenoxypropanoate herbicides to single-enantiomer products or introduced completely novel single-enantiomer compounds, both approaches creating a growing market for (S)-CPA.

Synthetic approaches to phenoxypropanoic acid herbicides not based on the use of (S)-CPA have also been proposed. These include direct microbiological resolution of the final racemic herbicide, which was demonstrated by ZENECA with fluazifop,[3] and also, more recently, by esterase hydrolysis.[4] However, these do not appear to compete economically with the (S)-CPA option, indicating the effectiveness and high productivity of the chemistry from (S)-CPA and the cost implication of yield loss when working with the final active molecule. This substantiates the general rule that resolutions should take place as early as possible to minimise the total quantity of material produced.

The herbicide market needs described above require (S)-CPA to be available at a low price, but with good chemical and enantiomer purity (>92% ee) and on a scale of several thousand tonnes per annum.

## 10.3 (S)-2-CHLOROPROPANOIC ACID SYNTHESIS OPTIONS

The principal route options to (S)-CPA have all included a biological stage. The oldest established route is based upon the fermentative production of (R)-lactic acid from glucose using selected strains of *Lactobacillus*. Although the (R)-isomer is not considered to be the natural lactic isomer, bacteria which are able to produce this isomer almost exclusively are now well characterised.[5,6] However, they do require a much higher degree of fermentation control than is common with *Lactobacillus* species, in order to prevent contamination with (S)-isomer producing species. Conversion of (R)-lactic acid to (S)-CPA can be done by various chemical methods, e.g. esterification followed by chlorination with thionyl chloride.

The second main synthesis option utilises the now well known ability of esterase and lipase enzymes to catalyse the selective hydrolysis or transesterification of esters of acids such as CPA. The early work on CPA resolution with esterases was done by Klibanov and co-workers in the USA[7] simultaneously with studies by the Stauffer Chemical Company.[8] However, extended reaction times, a carefully controlled ratio of organic and aqueous phases and a degree of conversion in excess of 60% were necessary to obtain good enantiomer purity. This could be a result of the reaction taking place at a site distant from the chiral centre on a small molecule. More recently this reaction has been further developed by BASF (using the isobutyl ester of CPA and an immobilised *Pseudomonas* esterase[9]), Chemie Linz[10] and others[11] (Figure 2).

Other routes have been investigated in the laboratory, including stereospecific enzymic hydrolysis of racemic 2-chloropropionamide,[12] hydroformylation with

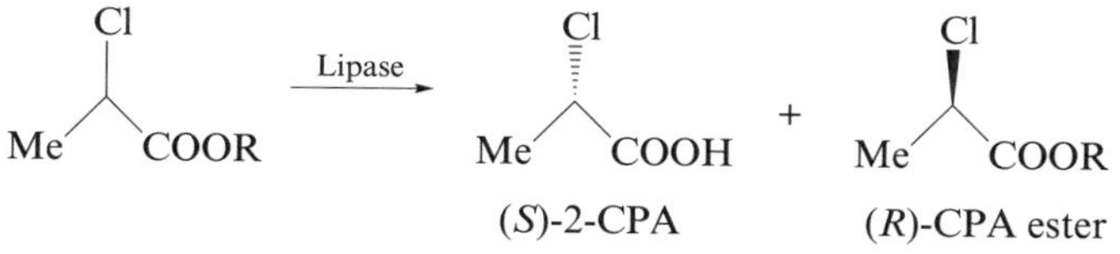

# INDUSTRIAL MANUFACTURE OF (S)-2-CHLOROPROPANOIC ACID

```
      Cl                    Cl                        Cl
      |          Lipase     ⋮                         |
   Me–C–COOR    ────────►  Me–C–COOH         +     Me–C–COOR
                              (S)-2-CPA               (R)-CPA ester

   R = isobutyl; lipase from Pseudomonas DSM8246
```

**Figure 2.** Proposed use of ester hydrolysis as a route to (S)-CPA

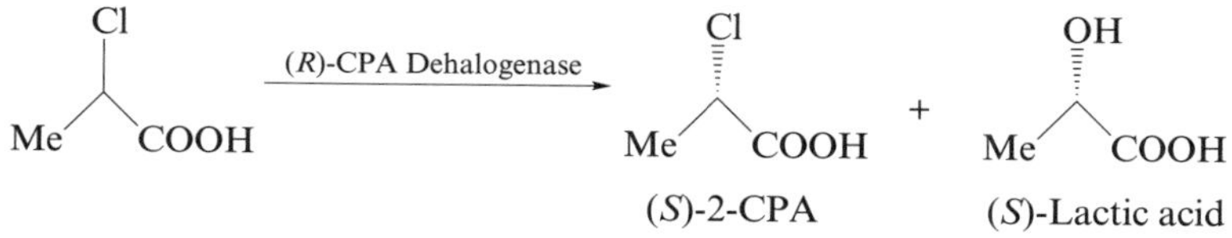

```
      Cl                              Cl                      OH
      |       (R)-CPA Dehalogenase    ⋮                       ⋮
   Me–C–COOH  ───────────────────►  Me–C–COOH       +      Me–C–COOH
                                      (S)-2-CPA               (S)-Lactic acid
```

**Figure 3.** Proposed use of stereoselective enzyme-catalysed dehalogenation as a route to (S)-CPA

metal–ligand complexes [13] and diastereoselective halogenation of silyl ketene acetals.[14]

The development of a ZENECA process was aimed at overcoming the performance shortcomings indicated above with the lactic acid and esterase routes, and achieving a cost base in advance of other more speculative alternatives. The low cost of racemic CPA made resolution approaches potentially economic, and novel enzyme-catalysed dehalogenation, where the reaction would occur directly at the chiral centre and thus, potentially, offer high enantiomer specificity, was recognised by ZENECA as an innovative approach. The principle was to use a hydrolytic, non-cofactor-requiring enzyme that catalysed the removal of chlorine from (R)-CPA but not from (S)-CPA. Thus, from racemic CPA, reaction products would include unreacted (S)-CPA, which should be readily separated from the expected hydroxy acid (lactic acid) co-product of resolution (Figure 3).[15]

## 10.4 THE ZENECA APPROACH TO (S)-2-CHLOROPROPANOIC ACID

### 10.4.1 BIOLOGICAL DEHALOGENATION

Enzymes that catalyse the dehalogenation of molecules are widespread in nature and occur in both prokaryotic and eukaryotic organisms. Microbes that live in environments contaminated with halogenated compounds, particularly *Pseudomonas* bacteria, are good sources of enzymes that catalyse the elimination of halogen from organic molecules and many organisms are able to use halogenated aliphatic and aromatic compounds as sole sources of carbon and energy for growth. Many biochemical studies on CPA, 2,2-dichloropropanoate and monochloroacetate biodegradation[16,17] have demonstrated that the initial attack on the molecule is catalysed by dehalogenase enzymes in what appears to be a hydrolytic removal of halogen from the C-2 position yielding the corresponding hydroxy acid. Many dehalogenases can be present in single species of bacteria, and these can have different substrate and enantiomer specificities.[18] At the start of the ZENECA programme, several dehalogenases had been reported that were specific for the (S)-enantiomer of CPA, with no activity being shown against the (R)-enantiomer.[19,20] However, no dehalogenase had been reported that was specific for the (R)-enantiomer of CPA. From an industrial perspective, if such a novel enzyme could be isolated and its effectiveness and commercial value demonstrated, this would give rise to a sound intellectual property position.

### 10.4.2 IDENTIFICATION OF AN (R)-2-CHLOROPROPANOIC ACID DEHALOGENASE

The approach adopted to find a specific (R)-CPA dehalogenase was to screen bacteria isolated from soil sites contaminated with chloro acids, such as areas where agrochemicals were manufactured or applied. Although many strains were found that could grow on (R)-CPA but not (S)-CPA, suggesting the presence of an (R)-CPA-specific dehalogenase, in practice these invariably turned out to be deficient in the metabolism of the (R)-lactic acid isomer (the expected product of CPA dehalogenation), rather than having a CPA dehalogenase specificity. This probably indicates the selective disadvantage to an organism of being unable to dehalogenate an isomer of a toxic compound such as CPA. Furthermore, it illustrates the deficiency of microbial selection as a basis for identifying new enzymes. However, several racemic CPA-utilising strains displayed differential growth rates on (R)- and (S)-CPA, which suggested that these strains might contain dehalogenases of different specificity. Dehalogenases in these strains were readily separated by polyacrylamide gel electrophoresis coupled with an enzyme activity stain[21] and, on examination, these were found to contain multiple dehalogenases of different specificities, including the required (R)-CPA specificity (Table 1). Thus, *Pseudomonas putida* AJ1, renamed NCIMB 12018, contained two specific dehalogenases, one specific for (R)-CPA and the other for (S)-CPA.

Table 1  Specificity of dehalogenases present in CPA-degrading microorganisms isolated by ZENECA

| Organism | CPA Enantiomer specificity | |
|---|---|---|
| | Dehalogenase 1 | Dehalogenase 2 |
| *P. putida* NCIMB 12018 | (R)-CPA | (S)-CPA |
| *P. fluorescens* NCIMB 12159 | (R)-CPA | (S)-CPA |
| NCIMB 12158 | (R)-CPA | (R/S)-CPA |
| NCIMB 12160 | (R)-CPA | (R/S)-CPA |
| NCIMB 12161 | (R)-CPA | (R/S)-CPA |

## 10.4.3 PROPERTIES OF (R)-2-CHLOROPROPANOIC ACID DEHALOGENASE AND *PSEUDOMONAS PUTIDA* NCIMB 12018

*P. putida* NCIMB 12018 was found to be a robust organism able to grow in the presence of relatively high concentrations (>100 mM) of CPA in minimal growth media. The strain also demonstrated rapid growth on simple, cheap carbon substrates such as glucose. As a *Pseudomonas* species it would be amenable to genetic manipulation by either mutation or molecular approaches. No toxicological problems were found with the strain, so it provided a good basis for development of a commercialisable fermentation process for production of the enzyme or whole-cell microbial biocatalyst.

The (R)-CPA dehalogenase was readily separated from the (S)-CPA dehalogenase by ion-exchange or molecular sieve chromatography, allowing examination of its catalytic properties, which was crucial to determining whether it could form the basis of a bio-transformation process. The enzyme was a tetramer of a molecular weight of 135 000 Da. Temperature and pH optima were in line with other dehalogenases and, most important, the enzyme stability was good under laboratory conditions at 30 °C and pH 7–8. At alkaline pH CPA rapidly racemises, so it was important to use a dehalogenase that retained high specific activity at near neutral pH; this was the case. The product of dehalogenation was (S)-lactic acid, which indicated a reaction with inversion of configuration. Activity with (S)-CPA as potential substrate was undetectable even in the presence of >0.2 M (S)-CPA. No significant inhibition of the enzyme was found in the presence of (S)-lactic acid, (S)-CPA or chloride ions (the products of the reaction) at concentrations in excess of 0.5 M. The values of $K_{cat}$ (15 s$^{-1}$) and, most importantly, $K_m$ (1 mM) were acceptable in terms of achieving a fast and complete removal of (R)-CPA from the reaction, and thus high enantiomeric purity of the (S)-CPA product. A low $K_m$ was an essential prerequisite for this, so that activity remained high at (R)-CPA concentrations of less than 1 g l$^{-1}$ which are typical of the end of a resolution bio-transformation.

In summary, both the host microbial strain and the basic properties of the enzyme indicated that, in principle, the enzyme and organism could be made in bulk, and that a high-purity (S)-CPA product could be made in an intensive reaction, with acceptable plant and capital implications for a multi-thousand tonne per annum facility. The realisation of this potential, however, required a substantial and multi-functional development programme.

## 10.4.4 DOWNSTREAM PROCESSING REQUIREMENTS

Separation problems can often arise with bio-transformation reactions owing to the presence of proteins and other biological materials which can interfere with solvent extraction and crystallisation processes. Thus, in addition to the fermentation and bio-transformation stages of the proposed (S)-CPA process, the principle of separating pure (S)-CPA from the reaction mixture also needed to be established. Although the product (S)-CPA is unstable at high pH, it has good stability at low pH in its un-ionised form. Therefore, at low pH, solvent extraction was applicable, and a carefully chosen solvent and operating conditions could separate (S)-CPA from the other major organic component, lactic acid. At low pH, proteinaceous material precipitated in a filterable form which was readily removed from the product stream. This offered flexibility in the nature of the biocatalyst, and implied that a whole cell process, where significant amounts of contaminating biological material would be present, was a viable option.

It was confirmed that all stages of a required process were feasible and that, provided appropriate plant volumetric and biocatalyst productivities could be achieved and the overall yield target met, a cost-effective process was attainable.

### 10.4.5 PROCESS DEVELOPMENT OVERVIEW

The key to success in bio-transformation technology is the process of translation of an embryonic laboratory procedure into a cost-effective, reliable and robust plant-scale operation in an acceptable time-scale, and with a resource appropriate to the opportunity. Many options needed to be considered for the (S)-CPA process, including isolated enzyme versus whole microbial cells, immobilised versus free cells or enzyme and batch versus continuous reaction for fermentation, bio-transformation and downstream processing stages. Making correct choices at an early stage in the development process for a bio-transformation is critical to economic success, and the learning required has become a key part of the intellectual property of ZENECA and other successful biotechnology businesses.

For simplicity and speed of scale-up, whole-cell systems offer several advantages over isolated enzymes. This is because several process steps, needed to prepare a cell free enzyme preparation, can be eliminated. *P. putida* NCIMB 12018 could not be used directly in a process as a whole-cell system owing to the presence of an (S)-CPA dehalogenase in the strain, which had first to be removed or inactivated. In the initial development work this was done by chemical mutagenesis, which proved effective at completely eliminating the unwanted dehalogenase activity, and resulted in strain NCIMB 12018-23. No other enzyme present in this strain interfered with the (R)-CPA dehalogenase or attacked the (S)-CPA product. (This can often be an issue with whole-cell systems; an isolated enzyme system is often required to overcome the problem.) Therefore, for the (S)-CPA system, a whole-cell approach was envisaged as being the most cost effective, with the shortest development time-scale, and consequently this approach was followed.

Early fermentation work concentrated on batch growth, although it was intended eventually to establish a continuous fermentation process and so derive the productivity benefits that are possible with this technology, which is well established in ZENECA. The dehalogenase enzyme was not constitutively expressed and substrate induction was required. Both isomers of CPA would induce (R)-CPA dehalogenase activity in *P. putida* NCIMB 12018-23. However,

# INDUSTRIAL MANUFACTURE OF (S)-2-CHLOROPROPANOIC ACID

(S)-CPA was relatively expensive to use in this way, and provided significant selection pressure on the mutant organism to revert to its former wild-type state. (R)-CPA, also expensive, was rapidly metabolised and had to be used at a concentration in excess of 50 mM to achieve good induction at target cell dry weights of >30 g l$^{-1}$. At this level, the resultant concentration of free chloride ion following dehalogenation causes corrosion problems with stainless steel, so specialist fermentation equipment would have been required. Racemic CPA was also an effective inducer but, like (S)-CPA, provided pressure on the culture to select for mutant revertants to wild type, and hence limited the options for fermentation on a longer time-scale.

Although a strain constitutive for (R)-CPA dehalogenase could have been constructed, it was found that some chlorinated acids, such as dichloropropanoic acid (DCP) and trichloroacetic acid (TCA), were not dechlorinated, but would act as powerful dehalogenase inducers at concentrations well below 1 mM. Furthermore, compounds such as TCA did not give selection pressure for strains reverting to (S)-CPA dehalogenase production. A low level of TCA or DCP, included in a sugar-based growth medium, allowed high fermentation culture dry weight of *P. putida* NCIMB 12018-23 with high dehalogenase activity. Over extended fermentation times typical of continuous fermentation mode, however, use of DCP resulted in a dramatic and unexpected total loss of the cells ability to make dehalogenase enzyme (Figure 4), possibly owing to the cell adapting to prevent uptake of the toxic and unmetabolisible DCP.

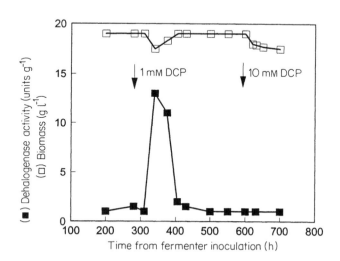

**Figure 4** Profile of a continuous fermentation with the (S)-CPA production strain using the non-metabolised dichloropropanoate as dehalogenase inducer. The continuous fermentation was established in the steady state prior to introduction of dichloropropanoate (DCA) at 1 mM concentration at 300 h. This was increased to 10 mM at 590 h

The ability of *Pseudomonas* species to display novel methods of regulatory control over dehalogenases has been reported previously,[22] and this may be another example. The effect was observed with all non-metabolisable chlorinated acids tested, which precluded use of this type of enzyme induction system for continuous fermentation operation.

To overcome these problems, an alternative inducer, monochloroacetic acid, was used. This compound is slowly dechlorinated and metabolised by *P. putida* NCIMB 12018 strains, and could induce dehalogenase activity at an acceptable concentration (<50 mM) from a metallurgical perspective. Furthermore, it also provided the organism with an additional potentially utilisable carbon source, glycolate, arising from dechlorination. As an achiral compound, monochloroacetic acid was fully dechlorinated by the (*R*)-CPA dehalogenase and provided no selection pressure for mutant reversion; these properties combined allowed for long-term continuous fermentation (Figure 5).

The cells of *P. putida* NCIMB 12018-23 could be used directly in a bio-transformation process. In the wet form, however, the stability of the intact cells as a bio-catalyst was poor, and a method for converting microbial cells into a stable biocatalyst was required. This is a situation where immobilisation may be applied. However, technology for drying cells and retaining high enzyme activity and viability had been researched by ZENECA, and a proprietary cell drying process was developed and applied to the dehalogenase-containing microorganism. This gave >90% retention of the dehalogenase enzyme activity and allowed for storage of a solid biocatalyst for over 12 months without significant loss of

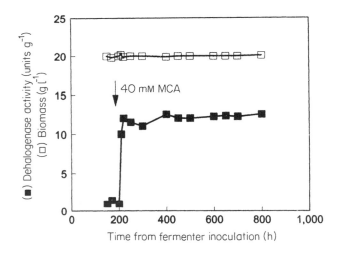

**Figure 5** Profile of a continuous fermentation with the (*S*)-CPA production strain using monochloroacetate as dehalogenase inducer. The continuous fermentation was established in the steady state prior to introduction of monochloroacetate (MCA) at 40 mM concentration at 190 h

# INDUSTRIAL MANUFACTURE OF (S)-2-CHLOROPROPANOIC ACID

activity. The adopted strategy was, therefore, to uncouple fermentation and bio-transformation reactions and site them independently.

In order to achieve an (S)-CPA product of high enantiomeric purity, a fed-batch rather than continuous bio-transformation process was developed. Optimisation of the process required understanding the effect of, and then controlling, several variables, including temperature, pH and alkali addition, standing (R)-CPA concentration, feedstock quality, mixing regime, oxygen tension and buffering capacity of the reaction medium. The importance of these parameters is illustrated by the effect of pH on the reaction; the activity and stability of the biocatalyst are affected differently, and subtle control of pH is required (Figure 6).

By this approach, a robust, intensive process was developed which could be used on a tonne scale in routine operation. The (S)-CPA product was of high enantiomer purity (90–99% ee) which could be tailored to meet the market need by adjustment of the bio-transformation cycle time; for example, the pharmaceutical opportunity demands a higher enantiomeric purity product, but can tolerate higher cost than would be the case for material for an agrochemical outlet, thus allowing longer cycle times to be used. A downstream processing route based on solvent extraction gave a high recovery yield with no racemisation of (S)-CPA, and this element of the process could be operated continuously utilising established ZENECA technology.

The final process was established 18 months after invention of the concept, and utilised the fermentation and biocatalyst preparation capacity of ZENECA BioProducts on Teesside, UK, and the bio-transformation reaction and downstream processing assets of ZENECA Fine Chemicals at Huddersfield, UK. The overall process, shown diagrammatically in Figure 7, has a capacity for multi-thousand tonne production.

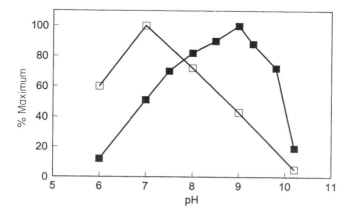

**Figure 6** Effect of pH on the activity and stability of the (S)-CPA biocatalyst. (■) Dehalogenase activity; (□) dehalogenase stability. The relative stability was determined by measuring the half-life of the biocatalyst held at the appropriate pH and 30 °C

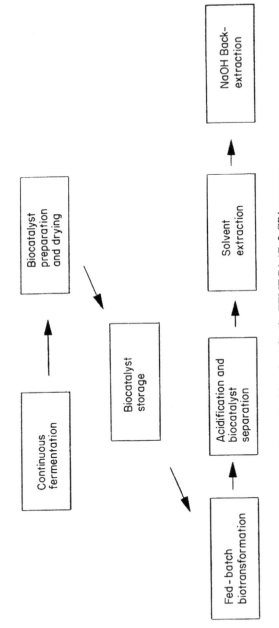

**Figure 7** Flowsheet for the ZENECA (*S*)-2-CPA process

# INDUSTRIAL MANUFACTURE OF (S)-2-CHLOROPROPANOIC ACID

## 10.4.6 DEVELOPMENTS THROUGH MOLECULAR BIOLOGY

Although the initial *Pseudomonas putida* NCIMB 12018 biocatalyst was derived by mutagenesis, it was recognised that an approach which utilised molecular biology techniques offered the potential for achieving step changes in performance, albeit over a longer time-scale, as little was known of the genetics of this particular system. Using sequence information derived from a sample of purified (R)-CPA dehalogenase, the Had-D gene encoding the enzyme was identified, cloned and expressed in a strain of *Escherichia coli*.[23] Manipulation utilising a variety of proprietary ZENECA plasmid systems and constructs led to development of strains of both *E. coli* and *P. putida* which produced elevated levels of active (R)-CPA dehalogenase in minimal growth media over many generations. By utilising molecular biology and with a good understanding of *Pseudomonas* physiology, a *P. putida* derivative and strain-selective growth environment were established which enabled the use of continuous fermentation culture for many hundreds of hours, maintaining a high level of (R)-CPA dehalogenase enzyme expression. The performance of this strain in biotransformation and downstream processing was comparable with that of the mutant, *P. putida* NCIMB 12018-23, but with much higher biocatalyst efficiency and productivity. This strain and its derivatives have been in production at ZENECA since late 1991.

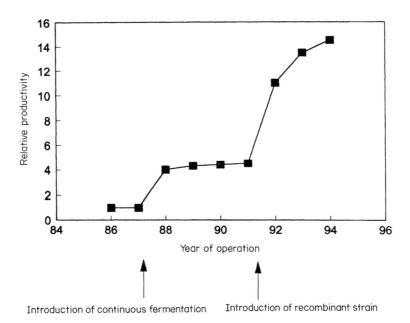

**Figure 8** Development in fermenter productivity for dehalogenase enzyme, 1986–94, indicating the impact of continuous operation of a recombinant strain

The molecular studies of the (R)-CPA dehalogenase have also served to confirm the novelty of this particular enzyme. No significant sequence homology exists between this enzyme and the (S)-CPA dehalogenase also present in *P. putida* NCIMB 12018,[24] or with other dehalogenases that have been sequenced.[25] This seems to rule out a common evolutionary heritage, an aspect of academic interest on which more light may be shed as data on the active sites of these enzymes is accumulated.

The impact of molecular biology and continuous fermentation on this process can be judged by examination of the overall fermenter productivity, measured in units of dehalogenase enzyme made per hour of fermenter occupation. The 20-fold increase (Figure 8) from an already relatively intensive process with a high natural level of dehalogenase within *P. putida* NCIMB 12018 has made a significant impact upon production economics, and has resulted in even lower cost, commodity herbicides being cost-effectively switched to single enantiomer forms.

### 10.4.7 DEHALOGENATION AS A GENERAL ROUTE TO CHIRAL MOLECULES

The (R)-CPA dehalogenase was examined for its ability to generate products from a wide range of racemic and non-chiral halogenated compounds. Substrate specificity was very limited (Table 2) and only compounds with a carbon chain length of two to four, and with an acid at C-1 and a single chlorine or bromine substitution at C-2, were effective substrates. Additional halogen substitution on C-3 was also accepted by the dehalogenase. Where relevant, the enantiomeric purity of products was ascertained and without exception, the dehalogenase demonstrated single-enantiomer specificity.

Hence the (R)-CPA dehalogenase, whilst able to provide for the production of very high enantiomeric purity products, was limited in the range of substrates that could be utilised. The system does provide, *inter alia*, for high-purity (S)-2-bromopropanoic acid and the other acids shown in Figure 9, which have several potential applications in drug development programmes.

**Table 2** Substrate range of the (S)-CPA dehalogenase from *P. putida* NCIMB 12018

| |
|---|
| 2-Chloropropanoate |
| 2-Bromopropanoate |
| Dibromoacetate |
| 2,3-Dibromopropanoate |
| Iodoacetate |
| Bromomalonate |
| Chloroacetate |
| 2,3-Dichloropropanoate |
| 2-Bromo,3-hydroxypropanoate |
| Dichloroacetate |

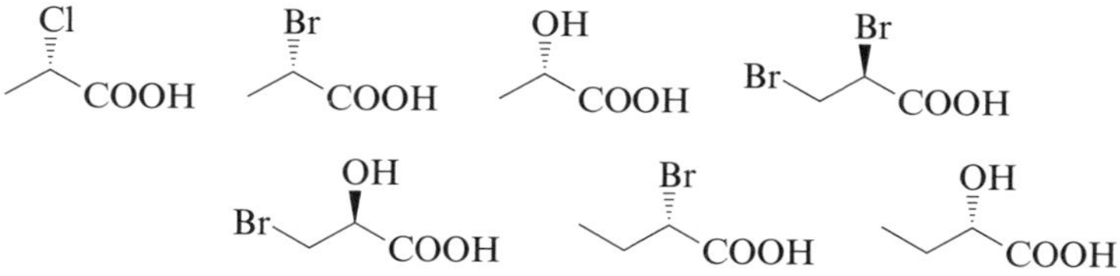

# INDUSTRIAL MANUFACTURE OF (S)-2-CHLOROPROPANOIC ACID

**Figure 9** Chiral products derived by use of the ZENECA (S)-CPA dehalogenase technology

When 2,3-dibromopropanoic acid is used as substrate, the product is (S)-2,3-dibromopropanoate, an unexpected stereochemical outcome.[26] Studies with the (S)-CPA dehalogenase from *P. putida* NCIMB 12018 established a similar substrate profile for this complementary enzyme.

Hence, although dehalogenases can clearly be used to very good effect to produce single-isomer products on a very large commercial scale, there is no evidence to show the broad specificity and utility that have become characteristic of esterase resolution systems. Consequently, current dehalogenases are unlikely to find wide utility. However, they will find application where individual chemicals in development command sufficient value to justify a specific application programme. Exceptions to this situation will arise when the chloro- and bromopropanoic derivatives provide appropriate chiral building blocks.

## 10.5 CONCLUSIONS

The ZENECA development of a bio-transformation process for (S)-CPA has clearly demonstrated that these types of enzyme reactions can be scaled up and operated economically on commercial plant on a >1000 tpa scale. In order to realise the full economic benefits that biotechnology can bring to chiral chemistry, a holistic view of all elements of the process must be taken. In developing new technologies of this sort, options in one aspect of the process, such as the genetic make-up of the microbial strain, can provide ways to address problems in another distant aspect, e.g. separation problems downstream of the bio-transformation. The essence of success in this, as in so many applications of bio-technology, is a multi-disciplinary technical and commercial project team, focused on a common goal.

That the development time-scale for the (S)-CPA process was relatively long is, perhaps, not surprising for a novel technology. Subsequent bio-transformations developed by ZENECA for chiral molecules have benefited from the (S)-CPA learning experience, and the development time-scale for new bio-transformations has been reduced dramatically. Time-scale has now become possibly the most critical aspect in meeting the needs of the pharmaceutical industry, and what used to be a major constraint to the exploitation of biotechnology in this area has now largely been overcome. The key determinant of development timescale now is generally not identification of an appropriate enzyme, but rather the speed of scale-up reflected in the ability to mobilise a cohesive team covering molecular biology, microbiology, enzyme and fermentation technology, chemistry and engineering, and direct it at the key issues while drawing on experience to make early choices from process options. The importance of integrated process development is the key lesson for future bio-transformation developments.

## 10.6 ACKNOWLEDGEMENTS

The development of a successful (S)-CPA process would not have been possible without the strong support of ZENECA business and corporate management, who recognised the long-term value of this aspect of biotechnology. Most important, however, it is the R&D staff of ZENECA BioProducts, ZENECA Fine Chemicals, the ZENECA Fine Chemicals Manufacturing Organisation and ZENECA Pharmaceuticals whose combined expertise made this bio-transformation concept a reality.

## 10.7 REFERENCES

1. Blacker, A. J., ZENECA FCMO, unpublished information.
2. Worthing, C. R. (ed.) *The Pesticide Manual*, 7th edn, British Crop Protection Council, 1983.
3. ZENECA Agrochemicals, unpublished information.
4. Smeets, J. W. H., and Kieboom, A. P. G., *Recl. Trav. Chim. Pays-Bas*, **111**, 490 (1992).
5. Rhône Poulenc, Eur. *Pat.*, EP 266258, 1988.
6. Daicel Chem. Ind., *Eur Pat.*, EP 190 770, 1986.
7. Cambou, B., and Klibanov, A. M., *Appl. Biochem. Biotechnol.*, **9**, 255 (1984).
8. Stauffer, *US Pat.* 4 613 690, 1986.
9. Hansen, H., Ladner, W., Rettenmaier, H., and Zipperer, B. (BASF), *Ger. Pat.*, DE 4 328 231, 1993.
10. Banko, G., Buchner, M., Estermann, R., and Mayerhofer, H. (Chemie-Linz), *Ger. Pat.*, DE 4 117 255, 1991.
11. Bodnar, J., Gubicza, L., and Szabo, L. P., *J. Mol. Catal.*, **61**, 353 (1990).
12. Baioru, K. K., *Jpn. Pat.*, JP 5 049 498, 1991
13. Babin, J. E., and Whiteker, G. T. (Union Carbide), *Eur. Pat.*, EP 600 020, 1993.
14. Duhamel, L., Angibaud, P., Desmurs, J. R., and Valnot, J. Y., *Synlett*, **11**, 807 (1991).
15. Taylor, S.C. (ZENECA), *US Pat.*, 4 758 518, 1988.
16. Slater, H., Lovatt, D., Weightman, A. J., Senior, E., and Bull, A. T., *J. Gen. Microbiol.*, **114**, 125 (1979).
17. Berry, E. K. M., Allison, N., Skinner, A. J., and Cooper, R. A., *J. Gen. Microbiol.*, **110**, 39 (1979).
18. Weightman, A. J., Weightman, A. L., and Slater, J. H., *J. Gen. Microbiol.*, **128**, 1755 (1982).
19. Little, M., and Williams, P. A., *Eur. J. Biochem.*, **21**, 99 (1971).

20. Motosugi, K., Esaki, N., and Soda, K., *Agric. Biol. Chem.*, **46**, 837 (1982).
21. Weightman, A., and Slater, J. H., *J. Gen. Microbiol.*, **121**, 187 (1980).
22. Weightman, A., Weightman, A.L., and Slater, J. H., *Appl. Environ. Microbiol.*, **49**, 1494 (1985).
23. Barth, P., Bolton, L., and Thomson, J. C., *J. Bacteriol.*, **174**, 2612 (1992).
24. Jones, D. H. A., Barth, P., Byrom, D. and Thomas, C. M., *J. Gen. Microbiol.*, **138**, 675 (1992).
25. Schneider, B., Muller, R., Frank, R., and Lingens, F., *J. Bacteriol.*, **173**, 1530 (1991).
26. Crosby, J., ZENECA Pharmaceuticals, unpublished information.

# 11 Development of a Full-scale Process for a Chiral Pyrrolidine

## H.-J. FEDERSEL
*Astra Production Chemicals AB, Sodertalje, Sweden*

| | | |
|---|---|---|
| 11.1 | Introduction | 225 |
| 11.2 | Background and Initiation of the Development Project | 227 |
| 11.3 | Route Identification and Evaluation | 229 |
| 11.4 | Experimental Phase: Comparative Studies and Initial Method Development as a Basis for Route Selection | 233 |
| | 11.4.1 Route A: Proline → 1-Ethylproline Ethyl Ester → 1-Ethylprolinamide (5) | 233 |
| | 11.4.2 Route B: Proline → Proline Ethyl Ester → Prolinamide → 1-Ethylprolinamide (5) | 234 |
| | 11.4.3 Route C: Proline → Proline Ethyl Ester → 1-Ethylproline Ethyl Ester → 1-Ethylprolinamide (5) | 235 |
| | 11.4.4 Route D: Proline → 1-Ethylproline → 1-Ethylproline Ethyl Ester → 1-Ethylprolinamide (5) | 235 |
| | 11.4.5 Final Step in Routes A–D: 1-Ethylprolinamide (5) → (S)-2-Aminomethyl-1-ethylpyrrolidine (2) | 236 |
| | 11.4.6 Rational Route Selection | 236 |
| 11.5 | Development and Scale-up Phase: Streamlining and Optimising the Process | 237 |
| 11.6 | The Commercial Manufacturing Process | 240 |
| 11.7 | Summary and Conclusions | 240 |
| 11.8 | Acknowledgements | 242 |
| 11.9 | References | 242 |

## 11.1 INTRODUCTION

The first observation that enantiomers of chiral molecules can exhibit biologically different properties dates back well over a century with the report in 1886 by Piutti,[1] who found (*R*)-asparagine to be sweet and the (*S*)-form to display an insipid character, but it is only in the last 30 years that this issue has become of major industrial significance, especially in the pharmaceutical area.[2] The most important reason for this is probably the gradual accumulation of data indicating that molecular stereoisomerism has a profound impact on the nature of the

*Chirality in Industry II.* Edited by A. N. Collins, G. N. Sheldrake and J. Crosby
© 1997 John Wiley & Sons Ltd

response in living systems; the development of stereospecific analytical techniques such as chiral chromatography, which not only allow a very precise determination of absolute stereochemical purity in isolated compounds but also offer the ability to monitor the fate of stereoisomeric molecular species *in vivo*, is another strongly contributing factor. An illustrative example is the different therapeutic profiles of dextropropoxyphene; the (2$R$,3$S$)-isomer is used as an analgesic, and its enantiomer levopropoxyphene (2$S$,3$R$) exhibits antitussive activity.[3]

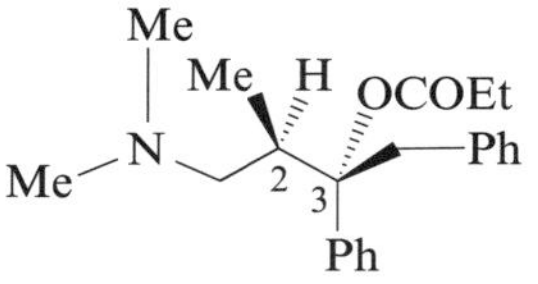

Dextropropoxyphene (2$R$,3$S$) 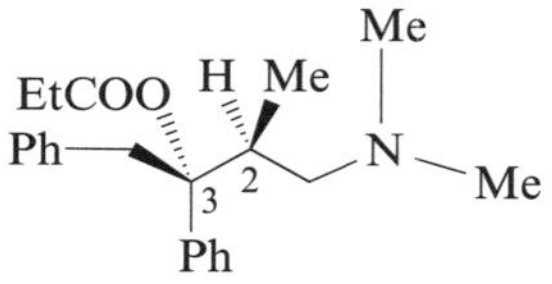          Levopropoxyphene (2$S$,3$R$)

Besides differing in a 'positive' manner in their pharmacological profiles, as in the previous case, often one of the enantiomers shows a toxic side-effect of some kind, or a lower potency, making its presence undesirable or, at best, without any appreciable value. This situation can be illustrated by dopa (3,4-dihydroxyphenylalanine) and penicillamine; the ($S$)- and ($R$)-enantiomers of the former display an anti-Parkinson effect and granulocytopenia (a severe eye disease), respectively,[4] whereas those of the latter are antiarthritic ($S$) and mutagenic ($R$).[5]

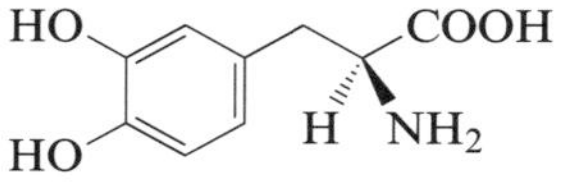

($S$): anti-Parkinson 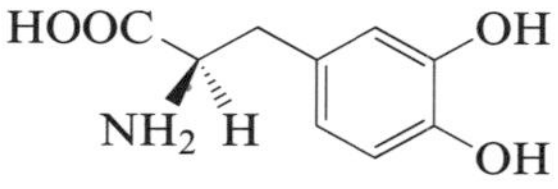          ($R$): causes granulocytopenia

Dopa

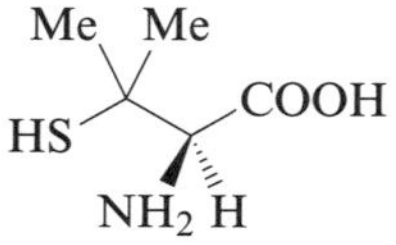

($S$): antiarthritic 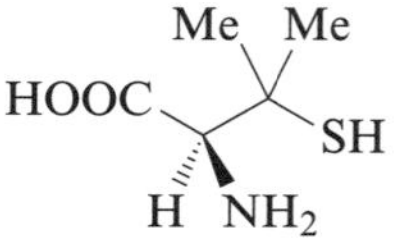          ($R$): mutagenic

Penicillamine

Given this background, the current trend to focus on a single stereoisomer, not only in the development of new drugs[6] but also to an increasing extent in the agrochemicals area,[7] is inevitable. The need to supply kilogram quantities of these materials to conduct extended toxicological and clinical (or field) trials throughout the entire R&D programme, has put a strong emphasis on the scale-up of single-enantiomer materials.[8] It should be pointed out, however, that racemates are not ruled out *a priori*, since instances of beneficial synergistic medicinal effects are known.[9] This is a view taken by representatives of both 'sides,' industry[10] and the various regulatory authorities[11] (see Chapter 2 for a more detailed discussion). The unbiased assessment of the three key parameters *quality*, *safety*, and *efficacy* should be adopted to enable a scientifically sound decision to be taken in favour of the use of either a single enantiomer or a racemate.[12]

## 11.2 BACKGROUND AND INITIATION OF THE DEVELOPMENT PROJECT

In the late 1960s, it was found that the racemic compound sulpiride exhibited pronounced antipsychotic activity.[13] Further studies revealed that a whole family of compounds belonging to the so-called benzamide group displayed an interesting pharmacological profile with regard to activities in the central nervous system (CNS). Screening and evaluation eventually led Astra to choose a number of

|  | $R^1$ | $R^2$ | $R^3$ |
|---|---|---|---|
| Remoxipride: | Me | Br | H |
| Raclopride: | H | Cl | Cl |

potent and particularly promising molecules, such as remoxipride and raclopride, as candidate drugs. The prime target was to develop these dopamine $D_2$-receptor antagonists for the treatment of schizophrenia.[14]

This class of products is composed of a variably polyfunctionalised benzoic acid moiety attached via an amide linkage to a common structural element comprising a chiral $N^1$-substituted pyrrolidinamine unit. A fairly broad range of different substituents at the pyrrolidine nitrogen has been investigated, but it seems that an ethyl group confers a particularly beneficial combination of properties with regard to key pharmacological parameters, such as efficient receptor binding.

The structural subunits **1** and **2** are easily identifiable as the penultimate building blocks to be combined in a condensation reaction to afford the desired amide products. This convergent synthetic strategy offers the advantage of allowing both moieties, the benzoic acid **1** and the amine **2**, to be assembled in separate reaction sequences.

The aromatic portion can be generated in a fairly straightforward manner using various 'classical' substitution reactions on readily available, inexpensive starting materials, such as 2,6-dimethoxybenzoic acid.[15] However, synthesis of compound **2** is considerably more complex owing to the need to obtain it enantiomerically pure, and to the many routes which can be envisaged for the synthesis. Initially, the pilot-scale demand for this pivotal intermediate was satisfied by a 'classical' resolution of the racemic material (**3**), available on the market in bulk quantities and accessible in a number of ways: (i) from N-ethyl-2-pyrrolidinone via catalytic hydrogenation of the 2-nitromethylene derivative,[16] (ii) from 2-substituted tetrahydrofurans, e.g. the chloromethyl compound, via ring opening and reaction with an appropriate alkyl amine,[17] and (iii) via ring contraction of 3-

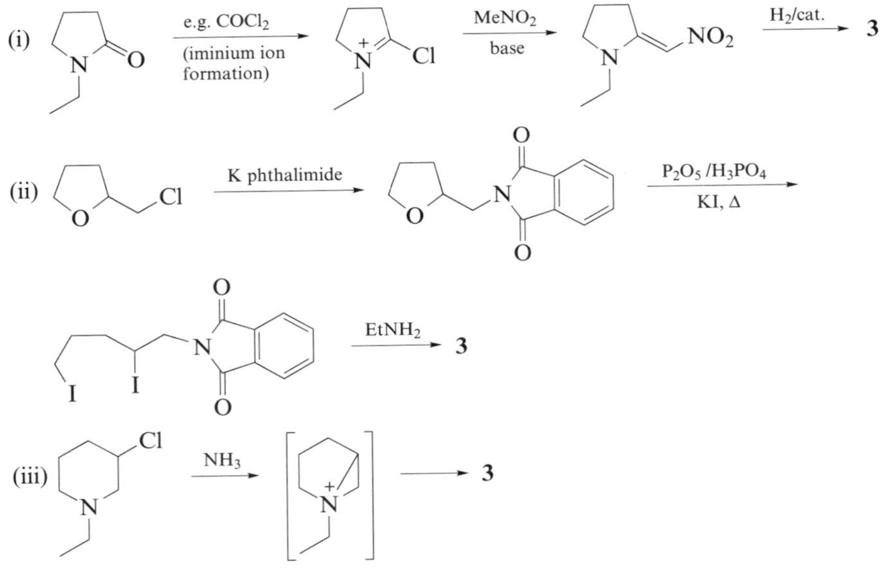

**Scheme 1** *Non*-stereospecific syntheses generating racemic pyrrolidinamine (**3**)

halopiperidines upon treatment with ammonia (or an appropriate amine, e.g. benzylamine)[18] (Scheme 1).

On treating racemic amine **3** with expensive, unnatural D-(−)-tartaric acid, a diastereoisomeric 1:2 tartrate salt precipitated which, after neutralisation with base, liberated the desired (*S*)-enantiomer of **2** in ca 90% ee and an overall yield of ca 25% (i.e. 50% of theory). An economic improvement was achieved by par-

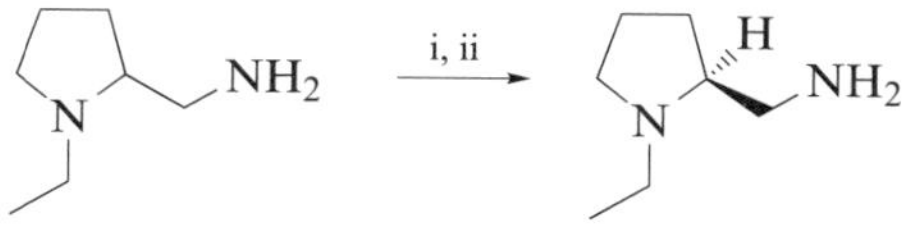

# FULL-SCALE PROCESS FOR A CHIRAL PYRROLIDINE

25% yield (from racemate)
>90% ee

**Scheme 2** Resolution of racemic pyrrolidinamine (**3**) to its (*S*)-enantiomer (**2**) using D- and L-tartaric acid. (i) (a) L-(+)-tartaric acid (*ca* 1.7 equiv.); (b) OH⁻ [liberation of enriched (*S*)-amine]. (ii) (a) D-(−)-tartaric acid (*ca* 1.8 equiv.); (b) recrystallisation (EtOH/H$_2$O); (c) OH⁻ (liberation of free base)

tially removing the (*R*)- isomer by crystallisation with natural L-(+)-tartaric acid, leaving an (*S*)-enriched mother liquor of the free base (after basification), which was then treated with D-tartaric acid, (Scheme 2). The yield and stereochemical purity were not affected, but the cost situation was improved, since the consumption of expensive D-tartaric acid was reduced considerably.

Evaluation of the potential for the procedure outlined in Scheme 2 revealed a number of weaknesses, besides the possibility of being involved in an interference situation with existing patents.[19] The main identified drawbacks were thus:

- the lack of an obvious method for racemising the unwanted (*R*)-enantiomer remaining in the mother liquor, which results in a 50% loss of expensive starting material;
- low overall yield of the required (*S*)-amine which, despite efforts, could not be improved significantly beyond 50% of theory;
- large usage of expensive D-tartaric acid;
- the need to operate an in-process recovery loop to recycle the D-tartaric acid;
- stability problems in the crystallisation leading to an uncontrolled, scale- and time-dependent precipitation of the tartrate salt of the undesired enantiomer, which was difficult to monitor. In the interval between crystallisation and completion of the product isolation by centrifugation (time consuming on a large scale), deterioration of the optical purity occurred; all efforts to improve the situation, such as changing the solvent composition or modifying the temperature (cooling profile; end-point value), were unsuccessful.

Given this background, the decision was taken to initiate a programme to identify alternative routes to **2**, the outcome of which is now presented[20].

## 11.3 ROUTE IDENTIFICATION AND EVALUATION

A particularly attractive option, which entirely eliminates the need to conduct a resolution, would be to use a building block with the required configuration from the ever-growing pool of chiral compounds, which could then be transformed into the target molecule with careful control of the stereochemistry. In this case, natural (S)-proline (4) is the obvious starting material. While its availability on a large scale in high stereochemical purity [bulk quantities of (4) at > 99% *ee*] and reasonable price is of paramount importance, the close structural relationship to the final product is also advantageous in order to minimise the number of large-scale synthetic operations.

(S)-Proline (4)

(S)-Proline displays a number of structurally important elements relevant in the present case, notably the pyrrolidine nucleus and a $C_1$-unit (carboxy group) attached to the stereogenic centre (C-2) bearing the correct configuration with respect to the target molecule. In fact, any sequence which incorporates a formal transformation of the carboxy function into an aminomethyl substituent, and the coupling of an ethyl group on to the ring nitrogen, while keeping the stereochemistry intact, would have satisfied the primary objectives for a new route. The real choice, however, between the different options was to a large extent directed by secondary, process-oriented parameters such as yields, ease of operation, product quality and reaction conditions. Patentability would also increase the value of a given methodology in that an exclusive entry to a specific product is guaranteed for some time. This is especially so for entities such as the pyrrolidine 2, which is useful not only for the purposes described here, but equally in the synthesis of a broad range of analogues.

A literature survey revealed that several sequences where the stereochemistry is retained have indeed been patented, two starting from (S)-proline[21,22] and others starting with either (S)-prolinol or (S)-glutamic acid.[21] These are compared in Scheme 3.

Their common feature is a linear sequence which, in the case of proline, requires four separate transformations, whilst the other approaches require up to eight steps. More specifically, the proline routes comprise either (i) an esterification, amide formation, pyrrolidine N-acetylation and a final reduction of both carbonyl functionalities[21] or (ii) an initial N-acetylation followed by a reduction to N-ethylprolinol, chlorination of the alcohol and finally an ammonia substitution.[22] The level of detail given in these patented methods is rather superficial with respect to crucial process information. In the former sequence, there are no yield data, stereochemical purities being reported only as optical rotations, and the latter quotes a 35% overall yield of a crude material (2) used 'as is' without specifying any purity.

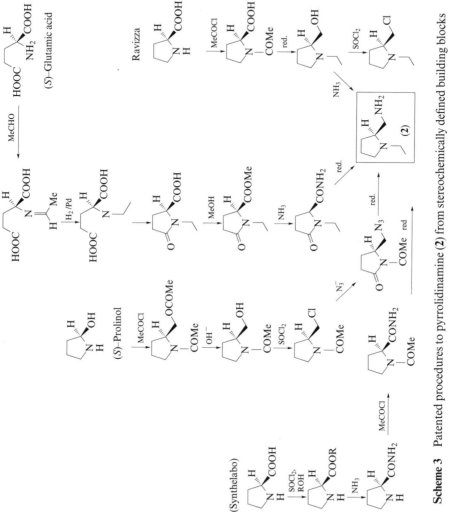

**Scheme 3** Patented procedures to pyrrolidinamine (**2**) from stereochemically defined building blocks

Based on what was known in this area at the time, the four routes shown in Scheme 4 were designed as potential pathways for transforming (*S*)-proline in a

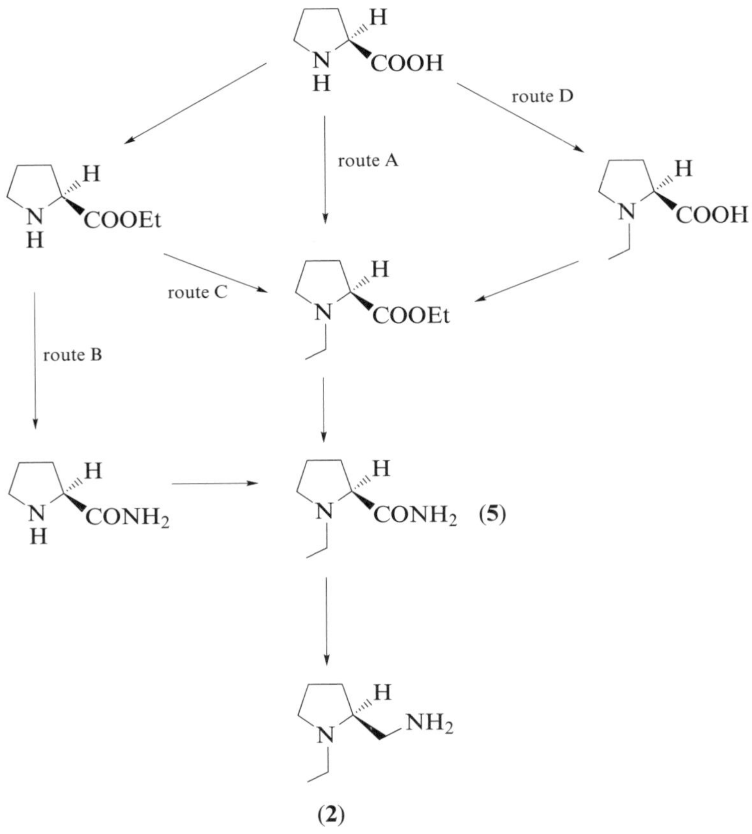

**Scheme 4** Short and stereoconservative Astra patented[23] syntheses of pyrrolidinamine (**2**) starting from (*S*)-proline

concise and straightforward manner into the desired pyrrolidinamine (**2**).[23]

A quick examination of the different routes reveals that each sequence is composed of a set of fairly standard reaction types, an esterification (or *O*-alkylation), an *N*-alkylation, an ester-to-amide transformation and an amide reduction. Although differing in the order of the required steps, all four routes involve a common penultimate intermediate, (*S*)-1-ethylprolinamide (**5**), which clearly becomes a key compound. As will be shown below, it is the fine-tuning of the individual steps which guarantees both a high overall yield and more or less complete control over stereochemistry, eventually ensuring that the desired product can be produced economically and in good quality.

## 11.4 EXPERIMENTAL PHASE: COMPARATIVE STUDIES AND INITIAL METHOD DEVELOPMENT AS A BASIS FOR ROUTE SELECTION

The first undertaking was to assess the strengths and weaknesses of each sequence outlined in Scheme 4, attempting to identify quickly the most promising in order to direct all efforts towards its further development. The synthetic approaches are detailed individually up to amide 5, followed by a separate discussion of the common final reductive step to target amine 2. The results obtained will be presented in a fairly concise format.[24] Factors which received the highest priority during this screening phase were the overall isolated chemical yield, the technical feasibility of a successful scale-up and the stereochemical quality of the product. Of these, the last was given particular attention because the optical purity had to be retained at a very high level to avoid the need for any additional resolution. Monitoring of optical quality was, if possible, performed at each intermediate stage using mainly chiral gas chromatography (on a Chirasil-Val column) and, occasionally, NMR spectroscopy.

### 11.4.1 ROUTE A: PROLINE → 1-ETHYLPROLINE ETHYL ESTER → 1-ETHYLPROLINAMIDE (5)

Here the first step is a novel and very attractive $O,N$-dialkylation which converts the starting proline into the $N$-alkylated ester in a one-pot procedure. A number of different conditions were investigated systematically, such as (i) type of solvent; (ii) reaction temperature; (iii) bromo- vs iodoethane; (v) time of addition of halide; (v) stoichiometric ratio of ethyl halide/proline; (vi) presence and type of base; and (vii) overall reaction time. The best yields obtained were in the range of 65–70%, and afforded a crude material with a purity of about 90% (by GC). The major by-product (6–8%) was the diketopiperazine 6, formed in a cyclodimerisation of the intermediate proline ester, a common reaction of α-amino acids (Scheme 5).

In common with the sluggishness frequently encountered when attempting to conduct ester-to-amide conversions, this transformation required rather forcing conditions (prolonged reaction times at elevated temperatures), which meant 1–2 days at about 50 °C when running in a methanolic ammonia (fortunately without causing any appreciable racemisation). Possibilities for improving this process have been investigated in a comparative study[25] where different catalysts were evaluated according to their abilities to effect an acceleration in the formation of amide 5 from the corresponding ethyl ester. Cyanide ion (added in *ca* 10 mol% as sodium cyanide) was significantly better than alternative, well known catalysts such as imidazole, DMAP, 2-hydroxypyridine and potassium iodide. Applying this cyanide method to the current reaction (45 °C/40 h), the crude target compound was obtained in 96% yield at a purity of 98%.

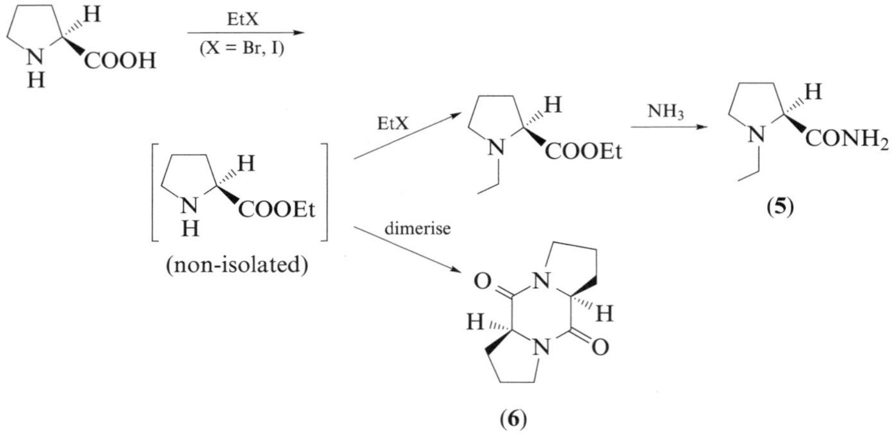

**Scheme 5** Conversion of (S)-proline to 1-ethylprolinamide according to route A and formation of piperazine by-product (**6**)

The immediate appeal of route A is its brevity (overall only three steps vs four in routes B–D; see Scheme 4). The main drawbacks are the requirement to handle a high-boiling point, water-miscible solvent such as DMF or DMSO, and the fact that the isolated yields in the first step are barely acceptable. Also, the use of cyanide in the amide-forming step would give rise to safety and hazard concerns and call for special handling precautions.

### 11.4.2 ROUTE B: PROLINE → PROLINE ETHYL ESTER → PROLINAMIDE →1-ETHYLPROLINAMIDE (5)

A standard one-pot procedure was applied to convert the starting material into its ethyl ester (as the hydrochloride salt) via the *in situ* formation of the acid chloride, a reaction which proceeded smoothly and in almost quantitative yield.[26] The next transformation was conducted on the free base of the ester using methanolic ammonia at an elevated temperature and in a sealed reaction vessel, which rapidly caused a transesterification to the methyl ester, analogous to the comparable step in route A,[25] before giving the amide in an excellent yield of ca 90%.[27] Finally, the *N*-alkylation was performed in a heterogeneous system at 30 °C over 24 h with ethanol as solvent and a slight excess of solid potassium carbonate as the main basic component in conjunction with catalytic amounts of triethylamine. Using bromoethane as the alkyl source resulted in yields of about 65%. The sequence is outlined in Scheme 6.

Overall, this sequence was judged, from both the chemistry and the process operability viewpoints, to possess good potential for scale-up. The reactions seemed to be robust and yields were high in spite of the early stage of development.

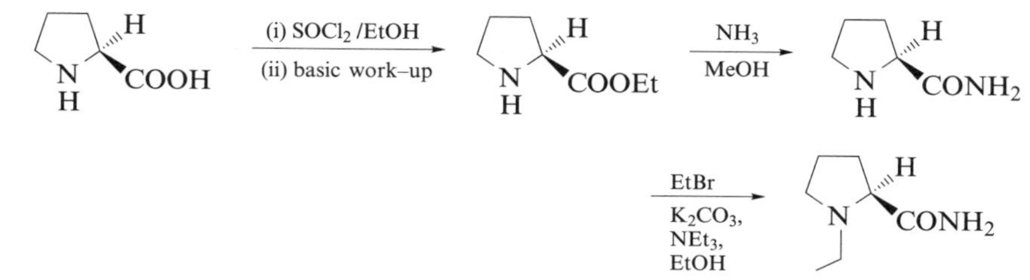

# FULL-SCALE PROCESS FOR A CHIRAL PYRROLIDINE

**Scheme 6** Conversion of proline to the amide **5** according to route B

### 11.4.3. ROUTE C: PROLINE → PROLINE ETHYL ESTER → 1-ETHYLPROLINE ETHYL ESTER → 1-ETHYLPROLINAMIDE (5)

This route is closely related to route A, with the exception that the first two steps are carried out separately and not in a one-pot fashion. This stepwise sequence had a yield benefit since each reaction could be performed with isolated product outputs of well above 90%. With respect to the final amide formation, the same limitations prevail as were outlined previously, meaning that a cyanide-catalysed process would probably be required. For the entire reaction sequence see Scheme 7.

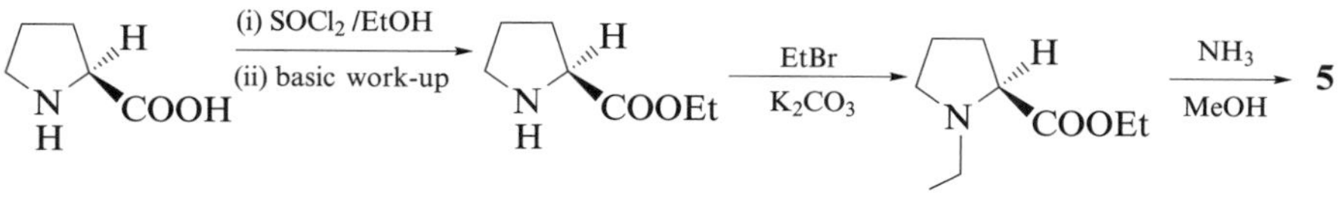

**Scheme 7** Formation of 1-ethylprolinamide (**5**) from proline using route C.

### 11.4.4 ROUTE D: PROLINE → 1-ETHYLPROLINE → 1-ETHYL-PROLINE ETHYL ESTER → 1-ETHYLPROLINAMIDE (5)

This inverted version of route C, with respect to the first two steps, required a selective $N$-alkylation, which could be effected using acetaldehyde under reductive conditions (e.g. NaBH$_4$, H$_2$/Pd); however, the yields could barely be increased beyond 50%.[28] Also, attempts to conduct this transformation in a direct fashion with bromoethane did not offer any appreciable yield advantages. The subsequent esterification and amide formation were carried out as described previously, and Scheme 8 displays the synthesis step by step.

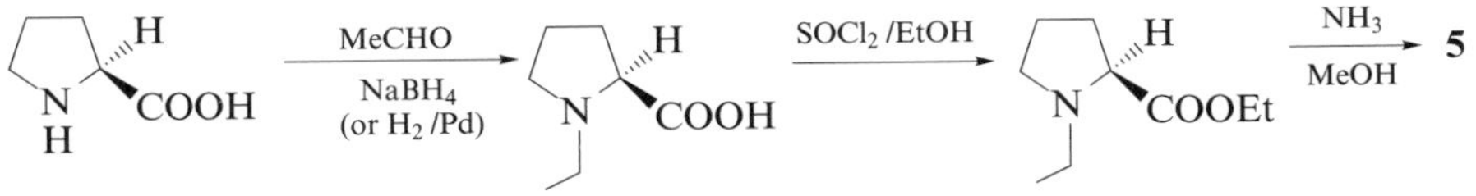

**Scheme 8** Route D as a means to generate 1-ethylprolinamide (**5**)

### 11.4.5 FINAL STEP IN ROUTES A–D: 1-ETHYLPROLINAMIDE (5) → (S)-2-AMINOMETHYL-1-ETHYLPYRROLIDINE (2)

Several methods exist for the reduction of an amide to an amine, and a number of them were evaluated. The more common methods, such as using $LiAlH_4$ and $NaBH_4$, readily gave the desired product in yields varying from 60 to 70%. Using sodium bis(2-methoxyethoxy)aluminium dihydride (available as a 70% solution in toluene under the commercial names Red-Al or Vitride), the yield could be increased well above 80%. Apart from the significantly better yields, two further advantages were a better reproducibility in combination with a slightly higher optical quality of the isolated amine (**2**), and the ease of use of the latter as a solution on the large scale compared with the other reagents which would be handled as solids. Scheme 9 summarises the chemistry.

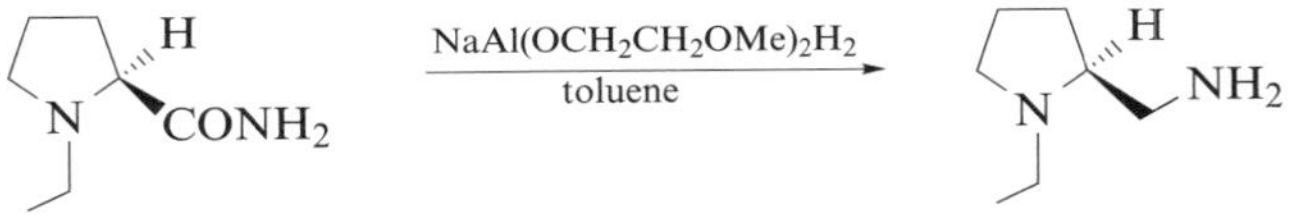

**Scheme 9** Reduction of amide **5** to target amine **2**

### 11.4.6 RATIONAL ROUTE SELECTION

The results obtained from the approaches described above were subjected to detailed side-by-side comparison, in addition to scrutiny against the overall project goals. Certain key parameters, as mentioned earlier, were rated as being more critical than others and had more impact on the final choice. Our ultimate choice of route B can be rationalised as follows (most important factors first):

- a high probability that a substantial increase in the chemical yield would be obtained, particularly in comparison with routes A and D;
- the ester, being the stereochemically most sensitive intermediate, can be transformed into the amide under milder conditions before attaching the *N*-substituent, which substantially decreases the risk of an uncontrolled loss of optical purity — a feature unique to route B;

- no requirement to use solvents such as DMF or DMSO (cf. route A), nor any need for a cyanide catalyst in the amide formation (cf. routes A, C and D);
- a technically straightforward process not requiring any extreme reaction conditions, performable with standard equipment using common unit operations;
- no obvious scaleability problems are envisaged;
- bulk availability of all components (starting materials, reagents, solvents);
- with proper precautions, the handling of the required chemicals and process streams is not liable to cause any severe environmental impact, nor would any major safety hazards (e.g. extreme exothermicities) be expected for the reactions and work-up procedures involved;
- a highly competitive production cost.

## 11.5 DEVELOPMENT AND SCALE-UP PHASE: STREAMLINING AND OPTIMISING THE PROCESS

Efforts were now focused on systematic process development involving both laboratory-scale experiments and scaled-up trial runs in the pilot plant and the production facility. A first step towards full-scale manufacture was taken fairly early using a method based on the preliminary laboratory procedure from the initial selection phase, i.e. route B. The primary purpose of this experimental manufacture was to gather process data regarding technical and operational feasibility, yields, product quality, impurity profile and cycle times, and to gain some experience for the process operators. The outcomes of this first pilot batch, and two subsequent repeats, all conducted more or less consecutively over a limited period (only 4 months), are summarised in Table 1.

A number of conclusions were drawn from the results: (i) the two larger runs proceeded in a fairly consistent manner, both with regard to overall yield and the performance in each single step, (ii) a severe yield penalty was incurred in the larger scale esterifications compared with the smaller reaction, whereas both the $N$-alkylation and the reduction worked significantly better (batch Nos II and III

**Table 1.** Compilation of key data from first pilot batches performed according to route B

| Batch No. | Relative scale/ proline amount kg (mol) | Stepwise yield, corrected for assay (%) | | | | Overall yield (%) | Stereochemical purity [% $S$ (GC)] |
|---|---|---|---|---|---|---|---|
| | | Esterification[a] | Amide formation | $N$-Alkylation | Reduction | | |
| I | 8.6(74.8) | 90.2 | ~97 | 61.1 | 79.5 | 42.5 | 99.1 |
| II | 100(868.6) | 70.6 | 97.5 | 88.5 | 91.8 | 55.9 | 96.3 |
| III | 100(868.6) | 76.1 | 95.2 | 83.0 | 88.0 | 52.9 | 97.8 |

[a] Synthesis of the methyl ester.

vs I); and (iii) there was a noticeable decrease in stereochemical purity on going to the larger scale, albeit to a level which still matched the quality requirement of ≥95% (S)-isomer. These findings prompted a careful examination of the manufacturing process, focusing on the areas where shortcomings were observed compared with the laboratory method.

The drastic decrease in yield in the ester step was of particular concern, since it was felt to be technically rather simple with an expected product conversion close to quantitative. On examining the reaction and work-up to the free ester base, it became evident that the problem was related to the stability of the product; a neat sample at ambient temperature (ca 25 °C) showed a ca 4.6% loss over 24 h, with formation of the piperazinedione **6** through an ester dimerisation.

Fortunately, the situation could be greatly improved by merely switching to the ethyl ester, which is almost twice as stable. The importance of this effect was demonstrated in a dramatic fashion when performing a subsequent first full production size batch in the 4000 l equipment. Thus, when running the ester step with 300 kg (2606 mol) of proline according to the pilot method used earlier and only replacing methanol by ethanol, the gain in yield was in the range 20–25%. The remainder of the sequence was conducted with only very minor modifications (e.g. relative amounts of solvents) with individual steps showing an acceptable to high degree of reproducibility (step 2, 95%; step 3, 75%; step 4, 90%) considering the fairly early stage of process development, resulting in an overall yield of ca 62% and an excellent stereochemical purity of ca 98% (S)-isomer.

Once the successful scale-up had been demonstrated, full attention was given to a stepwise optimisation with the goal of obtaining a streamlined commercial process. The immediate targets in this respect were to minimise the input of chemicals (building blocks, reagents and solvents) in order both to reduce environmental impact and to ensure a high process capacity.

In the step 1 esterification all of the conditions were challenged, eventually leading to the selection of a number of parameters which were examined more closely using a factorial design.[29] Substantial changes were made to the literature procedure: the excess of $SOCl_2$ was reduced (from 1.77 to 1.05 equiv.), as was the temperature (from 60 to 40 °C), time (from 4 to 1 h) and solvent usage (from 6 to 2 kg ethanol per kg proline) without a yield penalty. Interestingly, it was possible to trace the initiation of the formation of a by-product contaminating the target amine (**2**) [its *N-methyl* analogue (**7**)] to this first step. In the latter part of the synthesis it is transformed into the undesired methyl derivative of the final drug substance. The mechanism can be explained by the generation of diethyl sulphite from ethanol and thionyl chloride, a side reaction which occurs even if $SOCl_2$ is added on an equimolar basis relative to proline. This high-boiling impurity (lit.[30] b.p. 157 °C/768 mmHg) undergoes transesterification with the abundantly available MeOH to dimethyl sulphite, a potent methylating agent which results in the formation of 1-methylprolinamide (**8**), as depicted in Scheme 10.

Under the reaction conditions, the ester hydrochloride is formed as the primary product and is subsequently neutralised to the free base, preferably using

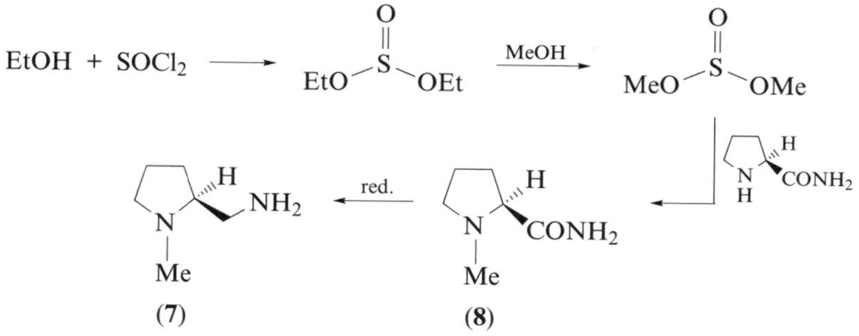

# FULL-SCALE PROCESS FOR A CHIRAL PYRROLIDINE

**Scheme 10** Formation of *N*-methylpyrrolidinamine (**7**) due to a by-product, diethyl sulphite, from the ester step

aqueous ammonia. Since the free base of the ester is somewhat volatile (b.p. 82–83 °C/22 mmHg), a low-boiling extraction solvent, dichloromethane, had to be used to prevent losses during concentration.

The amide-forming step is the stereochemically most sensitive part of the sequence, requiring a method which minimises racemisation. It was found that reaction temperature is important, 50 °C being established as the upper limit, at which point a 24 h reaction time was required. Although a lower temperature reduces the risk of racemisation, it would extend the reaction time unacceptably. Attempts to replace methanol with ethanol as solvent, and eliminate the trans-esterification, resulted in a drastically decreased reaction rate (50% conversion after 39 h).

The first scale-up batches in the pilot plant revealed the presence of a new impurity at the amide stage, which had not been observed in the laboratory. The cause was identified as dichloromethane, which remains in small amounts in the oily proline ester from the previous stage. This solvent residue, cf. the diethyl sulphite, acts as an alkylating agent under the rather forcing conditions of step 2 forming the novel pyrroloimidazolone **9**. The proposed mechanism for its formation is shown in Scheme 11.[31] This side reaction was easy to avoid by ensuring the complete removal of dichloromethane in a co-evaporation with ethanol during work-up.

Also, in the third step, an experimental plan based on factorial design was deployed in order to monitor the influence of a number of parameters: amounts of potassium carbonate and bromoethane, in addition to the charging time for the latter, the need to add triethylamine and sodium iodide individually and the reaction temperature and time. The results from this study were used to set values for parameters such as a reaction temperature (30 °C), reaction time (20 h) and bromoethane charging time 0.5 h. An additional simplex optimisation was performed in order to determine the relative amounts of potassium carbonate and bromoethane, which turned out to be in the range 2.5–3.0 equiv. at a reaction concentration of 5.5 kg EtOH per kg prolinamide. If, however, the concentration was increased to 2.4 kg kg$^{-1}$, an increase in yield of *ca* 5% was obtained, whilst at

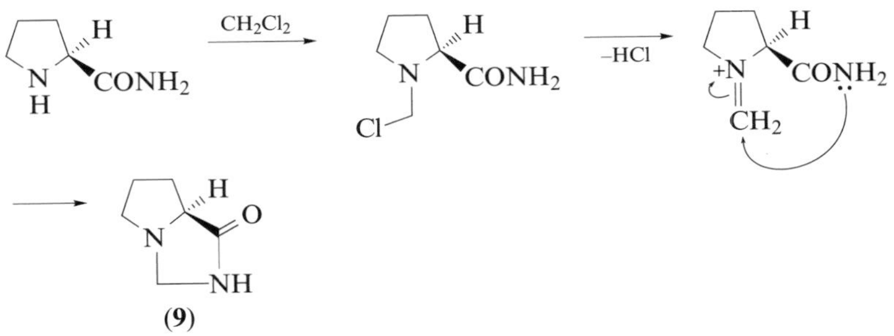

**Scheme 11** Proposed generation of pyrrolo[1,2-c]imidazolone (**9**) from prolinamide and dichloromethane.

the same time allowing the excess of potassium carbonate and bromoethane to be substantially reduced (to 1.8 and 2.2 equiv., respectively).

More detailed investigation of the reduction step revealed that both the temperature and time for the reaction were important with regard to the stereochemical quality of the product. Thus, the addition of the reducing agent had to be performed at less than 20 °C, whereafter the reaction mixture was cautiously heated to 40 °C. This procedure guarantees as short a reaction time as possible (in the range 1–2 h). Finally, during work-up, it is essential to keep the amount of water to a minimum owing to the extremely high solubility of the product amine in aqueous systems.

## 11.6 THE COMMERCIAL MANUFACTURING PROCESS

The implementation of the improvements and modifications outlined above in successive full-scale trial batches resulted in a procedure which was highly optimised and capable of delivering the required quantity of the target amine **2** with the required quality. Scheme 12 summarises the commercial production in a sequential flow chart indicating unit operations and key process parameters.

## 11.7 SUMMARY AND CONCLUSIONS

Driven by the need to establish an economically viable process for the synthesis of a chiral aminomethylpyrrolidine compound in high enantiomeric purity, four closely related and patentable routes were identified, all starting from (*S*)-proline. Careful comparative studies and evaluations led to the selection of the process consisting of the four sequential steps esterification, amide formation, *N*-alkylation and reduction as offering the best potential for further development.

Chemical and technical optimisations, coupled with scale-up trials, furnished

# FULL-SCALE PROCESS FOR A CHIRAL PYRROLIDINE

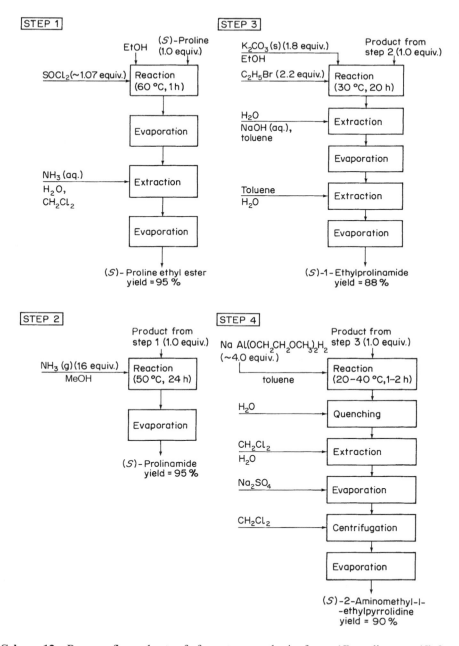

**Scheme 12** Process flow chart of four-step synthesis from (S)-proline to (S)-2-aminomethyl-1-ethylpyrrolidine (**2**)

the final commercial production method for a 4000 l reactor, in which 300 kg (*ca* 2600 mol) of (S)-proline can be converted into *ca* 240 kg of the target compound, representing an overall yield of 71%.

It is hoped that this account of an authentic development case has emphasised some of the more important elements of the development of a synthesis from its initially rather rudimentary state in the early laboratory phase to a fully controlled and optimised commercial manufacturing procedure. The chiral nature of the product does not by itself infer any unique feature to the methodology, but may add constraints, e.g. avoiding racemisation. It therefore underscores the importance of following a rational approach and exerting the highest possible degree of control over the different stages and operations.

## 11.8 ACKNOWLEDGEMENTS

The work reported in this chapter reflects the skill, commitment and endeavour of a combined task force and project team whose foremost members were my close collaborators Erik Könberg and Lars Lilljequist of the Chemical Process Development Laboratory and Dr Thomas Högberg, Dr Sten Rämsby and Peter Ström, all at that time belonging to the Medicinal Chemistry Department of the CNS research unit within Astra.

## 11.9 REFERENCES

1. Piutti, A., *C. R. Acad Sci.*, **103**, 134 (1886).
2. Selected reading: (a) Ariëns, E. J., *Eur. J. Clin. Pharmacol.*, **26**, 663 (1984); (b) Jamali, F., Mehvar, R., and Pasutto, F. M., *J. Pharm. Sci.*, **78**, 695 (1989); (c) Crossley, R., *Tetrahedron*, **48**, 8155 (1992); (d) Stinson, S. C., *Chem. Eng. News*, **70**, (39), 46 (1992); (e) Testa, B., Carrupt, P.-A., and Gal, J., *Chirality*, **5**, 105 (1993); (f) Waldeck, B., *Chirality*, **5**, 350 (1993); (g) Batra, S., Seth, M., and Bhaduri, A. P., *Prog. Drug Res.*, **41**, 191 (1993).
3. Drayer, D. E., *Clin. Pharm. Ther.*, **40**, 125 (1986).
4. Cotzias, G. C., Papavasiliou, P. S., and Gellene, R., *N. Engl. J. Med.*, **280**(7), 337 (1969).
5. Glatt, H., and Oesch, F., *Biochem. Pharmacol.*, **34**, 3725 (1985).
6. Millership, J. S., and Fitzpatrick, A., *Chirality*, **5**, 573 (1993).
7. Ramos Tombo, G. M., and Bellus, D., *Angew. Chem., Int. Ed. Engl.*, **30**, 1193(1991).
8. (a) Sheldon, R., *Chem. Ind. (London)*, 212 (1990); (b) Crosby, J., *Tetrahedron*, **47**, 4789 (1991); (c) Collins, A. N., Sheldrake, G. N., and Crosby, J. (Eds), *Chirality in Industry. The Commercial Manufacture and Applications of Optically Active Compounds*, Wiley, Chichester (1992); (d) Nugent, W. A., RajanBabu, T. V., and Burk, M. J., *Science*, **259**, 479 (1993); (e) Sheldon, R., *Chirotechnology. Industrial Synthesis of Optically Active Compounds*, Marcel Dekker, New York (1993); (f) Federsel, H.-J., *Chem. Unserer Zeit*, **27**(2), 78 (1993); (g) Federsel, H.-J., *CHEMTECH*, **23**(12), 24 (1993); (h) Kotha, S., *Tetrahedron*, **50**, 3639 (1994); (i) Federsel, H.-J., *Endeavour*, **18**(4), 164 (1994).
9. Testa, B., and Trager, W. F., *Chirality*, **2**, 129 (1990).
10. Cayen, M. N., *Chirality*, **3**, 94 (1991).
11. (a) Swedish Medical Products Agency, *Reg. Affairs J.*, **2**(1), 2 (1991); (b) Madden, S.,

Reg. Affairs J., **2**(1), 20 (1991). (c) Allen, M. E., Reg. Affairs J., **2**(2), 93 (1991); (d) Shindo, H., and Caldwell, J., Chirality, **3**, 91 (1991); (e) Hutt, A. J., Chirality, **3**, 161 (1991); (f) Rauws, A. G., and Groen, K., Chirality, **6**, 72 (1994), and references cited therein.
12. For a summary of recommendations, see Gross, M., Cartwright, A., Campbell, B., Bolton, R., Holmes, K., Kirkland, K., Salmonson, T., and Robert, J.-L., Drug Inf. J., **27**, 453 (1993).
13. Borenstein, P., Champion, C., Cujo, P., and Olivenstein, C., Ann. Méd-Psychol., **126**, 90 (1968).
14. (a) Högberg, T., Rämsby, S., Ögren, S.-O., and Norinder, U., Acta Pharm. Suec., **24**, 289 (1987); (b) Högberg, T., Drugs Future, **16**, 333 (1991), and references cited therein; (c) Eriksson, L., Prog. Neuro-Psychopharmacol. Biol. Psychiatry, **18**, 619 (1994). (d) Conley, R., Nguyen, J. A., Hain, R., and Tamminga, C., Psychopharmacol. Bull., **30**, 75 (1994).
15. Florvall L., and Ögren, S.-O., J. Med. Chem., **25**, 1280 (1982).
16. (a) Kamiya T., and Hashimoto, M. Société d'Etudes Scientifiques et Industrielles de l'Ile-de-France), Ger. Offen., 1 941 536, 1970; Chem. Abstr., **72**, 100497x (1970); Ger. offen., 1 966 195, 1972; Chem. Abstr., **77**, 48223s (1972); (b) Oto, K., Ichikawa, E., Tamazawa, K., and Takahashi, K., (Yamanouchi Pharmaceutical Co.) Ger. Offen., 2 635 587, 1977; Chem. Abstr., **87**, 5801t (1977); (c) Valenta, V., and Protiva, M., Collect. Czech. Chem. Commun., **52**, 2095 (1987).
17. (a) Fukuzawa, S., Imamura, T., Kakehi, M., and Watanabe, M., (Fujisawa Pharmaceutical Co.), S. Afr. Pat., 69 00 983, 1969; Chem. Abstr., **72**, 90273e (1970); (b) Rossier, J. P., Swiss Pat., 544 754, 1974; Chem. Abstr., **80**, 82643a (1974); (c) Fr. Demande, 2 229 693, 1974; Chem. Abstr., **83**, 43181r (1975).
18. (a) Société d'Etudes Scientifiques et Industrielles de l'Ile-de-France, Fr. Demande, 1 580 165, 1969; Chem. Abstr., **72**, 132509u (1970); (b) Rossier, J. P., Swiss Pat., 544 753, 1974; Chem. Abstr., **80**, 82640x (1974); (c) Fr. Demande, 2 229 692, 1974; Chem. Abstr., **83**, 43180q (1975); (d) Podesva, C., Scott, W. T., and Navratil, M. (Delmar Chemicals), Can. Pat., 1 013 750, 1977; Chem. Abstr., **88**, 22611u (1978); (e) Reitsema, R. H., J. Am. Chem. Soc., **71**, 2041 (1949).
19. Bulteau, G. (Société d'Etudes Scientifiques et Industrielles de l'Ile-de-France), S. Afr. Pat., 68 02 593, 1968; Chem. Abstr., **71**, 30354b (1969).
20. For a brief account of the development work, see Federsel, H.-J., Celgene Chiral Rep., **1**(1), 1 (1993).
21. Kaplan, J. P., Najer, H., and Obitz, D. C. L. (Synthelabo), Ger. Offen, 2 735 036, 1978; Chem. Abstr., **88**, 152414t (1978).
22. Mauri, F. (Ravizza SpA), Ger. Offen., 2 903 891, 1979; Chem. Abstr., **91**, 211259h (1979).
23. Federsel, H.-J., Högberg, T., Rämsby, S., and Ström, P. (Astra Läkemedel AB), PCT Int. Pat. Appl., WO 87 07 271, 1987; Chem. Abstr., **109**, 22839b (1988); in accepted form as US Pat., 5 300 660, 1994.
24. For a full paper discussing these reactions in greater detail, see Högberg, T., Rämsby, S., and Ström, P., Acta Chem. Scand., **43**, 660 (1989).
25. Högberg, T., Ström, P., Ebner, M., and Rämsby, S., J. Org. Chem., **52**, 2033 (1987).
26. (a) Deimer, K.-H., in Methoden der Organischen Chemie (Houben–Weyl), 4th edn, Georg Thieme, Stuttgart, 1974, Vol. XV/1, pp 315ff; (b) for a recent corresponding synthetic description of the analogous methyl ester, see Elliott, R. L., Kopecka, H., Lin, N.-H., He, Y., and Garvey, D. S., Synthesis, 772 (1995).
27. In close agreement with Flouret, G., Morgan, R., Gendrich, R., Wilber, J., and Seibel, M., J. Med. Chem., **16**, 1137 (1973).
28. Methodology essentially according to Möhrle, H., and Sieker, K., Arch. Pharm. (Weinheim), **309**, 380 (1976).
29. For a description of this methodology, see for example Carlson, R., Design and

*Optimisation in Organic Synthesis*, Elsevier, Amsterdam, 1992.
30. Weast, R. C. (Ed.), *Handbook of Chemistry and Physics*, 60th edn, CRC Press, Boca Raton, FL, 1979, p. C-508.
31. Federsel, H.-J., Könberg, E., Lilljequist, L., and Swahn, B.-M., *J. Org. Chem.*, **55**, 2254 (1990).

# 12 Development of a Multi-stage Chemical and Biological Process for an Optically Active Intermediate for an Anti-glaucoma Drug

**A. J. BLACKER**
*ZENECA Process Technology Department, Huddersfield, UK*

**and**

**R. A. HOLT**
*ZENECA Life Science Molecules, Billingham, UK*

| | | |
|---|---|---|
| 12.1 | Introduction | 245 |
| 12.2 | Development of the Multi-stage Chemical and Biological Process | 248 |
| | 12.2.1 Synthesis of Methyl (*R*)-3-hydroxybutyrate | 248 |
| | 12.2.2 Formation of Methyl (*S*)-3-(2-thienylthio)butyrate | 249 |
| | 12.2.3 Hydrolysis of Methyl (*S*)-3-(2-thienylthio)butyrate | 250 |
| | 12.2.4 Synthesis of the Ketosulphide and Ketosulphone | 250 |
| | 12.2.5 Development of an Asymmetric Bioreduction of the Ketosulphone | 253 |
| 12.3 | Conclusion | 260 |
| 12.4 | Acknowledgements | 260 |
| 12.5 | References | 260 |

## 12.1 INTRODUCTION

Glaucoma is a disease of the eye characterised by increased intraocular pressure which results in defects in the field of vision. If left untreated, the disease can cause irreversible damage to the optic nerve, eventually leading to blindness. The most common form of the disease is open-angle or chronic glaucoma, in which the trabecula network which drains aqueous humour from the eye is obstructed whilst inflow of aqueous humour is unrestricted.

There are a number of drug treatments for glaucoma which act by opening the trabecula network, including pilocarpine and physostigmine. Adrenaline has a similar mode of action but also reduces aqueous humour formation, and $\beta$-blockers, which have also been used successfully in the treatment of glaucoma,

---

*Chirality in Industry II.* Edited by A.N. Collins, G. N. Sheldrake and J. Crosby
© 1997 John Wiley & Sons Ltd.

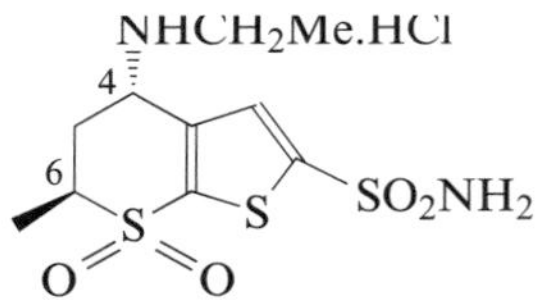

**Figure 1** Structure of MK-0507 (trade name Trusopt)

also act by reducing aqueous humour formation. Unfortunately, all of these treatments have side effects which, in certain instances, can preclude their use. A further class of compounds useful in reducing intraocular pressure are the carbonic anhydrase inhibitors (CAIs). These act by reducing the levels of hydrogencarbonate in the aqueous humour, which has the effect of reducing the inflow of water into the eye. Until recently, CAIs were available only as orally administered drugs and, consequently, suffered from a number of systemic side effects. Recently, however, Merck have discovered a new range of CAIs which, by virtue of being water soluble, can be applied topically, which avoids most of the side effects associated with systemic administration.[1] One such compound, designated MK-0507, has recently been launched by Merck under the trade name Trusopt (Figure 1).

MK-0507 contains two chiral centres with defined absolute configuration and Merck have recently published a synthetic route (Scheme 1).[2] Incorporation of the chiral methyl group at C-6 was accomplished by tosylation of methyl 3-($R$)-hydroxybutyrate followed by $S_N2$ displacement with 2-(lithiomercapto)thiophene. A major consideration for Merck in developing a manufacturing route to MK-0507 was how to obtain the desired stereochemistry at C-4. Following a full analysis of the options, they opted for a route which introduced the acetamide through a Ritter reaction (Scheme 1). This Ritter reaction is unusual in that it proceeds predominantly with retention of configuration when the hydroxysulphone reactant has the (4$S$,6$S$)-configuration (a full discussion of the proposed mechanism is given in refs 2 and 3).

The stereoselectivity of the Ritter reaction required that the hydroxysulphone be synthesised in the (4$S$,6$S$)-configuration. Unfortunately, reduction of the ketosulphone or ketosulphide by borane–THF or lithium aluminium hydride, respectively, resulted in the formation of predominantly the (4$R$,6$S$)-configuration (>95%). Although the hydroxysulphide could be epimerised using sulphuric acid, the 4$S$:4$R$ ratio obtained was only 76:24.[2] The most elegant solution to this problem is the asymmetric reduction of the keto group, and ZENECA's approach was to identify a system capable of achieving this with the desired selectivity. Since this is a redox reaction, a whole-cell bio-transformation system was employed to avoid the need for enzyme purification and allow the *in vivo* recycling of the nicotinamide cofactor involved in the reduction. Consideration of the water solubilities of the ketosulphide and ketosulphone suggested the latter as the preferred substrate for the biological reaction. The manufacturing

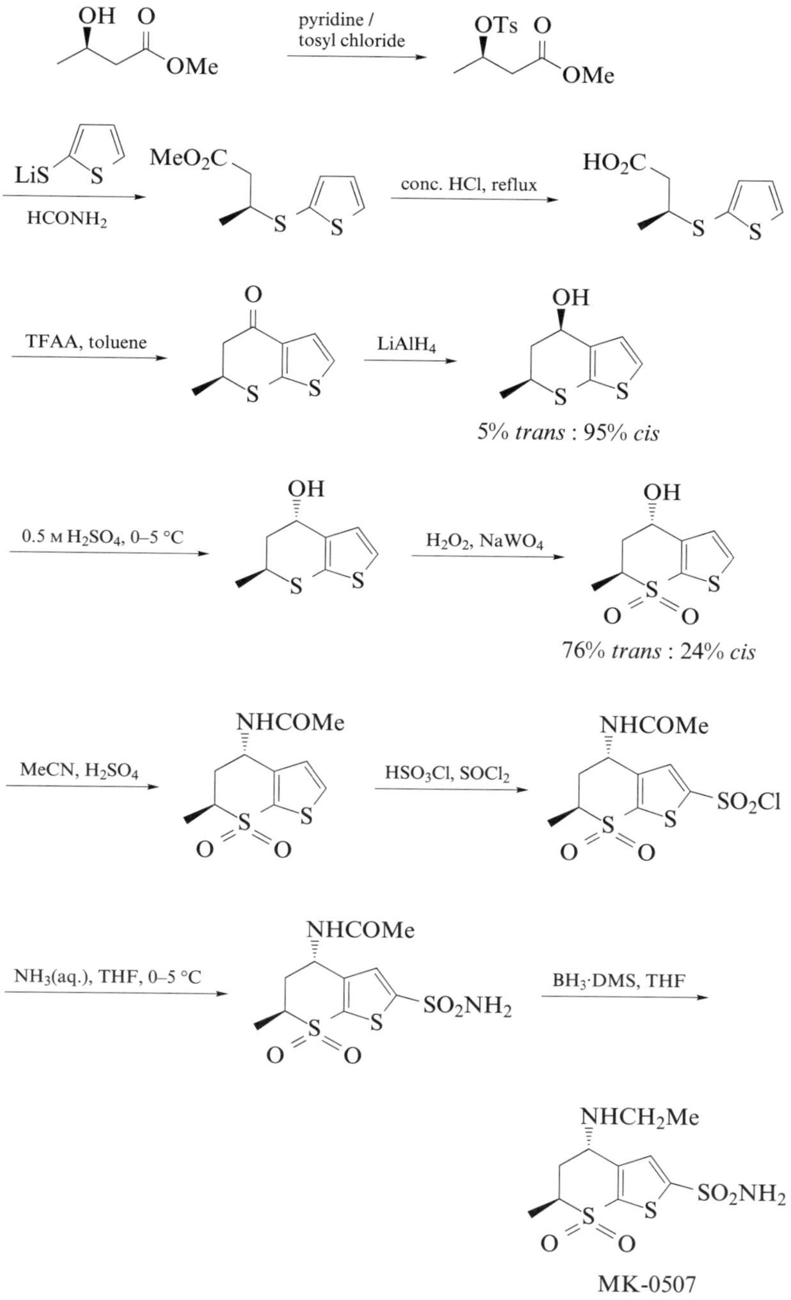

# OPTICALLY ACTIVE INTERMEDIATE FOR ANTI-GLAUCOMA DRUG

**Scheme 1** After ref. 2

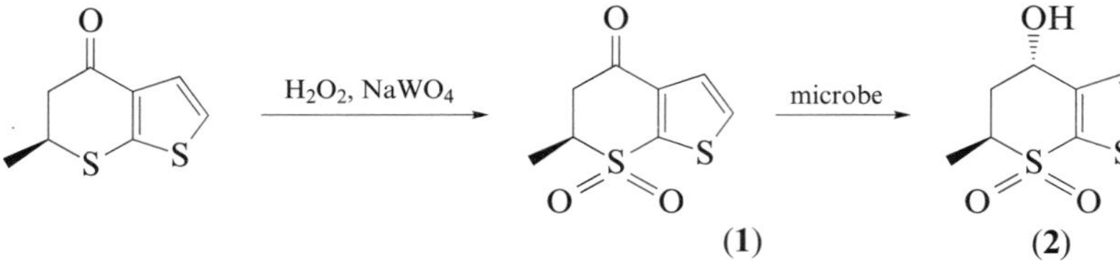

**Scheme 2**

route employed by ZENECA for the synthesis of the hydroxysulphone (**2**) is shown in Scheme 2, and a more detailed description of the process and its development is given below.

## 12.2 DEVELOPMENT OF THE MULTI-STAGE CHEMICAL AND BIOLOGICAL PROCESS

### 12.2.1 SYNTHESIS OF METHYL (R)-3-HYDROXYBUTYRATE

There are currently two commercial sources of (R)-3-hydroxybutyrate esters. The first is by depolymerisation of the microbially produced polymer poly[(R)-3-hydroxybutyrate] and the second via asymmetric hydrogenation of acetoacetate esters using Group VIII transition metal catalysts with optically active ligands. These processes are described in more detail below.

Biopol™ is a biodegradable plastic made by ZENECA. The name covers the generic family of polyhydroxyalkanoates, which are natural polymers produced by some microorganisms as storage compounds when carbon is abundant but growth is not possible owing to the limited supply of a nutrient, such as nitrogen. The polymers are intracellular inclusion bodies made from glucose and carboxylic acids. Polymer granules are isolated from the fermented microorganism, first by cell lysis, and then from the cellular debris by washing. These granules may, if necessary, be further purified before plasticising. Copolymers of (R)-3-hydroxybutyrate and (R)-3-hydroxyvalerate are preferred for the manufacture of Biopol used in biodegradable plastics, whereas it is the homopolymer that is used by ZENECA to produce methyl (R)-3-hydroxybutyrate for MK-0507.

The homopolymer has a molecular weight that may be varied depending on cellular growth conditions, but generally lies between 400 and 600 kDa. It is an optically active polyester that is insoluble in water and moderately soluble in chlorinated hydrocarbons. One of the problems in producing the monomer by saponification of the polymer is the insolubility of the latter [despite the high aqueous solubility of (R)-3-hydroxybutyric acid], and another is the formation of significant amounts of the by-product methylcrotonate. Processes have been developed for hydrolyses and transesterifications in chlorinated solvents,[4] but for manufacturing these suffer from problems of low reaction rates, complex work-ups and undesirable solvents.

# OPTICALLY ACTIVE INTERMEDIATE FOR ANTI-GLAUCOMA DRUG

An improved process operated by ZENECA to produce methyl (R)-3-hydroxybutyrate involves transesterification of the polymer in hot, acidic, methanolic medium. In this manner the insoluble polymer may be transesterified with good conversions (Scheme 3). After work-up, the desired product is recovered by fractional distillation under reduced pressure, which gives the methyl (R)-3-hydroxybutyrate as a clear, colourless liquid, with a chemical purity of >98% and of >99.5% ee. The process has been operated routinely on a multi-tonne scale.

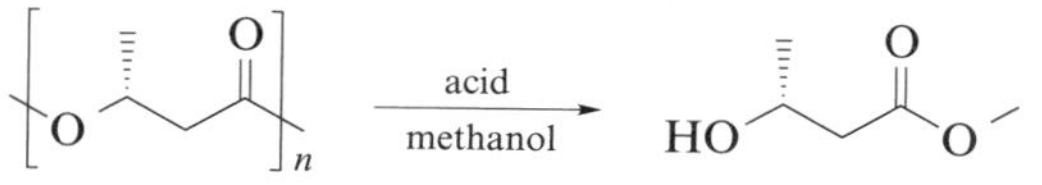

**Scheme 3**

Whilst methyl (R)-3-hydroxybutyrate is used for the manufacture of an intermediate to MK-0507, it may also be used as a starting material for syntheses of other optically active molecules. Much of the chemistry of the 3-hydroxybutyrates has been elegantly elaborated by Seebach's group;[4] examples are its conversion to (R)-butane-1,3-diol and (R)-3-hydroxybutyrolactone.[5] Should it be required, a simple method for inverting the stereochemistry is given in ref. 6.

Optically active hydroxybutyrates may also be made by asymmetric hydrogenation of acetoacetate esters using catalysts that employ rhodium or ruthenium complexed to optically active phosphine-based ligands. An example of such a catalyst is cycloocta-1,4-dienyl-2,2'-bis(diphenylphosphino)-1,1'-binaphthylruthenium, Ru(COD)(BINAP) (Scheme 4). This catalyst was invented by Noyori and co-workers, developed by Takasago Co. (Japan),[7] and is used to produce both (R)-and (S)-enantiomers of 3-hydroxybutyrate esters, albeit with a lower optical purity than the biologically derived material.

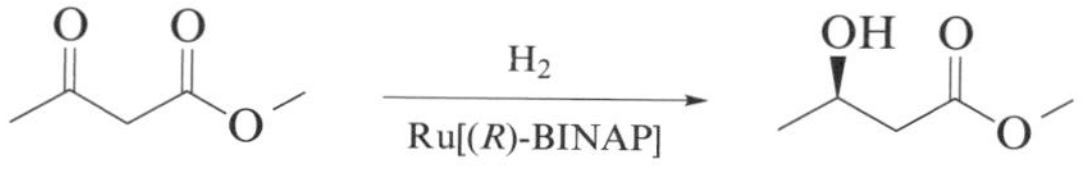

**Scheme 4**

## 12.2.2 FORMATION OF METHYL (S)-3-(2-THIENYLTHIO)BUTYRATE

Starting from methyl 3-(R)-hydroxybutyrate, the process developed by Merck involves activation of the 3-hydroxy group by tosylation in pyridine.[2] After isolating this product by crystallisation from an aqueous drown-out, the sulphonyl ester is treated with 2-lithiomercaptothiophene generated in situ by reaction of thiophene, butyllithium and sulphur. The optical activity of the product methyl

(S)-3-(2-thienylthio)butyrate indicates a clean inversion of stereochemistry. The reaction is complex, and requires careful control of stoichiometries and other parameters to minimise the formation of by-products.

### 12.2.3 HYDROLYSIS OF METHYL (S)-3-(2-THIENYLTHIO)BUTYRATE

The conditions needed to effect this reaction are fairly harsh: 6M HCl at reflux for several hours. Unfortunately, this procedure favours an undesirable rearrangement of the product (S)-3-(2-thienylthio)butyric acid to give the regioisomeric product (S)-3-(3-thienylthio)butyric acid. A rearrangement of this type has been reported to occur in an analogous system which, according to one hypothesis, passes through an unstable *ipso* compound (Scheme 5).[8] The regioisomer is virtually inseparable from the desired product and may be carried through the subsequent reactions to the final product, so it constitutes an important and undesirable impurity. Careful process optimisation has minimised formation of the regioisomer.

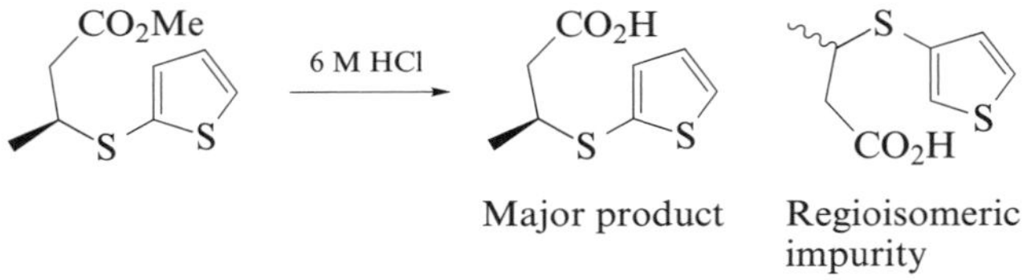

**Scheme 5**

### 12.2.4 SYNTHESIS OF THE KETOSULPHIDE AND KETOSULPHONE

The cyclisation procedure employed by ZENECA is essentially that developed by Merck, using trifluoroacetic anhydride, which was shown to be superior to standard Friedel–Crafts cyclisations.[2,9,10] It appears that use of trihaloacetic anhydrides has a further, elegant advantage over the Friedel–Crafts acylation, but before describing this, it is of interest to note another method of oxidation introduced by the Merck chemists.

The ketosulphide, in a toluene solution is oxidised to ketosulphone using hydrogen peroxide with a sodium tungstate catalyst. The reaction is highly exothermic and is, therefore, operated by controlled addition of the peroxide. The tungstate catalyst is highly active, and the peroxide is consumed as added, preventing an undesirable accumulation. Moreover, the sulphide is oxidised rapidly to sulphone, and the intermediate sulphoxide is barely observed. The use of toluene is advantageous over mixed solvents since it reduces costs and simplifies the work-up, which is carried out by separating the biphasic reaction mixture and drowning the aqueous phase into a solution of sodium hydrogensulphite. After washing and concentrating the toluene solution, the product can be isolated by crystallisation from mixtures of toluene and methanol.

In an attempt to improve this process still further, it was noted that the trifluoroacetic acid generated in the cyclisation reaction could, potentially, be used to generate trifluoroperacetic acid, an oxidation catalyst for the formation of ketosulphone. Indeed, it was demonstrated that the reaction mixture at the end of the cyclisation reaction could be used directly for the oxidation simply by addition of hydrogen peroxide (Scheme 6).[11] Initial experiments which gave ketosulphone of

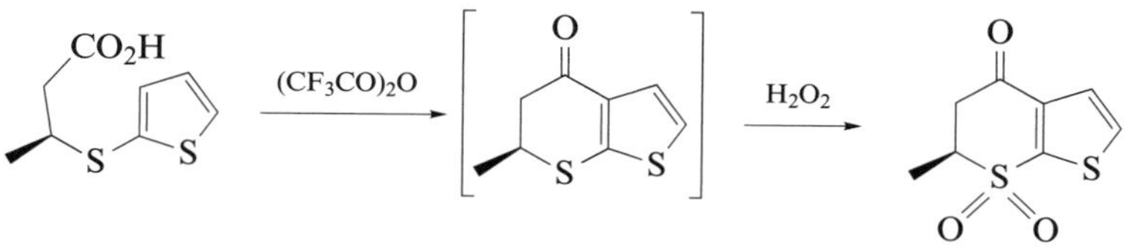

**Scheme 6**

high enantiomeric excess were carried out in the laboratory by relatively fast additions of hydrogen peroxide. However, as with the tungstate-catalysed oxidation, the heat of reaction was considerable, and it was necessary to slow the rate of hydrogen peroxide addition on a larger scale. Unfortunately, in simulating this in the laboratory we observed that the enantiomeric excess of the ketosulphone dropped dramatically from 98% to about 60%. On further investigation it was found that using slow addition of hydrogen peroxide allowed the build-up of the intermediate ketosulphoxide (Figure 2). Moreover, we observed that both *cis*- and *trans*-(6S)-methyl sulphoxides were formed and consumed at different rates. Acid-catalysed racemisation of sulphoxides has been reported,[12] and one of the mechanisms proposed is by sulphenolisation and consequent loss of the chiral centre at C-6 (Scheme 7). Although it might have been possible to increase the rate of peroxide addition on the plant to reduce racemisation, this would have required a far greater cooling capacity than was available, hence the choice of the sodium tungstate-catalysed oxidation.

Although it was recognised at an early stage that the ketosulphone was base sensitive, an unexpected observation was made during process development. In the course of the work-up of the oxidation reaction, an aqueous sodium hydrogensulphite wash is used to destroy residual peroxides. The separation of both phases is not always clean, and this can result in a small carry-over of hydrogensulphite solution into the organic phase. After atmospheric azeotropic drying of the organic solution, and upon cooling, the ketosulphone crystallises and precipitation of insoluble sodium hydrogensulphite is observed. When this mixture is redissolved in a polar solvent, such as boiling methanol, decomposition of the

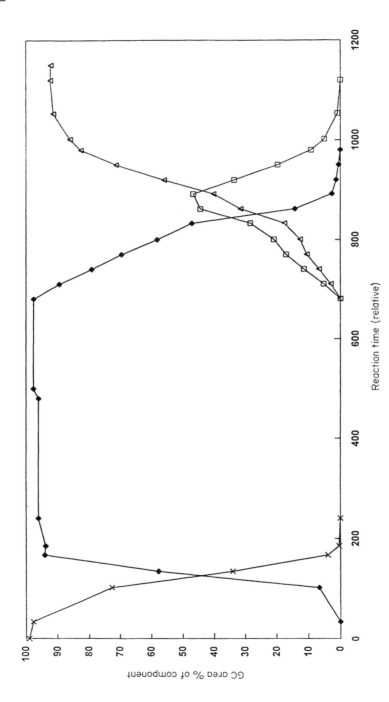

**Figure 2** Reaction profile for telescoped cyclisation–oxidation reaction (♦) ketosulphide; (□) ketosulphoxide, (△) ketosulphone; (×)(*S*)-3-(2-thienylthio)butyric acid

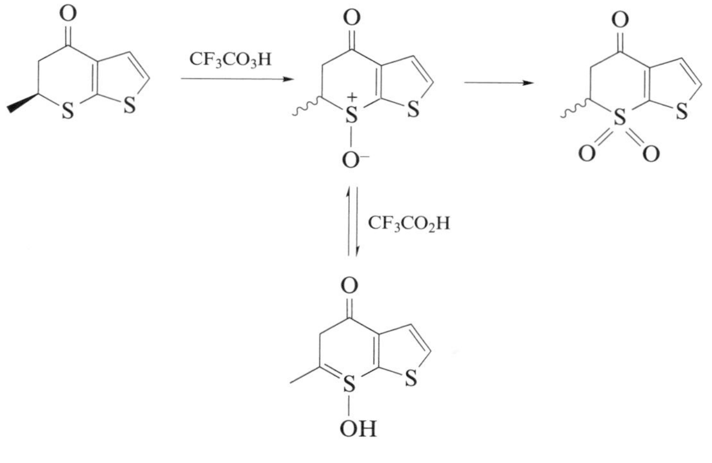

**Scheme 7**

sodium hydrogensulphite to sulphur dioxide and sodium hydroxide occurs. The latter reagent is then able to slowly racemise the ketosulphone. Clearly, this problem may be easily overcome by efficient washing of the product with water prior to the solvent drying step, but the episode serves to illustrate the importance of a thorough understanding of the process during the development stage.

## 12.2.5 DEVELOPMENT OF AN ASYMMETRIC BIOREDUCTION OF THE KETOSULPHONE

The use of microorganisms to carry out asymmetric reductions of ketones is well documented and many reactions have been described in the literature.[13] Most of these reactions have been carried out with strains of the yeast *Saccharomyces cerevisiae* and, although with certain compounds good stereoselectivity has been observed, this is by no means always the case. ZENECA's approach to the reduction of the ketosulphone began with a screen to identify potential microorganisms.

Initial studies were carried out using a range of carefully selected microorganisms (bacteria, yeasts and fungi) which were chosen on the basis of previous experience. Each microorganism was grown under the appropriate conditions, recovered by centrifugation and resuspended in a phosphate buffer medium at pH 7.0. To the cell suspension was added ketosulphone plus an appropriate carbon source to supply the microorganisms with reducing power. After incubation, the reaction mixtures were analysed for the presence of hydroxysulphone (**2**). As the hydroxysulphone is diastereoisomeric, it was possible to determine the ratio of *trans*-to *cis*-hydroxysulphone using conventional, non-chiral HPLC. From the results in Tables 1–3, it can be seen that a range of selectivities were obtained depending on the microorganism used.

**Table 1** Reduction of ketosulphone (2) by bacteria

| Organism | Conversion (%) | trans-Alcohol (%) |
|---|---|---|
| Lactobacillus salivarius | 96.9 | 69.6 |
| L. buchneri | 36 | 33.9 |
| L. fermentum | 84.3 | 54.2 |
| L. brevis | 76.6 | 20.9 |
| L. plantarum | 6.1 | 91.3 |
| Streptococcus faecium | 4.9 | 3.9 |
| Streptomyces lividans | 61 | 26 |
| Arthrobacter petroleophagus | 99.5 | 42.3 |
| Gluconobacter oxydans | 86.0 | 10.2 |
| Corynebacterium species | 99.6 | 27 |
| C. paurometabolum | 97.0 | 20 |
| Mycobacterium M156 | 100 | 40 |
| M. smegmatis | 99.4 | 53.6 |
| Pseudomonas putida UV4 | 0 | – |
| Ps. putida | 20 | 12 |
| Ps. oleovarans | 48 | 74 |
| Ps. species LC2 | 41 | 77 |
| Ps. aeruginosa | 80 | 17.4 |
| Proteus vulgaris X19 | 24 | 13 |
| P. mirabilis | 26 | 13 |
| Xanthomonas campestris | 69 | 34 |
| Alcaligenes eutrophus | 98.4 | 43.5 |
| Clostridium tyrobutyricum | 91.8 | 40.0 |
| C. tyrobutyricum | 24.3 | 5.9 |
| C. pasteurianum | 60.2 | 36.4 |
| Nocardia species | 84.2 | 34.9 |

**Table 2** Reduction of ketosulphone (1) by yeast

| Organism | Conversion (%) | trans-Alcohol (%) |
|---|---|---|
| Candida chalmersii | 37.3 | 42.1 |
| C. diddensiae | 66.7 | 52.2 |
| C. lipolytica | 95 | 35 |
| C. utilis | 32.9 | 51.4 |
| C. oleophilia | 1.2 | 9.9 |
| Geotrichum candidum | 99.3 | 15 |
| G. candidum | 95.7 | 31.6 |
| Hansenula anomala | 99.8 | 51.2 |
| Pichia farinosa | 13 | 45.2 |
| P. farinosa | 12.3 | 51.6 |
| P. haplophila | 97.7 | 68.2 |
| P. pastoris | 15.1 | 41 |
| P. trehalophilia | 98.8 | 36.6 |
| Saccharomyces cerevisiae | 21.4 | 15.4 |
| S. cerevisiae | 24 | 15.0 |
| S. cerevisiae | 8 | 35 |

(*continued*)

## Table 2 (continued)

| Organism | Conversion (%) | trans-Alcohol (%) |
|---|---|---|
| S. cerevisiae | 40.4 | 40.2 |
| Zygosaccharomyces rouxii | 11.1 | 45 |
| Torulaspora hansenii | 10.8 | 50.3 |
| Pichia capsulata (Torulopsis molischiana) | 99.8 | 10.3 |

## Table 3  Reduction of ketosulphone by fungi

| Organism | Conversion (%) | trans-Alcohol (%) |
|---|---|---|
| Aspergillus flavus | 5 | 68 |
| Beauveria bassiana | 46 | 21 |
| B. bassiana | 41 | 53 |
| B. brogniartii | 49 | 26 |
| Fusarium graminearum | 73.5 | 56.9 |
| Neurospora crassa | 87 | 71 |
| Penicillium species | 16.9 | 39.5 |

A number of organisms reduced the ketosulphone in almost quantitative yield, but did not simultaneously demonstrate the required stereoselectivity under the chosen reaction conditions. One microorganism which was notable for the selectivity of the reduction was *Lactobacillus plantarum*, but the extent of reduction was low. Given the need for both high stereoselectivity and high yield, further investigations and development work were required. Initially, studies were restricted to *L. plantarum* in an attempt to understand better how variations in the growth and bio-transformation conditions would affect the yield and stereoselectivity of the reaction.

A reaction was carried out which was sampled at intervals with measurement of key variables (Table 4). It is interesting that the glucose provided was almost entirely consumed during the first 2 h of the reaction, being converted nearly quantitatively into racemic lactic acid, the normal product of *L. plantarum* fermentation. Coincident with the formation of lactic acid was a rapid fall in pH.

## Table 4  Time course of *Lactobacillus plantarum*-catalysed reduction of ketosulphone (1)

| Time (h) | pH | Glucose ($g\,l^{-1}$) | Lactic acid ($g\,l^{-1}$) | Ketosulphone reduction (%) | trans-Diastereo-isomers (%) |
|---|---|---|---|---|---|
| 0 | 7.0 | 10 | 0 | 0 | – |
| 2 | 5.0 | 0.1 | 9 | 13 | 91 |
| 8 | 5.3 | nd[a] | 9 | 18 | 90 |
| 20 | 5.3 | nd[a] | 9 | 27 | 89 |

[a]Not detectable.

**Table 5** Effect of glucose concentration on the extent and selectivity of ketosulphone reduction by *Lactobacillus plantarum*

| Glucose concentration ($gl^{-1}$) | Time (h) | pH | Ketosulphone reduction (%) | *trans*-Diastereo-isomers (%) |
|---|---|---|---|---|
| 0 | 2 | 6.9 | 1.2 | 61 |
|  | 4 | 6.8 | 1.4 | 69 |
|  | 20 | 6.5 | 2.4 | 55 |
| 10 | 2 | 4.6 | 5 | 88 |
|  | 4 | 4.8 | 6 | 91 |
|  | 20 | 5.0 | 12 | 89 |
| 50 | 2 | 3.6 | 25 | 90 |
|  | 4 | 3.5 | 36 | 92 |
|  | 20 | 3.6 | 36 | 92 |

The highest rate of ketosulphone reduction was seen during the first 2 h of the reaction, which suggested that glucose exhaustion, or a fall in pH, could be responsible for the slowing of the reaction, and these two variables were therefore examined.

In order to study the effect of glucose concentration, a batch of *L. plantarum* was grown, resuspended in buffer (pH 7.0) and supplemented with different amounts of glucose. Table 5 shows that glucose is necessary for any significant reduction of ketosulphone to occur; this is not surprising, as glucose is required to recycle the reductant, NAD(P)H, which is present in the cell in only catalytic amounts. In the absence of added glucose the pH of the reaction medium stayed close to its initial value, as little lactic acid was formed, but the hydroxysulphone which was formed was rich in the *cis* diastereoisomer.

In order to study the effect of pH on the reduction of ketosulphone by *L. plantarum*, a series of experiments was carried out in which the reaction pH was maintained at a predetermined value by the automatic addition of sodium hydroxide to neutralise the lactic acid formed during metabolism of glucose. Two important observations were made during these experiments: neutral pH was favoured over acidic pH as far as yield was concerned, but the opposite was true for the formation of the *trans* diastereoisomer (Table 6). There were two possible explanations: first, *L. plantarum* contains more than one enzyme system capable of reducing ketosulphone (these systems having opposite stereoselectivities and different sensitivities to pH); second, the ketosulphone or hydroxysulphone may be undergoing racemisation under the reaction conditions. Since the ketosulphone used in these experiments has a 6$S$:6$R$ ratio of greater than 99:1, it was possible to test the racemisation hypothesis by measuring the ratio of the four diastereoisomers of the hydroxysulphone produced by *L. plantarum*. It was apparent from these measurements that racemisation at the C-6 position was occurring (Figure 3). Since previous experiments had indicated a strong influence of pH on the stereoselectivity of the reduction reaction, the effect of pH on the

# OPTICALLY ACTIVE INTERMEDIATE FOR ANTI-GLAUCOMA DRUG

**Table 6** Effect of pH on the stereochemical outcome of ketosulphone reduction by *Lactobacillus plantarum*

| Reaction pH | Time (h) | Ketosulphone reduction (%) | *trans*-Diastereoisomers (%) |
|---|---|---|---|
| 4.0 | 2 | 23 | 87 |
|  | 22 | 46 | 81 |
| 5.0 | 1 | 28 | 82 |
|  | 19 | 60 | 72 |
| 7.2 | 2 | 79 | 44 |
|  | 15 | 80 | 45 |

optical purity of the ketosulphone was examined. Ketosulphone was suspended in aqueous buffers of various pH values and stirred; the mixtures were extracted at intervals and the optical purity was measured by HPLC with a chiral stationary phase. The extreme optical lability of the ketosulphone can be seen clearly in Figure 4. Racemisation appears to occur through a reverse Michael reaction following loss of an acidic hydrogen at C-5 (Scheme 8); this has been demonstrated by deuterium exchange NMR experiments.[2]

The requirement for an acidic pH to avoid ketosulphone racemisation was clearly incompatible with the requirements of *L. plantarum*, which only gave acceptable yields when the pH was maintained near neutral. Consequently, it was

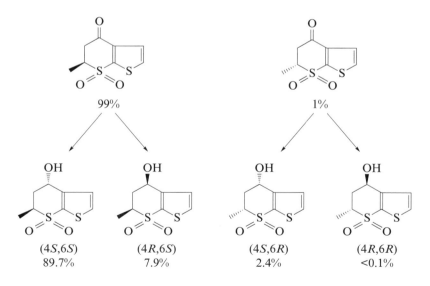

**Figure 3** Ratio of hydroxysulphone diastereoisomers obtained from an *L. plantarum*-catalysed reaction. Initial pH 7.0; no pH control

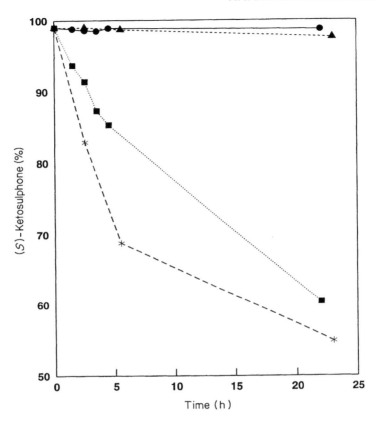

**Figure 4** Optical stability of ketosulphone over a range of pH values: (●) 4.3; (▲) 5.0; (■) 6.5; (×) 9.0

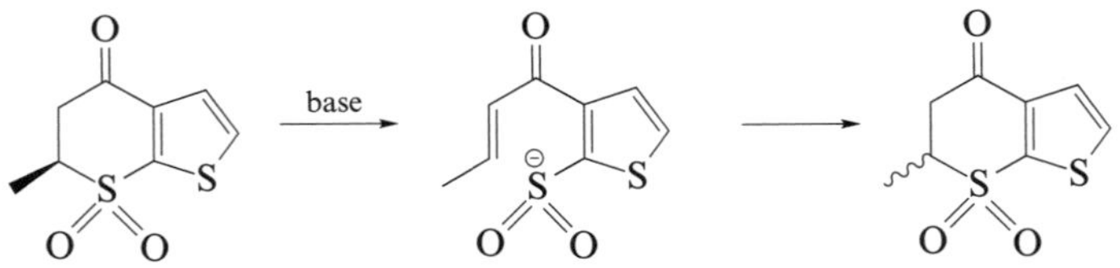

**Scheme 8**

necessary to identify an alternative microorganism demonstrating the desired selectivity coupled with the ability to tolerate pH values below 5.0.

From the results of the initial screening work, the fungus *Neurospora crassa* appeared worthy of closer investigation, having given a yield of 87% with a

# OPTICALLY ACTIVE INTERMEDIATE FOR ANTI-GLAUCOMA DRUG

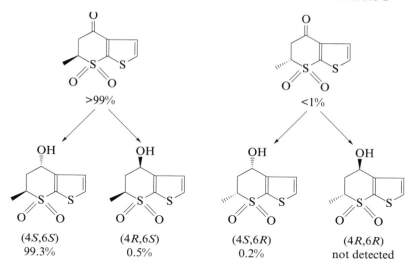

**Figure 5** Ratio of hydroxysulphone diastereoisomers obtained from an *N. crassa*-catalysed reaction. Reaction controlled at pH 4.3

*trans:cis* hydroxysulphone ratio of 71:29. One of the advantages of *N. crassa* is its ability to grow at low pH values, which avoids the problem of ketosulphone racemisation. When tested under these conditions *N. crassa* provided hydroxysulphone of very high optical purity (Figure 5). On the basis of these results and our previous experience of developing processes based on the submerged culture of filamentous fungi, *N. crassa* was chosen as the prime candidate for the rapid development of a full-scale process. Optimisation of growth and bio-transformation variables such as culture medium, temperature and, most important, pH were carried out to define the optimum growth and bio-transformation conditions for application at plant scale.

The final process involves the growth of *N. crassa* in a fully controlled fermenter. At a specific stage in the growth the bio-transformation is initiated by the slow addition of ketosulphone to the fermenter. The rate of ketosulphone addition is matched to the rate of reduction to hydroxysulphone so that ketosulphone does not accumulate in the reactor, thus minimising the possibility of ketosulphone racemisation. On completion of the bio-transformation, the fungus is removed and the bio-transformation broth is extracted with organic solvent. Finally, the hydroxysulphone is recovered by crystallisation from the solvent phase. On a plant scale this process provides hydroxysulphone in > 80% isolated yield (based on ketosulphone) with high chemical purity (>99%) and high optical purity [>98% of the (4*S*,6*S*)-hydroxysulphone and <0.5% of (6*R*)-diastereoisomers].

The highly stereospecific dehydrogenase responsible for the reduction of the ketosulphone is found in the cytoplasmic fraction of *N. crassa* cell extracts and is NADPH-specific. The enzyme has been purified using a two-stage procedure involving ion-exchange and affinity chromatography and the enzyme's *N*-terminus has been sequenced (38 amino acid residues) as a first step towards isolating and cloning the gene (work now being actively pursued by ZENECA).

## 12.3 CONCLUSION

This process demonstrates that complex, optically active compounds can be efficiently synthesised on a plant scale using bio-transformations. The application of bio-transformation technology was particularly vital to the success of this project, both chiral centres being derived through biological reactions. Of note was the availability of enantiopure methyl (*R*)-3-hydroxybutyrate from the bacterial polymer; although this may be considered a natural product, the chirality of this polymer is derived through asymmetric reduction of a ketone [acetoacetyl-CoA to (*R*)-3-hydroxybutyryl-CoA to poly (*R*)-3-hydroxybutyrate]. The novel,[14] highly stereoselective reduction of the ketosulphone was also vital to the success of the project, and marks a step forward in the development of industrial bio-transformations from resolutions (such as those catalysed by hydrolase enzymes) towards more complex, cofactor-requiring, synthetic reactions. Equally important was the development of processes that would avoid racemisation of the chiral centres, particularly at C-6, during the oxidation of the ketosulphide and during the biological reduction of the ketone to form the hydroxysulphone.

## 12.4 ACKNOWLEDGEMENTS

The authors thank all the members of the ZENECA project team, especially Dr D. J. Moody. Special thanks are also due to Dr E. J. J. Grabowski, Dr D. J. Mathre and Dr D. Matsumoto of Merck for useful discussions.

## 12.5 REFERENCES

1. (a) Baldwin, J. J., Ponticello, G. S., and Christy, M. E. (Merck) *Eur. Pat.*, EP 0 296 879; (b) Ponticello, G. S., Freedman, M. B., Habecker, C. N., Lyle, P. A., Schwam, H., Varga, S. L., Christy, M. E., Randall, W. C., and Baldwin, J. J., *J. Med. Chem.*, **30**, 591 (1987); (c) Baldwin, J. J., Ponticello, G. S., Anderson, P. S., Christy, M. E., Murcko, M. A., Randall, W. C., Schwam, H., Sugrue, M. F., Springer, J. P., Gautheron, P., Grove, J., Mallorga, P., Viader, M.-P., McKeever, B. M., and Navia, M.-A., *J. Med. Chem.*, **32**, 2510 (1989).
2. Blacklock, T. J., Sohar, P., Butcher, J. W., Lamanec, T., and Grabowski, E. J. J., *J. Org. Chem.*, **58**, 1672 (1993).

3. Sohar, P., Mathre, D. J., and Blacklock, T. J., *Eur. Pat.*, 0 617 037.
4. (a) Seebach, D. Z., and Zuger, M., *Helv. Chim. Acta*, **65**, 495 (1982); (b) Seebach, D., Beck, A. K., Breitschuh, R., and Job, K., *Org. Synth.*, **71**, 39 (1993); (c) *Eur. Pat.* EP 0 043 620 and EP 0 320 046.
5. (a) Fräter, G., *Helv. Chim. Acta*, **62**, 2825 (1979); (b) Seebach, D., and Zuger, M. F., *Tetrahedron Lett.*, **25**, 2747 (1985) (c) Kramer, A., and Pfanler, H., *Helv. Chim. Acta*, **65**, 293 (1982); (d) Seebach, D., Roggo, S., and Zimmermann, J., in *Stereochemistry of Organic and Bioorganic Transformations* (ed. W. Bartmann, and K. B., Sharpless), Proceedings of the Seventh Workshop Conference Hoechst, Verlag Chemie, Weinheim, 1986, and New York, 1987.
6. (a) Cainelli, G., Manescalchi, F., Martelli, G., Panunzio, M., and Plessi, L., *Tetrahedron Lett.*, **26**, 3369 (1985); (b) Griesbeck, A., and Seebach, D., *Helv. Chim. Acta*, **70** , 1320 (1987); (c) Breitschuh, R., and Seebach, D., *Chimia*, **44**, 216 (1990).
7. (a) Noyori, R., and Kitamura, M., in *Modern Synthetic Methods* (ed. R. Scheffold), Springer, Heidelberg, 1989, Vol. 5, p. 115; (b) Noyori, R., *Asymmetric Catalysis in Organic Synthesis*, Wiley, New York, 1994; (c) Akutagawa, S., in *Chirality in Industry* (ed. A. N. Collins, G. N. Sheldrake, and J. Crosby), Wiley, Chichester, 1992, p. 325; (d) Kitamura, M., Tokunaga, M., Ohkuma, T., and Noyori, R., *Org. Synth.*, **71**, 1 (1990).
8. (a) Gronowitz, S., and Pinchas, M., *Acta Chem. Scand.*, **16**, 155 (1962); (b) De Sales, J., Greenhouse, R., and Muchowski, J. M., *J. Org. Chem.*, **47**, 3668 (1982).
9. Shinkai, I., *J. Heterocycl. Chem.*, **29**, 627 (1992).
10. Jones, T. K., Mohan, J. J., Xavier, L. C., Blacklock, T. J., Mathre, D. J., Sohar, P., Turner-Jones, E. T., Reamer, R. A., Roberts, F. E., and Grabowski, E. J. J., *J. Org. Chem.*, **56**, 763 (1991).
11. Merck ZENECA, *PCT Int. Pat. Appl.*, 60/002 890, 1995.
12. House, H. O., *Modern Synthetic Reactions*, Benjamin, New York, 1972, p. 726.
13. (a) Sih, C. J., and Chen, C.-S., *Angew. Chem., Int. Ed. Engl.*, **23**, 570 (1984); (b) Servi, S., *Synthesis*, 1 (1990).
14. Holt, R. A., and Rigby, S. R., *PCT Int. Pat. Appl.*, 93/01 776.

# 13 Synthesis and Application of Chiral Liquid Crystals

D. PAULUTH and A. E. F. WÄCHTLER
*H. Merck K GaA, Darmstadt, Germany*

| | | |
|---|---|---|
| 13.1 | Introduction | 263 |
| 13.2 | Properties and Applications of Liquid Crystals | 264 |
| | 13.2.1 Properties of Thermotropic Liquid Crystals | 264 |
| | 13.2.2 The Cholesteric Phase and the TN Display | 266 |
| | 13.2.3 The Chiral Smectic C Phase and the FLC Display | 267 |
| | 13.2.4 Properties of Thermochromic Liquid Crystals | 272 |
| 13.3 | Requirements for Commercial Liquid Crystals | 273 |
| | 13.3.1 Nematic Mixtures | 273 |
| | 13.3.2 Dopants for Nematic Mixtures | 275 |
| | 13.3.3 Ferroelectric Mixtures | 277 |
| | 13.3.4 Ferroelectric Dopants | 277 |
| | 13.3.5 Thermochromic Liquid Crystals | 279 |
| 13.4 | Synthesis of Chiral Liquid Crystals | 280 |
| | 13.4.1 Chiral Building Blocks for Dopants Used in Nematic Mixtures | 280 |
| | 13.4.2 Chiral Dopants for Nematic Mixtures | 280 |
| | 13.4.3 Chiral Building Blocks for Ferroelectric Dopants | 282 |
| | 13.4.4 Ferroelectric Dopants | 283 |
| | 13.4.5 Thermochromic Liquid Crystals | 283 |
| | 13.4.6 Optical Purity | 283 |
| 13.5 | Trends and Outlook | 284 |
| 13.6 | Acknowledgements | 285 |
| 13.7 | Notes and References | 285 |

## 13.1 INTRODUCTION

Liquid crystalline behaviour was first described by the Austrian botanist Friedrich Reinitzer more than 100 years ago.[1] He found that on heating optically active cholesteryl benzoate there was no direct transition from the crystalline state to the liquid state. In a well defined temperature range above the melting point, a hitherto unknown, strangely coloured, turbid liquid phase existed,

*Chirality in Industry II*. Edited by A. N. Collins, G. N. Sheldrake and J. Crosby
© 1997 John Wiley & Sons Ltd

which changed into a clear and colourless ordinary liquid phase on raising the temperature above the 'clearing point.' Since that time, similar temperature-dependent (thermotropic) liquid crystalline behaviour has been found for numerous anisotropic molecules, which are either rod- or disc-shaped. This chapter will deal only with rod-like molecules, because these have been of greater industrial importance to date.

## 13.2 PROPERTIES AND APPLICATIONS OF LIQUID CRYSTALS

### 13.2.1 PROPERTIES OF THERMOTROPIC LIQUID CRYSTALS

In the crystalline state, molecules are held together by strong intermolecular forces. They are located at defined positions in the crystal lattice, giving rise to a long-range positional and orientational order. In contrast to the highly ordered arrangement in the crystalline state, there is neither positional nor orientational order in the liquid state; the molecules move randomly and there is no net intermolecular alignment. Positional and orientational order are destroyed by heating an ordinary substance from the crystalline state to the liquid state. For 'thermotropic' liquid crystalline compounds this is a stepwise process. At the melting point the positional order is lost and, at the 'clearing point,' the orientational order vanishes.

There are different types of liquid crystalline phases (mesophases), classified according to their structural features, which reflect their degree of order. We shall confine our discussion to the nematic (N), the smectic A ($S_A$) and the smectic C ($S_C$) phases. The least ordered is the nematic phase. In a similar manner to the liquid phase, the molecules in the nematic phase move almost randomly, but the long axes of the rod-like molecules tend to be oriented in a certain direction defined by the director **n** (physically there is no difference between $+\mathbf{n}$ and $-\mathbf{n}$). The order in the 'nematic phase' is described by the order parameter $S$, with $S = 1$ in the ideal case with parallel alignment of all molecules and $S = 0$ in the non-ordered isotropic phase.[2] Typical values for the order parameter $S$ are in the range between 0.3 and 0.7 at temperatures not too close to the clearing point (Figure 1).

The translational freedom of molecules in the nematic phase is restricted in the more ordered 'smectic phases.' For example, in the $S_A$ phase the molecules are arranged in layers with their long axes perpendicular to the planes of the layer or, in other words, with the director parallel to the normal of the layers. Within a given layer the parallel-aligned molecules move randomly, and there is a statistical exchange of molecules between adjacent layers. The structure of the $S_C$ phase is closely related to the $S_A$ phase, the only difference being the tilted director (tilt angle $\vartheta$) with respect to the normal of the layer's surface (Figure 2).

On decreasing the temperature, the phases are observed in the following sequence: liquid phase, [clearing point], N, $S_A$, $S_C$, $S_B$, $S_I$, $S_F$, higher ordered smectic phases, [melting point], crystalline phase. In practice, only a selection of phas-

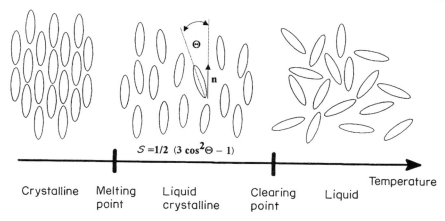

**Figure 1** Molecular order in the crystalline, liquid crystalline and liquid states. The order parameter $S$ is defined by the angle $\Theta$ between the molecular long axis and the director **n**

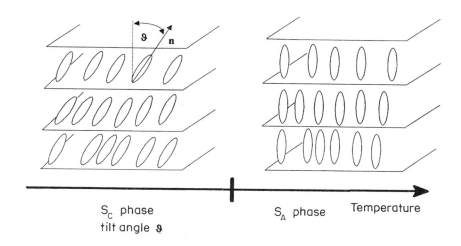

**Figure 2** Schematic drawing of the $S_C$ and $S_A$ phases. In the $S_C$ phase the director is tilted by an angle $\vartheta$ relative to the normal of the layers

es is found for a given liquid crystalline system, but the phase sequence will not be inverted. While it is difficult to supercool below the clearing point, it is possible to find liquid crystalline phases at temperatures below the melting point (metastable monotropic liquid crystalline phases).

As a consequence of both orientational order and short-range positional order, physical properties in the liquid crystalline phases must be described as vectors, although they can be described as scalars in the isotropic liquid phase.

This means that physical properties in the direction parallel to the director, or perpendicular to the director of the liquid crystalline system, are different.

For technical applications, the most important anisotropic parameters are the refractive indices ($n$) and the dielectric constants ($\varepsilon$). The birefringence ($\Delta n$) of a liquid crystal is the difference between the ordinary refractive index $n_o$ (light propagation perpendicular to the director) and the extraordinary refractive index $n_e$ (light propagation parallel to the director):

$$\Delta n = n_e - n_o > 0 \qquad (1)$$

In a similar way, the dielectric anisotropy ($\Delta \varepsilon$) is described by the dielectic vectors parallel ($\varepsilon_\parallel$) and perpendicular ($\varepsilon_\perp$) to the director according to the equation

$$\Delta \varepsilon = \varepsilon_\parallel - \varepsilon_\perp \qquad (2)$$

Depending on the molecular structure, the dielectric anisotropy can be positive (molecular dipole parallel to the long axis of the molecule) or negative (polar substituent perpendicular to the long axis of the molecule).

### 13.2.2 THE CHOLESTERIC PHASE AND THE TN DISPLAY

A special variety of the nematic phase is the 'cholesteric phase' (Ch, chiral nematic phase), in which the liquid crystalline phase is made up of optically active molecules (e.g. cholesteryl benzoate) or a nematic phase is doped with optically active molecules. Looking at one hypothetical layer of the cholesteric phase (Figure 3), all molecules are aligned parallel to the director lying in the plane of the layer. In a second adjacent layer (parallel to the first) the same situation exists, but the director has changed its direction within the plane of the layer. From one layer to the next the director rotates gradually about an axis which is identical with the normal of the layers.

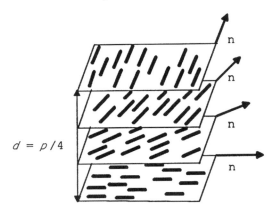

**Figure 3** Schematic drawing of the Ch phase showing the precession of the director around the normal of the layers of the Ch phase

# SYNTHESIS AND APPLICATION OF CHIRAL LIQUID CRYSTALS

The pitch, $p$, is the distance for one full revolution of the director, and its magnitude depends on the structure of the optically active molecules. If the cholesteric phase is induced by an optically active dopant, the pitch varies linearly with the concentration ($c$). This property of the dopant is termed the 'helical twisting power' (HTP):[2]

$$\text{HTP} = (pc)^{-1} \qquad (3)$$

From a consideration of its structural features, the cholesteric phase can be interpreted as a helically twisted nematic phase. The interaction of polarised light with a helically twisted nematic phase controlled by an electrical field is the basic principle of the most widely used liquid crystal electrooptic effect in display technology,[3] and constitutes the most important technical application of liquid crystals. The plane of linearly polarised light passing through a helically twisted nematic phase of thickness $d = p/4$ along the helical axis follows the rotation of the director. Consequently, the planes of polarisation of the entering and the leaving light are perpendicular to each other. A device with a dielectrically positive liquid crystal sandwiched between two glass plates which are coated with alignment layers to induce the director to be parallel to the surface (homogeneous alignment), and with the direction of the two alignments twisted by 90°, transmits light if it is viewed between crossed polarisers (Figure 4a). As nematic liquid crystals with positive dielectric anisotropy orient their molecular long axes parallel to an electric field, the arrangement of the liquid crystals in this device can be changed by applying a voltage across two transparent electrodes on the glass plates. In this case the director of the liquid crystals switches to a position perpendicular to the glass plates (homeotropic alignment) and the entering polarised light does not change the direction of its plane of polarisation by passing through the liquid crystal parallel to the director (Figure 4b).

On switching off the electrical field, the original helically twisted arrangement of the liquid crystal is restored by elastic forces resulting from the interaction of the liquid crystal with the surfaces of the alignment layers. The twisted nematic arrangement in this type of electrooptic cell (TN cell) is structurally equivalent to a cholesteric phase.

## 13.2.3 THE CHIRAL SMECTIC C PHASE AND THE FLC DISPLAY

The structure of the $S_A$ phase is the same for optically active and for achiral molecules but, by analogy with the structural features of the nematic and cholesteric phase, there is a chiral smectic C ($S_C^*$) phase, which differs from the $S_C$ phase only in its helical structure. From one layer to the next the tilted director rotates gradually around an axis perpendicular to the layer, thus forming a helix with a pitch $p$ (Figure 5).

Symmetry considerations led Meyer et al[4]. to the conclusion that smectic C liquid crystals composed of optically active molecules must have a spontaneous

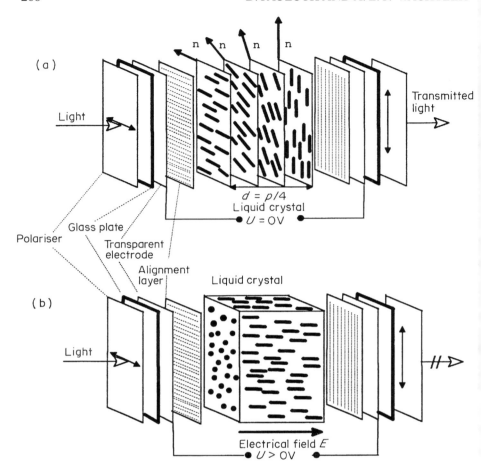

**Figure 4** The TN cell with crossed polarisers is the most widely used display. (a) The bright 'off' state with no voltage applied; (b) the dark 'on' state

electrical polarisation characteristic of a ferroelectric state. This was proved by p-decyloxybenzylidene p'-amino-2-methylbutylcinnamate (DOBAMBC), a material designed and synthesised to exhibit the above-mentioned properties (Figure 6).

Any macroscopic physical property of a material must include all symmetry elements given by the structure of the material; thus ferroelectricity is compatible only with an $S_C^*$ phase and not with an $S_C$ phase. Considering the local symmetry of an $S_C$ layer averaged over time, we find $C_{2h}$ symmetry with a twofold rotation axis $C_2$, a mirror plane $\sigma_h$ and a centre of inversion $i$ (Figure 7). These symmetry elements do not allow the existence of a macroscopic vectorial physical property such as the spontaneous polarisation P ($p_x, p_y, p_z$).

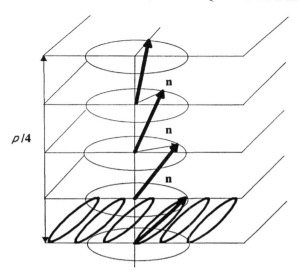

**Figure 5** Schematic drawing of the $S_C^*$ phase showing the precession of the director around the normal of the layers of the $S_C^*$ phase

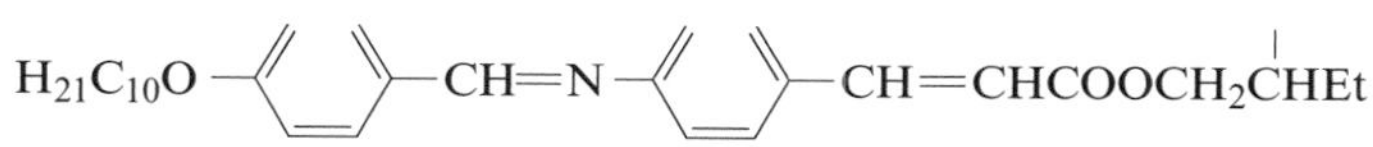

$$H_{21}C_{10}O-\langle\!\!\bigcirc\!\!\rangle-CH=N-\langle\!\!\bigcirc\!\!\rangle-CH=CHCOOCH_2\overset{|}{C}HEt$$

K76 $S_I^*$(63) $S_C^*$ 95 $S_A^{117I^-}$          $P_S$–3nC cm$^{-2}$

**Figure 6** Chemical structure, phase range and spontaneous polarisation of the first reported ferroelectric $S_C^*$ compound[4,18]

$$\text{point group } C_{2h}: \quad \mathbf{P}(p_x, p_y, p_z) \xrightarrow{C_2} \mathbf{P}'(-p_x, p_y, -p_z) \quad (4)$$

$$\mathbf{P}(p_x, p_y, p_z) \xrightarrow{\sigma_h} \mathbf{P}'(p_x, -p_y, p_z) \quad (5)$$

$$\mathbf{P}(p_x, p_y, p_z) \xrightarrow{i} \mathbf{P}'(-p_x, -p_y, -p_z) \quad (6)$$

$$\implies p_x = p_y = p_z = 0$$

**Figure 7** Allowed symmetry operations for the $S_C$ phase (point group $C_{2h}$)

In an $S_C^*$ layer composed of optically active molecules, there is no mirror plane and no centre of inversion because these symmetry elements are incompatible with optically active molecules. The remaining twofold $C_2$ axis allows the existence of a permanent dipole moment parallel to this axis (parallel to the $y$-coordinate) and perpendicular to the director **n** (Figure 8), because the compo-

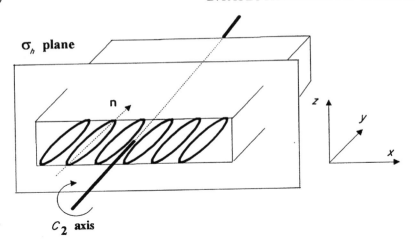

**Figure 8** Schematic drawing showing the tilted molecules within one $S_C$ layer. The molecules are lying in the $\sigma_h$ plane with the $C_2$ axis being perpendicular to the $\sigma_h$ plane

nent $p_y$ of the spontaneous polarisation **P** remains unchanged upon rotation around C2:

$$\text{point group } C_2: \mathbf{P}\,(p_x,p_y,p_z) \xrightarrow{C_2} \mathbf{P}'\,(-p_x,p_y,-p_z) \quad (7)$$
$$\Rightarrow p_x = p_z = 0;\, p_y \text{ may be different from } 0$$

As a consequence of the helically twisted arrangement of the layers in the $S_C^*$ phase, the direction of the polarisation rotates with the helix and the polarisation

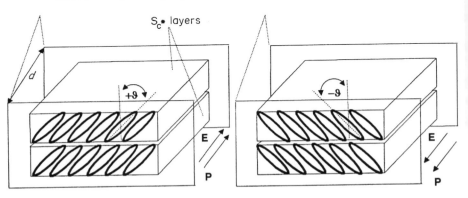

**Figure 9** Operating principle of the SSFLC cell. E, electric field; P, spontaneous polarisation

# SYNTHESIS AND APPLICATION OF CHIRAL LIQUID CRYSTALS

is averaged to zero in the bulk. By unwinding the helix by surface action, Clark and Lagerwall invented a device showing a bistable electrooptic effect.[5] In this surface-stabilised ferroelectric liquid crystal (SSFLC) cell (Figure 9), the smectic layers are arranged perpendicular to the cell glass plates. By interaction with the surface, the rod-like molecules are constrained to a position parallel to the glass plates, thus preventing the $S_C^*$ phase from forming a helical structure.

By applying an electric field **E**, the spontaneous polarisation **P** is adjusted parallel to **E**. Reversing the direction of **E** causes a reversal of **P** by rotation of the molecules on a cone around the layer's normal (Figure 10). If the electric field is switched off, the positions of the molecules are retained. In this way, an SSFLC cell can be switched between two stable states by changing the polarity of an electrical pulse. Given the condition that the tilt angle ϑ is 22.5° and

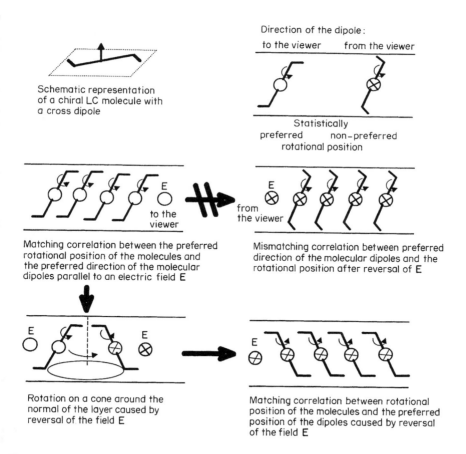

**Figure 10** Schematic drawing showing the switching of a chiral material in the $S_C^*$ phase on the molecular level

$$d\Delta n = \lambda/2 \tag{8}$$

(where $d$ = cell gap, $\Delta n$ = birefringence, and $\lambda$ = wavelength of the light), one of these two states transmits light and the other absorbs light when observed through crossed polarisers.

In order to understand the switching process on the molecular level, we have to look at the rotational motions of the molecules. In the isotropic, nematic and $S_A$ phases the molecules rotate freely around their molecular long axis. Statistically (averaged over time) even molecules with $C_1$ symmetry can be regarded as having an axial symmetry around their molecular long axes. In the tilted $S_C$ phase the situation is different. The rotation around the molecular long axes is unbalanced, giving rise to statistically preferred or non-preferred rotational positions of the molecules. This means that during one revolution a molecule remains for a longer time in the preferred position than in the other positions (Figure 10).

In an electric field **E**, molecules align themselves in such a way that, in the statistically preferred position, their dipoles are parallel to **E**. A reversal of the electric field causes a mismatch between the preferred direction of the dipoles and the preferred rotational positions of the molecules. Excluding an intramolecular rotation of the molecular dipole by rigidly attaching the dipole to the rest of the molecule, this mismatch situation can be overcome only by rotation of the molecules on a cone around the normal of the smectic layer (Figure 10).

Compared with the TN cell, this special switching mode of the SSFLC cell is much faster because, in the switching process, the extent of reorganisation of the $S_C$ phase is smaller than that of the liquid crystalline phase in the TN cell. Additionally, the optical performance of the SSFLC cell depends less on the viewing angle. The prominent feature of the SSFLC display is bistability, offering data storage without power consumption. A problem arising from bistability is the difficulty of realising states between black and white (grey levels). Drawbacks of the SSFLC displays are higher power consumption compared with other advanced LC displays and the technological problems associated with the large-scale manufacture of displays with a very small cell gap of 2–3 $\mu$m (which is necessary in order to unwind the helix of the $S_C^*$ phase).

## 13.2.4 PROPERTIES OF THERMOCHROMIC LIQUID CRYSTALS

If circularly polarised light enters a cholesteric phase, and the wavelength of the light is of the order of the pitch, left-handed circularly polarised light can follow a right-handed helical path and the right-handed circularly polarised light is reflected. The corresponding effect is found for a left-hand helix.[6] As a consequence, a cholesteric phase can reflect light of an appropriate wavelength depending on pitch and handedness of the helix selectively, giving rise to highly iridescent colours.[7] The pitch of the chiral nematic phase usually shows a weak temperature dependence.[7] This results in a shift of the reflected colour from the

# SYNTHESIS AND APPLICATION OF CHIRAL LIQUID CRYSTALS

blue to the red limit of the visible spectrum with decreasing temperature.[8] Such thermochromic liquid crystals find applications in temperature-indicating devices, inks and other areas.

## 13.3 REQUIREMENTS FOR COMMERCIAL LIQUID CRYSTAL MIXTURES

### 13.3.1 NEMATIC MIXTURES

Liquid crystals are organic molecules consisting of a rod-like rigid core and two terminal substituents ($R^1$ and $R^2$). Twenty-five years of industrial research have shown that six membered carbo- and heterocycles (A) are the most efficient building blocks for construction of the rigid core. These building blocks are either linked by a single bond, or by groups such as ethylene or ester (Z). The terminal substituents are generally unbranched alkyl and alkoxy groups or polar substituents, such as nitrile or a fluorine atom (Figure 11).

Displays for different applications require different liquid crystal properties. The common requirements are long-term stability, wide phase range and low viscosity, and important parameters are clearing point, dielectric and optical anisotropy and ohmic resistance. Not all of these requirements can be fulfilled by a single compound, and mixtures of, typically, 10–20 compounds are required (Figure 12).

For a uniform optical performance of a display based on the TN effect, it is necessary to have a uniform chirality of the helical twist of the liquid crystal.

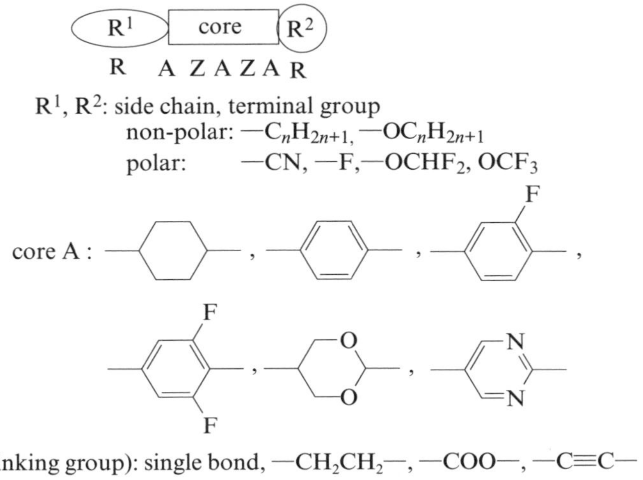

R A Z A Z A R

$R^1$, $R^2$: side chain, terminal group
  non-polar: —$C_nH_{2n+1}$, —$OC_nH_{2n+1}$
  polar: —CN, —F, —$OCHF_2$, $OCF_3$

Z (linking group): single bond, —$CH_2CH_2$—, —COO—, —C≡C—

**Figure 11** General formula of a rod-like liquid crystal. Ring systems for construction of the core and linking groups are limited here to those of commercial importance

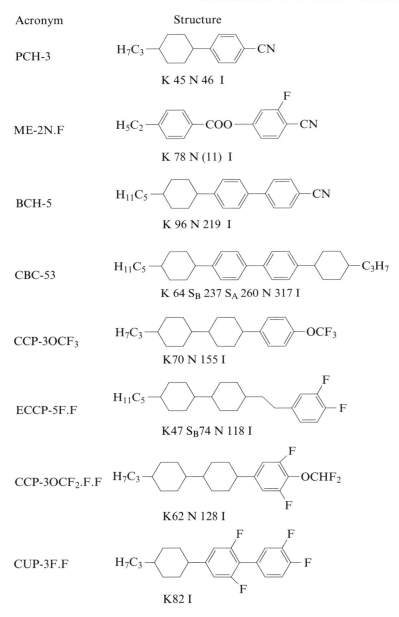

**Figure 12** The Merck Group produces more than 100 different compounds for application in nematic mixtures. The compounds shown here demonstrate different core structures and linking and terminal groups. Compounds with a cyano group are used in TN and STN mixtures. Compounds with fluorine atoms and SFM groups are mainly used in TFT mixtures

While the twist is generated by the interaction between the alignment layers of the cell, the uniform handedness of the twist is achieved by the addition of a chiral dopant. This dopant should not affect either the phase range or the viscosity of the mixture. For practical reasons, the concentration should not be significantly lower than 0.1%. Additionally, the pitch of the mixture has to fit the cell gap of the display. A consequence of these factors is that the HTP of the dopants should be of the order of 10 $\mu m^{-1}$.

Further development of the TN cell has resulted in the more advanced 'super twisted nematic' (STN) cell,[9a] requiring shorter pitch values for the mixture. Fine tuning of the pitch is achieved more readily by the use of a dopant with a larger HTP rather than increasing the dopant concentration.[9b]

Another approach to liquid crystal cells for high information content displays is 'thin film transistor' (TFT) technology. The TFT display differs from the conventional TN display only by integration of thin-film transistors into the transparent electrodes. This technology requires liquid crystals with superior stability and a larger resistivity than conventional TN technology. Generally, liquid crystals made from building blocks such as heterocycles and ester groups are less stable than those built up from six-membered carbocycles linked by single bonds or ethylene groups. All building blocks with a high affinity for ions result in materials with low resistivities. In particular, structural elements such as cyano groups must be avoided in materials for TFT displays. In this case polarity is introduced by fluorine atoms or multiply fluorinated groups such as $OCF_3$ or $OCHF_2$ ['super fluorinated materials', (SFM)].[10]

## 13.3.2 DOPANTS FOR NEMATIC MIXTURES

Initially, readily available cholesterol esters such as cholesteryl nonanoate (CN) were used as chiral dopants. The poor stability of these compounds to ultraviolet light[8] resulted in the search for more robust dopants. An additional target was to develop dopants with a less complex structure than steroids.

In the 1970s, liquid crystal core structures were mainly of the ester type, and an obvious approach to these was to link typical rod-like building blocks with commercially available chiral alcohols by an ester group. This resulted in the development of S-811, which requires (S)-octan-2-ol, and CB-15 with opposite handedness. As a consequence of the high price and the limited availability of (S)-octan-2-ol, an in-house synthesis was developed according to the method of Kenyon.[11] As this approach was based on resolution, (R)-octan-2-ol also became available, and R-811, a dopant having opposite handedness, but with all other properties being the same, was synthesised.

The introduction of the STN cell resulted in the need for a dopant with higher HTP, for example (R)- and (S)-1011.[9b] These new ester type dopants are based on (R)- and (S)-1-phenylethane-1,2-diol. The rapidly growing importance of TFT-TN displays in the last five years has created a need for a dopant with HTP comparable with that of S-811 but with improved stability and resistivity. This was

| Acronym | Structure | Phase range | HTP ($\mu m^{-1}$) ZLI-1132 | HTP ($\mu m^{-1}$) MLC-6012 |
|---|---|---|---|---|
| CN | 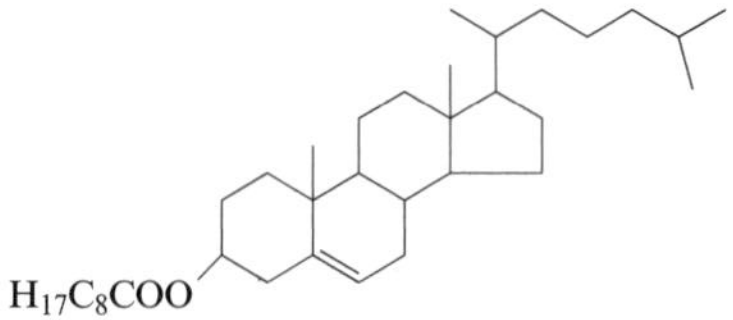 $H_{17}C_8COO$– (cholesteryl) | K 26 Ch 41 I | −4.4 | |
| S-811 | 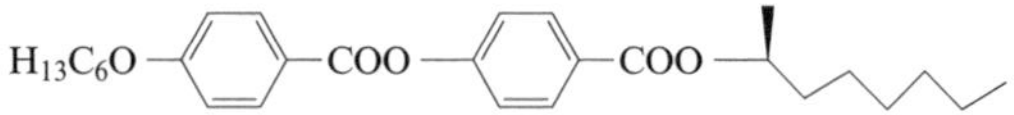 $H_{13}C_6O$–C$_6$H$_4$–COO–C$_6$H$_4$–COO–CH(CH$_3$)C$_6$H$_{13}$ | K 48 I | −13.3 | −14.0 |
| R-811 | Enantiomer to S-811 | | +13.3 | +14.0 |
| C-15 | 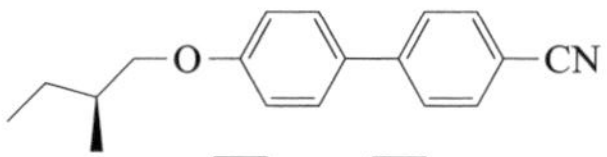 (cyanobiphenyl-O-CH$_2$-CH(Et)-C$_2$H$_5$) | K 154 I | −1.6 | |
| CB-15 | 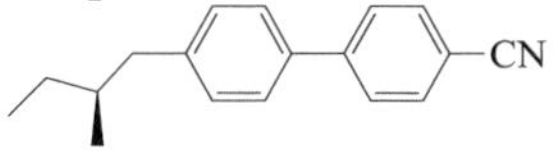 (cyanobiphenyl-CH$_2$-CH(Et)-C$_2$H$_5$) | K 4 Ch (−30) I | +9.6 | |
| S-1011 | 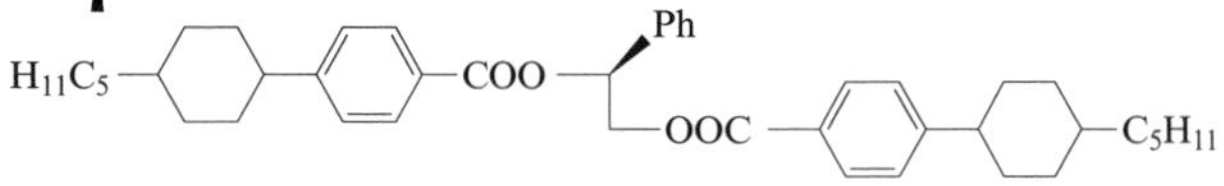 $H_{11}C_5$–Cy–C$_6$H$_4$–COO–CH(Ph)–OOC–C$_6$H$_4$–Cy–C$_5$H$_{11}$ | K 134 I | −39.5 | |
| R-1011 | Enantiomer of S-1011 | | +39.5 | |
| S-2011 | 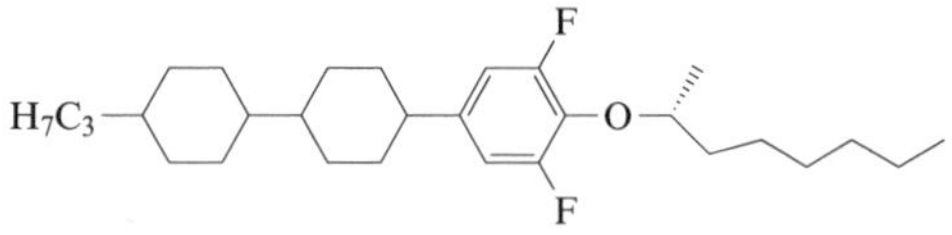 $H_7C_3$–Cy–Cy–C$_6$H$_2$F$_2$–O–CH(CH$_3$)C$_6$H$_{13}$ | K 19 S$_B$ 45 Ch 72 I | | −13.6 |
| R-2011 | Enantiomer of S-2011 | | | +13.6 |

**Figure 13** Commercial dopants for nematic mixtures. HTP values are calculated (equation 3) from pitch values determined in a Grandjean

# SYNTHESIS AND APPLICATION OF CHIRAL LIQUID CRYSTALS

achieved by combining core structures developed for SFM materials with enantiopure octan-2-ol via an ether linkage (Figure 13).

### 13.3.3 FERROELECTRIC MIXTURES

Much research effort on non-chiral $S_C$ and chiral $S_C^*$ materials[12] in the last 10 years has resulted in several commercial ferroelectric mixtures.[13] Whereas the nematic materials are mainly based on cyclohexylphenyl and bicyclohexylphenyl core structures, with a short alkyl chain on the cyclohexane ring and a polar substituent at the phenyl ring, $S_C$ materials are based on biphenyl and analogous heterocyclic core structures, with two long-chain alkyl or alkoxy groups (Figure 14). The $S_C$ phase is converted into an $S_C^*$ phase by the addition of a chiral dopant. This dopant must not affect the phase range ($S_C$ or $S_C^*$ respectively) of the base mixture, or interfere with the unwinding of the $S_C^*$ helix in the SSFLC cell by a large HTP. High values of the spontaneous polarisation are required to switch the mixture.[14]

### 13.3.4 FERROELECTRIC DOPANTS

The first optically active ferroelectric $S_C^*$ compound was synthesised in 1974 (Figure 6),[4] and industrial research on $S_C^*$ materials began ten years later. It became obvious that the optically active materials are of much greater importance for smectic formulations than for nematic mixtures. In addition to the more complex properties of the latter, the fact that they are used in concentrations up to 10% contributes to their economic importance. Consequently, building blocks

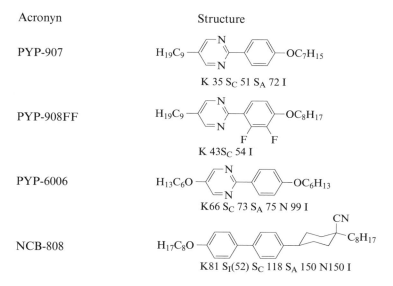

**Figure 14** Overview of compounds used in commercial FLC mixtures[13]

from the chiral pool, and derivatives thereof, were screened intensively. Much work was devoted to terpenes and terpene derivatives,[15] amino acid derivatives[16] and lactic acid derivatives.[17] A general disadvantage of terpene-derived compounds is that these materials offer very low values for the spontaneous polarisation. During the systematic variation of the substituents it was found that only compounds with a strong dipole at the chiral centre result in large values for the spontaneous polarisation. For a given structure, spontaneous polarisation increases when the chiral centre is moved closer to the core.[18] Thus, high values of spontaneous polarisation are found for materials with the chiral centre and the cross-dipole coupled rigidly to the core. This was effected by using optically active α-chloro acid esters, ethers of lactic acid esters and ethers of optically active cyanohydrins. Later, optically active 2-fluoroalkanols[19a] were identified as allowing the synthesis of dopants with a wide phase range and high spontaneous polarisation[19b] (Figure 15).

As in the case of the nematic dopants, narrowing of the phase range of the base mixture can be avoided by using the same core structure for dopants and

| Code No. | Structure | HTP ($\mu m^{-1}$) | $P_s$ (nC cm$^{-2}$) |
| --- | --- | --- | --- |
| IS-2175 | 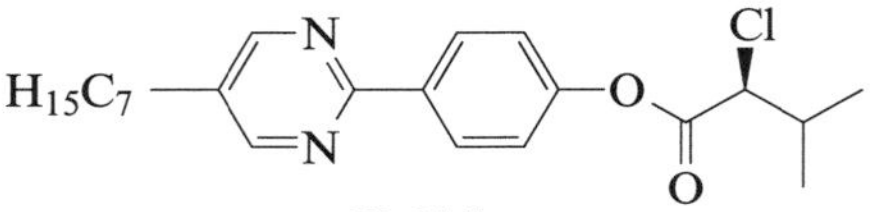 K 64 I | +0.3 | 14.3 |
| IS-2874 | 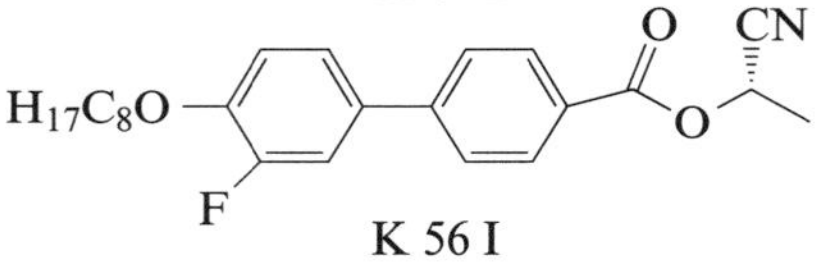 K 56 I | +2.1 | 31.3 |
| IS-2875 | 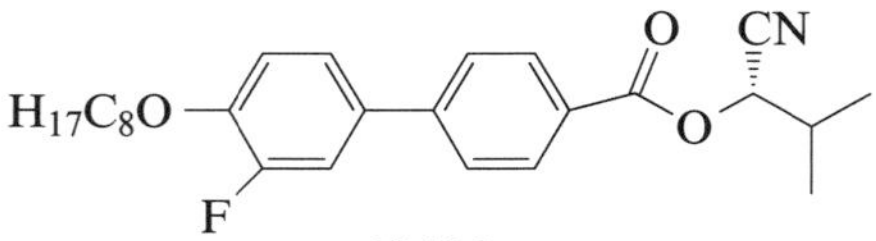 K 49 I | +5.9 | 36.5 |
| IS-3512 | 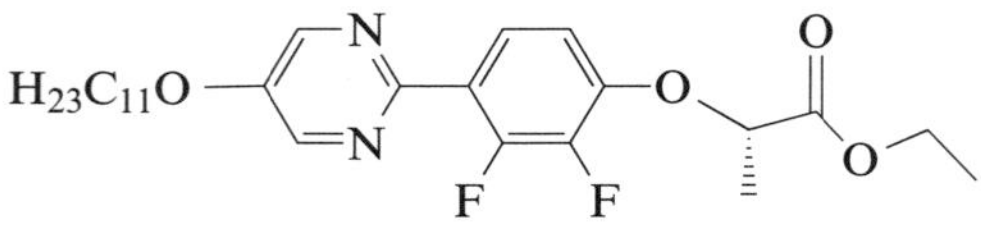 K 38 I | +0.2 | 15.3 |
| IS-4006 | 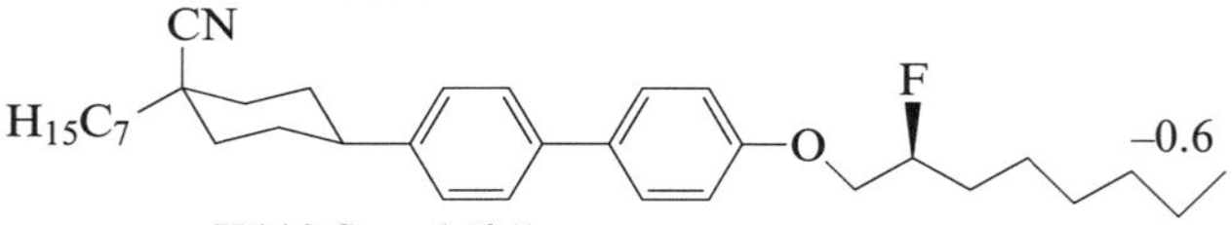 K 110 Sm$_A$ 159 I | −0.6 | −19.5 |

**Figure 15** Dopants developed by Merck for FLC mixtures[13]

base materials. Ideally, dopants for ferroelectric mixtures should combine a broad $S_C^*$ phase and a high value for spontaneous polarisation, but this combination of properties is still difficult to realise in practice.

### 13.3.5 THERMOCHROMIC LIQUID CRYSTALS

Thermochromic liquid crystals find many applications, mainly in the areas of thermometry, medical thermography, non-destructive testing, radiation sensing and colours/inks.[7] Common requirements for thermochromic liquid crystals are stability and low melting points, and commercial compounds with these features have been developed using benzene and biphenyl core structures. Their properties can be tuned by variation of the length of the non-chiral alkyl and alkoxy side chains (Figure 16). The compounds are commercially available in different formulations including mixtures of thermochromic compounds, microencapsulated thermochromic materials and combinations of micro-capsules with various inks and surface coatings.

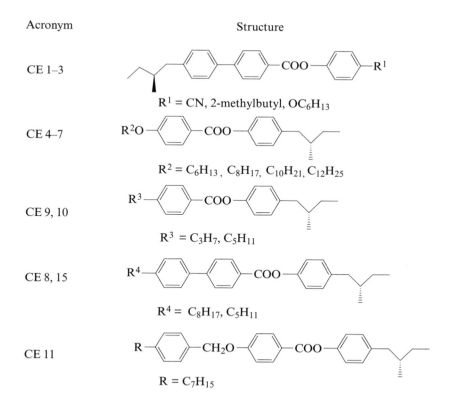

**Figure 16** Commercially available thermochromic liquid crystals from Merck

## 13.4 SYNTHESIS OF CHIRAL LIQUID CRYSTALS

### 13.4.1 CHIRAL BUILDING BLOCKS FOR DOPANTS USED IN NEMATIC MIXTURES

The world market for nematic liquid crystal mixtures in 1994 was of the order of 16.5 tons.[20] Bearing in mind the typical dopant concentration of 0.1%, it is clear that the dopants are produced on a small scale. As noted previously, the required chiral building blocks are cholesterol, (S)-2-methylbutanol, (R)- and (S)-octan-2-ol and (R)- and (S)-1-phenylethane-1,2-diol. Cholesterol is found in animal tissue and (S)-2-methylbutanol is a by-product formed in alcoholic fermentation. Both of these naturally occurring optically active alcohols have been available in commercial quantities for a long time. As already noted, (R)- and (S)-octan-2-ol are prepared by classical resolution, which is effected with brucine via the octyl hydrogenphthalate.[11] Although this method is convenient, great care has to be taken in handling the highly toxic brucine which, owing to its high price, must be recycled.

Many other approaches have now been developed, mainly based on enzymatic and microbiological methods,[21] to prepare the necessary enantiopure secondary alcohols. Today, both enantiomers of octan-2-ol and 1-phenylethane-1,2-diol are readily available in larger quantities, making small-scale in-house production unattractive.

### 13.4.2 CHIRAL DOPANTS FOR NEMATIC MIXTURES

The nematic dopants can be classified according to the linking group between the liquid crystal core structure and the chiral building block into esters, alkylated aromatics and ethers. Starting from optically active alcohols, the esterification processes can be carried out under standard conditions, since the chiral centre is not involved in the reaction. As an example, the synthesis of R-1011 is shown in Figure 17.[22]

Direct attachment of a branched chiral side chain to an aromatic core can be achieved by metal-catalysed cross-coupling of an aryl Grignard reagent and an

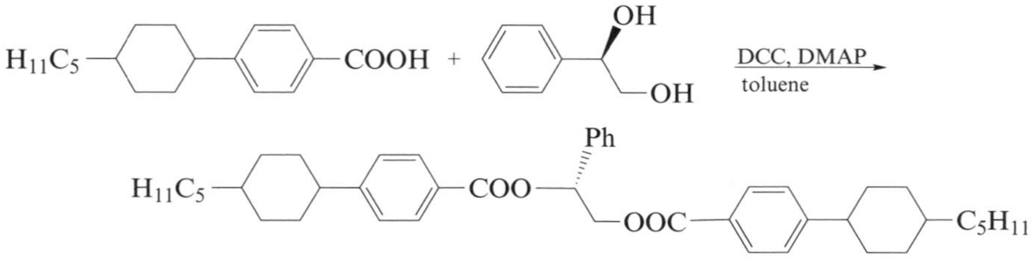

**Figure 17** Synthesis of R-1011

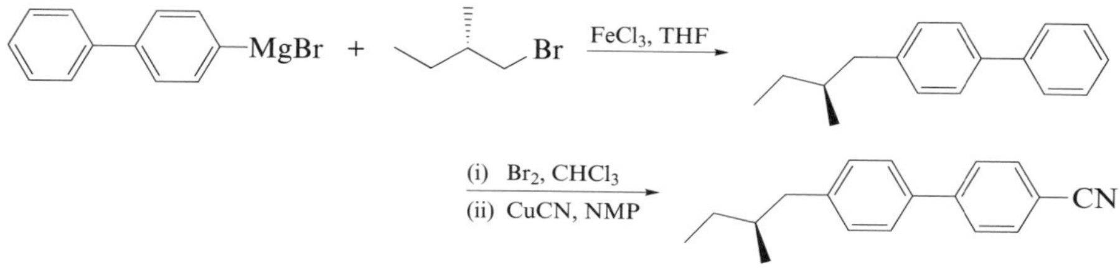

**Figure 18** Synthesis of CB-15

alkyl bromide, exemplified in the synthesis of CB-15[23] (Figure 18). The ethers R- and S-2011 are prepared from 4-[*trans*-4-(*trans*-4-propylcyclohexyl)cyclohexyl]-2,6-difluorophenol via the Mitsunobu reaction with inversion of the chiral centre[24] of the corresponding enantiopure octan-2-ol (Figure 19).[25] For safety reasons, diisopropyl azodicarboxylate (DIAD) is used instead of the explosive diethyl azodicarboxylate (DEAD).[26]

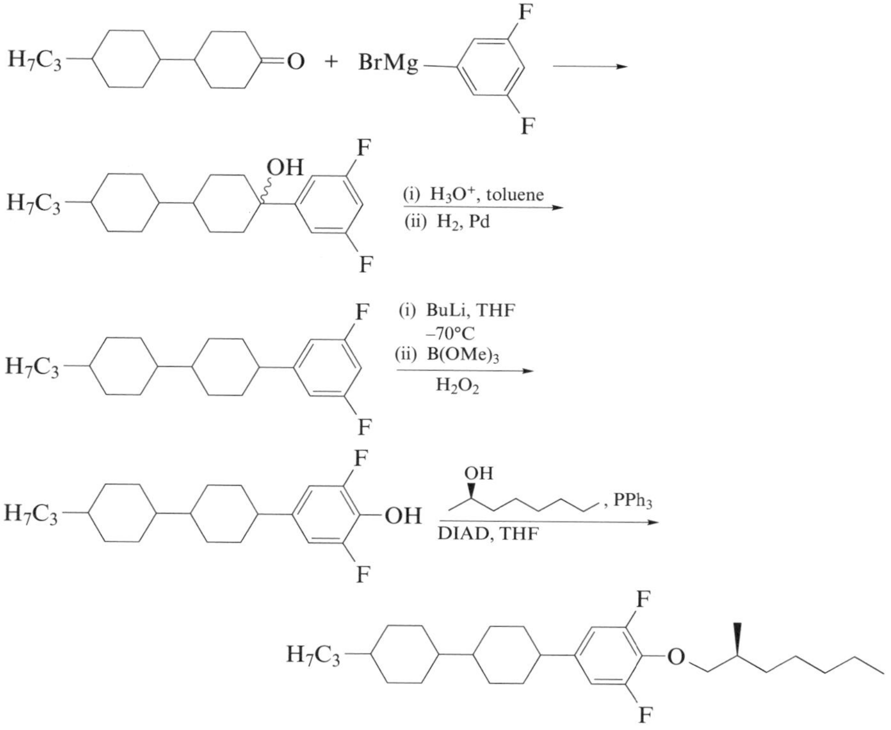

**Figure 19** Synthesis of R-2011

### 13.4.3 CHIRAL BUILDING BLOCKS FOR FERROELECTRIC DOPANTS

The first-generation dopant for commercial ferroelectric mixtures was a derivative of the optically active α-chloroacid derived from L-valine by diazotisation in the presence of HCl[27] (Figure 20). The α-chloro acids were succeeded by derivatives of optically active α-hydroxy acids. The latest generation of dopants are derivatives of optically active 2-fluoroalkan-2-ols which are readily available from commercial optically active 1,2-epoxyalkanes, (see also Volume I, Chapter 7).

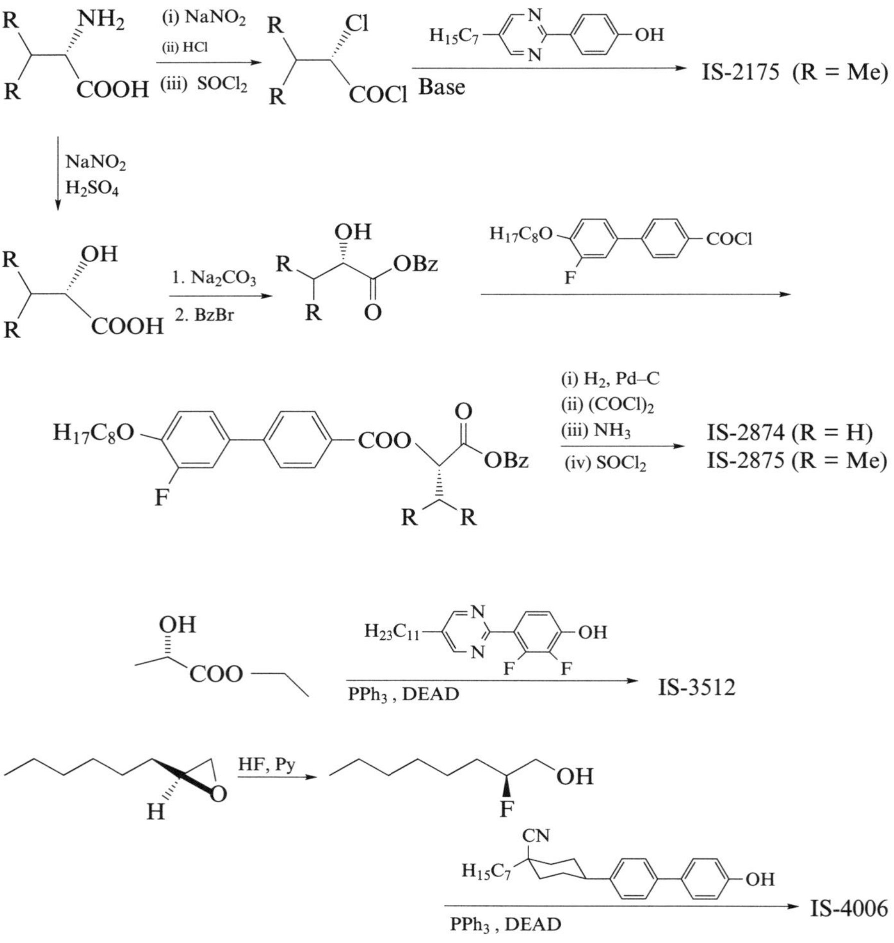

**Figure 20** Synthesis of dopants for $S_C$ mixtures from commercial enantiopure building blocks. The synthesis of the liquid crystal intermediates is described in the references

# SYNTHESIS AND APPLICATION OF CHIRAL LIQUID CRYSTALS

## 13.4.4 FERROELECTRIC DOPANTS

As in the case of the nematic dopants, the ferroelectric dopants can be classified according to the linking group into esters and ethers. The ester IS-2175 is made by esterification of enantiopure (S)-2-chloro-3-methylbutanoic acid chloride with 4-(5-heptyl-1,3-pyrimid-2-yl)phenol under basic conditions (Figure 20). The syntheses of the cyanohydrin esters IS-2874 and IS-2875[28] are more difficult. To make IS-2875, L-valine is transformed, with retention of configuration, into the corresponding optically active α-hydroxy acid, which, as the sodium salt, is treated with benzyl bromide to give the benzylic ester. The hydroxy group of this intermediate is esterified with 3′-fluoro-4′-octyloxybiphenyl-4-carboxylic acid chloride. The benzylic ester is cleaved by hydrogenolysis, and the resulting acid is treated successively with oxalic acid chloride, ammonia and thionyl chloride to produce IS-2875 (Figure 20). Starting from lactic acid, the dopant IS-2874 is made using an analogous reaction sequence.

The ether-type compounds IS-3512 and IS-4006[29] are prepared via the Mitsunobu reaction of 4-(5-undecanyl-1,3-pyrimid-2-yl)-2,3-difluorophenol with ethyl (S)-lactate and reaction of 4′-(4-cyano-4-*trans*-heptylcyclohexyl)biphenyl-4-ol with (S)-2-fluorooctanol, respectively. The latter compound is available from (R)-1,2-epoxyoctane by treatment with hydrogen fluoride–pyridine complex[19a] (Figure 20).

## 13.4.5 THERMOCHROMIC LIQUID CRYSTALS

With a demand of the order of 1 tpa, chiral compounds for thermochromics are produced on a larger scale than chiral dopants for display application. The products described here compete with inexpensive cholesterol esters and, for economic reasons, they are all based on two key optically active intermediates derived from cheap (S)-2-methylbutan-1-ol. These intermediates are combined with building blocks from the production of thermotropic liquid crystals.

The first key intermediate is (S)-4-(2-methylbutyl)-biphenyl-4′-carboxylic acid prepared from CB-15 by hydrolysis of the nitrile group under basic conditions. Treatment with thionyl chloride yields the corresponding acid chloride, which is then esterified with 4-alkylphenols to give the thermochromic liquid crystals CE 1 to 3[23] (Figure 21).

The second key intermediate is (S)-4-(2-methylbutyl)phenol, which is synthesised by a metal-catalysed cross-coupling reaction between 4-methoxyphenylmagnesium bromide and (S)-1-bromo-2-methylbutane. The methoxy group is then cleaved with hydrogen bromide to yield (S)-4-(2-methylbutyl)phenol. Esterification of this phenol yields the thermochromic liquid crystals CE 4 to 15.[23]

## 13.4.6 OPTICAL PURITY

The optical purities of the chiral building blocks used in the synthesis of the dopants are listed in Table 1. As noted above, the dopants are used to induce a

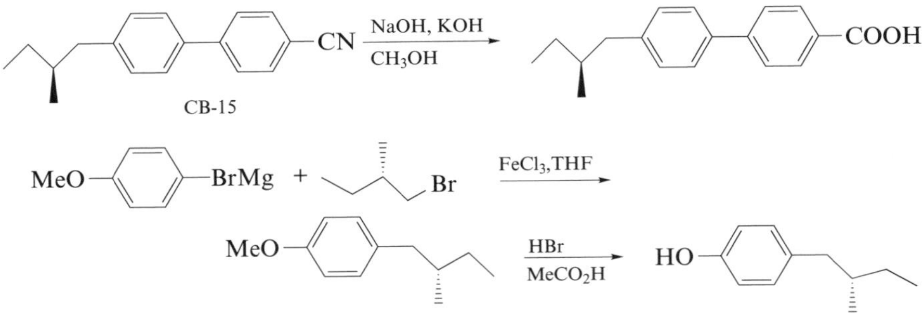

**Figure 21** Synthesis of the key chiral intermediates for thermochromic liquid crystal

**Table 1** The *ee* values of chiral building blocks used in the synthesis of dopants

| Compound | ee(%) | Ref. |
|---|---|---|
| (S)-2-Methylbutanol | 80 | 12 |
| (R)-Octan-2-ol | > 99 | 30 |
| (S)-Octan-2-ol | > 99 | 30 |
| (S)-Valine | > 99 | 30 |
| (S)-2-Chloro-3-methylbutanoic acid | 98 | 27 |
| Ethyl (S)-lactate | > 98 | 30 |

specific pitch in nematic mixtures or a specific spontaneous polarisation in ferroelectric mixtures. Consequently, the HTP and the spontaneous polarisation are used in the characterisation of the dopants.

## 13.5 TRENDS AND OUTLOOK

The world market for liquid crystal displays is expected to grow from US$5 billion in 1994 to $12 billion in 2000, with the largest growth expected in the area of TFT displays.[31] The corresponding market for liquid crystal mixtures will grow in this period from 16.5 tons ($89 million) to 27.3 tons ($179 million).[20] The market is demanding larger displays with improved optical performance, and reduced power consumption for portable devices. These targets will be approached both by display development by the use of new driver chips and also by the development of new TFT modes[32] and technologies.[33] Large improvements of materials performance are necessary, requiring mainly the development of new polar materials with low viscosity and improved electrical purity. Another target is the development of systems prepared from prepolymers/polymers and liquid crystals in which conventional orientation layers are replaced by molecular networks. The development of new TFT modes might also require dopants with properties fine-tuned appropriately.

The future of ferroelectric liquid crystals will depend heavily on the introduction and success of the first commercially available SSFLC display in 1995.[34]

## 13.6 ACKNOWLEDGEMENTS

The authors would like to dedicate this chapter to Professor Dr H. C. Michael Hanack on the occasion of his 65th birthday. The authors are grateful to Dr M. Bremer, Dr D. Coates, Dr J. Krause and Mr V. Reiffenrath for helpful discussions and Ms B. Nipkow for preparing the manuscript.

## 13.7 NOTES AND REFERENCES

1. Reinitzer, F., *Monatsh. Chem.*, 9, 421 (1888).
2. Vertogen, G., and de Jeu, W. H., *Thermotropic Liquid Crystals, Fundamentals*, Springer, Berlin, 1988.
3. Schadt, M., and Helfrich, W., *Appl. Phys. Lett.*, 18, 127 (1971).
4. Meyer, R. B., Liebert, L., Strzelecki, L., and Keller, P., *J. Phys. Lett.*, 36, 4 (1975).
5. (a) Clark, N. A., and Lagerwall, S. T., *Appl. Phys. Lett.*, 36, 899 (1980); (b) Lagerwall, S.T., and Dahl., I., *Mol. Cryst., Liq. Cryst.* 14, 151 (1984).
6. Bouligand, Y., *J. Phys.*, 34, 603 (1973).
7. McDonnell, D. G., in *Thermotropic Liquid Crystals*, (ed. G. W. Gray), Wiley, Chichester, 1987, Vol. 22; Sage, I., in *Liquid Crystals Applications and Uses* (ed. B. Bahadur), World Scientific, Singapore, 1992, Vol. 3.
8. Parsley, M., in *Licritherm* (ed. I. C. Sage and K. G. Archer), BDH Poole, 2nd edn, Sage, I., in *Ullmann's Encyclopedia of Industrial Chemistry*, VCH, Weinheim, 1990, Vol. A15.
9. (a) Waters, C. M., and Raynes, E. P. (Ministry of Defence), *Br. Pat.* 2 123 163; (b) Heppke, G., Lötzsch, D., and Oestreicher, F., *Z. Naturforsch., Teil A.*, 41, 1214 (1986); Hochgesang, R., Plach, H. J., and Reiffenrath, V., *Produktinformation*, Merck, Darmstadt, 1983; Heppke, G., and Oestreicher, F. (Merck), *Eur. Pat.*, EP 0 168 043 1990.
10. Eidenschink, R., and Pohl, L. (Merck), *Eur. Pat.*, EP 0 051 738, 1984; Kurmeier, H. A., Scheuble, B., Poetsch, E., and Finkenzeller, U. (Merck), *Eur. Pat.*, EP 0 334 911 1, 1993; Reiffenrath, V., Kurmeier, H. A., Poetsch, E., Plach, H. J., and Finkenzeller, U. (Merck) *PCT Int. Pat Appl.*, WO 91/03450, 1991.
11. Kenyon, J., *Org. Synth., Coll. Vol.*, 1, 418.
12. Goodby, J. W., Blinc, R., Clark, N. A., Lagerwall, S. T., Osipov, M. A., Pikin, S. A., Sakurai, T., Yoshino, K., and Zeks, B., *Ferroelectric Liquid Crystals, Principles, Properties and Applications*, Gordon and Breach, New York, 1991.
13. Merck has terminated all FLC activities and has sold its patent portfolio to Hoechst (February 95). Correspondingly, Hoechst has transferred all patent rights for nematic liquid crystals to Merck.
14. Finkenzeller, U., Pausch, A. E., Poetsch, E., and Suermann, J., *Kontakte (Darmstadt)*, 2, 3 (1993).
15. Coates, D., Gray, G. W., Lacey, D., Young, D. J. S., Toyne, K. J., and Bone, M. F. (Ministry of Defence), *PCT Int. Pat. Appl.*, WO 86/04328, 1986.

16. Bahr, C., and Heppke, G. *Mol. Cryst. Liq. Cryst. Lett.*, **4**, 31 (1986); Demus, D., Zaschke, H., Weissflog, W., Mohr, K., Köhler, S., and Worm, K. (VEB Werk), *Ger. Pat. Appl.*, DE 362 796 4, 1985; Eidenschink, R., Escher, C., Geelhaar, T., Hittich, R., Kurmeier, H. A., Pauluth, D., and Wächtler, A.E.F (Merck), *Eur. Pat.* EP 0 257 049, 1990.
17. Sage, I., Jenner, J., Kurmeier, H. A., Pauluth, D., Escher, K., and Poetsch, E. (Ministry of Defence), *Eur. Pat.*, EP 0 263 843, 1986; Geelhaar, T., Kurmeier, H. A., and Wächtler, A.E.F., paper presented at the 12th ILCC, Freiburg, 1988.
18. Walba, D. M., Slater, S. C., Thurmes, W. N., Clark, N. A., Handschy, M. A., and Supon, F., *J. Am. Chem. Soc.*, **108**, 5210 (1986).
19. (a) Nohira, H., Kamei, M., Nakamura, S., Yoshinaga, K., and Kai, M. (Canon KK), *Jpn. Pat.*, JP 62 093 248, 1985; (b) Pausch, A.E., Geelhaar, T., Poetsch, E., and Finkenzeller, U., in *Liquid Crystal Materials, Devices and Applications* (ed. P. S. Drzaic and E. Uci), *Proc. SPIE*, 1665, **128** (1992).
20. Mentley, D. E., and Castellano, J. A., *Liquid Crystal Display Manufacturing, Materials and Equipment*, Stanford Resources, San Jose, CA, 1994.
21. Kawashima, M., (Kankyo Kagaku Sentaa KK)., *Jpn. Pat.*, JP 05 317 090, 1992; Wong, C. H., Wang, Y. F., Hennen, W. I, and Babiak, K. (Searle), *Eur. Pat.*, 0 357 009, 1989; Nishio, T., Seto, K., Ohashi, M., Terao, Y., Tsuji, K., and Achinami, K. (Sapporo Breweries), *Jpn. Pat.* JP 02 142 495, 1988; Murakami, N., and Hara, S., (Idemitsu Kosan), *Jpn. Pat.*, JP 01 257 484; Barton, P., and Page, M. I. (ICI), *Int. Pat.*, 93/02207, 1993; Ogura, M., Shiraishi, T., Takahashi, H., and Hasegawa, J. (Kanegafuchi), *Eur. Pat.*, EP 0 317 998, 1988; Suzuki, G., and Murakami, N. (Idemitsu Kosan), *Jpn. Pat.*, JP 63 042 700, 1986.
22. Heppke, G., and Oestreicher, F., (Merck), *Eur. Pat.*, EP 0 168 043, 1990.
23. Gray, G. W., and McDonnell, D. C. (Ministry of Defence), *Br. Pat.*, 1 592 161, 1977.
24. Bittner, S., and Assaf, Y. *Chem. Ind.* (London), 281 (1975); Mitsunobu, O., *Synthesis*, 1 (1981).
25. Pauluth, D., and Plach, H. J., (Merck) *Ger. Pat. Appl.*, DE 4 322 905, 1993.
26. Camp, D., and Jenkins, I. D., *J. Org. Chem.*, **54**, 3045 (1989).
27. Fu, S.-C., Birnbaum, S. M., and Greenstein, J. P., *J. Am. Chem. Soc.*, **76**, 6054 (1954); Koppenhöfer, B., personal communication.
28. Coates, D., Greenfield, S., and Sage, I. C. (Ministry of Defence), *US Pat.*, 5 230 830, 1992.
29. Wächtler, A. E. F., Geelhaar, T., Hittich, R., Bartmann, E., Krause, J., and Poetsch, E., (Merck), *PCT Int. Pat Appl.* WO 90/13 611, 1990.
30. *Chemika–BioChemika Catalogue*, Fluka, Buchs, 1991, pp. 1044, 1375.
31. Trish, C., Flat Panel Display Industry Overview in *Technology and Application Trends*, (ed. J. A. Castellano), Stanford Resources., San Jose, CA, 1993: worldwide LCD market opportunity past present and future, flat panel display industry overview.
32. Shimada, S. (Sharp KK), *Eur. Pat. Appl.*, EP 0 626 607, 1994.
33. Kiefer, R., Weber, B., Windscheid, F., and Baur, G., presentation at the 12th International Display Research Conference, Hiroshima, Japan, October 1992: in-plane switching of nematic liquid crystals.
34. The start of the production of 15 inch SSFLC displays at 5000 per month by Canon in the spring of 1995 was announced by, amongst others, Nikkon Kogyo on 19 November 1994 and by Handontai Sangyo on 21 December 1994.

# 14 Bio-transformation in the Production of L-Carnitine

Th. P. ZIMMERMANN, K. T. ROBINS,* J. WERLEN and F. W. J. M. M. HOEKS

*LONZA AG, Visp, Switzerland*

*In memory of Hans G. Kulla*

| | | |
|---|---|---|
| 14.1 | Introduction | 287 |
| 14.2 | Strain Isolation | 288 |
| 14.3 | Butyrobetaine/L-Carnitine Metabolism | 288 |
| 14.4 | Transport | 292 |
| 14.5 | Genetic Background | 293 |
| 14.6 | The Bio-transformation Strain | 294 |
| 14.7 | The Recombinant Bio-transformation Strain | 296 |
| 14.8 | Process Design | 298 |
| 14.9 | Kinetic Analysis | 300 |
| 14.10 | Raw Materials Aspects | 303 |
| 14.11 | Conclusion | 304 |
| 14.12 | References | 304 |

## 14.1 INTRODUCTION

L-Carnitine, (R)-3-hydroxy-4-(trimethylammonio)butanoate, belongs to a group of vitamin-like nutrients. Also known as vitamin $B_T$, it is synthesised in the human liver from lysine and methionine and functions in the transport of activated fatty acids over the mitochondrial inner membrane in eukaryotes. L-carnitine plays an essential role in the β-oxidation of long-chain fatty acids by mitochondria; skeletal and cardiac muscle cells rely upon this mechanism as a source of metabolic energy. Insufficient L-carnitine is produced in infants,[1] adolescents and adults under certain physiological conditions, and a dietary deficiency of lysine or methionine can lead to an L-carnitine deficiency. This condition is associated with muscle weakness, fatigue and elevated levels of triglycerides in the blood. The role of L-carnitine in infant, health, sport and geriatric nutrition is receiving increasing interest, and applications in the medical field have been reviewed and discussed recently.[2] L-Carnitine is considered to be an essential nutrient for

---

*K. T. R. was not directly involved with the invention covered by the patent applications.

*Chirality in Industry II.* Edited by A. N. Collins, G. N. Sheldrake and J. Crosby
© 1997 John Wiley & Sons Ltd

infants because endogenous biosynthesis is not adequate to meet infant needs.[1] Infants fed unsupplemented soya-based formulae receive virtually no dietary L-carnitine until other foods are introduced. It has been demonstrated that supplementation of the diet of healthy people can improve physical performance,[3] and the beneficial effect of L-carnitine for the heart has led to its use in geriatric nutrition. The use of L-carnitine also has a large commercial potential in the animal feed sector; improved growth rates and improved feed efficiencies have been reported for various species.[4,5]

The metabolically active form of carnitine is L-carnitine. D-Carnitine competes with L-carnitine for active transport into the eukaryotic cell,[1] and it has been shown in fish cultures that D-carnitine causes slower growth.[6] For this reason, the pure L-enantiomer is required for use in the medical, nutritional and feed applications of carnitine.[7] LONZA has developed a new bio-transformation route from the achiral compound 4-butyrobetaine, which gives L-carnitine in 99.5% yield and 100% *ee*. Other possibilities for the bio-technological production of L-carnitine[8] which involve resolution of racemic carnitine and its precursors have an inherent yield disadvantage and afford less than 100% *ee*. As at least 50% of this yield loss results from complete degradation of the D-isomer, the recycling of D-carnitine via chemical racemisation should, in principle, offset the yield loss, but a chemical resolution process developed at LONZA turned out to be uncompetitive.[7]

## 14.2 STRAIN ISOLATION

Betaines of the type shown in Figure 1 occur widely in nature. One of their functions is as osmoregulatory substances in plants and bacteria. We therefore expected that a microorganism capable of enantioselective synthesis of L-carnitine from 4-butyrobetaine or crotonobetaine could be isolated from soil samples using a selective medium. Strains isolated in this way were tested for their ability to use 4-butyrobetaine, crotonobetaine and L-carnitine as the sole carbon, nitrogen and energy source under aerobic conditions. An important selection criterion was the inability of the strains to utilise D-carnitine.

One of these strains, HK4, was chosen for further work. This strain was phenotypically characterised (Table 1) but, despite extensive studies, it could not be assigned to one genus. It was found to be related taxonomically to the common soil microorganisms *Agrobacterium* and *Rhizobium*, a close relationship to *Rhizobium meliloti* being established using the UPGMA method[9] and by the high genetic homology that was found between the two microorganisms. The strain was assumed to belong to a new, as yet undefined, genus.[10,11,12]

## 14.3 BUTYROBETAINE/L-CARNITINE METABOLISM

There are several known pathways for the metabolism of L-carnitine in microorganisms, and an excellent review was given by Jung *et al*.[8] It appeared that the

# BIO-TRANSFORMATION IN THE PRODUCTION OF L-CARNITINE

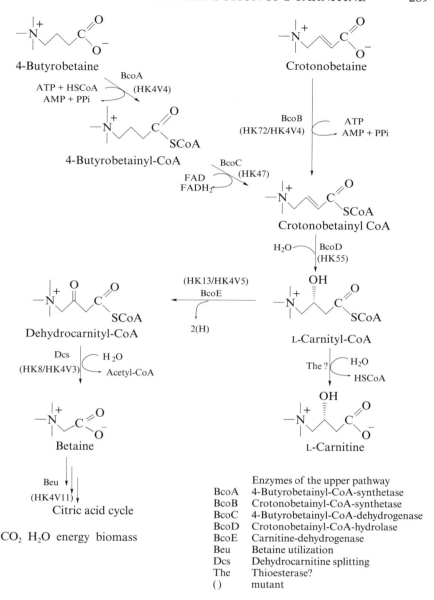

**Figure 1** Catabolic pathway of betaine derivatives in the strain HK4. The first steps of the degradation of 4-butyrobetaine and/or crotonobetaine via L-carnityl-CoA to dehydrocarnityl-CoA are under coordinated regulation and described as the upper pathway of the catabolism. The further mineralisation of dehydrocarnityl-CoA is independent of the upper pathway on both a genetic and a biochemical level and is therefore termed the lower pathway of the catabolism. For each step frame shift or transposon insertion mutants were obtained as indicated in the scheme (compare Figure 2 and Table 2)

**Table 1** Taxonomic description of the strain HK4 and composition of the nutrient medium[a]

| Form of the cell: | | | | | Formation of acids (OF-test) from: | |
|---|---|---|---|---|---|---|
| Shape | Rods partly pleomorphic | | | | Glucose aerobic | – |
| Length | 1–2 μm | | | | Anaerobic | – |
| Width | 0.5–0.8 μm | | | | Fructose aerobic | – |
| Mobility | + | | | | ASS glucose | + |
| Flagellata | Peritric | | | | Xylose | + |
| Gram reaction | – | | | | Trehalose | + |
| Spores | – | | | | Ethanol | – |
| Formation of poly-β-hydroxybutyrate | – | | | | Gas formation from glucose | – |
| Oxidase | + | | | | ONPG | + |
| Catalase | + | | | | Arginine dehydrolase | – |
| | | | | | Lysine decarboxylase | – |
| Growth: | | | | | Phenylalanine deaminase | – |
| Anaerobic | – | | | | Ornithine decarboxylase | – |
| 30 °C | + | | | | $H_2S$ | – |
| 37 °C | – | | | | Voges–Proskauer | – |
| MacConkey agar | + | | | | Indole | – |
| SS agar | – | | | | Nitrite from nitrate | + |
| Cetrimide agar | – | | | | Denitrification | + |
| | | | | | Formation of lavan | – |
| Utilisation of substrate: | | | | | Lecithinase | – |
| Acetate | – | | | | Urease | + |
| Citrate | – | | | | | |
| Malonate | – | | | | Decomposition of: | |
| Glycine | – | | | | Starch | – |
| Norleucine | – | | | | Gelatin | – |
| Xylose | + | | | | Casein | – |
| Fructose | + | | | | Tyrosine | – |
| Glucose | + | | | | Tween 80 | – |
| Autotrophic growth with $H_2$ | – | | | | DNA | + |
| 3-Ketolactose | – | | | | Aesculin | + |
| | | | | | | |
| Growth on: | | | | | Formation of pigment: | |
| 4-Butyrobetaine | (–) | + | (–) | | Not diffusing | – |
| Crotonobetaine | (–) | + | (–) | | Diffusing | – |
| L-Carnitine | (–) | + | (–) | | Fluorescing | – |
| Betaine | (+) | + | (+) | | | |
| Choline | (+) | + | (+) | | | |
| Dimethylglycine | (+) | + | (+) | | | |
| Sarcosine | (+) | + | (+) | | | |
| L-Glutamate and 4-butyrobetaine | (+/–) | +/– | (+) | | | |
| L-Glutamate and crotonobetaine | (+/–) | +/– | (+) | | | |
| L-glutamate and L-carnitine | (+/–) | +/– | (+) | | | |

(*continued*)

## Table 1 (continued)

**Composition of the nutrient medium:**

| | | | |
|---|---|---|---|
| L-Glutamate | 2 g | *Trace elements solution:* | |
| Betaine | 2 g | $ZnSO_4 \cdot 7H_2O$ | 100 mg |
| Crotonobetaine | 2 g | $MnCl_2 \cdot 4H_2O$ | 30 mg |
| Buffer solution | 100 ml | $H_3BO_3$ | 300 mg |
| Mg–Ca–Fe solution | 25 ml | $CoCl_2 \cdot 6H_2O$ | 200 mg |
| Trace elements solution | 1 ml | $CuCl_2 \cdot 2H_2O$ | 10 mg |
| Vitamin solution | 1 ml | $NiCl_2 \cdot 6H_2O$ | 22 mg |
| With water to | 1 ℓ | $NaMoO_4 \cdot 2H_2O$ | 30 mg |
| | | With water to | 1 ℓ |
| *Buffer solution:* | | | |
| $Na_2SO_4$ | 1 g | *Vitamin solution:* | |
| $Na_2HPO_4 \cdot 2H_2O$ | 25.08 g | Pyridoxal·HCl | 10 mg |
| $KH_2PO_4$ | 10 g | Riboflavin | 5 mg |
| NaCl | 30 g | Nicotinamide | 5 mg |
| With water to | 1 ℓ | Thiamine·HCl | 5 mg |
| | | Biotin | 2 mg |
| *Mg–Ca–Fe solution:* | | Sodium pantothenate | 5 mg |
| $MgCl_2 \cdot 6H_2O$ | 16 g | *p*-Aminobenzoic acid | 5 mg |
| $CaCl_2 \cdot 2H_2O$ | 0.58 g | Folic acid | 2 mg |
| $FeCl_3 \cdot 6H_2O$ | 0.032 g | Vitamin $B_{12}$ | 5 mg |
| With water to | 1 ℓ | With water to | 1 ℓ |

[a] The strain is deposited at the German Culture Collection under the registration number DSM2938. The phenotypical changes of the mutant strains HK13 (DSM2903) and HK1349 (DSM3944) are given in parentheses.

conversion of 4-butyrobetaine into L-carnitine in HK4 did not follow one of the known pathways.[10] Small, but measurable, amounts of L-carnitine were formed and excreted into the medium during growth on crotonobetaine and 4-butyrobetaine.

A lag phase was observed when the cells were grown on succinate or glucose and then transferred to betaine derivatives as the carbon source. In contrast, no lag phase occurred when the cells were grown on one betaine derivative and then transferred to a medium containing a different betaine derivative, which indicated a close coupling of the metabolism of 4-butyrobetaine, crotonobetaine and L-carnitine, and that the degradation of these substrates occurred under coordinated induction. The biochemical evidence showed that the metabolism of the betaine derivatives was analogous to the β-oxidation of fatty acids (Figure 1). The activities of each catabolic step could be measured *in vitro* with a direct enzymatic assay or by product formation, and the requirement for coenzyme A, FAD and ATP was also demonstrated. The enzymes of the pathway were enriched by chromatography and gel filtration for the biochemical assays, but not purified because of their instability (Table 2). 4-Butyrobetainyl-CoA is a key intermediate of this pathway and could be isolated from *in vitro* assays using ion-exchange chromatography (its structure being confirmed by $^1$H NMR spectroscopy). The CoA-synthetase and the hydrolase appear to form an enzymatic complex that

**Table 2** Biochemical, phenotypical and genotypical features of HK4 and selected mutants[a]

| Strain | Phenotype | | | | Activity | | | | | | DNA-fragment Complementing | Mutated gene |
|---|---|---|---|---|---|---|---|---|---|---|---|---|
| | 4-bb | cb | L-c | b | BcoT | BcoB | BcoA | BcoC | BcoD | BcoE | | |
| HK4 | + | + | + | + | + | + | + | + | + | + | nt. | Wild-type |
| HK13 | − | − | − | + | + | + | + | + | + | − | 1 | bcoE |
| HK1349 | − | − | − | + | + | + | + | + | + | − | 1 | bcoE |
| HK47 | − | + | + | + | + | + | + | − | + | + | 1 | bcoC |
| HK55 | − | − | + | + | +/− | + | + | + | − | + | 1 | bcoD |
| HK72 | + | − | + | + | + | − | + | + | + | + | 1 | bcoB |
| HK4V4* | − | − | + | + | − | − | − | + | − | + | 1 | bcoA/B |
| HK4V3* | − | − | − | + | + | + | + | + | + | + | 2 | dcs |
| HK4V5* | − | − | − | + | − | − | − | − | − | − | 1 | bcoE |
| HK4V11* | − | − | − | − | + | + | + | + | + | + | 3 | beu |

[a] The phenotype characterises the use of betaine derivatives as growth substrates, e.g. 4-bb = 4-butyrobetaine, cb = crotonobetaine, L-c = L-carnitine, b = betaine. The listed enzyme activities correspond to the catalytic steps indicated in Figure 1. The gene loci were identified by the ability of cosmid gene bank clones to restore a wild-type function to the mutant (Figure 2). Asterisks indicate transposon insertion mutants

can be easily separated from the 4-butyrobetainyl-CoA-dehydrogenase. Both CoA-synthetase activities have been assigned to the same protein, but they can be distinguished from one another by mutation (Table 2). Mutants for each metabolic step were obtained using the frame shift mutagen Acridin ICR191[13] followed by conventional counter-selection.

Further mutants were obtained by Tn5 transposon insertion mutagenesis[14,15] (Table 2), and more detailed information about the butyrobetaine-L-carnitine metabolism in HK4 was obtained using these strains. The strain HK13, a mutant of HK4 with an irreversibly deactivated carnitine dehydrogenase, can transform 4-butyrobetaine and/or crotonobetaine to L-carnitine, the latter being excreted quantitatively into the medium, possibly by an 'overflow' mechanism. This efficient L-carnitine producer was used in $^{18}$O-incorporation studies which confirmed that the origin of the O-atom in L-carnitine is water, and not molecular oxygen.[10]

## 14.4 TRANSPORT

A specific transport system is responsible for the uptake and excretion of betaines by the HK4 strain. It is dependent on a specific periplasmic binding protein, needs energy in the form of ATP and is also induced by the betaine derivatives.[16-20] This 4-butyrobetaine transport system was studied using the mutant *Agrobacterium/Rhizobium* HK47 (DSM2938), which had lost the ability to form crotonobetaine from 4-butyrobetaine. We found that the transport system obeyed Michaelis–Menten kinetics;[16] the saturation constant was very low (0.5 $\mu$M), which should, in theory, allow the complete conversion of 4-butyrobetaine. However, the maximum transport velocity for 4-butyrobetaine measured in the transport assay was relatively low at 0.0055 mol C mol$^{-1}$ h$^{-1}$,[16] indicating that the transport system should be rate-limiting in the production of L-carnitine. In practice this was not the case, which implies that another transport phenomenon for 4-butyrobetaine must be involved. Passive diffusion is the most likely, considering the high 4-butyrobetaine concentrations required for fast production.

## 14.5 GENETIC BACKGROUND

The DNA regions coding for the butyrobetaine-L-carnitine pathway in the bacterial genome were identified by complementation and hybridisation experiments.[21] An HK4 cosmid clone gene bank was prepared and screened for the enzymes of the metabolic pathway. This was achieved by conjugative transfer of the cosmid clones bearing the HK4 DNA fragments into HK4 mutants (Table 2). The appropriate clones that complemented the specific mutation in the HK4 mutants were then isolated. The identity of these DNA fragments was further confirmed by Southern blot hybridisation of the isolated clones with transposon Tn5-labelled and cloned DNA fragments of the transposon insertion mutants.

Four different DNA fragments (Figure 2, fragments 1-4) implicated in the complete mineralisation of 4-butyrobetaine or crotonobetaine via L-carnitine and betaine were identified. The first enzymes of the upper pathway leading to dehydrocarnityl-CoA, i.e. 4-butyrobetaine-CoA- and crotonobetaine-CoA-synthetase, 4-butyrobetaine-CoA-dehydrogenase, crotonobetaine-CoA-hydrolase and carnitine-dehydrogenase are coded in one transcription unit as an oper-

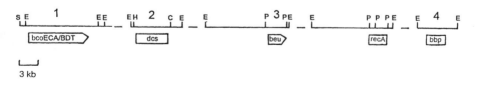

**Figure 2** DNA fragments from an HK4 cosmid gene bank encoding the identified gene loci of 4-butyrobetaine catabolism. The DNA from HK4 was extracted, cut into fragments with the restriction enzyme EcoR1 and ligated into the similarly cut and dephosphorylated mobilisable cosmid vector pVK100.[33] E. coli S17-1[34] was transfected by lambda phages after in vitro packaging of the recombinant cosmids.[22] The cosmid vector contains genes coding for tetracycline and neomycin resistances and therefore cosmid carrying cells were easy to select for by their ability to grow in the presence of selectable amounts of the antibiotics (25 $\mu$g ml$^{-1}$). The gene library contained statistically about 5500 individual clones. S17-1 can mobilise pVK100 hybrids because of its chromosomally integrated R-plasmid coding for transfer functions. The specific cosmid clones bearing the desired genes of the upper pathway of the 4-butyrobetaine catabolism could be screened for after conjugation between the gene bank E. coli donors and the receptor HK mutant strains. The restoration of the genetic defects of the different HK mutants by a cosmid clone led to the isolation and identification of the desired gene loci. After subcloning of smaller DNA fragments of the identified gene loci and further in vivo complementations of HK mutants the genes of the upper pathway could be localised on the cloned DNA more precisely (Figure 3)

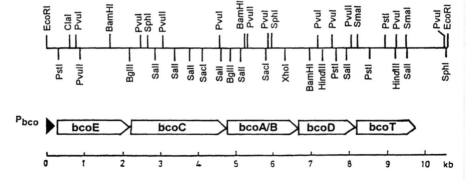

**Figure 3** Restriction map of the DNA fragment encoding the catabolic enzymes of the upper pathway and a transport protein. This DNA fragment could be isolated from an HK4 gene bank clone and the different genes were identified via *in vivo* complementation of HK mutants with subfragments of the DNA

on: *bcoECA/BD* (Figures 2 and 3). In view of the fact that an active 4-butyrobetaine uptake can no longer be measured in the transposon insertion mutants (and that it is known that transposon insertion has a polar effect on the transcription of the following genes in a gene cluster), it was assumed that at least one transport component (*bcoT*) is also coded for on the DNA region following the gene cluster for the catalytic protein.

It was also observed that the transport could be restored in the Tn insertion mutants by the same gene bank clone that complemented the catalytic mutations of the upper pathway. The fact that 4-butyrobetaine, crotonobetaine and L-carnitine induce both the transport and the catalytic activities of the enzymes of the upper pathway indicates that the associated genes are on one operon. The catalytic activities responsible for the splitting of dehydrocarnitine (Dcs) and the utilisation of betaine (Beu) that follow in the further degradation of dehydrocarnityl-CoA (the lower pathway of 4-butyrobetaine catabolism) are coded by separate DNA regions of the HK4 genome, and mutations in these catalytic steps were complemented by other cosmid clones (Table 2). A further gene locus *bbp*, responsible for the butyrobetaine binding protein (Figure 2), was also located on the HK4 genome by Southern blot hybridisation with mixed DNA oligomers of the *N*-terminus of the purified and sequenced Bbp [17].

## 14.6 THE BIO-TRANSFORMATION STRAIN

As already mentioned, the production strain for the fermentative production of L-carnitine was obtained by irreversible inactivation of the carnitine dehydrogenase in HK4 by frame shift mutation.[10–12] This strain, HK13 (Table 2, Figure 1) converts 4-butyrobetaine and crotonobetaine quantitatively into L-carnitine with an enantiomeric purity of 100% and excretes it into the medium. Cofactor regeneration and educt/product transport by the cell during the formation of L-carnitine require large amounts of energy, and occur only in growing cells. Cells in the maintenance state, which is characterised by high metabolic activities at very low growth rates, are ideal for the L-carnitine production, giving nearly complete conversion of the educt. The growth of HK13 is inhibited by product concentrations higher than 3%, although resistant production mutants that tolerate much higher product concentrations (>7%) were isolated from a strain development programme. The improved tolerance to high concentrations of L-carnitine is associated with a reduced uptake affinity for the educt, 4-butyrobetaine, and other modified traits in the HK strain. This can be clearly seen in Table 1 and

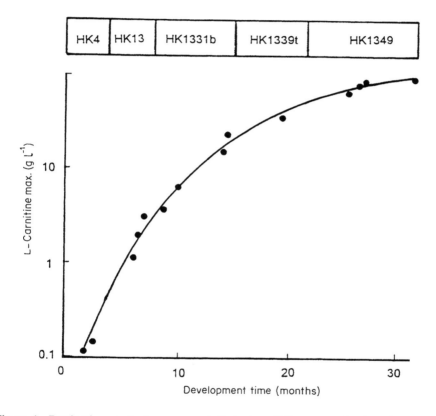

**Figure 4** Production strain improvement. It took *ca* 2.5 years to develop the system ready for scale-up.[10] The programme started with the isolation of a microorganism, HK4, from a soil sample. A strain improvement programme, which aimed to isolate mutants that were resistant to high concentrations of L-carnitine, but still retained a high affinity for 4-butyrobetaine uptake, resulted ultimately in the isolation of the production strain, HK1349.

Figure 4. In the production strain, HK1349, isolated from the programme, a balance between an acceptable tolerance to high concentrations of L-carnitine and sufficient uptake affinity for 4-butyrobetaine was achieved.

## 14.7 THE RECOMBINANT BIO-TRANSFORMATION STRAIN

A recombinant production strain was obtained by cloning the upper pathway gene operon bearing the genes responsible for the conversion of 4-butyrobetaine to L-carnitine (Figures 3 and 5). This strain displays an increased productivity with respect to the mutant HK1349 used in the current production process (Table 3).

The advantages of the recombinant strain are the reduced volume and reduced time-scale needed for the production of L-carnitine.[21] The butyrobetaine-L-carnitine operon *bcoE'CA/BDT* with the deactivated carnitine dehydrogenase gene was isolated from a HK1349 gene bank and cloned into different broad host range vectors with varying copy numbers. The operon was still under the control of the natural promotor ($P_{bco}$), and the increased gene dosage led to increased

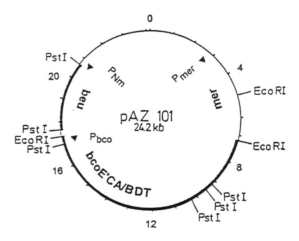

**Figure 5** The recombinant hybrid plasmid pAZ101. The mobilisable broad host range vector pME285[35] was used for the construction of a recombinant hybrid plasmid containing the upper pathway operon of the 4-butyrobetaine catabolism from HK1349 instead of the operon from HK4, because of the carnitine dehydrogenase mutation (bcoE') within the cloned 10.6 kb EcoR1 DNA fragment. Furthermore, the betaine utilisation encoding gene locus *beu* was used to create a selectable gene marker for this plasmid in the *beu* negative deletion mutant HK1349.4. Therefore, a 3.0 kb PstI HK4 DNA fragment was cloned into the newly created PstI site of the vector under the control of the promotor $P_{Nm}$. Another selectable marker that could be used in the genetic experiments is presented by the vectors *mer* locus transmitting mercury resistance to the plasmids host

Table 3  Bio-transformations with HK1349 and its recombinant descendants[a]

| Strain | Medium MM + glycerol | Specific activity (mmol l$^{-1}$h$^{-1}$ OD$^{-1}$) |
| --- | --- | --- |
| HK1349 | + Betaine | 0.24 |
| HK1349/pVK100q | + Betaine | 0.36 |
| HK1349/pAZ7 | + Betaine | 0.49 |
| HK1349 | + L-Glutamate | 0.14 |
| HK1349.4 | + L-Glutamate | 0.11 |
| HK1349.4/pLO41 | + L-Glutamate | 0.29 |
| HK1349.4 | + Ammonia | 0.12 |
| HK1349.4/pVK1011 | + Betaine | 0.54 |
| HK1349.4/pAZ7::beu | + Betaine | 0.47 |
| HK1349.4/pAZ101 | + Betaine | 1.08 |

[a] The bio-transformations for the comparison and estimation of L-carnitine productivities with different HK strains were carried out in shake flasks. The nitrogen-free mineral salt medium MM[36] contained 0.4% (w/v) glycerol as carbon source, 0.2% (w/v) 4-butyrobetaine as the educt for the bio-transformation and 0.2 % (w/v) betaine was added as the nitrogen source; 0.2% (w/v) L-glutamate was used for the betaine-negative strains as the nitrogen source and additional carbon source. The formation of L-carnitine from 4-butyrobetaine was measured by the DTNB (5,5'-dithiobis-2-nitrobenzoate) method described by Bergmeyer[37]

productivity in the biocatalyst. Plasmid stabilisation using an antibiotic selection was not attempted for the recombinant bio-transformation process, as the use of antibiotics, e.g. tetracycline, is known to lead to the loss of the plasmid through segregation and also to growth disturbances.[22,23] Antibiotics are, in any case, not desirable in a large-scale fermentation, so an antibiotic-free selection was developed. The gene *beu,* which is necessary for the utilisation of betaine, was completely deleted from the bacterial chromosome and transferred to a plasmid under the control of the neomycin phosphotransferase promotor $P_{Nm}$ (Figure 5). This selectable plasmid also contains the genes encoding for the upper 4-butyrobetaine–L-carnitine pathway (*bco*). Plasmid stabilisation was now achieved by growth on betaine as the sole nitrogen and/or carbon source.[22]

Another problem that had to be overcome with the recombinant production strain was the tendency of the recombinant plasmid to integrate into the chromosome, owing to homologous recombination between the chromosomal and plasmid encoded operon of the catalytic proteins (BcoE'CA/BDT). One way to reduce this tendency is to introduce a *recA* mutation. The *recA* gene plays an important role in the bacterial repair mechanism of naturally occurring mutations which are otherwise frequently fatal.[24] However, the use of a *recA* minus strain is desirable from a biosafety point of view, because it dramatically reduces the survival rate of the strain outside of the fermenter environment.[24] Another solution to the problem of recombination in the HK strain would be the complete deletion of the homologous DNA from the bacterial chromosome. Yet another possible solution, cloning of the *bco* operon into another host, was not considered because the transport system and the cofactor background in the HK strain are important elements of the L-carnitine production system.

## 14.8 PROCESS DESIGN

The economics of the bio-technological production of fine chemicals are influenced not only by the bio-process but also by the downstream processing and product isolation. Despite this fact, bioprocess technology often concentrates on the bio-process, with the risk that much effort is put into the production of an aqueous product solution which may be unsuitable for downstream processing. This can result in the loss of product during isolation, or poor product quality which can only be improved by numerous costly unit operations. Consequently, the development and optimisation of a bio-process and the subsequent downstream processing (in addition to the upstream processing) should be done in an integrated way which leads to a cost-optimised production process.

There are several options for the process design of the bio-transformation for the production of L-carnitine, e.g. one-stage or multi-stage chemostat with cell recycling or fed batch. The first reported process for the bio-transformation of

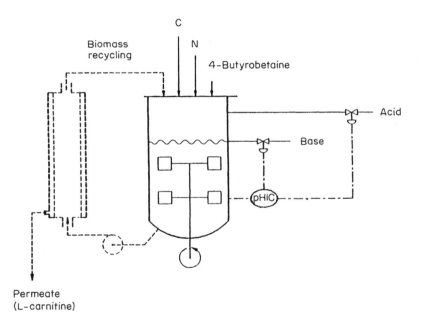

**Figure 6** Scheme of the fed-batch bio-transformation process showing the additions of C and N sources, 4-butyrobetaine, acid and base. The dotted line indicates the modifications of the fed-batch system that are necessary for the one-stage continuous process with cell recycling. For the continuous system with cell recycling, the following data have been reported:[25] flow rate ($Q$) = 32 l h$^{-1}$; educt concentration at the inlet of bioreactor ($c_s$,0) = 60 g l$^{-1}$ = 0.41 mol l$^{-1}$; bioreactor volume ($V_{cc}$) = 360 l; biomass concentration ($c_{x,cc}$) = 40 g l$^{-1}$ as dry matter; product concentration at the outlet of bioreactor ($c_{p,cc}$) = 60 g l$^{-1}$ = 0.37 mol l$^{-1}$; educt concentration at the outlet of bioreactor ($c_{s,cc}$) = 4.6 g l$^{-1}$ = 0.032 mol l$^{-1}$.

# BIO-TRANSFORMATION IN THE PRODUCTION OF L-CARNITINE

chemically synthesised 4-butyrobetaine into L-carnitine consisted of a one-stage chemostat with cell recycling.[10] This process was successfully scaled up to 2000 l,[25] and the phenomenon of maintenance state production was optimally exploited. The process configuration and the cultivation of the strain are shown in Figure 6 and have been described previously.[10,16,25] Betaine was used as the carbon, energy and nitrogen source. The separation of the biomass after the bio-transformation by micro- or ultrafiltration is the first step in the product isolation, which is integrated with the biotransformation in the continuous process (Figure 6). The biomass separation must be reliable and operated aseptically, which is difficult on a large scale, and ceramic cross-flow filters proved to be most suitable for this continuous bio-transformation process.[25]

The downstream processing for L-carnitine is a purification of the aqueous solution after micro- or ultrafiltration. L-Carnitine is highly soluble in water at low, neutral and high pH. Butyrobetaine (educt) and L-carnitine have very similar physio-chemical properties which make their separation from an aqueous solution difficult, so no selective recovery method for L-carnitine is available.

The product isolation consisted of several unit operations which are shown schematically in Figure 7. The continuous bio-transformation process with cell

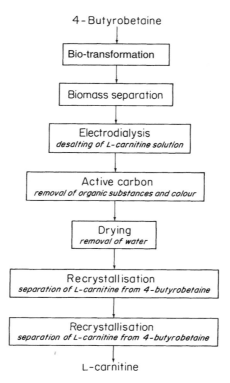

**Figure 7** Unit operations for the production of L-carnitine from 4-butyrobetaine

recycling has a high volumetric productivity (130 g l$^{-1}$ d$^{-1}$) in the steady state.[25] However, the kinetics of the bio-transformation process are such that this is only possible when the concentration of 4-butyrobetaine is relatively high (5 g l$^{-1}$). This results in a product solution containing a mixture of approximately 92% L-carnitine and 8% 4-butyrobetaine,[10] so two recrystallisations from isobutanol are necessary after the continuous biotransformation process in order to obtain the L-carnitine in the purity required by the market (>99%). These recrystallisations are an expensive part of the work-up of L-carnitine.[25]

A fed-batch system has also been developed (Figure 6) in which the volumetric productivity (30 g l$^{-1}$d$^{-1}$) was much less than the one-stage chemostat with cell recycling. However, the bioconversion yield was >99%,[25] which simplified the downstream processing.

## 14.9 KINETIC ANALYSIS

A kinetic analysis of the bio-process was carried out in order to define an optimum design with respect to productivity, downstream processing and costs. For complex biochemical multi-enzyme and multi-component reaction systems with microorganisms, Blackman kinetics is a simple model of the reaction rate as a function of the concentration of a substrate or reaction component.[26-29] In this technique it is assumed that the substrate is rate limiting below a critical concentration (Figure 8). Above the critical concentration other components and/or reaction mechanisms are rate limiting, because the substrate uptake system is saturated. Below the critical concentration, the reaction kinetics approximate to those of a first-order reaction.

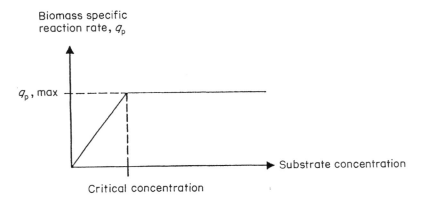

**Figure 8** Blackman kinetics: specific product formation rate of whole cells as a function of the substrate concentration

The continuous system with cell recycling (single- or multi-stage) and the fed-batch system were analysed and compared using the kinetic data and concentrations from the continuous process (Figure 6). To derive the volumetric productivity of the fed-batch biotransformation the four periods of the process cycle time must be taken into account (Figure 9).

As can be seen from Table 4, a one-stage continuous bio-process with cell recycling has a high volumetric productivity, but also a much higher residual

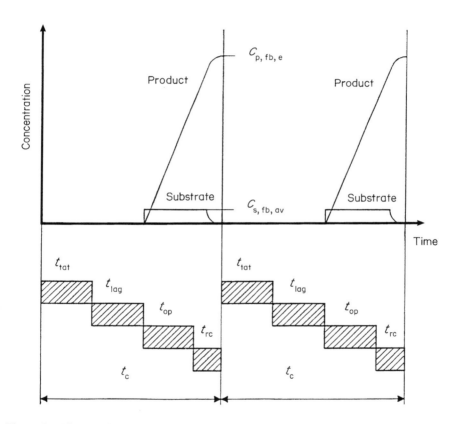

**Figure 9** Time periods in the fed-batch bio-transformation of 4-butyrobetaine into L-carnitine showing substrate and product concentrations as a function of time. The duration of both the period with fast production and the period of the residual substrate conversion was calculated:[29] $t_{tat}$ = turnaround time in which the bioreactor is emptied, cleaned, filled, sterilised, etc; it is usually 15–25 h, and here 15 h will be assumed; $t_{lag}$ = time taken after inoculation of the bioreactor for the production to start; growth can take 10–30 h,[30] and here 20 h will be assumed; $t_{op}$ = in this period, the optimum conditions for production are fulfilled,[25] and have $t_{op}$ = 11.5 h; $t_{rc}$ = in this period, when there is no feed, the remainder of the substrate 4-butyrobetaine in the reactor is converted into L-carnitine.[27] $t_{rc}$ = 2.8 h; $t_c$ = total cycle time; $c_p$, fb, e = final product concentration in the fed-batch process; $c_s$, fb, av = average educt concentration in the fed-batch process

**Table 4** Comparison of continuous processes with cell recycling and a fed-batch process for the bio-transformation of 4-butyrobetaine into L-carnitine[a]

| Parameter | One-stage continuous process with cell recycling | | Fed-batch process | Two-stage continuous process with cell recycling |
|---|---|---|---|---|
| Volumetric productivity (mol l$^{-1}$) | 0.034 | 0.0019 | 0.0075 | 0.0155 |
| Bioconversion yield (as molar % of substrate) | 91 | 99.5 | 99.5 | 99.5 |
| Cost estimation for the bio-transformation and the isolation of L-carnitine (arbitrary units) | 100 | 105 | 60 | 75 |

[a] The volumetric productivities and the yields were calculated as in ref. 29.

substrate level, than a fed-batch process. Consequently, the cost of the downstream processing is disproportionally high for the continuous process owing to the necessary separation of L-carnitine and 4-butyrobetaine. In the fed-batch process, 99.5% of 4-butyrobetaine is converted into L-carnitine.[10,25] The continuous process can be adapted so that the conversion of 4-butyrobetaine into L-carnitine is as high as the fed-batch process, but the resulting productivity would

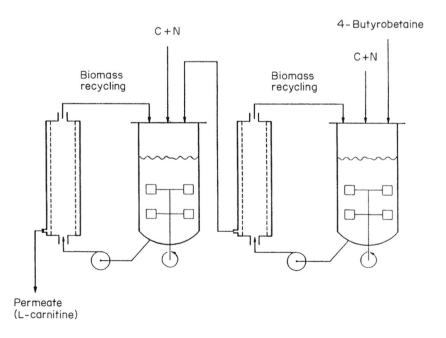

**Figure 10** Schematic presentation of a two-stage continuous process with cell recycling for the bio-transformation of 4-butyrobetaine into L-carnitine

be less by a factor of 18. This loss in productivity nullifies the savings in the downstream processing (Table 4).

The low product to substrate ratio of the continuous process can also be improved with a multi-stage reactor system (Figure 10). The volumetric productivity of the multi-stage reactor system is twice as high as that for the fed-batch process (Table 4). Despite this, the system would be unattractive on a production scale because of the technical difficulties associated with a complex plant with two bioreactors in series (and with two cell-recycling systems), which has to be operated in a continuous aseptic manner. The technical difficulties of a one-stage system have been reported before.[25] The two-stage continuous process would undoubtedly require a relatively high investment per unit working volume, as reflected in the estimated production costs (Table 4).

Given that the fed-batch process is economically the most favourable (Table 4), LONZA decided to produce L-carnitine in a multi-purpose 15 m$^3$ fed-batch plant adapted to the specific needs in the Czech Republic.[31] This process gives a product solution comprising >99% L-carnitine and <1% 4-butyrobetaine. After electrodialysis and active carbon treatment, this solution is dried to a product which meets the required specification. Consequently, the previously described recrystallisations with isobutanol could be omitted, with a large economic advantage.[25] In 1996, the fed-batch process was scaled up to 50 m$^3$ without altering the yield and quality of the product.

## 14.10 RAW MATERIALS ASPECTS

Other criteria in process optimisation are the media composition and the quality of the raw materials, and in our case these had important consequences for the downstream processing. Originally, betaine was used as sole carbon and energy source for the microorganism, leading to the production of ammonia, which had to be neutralised with phosphoric acid. The formation of ammonium phosphate led to complications both in the biomass separation and the electrodialysis steps in the downstream processing.

The flux rate through the ceramic filters of the biomass separation system appeared to be twice as high when glycerol was used as carbon and energy source instead of betaine (0.15 compared with 0.075 m h$^{-1}$ bar$^{-1}$). The reduced flux rate when betaine was used is thought to be caused by the ionic strength of 0.11 that resulted from the ammonium phosphate formation. It has been reported that an ionic strength of this magnitude reduces the flux rate through membranes substantially;[32] ammonium phosphate was not formed when glycerol was used.

The above-mentioned formation of ammonium phosphate when betaine was used as sole carbon source was also disadvantageous in the electrodialysis. There were more salts to be removed from the L-carnitine solution and the duration of the electrodialysis step affected the manufacturing costs of L-carnitine. A longer electrodialysis period implies a higher operating cost and higher product losses. For this reason, glucose or glycerol as carbon and energy sources in the bio-transformation are preferred over betaine. Betaine was retained as a nitrogen source owing to its postulated role in the interrupted catabolic pathway.

Replacement of phosphoric acid with sulphuric acid resulted in a more efficient electrodialysis, as sulphate removal was 30% faster than phosphate removal. The colour of the L-carnitine solution after the bio-transformation process depended on the quality and nature of the raw materials used in the bio-transformation process. Technical-grade betaine, although cheaper, resulted in a coloured product. A white L-carnitine product could be obtained if a pure grade of betaine and glycerol was used as the substrate in the bio-transformation. This made the subsequent active carbon treatment effective, with the extra cost of pure betaine offset by savings in the downstream processing.

## 14.11 CONCLUSION

The development of the L-carnitine process illustrates the importance of a multi-disciplinary approach. The isolation of the microorganism, strain development and biochemical and genetic studies resulted in a comprehensive understanding of L-carnitine metabolism at a cellular level. This led to strains which could produce high levels of optically pure L-carnitine. The exploitation of the maintenance state of this strain allowed the successful development of the single-stage continuous culture with a cell recycling process. Although this system had a high specific productivity, a drawback was the mixture of product and substrate in the product stream. Costly recrystallisations were required to separate the product from the substrate.

The 'black box' approach (system modelling based on Blackman's kinetics) allowed an evaluation and comparison of other process options, e.g. multi-stage continuous culture with cell recycling and fed-batch processes, with the single-stage continuous culture and cell recycling system. From this analysis LONZA opted for a fed-batch process, in which a lower productivity was compensated for by cheaper and simplified downstream processing and lower investment costs. A careful choice of raw materials for the bio-transformation resulted in a reduced number of unit operations during purification of the product.

## 14.12 REFERENCES

1. Scholte, H. R., and De Jonge, P. C., in *Carnitin in der Medizin*, (ed.R. Gitzelmann, K. Baerlocher and B. Steinmann), Schattauer, Stuttgart, 1987, p. 22.
2. Ferrari, R., Di Mauro, S., and Sherwood, G., *L-Carnitine and its Role in Medicine from Function to Therapy*, Academic Press, London, 1992.
3. Wagenmakers, A. J. M., *Med. Sport Sci.*, **32**, 110 (1991).
4. Carter, A. L., *Current Concepts in Carnitine Research*, CRC Press, Boca Raton, FL, 1992.

5. Torreele, E., Sluiszen, A. van der, and Verreth, J., *Br. J. Nutr.*, **69**, 289 (1993).
6. Santulli, A., and D'Amelio, V., *J. Fish. Biol.*, **28**, 81 (1986).
7. Voeffray, R., Perlberger, J. C., Tenud, L., and Gosteli, J., *Helv. Chim. Acta*, **70**, 2058 (1987).
8. Jung, H., Jung, K., and Kleber, H. P., *Adv. Biochem. Eng. Biotechnol.*, **50**, 21 (1993).
9. Sneath and Sokal (Eds), *Numerical Taxonomy*, Freeman and San Francisco, 1973, p. 230.
10. Kulla, H., *Chimia*, **45**, 81 (1991).
11. Kulla, H., and Lehky, P., *Eur. Pat.*, 01 58 194, 1985; *US Pat.*, 5 187 093, 1993.
12. Kulla, H., Lehky, P., and Squaratti, A., *Eur. Pat.*, 0 195 944, 1986; *US Pat.*, 4 708 936, 1987.
13. Miller, J. H., *Experiments in Molecular Genetics*, Cold Spring Harbor Laboratory Press, Cold Spring Harbor, NY, 1972.
14. Rella, M., Mercenier, A., and Haas, D., *Gene*, **33**, 293 (1985).
15. Simon, R., O'Connell, M., Labes, M., and Pühler, A., *Methods. Enzymol.*, **118**, 640 (1986).
16. Nobile, S., and Deshusses, J., *J. Bacteriol.*, **168**, 780 (1986).
17. Nobile, S., Baccino, D., Takagi, T., and Deshusses, J., *J. Bacteriol.*, **170**, 5236 (1988).
18. Nobile, S., Baccino, D., and Deshusses, J., *FEBS Lett.*, **233**, 325 (1988).
19. Nobile, S., and Deshusses, J., *Biochimie*, **70**, 1411 (1988).
20. Nobile, S., and Deshusses, J., *J. Chromatogr.*, **449**, 331 (1988).
21. Zimmermann, Th. P., and Werlen, J., PCT Int. Pat. Appl., WO 95/10613, 1995.
22. Zimmermann, Th. P.,Boraschi, C., Burgdorf, K., and Caubere, C., *Eur. Pat. Appl.*, EP 543 344, 1992.
23. Lee, S.W., and Edlin, G., *Gene*, **39**, 173 (1985).
24. Selbitschka, W., Arnold, W., Priefer, U. B., Rottschäfer, Th., Schmidt, M., Simon, R., and Pühler, A., *Mol. Gen. Genet.*, **229**, 86 (1991).
25. Hoeks, F. W. J. M. M., Kulla, H., and Meyer, H. P., *J. Biotechnol.*, **22**, 117 (1992).
26. Atkinson, B., and Mavituna, F., *Biochemical Engineering and Biotechnology Handbook*, Nature Press, New York, 1983.
27. Dabes, J. N., Finn, R. F., and Wilke, C. R., *Biotechnol. Bioeng.*, **15**, 1159 (1973).
28. Roels, J. A., *Energetics and Kinetics in Biotechnology*, Elsevier, Amsterdam, 1983.
29. Hoeks, F. W. J. M. M., Mühle, J., Böhlen, E., and Psenicka, I., *Chem. Eng. J.*, **61**, 53 (1996).
30. Heijnen, J. J., Van Scheltinga, A. H. T., and Straathof, A. J., *J. Biotechnol.*, **22**, 3 (1992).
31. Meyer, H. P., *Chimia*, **47**, 59 (1993).
32. Fane, A. G., Fell, C. J. D., Hodgson, P. H., Leslie, G., and Marshall, K. C., *Filtr. Sep.*, **28**, 332 (1991).
33. Knauf, V. C., and Nester, E. W., *Plasmid*, **8**, 45 (1982).
34. Simon, R., Priefer, U., and Pühler, A., *Bio/Technology*, **1**, 784 (1983).
35. Itoh, Y., and Haas, D., *Gene*, **36**, 27 (1985).
36. Kulla, H. G., Klausener, F., Meyer, U., Lüdeke, B., and Leisinger, Th., *Arch. Microbiol.*, **135**, 1 (1983).
37. Bergmeyer, H. U., *Methoden der Enzymatischen Analyse*, Verlag Chemie, Weinheim, 3rd edn., 1974.

# ASYMMETRIC SYNTHESES
# BY CHEMICAL METHODS

# 15 The Asymmetric Hydrocyanation of Vinylarenes

**A. L. CASALNUOVO**
*DuPont Agricultural Products, Newark, DE, USA*

**and**

**T. V. RAJANBABU**
*Ohio State University, Columbus, OH, USA*

| | | |
|---|---|---|
| 15.1 | Introduction and Background . . . . . . . . . . . . . . . . . . . . . | 309 |
| | 15.1.1 Nickel-catalysed Hydrocyanation of Alkenes . . . . . . . | 310 |
| | 15.1.2 Asymmetric Hydrocyanation of Norbornene Derivatives | 312 |
| | 15.1.3 Asymmetric Hydrocyanation of Vinylarenes with Non-carbohydrate Phosphorus Ligands . . . . . . . . . | 312 |
| 15.2 | Asymmetric Hydrocyanation with Carbohydrate-derived Phosphorus Ligands . . . . . . . . . . . . . . . . . . . . . . . . . | 315 |
| | 15.2.1 Carbohydrate Ligand Syntheses . . . . . . . . . . . . . | 315 |
| | 15.2.2 Hydrocyanation and Analytical Protocol . . . . . . . . | 317 |
| | 15.2.3 Trials of Asymmetric Hydrocyanation of Vinylarenes . . . | 318 |
| |     15.2.3.1 Steric and Electronic Effects of Phosphorus Substituents . . . . . . . . . . . . . . . . . . | 318 |
| |     15.2.3.2 Site of Phosphorus Substitution . . . . . . . . | 322 |
| |     15.2.3.3 Effect of Electronic Asymmetry . . . . . . . . | 324 |
| |     15.2.3.4 Effect of Other Parameters . . . . . . . . . . | 326 |
| | 15.2.4 Asymmetric Hydrocyanation Mechanism . . . . . . . . | 327 |
| | 15.2.5 Laboratory Preparations of Naproxen and Ibuprofen Precursors . . . . . . . . . . . . . . . . . . . . . . . | 331 |
| | 15.2.6 Asymmetric Hydrocyanation as a Commercial Route to Naproxen . . . . . . . . . . . . . . . . . . . . . . . | 331 |
| 15.3 | Acknowledgements . . . . . . . . . . . . . . . . . . . . . . . | 332 |
| 15.4 | Notes and References . . . . . . . . . . . . . . . . . . . . . | 332 |

## 15.1 INTRODUCTION AND BACKGROUND

The asymmetric, catalytic hydrocyanation reaction is a potentially competitive route to the commercial manufacture of a diverse array of optically active compounds because hydrogen cyanide is a low-cost feedstock, and the cyano group

---

*Chirality in Industry II.* Edited by A. N. Collins, G. N. Sheldrake and J. Crosby
© 1997 John Wiley & Sons Ltd

can be readily transformed into a variety of other functional groups.[1-3] Thus, chiral nitriles, amines, aldehydes, carboxylic acids, esters and imidates are all potential targets. Even so, the only commercial process based on an asymmetric, catalytic hydrocyanation reaction appears to be in the manufacture of a pyrethroid insecticide.[4] This process utilises a cyclic dipeptide catalyst (**1**) to effect the asymmetric addition of hydrogen cyanide to an aromatic aldehyde.

(**1**)

Highly enantioselective hydrocyanations of carbonyl groups have been achieved in the laboratory using both chemical and biochemical catalysts.[2,5,6] In contrast, there are few good examples of either base-catalysed or transition metal-catalysed asymmetric hydrocyanations of alkenes.[7,8] DuPont's discovery and commercialisation of the nickel-catalysed hydrocyanation of alkenes has made the development of an asymmetric version of this reaction a natural research target,[1,3] and towards this end we began a programme that utilised the asymmetric hydrocyanation of vinylarenes as a prototypical reaction.[8] One of the principal, and unanticipated, lessons learned in this exploratory programme was the remarkable impact of subtle changes in the electronic properties of the catalyst on the reaction enantioselectivity. In this chapter we describe the development of this catalyst system and its potential application to the synthesis of the anti-inflammatory agent naproxen.

### 15.1.1 NICKEL-CATALYSED HYDROCYANATION OF ALKENES

The homogeneous, transition metal-catalysed hydrocyanation of alkenes, generically illustrated in equation 1, has been extensively reported in the literature.[1,3] Although a comprehensive review of this reaction is beyond the scope of this chapter, a few relevant details are given here as background.

$$R^1R^2C=CR^3R^4 + HCN \xrightarrow{[M]} R^1R^2HCC(CN)R^3R^4 \qquad (1)$$

A number of transition metal complexes have been reported to catalyse this reaction, although zero-valent nickel phosphite catalysts, $Ni[P(OR)_3]_x$ ($x = 3, 4$), have generally proved to be the most effective. Using triarylphosphite nickel catalysts, for example, DuPont produces over one billion pounds of adiponitrile from butadiene annually (equation 2).

$$\diagup\!\!\!\diagdown + 2HCN \xrightarrow{Ni[P(OAr)_3]_4} NC\diagup\!\!\!\diagdown\!\!\!\diagup CN \qquad (2)$$

With some notable exceptions,[1] these catalysts tolerate a broad range of substrates and functional groups. Thus alkynes, dienes, internal and terminal alkenes, and vinylarenes are all suitable, although for certain substrate classes, particularly isolated alkenes, catalytic amounts of Lewis acids (e.g. $ZnCl_2$) are necessary to achieve acceptable catalyst turnover.

The regioselectivity of the hydrocyanation depends on several factors, including the substrate type, ligand and Lewis acid. For isolated terminal alkenes, linear hydrocyanation products usually predominate. Even internal, isolated alkenes will give rise to a preponderance of the linear nitrile, provided alkene isomerisation is possible through a hydride insertion/$\beta$-hydride elimination mechanism (Scheme 1). Lewis acids tend to increase the rate of alkene isomerisation and thus promote the formation of linear nitriles from internal alkenes.

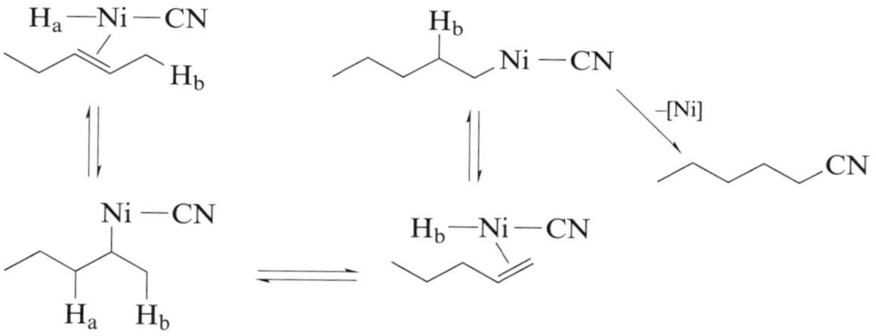

**Scheme 1**

The hydrocyanation of conjugated dienes occurs via the formation of stable $\eta^3$-allyl intermediates.[1,9,10] For example, the $\eta^3$-allyl intermediate **2** (equation 3) has been observed by [31]P NMR spectroscopy in the $Ni[P(OEt)_3]_4$ and $Ni[P(O\text{-}o\text{-}tolyl)_3]_3$-catalysed hydrocyanation of buta-1,3-diene. Similar allylic interactions have been invoked to explain the predominance of the branched nitrile in the $Ni[(P(O\text{-}o\text{-}tolyl)_3]_3$-catalysed hydrocyanation of styrene (Scheme 2).[1]

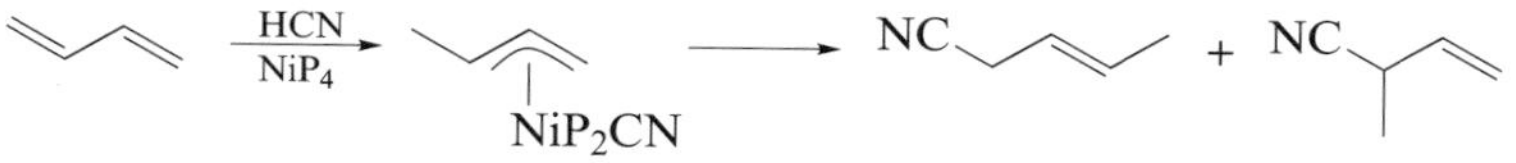

$$P = P(O\text{-}o\text{-}tolyl)_3 \quad (3)$$

One of the primary deactivation pathways of most transition metal hydrocyanation catalysts is metal oxidation by hydrogen cyanide. In the case of zerovalent nickel catalysts, the net redox process is given by equation 4. The overall oxidation rate is thought to be second order with respect to hydrogen cyanide in most cases, whereas the alkene hydrocyanation rate is usually first order. Thus, maintaining a low hydrogen cyanide concentration in the reaction is usually necessary to obtain good catalyst turnover.

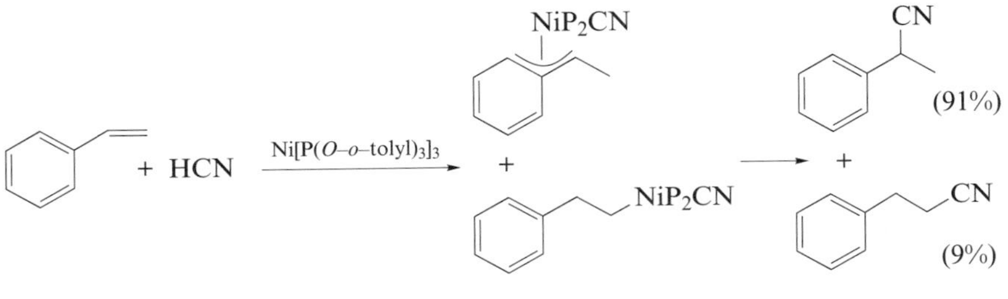

## Scheme 2

[Scheme 2: styrene + HCN with Ni[P(O-o-tolyl)₃]₃ catalyst gives branched NiP₂CN intermediate and linear NiP₂CN intermediate, leading to PhCH(CN)Me (91%) and PhCH₂CH₂CN (9%)]

$$2\text{ HCN} + \text{NiL}_n \longrightarrow \text{Ni(CN)}_2 + \text{H}_2 + n\text{ L} \qquad (4)$$

### 15.1.2 ASYMMETRIC HYDROCYANATION OF NORBORNENE DERIVATIVES

By comparison with the asymmetric hydrogenation of alkenes, the catalytic, asymmetric hydrocyanation of alkenes is still in its infancy. In fact, there have been only a few published accounts of the catalytic asymmetric hydrocyanation of an alkene since the first report appeared in 1982.[7,8] The earlier work, summarised in Table 1, was focused almost exclusively on the asymmetric hydrocyanation of norbornene or norbornene derivatives (equation 5).

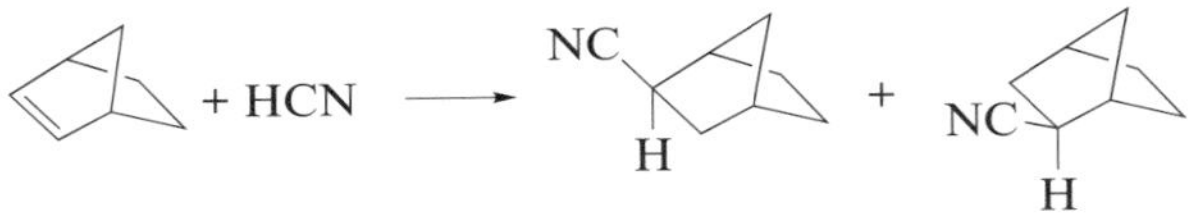

[Equation 5: norbornene + HCN → exo-2-cyanonorbornane enantiomers]  (5)

Although these reactions were completely facially selective (ca 100% exo nitrile), they met with only modest success in terms of reaction enantioselectivity and yield. For example, the highest ee reported for this class of substrate was 38–40% (Table 1; entries 9 and 11) using either Pd(BINAP)₂ or the binaphthol-derived phosphite complex represented as Ni(**B**)₂.

### 15.1.3 ASYMMETRIC HYDROCYANATION OF VINYLARENES WITH NON-CARBOHYDRATE PHOSPHORUS LIGANDS

$$\text{ArCH}=\text{CH}_2 + \text{HCN} \longrightarrow \underset{\text{2-Arylpropionitrile}}{\text{ArCH(CN)Me}} \longrightarrow \underset{\text{2-Arylpropionic acid}}{\text{ArCH(CO}_2\text{H)Me}} \qquad (6)$$

2-Arylpropionitriles constitute a more interesting target for the asymmetric hydrocyanation reaction because they are precursors to chiral 2-arylpropionic acids; a class of commerically important anti-inflammatory drugs (equation 6).[11–13] In general, the (S)-enantiomers of these acids have a greater physiological activity and, in the case of the 6-methoxynaphthalene derivative, naproxen, only

ASYMMETRIC HYDROCYANATION OF VINYLARENES

**Table 1** Asymmetric hydrocyanation of norbornene and derivatives

| Entry | Catalyst[a] | Substrate | Yield (%) | Temperature (°C) | ee (%) | Ref. |
|---|---|---|---|---|---|---|
| 1 | Ni(DIOP)$_2$ | Norbornene | 20 | 80 | 6[b] | 7a |
| 2 | Ni(DIOP)$_2$ | Norbornadiene | 78 | 120 | 8 | 7a |
| 3 | Ni(DIOP)$_2$ | Benzonorbornadiene | 85 | 120 | 16 | 7a |
| 4 | Pd(DIOP)$_2$ | Norbornene | 94 | 80 | 13 | 7c |
| 5 | Pd(DIOP)$_2$ | Norbornadiene | 83 | 120 | 17 | 7a |
| 6 | Pd(DIOP)$_2$ | Benzonorbornadiene | 83 | 120 | 13 | 7a |
| 7 | Pd[(−)-ARSOP)]$_2$ | Norbornene | <5 | 120 | 5[b] | 7a |
| 8 | Pd(DIPHIN)$_2$ | Norbornene | 10 | 120 | 3[b] | 7b |
| 9 | Pd(BINAP)$_2$ | Norbornene | 6 | 120 | 40 | 7c |
| 10 | Pd(BPPM) | Norbornene | 13 | 80 | 25 | 7c |
| 11 | Ni(B)$_2$ + BPh$_3$ | Norbornene | 58 | 100 | 38 | 7d |
| 12 | Ni(A)$_2$ | Norbornene | 16 | 150 | 10 | 7d |

[a] Ligand abbreviations:

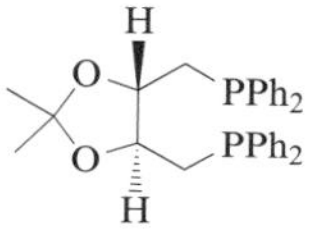

(+)-DIOP 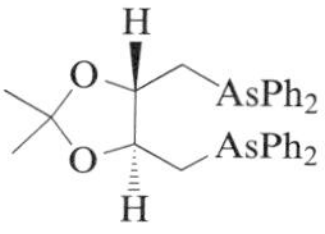    (+)-ARSOP 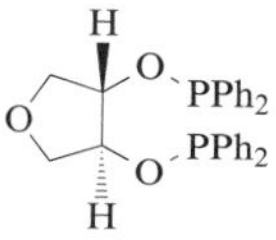    (+)-DIPHIN 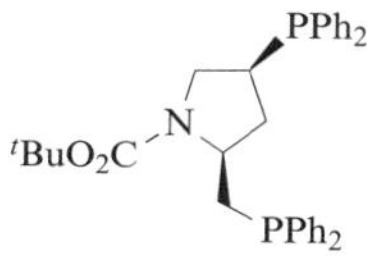    BPPM

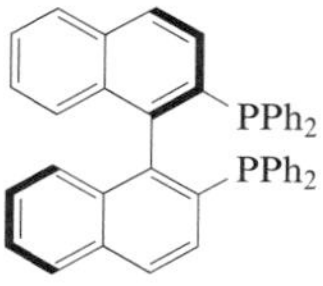

BINAP 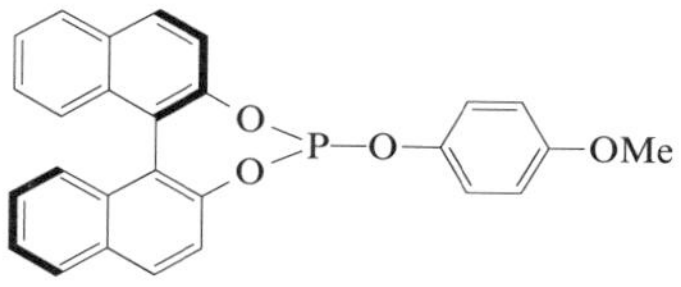    A

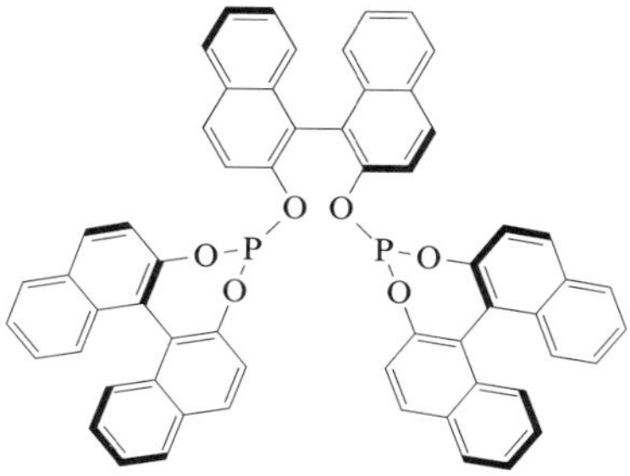

B

[b] ee Values have been corrected from the original report (see ref. 7c).

the pure (S)-(+)-enantiomer has been licensed for sale (see also Volume I, Chapters 15, and this volume Chapter 3). From a regioselectivity standpoint, the starting prochiral vinylarenes are also attractive substrates for the asymmetric hydrocyanation reaction because an overall Markovnikov addition is expected.

In this regard, Nugent[14] and McKinney first reported the efficient synthesis of racemic ibuprofen and naproxen via the Ni[P(O-p-tolyl)$_3$]$_4$-catalysed hydrocyanation of the corresponding vinylarene precursors (equations 7 and 8). Styrene derivatives, such as the ibuprofen precursor p-isobutylstyrene, suffered from competitive oligomerisation. In the latter case, the addition of Lewis acids decreased the amount of alkene oligimerisation but increased the amount of linear nitrile (equation 8).

In what probably constitutes the earliest attempt at the asymmetric hydrocyanation of a vinylarene, Gosser obtained a slight asymmetric induction (ca 10% ee) in the nickel-DIOP-catalysed hydrocyanation of styrene.[15] Later, Nugent and McKinney used various chiral phosphorus ligands to explore the asymmetric hydrocyanation of 6-methoxy-2-vinylnaphthalene (MVN) and 2-vinylnaphthalene (VN). They obtained modest enantioselectivities, with the highest ee (25%, 2-vinylnaphthalene) obtained using the dimethyl tartrate-derived phosphinite ligand **3**.[16]

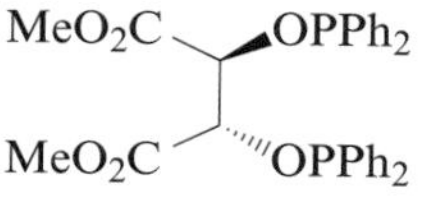

$$\text{MeO}_2\text{C}\diagup\diagdown\text{OPPh}_2$$
$$\text{MeO}_2\text{C}\diagdown\diagup\text{OPPh}_2$$

(3)

## 15.2 ASYMMETRIC HYDROCYANATION WITH CARBOHYDRATE-DERIVED PHOSPHORUS LIGANDS

The catalyst characteristics which control enantioselectivity in asymmetric catalysis can be subtle and complex, particularly in multi-step reaction mechanisms such as the nickel-catalysed hydrocyanation reaction[1] or the rhodium-catalysed asymmetric hydrogenation of alkenes.[17] In such cases, the *a priori* design of suitable chiral ligands is often unproductive and a more empirical approach may be required. Carbohydrates provide many accessible, unique and structurally diverse frameworks for building chiral phosphorus ligands. Indeed, certain glucose-based diphenylphosphinite ligands, such as compound **4**, have been used successfully in the asymmetric hydrogenation of alkenes.[18]

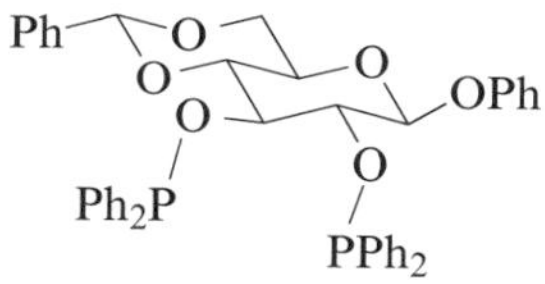

(4)

Furthermore, the existence of well developed, carbohydrate synthetic methodologies allows a modular approach to ligand construction and so simplifies the systematic modification of the ligand. For example, several possible sites for modifying a carbohydrate-derived bidentate phosphorus ligand are illustrated in Figure 1. Variation of these sites using standard synthetic methods leads to a number of potentially useful chiral phosphorus ligands, a few of which are shown Figure 1. With this strategy we have synthesised over 100 carbohydrate-based phosphorus ligands and used them to evaluate systematically the asymmetric hydrocyanation of vinylarenes. Ultimately, this empirical approach to ligand design pointed the way to the discovery of more enantioselective hydrocyanation catalysts and, perhaps more importantly, to some useful insights into electronic effects in asymmetric catalysis.[8]

### 15.2.1 CARBOHYDRATE LIGAND SYNTHESES[8,18]

In addition to the synthesis of the carbohydrate fragment, the synthesis of the carbohydrate ligands used in these studies generally requires the reaction of

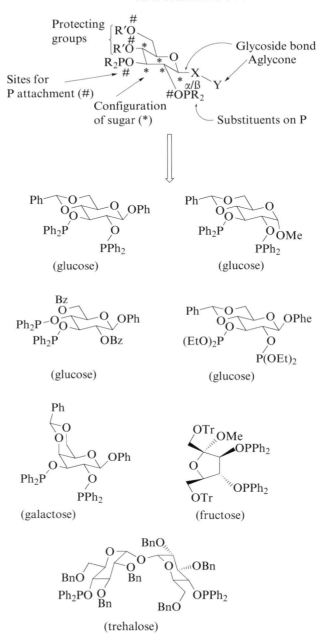

**Figure 1** Modifiable sites on a carbohydrate-derived phosphorus ligand and examples of accessible ligands. Bz = benzoyl; Bn = benzyl; Tr = triphenylmethyl. The parent sugars are indicated in parentheses

# ASYMMETRIC HYDROCYANATION OF VINYLARENES

R$_2$PCl with the hydroxy groups of a suitably protected carbohydrate in the presence of a base. Commercially unavailable R$_2$PCl reagents can be prepared by displacement of the chlorine atom of PCl groups with the appropriate Grignard reagents or alcohols. An example of a typical ligand preparation is shown in Scheme 3. The synthesis of diarylchlorophosphines is shown in Scheme 4.

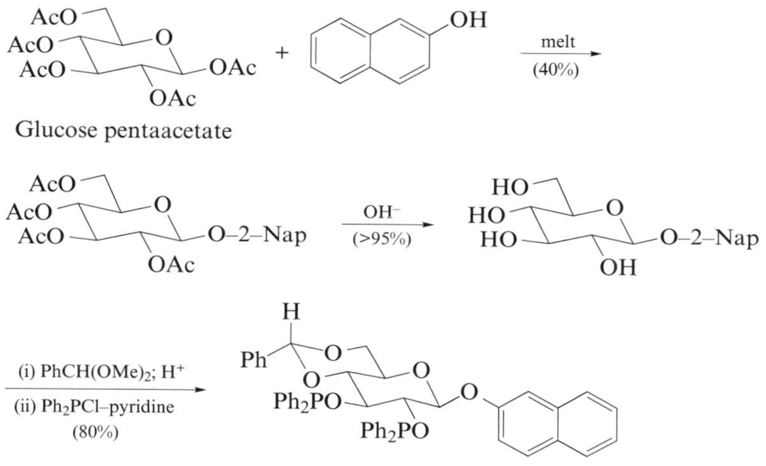

**Scheme 3**

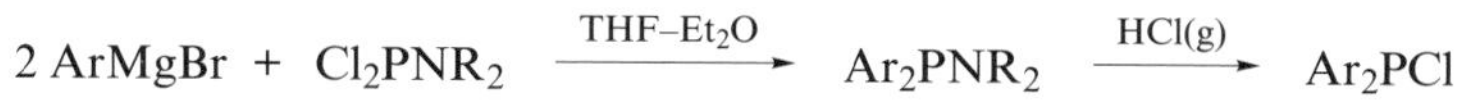

**Scheme 4**

## 15.2.2 HYDROCYANATION AND ANALYTICAL PROTOCOL

Hydrocyanation catalysts are conveniently prepared by mixing commercially available Ni(COD)$_2$ (COD = cycloocta-1,5-diene) with solutions of the chiral phosphorus ligand (P) at room temperature under nitrogen. In most cases, the formation of the typical catalyst precursors NiP$_2$(COD) or NiP$_4$ occurs rapidly at this temperature and any unreacted Ni(COD)$_2$ is catalytically inactive. For trial reactions, HCN is usually added dropwise to a mixture of the vinylarene and catalyst because this procedure minimises oxidation of the catalyst by HCN. Conventional GC methods can be used to evaluate reaction yields and regioselectivity. Chiral HPLC has proved to be the most useful and reliable method for measuring the enantiomeric purities of the 2-arylpropionitriles.

### 15.2.3 TRIALS OF THE ASYMMETRIC HYDROCYANATION OF VINYLARENES

#### 15.2.3.1 Steric and Electronic Effects of Phosphorus Substituents

$$\text{MVN} + \text{HCN} \xrightarrow[\text{L}]{\text{Ni(COD)}_2} \text{MeO-naphthyl-CH(CN)CH}_3 \quad (9)$$

Trial studies on the asymmetric hydrocyanation of 6-methoxy-2-vinylnaphthalene (MVN, equation 9) and 2-vinylnaphthalene (VN) were initially carried out using various $O$-substituted, glucose-derived phosphite [P(OR)$_3$] and phosphinite [PR$_2$(OR'), R' = glucose fragment] ligands. For both types of ligands, the results pointed to the important effect of the phosphorus substituents on the reaction enantioselectivity. In the case of the β-phenyl glucophosphite ligands of type **5** [R = P(OR')$_2$], the data suggested that the size and orientation of the phosphorus alkoxy groups were the primary factors influencing enantioselectivity. As shown in Table 2, the *ee*s ranged from 13% ($R$) to 60% ($S$), with the largest enantioselectivities (entries 4 and 6) obtained when chiral diols were used in constructing the phosphite group. A comparison of these results with those obtained with the phosphinite ligands showed that this glucose moiety imposes a transition state arrangement which favours the formation of ($S$)-nitrile; the presence of other chiral centres near phosphorus can either enhance or counteract this effect (compare entry 4 with 5 and entry 6 with 7).

(5)

In contrast to the glucophosphite ligands, a similar trial study of the asymmetric hydrocyanation of MVN and VN using $o$-, $m$- or $p$-substituted glucodiarylphosphinite ligands (**5**, R = PAr$_2$) led to the unexpected discovery that the electronic characteristics of the phosphorus–aryl substituents have a much greater impact on the enantioselectivity than the inherent size of the substituent. Only a few instances of such ligand electronic effects have been reported in the field of asymmetric catalysis, the principal examples being Jacobsen's asymmetric alkene epoxidation reaction and Nishiyama's ketone hydrosilylation reaction.[19] This electronic effect can be seen clearly from the result obtained using the $m$-disubstituted ligands (**5a–d**), as shown in Table 3. Here the *ee*s increased dramatically (from 16% to 78% for MVN) as the electron-withdrawing power of phosphorus–aryl substituents increased (for **5d–5a**, $\sigma_m = -0.07, 0, 0.34, 0.43$,

## ASYMMETRIC HYDROCYANATION OF VINYLARENES

**Table 2** Asymmetric hydrocyanation of VN using glucophosphite ligands (5)

| Entry | R | ee % |
|---|---|---|
| 1 | (EtO)$_2$P— | 18 (S) |
| 2 | (3-CF$_3$-C$_6$H$_4$-O)$_2$P— | 35 (S) |
| 3 | (2-Me-C$_6$H$_4$-O)$_2$P— | 39 (S) |
| 4 | (R)-BINOL-O$_2$P— | 49 (S) |
| 5 | (S)-BINOL-O$_2$P— | 13 (R) |
| 6 | (S,S)-diphenyl-dioxaphospholane | 60 (S) |
| 7 | (R,R)-diphenyl-dioxaphospholane | 26 (S) |

5a–d

Ar =
(a) 3,5-(CF$_3$)$_2$-C$_6$H$_3$
(b) 3,5-F$_2$-C$_6$H$_3$
(c) C$_6$H$_5$
(d) 3,5-Me$_2$-C$_6$H$_3$

**Table 3** Hydrocyanation of MVN and VN using ligands **5a–d**[a]

| Substrate | ee (%) | | | |
|---|---|---|---|---|
| | 5a | 5b | 5c | 5d |
| MeO-naphthyl-vinyl (MVN) | 85–91[b] | 78[c] | 35[d] | 16[d] |
| naphthyl-vinyl (VN) | 77 | 75 | 46[d] | 25[d] |

[a] 0.10–0.20 M alkene, 1.0–5.0 mol% Ni(COD)$_2$/**5** in hexane.
[b] 91% at 0°C in heptane.
[c] Toluene.
[d] Benzene.

**Table 4** Effect of other aryl substituents on the ee for the hydrocyanation of MVN

| Aryl substituents in **5** | ee (%)[a] |
|---|---|
| 3,4,5-F$_3$ | 61 |
| 3,5-Cl$_2$ | 64 |
| 4-CF$_3$ | 65 |
| 4-F | 38 |
| 2-Me | 0 |
| 3,5-(TMS)$_2$ | 16 |
| 2,5-Me | 9 |
| 4-MeO | 17 |

[a] 0.10–0.20 M MVN, 1–6 mol% Ni(COD)$_2$/**5**, hexane

respectively). A study of other phosphorus–aryl substituents (Table 4) also showed that the highest ees were generally obtained with the more electron-withdrawing substituents (although a strict correlation between enantioselectivity and the substituent σ values was not observed, cf. 4-CF$_3$ vs 3,5-Cl$_2$).

The asymmetric hydrocyanation of other vinylarenes showed a similar response to the electron-withdrawing character of the phosphorus–aryl substituent. For example, the results from the use of ligands **5a–5c** in the asymmetric hydrocyanation of a series of p-substituted styrene derivatives are shown in Table 5. In every case the highest enantioselectivities were obtained with the most electron-withdrawing trifluoromethyl derivative (**5a**). Even so, the enantioselectivities showed a strong dependence on the substrate (ee range: 14–70% for **5a**) and some of the data suggests that the electronic nature of the *substrate* is also a major factor (compare entries 1 and 9).

Studies with other ligand frameworks (see Table 6), both carbohydrate and non-carbohydrate, support the notion that the electronic enhancement of the enantioselectivity is a general feature of the diaryl phosphinite-catalysed asym-

ASYMMETRIC HYDROCYANATION OF VINYLARENES

Table 5. Hydrocyanation of 4-substituted styrene derivatives[a]

| Entry | 4-Substituent | ee (%) | | |
|---|---|---|---|---|
| | | 5a | 5b | 5c |
| 1 | Me | 70 | 47 | 1 |
| 2 | Ph | 68 | 41 | 8 |
| 3 | $Me_2C=CH$ | 63 | – | – |
| 4 | PhO | 60 | 38 | 7 |
| 5 | $Me_2CHCH_2$ | 56 | 38 | 6 |
| 6 | MeO | 52 | 39 | 6 |
| 7 | Cl | 40 | – | – |
| 8 | F | 28 | 15 | 4 |
| 9 | $CF_3$ | 14 | 9 | 1 |

[a] 0.65 mmol alkene, 0.020 mmol $Ni(COD)_2$, 0.020 mmol ligand, 0.65 mmol HCN in 5 ml of hexane after 24 h.

Table 6. Enantioselectivities (% ee) of MVN hydrocyanation using other carbohydrate and non-carbohydrate ligands

| Entry | Ligand | p-Aryl | |
|---|---|---|---|
| | | Ph | $3,5-(CF_3)_2C_6H_3$ |
| 1 | (furanose with OTr, OMe, OPAr$_2$, OPAr$_2$, OTr) | 30 (or 43) | 56 |
| 2 | (PhN succinimide-type with OPAr$_2$, OPAr$_2$) | 54 | 72 |
| 3 | (cyclohexane with OPAr$_2$, OPAr$_2$) | 26 | 35 |

metric hydrocyanation reaction. However, other subtle catalyst features must be important in determining the overall enantioselectivity because the enantioselectivity does not always correlate directly with substituent $\sigma$ values and the magnitude of the effect varies considerably from ligand system to ligand system.

### 15.2.3.2 Site of Phosphorus Substitution

One of the frequently cited drawbacks of using chiral catalysts derived from nature's chiral pool is the unavailability of both catalyst enantiomers. This is a particularly pertinent criticism for many of the unnatural carbohydrate derivatives. From the perspective of making precursors to profen drugs such, as naproxen, it was fortunate that the 2,3-O-bis(diarylphosphino)glucopyranoside ligands, such as **5a**, gave the desired (S)-nitrile as the predominant enantiomer. However, if the other nitrile enantiomer had been needed, the results from an investigation on the nickel-catalysed asymmetric cross-coupling reaction[20] suggested a possible solution. In a prototypical cross-coupling reaction, shown in Figure 2, the cross-coupled product's sense of chirality changed from (S) to (R)-when a 3,4-O-disubstituted glucophosphinite was used instead of a 2,3-O-disubstituted glucophosphinite. In accordance with this observation, the asymmetric hydrocyanation of MVN using 3,4-O-disubstituted glucophosphinite ligands gave the (R)-nitrile as the major enantiomer, although the magnitude of the ee was much less than in the 2,3-system.

Another clue was found when reactions conducted with 2,3-O-disubstituted α-methylfructofuranoside ligands, (**6**) also gave an excess of the (R)-nitrile. A summary of these results and those obtained from other ligand frameworks is given in Table 7. Inspection of these ligand structures suggests that the local chirality defined by the phosphorus-substituted diol controls the sense of asymmetric induction. In this approximation, illustrated in Figure 3, groups adjacent to the P—O—C—C—O—P linkage are ignored. Thus, the 2,3- and 3,4-O-disubstituted

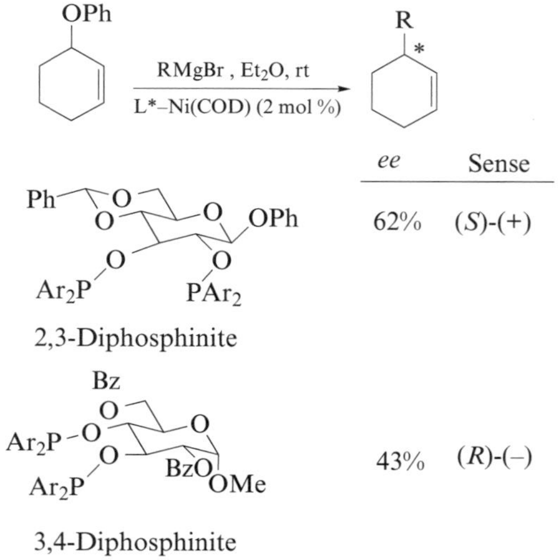

**Figure 2** Effect of the phosphorus substitution site on the major product enantiomer for an asymmetric cross-coupling reaction

ASYMMETRIC HYDROCYANATION OF VINYLARENES 323

(6)

**Table 7** Nitrile configuration for MVN hydrocyanation using various ligand systems

| Entry | Ligand | p-Aryl | ee(%) |
|---|---|---|---|
| 1 | (Bz-protected glucopyranose diphosphinite) | 3,5-$(CF_3)_2C_6H_3$ | 33 (R) |
| 2 | As above | 3,5-$Me_2C_6H_3$ | 13 (R) |
| 3 | (OTr, OMe fructofuranose diphosphinite) | 3,5-$(CF_3)_2C_6H_3$ | 56 (R) |
| 4 | As above | Ph | 30 (R) |
| 5 | (Ph-benzylidene OPh glucopyranose diphosphinite) | 3,5-$(CF_3)_2C_6H_3$ | 78 (S) |
| 6 | (trans-cyclohexane-1,2-diyl diphosphinite) | 3,5-$(CF_3)_2C_6H_3$ | 35 (S) |
| 7 | (N-phenyl succinimide-3,4-diyl diphosphinite) | 3,5-$(CF_3)_2C_6H_3$ | 72 (S) |

glucophosphinites can be viewed as near mirror images or 'quasi-enantiomers,' as can the 2,3-*O*-disubstituted glucophosphinites and 2,3-*O*-disubstituted fructophosphinites.

One explanation of the influence of the diol stereochemistry on the stereochemical outcome of the reaction is that the positions of the phosphorus aryls in

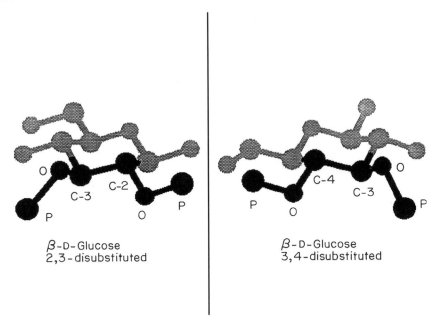

**Figure 3** 'Quasi-enantiomeric' relationship of 2,3- and 3,4-$O$-disubstituted glucophosphinites

the metal complex define the overall shape of the catalyst's chiral pocket.[21] We presume, therefore, that the approximate inversion of the diol stereochemistry effectively inverts the conformational positions of the phosphorus-aryl groups in the seven-membered chelate ring.

Because these quasi-enantiomers are not true mirror images, the magnitude of the *ee* may change, and further catalyst optimisation may be necessary. For example, the *ee* drops from 85% (*S*) to 56% (*R*) on going from the glucose-based ligand **5a** to the fructose based ligand **6a** [Ar = 3,5-$(CF_3)_2C_6H_3$]. Nevertheless, one should be able to predict which phosphorus-substitution sites are likely to produce a particular product enantiomer, and thus profit from the wide variety of possible carbohydrate precursors. As suggested by the initial cross-coupling experiment, one would also expect this principle to apply to other asymmetric, catalytic reactions. In fact, the use of the 2,3- and 3,4-$O$-disubstituted glucophosphinites ligands in the rhodium-catalysed asymmetric hydrogenation of alkenes allows the synthesis of either the (*R*) or (*S*) product enantiomers in very high *ee*.[22]

### 15.2.3.3 Effect of Electronic Asymmetry

$C_2$-Symmetric chiral ligands are often targeted in asymmetric catalysis because they reduce the number of competing diastereoisomeric intermediates in the catalytic cycle and are usually easier to synthesise. Nature, however, is replete with both chiral catalysts and chiral building blocks that lack a $C_2$ symmetry element. Although, strictly, the carbohydrate frameworks used in these hydrocyanation studies have $C_1$ symmetry, it can be argued, as we have done with the quasi-enantiomers, that the local chiral environment around the phosphorus atoms is essentially $C_2$. In principle, further steric or *electronic* differentiation of the two ligating phosphorus centres should provide an interesting test of the importance of $C_2$ symmetry in these ligand systems. Given the significant ligand electronic effects observed in asymmetric hydrogenation, we were particularly eager to explore the effects of ligand electronic asymmetry in this reaction.

Although there are synthetic obstacles to preparing bidentate diarylphosphinite ligands where the two phosphorus atoms have different aryl substituents, we were successful in preparing a series of such electronically asymmetric ligands using an α-methylfructofuranoside framework (**6'**). The results of using these ligands in the asymmetric hydrocyanation of MVN, summarised in Table 8, demonstrated that the enantioselectivity depended strongly on the ligand electronic asymmetry. In every case, the highest enantioselectivities were obtained when the phosphorus attached to the 4-hydroxy group carried a more electron-withdrawing aryl group than the phosphorus attached to the 3-hydroxy group. For example, the highest enantioselectivity (89%, entry 1) was obtained with 3,5-bis(trifluoromethyl)phenylphosphino groups at the 4-*O*-position and the corresponding diphenylphosphino groups at the 3-*O*-position. Simply

(**6'**)

Table 8   Effect of electronic asymmetry on MVN hydrocyanation using a fructofuranoside framework

| Entry | X | Y | ee (%) of (*R*)-nitrile |
|---|---|---|---|
| 1  | 3,5-(CF$_3$)$_2$ | H                 | 89 |
| 2  | 3,5-(CF$_3$)$_2$ | 4-F               | 88 |
| 3  | 3,5-(CF$_3$)$_2$ | 4-MeO             | 84 |
| 4  | 3,5-(CF$_3$)$_2$ | 3,5-(Me)$_2$      | 78 |
| 5  | H                | 3,5-(CF$_3$)$_2$  | 58 |
| 6  | 3,5-(CF$_3$)$_2$ | 3,5-(CF$_3$)$_2$  | 56 |
| 7  | 3,5-(CH$_3$)$_2$ | 3,5-(Me)$_2$      | 40 |
| 8  | 4-MeO            | 4-MeO             | 25 |
| 9  | 3,5-(CF$_3$)$_2$ | 3,5-F$_2$         | 78 |
| 10 | 3,5-F$_2$        | 3,5-(CF$_3$)$_2$  | 42 |
| 11 | 3,5-F$_2$        | 3,5-F$_2$         | 45 |

exchanging the positions of these groups or having both sites substituted with 3,5-bis(trifluoromethyl)phenylphosphino groups decreases the *ee* from 89% to 58% and 56%, respectively (entries 5 and 6). Similar changes in enantioselectivity were observed with the 3,5-difluorophenyl groups (entries 9–11).

There is also some evidence that electronic asymmetry can enhance enantioselectivity in other ligand systems. Diaryl phosphinites (**7**), derived from the $C_2$-symmetric tartranil framework, showed a similar, although less dramatic, effect. Thus, the highest *ee*, 77%, was obtained with the mixed phenyl–3,5-bis(trifluoromethyl)phenyl derivatives, whereas the $C_2$-symmetric phenyl and 3,5-bis(trifluoromethyl)phenyl derivatives gave *ee*s of 54% and 70%, respectively.

$$\underset{(\mathbf{7})}{\text{PhN}\begin{array}{c} O \\ \diagup \diagdown \\ \diagdown \diagup \\ O \end{array}\begin{array}{c} \text{\tiny{...}}OAr^1_2 \\ \\ OAr^2_2 \end{array}}$$

These results clearly demonstrate that the incorporation of electronically different phosphorus chelates can markedly enhance the enantioselectivity obtained with certain ligand arrangements. Thus, the selective manipulation of the electronic properties of the ligand on two different carbohydrate frameworks allows the preparation of either the (*R*)- or (*S*)-enantiomers of 2-(6-methoxy-2-naphthalene)propionitrile in high enantioselectivity [85% (*S*), 89% (*R*) at 25 °C].

### 15.2.3.4 Effect of Other Parameters

The glycoside bond is one of the most readily modified groups in carbohydrate synthesis. Unfortunately, in the case of the 2,3-*O*-disubstituted glucophosphinites the effect of this group on the enantioselectivity was very modest. For example, varying the glycoside groups in ligands of the general structure **8** (Table 9) led to a net change of only 10% in the *ee* for the hydrocyanation of VN. On the other hand, strong solvent effects were observed for the compound **5**-catalysed hydrocyanation of MVN, with the highest *ee*'s obtained in non-polar solvents. For example, *ee*s of 27%, 65%, 78% and 85% were obtained in MeCN, THF, benzene and hexane, respectively, in the **5a**-catalysed hydrocyanation of MVN. Modest increases in enantioselectivity have been observed at temperatures of 0–10 °C but these reactions have been accompanied by a much lower catalyst turnover.

(**8**)

# ASYMMETRIC HYDROCYANATION OF VINYLARENES

**Table 9** Effect of C-1 substituent in β-glucophosphinite (**8**) on the hydrocyanation of VN

| Entry | C-1 substituent | ee (%) (S) |
|---|---|---|
| 1 | OCH$_3$ | 45 |
| 2 | O-C$_6$H$_5$ (phenoxy) | 46 |
| 3 | O-(2-naphthyl) | 52 |
| 4 | O-(2-methylphenyl) | 53 |
| 5 | O-(2-trifluoromethylphenyl) | 55 |
| 6 | Ph | 49 |

## 15.2.4 ASYMMETRIC HYDROCYANATION MECHANISM[8c]

The mechanism proposed for the **5**–Ni(COD)$_2$-catalysed asymmetric hydrocyanation of MVN is shown in Scheme 5. Note that this simplified mechanism is meant to represent all of the diastereoisomers which result from the $C_1$ symmetry of the ligand. The catalytic cycle is based on a number of experimental observations for the **5**–Ni(COD)$_2$–MVN system and also on previously proposed hydrocyanation mechanisms. Several key observations regarding this mechanism include:

- the preparation of a series of other catalyst precursors showed that the active catalyst contains **5** and Ni in a 1:1 ratio;
- the ee is independent of conversion, the addition of other substrates and the addition of enantiomerically enriched nitriles;
- the only directly observed intermediate is the Ni(**5**)(COD) complex **9**, which forms rapidly upon mixing Ni(COD)$_2$ and glucophosphite ligand **5**; this same complex is the catalyst resting state in the absence of or at low concentrations of HCN.
- one other proposed intermediate, Ni(**5a**)(MVN) (**10**), can be independently prepared, and gives the same ee as the Ni(**5a**)(COD) catalyst;
- the only completely irreversible step is the final reductive elimination of the 2-arylpropionitrile from intermediate **13**.

Reaction kinetics for the Ni(**5a**)(COD)-catalysed hydrocyanation of MVN in toluene were zero order for both MVN and HCN at concentrations above 0.04 M in each reagent. This saturation behaviour implies a shift of the catalyst resting

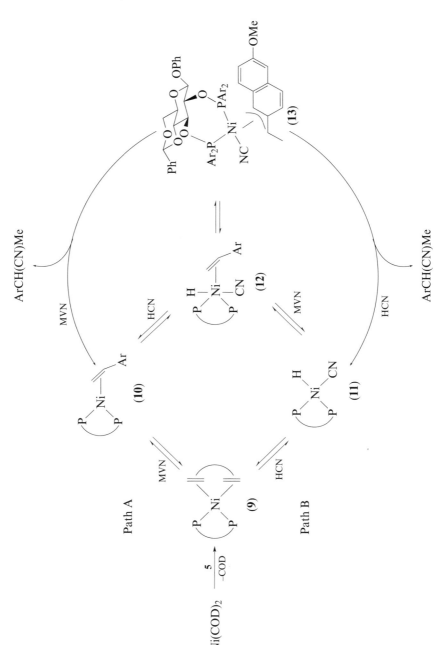

**Scheme 5** Proposed mechanism for the asymmetric hydrocyanation of MVN using **5**–Ni(COD)$_2$

state from Ni(**5a**)(COD) (**9**) to either complex **12** or complex **13**. Based on the known stability of allylic-type hydrocyanation intermediates and the complete regioselectivity of this reaction, we believe that complex **13** is the catalyst resting state under most hydrocyanation conditions. At room temperature under these saturation conditions, a maxium activity of about 2000 turnovers $h^{-1}$ (turnover = mol nitrile/mol Ni) was observed. Deactivation of the catalyst by HCN (equation 4) limits the catalyst lifetime to about 700–800 turnovers.

Two fundamental questions regarding this mechanism are, (i) when is the enantioselectivity determined, and (ii) how does the ligand electronic effect bring about an enhancement in the enantioselectivity? Our data suggest that the enantioselectivity is determined either in the formation of the $\eta^3$-benzylic intermediate **13**, or in the final reductive elimination of the product nitrile from this intermediate, and not in the initial coordination of MVN to form complex **10** or **12**. This conclusion is based upon the rapid exchange rates of free and complexed MVN in complex **10**, the very low ground-state differentiation of the *re* and *si* faces of MVN in this complex and deuterium labelling studies (equation 10), which suggested that the formation of intermediate **12** was also readily reversible.

$$\text{ArCH}=\text{CD}_2 + \text{HCN} \rightarrow \text{ArCH(CN)(CD}_x\text{H}_{3-x}) \quad (10)$$
$$\text{MVN-}d_2$$

In this latter deuterium-labelling study, terminally labelled MVN, MVN-$d_2$, was hydrocyanated using Ni(COD)$_2$ and the ligands **5a–d**. Analysis of the 2-arylpropionitrile product showed that the extent of deuterium scrambling decreased markedly with the more electron-withdrawing ligands **5a** and **5b** (Table 10). The decrease in deuterium scrambling is believed to be due to an increase in the rate of reductive elimination of nitrile from **13** ($k_2$, equation 11) relative to the rate of β-hydride elimination from **13** to form **12** ($k_{-1}$). This explanation is consistent with studies which suggest that less basic phosphorus ligands (i.e. more electron deficient) accelerate the rate of reductive elimination reactions. The electronic effect of the phosphorus aryl substituents on the nickel centre was further

**Table 10** Nitrile deuterium content from MVN-$d_2$ hydrocyanation[a]

| Ligand | Deuterium content (%) | | | | Conversion (%) |
|---|---|---|---|---|---|
| | $d_3$ | $d_2$ | $d_1$ | $d_0$ | |
| MVN-$d_2$ | | 92.3 | 7.2 | 0.5 | |
| **5a** | 1 | 94 | 5 | 0 | 76 |
| **5a**[b] | 2 | 90 | 8 | 0 | 87 |
| **5b** | 6 | 87 | 7 | 0 | 50 |
| **5c** | 31 | 50 | 15 | 3 | 69 |
| **5d** | 29 | 50 | 18 | 3 | 68 |

[a] 0.49 mmol MVN $d_2$, 0.015 mmol Ni(COD)$_2$, 0.015 mmol ligand, 0.50 mmol HCN in 5 ml of hexane.
[b] 0.25 mmol MVN-$d_2$, 0.18 mmol Ni(COD)$_2$, 0.015 mmol **5a**, 0.26 mmol HCN in 2 ml of hexane.

$$\text{Ni}(5)(\text{MVN})(\text{H})\text{CN} \underset{k_{-1}}{\overset{k_1}{\rightleftharpoons}} \text{Ni}(5)(\text{CN})(\eta^3\text{-benzyl}) \xrightarrow{k_2} \text{nitrile} + \text{Ni}(5) \quad (11)$$

(12)             (13)

demonstrated via IR analysis of a series of *m*- and *p*-substituted nickel carbonyl derivatives Ni(5)(CO)$_2$. In this study, a linear relationship was obtained from a plot of the CO stretching frequencies ($A_1$) vs the Hammett parameters $\sigma_m$ or $\sigma_p$, with an overall increase of 34 cm$^{-1}$ in $\nu_{\text{CO}}$ as the aryl substituent was varied from 4-CH$_3$O to 3,5-(CF$_3$)$_2$ (Figure 4).

If the primary effect of the electron-withdrawing phosphorus–aryl substituents is to accelerate the final reductive elimination, then the enhancement in enantioselectivity may result because one of the diastereoisomers of **13** [i.e. a diastereoisomer that produces the (*S*)-nitrile] is disproportionately affected. The precise factors favouring a particular diastereoisomer are at present unclear, but the results with the α-methylfructofuranoside ligand framework strongly suggest the importance of a *stereoelectronic* component. For example, the effect of electronic asymmetry may reflect the importance of a *trans* relationship between the $\eta^3$-aryl fragment and the phosphorus bearing the electron-withdrawing aryl groups in complex **14**.

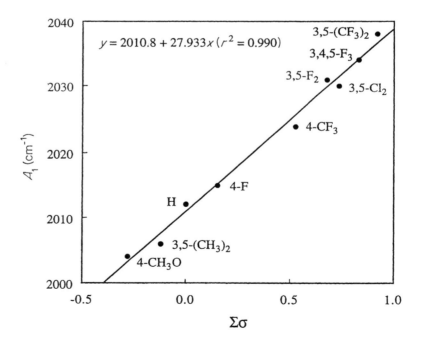

**Figure 4** Correlation of $\sigma_m$ and $\sigma_p$ with the $A_1$ stretching frequency in Ni(5)(CO)$_2$ complexes

# ASYMMETRIC HYDROCYANATION OF VINYLARENES

$Ar^1 = Ph$
$Ar^2 = 3,5\text{-}(CF_3)_2C_6H_3$

(14)

Although these explanations for the electronic effect are highly speculative, they provide a rationale for the ligand electronic effects observed in another asymmetric reaction: the rhodium-catalysed asymmetric hydrogenation of acetamido acrylates. Using similar diaryl phosphinite ligands, we found that electron-*donating* phosphorus aryl substituents dramatically enhanced the enantioselectivity of the reaction.[22] Here, the hydrogen oxidative addition rate is thought to play a key role in determining the reaction enantioselectivity,[17] and such oxidative addition reactions should be accelerated by more electron-donating phosphorus groups.

## 15.2.5 LABORATORY PREPARATIONS OF NAPROXEN AND IBUPROFEN PRECURSORS

The nickel catalyst derived from ligand **5a** is a practical laboratory reagent for the asymmetric hydrocyanation of vinylarenes. For example, the asymmetric hydrocyanation of MVN and *p*-isobutylstyrene has been carried out at room temperature on 5 and 10 g scales using 0.01 and 0.02 equiv. of the catalyst Ni(COD)$_2$-(**5a**), respectively. The resulting 2-arylpropionitriles were obtained in 96% (MVN) and 77% (*p*-isobutylstyrene) isolated yields with vinylarene conversions greater than 99%. In both cases, only the branched nitriles were detected and no yield losses due to vinylarene polymerisation were observed. Although the Ni(**5a**)(COD) catalyst is limited to about 700–800 turnovers in the asymmetric hydrocyanation of MVN, greater *ligand* utility can be obtained because deactivation of the catalyst occurs with dissociation of the free ligand (equation 4). Thus, we have obtained 4000–5000 turnovers based on ligand **5a** in the hydrocyanation of MVN without loss of enantioselectivity (84% *ee*, hexane–toluene) by periodically adding more Ni(COD)$_2$ to the reaction mixture.

## 15.2.6 ASYMMETRIC HYDROCYANATION AS A COMMERCIAL ROUTE TO NAPROXEN

Several potentially practical routes to naproxen have been reported and reviewed.[13] Of the current commercially practised methods, none is believed to utilise asymmetric catalytic methods. Although our work in this area did not focus on process and development issues for naproxen manufacture, several aspects of the Ni(5a)(COD)-catalysed hydrocyanation of MVN make it a feasible route for the commercial manufacture of naproxen: the regioselectivity is 100% and few, if any, by-products are produced; high yields, reasonable reaction rates and high levels of ligand utility can be obtained, and the ligand **5a** is readily accessible. Although the nitrile is not produced in >99% *ee*, it is enantiomerically enriched, and can be crystallised to afford optically pure nitrile. The hydrolysis of the nitrile to naproxen has some precedent in the related area of chiral cyanohydrins (see also Volume I, Chapter 14). For example, the hydrolysis of optically active cyanohydrins, such as 4-methoxymandelonitrile, to amino acids can be carried out without racemisation.

Some of the most pressing technical and economic issues concern the cost and handling of $Ni(COD)_2$. Unless ligand **5a** can be readily separated from the reaction mixture and recycled, $Ni(COD)_2$ (or an equivalent zero-valent nickel source) must be fed to the hydrocyanation reaction so that the use of **5a** can be minimised. Unfortunately $Ni(COD)_2$, currently available in only research quantities, is a relatively expensive complex that is also very air unstable and thermally sensitive. Other key issues for this process include the supply and safe handling of HCN, finding low-cost routes to MVN and the expense of incorporating trifluoromethyl groups into the ligand.

## 15.3 ACKNOWLEDGEMENTS

We gratefully acknowledge the seminal contributions of T. Ayers and T. Warren to the work reported here. We also thank M. Beattie, E. Duvall, G. Halliday, D. Johnson, S. Bernard and K. Messner for their invaluable technical assistance.

## 15.4 NOTES AND REFERENCES

1. Tolman, C. A., McKinney, R. J., Seidel, W. C., Druliner, J. D., and Stevens, W. R., *Adv. in Catal.*, **33**, 1 (1985).
2. Kruse, C. G., in *Chirality in Industry* (ed. A. N. Collins, G. N. Sheldrake and J. Crosby), Wiley, Chichester, 1992, p. 279.
3. Casalnuovo, A. L., McKinney, R. J., and Tolman, C. A., in *The Encyclopedia of Inorganic Chemistry*, Wiley, New York, 1994.
4. (a) Stoutamire, D. W., and Tieman, C. H. (Shell Oil Co.), *US Pat.*, 4 560 515, 1985; (b) Stoutamire, D. W., and Dong, W. (Shell Oil Co.), *US Pat.*, 4 594 196, 1986; (c) Stoutamire, D. W., and Dong, W. (Du Pont, *US Pat.*, 4 723 027, 1988.
5. Tanaka, K., Mori, A., and Inoue, S., *J. Org. Chem.*, **55**, 181 (1990).
6. North, M., *Synlett*, 807 (1993).
7. (a) Elmes, P. S., and Jackson, W. R., *Aust. J. Chem.*, **35**, 2041 (1982); (b) Jackson, W. R., and Lovel, C. G., *Aust. J. Chem.*, **35**, 2053 (1982); (c) Hodgson, M., Parker, D., Taylor, R. J., and Ferguson, G., *Organometallics*, **7**, 1761 (1988); (d) Baker, M. J., and Pringle, P. G., *J. Chem. Soc., Chem. Commun.*, 1292 (1991).

8. (a) Casalnuovo, A. L., and Rajanbabu, T. V., DuPont *US Pat.*, 5 175 335, 1992; (b) RajanBabu, T. V., and Casalnuovo, A. L., *J. Am. Chem. Soc.*, **114**, 6265 (1992); (c) Casalnuovo, A. L., RajanBabu, T. V., Ayers, T. A., and Warren, T. H., *J. Am. Chem. Soc.*, **116**, 9869 (1994); (d) RajanBabu, T. V., and Casalnuovo, A. L., *Pure Appl. Chem.*, **66**, 1535 (1994).
9. Tolman, C. A., Seidel, W. C., Druliner, J. D., and Domaille, P. J., *Organometallics*, **3**, 33 (1984).
10. Andell, O. S., and Bäckvall, J. E., *Organometallics*, **5**, 2350 (1986).
11. Rieu, J., Boucherle, A., Cousse, H., and Mouzin, G., *Tetrahedron*, **42**, 4095 (1986).
12. Sonawane, H. R., Bellur, N. S., Ahuja, J. R., and Kulkarni, D. G., *Tetrahedron: Asymmetry*, **3**, 163 (1992).
13. Giordano, C., Villa, M., and Panossian, S., in *Chirality in Industry* (ed. A. N. Collins, G. N. Sheldrake and J. Crosby), Wiley, Chichester, 1992, pp. 303–312.
14. Nugent, W. A., and McKinney, R. J., *J. Org. Chem.*, **50**, 5370 (1985).
15. See contributions of Gosser, L., in ref. 8a.
16. See contributions of McKinney, R., and Nugent, W., in ref. 8a.
17. Landis, C. R., and Halpern, J., *J. Am. Chem. Soc.*, **109**, 1746 (1987).
18. (a) Cullen, W. R., and Sugi, Y., *Tetrahedron Lett.*, 1635 (1978); (b) Selke, R., *React. Kinet. Catal. Lett.*, **10**, 135 (1979); (c) Jackson, R., and Thompson, D. J., *J. Organomet. Chem.*, **159**, C29-C31 (1978); (d) Yamashita, M., Kobayashi, M., Sugiura, M., Tsunekawa, K., Oshikawa, T., Inokawa, S., and Yamamoto, H., *Bull. Chem. Soc. Jpn.*, **59**, 175 (1986); (e) Habus, I., Raza, Z., and Sunjic, V., *J. Mol. Catal.*, **42**, 73 (1987); (f) Selke, R., Facklam, C., Foken, H. and Heller, D., *Tetrahedron: Asymmetry*, **4**, 369 (1993); (g) Selke, R., Schwarze, M., Baudisch, H., Grassert, I., Michalik, M., Oehme, G., and Stoll, N., *J. Mol. Catal.*, **84**, 223 (1993).
19. (a) Jacobsen, E. N., Zhang, W., and Guler, M. L., *J. Am. Chem. Soc.*, **113**, 6703 (1991); (b) Nishiyama, H., Yamaguchi, S., Kondo, M., and Itoh, K., *J. Org. Chem.*, **57**, 4306 (1992). Until recently (see ref. 22), only modest effects were observed in the asymmetric hydrogenation reaction. See (c) Morimoto, T., Chiba, M., and Achiwa, K., *Chem. Pharm. Bull.*, **41**, 1149 (1993); (d) Morimoto, T., Chiba, M. and Achiwa, K., *Chem. Pharm. Bull.*, **40**, 2894 (1992); (e) Hengartner, U., Valentine, D., Jr., Johnson, K. K., Larscheid, M. E., Pigott, F., Scheidl, F., Scott, J. W., Sun, R. C., Townsend, J. M., and Williams, T. H., *J. Org. Chem.*, **44**, 3741 (1979); (f) Werz, U., and Brune, H., *J. Organomet. Chem.*, **365**, 367 (1989).
20. RajanBabu, T. V., and Ayers, T. A., unpublished results. See also ref. 22.
21. (a) Knowles, W. S., *Acc. Chem. Res.*, **16**, 106 (1983); (b) Oliver, J. D., and Riley, D. P., *Organometallics*, **2**, 1032 (1983); (c) Brown, J. M., and Evans, P. L., *Tetrahedron*, **44**, 4905 (1988); (d) Pavlov, V. A., Klabumovskii, E. I., Struchkov, Y. T., Voloboev, A. A., and Yanovsky, A. I., *J. Mol. Catal.*, **44**, 217 (1988); (e) Bogdan, P. L., Irwin, J. J., and Bosnich, B., *Organometallics*, **8**, 1450 (1989); (f) Seebach, D., Plattner, D. A., Beck, A. K., Wang, Y. M., and Hunziker, D., *Helv. Chim. Acta*, **75**, 2171 (1992).
22. RajanBabu, T. V., Ayers, T. A., and Casalnuovo, A. L., *J. Am. Chem. Soc.*, **116**, 4101 (1994).

# 16 Enantioselective Protonation in Fragrance Synthesis

## C. FEHR
*Firmenich SA, Geneva, Switzerland*

| | | |
|---|---|---|
| 16.1 | Introduction | 335 |
| 16.2 | The Appeal of Enantioselective Protonation | 336 |
| 16.3 | (S)-α-Damascone | 337 |
| | 16.3.1 The Rose Ketones | 337 |
| | 16.3.2 Design of the Chiral Reagent | 339 |
| | 16.3.3 Interaction of Chiral Reagent, Counter Ion and Ligand | 339 |
| | 16.3.4 (S)-α-Damascone by Catalytic Enantioselective Protonation | 341 |
| | 16.3.5 Extension to the Syntheses of the Enantiomers of α- and γ-Cyclogeranic Acid, α- and γ-Cyclocitral and γ-Damascone | 343 |
| 16.4 | Enantioselective Synthesis of a New Tetralin Musk Perfume Component | 348 |
| 16.5 | Acknowledgements | 350 |
| 16.6 | Notes and References | 350 |

## 16.1 INTRODUCTION

The fragrance industry is not subjected to the same regulatory constraints as the pharmaceutical industries with respect to the enantiomeric purity of chiral compounds. However, very often one enantiomer shows superior odour properties compared with the other, justifying its synthetic development. As an example, the odour of (S)-α-damascone [(S)-**1**] is reminiscent of rose petals, whereas the (R)-enantiomer has a more pronounced apple note and is also responsible for an undesired cork odour. The difference in their odour threshold concentrations is also remarkable. Whereas the odour of the (S)-enantiomer is still perceived at 1.5 ppb, the (R)-antipode requires a 70-fold higher concentration for odour perception.[1] In the field of aromatic musk compounds, it has been established that the extremely strong and characteristic note of the racemic new Firmenich musk (±)-**2**[2] is almost entirely due to the (−)-enantiomer [(−)-**2**], the odour of which is extremely strong, both on the smelling strip and in perfume compositions.[3]

In this chapter, the syntheses of the fragrances (S)-(−)-**1** and (S,S)-(−)-**2**, together with related compounds are described, employing as the key step the promising new methodology of enantioselective protonation.[1,3–11]

*Chirality in Industry II.* Edited by A.N. Collins, G. N. Sheldrake and J. Crosby
© 1997 John Wiley & Sons Ltd.

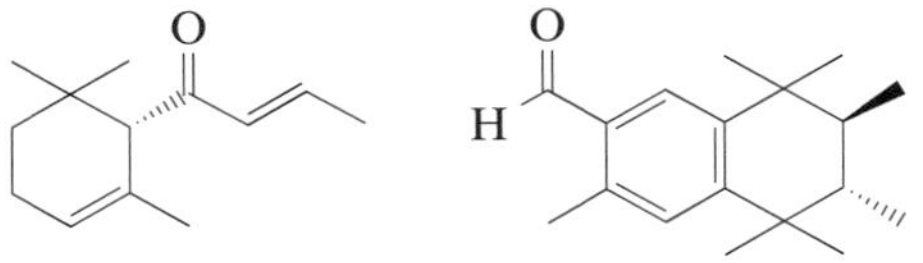

[(S)-(−)-**1**]   [(S,S)-(−)-**2**]

## 16.2 THE APPEAL OF ENANTIOSELECTIVE PROTONATION

Conceptually, enantioselective protonation is extremely simple. A prochiral enolate possesses two enantiotopic faces. Using a chiral, non-racemic proton source, the proton transfer to the enolate will be kinetically favoured either from the top face or from below, thus affording an (R)- or (S)-enantiomerically enriched carbonyl compound (Scheme 1). However, it should be stated immediately that a generally applicable chiral reagent is not yet available. In addition, the outcome of enantioselective protonation depends strongly on the structure and configuration of the enolate, the presence or absence of lithium salts (LiOR, LiNR$_2$, LiX) and the solvent. In spite of the complexity of the reaction, due to aggregated reactive species, application of enantioselective protonation is straightforward and very attractive. The enolate, obtained by deprotonation from the parent carbonyl compound, or directly from a synthetic reaction step, e.g. by addition of an organometallic reagent to a ketene or an enone, is 'quenched' with the chiral proton donor to afford the enantiomerically enriched carbonyl compound. After addition of water, the chiral reagent is regenerated to its protonated state and recovered by simple extraction.

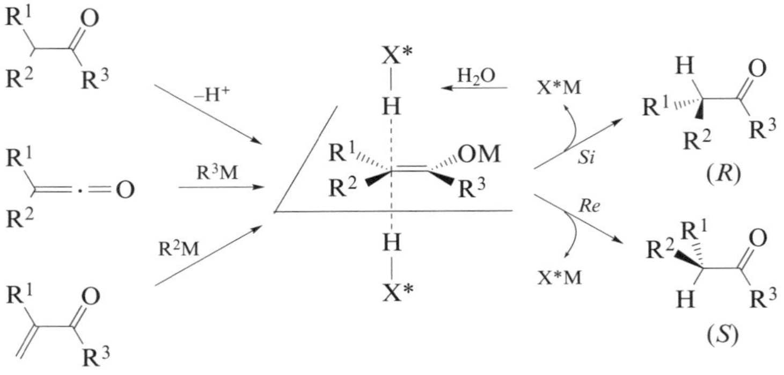

**Scheme 1** Enolate generation and enantioselective protonation

## 16.3 (S)-α-DAMASCONE

### 16.3.1 THE ROSE KETONES

In recent years the rose ketones, namely α-damascone [(±)-1], β-damascone (3), γ-damascone [(±)-10] and β-damascenone (4), have become important perfume components owing to their typical fruity–flowery scent and exceptional odour strength.[12] γ-Damascone, which has never been found in nature, has a slightly weaker odour, but nevertheless possesses a very elegant damascone-type note. Despite the similarity of their odours, each rose ketone has an intrinsic value, so it has been essential to develop efficient syntheses of each of these individual fragrance compounds. Although they are already effective at concentrations lower than 0.1%, their annual production is steadily increasing and has grown from 1 tonne in 1978 to 10 tonnes in 1988, of which α-damascone [(±)-1] represents the major part.

In the 1980s, the development of a unique reaction scheme for the synthesis of rose ketones (±)-1,3 and 4 starting from the corresponding, readily accessible, esters 5,6, and 7 constituted an important breakthrough. Whereas the normal allyl-Grignard reaction mainly leads to the tertiary alcohols via di-addition, the selectivity for mono-/di-addition could be completely reversed when the Grignard reaction was performed in the presence of LDA.[13] Under these conditions, the ketone formed *in situ* is protected from further reaction by deprotonation to the enolate and, after acid-catalysed isomerisation of the terminal double bond, the three rose ketones (±)-1,3 and 4 are isolated in high yield.

This methodology has been applied to other sterically hindered esters or amides and gives the best results with allylic Grignard reagents, as the product ketones are rapidly deprotonated.[14]

Evidently, for the synthesis of (S)-α-damascone [(S)-1] by enantioselective protonation, the regioisomeric, achiral ketone enolate 8 is required (compare Schemes 2 and 3), which can be obtained by Grignard reaction on the *deprotonated* ester 5.[15] This reaction, in which the ester enolate represents the electrophilic species, merits comment. Owing to the delocalisation of the negative charge over five centres, the carbonyl group possibly retains part of its electrophilic character. On the other hand, it is also known that ester enolates have only limited stability: elimination of LiOMe would then allow the Grignard reaction to occur via a ketene or ketenoid species.[16] Application of the same reaction conditions to methyl β-cyclogeranate (6) constitutes a highly efficient and direct synthesis of γ-damascone [(±)-10].[15] The addition of organometallic reagents to ketones also represents an interesting access to prochiral enolates, as exemplified by the synthesis of α-damascone [(±)-1] starting from the readily accessible and stable ketene 14[17] (see Scheme 5).

The aforementioned Grignard reactions, usually in combination with strong lithium bases, have now been developed into highly efficient industrial procedures.[18] In addition, the prochiral enolate intermediates 8 and 9, which afford

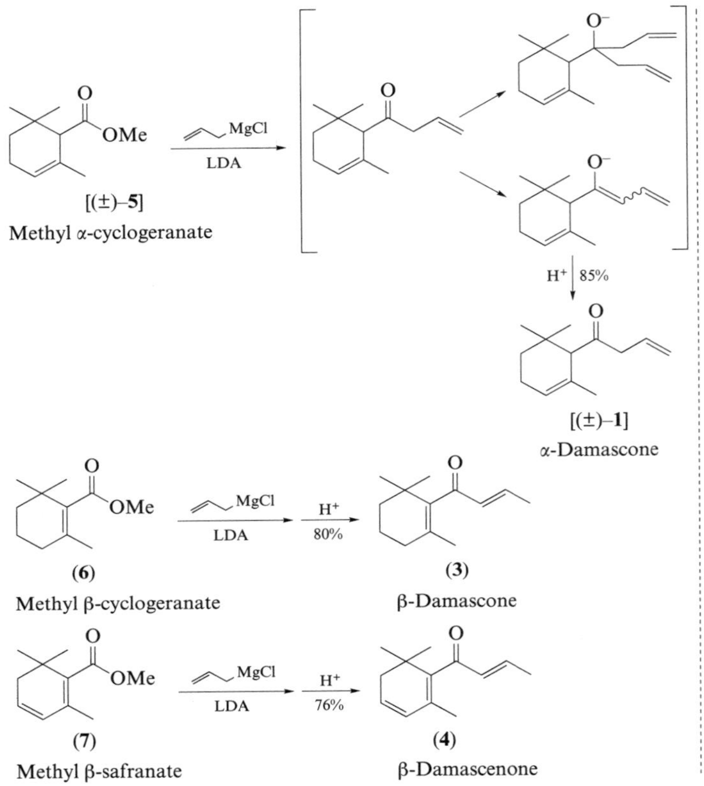
O
OMe
MgCl
LDA
[(±)–5]
Methyl α-cyclogeranate
O
O⁻
O⁻
H⁺
85%
O
[(±)–1]
α-Damascone
O
OMe
MgCl
LDA
H⁺
80%
O
(6)
Methyl β-cyclogeranate
(3)
β-Damascone
O
OMe
MgCl
LDA
H⁺
76%
O
(7)
Methyl β-safranate
(4)
β-Damascenone

| Selectivity ketone/alcohol | |
|---|---|
| With LDA | Without LDA |
| 98 : 2 | 28 : 72 |
| 90 : 10 | 15 : 85 |
| 96 : 4 | 15 : 85 |

**Scheme 2**

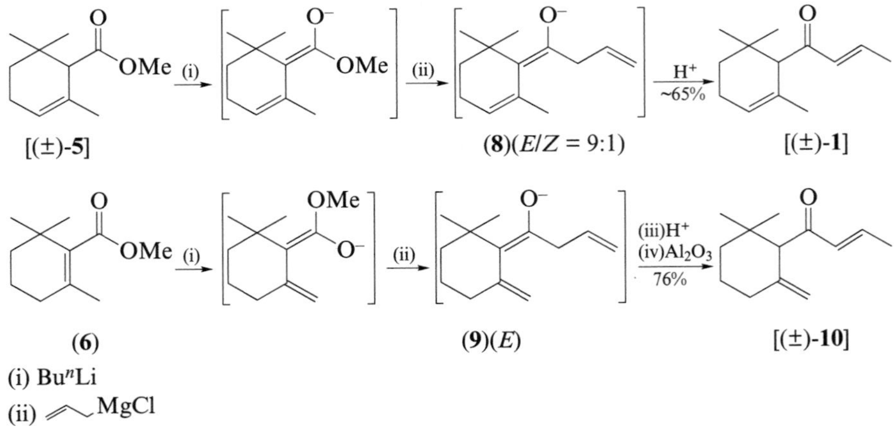
O
OMe
(i)
O⁻
OMe
(ii)
O⁻
H⁺
~65%
O
[(±)-5]
(8)(E/Z = 9:1)
[(±)-1]
O
OMe
(i)
OMe
O⁻
(ii)
O⁻
(iii)H⁺
(iv)Al₂O₃
76%
O
(6)
(9)(E)
[(±)-10]
(i) BuⁿLi
(ii) MgCl

**Scheme 3**

# ENANTIOSELECTIVE PROTONATION IN FRAGRANCE SYNTHESIS

racemic α- and γ-damascone [(±)-**1** and (±)-**10**] upon quenching with H$_2$O, offer the possibility of enantioselective protonation by a chiral proton source. Thus, a minimal change of reaction conditions should allow the production of enantio-enriched α- and γ-damascones.

## 16.3.2 DESIGN OF THE CHIRAL REAGENT

In the search for an efficient and synthetically useful chiral proton source, we were guided by the following criteria (Scheme 4). Ideally, the chiral reagent should be only weakly acidic to allow better transition-state discrimination. It should also contain electron-rich groups with coordination or chelation ability, which would enhance conformational rigidity in the transition state. For obvious reasons, the transferred proton should be located in the proximity of the stereogenic centre (within the 'chiral environment') and, further, the chiral reagent should be readily accessible in both enantiomeric forms and be easily recoverable. These criteria are fulfilled with the ephedrine derivatives **11** and **12**, of which only one enantiomer is illustrated. The imidazolidinone **11** has an N—H bond confined in a cyclic system, and the amino alcohol **12** can attain conformational rigidity through chelation. It should be added that (+)- and (−)-**11** are commercially available (Merck), and that (+)- and (−)-**12** are readily obtained by treatment of (+)- or (−)-ephedrine with acetone by catalytic hydrogenation.[19]

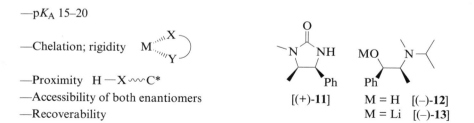

**Scheme 4** Criteria for chiral reagent

## 16.3.3 INTERACTION OF CHIRAL REAGENT, COUNTER ION AND LIGAND

In an initial experiment, the ester (±)-**5** was deprotonated with *n*-butyllithium and the resulting ester enolate was treated with allylmagnesium chloride to afford ketone enolate **8** (*E/Z* ≈ 9:1). Protonation of enolate **8** with ephedrine derivatives (+)-**11** or (+)-**12** at −50 to −10 °C and subsequent isomerisation of the terminal double bond afforded the desired product (*S*)-**1** with 58% and 70% *ee*, respectively (Scheme 5). The moderate chemical yield (60%) was due to competing nucleophilic attack of *n*-butyllithium on the ester function. Some additional

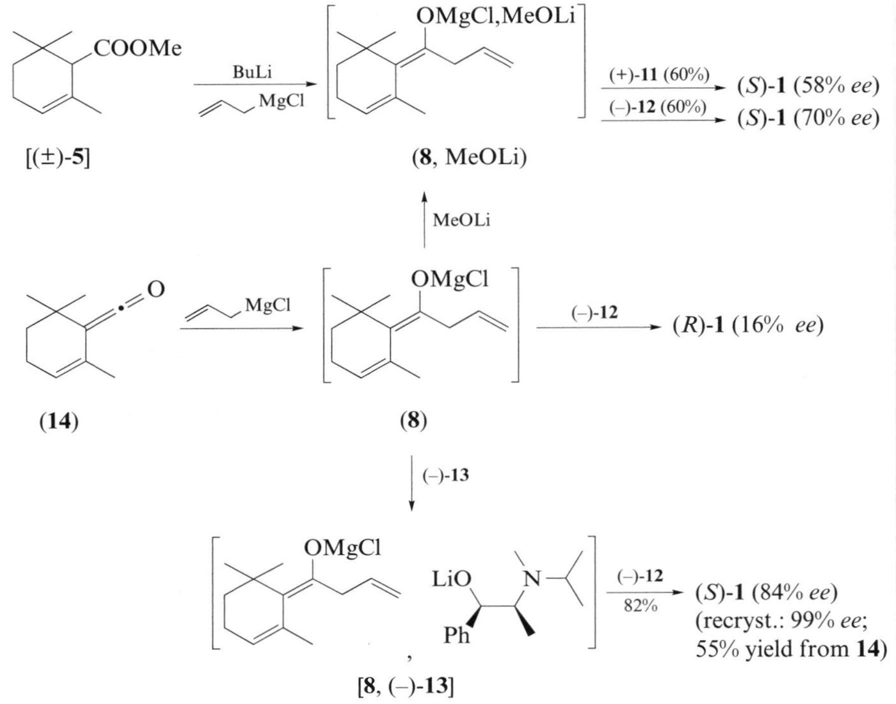

**Scheme 5**

screening experiments with differently *N*-substituted ephedrine derivatives revealed that inferior inductions were obtained both with less hindered (NHMe, NMe$_2$, NHPr$^i$) and more hindered [NMeCH(Et)$_2$] proton donors, which highlights the subtle balance between steric factors and ease of chelation. Therefore, the amino alcohol (−)-**12** [or (+)-**12**] was chosen as the standard chiral reagent for further investigations. This novel protonating agent is not only efficient, but also very practical, as it can be recovered almost quantitatively by a simple acid–base extraction.

To understand better the influence of the counter ion and ligand, and also to avoid the use of *n*-butyllithium, we next investigated the reaction of allylmagnesium chloride with ketene **14** (Scheme 5). The THF-solvated (but otherwise ligand-free) enolate **8** (*E/Z* ≈ 9:1) thus formed was protonated with amino alcohol (−)-**12**. Much to our surprise, protonation of the ligand- and lithium-free enolate **8** afforded (*R*)-**1** (16% *ee*) as the major enantiomer. On the other hand, addition of 1 equiv. of MeOLi prior to protonation, thus restoring the conditions present when starting from ester (±)-**5**, furnished (*S*)-**1** with 70% *ee*. Next, enolate **8** was treated with the chiral lithium alkoxide (−)-**13** derived from *N*-isopropylephedrine, and protonated with (−)-**12** to afford (*S*)-α-damascone with

84% *ee* in an improved yield of 82%.[6] This product could easily be purified further by crystallisation (99% *ee*; 55% yield from **14**), but for use as a fragrance building block, a 9:1 enantiomeric mixture was judged to be sufficient because of cost considerations.

At this stage of development, the major drawbacks of the (*S*)-α-damascone synthesis were the large amounts of chiral reagent employed [*ca* 3 kg of (−)-**12** for 1 kg of (*S*)-**1**], and the impractically low protonation temperature. In the industrial process which produced the first kilograms of (*S*)-**1**, the lithium alkoxide ligand (−)-**13** was replaced by Bu$^t$OLi (from Chemetal), and the protonation was effected at 0 °C with only minimal loss of enantioselectivity (79% *ee*).

### 16.3.4 (*S*)-α-DAMASCONE BY CATALYTIC ENANTIOSELECTIVE PROTONATION

As shown above (Scheme 5), protonation of the complex between magnesium enolate **8** and the lithium alkoxide (−)-**13** affords (*S*)-**1** with 84% *ee*, whereas the ligand- and lithium-free magnesium enolate affords (*R*)-**1** with a low *ee* of 16%. In order to examine the reactivity of the alkoxide-free lithium enolate, the magnesium enolate **8** was quenched with Me$_3$SiCl. Almost pure (*E*)-silyl enol ether **15** was readily obtained by fractional distillation. Treatment of this compound with methyllithium generated the lithium(*E*)-enolate **16** which, upon addition of (−)-*N*-isopropylephedrine [(−)-**12**] at +60 °C, underwent protonation with excellent enantioselectivity (94% *ee*).[10] Apparently, the lithium enolate does not require a lithium alkoxide ligand for selective protonation. Evidently, this reaction is only ligand-free and enolate-rich at low conversion, as 1 equiv. of the lithium alkoxide ligand (−)-**13** is generated during this process.

To test whether the enolate protonation is more enantioselective at low conversion, enolate **16** was treated with only 0.5 equiv. of (−)-**12** (−60 °C, 2 min) and the reaction mixture was quenched with Me$_3$SiCl (Scheme 6). Much to our surprise, no trace of enol silyl ether **15** was detected. Instead, approximately equal amounts of ketone (*S*)-**17** and enol silyl ether (*S*)-**19** of high enantiomeric purity (94% *ee*) were formed. This unique enantioselective protonation is thus catalytic: the protonated intermediate (*S*)-**17** makes available its most acidic hydrogen (C-2) for the protonation of (−)-**13** or, more likely, of a mixed aggregate of alkoxides **16** and (−)-**13**. This catalytic cycle is based on subtle kinetic differences for the proton transfer reactions between chiral reagents, enolate and non-inducing[20] proton donor. When the quantity of ligand (−)-**12** is lowered below *ca* 0.2 equiv., a marked decrease in reaction rate and selectivity was observed. Finally, for the synthesis of (*S*)-α-damascone [(*S*-**1**)], enolate **16** was treated with 0.3 equivalents of (−)-**12** (−60 °C, 10 min) and the intermediate enolate (*S*)-**18** was quenched with dilute HCl (Scheme 7). After isomerisation of the terminal double bond, (*S*)-**1** was obtained in 86% yield and 93% *ee*.[10] Ongoing experiments indicate that this reaction is not very temperature dependent, and we are confident that this elegant access to (*S*)-α-damascone is also the most efficient one.

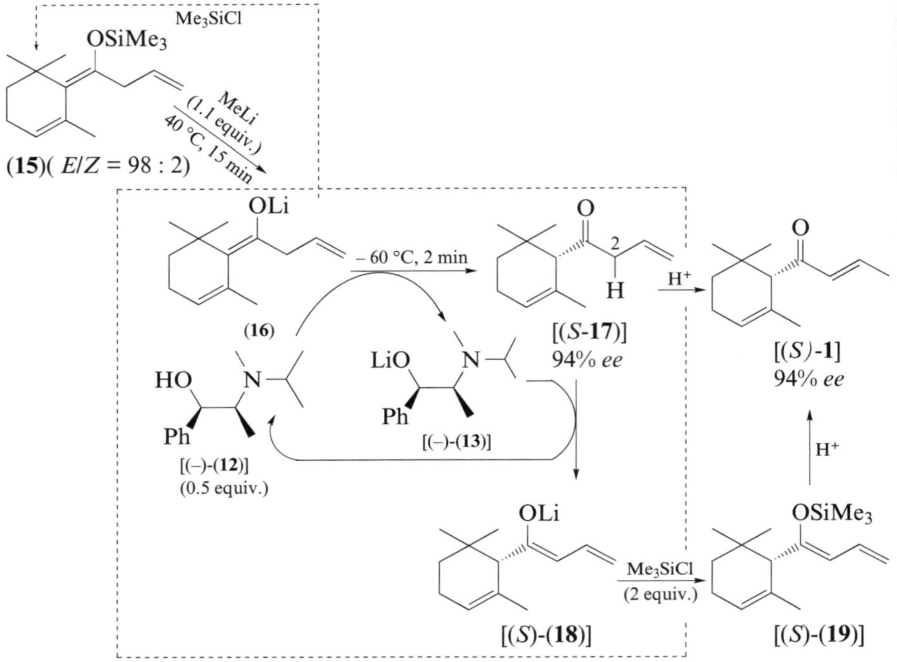

**Scheme 6**

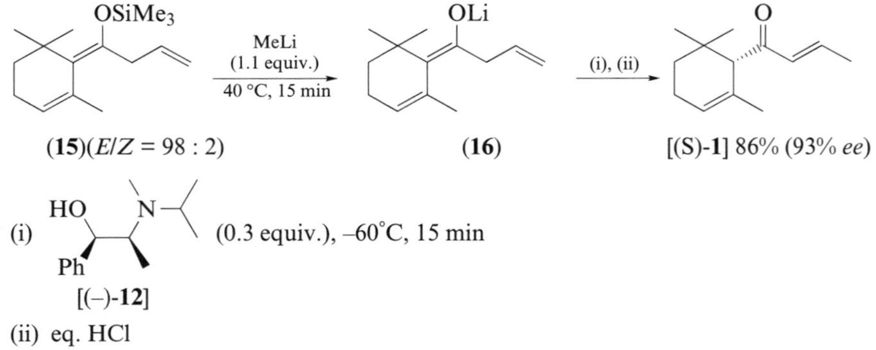

**Scheme 7**

The aforementioned case is, of course, very substrate specific, as the product ketone (S)-**17** is rapidly and exclusively deprotonated at C-2. For example, an analogous autocatalytic reaction is not possible with enolate **20**, as the acidity of the C-2 protons of (S)-21 is substantially weaker. To circumvent this problem, 1-phenylacetone has been used successfully as an external, achiral proton donor (Scheme 8). Thus, protonation of **20** (E/Z ≈ 97:3), readily obtained from ketene **14** and n-butyllithium, with equimolar amounts of (−)-**12** afforded butyl ketone (S)-**21** with 96% ee, and catalytic enantioselective protonation with 0.2 equiv. of (−)-**12**, followed by the addition of 0.85 equiv. of 1-phenylacetone, also furnished ketone (S)-**21** with high enantioselectivity (94% ee). With only 0.1 equiv. of (−)-**12** the ee decreased to 85%.[10]

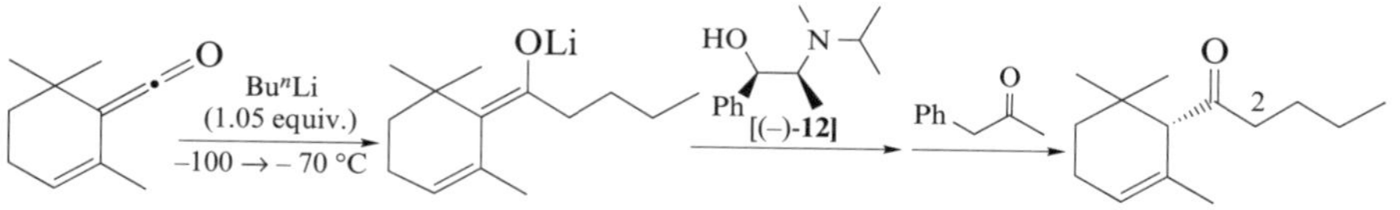

| (14) | (20)(E/Z = 97:3) | | | [(S)-21] |
|---|---|---|---|---|
| | | 1 equiv. | – | 96% ee (90%) |
| | | 0.2 equiv. | 0.85 equiv. | 94% ee (94%) |
| | | 0.1 equiv. | 0.95 equiv. | 85% ee |

**Scheme 8**

### 16.3.5 EXTENSION TO THE SYNTHESES OF THE ENANTIOMERS OF α- AND γ-CYCLOGERANIC ACID, α- AND γ-CYCLOCITRAL AND γ-DAMASCONE

The single enantiomers of α- and γ-cyclogeranic acids and esters represent versatile intermediates in the field of fragrance and medicinal chemistry, and the racemic α- and γ-methyl esters (±)-**5** and (±)-**33** are used as fragrance materials which complement the widely used rose ketones. Applying the aforementioned technology for enolate generation/protonation, methyl α-cyclogeranate [(±)-**5**] was deprotonated with n-butyllithium and the resulting ester enolate **24** was protonated with (−)-**12**. Unfortunately, the isolated methyl ester (S)-**5** showed a low enantiomeric excess of 36%. This marked difference with the related ketone enolates (see above) may be ascribed to an over-rapid protonation of the enolate (due to the high $pK_a$ of **5**), coupled with an insufficient structural differentiation of the enolate substituents on C-1 (OMe vs OLi) for efficient enantiotopic recognition. For this reason, the phenyl ester (±)-**22** and the phenyl thioester (±)-**23** were chosen as substrates. Indeed, protonation of enolates **25** and **26** with 2 equiv. of (−)-N-isopropylephedrine [(−)-**12**] at −100 °C, followed by gradual warming to −10 °C, afforded the corresponding esters with 77% and 99% ee, respectively (Scheme 9, Table 1).[7,9] Likewise, almost enantiomerically pure (R)-thioester [(R)-**23**] was obtained by protonation of enolate **26** with (+)-**12**. Application of the catalytic version on thioester enolate **25** is also possible: use of 0.5 equiv. of (−)-**12** and 1.55 equiv. of 1-phenylacetone gives almost the same result (98% ee) as

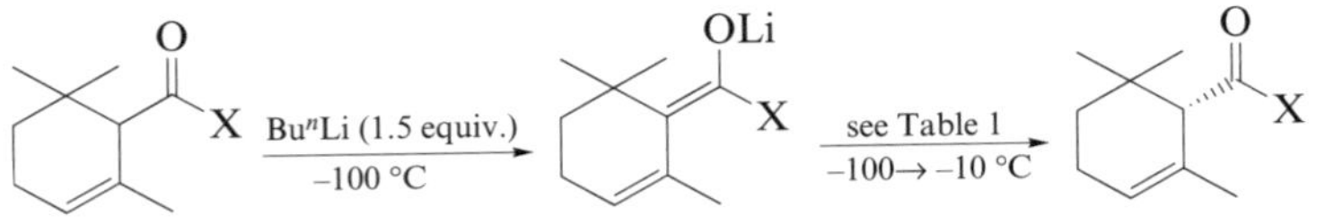

[(±)-5] X = OMe
[(±)-22] X = OPh
[(±)-23] X = SPh

(24)(Z)
(25)(Z)
(26)(Z)

[(S)-5]
[(S)-22]
[(S)-23]

**Scheme 9**

**Table 1** Protonation conditions for Scheme 9

| Substrate | Reagents (equiv.) | ee % | yield % |
|---|---|---|---|
| (±)-5 | (−)-**12** (2.0) | 36 (S) | – |
| (±)-22 | (−)-**12** (2.0) | 77 (S) | – |
| (±)-23 | (−)-**12** (2.0) | 99 (S) | 87 |
| (±)-23 | (+)-**12** (2.0) | 99 (R) | 87 |
| (±)-23 | (−)-**12** (0.5), 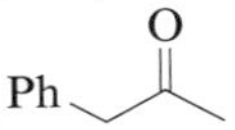 Ph-CO-CH₃ (1.55) | 98 (S) | 81 |
| (±)-23 | (−)-**12** (0.2), 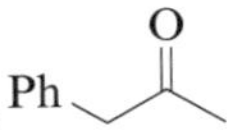 Ph-CO-CH₃ (1.85) | 81 (S) | |

2.0 equiv. of (−)-**12**. However, a substantial loss in selectivity is observed with smaller amounts of (−)-**12** [81% ee with 0.2 equiv. of (−)-**12**].

The foregoing protonations are performed at −100 °C because, above −80 °C, the thioester enolates eliminate LiSPh to afford ketene **14** (see Section 16.3.3). To circumvent this problem, we studied the enantioselective addition of aromatic thiols to ketene **14**[8,9] (Scheme 10). After some experimentation, it was found that slow addition of the chiral amino alcohol (−)-**12** to a mixture of ketene **14** and ArSLi at −60 °C leads to the thioesters (S)-**23** and (S)-**27** with 95 and 97% ee,

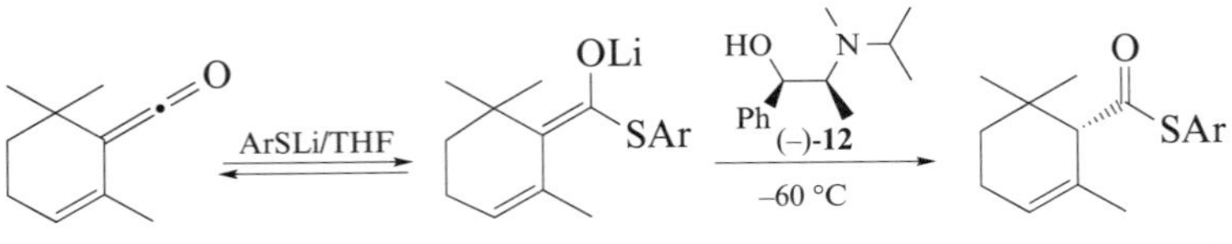

(14)  (26)

[(S)-23] Ar = Ph, 95% ee (80% yield)
[(S)-27] Ar = 4-ClPh, 97% ee (83% yield)

**Scheme 10**

respectively. Since, under these conditions, the equilibrium between ketene **14** and enolate **26** lies on the side of the former, the reaction rate is controlled by the addition of (−)-**12**. This allows rapid and irreversible C-protonation of the incipient enolate. In this manner, the concentration of (−)-**12** in the reaction medium is kept low, thus minimising the risk of proton exchange with the thiolate.

In the overall process, 1 equiv. of thiol is incorporated irreversibly into ketene **14**; however, the lithium ephedrate (−)-**13** is recovered unchanged. We next turned our attention, therefore, to the possibility of a catalytic process, in which the aromatic thiol serves as nucleophile *and* proton source, in the presence of catalytic amounts of lithium-ephedrate (−)-**13**. To allow this catalytic reaction to succeed, conditions had to be found which guaranteed the rapid consumption of the thiol, thus maintaining its concentration lower than that of the chiral lithium alkoxide (−)-**13**. The required rate acceleration could be achieved by working at higher temperatures and higher concentrations. Thus, continuous slow addition of ArSH to a reaction mixture containing 1 equiv. of ketene **14** and 0.05 equiv. of (−)-**13** at −27 °C furnished the thioesters (*S*)-**23** and (*S*)-**27** with excellent *ee*'s (89–90%) (Scheme 11). At lower temperatures the reaction becomes sluggish and at 0 °C competing addition of (−)-**13** to **14** is observed, leading to an accumulation of thiol and a corresponding drop in the *ee* of **23** (*ca* 50% *ee*).[8,9]

In comparison with the catalytic enantioselective enolate protonation, the thiol addition described here shows improved efficiency in terms of the overall selectivity and turnover of chiral reagent. This is undoubtedly due to the fact that the concentrations of achiral proton donor (ArSH) and enolate are very low throughout the course of the reaction.

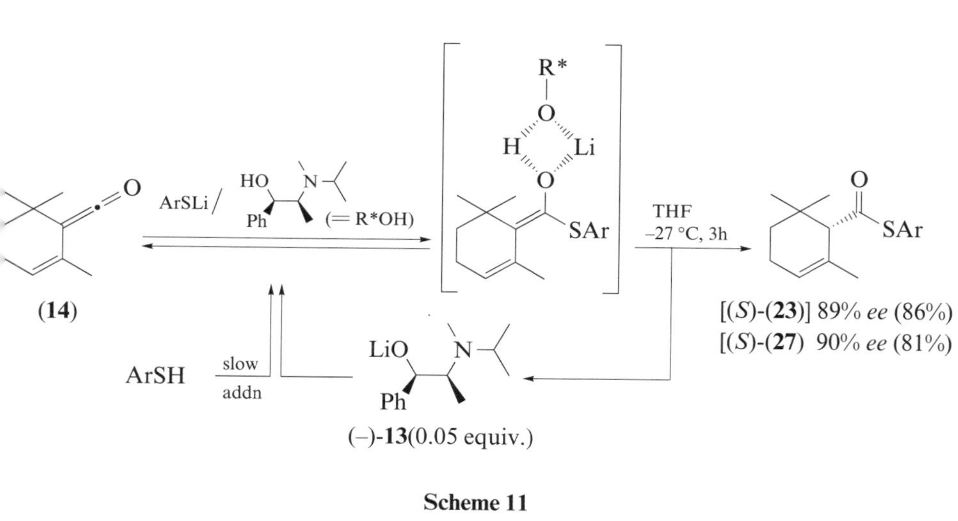

**Scheme 11**

The thiol esters (R)-**23** and (S)-**23** have been transformed readily into the corresponding methyl esters (R)-**5** and (S)-**5** [of which the (R)-enantiomer is the more valuable fragrance], the α-cyclocitrals (R)-**28** and (S)-**28** and the α-damascones (R)-**1** and (S)-**1**, all without racemisation.

[(S)-**5**]  R = OMe
[(S)-**28**] R = H
[(S)-**1**]  R = ⸻⁄⸺

Finally, the most recent extension of enantioselective protonation to the synthesis of the hitherto unknown enantiomers of γ-damascone [(R)- and (S)-**10**] and the γ-cyclogeranates is presented briefly for the (S)-series.[11] Following the synthetic route used for racemic γ-damascone [(±)-**10**] (Scheme 3), the prochiral mixed lithium–magnesium enolate complex **9** is protonated with (–)-**12**. Because this protonation is slow, and to minimise side reactions between enolate **9** and the product ketone such as double-bond isomerisation and racemisation, enolate **9** was added to excess (–)-**12** at 0 °C. Under these conditions and after treatment with $Al_2O_3$, a 3:1 mixture of (S)-**10** and (R)-**10** was isolated (Scheme 12). On the other hand, protonation of the magnesium-free lithium enolate **29** depends critically on the presence of the chiral lithium alkoxide (–)-**13**. The best results were obtained when a solution of **29** in THF was added to a 2:1 mixture of (–)-**13** and (–)-**12** (75% ee). By comparison, a 1:1 mixture of (–)-**13** and (–)-**12** gives 68% ee, and (–)-**12** alone gives 49% ee.

(S)-γ-Damascone of a higher enantiomeric purity was obtained by an LDA-mediated Grignard reaction on the corresponding (S)-γ-thioester (S)-**32**, which was prepared by the route illustrated in Scheme 13. Following the procedure for its positional isomer **23**, the α,β-unsaturated thioester **30** was deprotonated with excess LDA, and the resulting enolate **31** was protonated with an excess of (+)-amino alcohol (+)-**12**. It should be stressed that, unlike the case with thiol ester **23**, deprotonation of **30** leads to an enolate of opposite configuration (E), and that the protonation takes place with excellent, but *opposite*, selectivity. Here, use of the dextrorotatory reagent (+)-**12** allows the preparation of the (S)-thioester (S)-**32** with greater than 96% ee (Scheme 13). Unfortunately, competing γ-protonation also affords substantial amounts of starting ester **30**, thus rendering the conversion incomplete. The quantities of base and proton donor could be reduced with minimal loss of enantioselectivity, and we were pleased to find that the combination of phenylacetone and 0.5 equiv. of amino alcohol (+)-**12** also

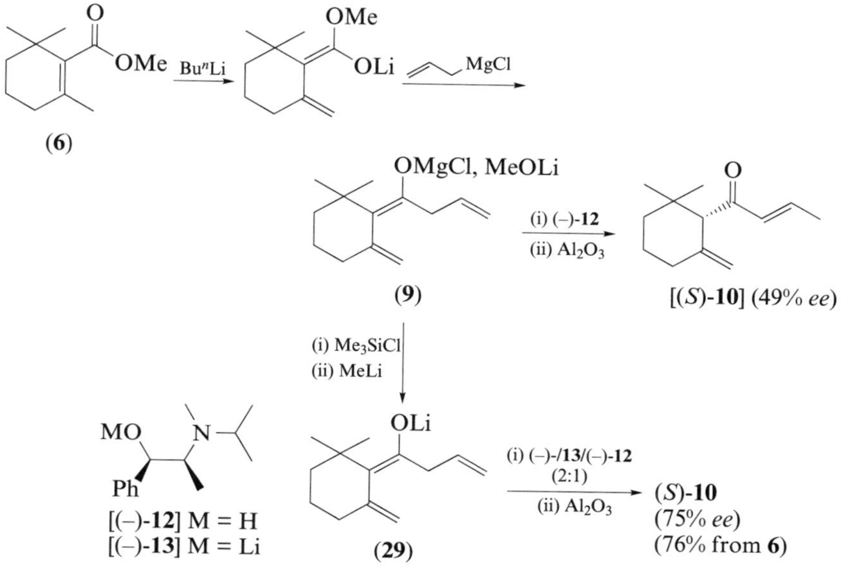

# ENANTIOSELECTIVE PROTONATION IN FRAGRANCE SYNTHESIS

**Scheme 12**

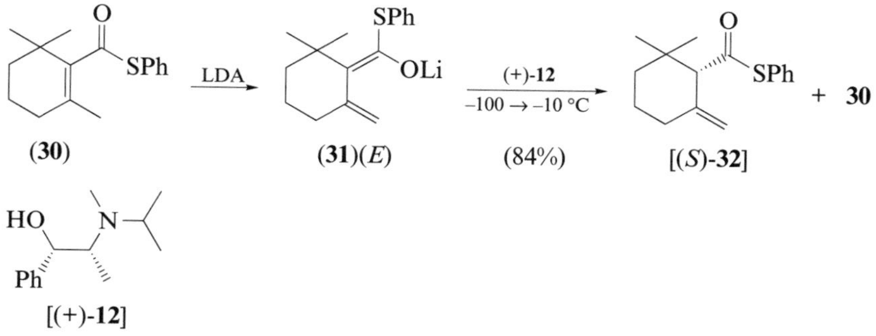

**Scheme 13**

| LDA (equiv.) | (+)-12 (equiv.) | 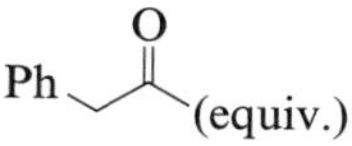 Ph-CO-CH₃ (equiv.) | (S)-32 | 32:30 |
|---|---|---|---|---|
| 3.0 | 4.0 | – | ≥96% ee | 53:47 |
| 1.5 | 2.0 | – | 94% ee | 55:45 |
| 1.5 | 0.5 | 1.5 | 88% ee | 45:55 |

resulted in a high enantiofacial discrimination, affording (*S*)-**32** with 88% *ee*. With the enantiomers of (*S*)-thioester **32** in hand, the enantiomers of methyl γ-cyclogeranate [(*S*)- and (*R*)-**33**], γ-damascone [(*S*)- and (*R*)-**10**] and γ-cyclocitral [(*S*)- and (*R*)-**34**] could thus be readily synthesised in high enantiomeric purity.

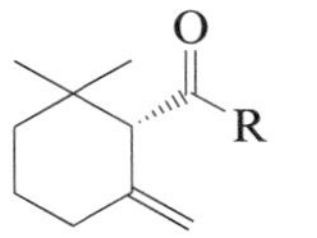


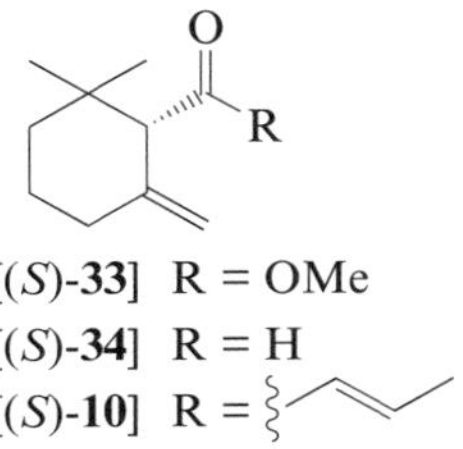

[(*S*)-**33**] R = OMe
[(*S*)-**34**] R = H
[(*S*)-**10**] R =

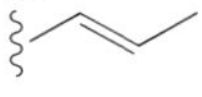

For both γ-damascone (**10**) and methyl γ-cyclogeranate (**33**), the (*S*)-enantiomer exhibits better odour properties, this difference being more pronounced in the latter case; whereas (*S*)-**33** excels with rich aromatic, herbal, floral, damascone-reminiscent notes, (*R*)-**33** has a predominant camphoraceous, cork, cellar note.

## 16.4 ENANTIOSELECTIVE SYNTHESIS OF A NEW TETRALIN MUSK PERFUME COMPONENT

Several hundred structurally related aromatic musk odourants of different odour strength are known, of which Tonalid [(±)-**35**], first commercialised by Polak Frutal Works, has long been considered the strongest. Its annual consumption is estimated at more than 3000 tonnes. Recently, we have discovered a new aldehyde musk, (±)-**2**, which, according to perfumery performance tests, is approximately ten times stronger than Tonalid.[2]

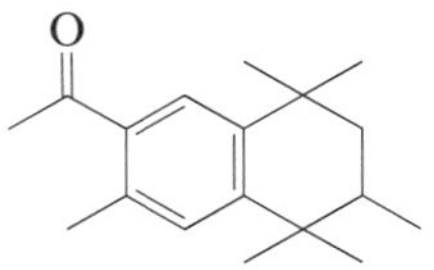


[(±)-**35**]
Tonalid® (*PFW*)
[=Fixolide® (*Givaudan*)]
>3000 tpa

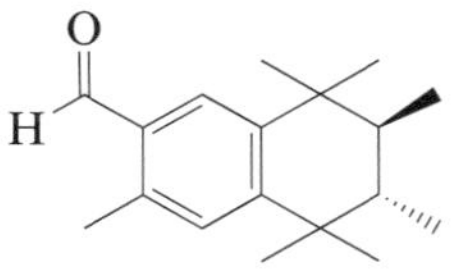


[(±)-**2**]
New *Firmenich* m usk
~10 × stronger than Tonalid®

The synthesis of (±)-**2** is outlined in Scheme 14. Friedel–Crafts alkylation of readily accessible *tert*-butyl ketone (±)-**36** with *o*-xylene furnishes ketone (±)-**37**, the reduction of which affords alcohol (±)-**38** as a 94:6 diastereoisomeric mixture. Under strongly acidic conditions (MsOH, $P_2O_5$ or $H_2SO_4$), a Wagner–Meerwein rearrangement generates a tertiary carbocation, which

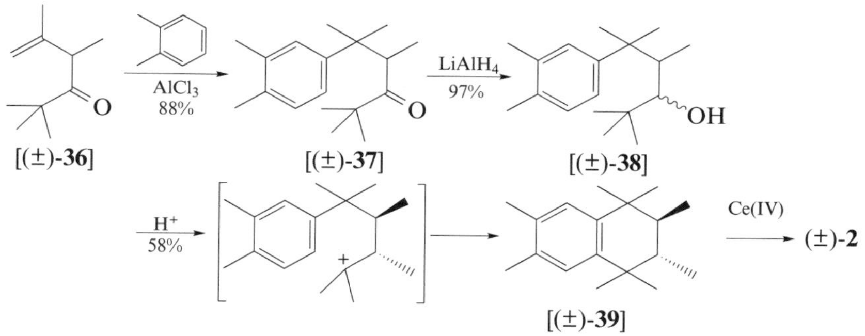

ENANTIOSELECTIVE PROTONATION IN FRAGRANCE SYNTHESIS 349

**Scheme 14**

undergoes cyclisation to furnish, after crystallisation, hydrocarbon (±)-**39** in 58% yield; the mother liquors contain about 15% of the diastereoisomeric *meso*-hydrocarbon. The final step involves Ce(IV)-mediated oxidation to the musk odourant (±)-**2**.

Subsequently, the two enantiomers of **2** were synthesised by a multi-step sequence including a classical resolution step,[21] thus allowing the evaluation of the fragrances of (*R,R*)-(+)- and (*S,S*)-(−)-**2** (see Introduction). In view of its clear superiority, it was decided to elaborate an enantioselective synthesis of (−)-**2**.[3]

If known enantioselective reactions are examined, it becomes evident that only a few are suited to the construction of sterically crowded chiral centres, and enantioselective protonation is often the method of choice. Our plan was to prepare ketone (*R*)-**36** by enantioselective protonation of either the (*Z*)- or (*E*)-enolate **40**, and to follow the efficient synthetic route shown in Scheme 14. In spite of the risk of racemisation, both in a Friedel–Crafts alkylation with (*R*)-**36** and, even more important, during acid-catalysed cyclisation (Scheme 14; **38** → **39**), this approach was considered the most interesting in view of its simplicity and similarity to the established route.

For the generation of enolate **40**, the deprotonation of both the β,γ-unsaturated enone (±)-**36** and of the readily accessible α,β-unsaturated enone **41** was examined. Whereas deprotonation of (±)-**36** with LDA afforded a 9:1 mixture of (*Z*)-**40** and (*E*)-**40**, chelation-controlled deprotonation of **41** generated (*Z*)-**40** with complete stereocontrol, thus enhancing the chances of an efficient enantioselective protonation (Scheme 15). Indeed, we were gratified to find that protonation of (*Z*)-**40** with (−)-*N*-isopropylephedrine [(−)-**12**] led to the desired ketone (*R*)-**36** with 90% *ee*. The site selectivity for protonation is also satisfactory, as only 31% of starting enone **41** was regenerated. Decreasing the amounts of LDA and (+)-**12** had only a minor effect on enantioselectivity (86% *ee*), but dramatically reversed the site selectivity for α- *vs* γ-protonation. As (*R*)-**36** is much more volatile than its isomer **41**, these compounds are easily separated by distillation. On the other hand, owing to the inertness of **41** to Friedel–Crafts condi-

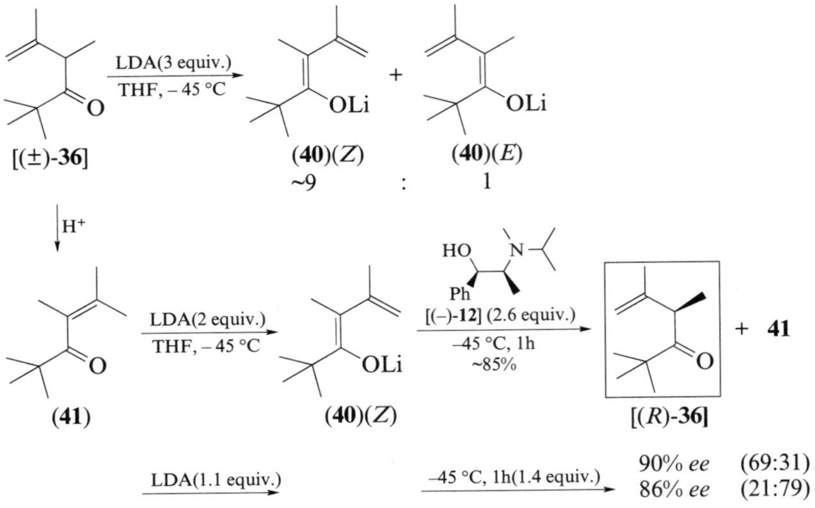

**Scheme 15**

tions, the mixture (R)-**36** and **41** can also be treated with o-xylene. To our satisfaction, the synthesis of (−)-**2** according to Scheme 14 could be achieved with minimal racemisation (<10%), and the organoleptic evaluation of a 9:1 mixture of (−)-**2** and (+)-**2** clearly confirmed the superiority of (−)-**2** over the racemic version.[3]

In view of these encouraging results, we are confident that the technique of enantioselective protonation will often be considered for future synthetic applications in industry.

## 16.5 ACKNOWLEDGEMENTS

The author gratefully acknowledges the postdoctoral contributions of Dr Isabelle Stempf and Dr Nathalie Chaptal-Gradoz and the excellent work of Mr José Galindo.

## 16.6 REFERENCES AND NOTES

1. Fehr, C., and Galindo, J., *J. Am. Chem. Soc.*, **110**, 6909 (1988); Fehr, C., and Galindo, J. (Firmenich), *Eur. Pat.*, EP 326 869, 1988.
2. Fehr, C., Galindo, J., Haubrichs, R., and Perret, R., *Helv. Chim. Acta*, **72**, 1537 (1989); Fehr, C., and Galindo, J. (Firmenich), *US Pat.*, 5 324 875, 1989; *Eur. Pat.*, EP 405 427, 1989.

3. Fehr, C., Delay, F., Blanc, P.-A., and Chaptal–Gradoz, N. (Firmenich), *Eur. Pat. Appl.*, 1994.
4. Reviews: Fehr, C., *Angew. Chem.*, **108**, 2726 (1996); *Angew. Chem. Int. Ed. Engl.*, **35**, 2566 (1996); Hünig, S., *Houben-Weyl, Methods of Organic Chemistry*, Vol. E 21 d (ed. Helmchen, G., Hoffmann, R.-W., Mulzer, J. and Schaumann, E.), Thieme, Stuttgart, 1995, p. 3851. Duhamel, L., Duhamel, P., Launay, J.-C., and Plaquevent, J.-C., *Bull. Soc. Chim. Fr.* II, 421 (1984) and ref. 5. Recent examples: Vedejs, E., and Lee, N., *J. Am. Chem. Soc.*, **117**, 891 (1995); Vedejs, E., Lee, N., Sakata, S. T., *J. Am. Chem. Soc.*, **116**, 2175 (1994); Ishihara, K., Kaneeda, M., and Yamamoto, H., *J. Am. Chem. Soc.*, **116**, 11179 (1994); Yanagisawa, A., Kuribayashi, T., Kikuchi, T., and Yamamoto, H., *Angew. Chem.*, **106**, 129 (1994); *Angew. Chem., Int. Ed. Engl.*, **33**, 107 (1994); Takeuchi, S., Ohira, A., Miyoshi, N., Mashio, H., and Ohgo, Y., *Tetrahedron: Asymmetry*, **5**, 1763 (1994); Gerlach, U., Haubenreich, T., Hünig, S., and Klaunzer, N., *Chem. Ber.*, **127**, 1989 (1994); Kosugi, H., Hoshino, K., and Uda, H., *Phosphorus Sulfur*, **95–96**, 401 (1994); Yanagisawa, A., Kikuchi, T., Watanabe, T., Kuribayashi, T., and Yamamoto, H. *Synlett*, 372 (1995); Aboulhoda, S. J., Létinois, S., Wilken, J., Reiners, I., Hénin, F., Martens, J., and Muzart, J., *Tetrahedron: Asymmetry*, **6**, 1865 (1995); Fuji, K., Kawabata, T., and Kuroda, A., *J. Org. Chem.*, **60**, 1914 (1995).
5. Fehr, C., *Chimia*, **45**, 253 (1991).
6. Fehr, C., and Guntern, O., *Helv. Chim. Acta*, **75**, 1023 (1992).
7. Fehr, C., Stempf, I., and Galindo, J., *Angew. Chem.*, **105**, 1091 (1993); *Angew. Chem., Int. Ed. Engl.*, **32**, 1042 (1993).
8. Fehr, C., Stempf, I., and Galindo, J., *Angew. Chem.*, **105**, 1093 (1993); *Angew. Chem., Int. Ed. Engl.*, **32**, l044 (1993).
9. Fehr, C., Stempf, I., and Galindo, J. (Firmenich), *Eur. Pat.*, EP 593 917, 1992.
10. Fehr, C., and Galindo, J., *Angew. Chem.*, **106**, 1967 (1994); *Angew. Chem., Int. Ed. Engl.*, **33**, 1888 (1994).
11. Fehr, C., and Galindo, J., *Helv. Chim. Acta*, **78**, 539 (1995).
12. Kastner, D., *Parfüm. Kosmetik*, **75**, 170 (1994).
13. Fehr, C., and Galindo, J., *Helv. Chim. Acta*, **69**, 228 (1986).
14. Fehr, C., Galindo, J., and Perret, R., *Helv. Chim. Acta*, **70**, 1745 (1987).
15. Fehr, C., and Galindo, J., *J. Org. Chem.*, **53**, 1828 (1988).
16. Häner, R., Laube, T., and Seebach, D., *J. Am. Chem. Soc.*, **107**, 5396 (1985).
17. Naef, F., and Decorzant, R., *Tetrahedron*, **42**, 3245 (1986).
18. *Perfum. Flavorist*, **19**, 59 (1994).
19. Adamski, R. J., and Numajiri, S., *US Pat.*, 3 860 651, 1969; *Chem. Abstr.*, **82**, 170 330t (1975).
20. (*S*)-**17** does not protonate **16** enantioselectively in the absence of (–)-**12**.
21. Fehr, C and Chaptal-Gradoz, N., unpublished work.

# 17 Thiamphenicol: a Manufacturing Process Involving a Double Inversion of Stereochemistry

**L. COPPI, C. GIORDANO, A. LONGONI and S. PANOSSIAN**
*Zambon Group SpA, Bresso (Milan), Italy*

| | | |
|---|---|---|
| 17.1 | Introduction | 353 |
| | 17.1.1  The Drug | 353 |
| | 17.1.2  The Market | 354 |
| 17.2. | Main Synthetic Approaches | 354 |
| | 17.2.1  Process 1 (from 4-Methylmercaptobenzaldehyde) | 355 |
| | 17.2.2  Process 2 (from 4-Methylsulphonylbenzaldehyde) | 358 |
| 17.3. | Conversion of (1S,2S)-2-Amino-1-[(4-methylthio)-phenyl]propane-1,3-diol into its Enantiomer | 358 |
| | 17.3.1  Degradation Route | 359 |
| | 17.3.2  Racemisation Route | 359 |
| | 17.3.3  Direct Conversion: the Present Zambon Industrial Process | 359 |
| 17.4. | References | 362 |

## 17.1 INTRODUCTION

### 17.1.1 THE DRUG

Thiamphenicol[1] is the international non-proprietary name for (1R,2R)-2-(dichloroacetamido)-1-(4-methylsulphonyl)phenylpropane-1,3-diol. The compound was described for the first time by Sterling Drug[2] in 1951, and is used as an antibiotic for both human and animal health. Thiamphenicol belongs to the family of the so-called 'amphenicols, which includes chloramphenicol[1b] and florfenicol.[1b] It was originally synthesised during a research programme aimed at the identification of compounds with a spectrum of activity similar to that of chloramphenicol, but with a better safety profile.

The configurational requirements for the antibacterial activity of the family are very specific, as shown by the fact that the D-*threo* configuration is essential. Thus, thiamphenicol possesses a fairly wide spectrum of antimicrobial activity against Gram-negative bacteria, whereas its enantiomer is not active. All compounds in this family have the same mechanism of antibacterial activity, which

*Chirality in Industry II.* Edited by A. N. Collins, G. N. Sheldrake and J. Crosby
© 1997 John Wiley & Sons Ltd

**Thiamphenicol** (structure: 4-(O₂S-Me)-C₆H₄-CH(OH)-CH(NHCOCHCl₂)-CH₂OH, with R,R stereochemistry)

**Chloramphenicol** (structure: 4-(O₂N)-C₆H₄-CH(OH)-CH(NHCOCHCl₂)-CH₂OH, with R,R stereochemistry)

**Florfenicol** (structure: 4-(O₂S-Me)-C₆H₄-CH(OH)-CH(NHCOCHCl₂)-CH₂F, with R,S stereochemistry)

involves the inhibition of peptidyl transferase following the interaction of the compounds with ribosomal sub-units of sensitive bacteria.

### 17.1.2 THE MARKET

Thiamphenicol and its derivatives[3] (thiamphenicol glycinate acetylcysteinate, thiamphenicol glycinate hydrochloride, thiamphenicol glycinate 4-hydroxyisophthalate) are marketed for both human and veterinary applications. The development of the veterinary application of thiamphenicol was undertaken in the early 1960s and the marketing of a different formulation began in Europe and Japan in the second half of the same decade.

The recent prohibition of chloramphenicol from major markets in the treatment of food-producing animals and the positive evaluation of the safety features of thiamphenicol (establishment of MRL according to the EEC Regulation 2377/90) have opened up new opportunities for the product. As a result of the growth of this veterinary application, the worldwide market for thiamphenicol is increasing and is currently estimated to be 100 Mtonne per/year. At present, bulk thiamphenicol is produced by the Zambon Group in Italy and by local producers in China. The average price of the bulk drug is about US$ 150 kg$^{-1}$.

## 17.2 MAIN SYNTHETIC APPROACHES

Owing to the presence of two non-equivalent stereogenic carbon centres, (1$R$,2$R$)-thiamphenicol is one of the four possible stereoisomers of 2-(dichloroacetamido)-1-[(4-methylsulphonyl)phenyl]propane-1,3-diol. Therefore, the target of any synthetic approach to thiamphenicol must be effective control of the relative (*threo* vs *erythro*) and absolute (1*R*,2*R* vs 1*S*,2*S*) configurations. In order to satisfy the above stereochemical requirements, all the practical synthetic routes to thiamphenicol follow a similar strategy consisting of the following steps:

(i) preparation of racemic *threo* intermediates;
(ii) classical resolution;
(iii) preparation of thiamphenicol from the 'correct' *threo* resolved intermediate.

To the best of our knowledge, no asymmetric synthesis has yet been found to be commercially viable on a manufacturing scale.

The only two synthetic methods applying the above strategy [(i)–(iii)] which have been shown to be useful for the production of commercial quantities of thiamphenicol are described in Schemes 1 and 2. In both processes, the relative *threo* configuration is determined in the first step, in which 4-methylmercaptobenzaldehyde (Process 1, Scheme 1) or 4-methylsulphonylbenzaldehyde (Process 2, Scheme 2) is condensed with glycine to give a *threo* racemic arylserine, virtually free from the *erythro* epimer. The optical resolution is performed at different stages and with different techniques:[4] entrainment resolution in Process 1 and diastereomeric salt formation in Process 2.

## 17.2.1 PROCESS 1 (FROM 4-METHYLMERCAPTOBEN-ZALDEHYDE)

Zambon has produced thiamphenicol for several decades using Process 1. Here, 2 mol of 4-methylmercaptobenzaldehyde are condensed with 1 mol of glycine in an aqueous alcoholic medium under basic conditions to provide a heterogeneous mixture of *threo* and *erythro* Schiff's bases 7 and 8. Prolonged stirring of the reaction mixture allows the equilibration between these compounds, causing the almost complete conversion of 8 into 7. The *threo* Schiff base obtained (7) is hydrolysed with hydrochloric acid to give, in 80% overall yield, *threo*-4-methylthiophenylserine (1).

(**7**) *threo* (2*S*, 3*R*) (2*R*, 3*S*)
(**8**) *erythro* (2*R*, 3*R*) (2*S*, 3*S*)

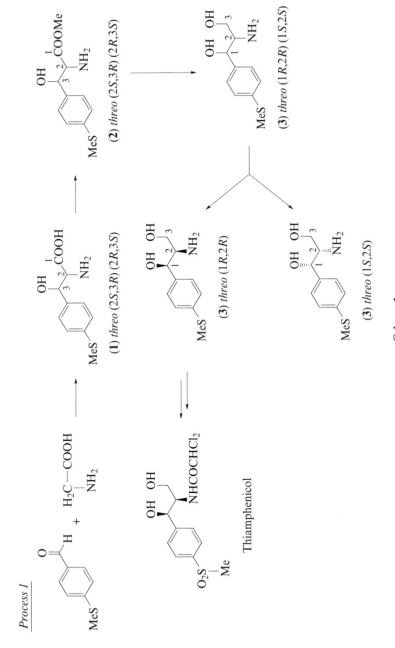

**Scheme 1**

*Process 2*

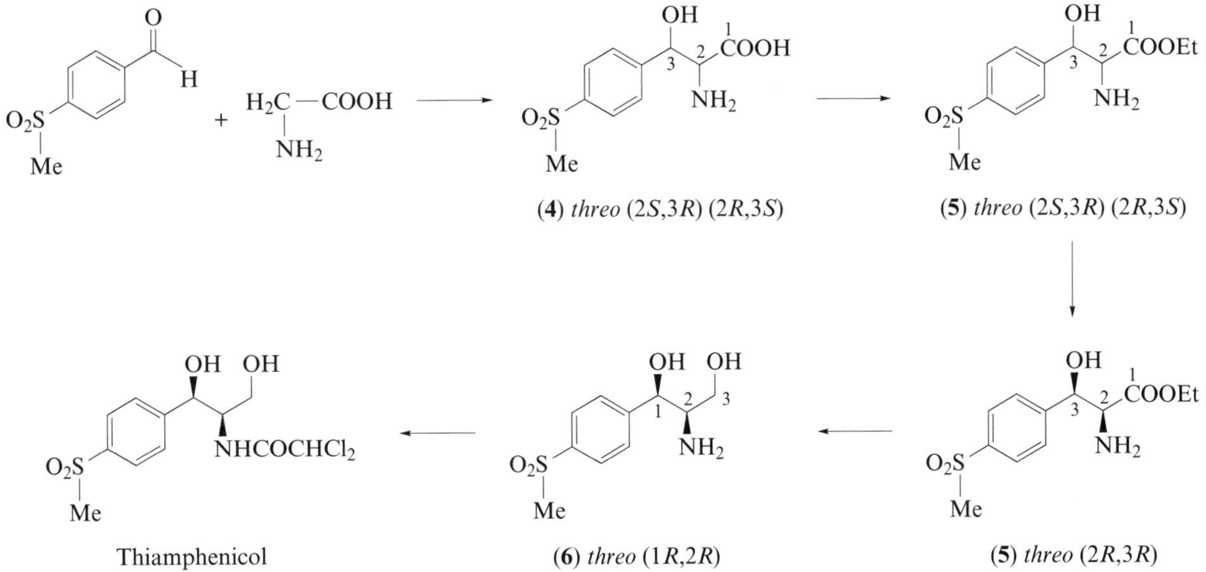

$\text{4-(MeSO}_2\text{)C}_6\text{H}_4\text{CHO}$ + H$_2$C(NH$_2$)—COOH ⟶ (**4**) *threo* (2S,3R) (2R,3S) ⟶ (**5**) *threo* (2S,3R) (2R,3S) ⟶ (**5**) *threo* (2R,3R) ⟶ (**6**) *threo* (1R,2R) ⟶ Thiamphenicol

**Scheme 2**

*threo*-4-Methylmercaptophenylserine (**1**) is transformed in almost quantitative yield into ester **2** by treatment with thionyl chloride in methanol. The methyl ester so obtained is subsequently reduced to the corresponding *threo*-2-amino-1-[(4-methylthio)pheny]propane-1,3-diol (**3**) by sodium borohydride in the presence of calcium chloride in about 90% yield. The entrainment resolution of the *threo* racemate (**3**),[5] carried out in aqueous acid, provides both enantiomers in enantiomerically pure form. The main features which make this entrainment resolution particularly suitable for industrial application are as follows:

—mild conditions (temperature range 25–50 °C);
—almost quantitative yield (high stability of products under operating conditions);
—high volume productivity;
—an exceedingly high number of cycles (>70);
—the enantiomers are collected as wet free base, which can be used directly in the next step.

The main drawback of this process is that the resolution is perfomed at a late stage and co-produces the 'incorrect' enantiomer, (1*S*,2*S*)-thiomicamine (**3**), whose chemical structure does not allow a straightforward recycle (see Section 17.3). The wet (1*R*,2*R*) enantiomer is converted into thiamphenicol by *N*-acylation followed by sulphur oxidation with hydrogen peroxide in the presence of sodium tungstate.

### 17.2.2 PROCESS 2 (FROM 4-METHYLSULPHONYL-BENZALDEHYDE)

We believe that Process 2 is used by Chinese producers, but only limited information is available. The racemic *threo*-4-methylsulphonylphenylserine (**4**) is converted into its ethyl ester,[5] which is resolved in alcoholic solution in high yield by treatment with the natural (L) isomer of tartaric acid.[6] The sequence to thiamphenicol is completed by converting the enantiomerically pure (2*S*,3*R*)-4-methylsulphonylphenylserine ethyl ester (**5**) into the corresponding aminopropanediol (1*R*,2*R*)-**6** followed by *N*-acylation.

Alternative methods have been described for the resolution of either arylserine[7,8] or aminopropanediol derivatives.[9–12] However, as far as we know, none of these has found industrial application.

## 17.3 CONVERSION OF (1*S*,2*S*)-2-AMINO-1-[4-METHYLTHIO)PHENYL]PROPANE-1,3-DIOL INTO ITS ENANTIOMER

Historically, Zambon has produced more than 1300 tonnes of thiamphenicol using Process 1 which, in turn, provided approximately 1500 tonnes of the by-product (1*S*,2*S*)-thiomicamine (**3**) (Scheme 1) of high chemical and enantiomeric purity. In the course of efforts to establish how to transform (1S,2S)-thiomicamine into a source of the desired (1R,2R)-isomer, three possible strategies were investigated:

—degradation of the molecule to a useful intermediate;
—racemisation (which ultimately requires a resolution);
—direct conversion.

### 17.3.1 DEGRADATION ROUTE

Compound (1S,2S)-**3** is transformed into 4-methylthiobenzaldehyde by treatment with an oxidising agent (e.g. sodium hypochlorite) followed by acid hydrolysis.[13]

### 17.3.2 RACEMISATION ROUTE

Zambon's strategy for racemisation does not involve degradation of the carbon skeleton.[14,15] The preferred approach[15] is shown in Scheme 3, and involves the following steps:

(i) protection of both the amino group and the primary hydroxy group as a 1,3-oxazoline followed by oxidative removal of the benzylic stereogenic centre;
(ii) spontaneous and rapid racemisation at the adjacent stereogenic carbon centre due to the acidity of the hydrogen $\alpha$ to the carbonyl group;
(iii) regeneration of the benzylic stereogenic centre by diastereoselective reduction of the carbonyl group under conditions which favour the *threo* form of the resulting alcohol;
(iv) hydrolytic removal of the protecting group to afford the *threo*-aminodiol.

Using the last strategy, a synthetic procedure has been developed which allows racemisation in four steps with a 50% overall yield.

### 17.3.3 DIRECT CONVERSION: THE PRESENT ZAMBON INDUSTRIAL PROCESS

The only direct conversion of (1S,2S)-thiomicamine (**3**) into the (1R,2R)-enantiomer is, to the best of our knowledge, that investigated and developed to an industrial scale by Zambon, as described below.

An efficient conversion of (1S,2S)-**3** into thiamphenicol is effected by sequential inversion of configuration at C-2 and C-1, in which each stereogenic centre controls the other and maintains its stereochemical integrity. The process[16] has five stages, starting from (1S,2S)-**3** (Scheme 4).

Step 1: condensation of (1S,2S)-**3** with acetone followed by N-acetylation provides *trans*-N-acetyl-1,3-oxazolidine (**9**) (80% yield).

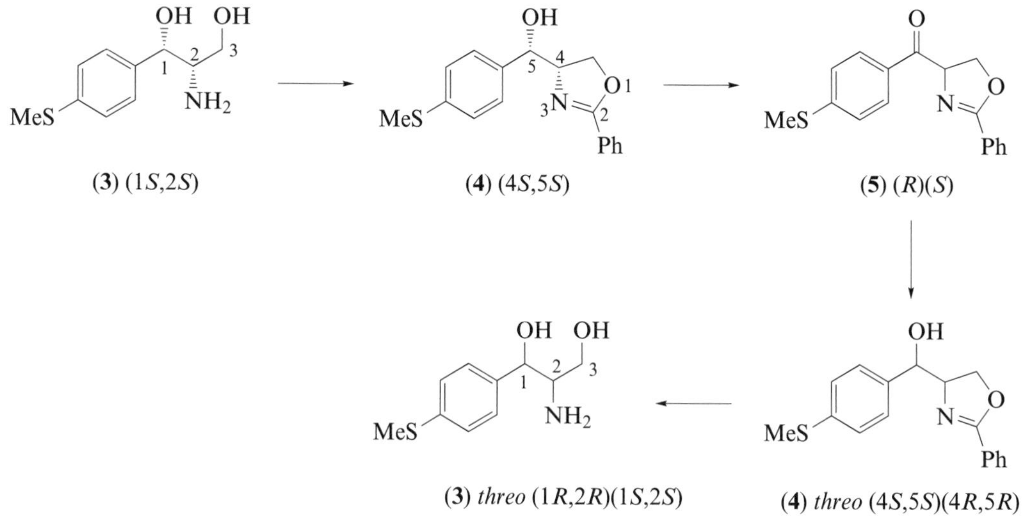

**Scheme 3**

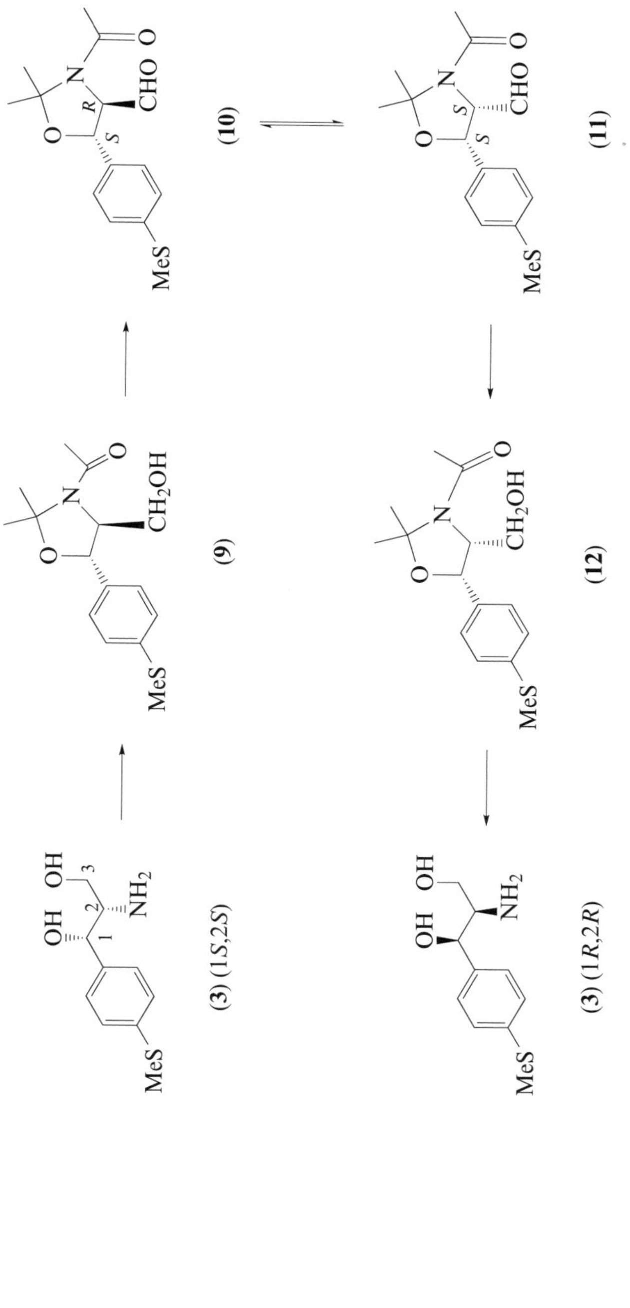

**Scheme 4**

Step 2: oxidation of the primary alcoholic function at room temperature to aldehyde **10** in 90% yield using DMSO and a Lewis acid.

Step 3: heating neat aldehyde **10** at 35 °C with a catalytic amount of DABCO, causing epimerisation to **11**, which crystallises from the medium, thus allowing virtually complete conversion of **10** into **11**.

Step 4: reduction of aldehyde epimer **11** providing oxazolidine alcohol **12** in almost quantitative yield.

Step 5: *O*-acetylation of alcohol **12**, followed by treatment with methanesulphonic acid and acetic anhydride, providing, after alkaline hydrolysis, crude (1*R*,2*R*)-**3**. Crystallisation then produces pure (1*R*,2*R*)-**3** in 86% yield based on aldehyde **11**.

The above chemical sequence is the result of 4 years of research; two years were needed to complete the scale-up of the process, which became an industrial reality in 1990.

The industrial process, which converts (1*S*,2*S*)-**3** into (1*R*,2*R*)-**3** in 55% yield, does not require any special apparatus or unusual operating conditions, and is carried out in standard stainless-steel and glass-lined reactors. The productivity of each stage is very high; stage 3 in particular is operated with a volumetric productivity of more than 90%, which allows the production of more than 1000 kg of aldehyde **11** in a 2000 l reactor.

This process has allowed Zambon to gain a competitive advantage in the thiampenicol market.

## 17.4 REFERENCES

1. (a) *The Merck Index*, XI Edn, No. 9230, p. 1445; (b) Elks, J., and Ganellin, C. R., *Dictionary of Drugs*, Chapman and Hall, London, 1990.
2. Sterling Drug, *Br. Pat.*, 745 900, 1951.
3. Kleeman, A., and Engel, J., *Pharmazeutische Wirkstoffe*, Georg Thieme Stuttgart, Vol. 5, 1982.
4. Jacques, J., Collet, A., and Wilen, S. H., *Enantiomers, Racemates and Resolutions*, Wiley, New York, 1981.
5. Parke, Davis and Co., *US Pat.*, 2 767 213, 1952.
6. Sumitomo Chemical, *Ger. Pat.*, 2 938 513, 1969.
7. Zambon Group, *Ital. Pat.*, 1 196 434, 1986.
8. Celgene, *Eur. Pat. Appl.*, EP 507 153, 1991.
9. E Gy.T. Gyogyszervegyeszeti Gyar, *Hung. Pat.* T3611, 1970; *Chem. Abstr.*, **77**, 5165g 1972.
10. Rebstock, M. C., and Bambas, L. L., *J. Am. Chem. Soc.*, **77**, 186 (1955).
11. Du Pont, *US Pat.*, 2 742 500, 1953.
12. Boehringer Mannheim, *Eur. Pat.*, EP 224 902, 1985.
13. Zambon Group, *Ital. Pat.*, 1 227 166, 1988.
14. Zambon Group, *Ital. Pat.*, 1 186 716, 1985.
15. Zambon Group, (a) *Ital. Pat.*, 1 223 563 1987; (b) Giordano, C., Cavicchioli, S., Levi, S., and Villa, M., *Tetrahedron Lett.*, **29**, 5561 (1988).
16. Zambon Group, (a) *Eur. Pat.*, 423 705, 1989; (b) Giordano, C., Cavicchioli, S., Levi, S., and Villa, M., *J. Org. Chem.*, **56**, 114 (1991).

# 18 Sharpless Asymmetric Epoxidation: Scale-up and Industrial Production

W. P. SHUM and M. J. CANNARSA
*ARCO Chemical Company, Newtown Square, PA, USA*

| | | |
|---|---|---|
| 18.1 | Introduction | 363 |
| 18.2 | Early Development | 366 |
| 18.3 | The Glycidol Process | 366 |
| 18.4 | Process Chemistry | 368 |
| 18.5 | Epoxy Alcohol Recovery Methods | 369 |
| 18.6 | Reduction of the Organic Hydroperoxide | 371 |
| 18.7 | Determination of the Optical Purities of Optically Active Epoxy Alcohols and Their Derivatives | 371 |
| 18.8 | Optical Purification of Optically Active Glycidol Derivatives | 373 |
| 18.9 | Synthetic Applications of Optically Active Epoxy Alcohols | 374 |
| 18.10 | Synthetic Applications of Aqueous Optically Active Glycidol | 376 |
| 18.11 | Conclusion | 377 |
| 18.12 | Acknowledgements | 377 |
| 18.13 | Notes and References | 377 |

## 18.1 INTRODUCTION

ARCO Chemical Company specialises in the development and commercialisation of olefin epoxidation using alkyl hydroperoxides, a technology which forms the basis of an international business in the manufacture and use of propylene oxide and its derivatives. The company constantly seeks additional applications of its expertise, and an excellent example of this is in the scale-up of the Sharpless asymmetric epoxidation (SAE) route to optically active epoxy alcohols (Figure 1).

Early attempts to prepare enantiopure epoxide intermediates started with optically active compounds available from nature. In this chiral pool approach, D-mannitol has been used for the preparation of optically active glycidol, isopropylideneglycerol and epichlorohydrin. The synthesis of (S)-isopropylideneglycerol from D-mannitol was reported as early as 1939[1] and, in the late 1970s, researchers at Merck disclosed in a patent[2] and papers[3,4] the synthesis and utilisation of optically active epichlorohydrin. They synthesised (R)- and (S)-epichlorohydrin starting with D-mannitol, in 6–8 steps and in 30% overall yield with an *ee* of >97%. The authors described several important chiral intermediates made accessible from a chiral $C_3$ synthon, and pointed out the general utility

*Chirality in Industry II.* Edited by A. N. Collins, G. N. Sheldrake and J. Crosby
© 1997 John Wiley & Sons Ltd

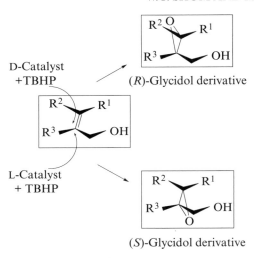

**Figure 1** Sharpless epoxidation chemistry

of the $C_3$ fragment in asymmetric synthesis. In the mid-1980s, important work was also carried out by Whitesides at Harvard University in developing the enzymatic resolution of glycidyl butyrate.[5-6] Porcine pancreatic lipase selectively hydrolysed the (S)-ester leaving the unreacted (R)-ester with 92% ee at 45% isolated yield. There are several other synthetic routes to chiral $C_3$ synthons based on the use of lipase enzymes and microorganisms.[7-11]

In the early 1980s, we became aware of the work of Sharpless, then at Stanford University, on the asymmetric epoxidation of allylic alcohols using alkyl hydroperoxides and titanium tartrate catalysts.[12] During the same period, single-enantiomer drug development began to generate interest, and we recognised this as an opportunity to extend our experience in epoxidation to the pharmaceutical market. Progress in asymmetric synthesis in academic and industrial laboratories, and also advances in enzymatic resolution, made single-enantiomer drug synthesis economically viable. However, a critical missing link in the utilisation of asymmetric synthesis was the ability to take new methodologies through the appropriate scale-up and engineering studies to develop commercial processes.

Discussions began between ARCO Chemical, Stanford University and Professor Sharpless in 1985, concerning possible licensing and commercial development of the SAE technology. While these discussions were taking place, a better understanding of the potential value of this technology was needed. In the absence of 'in-house' expertise of the pharmaceutical industry, several consultant studies were commissioned to provide an overview of potential markets. The results of these studies were sufficiently encouraging to motivate ARCO Chemical to enter a licensing agreement with Stanford. One attractive aspect of the technology is its broad applicability to a wide range of allylic alcohols. It was recognised that the three-carbon chiral epoxy alcohol building block would be a valuable intermediate for the production of a variety of pharmaceutical compounds (Figure 2).

This chapter does not review the literature pertaining to the discovery or mechanistic studies of the SAE, as these aspects have been covered previously in many excellent reviews, including several by Sharpless and co-workers.[13-14] Here, an overview is provided of the work required to commercialise the process, and

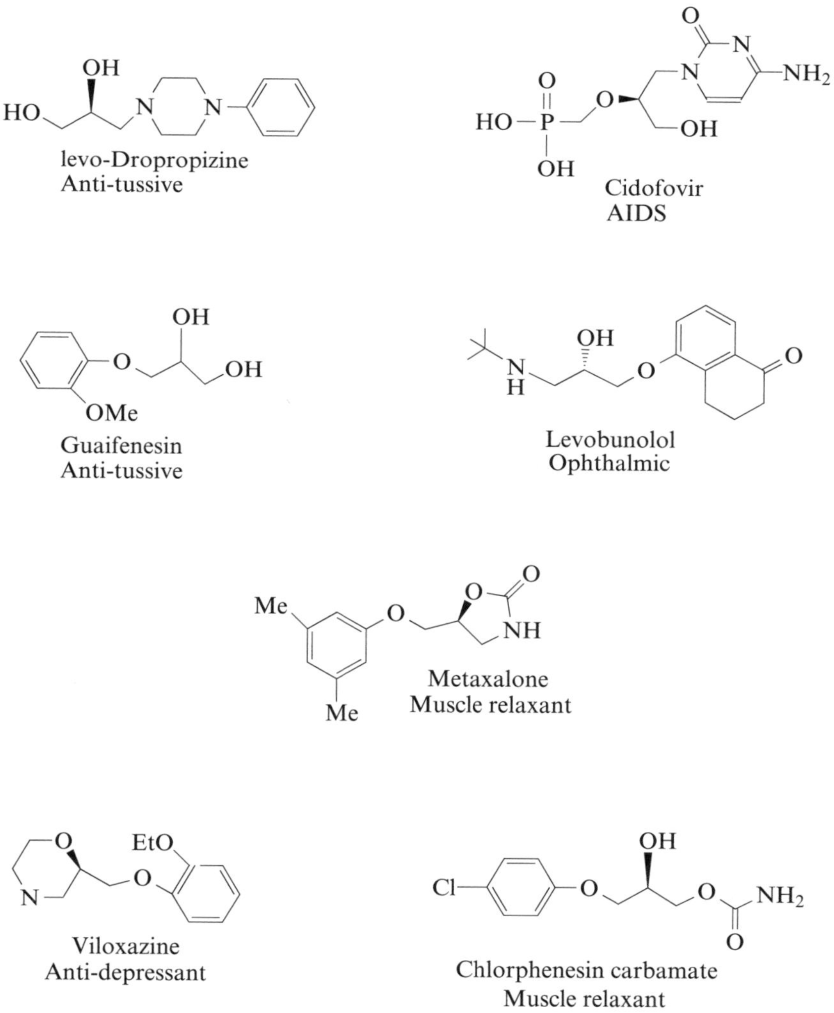

levo-Dropropizine
Anti-tussive

Cidofovir
AIDS

Guaifenesin
Anti-tussive

Levobunolol
Ophthalmic

Metaxalone
Muscle relaxant

Viloxazine
Anti-depressant

Chlorphenesin carbamate
Muscle relaxant

**Figure 2** Pharmaceutical compounds accessible from optically active glycidol

also some specific product development work. Also described is some important auxiliary work that was necessary, including the development of safety systems and analytical characterisation. Several important applications of the SAE in the pharmaceutical industry are covered in the final section.

## 18.2 EARLY DEVELOPMENT

The first paper on SAE, published in 1980,[12] described a process that utilised a stoichiometric quantity of titanium isopropoxide–diethyl tartrate in epoxidations using *tert*-butyl hydroperoxide. It is clear that it would be difficult, if not impossible, to practise this technology on a commercial scale owing to the large quantities of reagents required. For example, based on the original Sharpless papers, production of 100 kg of (*R*)-glycidol would have required 380 kg of titanium isopropoxide and 275 kg of D-(–)-diethyl tartrate. The cost of reagents, difficult product isolation and waste disposal made this technology unworkable.

While the license negotiations were taking place, Hanson and Sharpless published and patented an improved epoxidation process utilising molecular sieves.[15] This process allowed the titanium reagent and the tartrate ligand to be reduced to catalytic amounts. The same 100 kg batch of (*R*)-glycidol could now be prepared using only 20 kg of titanium isopropoxide and 15 kg of D-(–)-diisopropyl tartrate with the addition of 40 kg of relatively inexpensive and easily recycled molecular sieves. This improvement was the first in a series that led, ultimately, to a commercial process. Despite these improvements, the isolation of the water-soluble, unsubstituted (*R*)- and (*S*)-glycidols remained a significant challenge. These particular compounds had been identified from market research as highly desirable chiral starting materials. Several commercial products, including β-adrenergic blocking agents and anti-tussives had already been identified that could use (*R*)-glycidol in their synthesis. Therefore, although we knew that the water-soluble products would be the most difficult to process, especially in the presence of titanium alkoxides, these compounds were chosen for development work.

ARCO Chemical began work to commercialise the Sharpless laboratory-scale process in 1986. Several important process considerations are described below, including reagent concentrations and purity, reaction conditions and product isolation. In addition, critical safety considerations are discussed, including the handling of alkyl hydroperoxides on a large scale, the reduction of residual hydroperoxide and the isolation of a variety of reactive epoxy alcohol products.

## 18.3 THE GLYCIDOL PROCESS

Optically active glycidol was the first of several epoxy alcohols manufactured by ARCO Chemical in commercial quantities. The generalised process flow diagram for the epoxidation of allyl alcohol to provide optically active glycidol is shown

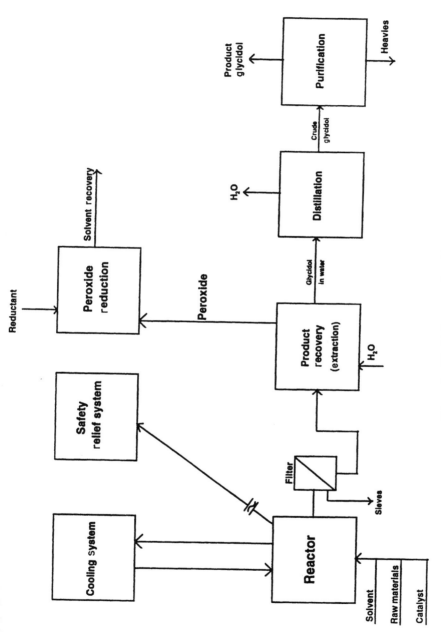

Figure 3   Glycidol process flow diagram

in Figure 3. Allyl alcohol, titanium tartrate catalyst, molecular sieves and solvent are present in the reactor, which is cooled to –20 °C. Cumene hydroperoxide is gradually added to the mixture, and the reaction exotherm is controlled by cooling. At the completion of the epoxidation, the molecular sieves are removed by filtration and the glycidol product is isolated by extraction into an aqueous phase. The deactivated catalyst, 2-phenylpropan-2-ol and residual cumene hydroperoxide remain in the organic solvent, which constitutes a waste stream which is treated with an appropriate reducing agent to remove the hydroperoxide before disposal. The glycidol product in the aqueous extract is concentrated by water removal and, finally, optically active glycidol of 88–90% ee is purified by vacuum distillation.

The isolation of optically active glycidol in high yield from the aqueous extract was one challenging aspect of the process development. Glycidol is prone to dimerisation and polymerisation at elevated temperatures, and these processes are sensitive to the pH of the medium.[16] The boiling point of glycidol at atmospheric pressure is reported to be 162 °C with decomposition[17], and so removal of water and high-boiling organic components requires distillation under carefully controlled conditions. In addition, trace alkali metal contamination can promote exothermic polymerisation of glycidol. Many safety systems were built into the distillation units to deal with these issues and, during process development, extensive accelerated rate calorimeter (ARC) evaluations on critical process streams were carried out to ensure safe plant operation.

## 18.4 PROCESS CHEMISTRY

In this section, catalyst chemistry related to the SAE process development is described. Although the chiral titanium tartrate catalyst formed *in situ* from titanium isopropoxide and the optically active dialkyltartrate ester provides predictable stereoselectivity in the asymmetric epoxidation of allylic alcohols, the optimal yield and enantiomeric purity of the epoxy alcohols are dependent on a number of process variables. In the SAE, a minimum amount of catalyst and solvent is required for the successful production of an epoxy alcohol of acceptable enantiomeric purity. The precipitous decrease in yield and enantiomeric purity of these epoxy alcohols observed in reactions with inadequate catalyst and solvent can be related to the proposed catalyst deactivation mechanism.[18] The Lewis acidic titanium(IV) metal centre is known to catalyse the nucleophilic addition of alcohol to an epoxy alcohol.[19] The diol made from this epoxide-opening reaction can act as a strong catalyst inhibitor. This detrimental effect on the epoxidation catalyst activity has been utilised as a safeguard against a run-away reaction in the scale-up of the SAE, in which propylene glycol is used as the catalyst inhibitor. In view of this chemistry, it is important to operate the batch SAE process under conditions that can suppress the epoxide-opening reactions of the epoxy alcohol product.

The dynamic titanium tartrate catalyst, with vacant sites available for ligand exchange, is also susceptible to hydrolytic degradation; sufficient molecular sieves must be used to adsorb trace water present in the solvent and the organic hydroperoxide. The enantioselectivity of the epoxidation is significantly reduced if trace water is present in the reaction medium. It is noteworthy that the catalyst used for the asymmetric oxidation of sulphides (see Chapter 19) is the SAE catalyst modified by the addition of water.[20] This suggests that the Sharpless titanium tartrate catalyst may undergo structural modification by water. Whether the reversible[21] hydrolytic degradation can be related to an equilibrium between a dimer and its monomer is open for debate in view of the conflicting proposals from Sharpless[22] and Corey.[23] From a practical standpoint, in the chiral glycidol process, which involves the SAE of the relatively inactive allyl alcohol, water must be rigorously eliminated from the reaction medium and cumene hydroperoxide must be used for optimal catalyst performance.

The effect of reactant purity on catalyst productivity and product purity is evaluated prior to production runs. It has been found that trace aldehyde impurities can have an adverse effect on catalyst performance. It is possible that aldehydes are oxidised to carboxylic acids which modify the titanium tartrate catalyst. This rationale is based on the reverse enantioselectivity observed for SAE using the slightly acidified phenethyl hydroperoxide.[24] Apart from the direct structural transformation of the catalyst, the carboxylic acid might catalyse the dehydration of alcohol to form water, which has already been demonstrated to have a detrimental effect on catalyst performance. In this context, the possibility of an acid-catalysed epoxide rearrangement to an aldehyde also has to be considered. Cascading side-reactions of this type would explain the continuous depletion of the epoxy alcohol product in the reaction mixture observed during storage, even at low temperatures.

## 18.5 EPOXY ALCOHOL RECOVERY METHODS

The inherent instability of the epoxy alcohol in the epoxidation reaction mixture imposes restrictions on the methodology used for product recovery in a production-scale process. The physical properties of the epoxy alcohol product and the organic hydroperoxide reactant must be considered and, as in many other homogeneous catalytic processes, the product recovery and purification steps require considerable process development. This is especially true because of the presence of unreacted organic hydroperoxide in the reaction effluent. The safe handling of the process streams containing hydroperoxide will be discussed in the following section. Product recovery methodologies that have been successfully commercialised can be conveniently categorised according to the physical properties of the epoxy alcohols.

Aqueous extraction works well for a number of water soluble epoxy alcohols such as glycidol (see Section 18.3) and 2- or 3-methylglycidol. In the glycidol case, where cumene hydroperoxide is used for the SAE, the use of this liquid–liquid extraction method can effectively remove the hydroperoxide and the allyl alcohol starting material from the product. The titanium tartrate catalyst is hydrolysed in the aqueous extraction process, and can also be effectively removed from the aqueous effluent containing the glycidol product. This method is suitable for epoxy alcohols such as the 2-substituted glycidols which are susceptible to decomposition at elevated temperatures in the presence of the titanium catalyst. Another advantage of this method is the possibility of direct utilisation of the aqueous glycidol effluent for the synthesis of a variety of synthetically useful glycidol derivatives.

Effective removal of the titanium tartrate catalyst and organic hydroperoxide from the product effluent prior to product isolation is necessary for the recovery of epoxy alcohols that are not water soluble. The titanium tartrate catalyst is hydrolysed, preferably by a caustic wash procedure under carefully controlled conditions to minimise epoxy alcohol loss. After phase separation, the organic hydroperoxide, typically *tert*-butyl hydroperoxide, present in the organic fraction with the epoxy alcohol product can be reduced by a variety of methods to *tert*-butyl alcohol, which can be easily removed by vacuum distillation. The isolated epoxy alcohol can be used without further purification, as in the case of *trans*-3-benzylglycidol for the synthesis of HIV protease inhibitor.[25] For aliphatic epoxy alcohols that are liquids, purification by fractional distillation can be carried out, as exemplified by the synthesis of *trans*-3-propylglycidol.

For solid epoxy alcohols of higher molecular weight, purification by recrystallisation can improve not only the chemical purity, but also the optical purity. This can significantly simplify the product isolation procedures, as illustrated in the synthesis of enantiopure *trans*-3-phenylglycidol.[26] In this process, the titanium tartrate catalyst was hydrolysed, and the residual *tert*-butyl hydroperoxide was removed by an azeotropic distillation method. The crude product that crystallised out from the process effluent was purified to >99% *ee* by recrystallisation.

There are a number of low-boiling epoxy alcohols that can be recovered by an azeotropic distillation approach. Glycidol is known to form azeotropic mixtures with a variety of solvents,[27] and it is possible to recover the glycidol made in the SAE by vacuum distillation with cumene as the azeotropic agent.[28] In one case, the unreacted cumene hydroperoxide was completely reduced prior to distillation. In another case, where cumene hydroperoxide was allowed to remain in the distillation mixture, the addition of excess propylene glycol was found to deactivate the titanium tartrate catalyst and no loss in the optical purity of glycidol was observed in the product recovery step. The azeotropic distillation methodology is an important alternative to the aqueous extraction procedure, especially for medium-chain aliphatic epoxy alcohols that have low solubility in water. The distillate containing the epoxy alcohols can be used without further purification for the synthesis of optically active epoxy alcohol derivatives, e.g. glycidyl arenesulphonates and trityl glycidol.

The recovery of epoxy alcohol would be greatly facilitated if a heterogeneous catalyst could be used for the SAE. A titanium-pillared montmorillonite has been used as a heterogeneous catalyst for the asymmetric epoxidation of allylic alcohols in the presence of an optically active tartrate ester.[29] In another approach, polymer-linked tartrate esters were used with titanium isopropoxide for the SAE of geraniol.[30] However, none of these heterogeneous catalysts gave satisfactory yields or optical purity, and research continues in this area.

## 18.6 REDUCTION OF THE ORGANIC HYDROPEROXIDE

The safe and complete removal of unreacted organic hydroperoxide is an important development issue in the scale-up of SAE. There are many methods for the effective removal of the hydroperoxide in the reaction effluent. The azeotropic distillation approach and the use of trialkyl phosphite or iron(II) sulphate heptahydrate as reducing agents have been described in the literature.[21] Selection of the best reducing agent for hydroperoxide removal varies with the epoxy alcohol process. In all cases, the reduction step has to be carefully integrated with the product recovery step in the process to ensure satisfactory isolated yield of the epoxy alcohol and compliance with waste disposal regulations. In an ideal situation, where the epoxy alcohol product can be separated from the unreacted hydroperoxide, the subsequent reduction step would be carried out with the more common reducing agents such as aqueous sodium hydrogen sulphite, which cannot be used satisfactorily in the presence of the epoxy alcohol product because of the high acidity of the reaction medium. The high acidity also catalyses dehydration of the alcohols to form olefins and high boiling point species on ageing. Some reduction processes also produce diperoxide (e.g. dicumyl peroxide, di-*tert*-butyl peroxide) by-products which must be taken into consideration.

During process development, the proportions of all of these side reactions had to be accurately determined to address the fate of all streams. In addition, the stability of the reaction effluents containing hydroperoxide at various stages of the process has also been adequately evaluated by a series of ARC tests to ensure safe process design for the epoxidation, product isolation, product purification and waste disposal steps.

## 18.7 DETERMINATION OF THE OPTICAL PURITIES OF OPTICALLY ACTIVE EPOXY ALCOHOLS AND THEIR DERIVATIVES

Both chemical and optical purity specifications must be addressed in the manufacture of optically active epoxy alcohols. Although the chemical purity of these compounds can be readily determined by methods such as GC, HPLC and NMR spectroscopy, the optical purity determination usually requires more elaborate method development on a case-by-case basis. In the early phase of our process research work, the traditional Mosher ester method[31] had been employed for the determination of *ee* values of optically active epoxy alcohols. This involves the derivatisation of the epoxy alcohols with the enantiopure α-methoxy-α-(trifluoromethyl)phenylacetyl chloride reagent (MTPA chloride) in the presence of triethylamine and 4-(dimethylamino)pyridine (DMAP) catalyst, followed by isolation of the diastereoisomeric Mosher esters of the epoxy alcohols by column chromatography or vacuum distillation. With care, this method is perfectly reliable for *ee* determination by the subsequent NMR analysis of the isolated Mosher esters. However, such a derivatisation method requires a synthetic procedure that is too time consuming to apply to the large number of product samples that must be analysed during process development work, so more convenient analytical methods were investigated

Capillary gas chromatography using liquid derivatives of α-, β- and γ-cyclodextrins as chiral stationary phases emerged as a powerful method for the separation of enantiomers in the late 1980s.[32,33] The application of these commercially available[34] cyclodextrin-based columns to the precise *ee* determination of many epoxy alcohols prepared by the SAE route was successfully accomplished.[35] Gas chromatographic separations of epoxy alcohols were carried out on a 30 m × 0.25 mm i.d. Chiraldex™ permethyl-*O*-hydroxypropyl-α-cyclodextrin A-PH column in the temperature range 40–80 °C. The three short-chain aliphatic epoxy alcohols, namely glycidol, 2-methylglycidol and *trans*-3-methylglycidol, can be resolved without derivatisation. Since these epoxy alcohols are water soluble, their enantiopurity during the course of the epoxidation can be readily followed by direct analysis of an aqueous extract of the reaction mixture. This A-PH column was found to be very resistant to hydrolysis, and it had been used extensively for the determination of enantiopurity of many aqueous glycidol samples in our process development work with no signs of column degradation. There have been several more recent reports[36–38] on the use of the α-cyclodextrin-based GC columns for the separation of glycidol enantiomers. The (*R*), (*S*) elution order is always the same on the α- and γ-cyclodextrin-based columns. Interestingly, a reversal of elution order has been observed on a commercially available Chiraldex™ B-TA column prepared from β-cyclodextrin.[38] This provides a means of avoiding the tailing problem encountered in the enantiopurity determination of highly enriched (*R*)-glycidol. However, the B-TA (trifluoroacetyl-derivatised) Chiraldex™ column is susceptible to hydrolytic degradation, so that regeneration with trifluoroacetic anhydride is frequently required for optimal separation of anhydrous glycidol. The long-chain aliphatic epoxy alcohols such as *trans*-3-ethylglycidol and *trans*-3-propylglycidol require derivatisation with trifluoroacetic anhydride for enantiomeric separation on the A-PH column.

For the determination of the enantiopurity of aromatic epoxy alcohols such as *trans*-3-phenylglycidol, 2-(2,4-difluoro)phenylglycidol and *trans*-3-benzylglycidol, the direct HPLC method is preferred. These compounds, without derivatisation, can be resolved on the commercially available[39] DAICEL™ chiral HPLC columns. This direct analytical method is well suited to the determination of both the enantiomeric purity and the diastereoisomeric purity of the epoxy alcohol products. For example, in the synthesis of *trans*-3-benzylglycidol by the SAE route, the reaction product typically contains 3–4% of *cis*-3-benzylglycidol that originates from the *cis*-4-phenylbut-2-en-1-ol impurity present in the *trans*-4-phenylbut-2-en-1-ol starting material. A Chiralpak® AD column has been used to resolve all four stereoisomers of 3-benzylglycidol in addition to the trace amount of *trans*- and *cis*-allylic alcohols and homoallylic alcohol typically present in the crude reaction product.[25] This chiral HPLC method allows a complete assay of the purity of the desired *trans*-3-benzylglycidol by a single analysis. In this case, *ee* values of 93% and 88% were obtained for *trans*- and *cis*-3-benzylglycidol, respectively. The slightly lower *ee* observed for the *cis*-epoxy alcohol is consistent with the general trends observed for the SAE in this class of *cis*-olefinic substrates.[13]

The chiral HPLC method is also very useful for the determination of the enantiopurity of many synthetically useful glycidol derivatives. We[40] and others[41–43] have published results on the enantiomeric separations of a variety of these derivatives including trityl glycidol and glycidyl arenesulphonates. The direct HPLC method is becoming the standard analytical method for *ee* determination because of the extra synthetic complexity involved in the traditional Mosher ester derivatisation method and the relatively small rotation values from polarimetric measurements. Furthermore, this convenient analytical method can provide the accuracy required for the construction of the melting point phase diagrams of these glycidol derivatives.

## 18.8 OPTICAL PURIFICATION OF OPTICALLY ACTIVE GLYCIDOL DERIVATIVES

The use of recrystallisation for the optical purification of chiral compounds is common practice in asymmetric synthesis, and it is possible to use this method to upgrade the enantiopurity of the epoxy alcohols obtained from SAE. A single recrystallisation from hexanes was claimed to be sufficient to improve the enantiopurity of *cis*-(2*S*,3*R*)-3-propylglycidol from 81% to 100% *ee*.[44] Another epoxy alcohol that can similarly be purified is *trans*-3-phenylglycidol, which was obtained in >99% *ee* after one recrystallisation.[21] Although the short-chain aliphatic epoxy alcohols cannot be enantiomerically enriched by this methodology, many of their synthetically useful derivatives are crystalline solids and therefore can be purified to higher enantiopurity by multiple recrystallisations.

We have investigated the optical purification of several important epoxy alcohol derivatives.[45] The melting point phase diagrams of these compounds were constructed to understand better their enantiomeric enrichment behaviours (Figure 4). Although the phase diagram of glycidyl 3-nitrobenzenesulphonate (GNBS) exhibits the distinctive V-shape of a conglomerate, optical purification of this compound above 97% *ee* by multiple recrystallisations is exceedingly diffi-

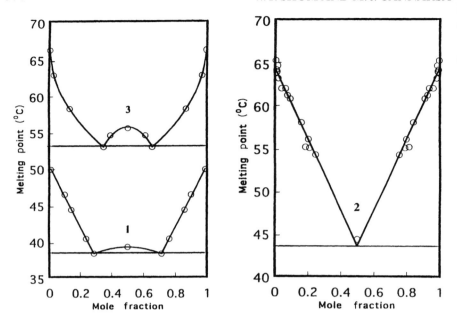

**Figure 4** Phase diagrams of glycidyl tosylate (1), glycidyl 3-nitrobenzenesulphonate (2) and *trans*-3-methylglycidyl tosylate (3)

cult. Preparation of enantiopure GNBS is impeded by the existence of 'terminal solid solutions.'[46] Recrystallisation of GNBS of low enantiomeric purity is further complicated by the formation of polymorphs, and preferential crystallisation of its seeded solutions did not afford an enantiopure product.[47] It is also difficult to obtain enrichment above 95% *ee* for glycidyl tosylate owing to solid solution formation.

In contrast to these two compounds, *trans*-3-methylglycidyl tosylate was found to be enriched to 99.3% from 89% *ee* by two recrystallisations from absolute ethanol. Although a perfect 100% *ee* cannot be achieved with most of these glycidyl arenesulphonates derived from SAE (since all appear to form solid solutions), many enantiopure pharmaceuticals have been successfully synthesised using the GNBS and glycidyl tosylate of 95% *ee* by recourse to further enrichment in the subsequent synthetic steps.

## 18.9 SYNTHETIC APPLICATIONS OF OPTICALLY ACTIVE EPOXY ALCOHOLS

Owing to their synthetically useful hydroxy and epoxy functionalities, optically active 2,3-epoxy alcohols prepared by the SAE route have found diverse applications as chiral building blocks in the syntheses of many pharmaceutical and agrochemical products.[48] Chiral glycidol is a particularly useful $C_3$ synthon that has been used for the syntheses of ß-adrenergic blocking agents[49] and related compounds,[50] alkyl glycerophospholipids,[51–53] acyclonucleotides,[54] monoamine oxidase inhibitors,[55] dropropizine,[56] guaifenesin,[57] and other pharmaceuticals. Applications of optically active 2,3-epoxy alcohols include the syntheses of HIV protease inhibitors,[58–59] AZT,[60] taxol side chain,[61] antifungal agents,[62–63] antibiotics,[64] vitamin $D_3$[65] and many other biologically active compounds. There are also reports on the use of these optically active epoxy alcohols for the syntheses of novel materials such as ferroelectric liquid crystals[66] and liquid crystalline polymers.[67] Since it is not possible to cover all aspects of these applications here, we have selected only two examples that serve to demonstrate the effectiveness of the Sharpless epoxidation route in the manufacture of chiral compounds.

The optically active aminoalkyl epoxide fragment (**4** in Figure 5) is an important intermediate for the syntheses of a variety of HIV protease inhibitors that are currently under active investigation.[58,68–69] This BOC-epoxide can be prepared very effectively by a multi-step process developed jointly by ARCO Chemical and SIPSY (Avrillé, France). Starting with a Wittig olefination reaction on phenylacetaldehyde (**1**), the desired (2R,3R)-trans-3-benzylglycidol (**3**) was obtained in three steps via the SAE on trans-4-phenylbut-2-en-1-ol (**2**).[25] This was followed by a series of three further steps to afford the enantiopure BOC-epoxide.[70] This

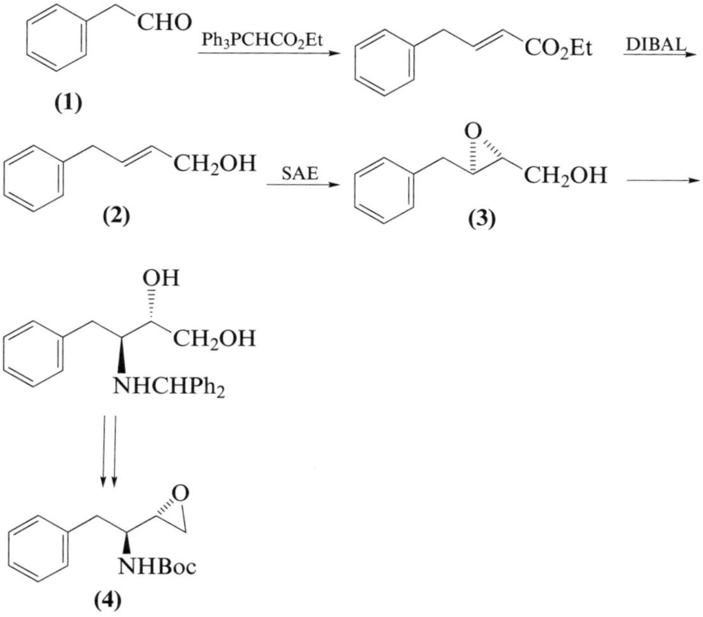

**Figure 5** Synthesis of aminoalkyl epoxide intermediate **4**

process has been carried out on pilot-plant scale to produce tens of kilogrammes of the BOC-epoxide for clinical trials of the HIV protease inhibitor.

Glycidol, with its dual functional groups, can readily be derivatised to many pharmaceutically important compounds. Guaifenesin, an expectorant that has been extensively used in anti-tussive formulations in its racemic form, can be synthesised in high yield and high enantiopurity by the direct nucleophilic addition of 2-methoxyphenol to optically active glycidol in the presence of triethylamine catalyst.[57] This route is more effective than other approaches reported in the literature, which include the use of optically active 1-tosyloxypropane-2,3-diol acetonide as the key intermediate,[71] the Sharpless asymmetric dihydroxylation of 2-methoxyphenoxy allyl ether[72] and the lipase-catalysed sequential esterification of racemic guaifenesin.[73] With the availability of enantiopure guaifenesin, several other guaifenesin-related drugs that have been used previously in racemic forms can be readily synthesised with high enantiomeric purity.[74] The list includes moprolol, methocarbamol, mephenoxalone, guafecainol and ranolazine. Furthermore, our simple epoxide-opening reaction provides a convenient means to obtain many enantiopure 3-aryloxypropane-1,2-diols of pharmaceutical and agrochemical importance.[75-77]

In the above examples, the two epoxy alcohols obtained from SAE are not enantiomerically pure: the enantiopurity of *trans*-3-benzylglycidol is 93% ee and that of glycidol is 90% ee. Both are liquids, and cannot be further enriched enantiomerically by traditional crystallisation methodology. Nevertheless, because of the highly stereoselective nature of the subsequent reactions on the epoxides and the crystallinity of the desired products in their solid-state, the BOC-epoxide and guaifenesin are ultimately isolated in >99% ee. Many other biologically active compounds that are synthesised from chiral $C_3$ synthons or by enzyme catalysis can also be efficiently prepared in enantiopure forms using the optically active glycidol derived from the SAE.

## 18.10 SYNTHETIC APPLICATIONS OF AQUEOUS OPTICALLY ACTIVE GLYCIDOL

Chiral glycidol made from the SAE can be effectively recovered in high purity by aqueous extraction because of its excellent solubility in water. This aqueous glycidol solution has itself been used for the syntheses of several glycidol derivatives. For example, an aqueous glycidol solution of about 20% was treated with 3-nitrobenzenesulfonyl chloride in the presence of triethylamine to afford a good yield of glycidyl 3-nitrobenzenesulphonate of >96% ee after recrystallisation from absolute ethanol.[78] The use of a phase-transfer catalyst was not necessary for this reaction. Other optically active compounds, such as glycidyl tosylate, levodropropizine and guaifenesin can similarly be prepared in good yield and in high enantiomeric purity by this approach. The aqueous glycidol used for these syntheses has an ee of 88–90%, which is substantially higher than the 60–65% ee of the aqueous (R)-glycidol typically obtained from the enzymatic resolution route.[5] The SAE is, therefore, preferred for the synthesis of compounds such as glycidyl tosylate that cannot readily be enriched enantiomerically by recrystallisation.

Derivative synthesis using the aqueous glycidol stream is an improvement over the *in situ* derivatisation method described by Sharpless and co-workers.[21] In the Sharpless method, the residual catalyst and alcohol derived from the hydroperoxide, and also other impurities, prevent a convenient purification of the products by crystallisation. In many cases chromatography is required to reach a purity sufficient for crystallisation. Therefore, aqueous extraction and subsequent derivatisation as described here constitute the preferred route on an industrial scale.

## 18.11   CONCLUSION

The list of available technologies for producing single enantiomer intermediates continues to grow, and optically active glycidol and derivatives produced by the SAE will continue to play an important role in the synthesis of many new drugs.

The most efficient implementation of the chiral epoxidation technology can sometimes be realised by carrying out several synthetic steps in a reaction scheme and providing an advanced intermediate to the customer. In some cases, isolation of the optically active epoxy alcohol is not required, and an efficient synthesis can proceed with the crude epoxy alcohol in a subsequent step; this can also mitigate some of the safety issues discussed above. In addition, trends in the pharmaceutical industry towards focus on drug discovery and marketing have shifted manufacturing towards third parties for the supply of key intermediates. As the expertise of ARCO Chemical is in epoxidation, the company has licensed the developed epoxidation technology to SIPSY[79], a supplier of key intermediates to the pharmaceutical industry. This technical partnership will continue to pursue the development and implementation of the Sharpless asymmetric epoxidation (SAE) technology.

## 18.12   ACKNOWLEDGEMENTS

We thank the following for their contributions: Jon R. Valbert, Harry Mazurek, Jian Chen, Angela M. Hughes, Ram Ananth, Carl E. Johnson, Sam Onyekere and Glenn E. Coughenour.

## 18.13   REFERENCES AND NOTES

1. Baer, E., and Fischer, H. O. L., *J. Biol. Chem.*, **128**, 463 (1939); *Chem. Abstr.*, **33**, 7276, (1939).

2. US Pat., 4 408 063, 1983.
3. Baldwin, J. J., Raab, A. W., Mensler, K., Arison, B. H., and McClure, D. E., *J. Org. Chem.*, **43**, 4876 (1978).
4. McClure, D. E., Arison, B. H., and Baldwin, J. J., *J. Am. Chem. Soc.*, **101**, 3666 (1979).
5. Ladner, W. E., and Whitesides, G. M., *J. Am. Chem. Soc.*, **106**, 7250 (1984).
6. US Pat., 4 732 853, 1988.
7. Wang, Y.-F., Lalonde, J. J., Momongan, M., Bergbreiter, D. E., and Wong, C. H., *J. Am. Chem. Soc.*, **110**, 7200 (1988).
8. Vanttinen, E., and Kanerva, L., *J. Chem. Soc., Perkin Trans.*, *1*, 3459 (1994).
9. Suzuki, T., and Kasai, N., *Bioorg. Med. Chem. Lett.*, **1**, 343 (1991).
10. US Pat., 5 246 843, 1993.
11. Eur. Pat., EP464 905, 1992.
12. Katsuki, T., and Sharpless, K. B., *J. Am. Chem. Soc.*, **102**, 5974 (1980).
13. Johnson, R. A., and Sharpless, K. B., in *Catalytic Asymmetric Synthesis*, (ed. J. Ojima), VCH, New York, 1993, p. 103.
14. Finn, M. G., and Sharpless, K. B., in *Asymmetric Synthesis*, (ed. J. D. Morrison), Academic Press, Orlando, FL, 1985, Vol. 5, 247.
15. Hanson, R. M., and Sharpless, K. B., *J. Org. Chem.*, **51**, 1922 (1985).
16. Rider, T. H., and Hill, A. J., *J. Am. Chem. Soc.*, **52**, 1521 (1930).
17. De Gegerfelt, M. H., *Bull. Soc. Chim. Fr.*, **23**, 160 (1875).
18. Ref. 14, p. 287.
19. Caron, M., and Sharpless, K. B., *J. Org. Chem.*, **50**, 1557 (1985).
20. Kagan, H. B., in *Catalytic Asymmetric Synthesis*, (ed. J. Ojima), VCH, New York, 1993, p. 203.
21. Gao, Y., Hanson, R. M., Klunder, J. M., Ko, S. Y., Masamune, H., and Sharpless, K. B., *J. Am. Chem. Soc.*, **109**, 5765 (1987).
22. See ref. 13, and references cited therein.
23. Corey, E. J., *J. Org. Chem.*, **55**, 1693 (1990).
24. US Pat., 4 764 628, 1988.
25. Shum, W. P., Chen, J., and Cannarsa, M. J., *Chirality*, **6**, 681 (1994).
26. US Pat., 4 935 101 (1990).
27. Kleemann, A., and Wagner, R. M., *Glycidol: Properties, Reactions, Applications*, Verlag Chemie, Heidelberg, 1981, p. 11.
28. US Pat., 5 288 882, 1994.
29. Choudary, B. M., Valli, V. L. K., and Prasad, D., *J. Chem. Soc., Chem. Commun.*, 1186 (1990).
30. Farrall, M. J., Alexis, M., and Trecarten, M., *Nouv. J. Chim.*, **7**, 449 (1983).
31. Dale, J. A., Dull, D. L., and Mosher, H. S., *J. Org. Chem.*, **34**, 2543 (1969).
32. Konig, W. A., Lutz, S., Wenz, G., Gorgen, G., Neumann, C., Gabler, A., and Boland, W., *Angew. Chem., Int. Ed. Engl.*, **28**, 178 (1989).
33. Li, W.-Y., Jin, H. L., and Armstrong, D. W., *J. Chromatogr.*, **509**, 303 (1990).
34. Available from Advanced Separation Technologies, Whippany, NJ, USA.
35. Dougherty, W., Liotta, F., Mondimore, D., and Shum, W., *Tetrahedron Lett.*, **31**, 4389 (1990).
36. Geerlof, A., van Tol, J. B. A., Jongejan, J. A., and Duine, J. A., *J. Chromatogr.*, **648**, 119 (1993).
37. Jin, Z., and Jin, H. L., *Chromatographia*, **38**, 22 (1994).
38. *A Guide to Using Cyclodextrin Bonded Phases for Gas Chromatography*, Advanced Separation Technologies, Whippany, NJ.
39. Available from Chiral Technologies, Exton, PA, USA.
40. Chen, J., and Shum, W., *Tetrahedron Lett.*, **34**, 7663 (1993).
41. Shaw, C. J., and Barton, D. L., *J. Pharm. Biomed. Anal.*, **793** (1991).

42. Kennedy, J. H., and Weigel, L. O., *Chirality*, **4**, 132 (1992).
43. Duchateau, A. L. L., Jacquemin, N. M. J., Straatman, H., and Noorduin, A. J., *J. Chromatogr.*, **637**, 29 (1993).
44. *Eur. Pat.*, EP583 591 1994.
45. Ananth, R., Chen, J., and Shum, W., *Tetrahedron: Asymmetry*, **6**, 317 (1995).
46. For discussions on terminal solid solutions, see Jacques, J., Collet, A., and Wilen, S. H., *Enantiomers, Racemates, and Resolutions*, Wiley, New York, 1981, pp. 126 and 424.
47. *Eur. Pat.*, EP441 471, 1991.
48. For a review on the synthetic applications of optically active glycidol and related 2,3-epoxy alcohols, see Hanson, R. M., *Chem. Rev.*, **91**, 437 (1991).
49. Klunder, J. M., Ko, S. Y., and Sharpless, K. B., *J. Org. Chem.*, **51**, 3710 (1986).
50. Barton, D. L., Press, J. B., Hajos, Z. G., and Sawyers, R. A., *Tetrahedron: Asymmetry*, **3**, 1189 (1992).
51. Burgos, C. E., Ayer, D. E., and Johnson, R. A., *J. Org. Chem.*, **52**, 4973 (1987).
52. Ali, S. and Bittman, R., *J. Org. Chem.*, **53**, 5547 (1988).
53. Thompson, D. H., Svendsen, C. B., Meglio, C. D., and Anderson, V. C., *J. Org. Chem.*, **59**, 2945 (1994).
54. Brodfuehrer, Howell, H. G., Sapino, C., Jr, and Vemishetti, P., *Tetrahedron Lett.*, **35**, 3243 (1994).
55. Gates, K. S., and Silverman, R. B., *J. Am. Chem. Soc.*, **112**, 9364 (1990).
56. Stamicarbon, *Eur. Pat.*, EP 0 349 066, 1990. Shum, W. P., and Chen, J., *Chirality*, **8**, 6 (1996).
57. Chen, J., and Shum, W., *Tetrahedron Lett.*, **36**, 2379 (1995).
58. Parkes, K. E. B., Bushnell, D. J., Crackett, P. H., Dunsdon, S. J., Freeman, A. C., Gunn, M. P. Hopkins, R. A., Lambert, R. W., Martin, J. A., Merrett, J. H., Redshaw, S., Spurden, W. C., and Thomas, G. J., *J. Org. Chem.*, **59**, 3656 (1994).
59. Askin, D., Eng, K. K., Rossen, K., Purick, R. M., Wells, K. M., Volante, R. P., and Reider, P. J., *Tetrahedron Lett.*, **35**, 673 (1994).
60. Jung, M. E., and Gardiner, J. M., *J. Org. Chem.*, **56**, 2614 (1991).
61. Bonini, C., and Righi, G., *J. Chem. Soc., Chem. Commun.*, 2767 (1994).
62. *Eur. Pat.*, EP 0 472 392, 1991.
63. *Eur. Pat.*, EP 0 539 938, 1992.
64. *US Pat.*, 5 352 832, 1994.
65. Okabe, M., and Sun, R.-C., *Tetrahedron Lett.*, **34**, 6533 (1993).
66. Walba, D. M., Vohra, R. T., Clark, N. A., Handschy, M. A., Xue, J., Parmar, D. S., Langerwall, S. T., and Skarp, K., *J. Am. Chem. Soc.*, **108**, 7424 (1986).
67. Taton, D., Le Borgne, A., Spassky, N., Friedrich, C., and Noel, C., *Polym. Adv. Technol.*, **5**, 203 (1994).
68. Kim, E. E., Baker, C. T., Dwyer, M. D., Murcko, M. A., Rao, B. G., Tung, R. D., and Navia, M. A., *J. Am. Chem. Soc.*, **117**, 1181 (1995).
69. Barrish, J. C., Gordon, E., Alam, M., Lin, P.-F., Bisacchi, G. S., Chen, P., Cheng, P. T. W., Fritz, A. W., Greytok, J. A., Hermsmeier, M. A., Humphreys, W. G., Lis, K. A., Marella, M. A., Merchant, Z., Mitt, T., Morrison, R. A., Obermeier, M. T., Pluscec, J., Skoog, M., Slusarchyk, W. A., Spergel, S. H., Stevenson, J. M., Sun, C., Sundeen, J. E., Taunk, P., Tino, J. A., Warrack, B. M., Colonno, R. J., and Zahler, R., *J. Med. Chem.*, **37**, 1758 (1994).
70. Castejon, P., Pasto, M., Moyano, A., Pericas, M. and Riera, A., *Tetrahedron Lett.*, **36**, 3019 (1995).
71. Nelson, W. L., Wennerstrom, J. E., and Sankar, S. R., *J. Org. Chem.*, **42**, 1006 (1977).
72. Wang, Z., Zhang, X., and Sharpless, K. B., *Tetrahedron Lett.*, **34**, 2267 (1993).
73. Theil, F., Weidner, J., Ballschuh, S., Kunath, A., and Schick, H., *J. Org. Chem.*, **59**, 388 (1994).

74. Shum, W., unpublished results.
75. Kleemann, A., Nygren, R., and Wagner, R., *Chem. Ztg.* **104**, 283 (1980).
76. Mazurek, H., in *Proceedings of International Conference on Pharmaceutical Ingredients and Intermediates*, Frankfurt, 1990.
77. Buser, H.-P., Spindler, F., *Tetrahedron: Asymmetry*, **4**, 2451 (1993).
78. *US Pat.*, 5252759,
79. *Chem. Mark. Rep.*, **247**, 24 April, 18 (1995).

# 19 Asymmetric Synthesis of Sulphoxides: Two Case Studies

**P. PITCHEN**
*Rhône-Poulenc Rorer Inc., Collegeville, PA, USA*

| | | |
|---|---|---|
| 19.1 | Introduction | 381 |
| 19.2 | Enantioselective Oxidation of Prochiral Sulphides | 382 |
| 19.3 | Synthesis of RP 73163 | 385 |
| 19.4 | Synthesis of RP 52891 | 386 |
| 19.5 | Conclusion | 388 |
| 19.6 | Acknowledgements | 389 |
| 19.7 | References | 389 |

## 19.1 INTRODUCTION

The concept of molecular chirality is more than a century old,[1] but it is only in the last 1–2 decades that the asymmetric synthesis of biologically active compounds has really boomed, prompted by the need for more active and selective drugs and rendered possible by the emergence of new analytical and synthetic methods.[2] The object of this chapter is to illustrate this recent trend with two examples: the potassium channel opener RP 52891[3] and the ACAT inhibitor RP 73163,[4] two drugs discovered and studied at the Franco-American pharmaceutical company Rhône-Poulenc Rorer (Figure 1).

Both of these molecules are chiral sulphoxides and, in each case, the scenario was the same: optimisation of lead compounds first led to the preselection of products containing sulphide moieties; it was then discovered that these sulphides were metabolised into the corresponding sulphoxides; the racemic

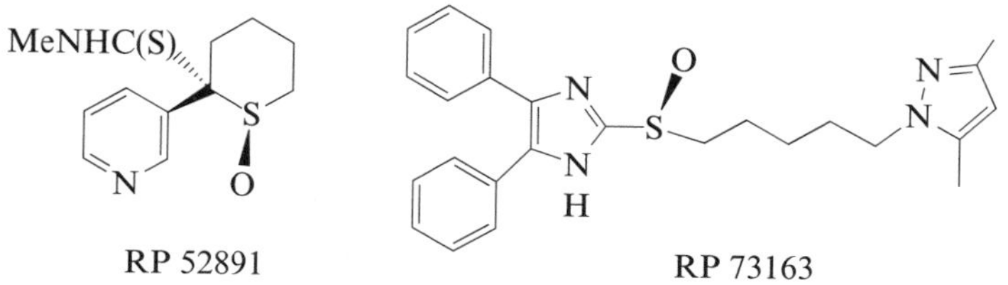

**Figure 1** Structure of potassium channel opener RP52891 and ACAT inhibitor RP73163

*Chirality in Industry II.* Edited by A. N. Collins, G. N. Sheldrake and J. Crosby
© 1997 John Wiley & Sons Ltd

sulphoxides were therefore synthesised, thus introducing a new asymmetric centre in the molecules, and resolved on a very small scale. In both cases, the enantiomeric sulphoxides were then subjected to biological testing, and the more active (or less toxic) enantiomer (the eutomer) was selected for further pharmacological and toxicological profiling. The possible development of these candidate compounds depended on the availability of asymmetric syntheses which could be rapidly and safely transferred into existing standard pilot plant equipment for the early manufacture of increasingly large quantities of enantiomerically pure (>99% *ee*) drug substance.

## 19.2 ENANTIOSELECTIVE OXIDATION OF PROCHIRAL SULPHIDES

The chiral pool strategy[2a] cannot really be applied to the syntheses of RP 52891 and RP 73163, as there are no readily available chiral sulphoxides which could be used directly as starting materials for the stereospecific syntheses of these compounds. On the other hand, there is an increasing number of chemical and biochemical methods for the enantioselective synthesis of chiral sulphoxides.[5] Most of these methods are, however, only suitable for laboratory use. One of the few possible candidates for scale-up is the asymmetric oxidation of prochiral sulphides using organic hydroperoxides in the presence of specific combinations of titanium isopropoxide and diethyl tartrate. This method relies on simple but key modifications to the well-known Sharpless epoxidation of allylic alcohols.

In its stoichiometric or catalytic versions, the Sharpless reaction is based on the use of soluble complexes made from equimolar amounts of diethyl tartrate (DET) and titanium isopropoxide under anhydrous conditions, at −20 °C, in an organic solvent such as dichloromethane or toluene.[6] The reaction has been successfully applied to the preparation of numerous enantiomerically pure epoxyalcohols, some of them to large-scale operations, as discussed in Chapter 18.

The Sharpless reagent, when applied to sulphides, leads to a mixture of the racemic sulphoxide and the corresponding sulphone. As discovered by 1983 by Kagan's group at Orsay,[7] the simple addition of 1 mol equiv. of water to the Sharpless reagent and/or the use of 2 (instead of 1) mol equiv. of DET per mole of titanium isopropoxide generates new soluble titanium complexes which show complementary chemical properties to those of the Sharpless reagent. Indeed, these new species are highly enantioselective for the oxidation of some suitably designed prochiral sulphides, but exhibit no enantioselectivity whatsoever in the epoxidation of allylic alcohols. Similar results were published by Modena and co-workers,[8] who used a larger excess of DET under anhydrous conditions. The first published procedures typically used *tert*-butyl hydroperoxide as oxygen donor and a stoichiometric amount of titanium diisopropoxide. It was then discovered that higher enantioselectivities could be achieved using cumene hydroperoxide, and that it was possible to reduce the amount of chiral complex to 50% (with respect to the sulphide) and even lower in some cases, with limited effect on the enantioselectivity of the reaction.[9] Figure 2 shows the results obtained in the case of the oxidation of methyl p-tolyl sulphide. The quoted enantiomeric excesses have been measured in our laboratories by chiral HPLC and might differ slightly from those given in the literature.

The structures of these new reagents have not yet been elucidated, but are clearly different from that of the Sharpless reagent. For instance, the infrared absorption dramatically changes when the DET:Ti ratio of an anhydrous mixture of DET and titanium isopropoxide in dichloromethane is increased from 1:1 (Sharpless reagent) to 2:1. The same phenomenon is observed on addition of 1mol equiv. of water to the Sharpless system.[7b] It must be noted that what is true for one sulphide is not necessarily true for all sulphides, so it is recommended that several combinations [Ti(OPr$^i$)$_4$/DET/H$_2$O, 1:2:1, 1:1:1, 1:2:0 or 1:>2:0, in stoichiometric or sub-stoichiometric quantity] are examined when considering the synthesis of a new chiral sulphoxide.

Differences between the sulphoxidation reaction and the Sharpless epoxidation arise in the preparation of the reagent and also from the fact that, unlike allylic alcohols,[10] sulphides cannot form ionic or hydrogen bonds with chiral titanium complexes. The chemoselectivity and stereoselectivity of the reaction are, consequently, mainly governed by steric effects and weak bond interactions. This explains the high substrate dependence of the reaction, a common feature of all biochemical and chemical methods used for the enantioselective oxidation of prochiral sulphides:[5] the more 'unsymmetrical' the prochiral sulphide, the higher is the enantioselectivity. For instance, most aryl methyl sulphides can now be oxidised in more than 90% ee, and some of them in more than 99% ee. This weak substrate–reagent interaction is also consistent with the fact that the Sharpless epoxidation can be seen as a more 'robust' process than the sulphoxidation reac-

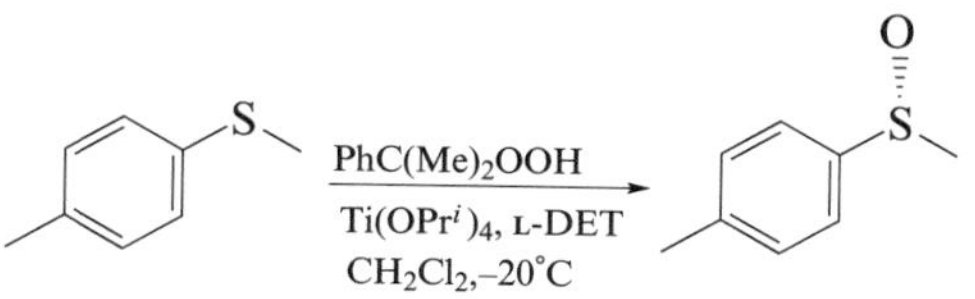

| | Mol equiv. | | | |
|---|---|---|---|---|
| | Ti(OPr$^i$)$_4$ | DET | | ee (%) |
| Sharpless reagent | 1 | 1 | Anhydrous | 0 |
| Modified Sharpless reagents | 1 | 1 | +1 mol equiv. H$_2$O | 96 |
| | 1 | 2 | Anhydrous | 89 |
| | 1 | 2 | +1 mol equiv. H$_2$O | 96 |

**Figure 2** Asymmetric oxidation of methyl p-tolyl sulphide by Sharpless' reagent

tion, where minor changes of the reaction conditions (solvent, tartrate, hydroperoxide, etc.) often result in decreased reactivity and enantioselectivity. There is a very good correlation between the absolute configuration of tartrate and sulphoxide when the sulphide bears two substituents of very different size, especially in the case of aryl methyl sulphides: L-DET leads to (*R*)-sulphoxides[5,7b] (Figure 3).

While the Sharpless epoxidation is now used for the bulk manufacture of several epoxy alcohols (see Chapter 18), the above asymmetric sulphoxidation reaction had, to our knowledge, never been used on a large scale prior to this work. We anticipated the suitability of this reaction for our syntheses for the following reasons:

(i) it does not require any special equipment and so could be easily implemented in standard pilot plant;
(ii) the raw materials are the same as those used in the Sharpless reaction and are available in large quantities from many suppliers;
(iii) it is not covered by a patent.

Possible areas for improvement prior to further scale-up and commercialisation included:

(i) the use of dichloromethane as solvent;
(ii) the relatively low throughput (10% w/v was a typical maximum concentration);
(iii) the need to use at least 20–50% of chiral complex with respect to the sulphide;
(iv) the fact that the occasionally required D-DET isomer is approximately 10 times more expensive than the natural L-DET.

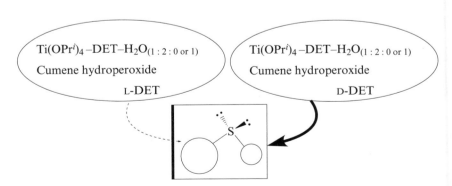

**Figure 3** Substrate–reagent orientation with unsymmetrical sulphides

## 19.3 SYNTHESIS OF RP 73163

RP 73163 is a hypocholesterolaemic agent which acts by inhibiting the enzyme ACAT. The opposite enantiomer of RP 73163 is inactive. This compound was first synthesised by oxidation of sulphide **1** using *m*-chloroperoxybenzoic acid, followed by resolution with semi-preparative high-performance liquid chromatography. Attempts to resolve the racemic sulphoxide using strong chiral acids failed. Indeed, owing to the presence of the sulphoxide group, the basicity of the imidazole ring is greatly reduced and the pyrazole ring is too remote from the chiral centre. The asymmetric oxidation of sulphide **1** using (+)-8,8-dichlorocamphorylsulphonyloxaziridine[11] in toluene at 40–50 °C afforded the desired enantiomer in 88–89.5% *ee*. However, this reaction could not be considered for the large-scale synthesis of RP 73163, mainly because of the low concentration required (2%) and the cost and availability of the chiral oxaziridine.[12] A first attempt to oxidise the sulphide **1** in dichloromethane at –20 °C using cumene hydroperoxide and the combination $Ti(OPr^i)_4$–DET–$H_2O$ (1:2:1) gave the racemic sulphoxide[13] (Figure 4).

The sulphide **1** is not a good substrate for the titanium-catalysed asymmetric sulphoxidation, as it bears two large substituents, one of them possessing a very flexible five-carbon chain, hence the decision to study the asymmetric oxidation of 4,5-diphenylimidazol-2-yl methyl sulphides, which could be used as intermediates in the synthesis of RP 73163.[12,13] Thus, the sulphide **2** was prepared in a one-pot process from commercially available 4,5-diphenylimidazole-2-thiol, and oxidised using the same reaction conditions as for sulphide **1** (Figure 5). In order

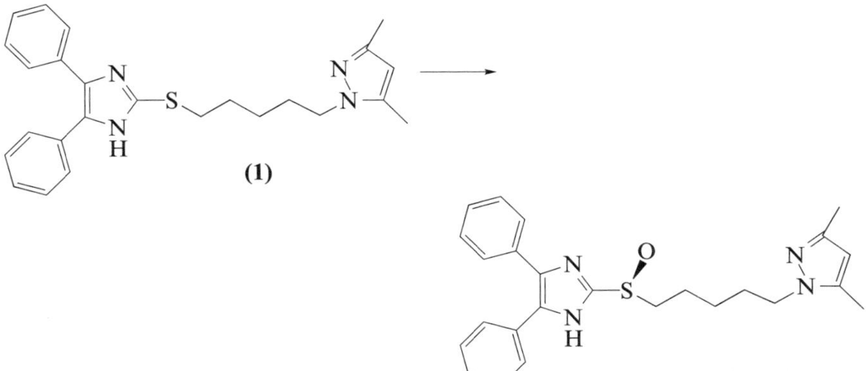

$Ti(OPr^i)_4$– DET– $H_2O$ (1 : 2 : 1)           0% *ee*
(+)-8,8-Dichlorocamphorylsulphonyloxaziridine: 88–89.5% *ee*

**Figure 4** Asymmetric epoxidation of sulphide **1**

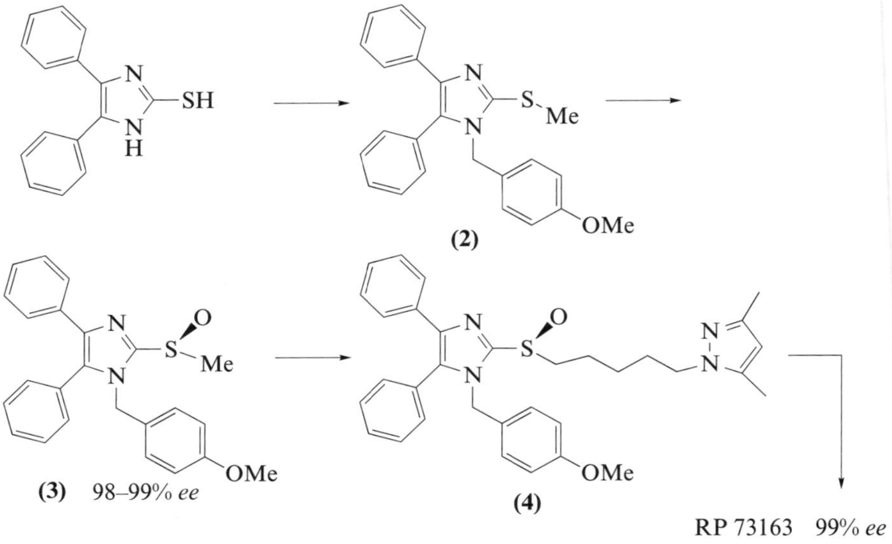

**Figure 5** An alternative route to RP 73163

to obtain the desired (S) absolute configuration, the unnatural D-DET had to be used.[7] The optimised reaction conditions required the use of cumene hydroperoxide as oxygen donor, 0.5 mol equiv. of titanium isopropoxide (with respect to the sulphide) and 1 equivalent of D-DET in anhydrous dichloromethane at −20 °C. The sulphoxide **3** was found to be surprisingly stable in acidic conditions in which other sulphoxides would racemise.[14] HCl(1 M) could therefore be used to wash out the titanium salts at the end of the reaction, and, after standard work-up, the sulphoxide **3** could be isolated as a crystalline material in 98–99% ee and 75% isolated yield.

Alkylation of sulphoxide **3** by 4-chloro-1-iodobutane and introduction of the pyrazole ring were combined in a one-pot process to give directly the intermediate **4**. Removal of the *p*-methoxybenzyl group was achieved without racemisation in refluxing trifluoroacetic acid in the presence of anisole. After a final recrystallisation stage, RP 73163 was obtained in >99% ee.[12,13]

## 19.4 THE SYNTHESIS OF RP 52891

RP 52891 is a potassium channel opener which has possible clinical indications in the treatment of, for instance, hypertension, irritable bladder, coronary artery and peripheral vascular disease and obstructive airway diseases.[3] The opposite enantiomer of RP 52891 is inactive.

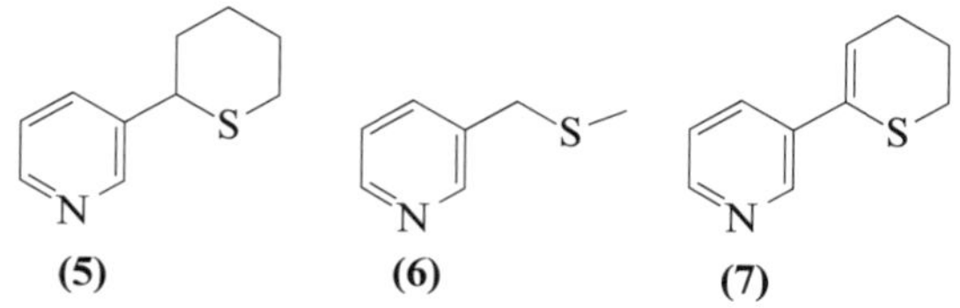

# ASYMMETRIC SYNTHESIS OF SULPHOXIDES

**Figure 6** Possible precursors to RP 52891

The first laboratory samples of RP 52891 were obtained by diastereoisomeric separation of synthetic intermediates and by chiral phase high-performance liquid chromatography.[3] As with RP 73163, several other asymmetric syntheses were then successfully studied in the laboratory and evaluated for future scale-up of RP 52891. Again, the method of choice for the introduction of the chiral sulphoxide moiety was the titanium-catalysed sulphoxidation reaction described above, and the plan was to design a substrate which could lead to the best enantioselectivity in this reaction (Figure 6).

The asymmetric oxidation of the sulphide **5** was obviously not the best strategy in this case, as it is a racemic mixture which could lead to four possible isomeric sulphoxides. As already mentioned, prochiral aryl methyl sulphides usually undergo asymmetric oxidation with good to high enantioselectivities. Thus, the sulphide **6** was subjected to the reaction conditions used with the sulphide **2** in the hope that it would then be possible to cyclise the resulting sulphoxide using an appropriate electrophile, and eventually produce RP 52891. However, the enantioselectivity of this oxidation never exceeded 20% ee. This was attributed to facile rotation about the CH$_2$—S bond. A more rigid substrate was therefore needed where the sp$^3$ nicotinic CH$_2$ group would be replaced by an sp$^2$ centre. To this end, the vinylic sulphide **7** was synthesised in three steps from methyl nicoti-

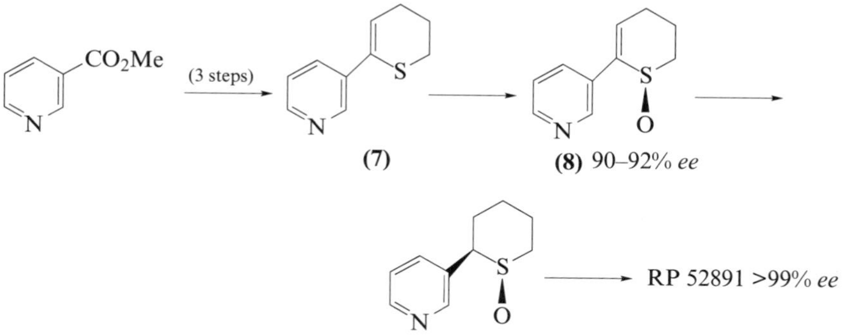

**Figure 7** Synthesis of RP 52891 from vinylic sulphide 7

nate, and oxidised under anhydrous conditions using L-DET as chiral ligand to give the desired (*R*)-sulphoxide (**8**) in 90–92% *ee*[15] (Figure 7).

The electron-deficient double bond of the sulphoxide **8** was reduced with sodium borohydride to afford sulphoxide **9**. In contrast to the intermediates **7** and **8**, the sulphoxide **9** is a highly crystalline material which can be recrystallised to give the desired (*R, R*) isomer in excellent enantiomeric, diastereoisomeric and chemical purity. Finally, diastereoselective alkylation with methyl isothiocyanate, followed by recrystallisation, gave RP 52891 in >99% *ee*.[15]

## 19.5 CONCLUSION

The strategy adopted in the syntheses of RP 52891 and RP 73163 was first to select an asymmetric oxidation method which would be suitable for rapid and safe scale-up, and then to design the rest of the synthesis around this key reaction. Because of the very limited number of asymmetric sulphoxidation methods potentially suitable for scale-up, the choice of the key reaction prevailed over the choice of the synthetic intermediate. The selected reaction was the enantioselective oxidation of prochiral sulphides by cumene hydroperoxide using soluble chiral complexes made from specific combinations of DET and titanium isopropoxide.[16] For better chiral recognition between the substrate and the chiral reagent, the prochiral sulphides had to be specifically designed with the sulphur atom in a very rigid and unsymmetrical environment. This strategy proved successful for both RP 52891 and RP 73163. Using the appropriate tartrate enantiomer, the sulphide intermediates then gave the corresponding sulphoxides in high enantiomeric purity and with the desired absolute configuration, thus allowing the preparations of multikilogram batches of RP 52891 and RP 73163 in >99% *ee* (Figure 8).

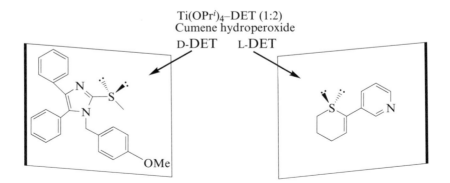

**Figure 8** Key asymmetric steps in the synthesis of drug candidates RP 52891 and RP 73163

## 19.6 ACKNOWLEDGEMENTS

The design and scale-up of the two syntheses presented in this chapter are the result of work done by the Process Chemistry team at the Dagenham Research Centre of Rhône-Poulenc Rorer in the UK. The author thanks all the members of this team for their constant enthusiasm and professionalism. The author also records his gratitude to Professor Henri Kagan for many fruitful discussions.

## 19.7 REFERENCES

1. See, for instance, Nasipuri, D., in *Stereochemistry of Organic Compounds*, Wiley, New York, 1991, and references cited therein.
2. (a) Crosby, J., in *Chirality in Industry* (ed.) A. N., Collins, G. N., Sheldrake, and J. Crosby, Wiley, Chichester, 1992, pp. 1–66; (b) For a recent article on the subject, see also Cadwell, J.; *Chem. Ind.*, (London) 175, 1995.
3. (a) Aloup, J. C., Farge, D., James, C., and Mondot, S., Cavero, I., *Drugs Future*, **15**, 1097 (1990); (b) Brown, T. J. Chapman, R. F., Cook, D. C., Hart, T. W., McLay, I. M., Jordan, R., Mason, J. S., Palfreyman, M. N., Walsh, R. J. A., Withnall, M. T., Aloup, J. C., Cavero, I., Farge, D., James, C., and Mondot, S. J., *Med. Chem.*, **35**, 3613 (1992).
4. Ashton, M. J., Bridge, A. W., Bush, R. C., Dron, D. I., Harris, N. V., Jones, G. D., Lythgoe, D. J., Ridell, D., and Smith, C., *Bioorg. Med. Chem. Lett.*, **2**, 375 (1992).
5. For a recent review, see Kagan, H. B, in *Catalytic Asymmetric Synthesis* (ed. I Ojima)., VCH, New York, 1993, Ch. 4.3, pp. 203–226
6. (a) Katsuki, T., Sharpless, K. B., and Rossiter, B. E., *J. Am. Chem. Soc.*, **102**, 5974 (1980); (b) Rossiter, B. E., Katsuki, T., and Sharpless, K. B., *J. Am. Chem. Soc.*, **103**, 464 (1981). For a review on the Sharpless epoxidation, see Rossiter, B. E., in *Asymmetric Synthesis* (ed.) J. D., Morrison, Academic Press, New York, 1985, Vol. 5, Ch. 7, pp. 193–246.
7. (a) Pitchen, P., and Kagan, H. B., *Tetrahedron Lett.*, **25**, 1049 (1984); (b) Pitchen, P., Deshmukh, M., Dunach, E., and Kagan, H. B. *J. Am. Chem. Soc.*, **106**, 8188, (1984); (c) Kagan, H. B., Dunach, E., Nemecek, C., Pitchen, P., Samuel, O., and Zhao, S. H., *Pure Appl. Chem.*, **57**, 1911 (1985).
8. Di Furia, F. Modena, G., and Seraglia, R., *Synthesis*, 325 (1984).
9. Zhao, S., Samuel, O., and Kagan, H. B., *Tetrahedron*, **43**, 5135 (1987).
10. (a) Burns, C. J., Matin, C. A., and Sharpless, K. B., *J. Org. Chem.*, **54**, 2826 (1989); (b) Jorgensen, K. A., *Tetrahedron: Asymmetry*, **2**, 515 (1991); (c) Potvin, P. G., and Bianchet, S., *J. Org. Chem.*, **57**, 6629 (1992).
11. Davis, Fig. A., and Sheppard, A. C., *Tetrahedron*, **45**, 5703 (1989).
12. Pitchen, P., *Chem. Ind.*, (London) 636 (1994).
13. Pitchen, P., France, C. J., McFarlane, I. M., Newton, C. G., and Thompson, D. M., *Tetrahedron Lett.*, **35**, 485 (1994).
14. See, for instance, (a) Mislow, K., Simmons, T., Melillo, J. T., Ternay, A. L., Jr, *J. Am. Chem. Soc.*, **86**, 1452 (1964); (b) Mislow, K., Green, M. M., Laue, P., Simmons, T., and Ternay, A. L., Jr, *J. Am. Chem. Soc.*, **87**, 1958 (1965); (c) Casarini, D., Foresti, E., Gasparrini, Fi., Lunazzi, L., Maciantelli, D., Misiti, D., and Villani, C., *J. Org. Chem.*, **58**, 5674 (1993).
15. Pitchen, P., paper submitted to *Proceedings of the Chiral 95 USA Symposium*, Boston, May 15–16, 1995.
16. For the latest improvements to this reaction see: (a) Brunel, J. M., Diter, P., Duetsch, M., Kagan, H. B., *J. Org. Chem.*, **60**, 8086 (1995); (b) Astra Aktiebolag, *Int. Pat. Appl.*, WO 96/02535.

# 20 Asymmetric Reduction of Prochiral Ketones

J.-C. CAILLE, M. BULLIARD and B. LABOUE
*SIPSY, Avrillé, France*

| | | |
|---|---|---|
| 20.1 | Introduction | 391 |
| 20.2 | The Stoichiometric Approach | 392 |
| 20.3 | The Catalytic Approach | 393 |
| 20.4 | Mechanistic Aspects | 394 |
| 20.5 | Preparation of the Catalyst | 395 |
| 20.6 | Asymmetric Reductions Catalysed by Oxazaborolines | 396 |
| 20.7 | Pharmaceutical Applications | 398 |
| 20.8 | Conclusion | 400 |
| 20.9 | References | 400 |

## 20.1 INTRODUCTION

Chirality continues to play an important role in the development of new pharmaceutical intermediates and of bulk drugs. Among the numerous techniques that industrial chemists now have at their disposal, asymmetric reduction and, in particular, reduction of unsymmetrical ketones to alcohols, has proved exceptionally useful over the last few years. The reduction is achieved by the overall addition of hydrogen to one face of the carbonyl group, which leads preferentially to the formation of one stereoisomer, as shown in Figure 1.

To achieve the chiral induction necessary for the selective preparation of one enantiomer, industrial chemists have modified the classical reagents with optically active ligands. The aim of this chapter is to survey some of the best practical methods for ketone reduction on an industrial scale, and to highlight in particu-

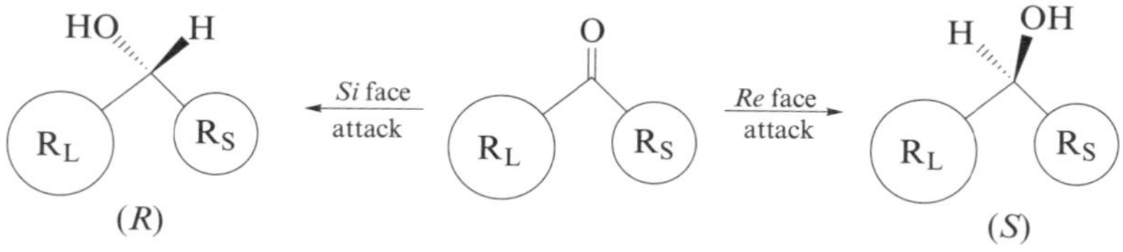

**Figure 1** Asymmetric reduction of prochiral ketones

*Chirality in Industry II.* Edited by A. N. Collins, G. N. Sheldrake and J. Crosby
© 1997 John Wiley & Sons Ltd

lar the use of oxazaborolidine reagents. We shall also illustrate the latter with some pharmaceutical targets which are currently prepared by this approach.

## 20.2 THE STOICHIOMETRIC APPROACH

The most obvious way to form a homochiral secondary alcohol from the corresponding ketone is to use a conventional reducing agent in a stoichiometric amount, associated with an optically pure chiral ligand. In this case the complex formed is able to deliver its hydride preferentially to one face. Among the reagents which illustrate this approach, those prepared from LiAlH$_4$ (LAH) are worthy of note. The Yamaguchi–Mosher reagent and BINAL-H have been most used over the last few years. Although these reagents have proved to be fairly useful for achieving good levels of enantiomeric excess, their manipulation is not straightforward. Both reagents usually require very low temperatures, sometimes below −78 °C to secure a good enantiomeric excess. Weigel[1] at Eli Lilly demonstrated the scope and limitation of the Yamagushi–Mosher reagent prepared from LAH and Chirald® [(2S,3R)-(+)4-dimethylamino-1,2-diphenyl-3-methylbutan-2-ol] or its enantiomer, *ent*-Chirald®, in the course of the synthesis of an analogue of the antidepressant fluoxetine (Prozac®) (Figure 2). (The merits of the technique for this particular compound are discussed in detail in Chapter 6.)

Another excellent way of producing enantiomerically pure secondary alcohols is to use chiral organoborane reagents. Following the pioneering work of Brown,

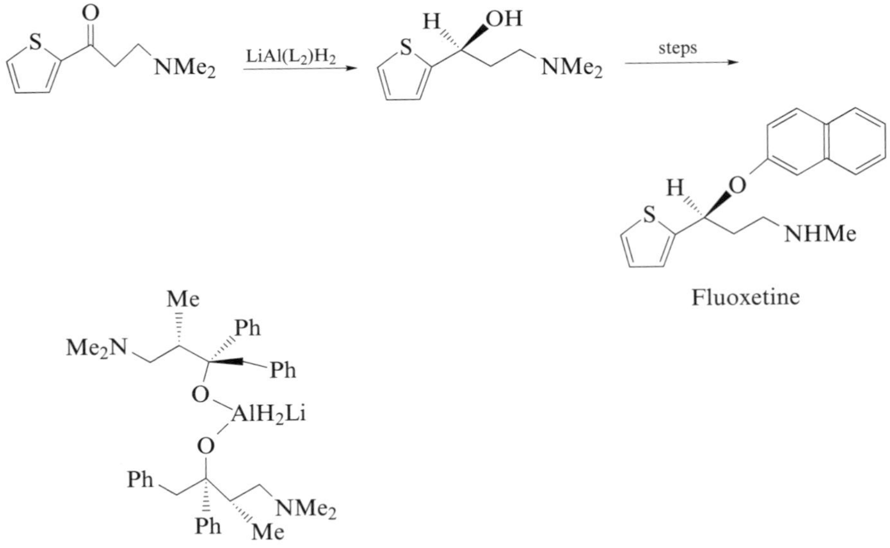

**Figure 2** Asymmetric synthesis of fluoxetine

# ASYMMETRIC REDUCTION OF PROCHIRAL KETONES

many such compounds have been reported, e.g., *B*-(3-pinyl)-9-borabicyclo-[3.3.1]nonane (Alpine-Borane) and chlorodiisopinocamphenylborane (Ipc$_2$BCl).[2] The latter is now commercially available from Callery Chemical and is a useful reagent achieving high chemical and optical yields. A good example of its efficacy was reported by Brown and co-workers during the preparation of the antipsychotic BMS 181100 from Bristol-Myers.

## 20.3 THE CATALYTIC APPROACH

All of the previously mentioned reagents, despite their effectiveness, have an important drawback. They are used in stoichiometric amounts and are, accordingly, often costly. Their use involves separation and purification steps which can become troublesome. Both academic and industrial chemists have searched for catalytic reagents to overcome these limitations and, on a small scale at least, the best technology to date is probably homogeneous catalytic hydrogenation. Numerous articles on asymmetric hydrogenation have appeared over the last few years, and we shall not discuss this area in detail here, because it appears that the method is currently of limited use on a preparative scale. While isolated enzymes and yeasts have been shown to be fairly useful in laboratory preparations, they are of limited use on a large scale owing to the high dilution required and troublesome work-ups.

Apart from these techniques, the most interesting reagents to be described in recent years are the so-called Corey reagents (chemzymes[3]). Capitalising on the excellent work of Itsuno and co-workers,[4] who first described the use of oxazaborolidine as a homochiral ligand, and of Kraatz[5], Corey was the first to report the enantioselective reduction of ketones to chiral secondary alcohols catalysed by oxazaborolidine (Figure 3).

Since then, numerous articles dealing with this versatile reagent have appeared in the literature. Many new catalysts have been designed, usually by replacing the diphenylprolinol moiety by other derivatives of various natural or unnatural amino acids. A large number of new oxazaborolidines have been reported which provide varying enantioselectivities in prochiral ketone reductions (we have counted more than 22 to date), but none has offered an improvement over the proline-derived oxazaborolidines. The only exception is the diphenyloxazaboroli-

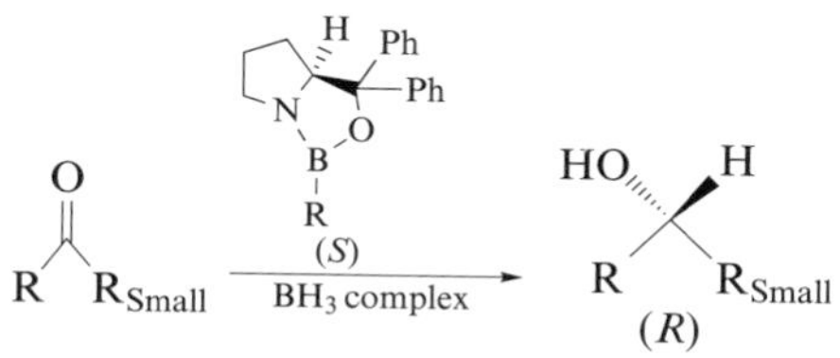

**Figure 3** Oxazaborolidine-catalysed reduction of prochiral ketones

dine reported by Quallich at Pfizer, which may be a good alternative since both enantiomers of *erythro*-aminodiphenylethanol are cheap and commercially available.[6]

On the industrial side, the discovery of this technique prompted many groups to investigate its versatility. At Merck, a group of chemists led by Mathre investigated the use of oxazaborolidine to prepare various optically pure pharmaceutical intermediates. In parallel with Corey's work, they studied the mechanistic aspects of this reaction and clearly demonstrated the pivotal role of the borane adduct CBS-B.[7] Their research culminated in the isolation of this stable complex described as a 'free-flowing crystalline solid,' and they were first to report its single-crystal X-ray structure.

## 20.4 MECHANISTIC ASPECTS

Several groups have attempted to rationalise the enantioselective reduction of prochiral ketones with this type of catalyst, Corey being the first to propose a catalytic scheme. The reduction occurs by coordination of the electron-deficient boron atom, through the lone pair *anti* to its larger substituent, with ketonic oxygen and hydrogen transfer from the $NBH_3^-$ unit to the carbonyl carbon via a six-membered cyclic transition state, following the path shown in Figure 4.

Shortly after his publication, Mathre at Merck in collaboration with Liotta's group at Emory University reconsidered this mechanism. They proposed an alternative 'chair transition state assembly' to explain the origin of enantioselec-

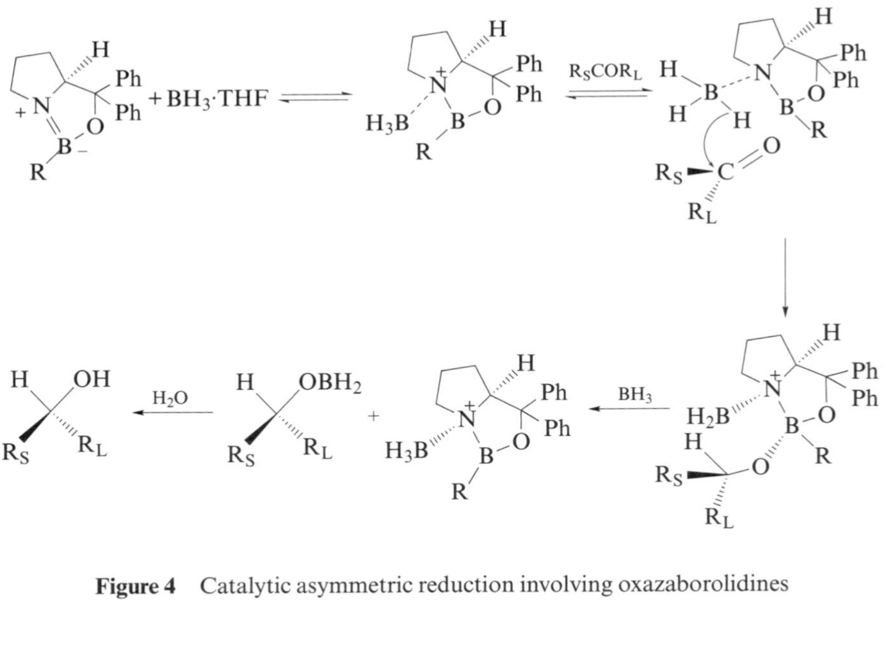

**Figure 4**  Catalytic asymmetric reduction involving oxazaborolidines

# ASYMMETRIC REDUCTION OF PROCHIRAL KETONES

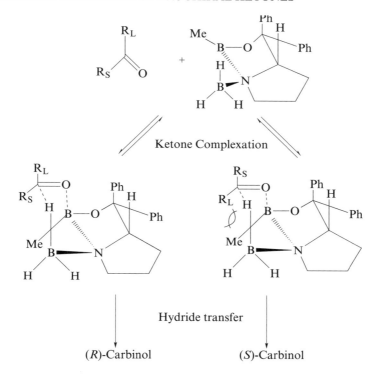

**Figure 5** 'Alternative' chair transition-state assembly

tivity observed in oxazaborolidine-catalysed reduction of prochiral ketones (see Figure 5). To achieve this goal, they elected to use the reaction surfaces generated using the MNDO Hamiltonian.[8] According to their model, the 1,3-diaxial interaction between the $R_L$ or $R_S$ ketonic substituent (L = large, S = small) and the oxazaborolidine alkyl group differentiates the two diastereoisomeric transition states leading to the major (R)- and minor (S)-isomers. This model, in our view, accounts much better for the role of the alkyl group borne by the boron atom, which had not been taken into consideration in the previous model.

Other interesting studies of the mechanism and the relationship of structure to enantiomeric excess have been published by Quallich et al.[9] in the case of a diphenylaminoethanol-derived catalyst.[9]

## 20.5 PREPARATION OF THE CATALYST

Two different approaches can be followed for the preparation and use of the catalyst. In the first, it is formed *in situ* by mixing the (R)- or (S)-diphenylprolinol

(DPP) and $BH_3$ complex. This route is advantageous because there is no need to use boronic acids (or boroxines) or to remove water to form the catalyst. Another possible way is to use preformed catalysts, some of which are commercially available from suppliers such as Callery Chemical. In this particular case, the instability of the catalyst over time may be a disadvantage. It is worth noting that investigators at Merck circumvented this drawback by making an air-stable complex called CBS-B. This complex is not yet commercially available.

DPP itself is now available on large scale, and can be prepared in several ways. The most straightforward route is based on an earlier publication by Corey: suitably protected (S)-proline is treated with phenylmagnesium halide, yielding the diphenylmethanol derivative. An alternative two-step enantioselective synthesis of DPP from proline, based on the addition of proline-N-carboxyanhydride to 3 equiv. of phenylmagnesium chloride, has been recently reported by the Merck group.[10] An elegant asymmetric preparation of DPP has been reported by Beak et al.[11] The enantioselective deprotonation of BOC-pyrrolidine is achieved with sec-butyllithium in the presence of sparteine as a chiral inducer. Subsequent quenching with benzophenone gives N-BOC-DPP in 70% yield and 99% ee after one recrystallization. It is worth noting that in order to cope with the rather high price of (R)-proline, Corey developed an alternative route to this ligand from racemic pyroglutamic acid.

The oxazaborolidine is finally made by refluxing DPP with a suitable alkyl-(aryl)boronic acid or, better, with the corresponding boroxine in toluene in the presence of molecular sieves (or by removing water by azeotropic distillation). According to the literature, methyloxazaborolidine (Me-CBS) can be either distilled or recrystallised. The key point is that the catalyst must be free from any trace of water or alkyl(aryl)boronic acid because these impurities decrease enantioselection.

## 20.6 ASYMMETRIC REDUCTIONS CATALYSED BY OXAZABOROLINES

The best reaction conditions depend on both the nature of the substrate and the catalyst. A wide range of conditions have been studied by Mathre's group at Merck, by Quallich and co-workers at Pfizer, by Stone at Sandoz (Basle) and by our group at SIPSY. Generally, the reaction must be carried out under anhydrous conditions, usually at room temperature. Temperature appears to be the most important parameter of the reaction and, although a clear temperature dependence has been demonstrated by Stone,[12] no good mechanistic explanations have been presented. These will probably follow shortly. The catalyst loading is typically 1–20%, and the DPP can be recovered at the end of the reaction and recycled in some cases. $BH_3$ complexes of THF, $Me_2S$ and 1,4-thioxane may be used, and diborane gas has also been successfully employed. Catecholborane is a suitable reagent for certain low-temperature reactions. The relationship between enantioselectivity and the nature of the substrate is highly predictable, and the best results are obtained with aryl ketones in which there is a significant size difference between the large and the small substituent, and in which the basicity of the ketone is favourable.

Besides the preparation of aryl alkyl carbinols, many other secondary alcohols have been prepared in good to excellent optical purity using this method, although improvements in the catalytic system still have to be made in order to obtain good results for all types of ketones, especially the problematic dialkyl ketones. Numerous complex target molecules have been prepared with the asymmetric reduction step as the key step. Some selected examples are shown in Figure 6.

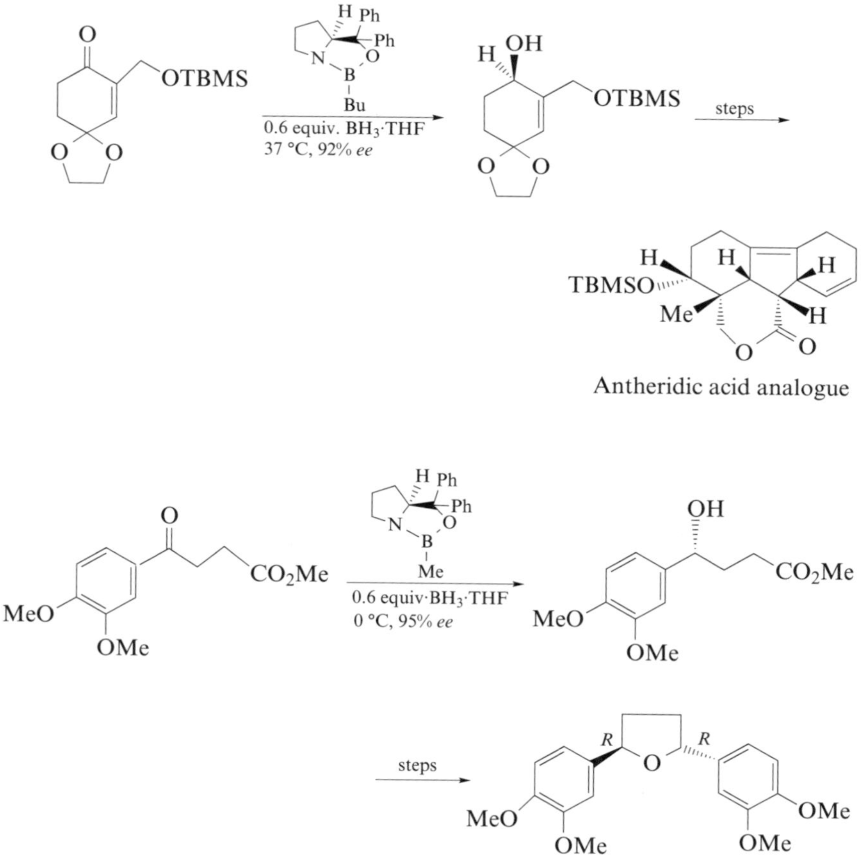

**Figure 6** Access to highly complex molecules by oxazaboroline-catalysed reductions

## 20.7 PHARMACEUTICAL APPLICATIONS

Numerous pharmaceutical intermediates and drug substances bear chiral secondary alcohol groups. We do not intend in this chapter to review all the target molecules that fall into this category, but rather to highlight the most important ones for which an oxazaborolidine-catalysed reduction has been used with success by industrial chemists. In some cases the processes have been commercialised. Corey was the first to become involved in the preparation of pharmaceutical molecules using this technology, and reported several preparations of important drugs. His first success was the synthesis of a pure enantiomer of fluoxetine, the Lilly antidepressant introduced under the brand name Prozac®.[13]

Shortly afterwards, he published the preparation of both enantiomers of isoproterenol, a β-adrenoreceptor agonist and denopamine, and a useful drug for congestive heart failure.[14] From a pharmaceutical point of view, the structure of isoproterenol is remarkable, since the same basic structure is also encountered in various drugs, such as those acting at the adrenoreceptor site. Therapeutic indications are broad, ranging from asthma ($\beta_2$-agonist) to obesity and diabetes ($\beta_3$-agonist). Halo aryl ketones have been shown to be useful intermediates to these drugs, giving rise, after reduction and basic cyclization, to the corresponding epoxystyrenes, which are important intermediates for the preparation of $\beta_3$-agonists (see Figure 7). The regiospecific ring opening is achieved with the appropriate amine to give an advanced intermediate possessing the correct configuration at the carbinol stereocentre. This compound is converted in a sequence of several steps into the active drug.

At SIPSY we have demonstrated the industrial feasibility of this approach in the production of hundreds of kilograms of related molecules.

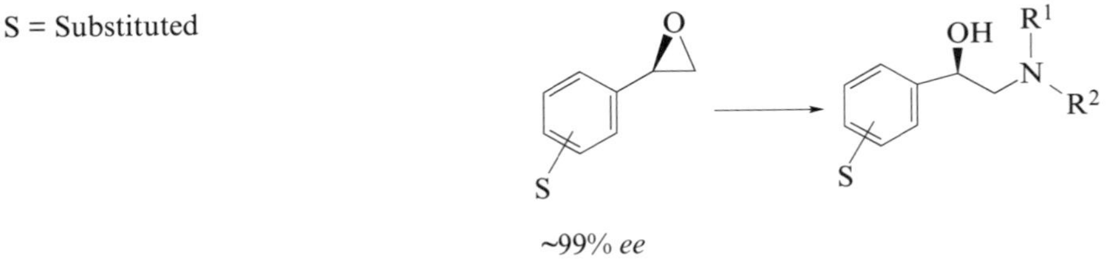

S = Substituted

~99% ee

**Figure 7**  Access to important $\beta_3$-agonist intermediates

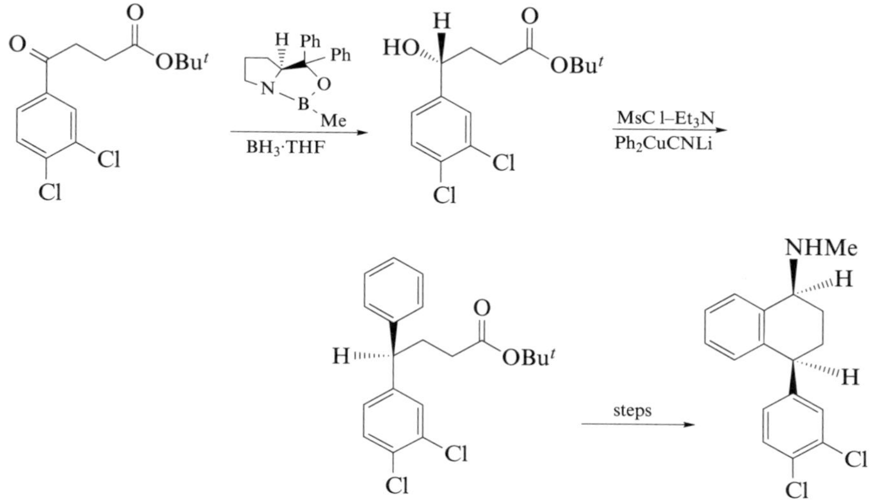

**Figure 8** Asymmetric synthesis of sertraline

Along the same lines, chemists at Merck successfully produced substantial quantities of various important pharmaceutical intermediates used in the preparation of drugs currently under development. Among these are MK-0417, a promising soluble carbonic anhydrase inhibitor useful in the treatment of glaucoma,[15] the LTD4 antagonist L-699392[16] and the anti-arrhythmic MK-499.[17] Equally successful has been the group of Quallich at Pfizer,[18] who have reported the preparation of a substituted dihydronaphthalenone, a pivotal intermediate in the preparation of the antidepressant sertraline (Lustral®).

In this synthesis (Figure 8), the newly created chiral centre subsequently determines the absolute stereochemistry of the final product. This illustrates that the use of Corey reduction is a valuable tool to introduce the chirality in a complex molecule owing to its high compability with other reactive centres. Quallich demonstrated that high enantiomeric excesses could also be obtained when other heteroatoms capable of coordinating borane were present in the starting material. Using an excess of borane to chelate/protect the nitrogen functions, a smooth reduction of these compounds could be effected.[19]

Reduction of alkyl aryl ketones is an equally valid route to optically pure secondary arylalkylamines, as researchers from Sandoz showed in an enantioselective synthesis of SDZ-ENA-713[20] (Figure 9).

The future developments of the Corey reduction will undoubtedly include finding ways to reduce effectively alkyl vinyl and dialkyl ketones. The first applications have already been demonstrated, e.g. the key reduction step of a 5-lipoxygenase inhibitor (Figure 10).

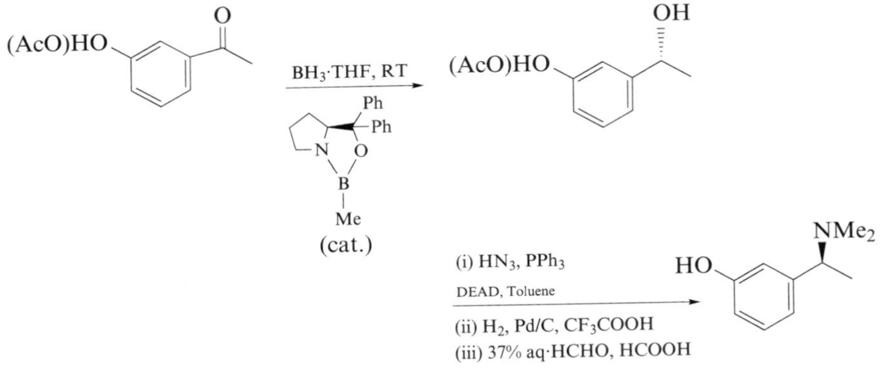

**Figure 9** Synthesis of SDZ–ENA–713 intermediate

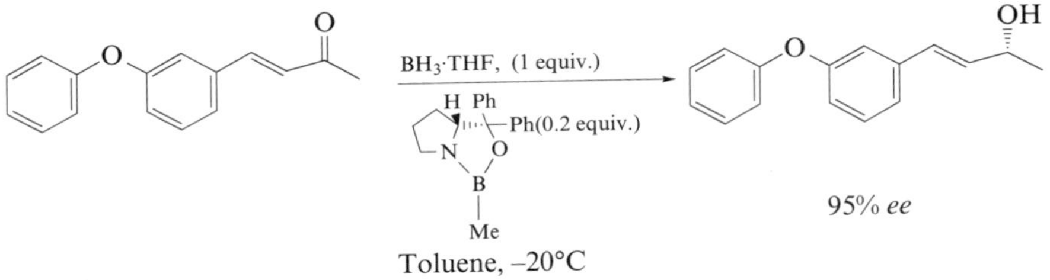

**Figure 10** Key step in the synthesis of a 5-lipoxygenase inhibitor intermediate

## 20.8 CONCLUSION

The representative examples described above are illustrative of the large amount of work going on in the area of oxazaboroline-catalysed reductions. We have demonstrated that asymmetric reduction of prochiral ketones is a powerful tool to produce important pharmaceutical intermediates, and it is also apparent that the discovery of oxazaborolidines has given rise to new insights into enantioselective reactions. Academic and industrial chemists are already looking beyond this type of reagent, and from this basis a new generation of efficient chiral catalysts is being developed.

## 20.9 REFERENCES

1. Weigel., L. O., *Tetrahedron Lett.*, 7101 (1990).
2. Brown, H. C., *Organic Synthesis via Boranes*, Wiley–Interscience, New York 1975.
3. Corey, E. J., Bakshi, R. K., and Shibata, S., *J. Am. Chem. Soc.*, **109**, 5551 (1987); Corey, E. J., (Harvard University), *US Pat.*, 4 943 635.

4. Itsuno, S., Hirao, A., Nakahama, S., and Yamazaki, N., *J. Chem. Soc., Perkin Trans. 1*, 1673 (1983); Itsuno, S., Nakano, M., Miyazaki, K., Masuda, H., Ito, K., Mhizao, A., and Nakahama, S., *J. Chem. Soc., Perkin Trans. 1*, 2039 (1985).
5. Kraatz, U. (Bayer), *Ger. Pat.*, DE 3 609 152.
6. Quallich, G. J., and Woodall, T. M., *Tetrahedron. Lett.*, **34**, 4145 (1993).
7. Thompson, A. S., Douglas, A. W., Hoogsteen, K., Carroll, J. D., Coreley, E. G., Grabowski, E. J. J., and Mathre, D. J., *J. Org. Chem.*, **58**, 2880 (1993).
8. Jones, D. K., Liotta, D. C., Shinkai, I., and Matthre, D. J., *J. Org. Chem.*, **58**, 799 (1993).
9. Quallich, G. J., Blake, J. F., and Woodall, T. M., *J. Am. Chem. Soc.*, **116**, 8516 (1994).
10. Blacklock, T. J., Jones, T. K., Matthre, D. J., and Xavier, L. C. (Merck), US *Pat.*, 5 039 802.
11. Beak, P., Kerrick, S. T., Wu, S., and Chu, J., *J. Am. Chem. Soc.*, **116**, 3231 (1994).
12. Stone, G. B., *Tetrahedron: Asymmetry*, **5**, 465 (1994).
13. Corey, E. J., and Reichard, G. A., *Tetrahedron Lett.*, **30**, 5207 (1989).
14. Corey, E. J., and Link, J. O., *J. Org. Chem.*, **56**, 442 (1991).
15. Jones, T. K., Mohan, J. J., Xavier, L. C. C., Blacklock, T. J., Mathre, D. J., Sohar, P., Tonner Jones, E. T., Reaner, R. A., Robers, F. E., and Grabardes, E. J. J., *J. Org. Chem.*, **56**, 763 (1991).
16. King, A. O., Corley, E. G., Anderson, R. K., Larsen, R. D., Verhoeven, T. R., and Reider, P. J., *J. Org. Chem.*, **58**, 3731 (1993).
17. Shi, Y. J., Cai, D., Dolling, U. M., Douglas, A. W., Tschaen, D. R., and Veroheven, T. R., *Tetrahedron. Lett.*, **35**, 6409 (1994).
18. Quallich, G. J., and Woodall, T. M., *Tetrahedron*, **48**, 10239 (1992).
19. Quallich, G. J., and Woodall, T. M., *Tetrahedron Lett.*, **34**, 785 (1993).
20. Chen, C.-P., Prasad, K., and Repic, O., *Tetrahedron Lett.*, **32**, 7175 (1991).

# Index

*(Entries in bold type are references to tables or figures)*

abzymes, 5, 9
*N*-acetyl amino acids, 197–200
D-(−)-(*S*)-acetyl-β-mercaptoisobutyric acid, 81
*N*-acetyl-L-methionine, 176
*N*-acetyl-phenylalanine, 176
acetylglutamic acid, 152
achiral assay, 13
acyclonucleotides, 375
α-acylase, 176
adrenaline, **93**, 245
β-adrenoreceptor agonist, 398
agrochemicals, 5, 8, 147–8, 209, 278, 283, 310
  market size, 1
Alcalase, 203
alkaline phosphatase, in lamivudine synthesis, 69
L-allylglycine, 200
alminoprofen, **20**
Alpine–Borane, 393
amino acid amides, resolution by aminoacylases, 197–200
amino acids,
  by membrane technologies, 7, 159–73
  liquid crystal applications, 277
  unnatural, 187–90, 194–203, 393
α-amino alcohols, 165
  as chiral proton source, 339–48
D-(+)-α-amino-ε-caprolactam, **87**, **93**
aminoacylases, 197–99
  combination with esterases, 200
(*S*)-(+)-2-aminobutan-1-ol, **87**, **93**
aminodiphenylethanol, 394
(+)-3-aminomethylpinane, **93**
(1*S*,2*S*)-(+)-*threo*-2-amino-1-phenyl-propane-1,3-diol, 85, **87**, 91, **93**
anisotropic liquid crystals, 264
anthenidic acid, analogue of, **397**
antihypertensive inhibitors, *see* renin inhibitors

antiviral drugs, 41–77
aristeromycin, **185**
(+)-ARSOP, **313**
3-aryloxy-1,2-propanediols, 376
2-arylpropionitriles, by asymmetric hydrocyanation, 312–3
(*R*)-asparagine, 154, 225
aspartame, 3
asymmetric catalysis, recent developments, 3
asymmetric dihydroxylation, Sharpless, 4
asymmetric epoxidation, 5
  Jacobsen, 4, 318
  Sharpless, 3, 6, 363–77, 382 (*see also* glycidol process)
asymmetric hydrocyanation, 309–32
  mechanism, 327–31
  in naproxen synthesis, 25, 331–2
  of norbornene derivatives, 312
asymmetric hydrogenation, *see* catalytic asymmetric hydrogenation
asymmetric ketene-imine Staüdinger reaction, 106
asymmetric oxidation, of sulphides, 381–7
asymmetric protonation, *see* enantioselective protonation
asymmetric reduction, 4, 9, 101–2 (*see also* catalytic asymmetric hydrogenation)
  of ketones, 101–2, 246–7, 253–60, 391–8
  with oxazaborolines, 4, 391–8
asymmetric transformation, *see* second order asymmetric transformations
autocatalysis, 341–2
azabicyclo[2.2.1]hept-5-en-3-one, **185**, 188
azlactones, dynamic resolution of, 201–3
AZT, 375

benoxaprofen, **20**
benzonorbornadiene, **313**
(*S*)-benzoyl-β-mercaptoisobutyric acid, 83
*N*-benzoylglycine, 203

D-(+)-N-benzyl-α-phenethylamine, 83
(2R,3R)-*trans*-3-benzylglycidol, 75, 370
*cis*-3-benzylglycidol, 370–2
beta-blockers, 160, **161**, 245, 375
BINAP, 3, **24**, 31, 313
  in asymmetric hydrocyanation, 312–3
  in asymmetric hydrogenation, 249
  in naproxen synthesis, 25
BINAP-LAH, 102
binaphthyl crown ethers, 167
1,1'-binapthol, resolution of, 165
biocatalysts, whole cell, *see*
  biotransformation, whole cell
Biopol, 248
bioresolution processes, whole-cell, *see*
  biotransformation, whole cell
biotechnology industry, development of, 9
biotransformation, 48
  continuous, 214–8
  fed-batch, 217, 302–3
  L-carnitine production, 287–304
  integration into process chemistry, 192, 194–200, 203–4, 298–304
  whole cell, 186–7, 211–4, 246, 253–60
birefringence, in liquid crystals, 266
Blackman kinetics, in L-carnitine production, 300–2
N-BOC-L-thiazolylalanine, (BOC-TAZ), 194–200
bornane-10,2-sultam, 3
(R)-(+)-bornylamine, **93**
bovine serum albumin, in enantioselective membranes, 160–61
brefeldin A, 192, **193**
(S)-1-bromo-2-methylbutane, 283–4
(S)-2-bromopropanoic acid, 208, 220
(–)-brucine, **87**, **93**, 280
(–)-bupivacaine, 2
bupranolol, **165**
(R)-1,3-butanediol, 249
2-butanol, resolution with enantioselective membranes, 161
butyrobetaine metabolism, 288–92

c-NECA, 188, **189**
C-3 synthons, 364, 376
(+)-calycanthine, **93**
camphorsulphonic acid, 84, 85, **87**, 95, 102
*Candida rugosa*, in ibuprofen resolution, 34
captopril, 81, **82**, 85
carbacephalosporin antibiotics, *see*
  β-lactam, antibiotics
carbinol, **395**
carbocyclic bromovinyldeoxyuridine, 52
carbocyclic nucleosides, 41–77, 184–90
carbohydrate-derived ligands, 315–7
carbonic anhydrase inhibitor, 246, 399
carbovir, 42–47, 185, 188, 192
carnitine dehydrogenase, 289–94
L-carnitine, 8
  metabolism, 288–92
  transport, 292–3
carprofen, **20**
catalytic antibodies, 5
catalytic asymmetric hydrogenation, 25, 249, 339
chemzymes, 393
chiral and non-chiral drugs, distribution, **2**
chiral auxilliaries, 3, 4, 26–7, 32, 102, 105, 111–2
chiral cyanohydrins, *see* cyanohydrins, optically-pure
chiral drugs, regulatory aspects, 11–17
chiral HPLC, 317
chiral intermediates, market size, 2
chiral inversion, *in vivo*, of ibuprofen, (S)-(+)-27
chiral nematic phase, 272–3
chiral pool, 3, 107–9, 230, 277, 382
chiral proton source, *see* enantioselective protonation
chiral selectors, in liquid-liquid extraction, 167, 173
chiral smectic C phase, 267–73
chiral solvents, 154
Chirald, 102, 392, **104**
Chiraldex, 372
chirality indices, 94
chloramphenicol, 353, **354**
(R)-3-chloro-1-phenylpropanol, 8
(S)-2-chloro-3-methylbutanoic acid, 283–4
α-chloroacids and esters, in liquid crystals, 278
chlorodiisopinocamphenylborane, 393
β-chloroisobutyric acid, 85
4-chlorophenylphosphoric acid, as resolving agent, 85, **87**
(R)-2-chloropropanoic acid, 220
(S)-2-chloropropanoic acid, 2, 151, 177, 207–21
  market requirements, 208–10
chlorphenesin, **365**
cholesteric phase, 266–7, 272

INDEX 405

cholesteryl benzoate, as chiral nematic phase, 263
cholesteryl-glutamate, 174
cholesterylnonanoate, 275
α-chymotrypsin, 196
cicloprofen, **20**
cidofovir, **365**
(–)-cinchonidine, 81, 82, **87, 93**
   in naproxen resolution, 22
classical resolution, *see* resolution, classical
CLECs, *see* crosslinked enzyme crystals
cliprofen, **20**
cloning, 260, 296–7
cofactor regeneration,
   in L-carnitine production, 295
   in membrane-assisted resolutions, 178
Committee for Proprietary Medicinal Products (CPMP), 12, 16
conglomerate, 15, **149**, 153, 373
Corey asymmetric reduction, *see* asymmetric reduction with oxazaborolidines
countercurrent adsorption, 7
(S)-CPA, *see* (S)-2-chloropropanoic acid
cross-linked enzyme crystals, (CLECs), 5, 34
crown ethers, chiral, 165
crystal packing, in diastereoisomeric salt pair, 89
crystal science techniques, for chiral compounds, 4, 6, 34–6, 119–55
crystal size distribution, 137–40
crystallisation,
   direct, *see* direct crystallisation
   evaporative, 154
   melt, 130
   preferential, 374
crystallisation-induced asymmetric transformations *see* second order asymmetric transformations
cyanohydrins, optically pure, 5, 278, 283
L-2-cyanophenylalanine, 200
cyclic phosphoric acids, as resolving agents, 84, 85, **86, 87**, 91
cyclocitrals, 346, 348
cyclodextrin-based GC, 372
cyclodextrins, **163**, 372
γ-cyclogeranates, 343, 346
(R)-(–)-1-cyclohexylethylamine, **87, 93**
cyclooxygenase, 19–21
(+)-(1R-cis)-4-cyclopentene-1,3-diol monoacetate, 47

cyhalothrin, 147–8
cytidine deaminase, in lamivudine synthesis, 69

γ-damascone, 346, 348
(S)-α-damascone, 335, 337
(S)-β-damascone, 337
dansyl-phenylalanine, resolution of, 165
Darvon alcohol, **93**
Darvon, 102
*p*-decyloxybenzylidene *p*'-amino-2-methyl-butylcinnamate, 268
dehalogenases, in resolution of 2-chloropropanoic acid, 211–4
(+)-dehydroabietylamine, **87, 93**
denopamine, 398
'designer' resolving agents, *see* resolving agents, rational design
desupersaturation, 140
dextropropoxyphene, **226**
L-(+)-2,4-diaminobutyric acid, **93**
diastereoisomeric salt formation, *see* resolution, classical
diastereomeric transition states, in asymmetric reduction with oxazaborolines, **395**
(S)-2,3-dibromopropanoate, 221
(1R,2R)-2-dichloroacetimido)-1-[(4-methylsulphonyl)phenyl]-1,3-propanediol, *see* thiamphenicol,
(+)-8,8-dichlorocamphorylsulphonyl oxaziridine, **385**
dichlorprop, **209**
dielectric constant, in liquid crystals, 266
diethyl tartrate, 366
   in enantioselective sulphide oxidation, 383–4
differential scanning calorimetry (DSC), in polymorph identification, 124–5, 151
2-(2,4-difluoro)phenylglycidol, 372
*cis*-dihydroxylation, of carbovir, 42–47
3,4-dihydroxyphenylalanine, *see* dopa
diisoamylborane, 58–59
D-(–)-diisopropyl tartrate, 366
diltiazem, 4, 176
(S)-dimethylcyclopropylcarboxamide, 5
N-(3,5-dinitrobenzoyl)leucine, 165
(+)-DIOP, **313**
(S)-diphenyl prolinol, **395**
diphenyloxazaborolidine, 393–4
diphenylphosphinite ligands, in asymmetric hydrocyanation, 315–332

# 406  INDEX

(+)-DIPHIN, **313**
direct crystallisation, 148–54
dispersion-free extraction, 170
display devices, 8, 263–87
$N$-$n$-dodecyl-L-hydroxyproline, in liquid-liquid extraction, 171
dopa, 226
dopamine $D_2$-receptor antagonists, 227
dopants for nematic mixtures, 275, 280–1
dropropizine, 375
drugs enantiopure, market size, 1
duloxetine, 101–4, **392**
dynamic resolution, *see* resolution, dynamic

(−)-eburnamonine, **93**
electrocarboxylation, route to ($S$)-(+)-ibuprofen, 29
electronic asymmetry, effect of in asymmetric hydrocyanation, 324–6
enantioselective membranes, *see* membranes, enantioselective
enantioselective nitrile hydrolysis, 5
enantioselective oxidation of prochiral sulphides, 382–4
enantioselective polymers, 158–62
enantioselective protonation, 32–3, 83, 335–49
enolates, enantioselective protonation of, 336–9
entrainment resolution, *see* resolution, by entrainment,
enzymatic kinetic resolution, 33–4, 69–70, 167, 175–8, 184–200
  of carbovir, 5, 42
  of ibuprofen esters, 33
  of phenyl-1,3-propanediol, 104
enzymes, "non-nameplate" applications, 5
ephedrine, 84, 85, **87**, **90**, **91**, **93**, **172**, **173**, 339
($S$)-epichlorohydrin, 363
Epivir, *see* lamivudine,
epoxides, chiral, by enzymatic resolution, 5
2,3-epoxy alcohols, *see* glycidols,
($S$)-1,2-epoxyalkanes, as building blocks for liquid crystal dopants, 282
($R$)-1,2-epoxyoctane, 283–4
esterase, 195–6, 210
  combination with aminoacylases, 200
($S$)-ethyl lactate, 283–4
*trans*-3-ethylglycidol, 372
($S$)-1-ethylprolinamide, 232, 234–5

eudismic ratio, 94
eutectic composition, in racemate separation, 6
  of naproxen, 24
  of ibuprofen, 34, 35
eutomer, 382
extremophiles, 6

FDA, 2, 11, 14,15, 27, 83
fed-batch biotransformation, *see* biotransformation, fed-batch
fenoprofen, **20**
fermentation, 252–60
  in CPA production, 213–7
ferrocenylethyl-thiophenol, resolution of, 163
ferroelectric dopants, 277–9, 283
ferroelectric liquid crystals, 375
ferroelectric mixtures, 277
flavours and fragrances, 8
florfenicol, 353, **354**
fluazifop
  racemic switch, 209
  resolution by esterase, 210
flunoxaprofen, **20**
2-fluoro-1-alkanols, as building blocks for liquid crystal dopants, 282
fluoroalkanols, in liquid crystals, 278
($S$)-2-fluorooctanol, 283–4
($S$)-fluoxetine (Prozac), 8, 104, **392**, 398
($S$)-(+)-flurbiprofen, **20**, 37, 208
fragrance synthesis, 335–49
fructose, phoshite ligands from, **316**
furcloprofen, **20**

galactose, phoshite ligands from, **316**
gas chromatography, chiral, 233
geraniol, Sharpless epoxidation of, 371
glucophosphinite ligands, in asymmetric hydrocyanation, 318–31
D-(+)-glucosamine, **93**
glucose, phoshite ligands from, **316**
($S$)-glutamic acid, 154, 230
glycerophospholipids, 375
glycidol process, 3, 363–77
  effect of reactant purity, 368
  recovery of epoxy alcohol, 369–71
  safety aspects, 366, 371
  scale-up, 368–9
($R$)-glycidol, 3, 364–71
  synthetic applications of, 374–6
glycidyl 3-nitrobenzenesulphonate

# INDEX

(GNBS), 373–4, 376
glycidyl arenesulphonates, by Sharpless epoxidation, 370, 372–3
glycidyl butyrate, resolution of, 176, 364
glycidyl tosylate, 376–7
guafecainol, 376
guafenesin, **365**, 375, 376–7

habit modifiers, 126–9, 154
hepatitis B, 66
herbicides, *see* agrochemicals
Herpes simplex, 42, 185
heterogeneous catalysis, 4
hexaprofen, **20**
hippuric acid, 203
HIV protease inhibitors, 4, 66–67, 185, 192, 370, 375
hollow fibre membranes, 7, 165, 170, 176–7
hollow fibre reactors, 34
homeotropic alignment, in liquid crystals, 267
hydroboration, stereoselective, 58–59, 66, 391–8
hydrocyanation, *see* asymmetric hydrocyanation
hydroformylation, route to (S)-(+)-ibuprofen, 29
hydroquinidine, **87**, **93**
4-*endo*-hydroxy-2-oxabicyclo[3.3.0]oct-7-en-3-one, 190–3
(R)-3-hydroxy-4-(trimethylammonio)-butanoate, *see* L-carnitine
α-hydroxyacids, as building blocks for liquid crystal dopants, 282
(R)-3-hydroxybutyric acid, 248–9, 260
(R)-3-hydroxybutyrolactone, 249
D-(–)-*p*-hydroxyphenylglycine,
    as resolving agent, 85
    resolution of, 148
(R)-3-hydroxyvalerate, 248
hypocholestemic agents, **193**, 385
(S)-(+)-ibuprofen, 2, 6, **20**, 27–37, **172**, **173**, 208, **314**
    by asymmetric hydrocyanation, 331–2
    racemic switch, 22
    resolution of esters by proteases, 177

immobilised enzymes, 5, 187–8
indoprofen, **20**
International Committee on Harmonisation (ICH), 11

(R)-(–)-isobornylamine, **93**
isoleucine, resolution of, 168
isoprofen, **20**
isopropyl D-tartrate, as chiral solvent, 154
N-isopropylephedrine, 340, **343**, 349
(S)-isopropylidene glycerol, **68**, 363
isoproterenol, 398

Jacobsen asymmetric epoxidation, *see* asymmetric epoxidation

ketone reduction, enantiospecific, *see* asymmetric reduction
(S)-ketoprofen, 2, **20**, 37
ketosulphones, asymmetric reduction of, 246–7, 253–60
kinetic analysis, in L-carnitine production, 300–2
kinetic resolution, *see* resolution, kinetic
kinetic resolutions, membrane-assisted, *see* resolution, membrane-assisted

β-lactam, antibiotics, 105, **106**
lactamase, 184–8
γ-lactams, 185–8
lactic acid esters, in liquid crystals, 278
(R)-lactic acid,
    as byproduct of dehalogenation, 213–5
    as precursor for (S)-2-chloropropanoic acid, 210
*Lactobacillus plantarum*, in asymmetric reduction, 255–7
lamivudine, **42**, 66–76
    resolution of, 69–70
    synthesis of racemate, 67–69
lasolocid A, 166, **167**
(+)-leucinamide, **93**
*tert*-leucine, resolution of, 171,
    by oxazolinone resolution, 201
    by aminoacylase based resolution, 199
levobunolol, **365**
levopropizine, **365**
levopropoxyphene, **226**
lipases, 176, 190–3, 364
    for guaifenesin resolution, 376
    for resolution of ibuprofen esters, 34
    in lamivudine synthesis, 71
    in oxazolinone resolution, 201–3
    *Mucor miehei*, 71
    *Pseudomonas*, 48
lipozyme, 202
liquid crystals, chiral, 263–84, 375

commercial requirements, 273–9
market, 280, 284
properties and applications, 263–84
synthesis, 280–4
liquid-liquid extraction, 168–71
liquid-membrane fractionation, 171–3
α-lobeline, **93**
loracarbef, 105–11
loxoprofen, **20**

mandelic acid, 96
as resolving agent, 84
D-mannitol, 363
mecoprop, racemic switch, 209
melting point phase diagram,
 ibuprofen, 34
 naproxen, 23
 epoxy alcohols, **374**
membrane-assisted separations, *see*
 resolutions, membrane-assisted)
membrane bioreactor, in carbocyclic
 nucleoside synthesis, **49**
membrane/separation technologies,
 6–8, 168–78
membranes, enantioselective, 158–62
membranes,
 hollow fibre, *see* hollow fibre
  membranes,
 selective liquid, 162–8
 ultrafiltration, 7, 158–62
(−)-menthol, 3, 72, 75
menthyl(−) laurate, 176
mephenoxalone, 376
metastable phase, 122,
 isolation of, 140–3
metastable zone width, **143**, 144
 measurement of, 134–9
 seeding within, 152
metaxalone, **365**
methocarbamol, 376
(S)-2-(methoxy)pyrrolidine (MMP), 111
α-methoxy-α-(trifluoromethyl)phenyl-
 acetyl chloride, (MTPA chloride),
 *see* Mosher's ester,
4-methoxymandelonitrile, 332
2-(6-methoxynaphthalene)propionitrile,
 326
*trans*-3-(4-methoxyphenyl)glycidic acid,
 methyl ester 176–7
methyl α-cyclogeranate, **338**, 343, 348
methyl β-safranate, **338**
methyl (R)-3-hydroxybutyrate, 246–8, 260

synthesis, 248–9
α-methyl fructofuranoside ligands, 322–5
methyl oxazaborolidine, 396
(2S)-(−)-methyl-1-butanol, 162
methyl-*p*-toyl sulphide, asymmetric
 oxidation of, 383–4
(R)-α-methylbenzylamine 56
(S)-α-methylbenzylamine, **87**, **93**
 in ibuprofen resolution, 34
 as chiral auxilliary, 111
 resolution of loracarbef intermediates,
  110
(S)-2-methylbutanol, 280, 283–4
(S)-4-(2-methylbutyl)-biphenyl-4'-
 carboxylic acid, 283–4
(S)-4-(2-methylbutyl)-phenol, 283–4
α-methylDOPA, 152
(+)-N-methylephedrine, **87**, **93**, 340
*trans*-3-methylglycidyl tosylate, 374
N-methylmorpholine-N-oxide, 46
*threo*-4-methylthiophenylserine, 358
mexoprofen, **21**
micellar-enhanced ultrafiltration (MEUF),
 158, 174
Michaelis-Menten kinetics, in betaine
 metabolism, 292
miroprofen, **21**
mirtazapine, **172**, **173**
molecular similarity model, 92
monoamine oxidase inhibitors, 375
moprolol, 376
Mosher's ester, 372
multi-stage reactors, in L-carnitine
 production, 302–3
mutagenesis, 219–20, 293
(−)-*cis*-myrtanylamine, **93**

(+)-1-(1-naphthyl)ethylamine, **93**
(S)-N-(1-naphthyl)leucine, 165
L-2-naphthylalanine, 200
(S)-(+)-naproxen, 6, 19, 22–27, 177, 208,
 310, 314,
 by classical resolution, 22
 by asymmetric hydrocyanation, 331–2
 Zambon process, 26
nematic mixtures, 273–6, 280–1
*Neurospora crassa*, in asymmetric
 reduction, 258–60
nickel-DIOP, in asymmetric
 hydrocyanation, 313–4
nicotine, **93**
nisoxetine, 104

INDEX
409

Nonsteroidal Anti-inflammatory Profen
  Drugs, (NSAIDs), 19–38
  mode of action, 19
  racemic switch, 22
nopol, **162**
noradrenaline, **93**
norbornadiene, 313
(*R*)-*endo*-norbornen-2-ol, **178**
norbornene derivatives, asymmetric
  hydrocyanation of, 312
norbornene, **313**
L-(−)-norephedrine, 75, **87**, **93**, **172**, **169**
  resolution by liquid-membrane
    fractionation, 173–4
NSAIDs *see* Nonsteroidal Anti-
  inflammatory Profen Drugs
nucleation inhibitors, 142, 154
  chiral, 148
nucleation, 134
nucleosides, *see* carbocyclic nucleosides
5′-nucleotidase, in lamivudine synthesis,
  69–70
number of transfer units (NTU), in
  liquid-liquid extraction, 169

(*S*)-2-octanol, 275, 280, **284**
olefin epoxidation, 363
organoborane reagents, chiral, 392–3
L-(+)-ornithine, **93**
oxazaborolidines, 4
  in asymmetric reduction of ketones,
    391–8
oxazolidinones, 3
oxprenolol, resolution with
  enantioselective membranes, 160–61
oxynitrilases, 5
oxyprenolol, **161**

paclobutrazol, **153**
penicillamine, **226**
penoxaprop, **209**
pertraction, 170
Pharmaceutical Manufacturers
  Association (PMA), 11
pharmacokinetic evaluation of
  enantiomers, 11, 13–14
phase transition temperature (PTT),
  132–3, 141
α-phenethylamine, 84, **87**
phenoxypropanoic acids, 208–9
(*S*)-1-phenyl-1,2-ethanediol, 275, 280
*trans*-4-phenyl-2-buten-1-ol, 75, 373

phenylalanine, 3, 160, 174
phenylalaninol, **93**
phenylethanesulphonic acid, as resolving
  agent, 148
β-phenylglucophosphite ligands, 318–26
β-phenylglyceric acid, 154
*trans*-3-phenylglycidol, 370, 372
(*R*)-phenylglycine, 165, 167, **172**
  in loracarbef synthesis, 107
D-(−)-phenylglycinol, **87**, **93**
(*S*)-4-(phenylmethyl)-2-oxazolidinone, 3
phosphinite ligands, asymmetric, 26, 31,
  314
physostygmine, 245
pilocarpine, 245
(−)-α-pinene, as chiral solvent, 154
(−)-β-pinene, in enantioselective
  membranes, 160
pirprofen, **21**
plasmid stabilisation, 296–7
poly(α-amino acids), 5, 159–60
poly(L-leucine), 5
poly(methyl-L-glutamate), in
  enantioselective membranes, 160
poly-L-glutamate, in enantioselective
  membranes, **159**
poly[(*R*)-3-hydroxybutyrate], 248, 260
polymorphic transformations, 125
polymorphs, 120–25, 373
  disappearing, 143
polysulphone UF membranes, 160
porcine pancreatic lipase, 364
potassium channel openers, 386
pranoprofen, **21**
preferential crystallisation, of epoxy
  alcohols, 374
Principle Component Analysis (PCA), 92
process design, L-carnitine, 298–304
Profens, *see* Nonsteroidal Anti-
  inflammatory. Profen Drugs,
(*S*)-proline, **82**, 96, 396
  as chiral building block, 230
D-proline, 96
proline-*N*-carboxanhydride, 396
(*S*)-prolinol, **93**
  as chiral building block, 230
propranolol, **165**, **172**, **173**
*N*-*n*-propyl-D-glucamine in naproxen
  resolution, 22
*trans*-3-propylglycidol, by Sharpless
  epoxidation, 370, 372
protizinic acid, **21**

protonation, enantioselective,
  see enantioselective protonation
pseudoephedrine, **87**, **93**, as resolving
  agent, 84
*Pseudomonas*
  esterase, in CPA resolution, 210
  lipase, in carbocyclic nucleoside
    synthesis, 48
  in lactam resolutions, 186–7
pyrethroid incecticides, 1, 147–8, 310
L-2-pyridylalanine, 200
pyroglutamic acid, 396
pyrrolidines, chiral, 225–40

(+)-quinidine, **87**, **93**
(−)-quinine, **87**, **93**

racemates *vs* single enantiomers, 11, 12,
    28–30, 66–7, 226, 382
racemic compounds, 15, 23, 34, 148–9, **150**
racemic mixture, *see* conglomerates
racemic switch, 2, 12, 14
  herbicides, 209
  ibuprofen, 22
racemisation, 256–60
  of enantiomers, *in vivo*, 13
  in thiamphenicol synthesis, 359
racemisation-recycle, 148, 153, 196, 358
  *in situ*, 94–6, 359–62
  in amino acid synthesis, 199
  in duloxetine synthesis, 103–5
  L-carnitine, 288, 298–304
raclopride, **227**
ranolazine, 376
recombinant biotransformation strain, in
    L-carnitine production, 296–7
recovery, of chiral proton source, 339
recycling *see* racemisation-recycle
reduction, enantioselective, *see*
    asymmetric reduction
refractive index, in liquid crystals, 266
regulatory agencies, policy on racemates, 1
remoxipride, **227**
renin inhibitors, 194
resolution,
  classical, 54–9, 75, 81–97, 99–115,
    144–8, 186, 228–9, 280, 349,
    355–8, 385
  enzymatic *see* enzymatic kinetic
    resolution
  'dynamic' of unnatural amino-acids,
    201–3

kinetic, enzymatic *see* enzymatic kinetic
  resolution
membrane-assisted, 48, 157–79
resolutions vs stereoselective synthesis, 27,
  99–115
resolutions, by fractional crystallisation of
  diasteriomeric salts – *see* classical
  resolution
  of Chirald, 102
  of Lamivudine, 69–71
  of naproxen, 22, 27
  of pyrrolidine amines, 228–9
  recent developments, 4
  efficiency, 4, **86**, **87**
  entrainment, 355
  of L-carnitine, 288
resolving agents,
  criteria, 84
  rational design, 81–97
reverse osmosis, 158–62
Ritter reaction, 246
rose ketones, 337, 349
Ru(COD)(BINAP), **249**

salbutamol, **172**, **173**
scale-up issues, 69–70, 75, 100, 102, 187–8,
    203–4, 214, 217–8, 226–7, 237–40,
    259, 298–303, 362, 384, 388, 398
second order asymmetric transformation,
    94–6, **148**, 153, 202, 359–62
selectively permeable membranes, *see*
    membranes, selectively permeable
seproxetine, 104
sequencing, 260
serine, 160
5-sertraline, lipoxygenase inhibitor,
    399–400
Sharpless asymmetric dihydroxylation, *see*
    asymmetric dihydroxylation
Sharpless asymmetric epoxidation (SAE),
    *see* asymmetric epoxidation
Sharpless epoxidation, in asymmetric
    sulphide oxidation, 6, 382–7
*E*-silyl enol ethers, enantioselective
    protonation of, 341–3
Simulated Moving Bed technology (SMB),
    7
sodium (*R,R*)-di-*p*-toluoyltartaric acid,
    (NaDTTA), 113
sodium erythorbate, in loracarbef
    synthesis, 107–9
solubility curves, measurement of, 131–3

# INDEX

solvates, 120–25
solvent, choice for crystallisations, 131
(−)-sparteine, **93**
spontaneous polarisation, in liquid crystals, 268, 278, 284
stereoselective assay, 15
stereoselective synthesis, *see* resolutions vs stereoselective synthesis
steroid 5-α-reductase inhibitors, 111–14
stilbene oxide, resolution with enantioselective membranes, **161**
strain isolation, 288
L-(−)-strychnine, **87**, **93**
sulphoxides, chiral, 3, 6, 383, 391–8
sulpiride, **227**
super twisted nematic cell, (STN), in liquid crystals, 275
supercritical fluid technology, 8
superfluorinated materials, (SFM), 275
suprofen, **21**
surface-stabilised ferroelectric liquid crystal (SSFLC) cell, **270**, 271–2, 277, 285
symmetrical countercurrent extraction, 7

Taft $\sigma$-values, 84, 320–1, **330**
tartaric acid, 84, **96**, 165, 169, 228–9, 358
tartaric esters, 7, 96, 154, 314, 366, 383–4
tartrate-derived phosphite ligands, 314
Taxol, 375
tazeprofen, **21**
terbutaline, **172**, **173**
terpenes, in liquid crystals, 277–8
tetralin musk perfume component, 348–50
tetriprofen, **21**
thermochromic liquid crystals, 272–3, 279, 283–4
thermogravimetric analysis (TGA), 122–3, 151
thermotropic liquid crystals, 264
thexylborane, 66
thiamphenicol,
derivatives, 354
manufacture, 353–62
market, 354
synthetic approaches, 354–8
($S$)-3-(2-thienylthio)butyrate, 249]–50
($1S,2S$)-thiomicamine, 358–9
tiaprofenic acid, **21**
tomoxetine, 104
Tonalid, 348
toxological assessment of enantiomers, 13
transition temperature, between crystal forms, 122
trityl glycidol, by Sharpless epoxidation, 370, 372–3
Trusopt, 246–8
tryptophan, **159**
tyrosine, 160

ultrafiltration membranes, *see* membranes, ultrafiltration
undesired enantiomers, FDA guidelines for, 11

L-valine, 168, 282, 283–4
*Varicella zoster*, 50
viloxazine, **365**
vinylarenes, asymmetric hydrocyanation of, 309–32
vitamin $B_T$, *see* L-carnitine
vitamin $D_3$, 375
volumetric productivity, in L-carnitine production, 300

whole cell biotransformation, *see* biotransformation, whole cell

X-ray diffraction (XRD), in crystal lattice determinations, 120–29, 133, 151
ximoprofen, **21**

zoliprofen, **21**